AF595065

Adsorbents for Water and Wastewater Treatment and Resource Recovery

Adsorbents for Water and Wastewater Treatment and Resource Recovery

Editors

Sílvia Santos
Cidália Botelho
Ariana Pintor

MDPI • Basel • Beijing • Wuhan • Barcelona • Belgrade • Manchester • Tokyo • Cluj • Tianjin

Editors
Sílvia Santos
Faculty of Engineering
University of Porto
Porto
Portugal

Cidália Botelho
Faculty of Engineering
University of Porto
Porto
Portugal

Ariana Pintor
Faculty of Engineering
University of Porto
Porto
Portugal

Editorial Office
MDPI
St. Alban-Anlage 66
4052 Basel, Switzerland

This is a reprint of articles from the Special Issue published online in the open access journal *Water* (ISSN 2073-4441) (available at: www.mdpi.com/journal/water/special_issues/adsorbents_water_wastewater_treatment_resource_recovery).

For citation purposes, cite each article independently as indicated on the article page online and as indicated below:

LastName, A.A.; LastName, B.B.; LastName, C.C. Article Title. *Journal Name* **Year**, *Volume Number*, Page Range.

ISBN 978-3-0365-4722-0 (Hbk)
ISBN 978-3-0365-4721-3 (PDF)

Contents

About the Editors

Sílvia Santos

Sílvia C. Santos is a Postdoctoral Researcher at LSRE-LCM (Laboratory of Separation and Reaction Engineering –Laboratory of Catalysis and Materials) and ALiCE (Associate Laboratory in Chemical Engineering), at the Faculty of Engineering of the University of Porto (FEUP). She obtained a PhD in Chemical and Biological Engineering in 2009 from FEUP. Her research interests include alternative adsorbents, plant-based coagulants, water decontamination, arsenic and antimony removal, textile wastewater treatment, and the recovery of precious metals. She has taught Bachelor's and Master's programmes of Environmental and Chemical Engineering, covering analytical methods, water quality and water treatment. S. Santos has been working on international projects related to surface water quality and modeling and sustainable processes for critical metals recovery from mining waste. S. Santos has published 28 papers and 3 chapters in international books (citations in 1649 documents; H-index of 18). Since 2018, she has been financed by the Portuguese national funding agency for science, research and technology (FCT), with a postdoctoral fellowship (SFRH/BPD/117387/2016).

Cidália Botelho

Cidália M. S. Botelho is an Assistant Professor at Faculdade de Engenharia—University of Porto (FEUP). She obtained her PhD (1998) in Chemical Engineering from FEUP, and Master's in Applied Chemistry from Instituto Superior Técnico—University of Lisbon. She teaches Environmental Chemistry, the Characterization and Monitoring of Water and Wastewater, Technologies and Systems of Water and Wastewater Treatment.

She is member of the research lab LSRE-LCM and of the associate lab ALiCE. Her main research interests are water chemistry; water quality and monitoring; water and wastewater treatment; and adsorption. In these areas, she has supervised 7 PhD theses (5 concluded), 35 master students and 5 postdoctoral researchers, and published 96 scientific papers, with 4061 citations and an H-index of 35 (Scopus, 6/2022), and 5 book chapters.

Ariana Pintor

Ariana Pintor is a junior researcher at the Laboratory of Separation and Reaction Engineering—Laboratory of Catalysis and Materials (LSRE-LCM) and the Associate Laboratory in Chemical Engineering (ALiCE) at the Faculty of Engineering of the University of Porto. She completed her PhD in Environmental Engineering at this institution in 2014. Since then, she has continued to pursue postdoctoral research on the development of natural adsorbents, especially cork-based materials, to remove various pollutants from water and wastewater. She has published 21 international peer-reviewed papers (H-index=14) and has supervised several undergraduate and graduate students on the topic of adsorption for environmental remediation.

Preface to "Adsorbents for Water and Wastewater Treatment and Resource Recovery"

Adsorption is a well-established operation used for water decontamination and the remediation of industrial effluents. It is also recognized as a key technology for recovering substances of economic interest or at risk of scarcity. Adsorption from aqueous solution has aroused great interest in the scientific community in recent decades. This interest has been driven by: (i) the increasing demanding standards of drinking water, irrigation water and discharge limits imposed by legislation; (ii) the recognition of new and emerging contaminants and their possible deleterious effects on human health and ecosystems; and (iii) the need to refine wastewater treatment to a level that allows its reuse in multiple applications. The new sustainability paradigm of the circular economy and the current context of promoting the efficient use of natural resources and energy are additional motivations for research on this topic. This Special Issue was open for submissions from January to November 2020. Submissions covering the latest findings and progress on the use of adsorbents for water and wastewater treatment and recovering substances were encouraged. In it are 21 papers (17 research articles and 4 reviews), published by researchers of 18 countries.

The removal of heavy metals from water has been addressed in several articles; some use biochars (Wang et al.; Guo et al.; Kim et al.; Zhu et al.), other bio-derived materials (Vázquez-Guerrero et al.; Hamoud et al., Zhang et al.) and industrially derived adsorbents (Kobayashi et al.; Liu et al.). Interesting contributions were also received on the removal of organoselenium compounds (Okonji et al.) by granular activated carbon and nano zerovalent iron, and on arsenic uptake (Ogata et al.) using complex metal hydroxides. Torrinha et al. presented a study on the uptake of gold from strong acidic solutions, aiming to recover the metal from secondary sources, such as hydrometallurgical liquors of e-waste. This Special Issue also contains several works addressing the removal of organic contaminants from water. Studies conducted by Angosto et al. showed the effective removal of diclofenac from water using agro-food waste and compared it with advanced oxidation techniques. Martín-Lara et al. presented a study on removing bisphenol A using commercial activated carbon and evaluated the effect of coexisting metal cations in solution. Turco et al. studied the removal of phenolics from olive mill wastewater by oxidized multiwalled carbon nanotubes entrapped in a microporous polymeric matrix of polydimethylsiloxane. Adewuyi reviewed the literature on biosorbent use for the removal of emerging pharmaceutical compounds from water. Zango et al. provided a state-of-the-art perspective on removing PAHs (polycyclic aromatic hydrocarbons) and phenolic compounds by carbon porous materials, porous polymers, mesoporous silica, and metal–organic frameworks. The research articles of Yu et al. and Kuang et al. cover the surfactant modifications of sepiolite and activated carbon to decolorize effluents containing mixed and cationic dyes, respectively. Srivatsav et al. presented a comprehensive review paper covering the decolorization of dye-containing wastewater using biochars. Frišták et al. introduced an additional adsorption application to this Special Issue, by presenting an interesting review on the use of biochars to remove microcystins and improve water quality.

This Special Issue expects to be a good reference material in the field of water decontamination and effluent treatment by adsorption. The guest editors would like to thank those who contributed with their research work or reviewed the manuscripts.

Sílvia Santos, Cidália Botelho, and Ariana Pintor

Editors

Review

The Use of Biochar and Pyrolysed Materials to Improve Water Quality through Microcystin Sorption Separation

Vladimír Frišták [1,*], H. Dail Laughinghouse IV [2] and Stephen M. Bell [3]

[1] Department of Chemistry, Trnava University in Trnava, 91843 Trnava, Slovakia
[2] Fort Lauderdale Research and Education Center, Agronomy Department, University of Florida/IFAS, Davie, FL 33314, USA; hlaughinghouse@ufl.edu
[3] Institute of Environmental Science and Technology (ICTA-UAB), Universitat Autónoma de Barcelona, 08193 Barcelona, Spain; Stephen.Bell@uab.cat
* Correspondence: fristak.vladimir.jr@gmail.com; Tel.: +421-33-592-1459

Received: 27 August 2020; Accepted: 13 October 2020; Published: 15 October 2020

Abstract: Harmful algal blooms have increased globally with warming of aquatic environments and increased eutrophication. Proliferation of cyanobacteria (blue-green algae) and the subsequent flux of toxic extracellular microcystins present threats to public and ecosystem health and challenges for remediation and management. Although methods exist, there is currently a need for more environmentally friendly and economically and technologically feasible sorbents. Biochar has been proposed in this regard because of its high porosity, chemical stability, and notable sorption efficiency for removing of cyanotoxins. In light of worsening cyanobacterial blooms and recent research advances, this review provides a timely assessment of microcystin removal strategies focusing on the most pertinent chemical and physical sorbent properties responsible for effective removal of various pollutants from wastewater, liquid wastes, and aqueous solutions. The pyrolysis process is then evaluated for the first time as a method for sorbent production for microcystin removal, considering the suitability and sorption efficiencies of pyrolysed materials and biochar. Inefficiencies and high costs of conventional methods can be avoided through the use of pyrolysis. The significant potential of biochar for microcystin removal is determined by feedstock type, pyrolysis conditions, and the physiochemical properties produced. This review informs future research and development of pyrolysed materials for the treatment of microcystin contaminated aquatic environments.

Keywords: biochar; microcystin; separation; water quality; cyanobacteria; algal bloom

1. Introduction

Recent studies indicate that cyanobacterial harmful algal blooms (cyanoHABs) are occurring more frequently in inland fresh waters (i.e., ponds, lakes, streams, and rivers) and estuaries worldwide [1–4]. Higher temperatures linked to global climate change and increased eutrophication have intensified cyanoHABs, posing significant environmental, social, and economic impacts partly due to the effects of toxins produced by several taxa [5–8]. Monitoring toxicity following such events in waterbodies can be complicated as cyanobacterial blooms contain complex mixtures of different classes of toxins [9–11]. Nevertheless, increased research efforts on the toxic impacts of these blooms under a changing climate and practical tools and methods for their management and remediation are clearly needed [4–14].

Microcystins (MCs) are a group of naturally occurring toxic secondary metabolites (cyanotoxins) produced by various genera of cyanobacteria [9–15]. They can be fatal to freshwater organisms even at very low concentrations [5–16] and have been classified by the International Agency for Research on Cancer (IARC) as possible human carcinogens [17]. To meet World Health Organization

(WHO) guidelines for maximum allowed surface water concentrations [18], there are several processes available for the removal of extracellular MCs. Powdered and granular activated carbon are commonly applied to adsorb MCs during treatment processes of drinking water because of their unique sorbent characteristics [19–21]. However, production costs for activated carbon and other materials (e.g., carbon nanotubes) have limited their potential for widespread application [19–22].

There is a current demand for more environmentally friendly and economically and technologically feasible sorbents for water quality remediation to be used in tandem with bloom prevention and risk reduction [23]. Pyrolysis, the thermochemical conversion of organic matter into char under anoxic or low-oxygen conditions, is a cheap and attractive method for the production of new sorption materials for water treatment purposes [24,25]. As a product of pyrolysis, biochar is of interest in this regard owing to the possibility of specifically designing, through feedstock and production process manipulation, more efficient sorption properties [26]. The high porosity, chemical stability, and significant sorption efficiency for removing cyanotoxins are among the central advantages of biochar [25]. In aqueous solutions, biochar and modified charcoal have provided promising results as sorbents [27–29], however, existing knowledge gaps have hindered widespread application in water treatment processes.

To address the need for an updated review of microcystin removal strategies, this paper first provides an overview of microcystin toxicity in natural water bodies followed by a discussion on conventional and alternative sorbents for microcystin separation. Here, this review focuses on the most pertinent chemical and physical sorbent properties responsible for effective removal of various pollutants from wastewater, liquid wastes, and aqueous solutions. Furthermore, an in-depth assessment of the pyrolysis process for sorbent production is given, providing the first comprehensive evaluation of biochar as a sorbent for microcystin removal. This review informs future research and development of pyrolysed materials for the treatment of microcystin in aquatic environments.

2. Microcystin Origin and Toxicity in Natural Water Bodies

Cyanobacteria are oxygenic photosynthetic bacteria that are prevalent in nutrient rich, warm, low turbulent freshwater bodies. When cyanobacteria die, their cell walls degrade, releasing intracellular toxins into the surrounding aquatic environment. MCs are a group of these naturally occurring toxins produced by various genera of cyanobacteria (e.g., *Anabaena*, *Anabaenopsis*, *Aphanocapsa*, *Cyanobium*, *Dolichospermum*, *Hapalosiphon*, *Limnothrix*, *Microcoleus*, *Microcystis*, *Nostoc*, *Oscillatoria*, *Phormidium*, *Planktothrix*, *Pseudanabaena*, *Sphaerospermopsis*, and *Synechocystis*). Their structures are extremely stable in water and can withstand chemical breakdown such as hydrolysis or oxidation. From a chemical point of view, MCs represent cyclic peptides composed of five common amino acids and pairs of L-amino acids as variants [30]. The common ones are D-erythro-β-methylaspartic acid, alanine, *N*-methyldehydroalanine, glutamic acid, and a unique amino acid called Adda (3-amino-9-methoxy-2,6,8-trimethyl-10-phenyldeca-4,6-diene acid). There have been more than 240 variants of MCs reported [31,32].

The main structural differences between microcystin variants arise from the substitution of single amino acids (Figure 1). In MC-LR, the most toxic and abundant microcystin, the two variable L-amino acids are L-arginine and L-leucine [33]. Trinchet et al. [34] and Roegner et al. [23] highlighted preferential organ uptake, and thus the importance of other congeners such as MC-RR (arginine at positions 2 and 4), MC-LA (leucine and alanine), MC-YR (tyrosine and arginine), MC-LF (leucine and phenylalanine), and MC-LW (leucine and tryptophan). The WHO established the guideline of a maximum recommended concentration of 1 μg/L for MC-LR in surface water. However, there remains a lack of data with respect to other structures and metabolites. According to the IARC, MCs are placed into group 2B as possible human carcinogens [17]. They are classified as tumour initiators with DNA-damaging effects [35]. These heptapeptides are produced via a nonribosomal biosynthesis pathway by large multienzyme complexes that include peptide synthetases, polyketide synthases, and tailoring enzymes [36].

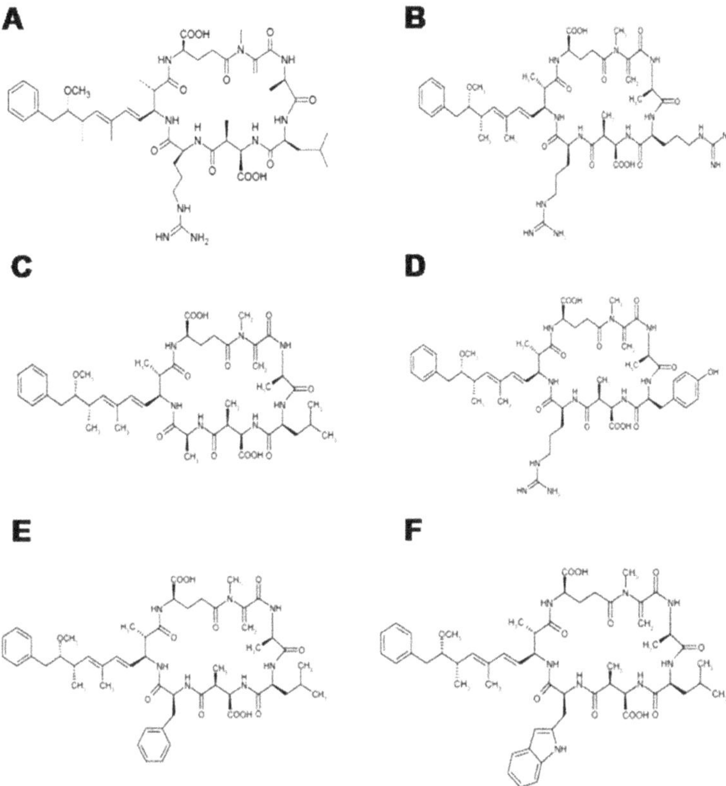

Figure 1. (**A**) Chemical structure of microcystin (MC)-LR, (**B**) microcystin-RR, (**C**) microcystin-LA, (**D**) microcystin-YR, (**E**) microcystin-LF, and (**F**) microcystin-LW.

In active and growing cyanobacteria, most MCs are cell-bound with intracellular concentrations of MCs several orders higher than concentrations of dissolved MCs in water. However, bloom senescence and cell lysis can cause the release of high quantities of MCs into the water. Mazur-Marzec et al. [37] and Jones et al. [38] reported that MC-LR can persist in aquatic for 1–3 months, but this is dependent on limnological and environmental parameters [39]. Based on acute toxicity, MC-LR is considered one of the most significant cyanobacterial toxins. As hepatotoxins, MCs mainly affect liver functioning, and can cause mortality and morbidity through necrosis and haemorrhage [40]. The hydrophilic characteristic of most MCs facilitates its uptake in vertebrates via active transport compared with cell membranes [41]. Membrane transport occurs mainly via organic anion transporting polypeptides [42].

Inside the hepatocytes, MCs inhibit eukaryotic protein serine/threonine phosphatases 1 and 2A [41], resulting in cytoskeletal disruption, rearrangement of associated filamentous actin, and morphological changes [23]. Frangez et al. [43] confirmed covalent binding of microcystin structure to protein phosphatases, leading to hyperphosphorylation of cytoskeletal proteins and the disruption of many cellular processes. Davis et al. [44] showed the effect of MC-LR on activation of the mitogen-activated protein kinase pathway. MCs induce apoptosis through free radical formation and mitochondrial alternations [23–45]. Human health problems are mostly related to chronic exposure to MCs via food and water consumption. As tumor-promoting toxins, MCs alter the regulation of phosphorylation and induce DNA damage in hepatocytes [46].

3. Conventional Strategies of Microcystin Removal

The removal or inactivation of MCs from water can be affected by several factors. The most crucial among these is if MCs are concentrated extracellularly or intracellularly. As mentioned, the majority (>95%) of MCs are found intracellularly during the growth stage of the bloom. However, when blooms collapse as a result of nutrient depletion, high temperatures, viral attack, and/or through the effect of chemical agents, intracellular MCs are released into the water as extracellular, increasing dissolved microcystin concentrations. The most common processes of extracellular MCs removal from drinking water are divided into three general steps: physical removal, biological inactivation, and chemical inactivation. Very often, these steps are cross-linked and supplied by processes such as sedimentation and flocculation. Physical removal includes the adsorption by effective sorbents (usually activated carbon) and the application of several filtration techniques [47,48]. Conventional sand filtration may be sufficient, but could also enhance the release of high concentrations of MCs into treated water [49,50]. Suboptimal set-up of filtration parameters, such as flow rate, can very often lead to enormous leaching of MCs from lysed microbial cells.

The main objective of the traditional sequence of coagulation–sedimentation–flocculation processes is to remove live algae, their residues, and organic and inorganic matter. A potential side-effect is cell lysis and subsequent increases in dissolved cyanotoxins' (e.g., MC) concentrations. The character and properties of cyanobacterial organic matter and toxic compounds decrease the efficacy of the conventional process for inorganic substances. Roegner et al. [23] described in detail the main disadvantages and pitfalls of the coagulation–sedimentation system when applied for MCs removal. Teixeira and Rosa [51] highlighted the significant effect of local water quality on MCs removal by this conventional method. Additionally, the quantity and quality of the chemical coagulant represents an important factor in the removal process. The application of flotation, a process whereby solids in suspension are recovered by their attachment to gas (usually air) bubbles, in combination with coagulation and flocculation, shows positive results in MCs separation [52]. The particles most effectively removed are in the size range of 10–200 μm. The particle-bubble aggregates that are formed have a density less than the suspension itself and they rise to the surface to be removed. Despite the high efficiency of this technology for algal residues removal, the question of extracellular toxins removal is still quite open. Therefore, the application of other methods is still needed.

From a treatment standpoint, MCs have three general areas subject to oxidation: the conjugated double bond in the Adda moiety; the single double bond in the Mdha moiety; and the side chain of amino acids [47]. Laszakovits and MacKay [53] described in detail the removal of cyanotoxins by potassium permanganate. However, this method represents a pH dependent and chemically demanding process. Several studies exist on verifying the efficiency of registered chemical algaecides and herbicides (especially copper-based and peroxide-based products) for bloom treatment and MC removal [54–56]. These compounds have been used successfully for decades, but initial cyanobacterial cell and toxin concentrations, algal species, and limnological characteristics can affect efficacy. Further, usage of waters (e.g., recreation vs. drinking water) and local regulations must be considered before application. Brooke et al. [57] showed reduced concentrations of MC-LR in water after ozone application. However, this pH dependent process is less effective under alkaline conditions. Chlorination can be viewed as another potential method to remove cyanotoxins from aqueous solutions [58]. Oxidation, chlorination, and ozonization can bring new risk such as secondary pollution from their by-products. Lawton et al. [59] showed the application of ultraviolet radiation as an effective, but expensive method to remove MCs from water.

Activated carbon produced from different precursor feedstocks such as coal or coconut represents the main sorption material used in separation processes of MCs from aqueous solutions [20–60]. The surface characteristics and pore structure of activated carbon determine its range of applicability in purification technologies of polluted water [61]. The activation process affects the adsorptive capacity, which can vary and only be effective for a specific group of contaminants. Activated carbon removes

MCs mainly via mechanisms of adsorption and intraparticular diffusion [62,63]. MCs primarily interact with carbon surfaces by hydrophobic forces and electrostatic interactions [64].

Generally, two basic types of activated carbon, powdered activated carbon and granular activated carbon, are applied in treatment processes for drinking water. The main difference in use is based on type and character of the contaminant (synthetic/natural, inorganic/organic). Porosity (total pore volume), porous character, and surface characteristics are crucial for efficiency of MCs sorption. Activated carbon provides a system of micropores with diameters of <2 nm, mesopores with diameters from 2 to 50 nm, and macropores at >50 nm. Microporous filtration materials restrict water flow and reduce contact between MCs and active surface sites, and thus can be considered as less effective in the removal process. On the other hand, MCs can be easily separated by activated carbon with high concentrations of mesopore structures [64].

Researchers have confirmed the different sorption efficiencies of activated carbons for MC-LR compared with MC-RR, MC-YR, and MC-LA. Mesoporous carbon materials with targeted functional groups represent an option in MCs separation from aqueous solutions [65]. Granular activated carbon is very often used in fixed beds as a filtration medium or adsorbent to remove fragile particles and chemicals under continual flow conditions [47]. However, Newcombie et al. [66] showed the inefficiency of granular activated carbon-based filters for microcystin separation compared with granular activated carbon-based adsorbents with proper regeneration or replacement, which can be potentially used as auxiliary barriers. Ho and Newcombie [67] found that powdered activated carbon can be easily bound up in the floc structure during the coagulation process, and thus can reduce the efficiency of MCs removal. Roegner et al. [23] showed the importance of the activated carbon's feedstock on the removal efficiency of MCs from potable water. The authors summarized that activated carbon produced from wood contains a system of mesopores and micropores, while coconut-based activated carbon contains just micropores, which can limit its MCs sorption application. Hena et al. [63] described the sorption process of MC-LR by activated carbon prepared from rubber wood sawdust as pH dependent with a maximal sorption capacity of 296 mg/g.

The effect of feedstock pH and the value of the point of zero charge (pHzpc) can significantly affect the adsorption efficiency of MCs [68]. Huang et al. [69] showed the relation between the pHzpc of activated carbon and sorption process of cyanotoxins. The higher pHzpc of the feedstock results in a neutral or positive charge of the produced activated carbon and higher adsorption potential for MCs [23]. The efficiency of activated carbon and activated carbon-based filters in separation of MCs from drinking water is considered sufficient (>99%) [70]. However, concentrations of cyanotoxins in water are potentially low and filters are not typically challenged with higher doses or the competitive effect of other toxic agents. Chlorine pre-treatment of potable water presents a problem for MCs sorption separation efficiency related to the reaction between residual chlorine and active carbon sites. Another problematic issue of causing a decrease in sorption separation with activated carbon application in water can be high concentrations of organic matter. Natural organic matter competes simultaneously with MCs for active sorption sites on surfaces of activated carbon.

Granular activated carbon can additionally be used as an ideal habitat for microbial growth thanks to its large surface area [47]. The growth of different microorganisms on activated carbon-based filters and biofilm formation can enhance the MCs removal owing to the combined effect of MCs biological degradation and sorption separation [69]. On the other hand, biofilm formation can decrease pore availability and thus reduce separation efficiency [23]. Drogui et al. [71] showed the lower efficiency of activated carbon-based filters after colonization by microbial communities compared with sterile filters during separation of MC-LR from water. The concerning literature discrepancies in the assessment of microbial colonization on MCs separation by activated carbon require additional and more detailed studies.

Numerous research studies refer to the decrease in accessibility and increase in mass production costs of activated carbon derived from typical feedstocks such as coal, coconut, or wood [19–22]. Zhi et al. [72] reported a price of 1650–9900 USD/t for activated carbon for pollutant removal, while Lou et al. [73] determined a range of 20,000–80,000 USD/t for carbon nanotubes. Based on these assessments, more economical and environmentally friendly sorbents are needed for water quality control.

4. Alternative Microcystin Separation Sorbents

Conventional technologies of MCs removal from water can be expensive and very often simply insufficient. More affordable methods with low technological requirements in water treatment are needed according to Gurbuz and Codd [50]. The authors highlighted a number of natural and artificial materials with extensive internal porosity for application in water purification. Despite possessing a lipophilic side chain of ADDA residue, molecules of microcystins are quite hydrophilic because of two carboxylic groups on its aspartic/glutamic acid side and one guanidine group on its arginine acid side [74]. Kim et al. [75] applied industrial waste biomass of *Escherichia coli* to develop biosorbents for MC-LR immobilization. The authors created biomass stabilization with polysulfone, coating the polysulfone-biomass composite with polyethylenimine, and subsequent decarboxylation of coated material. Such prepared materials can be recommended as highly efficient sorbents for MC-LR removal from aquatic environments with maximum sorption capacities of 1.2–1.9 mg/g at approximately pH 5. MC-LR can be effectively removed from aqueous solutions by sorbents prepared from *Moringa oleifera* Lam. seeds, as shown by Yasmin et al. [76]. The authors applied an acid pre-treatment of seed powder to increase pore size and volume. The sorption capacity of modified biosorbent for MC-LR was >92 mg/g. Sathya et al. [77] applied an exocellular polymer from *Enterobacter ludwigii* to remove microcystin-RR extracted from *Microcystis aeruginosa* (Kützing) Kützing. The maximum removal of MC-RR by the studied polymer was optimized by response surface methodology and achieved at a contact time of 5 h, pH 10, temperature of 30 °C, agitation speed of 150 rpm, and polymer dose of 4 mg/mL. This study confirmed the removal efficiency of MC-RR by exocellular polymer with a maximum sorption capacity of 23.76 ng/mg. Additionally, the authors showed a positive effect of Cu^{2+} ions on the MC-RR removal process by microbial polymeric material.

Application of native or dried biomass to separate MCs from aqueous solutions can cause secondary contamination by releasing other polymeric substances of heterogeneous composition into purified solution. The instability of biomass-derived sorbents represents a hazardous point that has to be mentioned [78]. Therefore, inorganic and thermochemically transformed materials are a more promising choice [79]. Morris et al. [80] showed high efficiency (>80%) of MC-LR separation from water by naturally-occurring clay particles. Laughinghouse et al. [81] found that lanthanum modified bentonite clay was able to sediment MC-LR at higher concentrations (around 500 μg/L) when compared with lower concentrations (50 and 100 μg/L). Adsorption of MC-LR onto nano-sized montmorillonite was investigated and optimized by Wang et al. [82]. Their work indicated a pH dependent process with the maximum of 0.186 mg/g reached at pH 2.96. Removal of MC-LR from model aqueous solutions by nano-sized montmorillonite represents a rapid process with several kinetic stages. Desorption efficiency of sorbed MC-LR (>75%) can be reached using 0.1 mol/L NaOH as the eluting media. Superabsorbent polymer composites appear to be a cost-effective solution for MC-LR adsorptive removal [83]. The porous network structure and the ionic functional groups of superabsorbent polymers ensure diffusion and binding of microcystin molecules. Wang et al. [83] showed the effective application of polyacrylamide/sodium alginate-modified montmorillonite to remove MC-LR from aqueous solution. Their study described sorption as a fast process, reaching equilibrium within 80 min of contact time. The maximum adsorption capacity of polyacrylamide/sodium alginate-modified montmorillonite for MC-LR was 32.66 mg/g. The authors also discovered that over 85% adsorption and 80% desorption could be achieved after five regeneration cycles. Recovery of studied microcystin reached more than 92% without ionic effect.

The sorption potential of pumice for MCs has also been described by Gurbuz and Codd [50]. Pumice is a porous material with high numbers of active sites that can be used as effective filtration material. The application of pumice in slow filtration processes to remove pathogens from irrigation water is a well-known method in horticulture [84]. Gurbuz and Codd [50] showed MC-LR adsorption on a pumice-based sorbent as a pH dependent process with maximal efficiency at pH 4. Experimental studies on the removal of MC-LR by peat showed the important role of this biomaterial in separation processes of cyanotoxins [85]. The maximum adsorption capacity of peat for MC-LR was found to be 0.26 mg/g at pH 3. Regeneration of peat-based sorbent using 2 mol/L NaOH showed a 94% efficiency in MC-LR desorption.

Iron oxides and hydroxides influence the mobility of organic and inorganic compounds in soils, sediments, and surface water. Therefore, Lee and Walker [86] studied the potential of maghemite nanoparticles to adsorb MC-LR from model solutions and confirmed a strong effect of pH with maximum removal efficiency at pH 3. They described the electrostatic interactions and hydrophobic forces as being crucial MC-LR sorption mechanisms. The adsorption process of microcystin onto maghemite nanoparticles is not significantly affected by the presence of river fulvic acid at lower concentrations (<2.5 mg/L). At higher concentrations of fulvic acid, the adsorption process of MC-LR by iron oxide nanoparticles decreased as a result of competition for limited active sorption sites. Along similar lines, Okupnik et al. [87] investigated the use of titanium dioxide nanoparticles for removal of MC-LR in environmental relevant concentrations from water samples. The authors tested different crystalline phases of nano-TiO_2 such as anatase, rutile, and an anatase–rutile mixture in comparison with a bulk TiO_2 counterpart. Sorption separation of MC-LR by titanium dioxide nanoparticles represents a complex mechanism, with chemisorption and sorbent particle size as the most influencing factors. Kinetic and equilibrium results of microcystin-LR and microcystin-RR removal by graphene oxide from water polluted by other environmental co-pollutants were described in a study by Pavagadhi et al. [61]. The tested graphene oxide showed significant adsorption capacity at 1.70 mg/g for MC-LR and 1.88 mg/g for MC-RR even in the presence of Mg^{2+}, Ca^{2+}, K^+, and Na^+ cations and NO^{2-}, NO^{3-}, Cl^-, $SO_4{}^{2-}$, and $PO_4{}^{3-}$ anions. The sorption kinetic experiments revealed that more than 90% removal of both microcystin variants is achieved within the first 5 min. New sorbents based on N-doped carbon xerogel for MC-LR adsorption were synthesized by Wu et al. [88]. Their results indicate fast microcystin uptake within 10 min and an adsorption capacity of 1.92 mg/g. The authors highlighted the efficiency of sorbent regeneration using 1 mol/L NaOH applied as the desorption agent.

5. New Sorption Materials Produced with Pyrolysis

Various feedstocks can be thermochemically converted into char using carbonization processes such as torrefaction, pyrolysis, hydrothermal carbonization, or gasification. Pyrolysis proceeds under strict anoxic or low oxygen conditions at temperatures from 350 to 1000 °C, and it is the main carbonization technique to produce conventional biochar with required characteristics [89]. During the pyrolysis process, solid (char), liquid (condensable crude oil), and gaseous (non-condensable syngas) products are formed. The solid products of hydrothermal carbonization, torrefaction, or gasification do not fulfil the main definition of biochar. According to Shaheen et al. [90], pyrolysis represents a less expensive and more robust technology involving thermochemical decomposition of organic matter compared with the other methods mentioned.

The physical and chemical characteristics of biochar are fully dependent on the feedstock properties and production system (Figure 2) [91]. The European Biochar Certificate outlines a wide range of biomass feedstocks approved for use in biochar production [89]. Biomass can be sourced through local waste collection services following waste separation (biodegradable wastes), garden waste (leaves, flowers, roots, hay, grass, prunings from trees, vines, bushes), agriculture and forestry (harvest leftovers, straw, husks and grain dust, seeds, plants, bark and chippings, sawdust, wood, wool), animal by-products (bones, manure, skins, hairs, feathers), material from food and confectionary production (leftovers from the production of canned food; residues from potato, rice, or corn starch production; residues from dairy processing; tobacco dust; slacks; ribs; tea; and coffee grounds), textiles (cellulose, cotton fibres, wool leftovers, hemp), or biogas plants (fermentation residues). Additionally, there exist several approved mineral-organic additives such as lime, bentonite, clay, loam, rock flour, or lignite to improve pyrolysis conditions and, subsequently, the quality of the produced biochar.

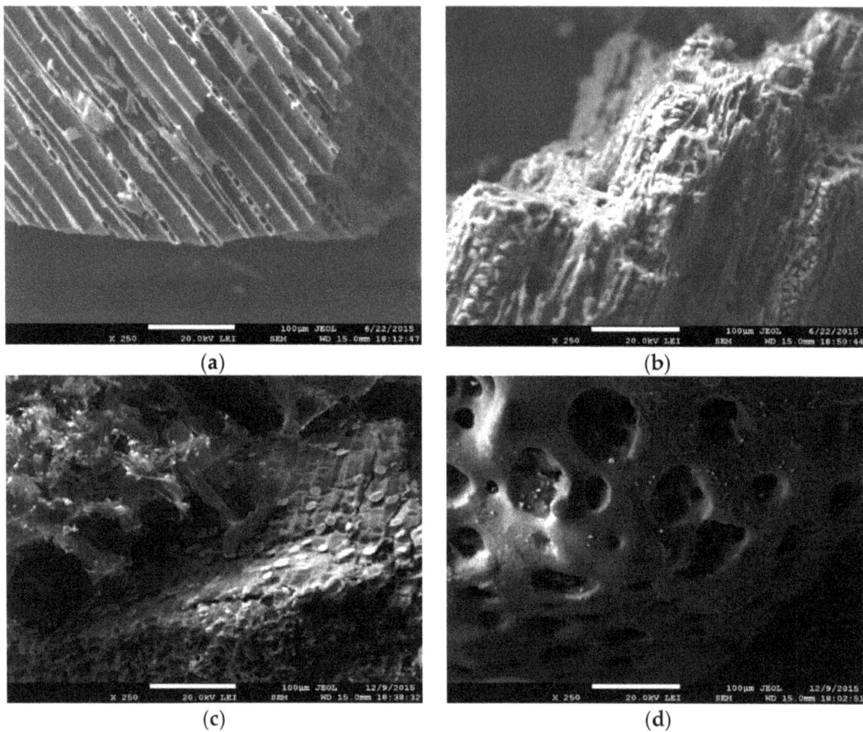

Figure 2. Scanning electron microscopy (SEM)—images of biochar derived from wood chips (**a**), garden green waste (**b**), corn cobs (**c**), and sewage sludge (**d**) at magnification 250× (with permission of authors—Micháleková-Richveisová et al. [91]).

Pyrolysis usually includes slow, fast, and flash processes [25]. Slow pyrolysis (conventional carbonization) is characterized by feedstock placement into a reactor and then initiating heating at a rate of about 0.1–1 °C/s until achieving a residence time from 20 min to several hours at temperatures of 400–700 °C. This process produces a solid char product rather that crude oil and syngas. While for fast and flash pyrolysis, addition of feedstock into the reactor occurs after pyrolysis temperature has been reached, with a significantly reduced residence time of several seconds to a few minutes at temperatures from 500 to 800 °C [92]. Fast pyrolysis produces a major bio-oil fraction and minor biochar and syngas by-products [93].

The fibrous biomass consists of cellulose, hemicelluloses, and lignin with smaller quantities of organic extractives and inorganic minerals. These constituents vary among different kinds of biomass. Cellulose as a linear condensation polymer of β-(1–4)-D-glucopyranose creates sheets of glucopyranose rings lying in a plane with successive sheets stacked on top of each other to form three-dimensional particles that aggregate into an elementary crystalline micro-fibril arrangement. This makes cellulose more thermal and chemically resistant compared with hemicellulose. Hardwood materials are rich in xylans and poor in glucomannans, while softwood feedstock contains more glucomannans compared with xylans [94]. Cellulose, hemicelluloses, and lignin have distinctive thermal decomposition behaviours that depend upon heating rates. Hemicellulose is the first to decompose, beginning at 220 °C and completed at 315 °C [94]. Cellulose does not start to decompose until about 315 °C.

Although lignin begins to decompose at 160 °C, it is a slow process extending to 850–900 °C. Pyrolysis of hemicellulose produces mainly no-condensable gases (CO, CO_2, H_2, and CH_4) and low molecular weight organic compounds such as carboxylic acids, alkanes, esters, and aldehydes [95]. Cellulose is a main source of considerable biochar product and the product of cellulose decomposition can vary depending on reaction conditions. The process of decomposition includes an exothermic and endothermic pathway. The exothermic pathway via anhydrocellulose yields char and non-condensable gases, while the endothermic pathway via levoglucosan is a devolatization process leading to tarry vapours or char. Mohan et al. [96] described the extent of secondary reactions affected by temperature, residence time, and catalysts occurring in feedstock. They yield a wide range of organic compounds such as carboxylic acids, ketones, aldehydes, and alcohols. The pyrolysis products of cellulose are mainly in the form of solid C, CO_2, and H_2O [97].

Many scientific studies focus on the utilisation of biochar and pyrolysis carbonaceous materials as effective sorbents in processes of organic and inorganic pollutant removal from aqueous solutions (Table 1). Several authors have confirmed different effects of physical and chemical properties on sorption efficiency. Therefore, the major material properties responsible for effective removal of various pollutants from wastewater, liquid wastes, and aqueous solutions are reviewed below.

Table 1. Removal of inorganic and organic pollutants from aqueous solutions by biochar.

Pollutant	Feedstock	Pyrolysis Temperature (°C)	Initial Concentration (mg/L)	Removal Efficiency (%)	Reference
copper	macroalga	500	10	80	[98]
cadmium	rape straw	600	20	100	[99]
lead	celery	500	400	98	[99]
zinc	wheat straw	650–700	100	100	[100]
chromium	rice husk	450–500	100	100	[101]
arsenic	sewage sludge	300	0.09	53	[102]
nickel	sewage sludge	600	30	26	[103]
tetracycline	sewage sludge	800	200	60	[104]
methylene blue	mangosteen peel	800	50	80	[105]
ibuprofen	wood chips	550	2	6	[106]
atrazine	cornstraw	500	30	100	[107]
sulfamethazine	cucumber	700	50	95	[108]

The area of internal and external surfaces, commonly called specific surface area (SSA), represents a crucial physical characteristic of pyrogenic carbonaceous material. Its value is the result of feedstock quality and production conditions (i.e., pyrolysis temperature, residence time, presence of catalysts, pre-treatment of feedstock or post-treatment of biochar). SSA determination for carbon-based sorbents involves mainly N_2 or CO_2 adsorptions followed by the Brunauer–Emmett–Teller (BET) equation application.

The method of SSA determination by liquid N_2 adsorption on biochar surface at a lower temperature (77 K) indicates the total specific surface area [109]. The method of CO_2 adsorption at a relatively high temperature (273 K) reflects the surface area of pores < 1.5 nm [110]. Micháleková-Richveisová et al. [91] and Dieguez-Alonso et al. [110] discussed the effect of pyrolysis temperature on porous structure formation and thus surface area development. Zama et al. [111] produced biochar from *Morus alba* at temperatures of 350 °C and 550 °C and confirmed more than 3.5 times higher SSA at the higher temperature. The effect of feedstock on SSA of biochar is also worth addressing. Jiang et al. [112] discussed the differences in surface areas of biochars prepared from pine wood and jarrah wood. Micháleková-Richveisová et al. [91] compared three biochars from different woody and agricultural wastes and found SSA values in the range of 16–26 m^2/g for the biochars with particle sizes of 0.5–1 mm. The effect of biochar post-treatment and feedstock pre-treatment can also significantly change SSA values of the resulting products [90].

Biochar pH ranges from 5 to 11 based on the feedstock and pyrolysis temperature [97]. There is a linear correlation between pH and pyrolysis temperature. Biochar produced at temperatures between 300 and 400 °C has a lower pH compared with biochar produced from the same feedstock at temperatures of 400–700 °C. Zhang et al. [29] explained the more alkaline pH of biochars produced at higher temperatures as a result of volatile organic compound losses and increases in basic cations. The lower pH values can be attributed to insufficient pyrolysis of material, biomass residues, and thus content with higher concentrations of phenolic and carboxylic functional groups. Elemental composition can be responsible for hydrophobicity and polarity of biochar, which can affect the contact of the material with the aqueous solution. The O/C molar ratio is used to estimate the hydrophilic nature of biochar and denote polar group concentrations in pyrogenic materials derived from carbohydrates in feedstock. The molar ratio of H/C can describe the degree of biochar carbonisation. Low values of H/C indicate a lower amount of non-carbonized biomass and thus a higher degree of carbonization [113].

The increased total content of elements such as K, Fe, Ca, Mg, and Mn can be attributed to higher concentrations of responding minerals, oxides, or oxohydroxides, which represent potential active sorption sites for pollutants removal [114]. Additionally, concentrations of surface functional groups such as carboxylic, hydroxyl, amino, phenolic, and lactic can be crucial for biochar assessment as effective sorption material. During pyrolysis under increasing temperature, most functional groups of feedstocks are lost. Kloss et al. [113] compared FT-IR spectra of biochars produced at different temperatures and found a decreased intensity of peaks corresponding to carboxylic and hydroxyl groups in samples produced at a higher temperature. Li et al. [115] applied ^{13}C NMR to investigate functional groups of rice straw and bran derived biochars produced at 100–800 °C. The authors confirmed the decreased content of aliphatic O-alkylated carbons and dominant aromatic structures in materials produced at temperatures > 300 °C.

Based on the type of contaminants, different mechanisms of sorption interactions can be proposed [116,117]. Sorption mechanisms of metals by biochar are fully dependent on the pH of the reaction solution, DOC, and competing elements concentrations [30]. Generally, complexation, cation exchange, precipitation, electrostatic interactions, and chemical reduction represent the main metal sorption mechanisms. However, the role of each mechanism is variable for each metal. Li et al. [115] described the target mechanisms of As, Cr, Pb, Cd, and Hg in detail. On the other hand, sorption of ionisable and ionic organic compounds such as pharmaceuticals and pesticides on biochar-based sorbents is a more complex process and the relative contribution of each sub-mechanism is dependent on the type of sorbate, pH, and sorbent properties [116]. These authors suggested and characterized the main active processes in sorption of organic molecules from liquid media as nonspecific van der Waals force (exactly London dispersion force), solvophobic effect, H-bond, charge assisted H-bond, electron donor-acceptor interactions, electrostatic (coulombic) interactions, ligand exchange, Lewis acid–base reactions, covalent bond, and oxidative coupling.

6. Biochar for Microcystin Sorption Separation

The study of biochar application in the removal of various organic structures has illustrated the potential of this pyrogenic material in separation processes. The biochar boom in environmental technologies and agronomy raised the question of potential risk evaluation for MCs interaction with this pyrogenic carbonaceous material present in sediment/water environments [74]. While at the same time, Zhang et al. [29], Li et al. [27], and Liu et al. [28] revealed the efficiency of modified charcoal and biochar utilization as potential sorbents of MC-LR from aqueous solutions. In each case, existing knowledge gaps on sorption behaviour of cyanotoxins in contact with biochar limit widespread usage of this engineered waste-derived sorbent in water purification processes. Sorbents derived from biochar exhibit different structure and properties compared with activated carbon. Incomplete carbonization at low temperatures in the range of 300–700 °C and the absence of activation processes can bring new challenges in the characterization and subsequent application of biochar sorbents. The effect of feedstock and mineral composition can be crucial for sorption properties of pyrogenic carbonaceous material [27].

The mechanisms of sorption interactions between microcystin and biochar can be characterized by several potential processes such as electrostatic and hydrogen bonding, ion exchange, complexation, mesopore filling, hydrophobic interactions, and interactions with aromatic carbon (such as $\pi \pm \pi$ EDA and $\pi - \pi$ stacking) (Figure 3). $\Pi \pm \pi$ EDA represents $\pi \pm \pi$ electron donor–acceptor interactions between protonated guanidine groups of MCs and π-electron rich graphene surfaces of biochar [28]. $\Pi - \pi$ stacking is the main interaction between the benzene ring of MCs and aromatic units on biochar surfaces [74]. As mentioned previously, molecules of MCs have a more hydrophilic character with reactive carboxyl or guanidine functional groups. Therefore, the dominant mechanism of MCs sorption can be often attributed to hydrophobic interactions [74]. Liu et al. [28] reported that the hydrophobic leucine unit in a MC-LR molecule tends to be attracted to hydrophobic surfaces of giant reed-derived biochar via a hydrophobic interaction. The authors showed that the positive correlation between pyrolysis temperature and hydrophobic interactions, thus hydrophobicity, is a driving force in the sorption of MC-LR. Teixidó et al. [118] described sorption removal of hydrophilic sulfamethazine by biochar with electron donor–acceptor interaction between protonated aniline ring and on π-electrons rich carbon surface as the predominant driving force. However, this kind of interaction is not a major force during MCs interactions with biochar-based sorbents. Isolated π electron systems in the Adda side chain represent small fractions of the hydrophilic macromolecule.

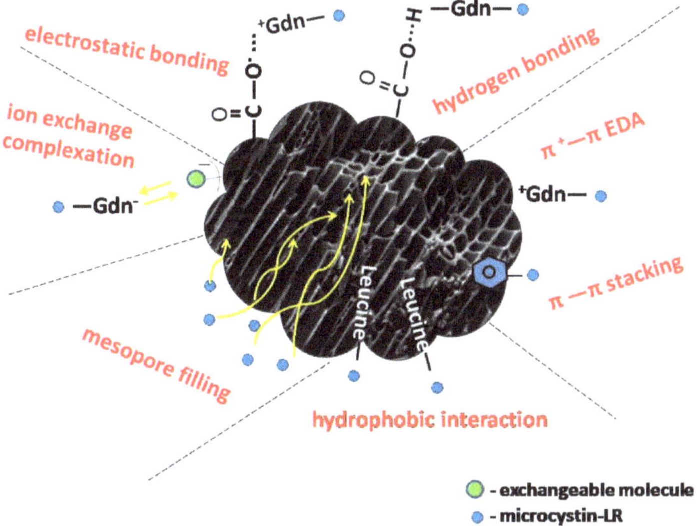

Figure 3. Suggested model of MC-LR sorption interactions with biochar-based sorbents.

On the other hand, Li et al. [74] showed how chemisorptions are the main force responsible for sorption of MC-LR molecules on wood-based biochar produced at 300 °C. Additionally, the authors confirmed physisorption and hydrogen bonding as the primary forces in sorption of MC-LR by the same biochar produced at 700 °C. One adsorbed MC-LR molecule can displace a considerable number of water molecules at the solid/liquid interface [119]. Thermodynamic calculations by Li et al. [74] demonstrated the ability of the MC-LR molecule to replace more water molecules on the biochar produced at lower temperatures. Li et al. [27] reported $\pi \pm \pi$ EDA interactions as the major driving forces of MC-LR sorption by biochars produced at 600 °C compared with biochars produced from the same feedstocks at 300 °C. Biochar produced at higher temperatures consists of highly aromatic structures providing larger π-electron-rich graphene sheets as electron donors.

The effect of porosity, as mentioned in Section 3, is crucial for sorption efficiency. According to Li et al. [27] and Zhang et al. [29], the three-dimensional size of the MC-LR molecule was determined to be 1.9 nm × 1.5 nm × 1.1 nm, with a maximum length of 2.94 nm and width of 2.55 nm. Therefore, micropores represent inaccessible spaces for MC-LR molecules [27]. Biochars with lower surface area and lower concentration of mesopores can be less effective in sorption removal of MC-LR [74]. Liu et al. [28] showed that giant reed derived biochars produced at higher temperatures (>500 °C) have predominantly mesoporous structures and are thus more suited and effective in the removal of MC-LR from aqueous solutions.

The pH values of biochar and surface point of zero charge are further driving forces in the sorption mechanism of MCs. Biochars with lower pH, or under lower pH environmental conditions, can still contain minor surfaces with positive charges, thus columbic attraction between anionic microcystin molecules and sorbent can still be considered. The pH of the aquatic environment can severely influence the sorption efficiency of MCs. Generally speaking, sorption of these microbial toxins decreases with increasing pH of the environment. Sorption processes of MCs are inhibited by electrostatic repulsion between their mono oxoanionic species induced by increasing pH (2–4) and higher negative charge of biochar surfaces at pH 3–8 [28].

Additionally, molecules of MCs tend to coil under more acidic conditions, and thus sorption can be more effective. Higher pH provides higher concentrations of hydrophilic microcystin molecules and thus a higher tendency of MCs to remain in the aqueous environment rather than to interact with hydrophobic biochar-based sorbents. Wu et al. [120] confirmed this trend in MC-LR sorption by natural sediments, and Sathishkumar et al. [85] for MC-LR sorption by peats-derived sorbents. Ionisable functional groups of MCs and biochar surfaces can be easily protonated or deprotonated at different pH values. According to Li et al. [74], adsorption of MC-LR on biochars is attributed to the binding of MC-LR- anions to biochars with pH-dependent charges in aquatic systems. Liu et al. [28] found the highest sorption efficiency of MC-LR removal by giant reed-derived biochars at pH 3.5.

The sorption separation of MCs by biochar can also be influenced by competitive substances such as phosphates, citric acid, oxalic acid, and tannic acid, but with negligible effects from gallic acid, nitrates, and chlorides [74]. These authors suggested anion exchange as a possible contributing factor to MC-LR adsorption. Liu et al. [28] investigated the contribution of mineral composition to the overall sorption of MC-LR by giant reed-derived biochar. Their reported results revealed that mineral removal by deashing treatment barely impacts the adsorption of MC-LR by biochars produced at temperatures < 400 °C, whereas it significantly reduces sorption of MCs by biochars produced at temperatures > 400 °C.

Li et al. [74] showed adsorption irreversibility of MC-LR sorbed to wood-derived biochar and increased accumulation of MC-LR in biochar-amended sediments. The authors confirmed the associated risk of the undetermined prolonged fate of MC-LR in sediments and thus in aquatic environments. However, the benefits of high porosity, chemical stability, and significant sorption efficiency for removal processes of cyanotoxins can help guide the optimization of pyrolysis processes and the creation of a new class of so-called engineered or designed sorbents for water purification.

7. Conclusions

Elevated cyanotoxin concentrations in freshwater systems and drinking water sources following toxic cyanobacterial blooms require remediation to reduce risks to human and ecosystem health. Conventional methods for microcystin removal are insufficient or, as in the case of activated carbon, too expensive for widespread adoption. Alternative methods exist and have shown interesting results, but require continued research efforts. This review illustrates the potential of the pyrolysis process to inexpensively transform organic materials into efficient sorbents. In particular, biochar has proven effective as a sorbent for MCs in water thanks to its high porosity, chemical stability, and notable sorption efficiency. The physiochemical properties of a specifically engineered biochar for MCs remediation can be optimized through manipulating feedstock and pyrolysis conditions. Future research and product development for MCs sorption separation from contaminated aquatic environments can consider pyrolysed materials as attractive candidates. This work supports the continued exploration of inexpensive and feasible water remediation sorbents produced through pyrolysis, but also demonstrates the need for further studies evaluating their effectiveness in the field under diverse conditions following harmful algal blooms worldwide.

Author Contributions: Conceptualization, V.F., H.D.L.IV, and S.M.B.; methodology, V.F. and H.D.L.IV; formal analysis, V.F. and S.M.B.; investigation, V.F., H.D.L.IV, and S.M.B.; writing—original draft preparation, V.F. and S.M.B.; writing—review and editing, V.F., H.D.L.IV, and S.M.B.; visualization, V.F.; supervision, V.F. and H.D.L.IV; project administration, V.F.; funding acquisition, V.F. All authors have read and agreed to the published version of the manuscript.

Funding: This research was funded by Scientific Grant Agency of the Ministry of Education, Science, Research, and Sport of the Slovak Republic, project number VEGA1/0178/20; and Trnava University in Trnava, project number 2/TU/2019 and 4/TU/2020.

Acknowledgments: Authors are thankful to David Erwin Berthold for technical support.

Conflicts of Interest: The authors declare no conflict of interest. The funders had no role in the design of the study; in the collection, analyses, or interpretation of data; in the writing of the manuscript; or in the decision to publish the results.

References

1. Griffith, A.W.; Gobler, C.J. Harmful algal blooms: A climate change co-stressor in marine and freshwater ecosystems. *Harmful Algae* **2020**, *91*, 101590. [CrossRef]
2. Hoagland, P.; Anderson, D.M.; Kaoru, Y.; White, A.W. The economic effects of harmful algal blooms in the United States: Estimates, assessment issues, and information needs. *Estuaries* **2002**, *25*, 819–837. [CrossRef]
3. Harke, M.J.; Steffen, M.M.; Gobler, C.J.; Otten, T.G.; Wilhelm, S.W.; Wood, S.A.; Paerl, H.W. A review of the global ecology, genomics, and biogeography of the toxic cyanobacterium, *Microcystis* spp. *Harmful Algae* **2016**, *54*, 4–20. [CrossRef]
4. Wood, S.A.; Kelly, L.; Bouma-Gregson, K.; Humbert, J.F.; Laughinghouse IV, H.D.; Lazorchak, J.; McAllister, T.; McQueen, A.; Pokrzywinski, K.; Puddick, J.; et al. Toxic benthic freshwater cyanobacterial proliferations: Challenges and solutions for enhancing knowledge and improving monitoring and mitigation. *Freshw. Biol.* **2020**, *65*, 1824–1842. [CrossRef]
5. Codd, G.A.; Morrison, L.F.; Metcalf, J.S. Cyanobacterial toxins: Risk management for health protection. *Toxicol. Appl. Pharmacol.* **2005**, *203*, 264–272. [CrossRef] [PubMed]
6. Dodds, W.K.; Bouska, W.W.; Eitzmann, J.L.; Pilger, T.J.; Pitts, K.L.; Riley, A.J.; Schloesser, J.T.; Thornbrugh, D.J. Eutrophication of U. S. freshwaters: Analysis of potential economic damages. *Environ. Sci. Technol.* **2009**, *43*, 12–19. [CrossRef] [PubMed]
7. Paerl, H.W.; Huisman, J. Climate change: A catalyst for global expansion of harmful cyanobacterial blooms. *Environ. Microbiol. Rep.* **2009**, *1*, 27–37. [CrossRef] [PubMed]
8. Almanza, V.; Pedreros, P.; Laughinghouse IV, H.D.; Félez, J.; Parra, O.; Azocar, M.; Urrutia, R. Association between trophic state, watershed use and blooms of cyanobacteria in south-central Chile. *Limnologica* **2019**, *75*, 30–41. [CrossRef]
9. Huisman, J.; Codd, G.A.; Paerl, H.W.; Ibelings, B.W.; Verspagen, J.M.H.; Visser, P.M. Cyanobacterial blooms. *Nat. Rev. Microbiol.* **2018**, *16*, 471–483. [CrossRef]
10. Kust, A.; Řeháková, K.; Vrba, J.; Maicher, V.; Mareš, J.; Hrouzek, P.; Chiriac, M.C.; Benedová, Z.; Tesařová, B.; Saurav, K. Insight into Unprecedented Diversity of Cyanopeptides in Eutrophic Ponds Using an MS/MS Networking Approach. *Toxins* **2020**, *12*, 561. [CrossRef]
11. Carvalho, L.R.; Pipole, F.; Werner, V.R.; Laughinghouse IV, H.D.; Rangel, M.; Konno, K.; Camargo, A.C.M.; Sant'Anna, C.L. A Toxic Cyanobacterial Bloom in an Urban Coastal Lake, Rio Grande do Sul, Southern Brazil. *Braz. J. Microbiol.* **2008**, *39*, 761–769. [CrossRef] [PubMed]
12. Burford, M.A.; Carey, C.C.; Hamilton, D.P.; Huisman, J.; Paerl, H.W.; Wood, S.A.; Wulff, A. Perspective: Advancing the research agenda for improving understanding of cyanobacteria in a future of global change. *Harmful Algae* **2020**, *91*, 101601. [CrossRef]
13. Wells, M.L.; Karlson, B.; Wulff, A.; Kudela, R.; Trick, C.; Asnaghi, V.; Berdalet, E.; Cochlan, W.; Davidson, K.; De Rijcke, M.; et al. Future HAB science: Directions and challenges in a changing climate. *Harmful Algae* **2020**, *91*, 101632. [CrossRef] [PubMed]

14. Paerl, H.W.; Gardner, W.S.; Havens, K.E.; Joyner, A.J.; McCarthy, M.J.; Newell, S.E.; Qin, B.; Scott, J.D. Mitigating cyanobacterial harmful algal blooms in aquatic ecosystems impacted by climate change and anthropogenic nutrients. *Harmful Algae* **2016**, *54*, 213–222. [CrossRef]
15. Dawson, R.M. The toxicology of microcystins. *Toxicon* **1998**, *36*, 953–962. [CrossRef]
16. Shahmohamadloo, R.S.; Poirier, D.G.; Ortiz Almirall, X.; Bhavsar, S.P.; Sibley, P.K. Assessing the toxicity of cell-bound microcystins on freshwater pelagic and benthic invertebrates. *Ecotoxicol. Environ. Saf.* **2020**, *188*, 109945. [CrossRef]
17. IARC. *Ingested Nitrate and Nitrite, and Cyanobacterial Peptide Toxins/IARC Monographs on the Evaluation of Carcinogenic Risks to Humans Volume 94*; IARC: Lyon, France, 2010.
18. WHO. *Guidelines for Drinking-Water Quality: Fourth Edition Incorporating the First Addendum*; WHO: Geneva, Switzerland, 2017.
19. Park, J.A.; Jung, S.M.; Choi, J.W.; Kim, J.H.; Hong, S.; Lee, S.H. Mesoporous carbon for efficient removal of microcystin-LR in drinking water sources, Nak-Dong River, South Korea: Application to a field-scale drinking water treatment plant. *Chemosphere* **2018**, *193*, 883–891. [CrossRef]
20. Ho, L.; Lambling, P.; Bustamante, H.; Duker, P.; Newcombe, G. Application of powdered activated carbon for adsorption of cylindrospermopsin and microcystin toxins from drinking water supplies. *Water Res.* **2011**, *45*, 2954–2964. [CrossRef]
21. Falconer, I.R.; Runnegar, M.T.C.; Buckley, T.; Huyn, L.; Bradshaw, P. Using activated carbon to remove toxicity from drinking water containing cyanobacterial blooms. *J. Am. Water Work. Assoc.* **1989**, *81*, 102–106. [CrossRef]
22. Mashile, P.P.; Mpupa, A.; Nomngongo, P.N. Adsorptive removal of microcystin-LR from surface and wastewater using tyre-based powdered activated carbon: Kinetics and isotherms. *Toxicon* **2018**, *145*, 25–31. [CrossRef]
23. Roegner, F.A.; Brena, B.; González-Sapienza, G.; Puschner, B. Microcystins in potable surface waters: Toxic effects and removal strategies. *J. Appl. Toxicol.* **2014**, *34*, 441–457. [CrossRef]
24. Basu, P. *Pyrolysis, in: Biomass Gasification, Pyrolysis and Torrefaction: Practical Design and Theory*; Elsevier: Amsterdam, The Netherlands, 2018; pp. 155–187.
25. Wang, J.; Wang, S. Preparation, modification and environmental application of biochar: A review. *J. Clean. Prod.* **2019**, *227*, 1002–1022. [CrossRef]
26. Ahmed, M.J.; Hameed, B.H. Insight into the co-pyrolysis of different blended feedstocks to biochar for the adsorption of organic and inorganic pollutants: A review. *J. Clean. Prod.* **2020**, *265*, 121762. [CrossRef]
27. Li, J.; Cao, L.; Yuan, Y.; Wang, R.; Wen, Y.; Man, J. Comparative studz for microcystin-LR sorption onto biochars produced from various plant-and animal-wastes at different pyrolysis temperatures: Influencing mechanisms of biochar properties. *Bioresour. Technol.* **2018**, *247*, 794–803. [CrossRef] [PubMed]
28. Liu, G.; Zheng, H.; Zhai, X.; Wang, Z. Characteristics and mechanisms of microcystin/LR adsorption by giant reed-derived biochars: Role of minerals, pores and functional groups. *J. Clean. Prod.* **2018**, *176*, 463–473. [CrossRef]
29. Zhang, H.; Zhu, G.; Jia, X.; Ding, Y.; Zhang, M.; Gao, Q.; Hu, C.; Xu, S. Removal of microcystin-LR from drinking water using a bamboo-based charcoal adsorbents modified with chitosan. *J. Environ. Sci.* **2011**, *23*, 1983–1988. [CrossRef]
30. Li, J.M.; Li, R.H.; Li, J. Current research scenario for microcystins biodegradation—A review on fundamental knowledge, application prospects and challenges. *Sci. Total Environ.* **2017**, *595*, 615–632. [CrossRef]
31. Spoof, L.; Catherine, A. Tables of microcystins and nodularins, Appendix 3. In *Handbook of Cyanobacterial Monitoring and Cyanotoxin Analysis*, 1st ed.; Meriluoto, J., Spoof, L., Cold, G.A., Eds.; John Wiley and Sons, Ltd.: Hoboken, NJ, USA, 2017.
32. Turner, A.D.; Dhanji-Rapkova, M.; OŃeill, A.; Coates, L.; Lewis, A.; Lewis, K. Analysis of microcystins in cyanobacterial blooms from freshwater bodies in England. *Toxins* **2018**, *10*, 39. [CrossRef] [PubMed]
33. Merel, S.; Clement, M.; Thomas, O. State of the art on cyanotoxins in water and their behaviour towards chlorine. *Toxicon* **2010**, *55*, 677–691. [CrossRef]
34. Trinchet, I.; Cadel-Six, S.; Djediat, C.; Marie, B.; Bernard, C.; Puiseux-Dao, S.; Krys, S.; Edery, M. Toxicity of harmful cyanobacterial blooms to bream and roach. *Toxicon* **2013**, *71*, 121–127. [CrossRef]

35. Humpage, R.; Hardy, S.J.; Moore, E.J.; Froscio, S.M.; Falconer, I.R. Microcystins (cyanobacterial toxins) in drinking water enhance the growth of aberrant cryptfoci in the mouse colon. *J. Toxicol. Environ. Health A* **2000**, *61*, 155–165. [PubMed]
36. Tillett, D.; Dittmann, E.; Erhard, M.; von Dohren, H.; Borner, T.; Neilan, B.A. Structural organization of microcystin biosynthesis in Microcystis aeruginosa PCC7806: An integrated peptide-polyketide synthetase systém. *Chem. Biol.* **2000**, *7*, 753–764. [CrossRef]
37. Mazur-Marzec, H.; Plinski, M. Do toxic cyanobacteria blooms pose a threat to the Baltic ecosystem? *Oceanologia* **2009**, *51*, 293–319. [CrossRef]
38. Jones, G.J.; Falconer, I.R.; Wilkins, R.M. Persistence of cyclic peptide toxins in dried Microcystis aeruginosa crusts from lake Mokoan, Australia. *Environ. Toxicol. Water Qual.* **1995**, *10*, 19–24. [CrossRef]
39. Schmidt, J.R.; Wilhelm, S.W.; Boyer, G.L. The Fate of Microcystins in the Environment and Challenges for Monitoring. *Toxins* **2014**, *6*, 3354–3387. [CrossRef]
40. Boopathi, T.; Ki, J.S. Impact of environmental factors on the regulation of cyanotoxin production. *Toxins* **2014**, *6*, 1951–1978. [CrossRef]
41. Žegura, B.; Štraser, A.; Filipič, M. Genotoxicity and potential carcinogenicity of cyanobacterial toxins–A review. *Mutat. Res. Rev. Mutat.* **2011**, *727*, 16–41. [CrossRef]
42. Fischer, A.; Hoeger, S.J.; Stemmer, K.; Feurstein, D.J.; Knobeloch, D.; Nussler, A.; Dietrich, D.R. The role of organic anion transporting polypeptides (OATPs/SLCOs) in the toxicity of different microcystin congeners in vitro: A comparison of primary human hepatocytes and OATP-transfected HEK293 cells. *Toxicol. Appl. Pharmacol.* **2010**, *245*, 9–20. [CrossRef]
43. Frangez, R.; Zuzek, M.C.; Mrkun, J.; Suput, D.; Sedmak, B.; Kosec, M. Microcystin-LR affects cytoskeleton and morphology of rabbit primary whole embryo cultured cells in vitro. *Toxicon* **2003**, *41*, 999–1005. [CrossRef]
44. Davis, M.A.; Chang, S.H.; Trump, B.F. Differential sensitivity of normal and H-ras Oncogene-transformed rat kidney epithelial cells to okadaic acid-induced apoptosis. *Toxicol. Appl. Pharmacol.* **1996**, *141*, 93–101. [CrossRef]
45. Ding, W.X.; Nam-Ong, C. Role of oxidative stress and mitochondrial changes in cyanobacteria-induced apoptosis and hepatotoxicity. *FEMS Microbiol. Lett.* **2003**, *220*, 1–7. [CrossRef]
46. Žegura, B.; Volcic, M.; Lah, T.T.; Filipic, M. Different sensitivities of human colon adenocarcinoma (CaCo-2), astrocytoma (IPDDC-A2) and lymphoblastoid (NCNC) cell lines to microcystin-LR induced reactive oxygen species and DNA damage. *Toxicon* **2008**, *52*, 518–525. [CrossRef] [PubMed]
47. Westrick, J.A.; Szlag, D.C.; Southwell, B.J.; Sinclair, J. A review of cyanobacteria and cyanotoxins removal/inactivation in drinking water treatment. *Anal. Bioanal. Chem.* **2010**, *397*, 1705–1714. [CrossRef] [PubMed]
48. Teixeira, M.R.; Rosa, M.J.; Sorlini, S.; Biasibetti, M.; Christophoridis, C. Removal of Cyanobacteria and Cyanotoxins by Conventional Physical-chemical Treatment. In *Water Treatment for Purification from Cyanobacteria and Cyanotoxins*; Hiskia, A.E., Triantis, T.M., Antoniou, M.G., Kaloudis, T., Dionysiou, D.D., Eds.; Wiley: Hoboken, NJ, USA, 2020; pp. 69–98.
49. Feitz, A.J.; Waite, T.D.; Jones, G.J.; Boyden, B.H.; Orr, P.T. Photocatalytic degradation of the blue green algal toxin microcystin-LR in a natural organic-aqueous matrix. *Environ. Sci. Technol.* **1999**, *33*, 243–249. [CrossRef]
50. Gurbuz, F.; Codd, G.A. Microcystin removal by a natural-occuring substance: Pumice. *Bull. Environ. Contam. Toxicol.* **2008**, *81*, 323–327. [CrossRef] [PubMed]
51. Teixeira, M.R.; Rosa, M.J. Comparing dissolved air flotation and conventional sedimentation to remove cyanobacterial cells of Microcystis aeruginosa Part II. The effect of water background organics. *Sep. Purif. Technol.* **2007**, *53*, 126–134. [CrossRef]
52. Crossley, I.A.; Valade, M.T. A review of the technological developments of dissolved air flotation. *J. Water Supply Res. Technol.* **2006**, *55*, 479–491. [CrossRef]
53. Laszakovits, J.; MacKay, A.A. Removal of cyanotoxins by potassium permanganate: Incorporating competition from natural water constituents. *Water Res.* **2019**, *155*, 86–95. [CrossRef]
54. Kinley-Baird, C.; Calomeni, A.; Berthold, D.E.; Lefler, F.W.; Barbosa, M.; Rodgers, J.H.; Laughinghouse, H.D., IV. Laboratory-scale evaluation of effectiveness for control of microcystin producing cyanobacteria from Lake Okeechobee, Florida (USA). *Ecotoxicol. Environ. Saf.* **2021**, *207*, 111233. [CrossRef]

55. Kinley, C.M.; Iwinski-Wood, K.J.; Geer, T.D.; Hendrikse, M.; McQueen, A.D.; Calomeni, A.J.; Liang, J.; Friesen, V.; Simair, M.C.; Rodgers, J.H. Microcystin-LR degradation following copper-based algaecide exposures. *Water Air Soil Pollut.* **2018**, *229*, 62. [CrossRef]
56. Jones, G.J.; Orr, P.T. Release and degradation of microcystin following algicide treatment of a *Microcystis aeruginosa* bloom in a recreational lake, as determined by HPLC and protein phosphatase inhibition assay. *Water Res.* **1994**, *28*, 871–876. [CrossRef]
57. Brooke, S.; Newcombe, G.; Nicholson, B.; Klass, G. Decrease in toxicity of microcystins LA and LR in drinking water by ozonation. *Toxicon* **2006**, *48*, 1054–1059. [CrossRef] [PubMed]
58. Ho, L.; Sawade, E.; Newcombe, G. Biological treatment options for cyanobacteria metabolite removal—A review. *Water Res.* **2012**, *46*, 1536–1548. [CrossRef] [PubMed]
59. Lawton, L.A.; Robertson, P.K.J. Physico-chemical treatment methods for the removal of microcystins (cyanobacterial hepatotoxins) from potable waters. *Chem. Soc. Rev.* **1999**, *28*, 217–224. [CrossRef]
60. Vlad, S.; Anderson, W.B.; Peldszus, S.; Huck, P.M. Removal of the cyanotoxin anatoxin-a by drinking water treatment processes: A review. *J. Water Health* **2014**, *12*, 601–617. [CrossRef] [PubMed]
61. Pavagadhi, S.; Tang, A.L.L.; Sathishkumar, M.; Loh, K.P.; Balasubramanian, R. Removal of microcystin-LR and microcystin-RR by graphene oxide: Adsorption and kinetic experiments. *Water Res.* **2013**, *47*, 4621–4629. [CrossRef] [PubMed]
62. Wang, X.L.; Xing, B.S. Sorption of organic contaminants by biopolymer-derived chars. *Environ. Sci. Technol.* **2007**, *41*, 8342–8348. [CrossRef]
63. Hena, S.; Ismail, N.; Isaam, A.M.; Ahmad, A.; Bhawani, S.A. Removal of microcystin-LR from agueous solutions using % burn-off activated carbon of waste wood material. *J. Water Supply Res. Technol.* **2014**, *63*, 332–341. [CrossRef]
64. Pendleton, P.; Schumann, R.; Wong, S.H. Microcystin adsorption by activated carbon. *J. Collod. Interf. Sci.* **2001**, *240*, 1–8. [CrossRef]
65. Teng, W.; Wu, Z.; Fan, J.; Zhang, W.X.; Zhao, D. Amino-functionalized ordered mesoporous carbon for the separation of toxic microcystin-LR. *J. Mat. Chem. A* **2015**, *3*, 19168. [CrossRef]
66. Newcombe, G.; Morrison, J.; Hepplewhite, C.; Knappe, D.R.U. Simultaneous adsorption of MIB and NOM onto activated carbon. II. Competitive effects. *Carbon* **2002**, *40*, 2147–2156. [CrossRef]
67. Ho, L.; Newcombe, G. Evaluating the adsorption of microcystin toxins using granular activated carbon (GAC). *J. Water Supply Res. Technol.* **2007**, *56*, 281–291. [CrossRef]
68. Hnatukova, P.; Kopecka, I.; Pivokonsky, M. Adsorption of cellular peptides of Microcystis aeruginosa and two herbicides onto activated carbon: Effect of surface charge and interactions. *Water Res.* **2011**, *45*, 3359–3368. [CrossRef] [PubMed]
69. Huang, W.J.; Cheng, B.L.; Cheng, Y.L. Adsorption of microcystin-LR by three types of activated carbon. *J. Hazard. Mater.* **2007**, *141*, 115–122. [CrossRef]
70. Campinas, M.; Rosa, M.J. Assessing PAC contribution to the NOM fouling control in PAC/UF systems. *Water Res.* **2010**, *44*, 1636–1644. [CrossRef] [PubMed]
71. Drogui, P.; Daghrir, R.; Simard, M.C.; Sauvageau, C.; Blais, J.F. Removal of microcystin-LR from spiked water using either activated carbon or anthracite as filter material. *Environ. Technol.* **2012**, *33*, 381–391. [CrossRef]
72. Zhi, M.; Yang, F.; Meng, F.; Li, M.; Manivannan, A.; Wu, N. Effects of pore structure on performance of an activated-carbon supercapacitor electrode recycled from scrap waste tires. *ACS Sustain. Chem. Eng.* **2014**, *2*, 1592–1598. [CrossRef]
73. Lou, F.; Zhou, H.; Tran, T.D.; Buan, M.E.M.; Vullum-Bruer, F.; Ronning, M.; Walmsley, J.C.; Chen, D. Coaxial carbon/metal oxide/aligned carbon nanotube arrays as high-performance anodes for lithium ion batteries. *ChemSusChem* **2014**, *7*, 1335–1346. [CrossRef]
74. Li, L.; Qiu, Y.; Huang, J.; Li, F.; Sheng, G.D. Mechanisms and factors influencing adsorption of microcystin-LR on biochars. *Water Air Soil Poll.* **2014**, *225*, 2220. [CrossRef]
75. Kim, S.; Yun, Y.S.; Choi, Y.E. Development of waste biomass based sorbent for removal of cyanotoxin microcystin-LR from aqueous phases. *Bioresource Technol.* **2018**, *247*, 690–696. [CrossRef]
76. Yasmin, R.; aftab, K.; Kashif, M. Removal of microcystin-LR from agueous solution using *Moringa oleifera* Lam. Seeds. *Water Sci. Technol.* **2019**, *79*, 104–113. [CrossRef] [PubMed]

77. Sathya, K.; Saranya, P.; Swarnalatha, S.; Mandal, A.; Sekaran, G. Removal of microcystin-RR, a membrane foulant using exocellular polymer from *Enterobacter ludwigii*: Kinetic and isotherm studies. *Desalination* **2015**, *369*, 175–187.
78. Park, J.H.; Wang, J.J.; Kim, S.H.; Kang, S.W.; Jeong, C.Y.; Jeon, J.R.; Park, H.P.; Cho, J.S.; Delaune, R.D.; Seo, D.C. Cadmium adsorption characteristics of biochars derived using various pine t tree residues and pyrolysis temperatures. *J. Colloid Interface Sci.* **2019**, *553*, 298–307. [CrossRef] [PubMed]
79. Frišták, V.; Pipíška, M.; Soja, G. Pyrolysis treatment of sewage sludge: A promising way to produce phosphorus fertilizer. *J. Clean. Prod.* **2018**, *172*, 1772–1778. [CrossRef]
80. Morris, R.J.; Williams, D.E.; Luu, H.A.; Holmes, C.F.; Andersen, R.J.; Calvert, S.E. The adsorption of microcystin-LR by natural clay particles. *Toxicon* **2000**, *38*, 303–308. [CrossRef]
81. Laughinghouse IV, H.D.; Lefler, F.W.; Berthold, D.E.; Bishop, W.M. Sorption of microcystin using lanthanum-modified bentonite clay. *J. Aquat. Plant Manag.* **2020**, *58*, 72–75.
82. Wang, Z.; Wang, C.; Wang, P.; Qian, J.; Hou, J.; Ao, Y. Process optimization for microcystin-LR adsorption onto nano-sized montmorillonite K10: Application of response surface methodology. *Water Air Soil Poll.* **2014**, *225*, 2124. [CrossRef]
83. Wang, Z.; Wang, C.; Wang, P.; Qian, J.; Hou, J.; Ao, Y. Response surface modelling and optimization of microcystin-LR remocal from agueous phase by polyacrylamide/sodium alginate-motmorillonite superabsorbent nanocomposite. *Desal. Eater Treat.* **2015**, *56*, 1121–1139. [CrossRef]
84. Akbal, F.O.; Akdemir, N.; Onar, A.N. FT-IR spectroscopic detection of pesticide after sorption onto modified pumice. *Talanta* **2000**, *53*, 131–135. [CrossRef]
85. Sathishkumar, M.; Pavagadhi, S.; Vijayaraghavan, K.; Balasubramanian, R.; Ong, S.L. Concomitant uptake of microcystin-LR and -RR by peat under various environmental conditions. *Chem. Eng. J.* **2011**, *172*, 754–762. [CrossRef]
86. Lee, J.; Walker, H.W. Adsorption of microczstin-LR onto iron oxide nanoparticles. *Colloid. Surface A* **2011**, *373*, 94–100. [CrossRef]
87. Okupnik, A.; Contardo-Jara, V.; Pflugmacher, S. Potential role of engineered nanoparticles as contaminant carriers in aquatic ecosystems: Estimating sorption processes of the cyanobacterial toxin microcystin-LR by TiO_2 nanoparticles. *Colloid. Surface A* **2015**, *481*, 460–467. [CrossRef]
88. Wu, L.; Lan, J.; Wang, S.; Zhu, J. Synthesis of N-doped carbon xerogel (N-CX) and its applications for adsorption removal of microcystin-LR. *Z. Phys. Chem.* **2017**, *231*, 1525–1541. [CrossRef]
89. EBC. *European Biochar Certificate-Guidelines for a Sustainable Production of Biochar*; EBC: Arbaz, Switzerland, 2015. [CrossRef]
90. Shaheen, S.M.; Niazi, N.K.; Hassan, N.E.E.; Bibi, I.; Wang, H.; Tsang, D.C.W.; Ok, Y.S.; Bolan, N.; Rinklebe, J. Wood-based biochar for the removal of potentially toxic elements in weter and wastewater: A critical review. *Int. Mater. Rev.* **2019**, *64*, 216–247. [CrossRef]
91. Micháleková-Richveisová, B.; Frišták, V.; Pipíška, M.; Duriška, L.; Moreno-Jimenez, E.; Soja, G. Iron-impregnated biochars as effective phosphate sorption materials. *Environ. Sci. Pollut. Res. Int.* **2017**, *24*, 463–475. [CrossRef]
92. Bridgwater, A.V. Review of fast pyrolysis of biomass and product upgrading. *Biomass Bioenergy* **2012**, *38*, 68–94. [CrossRef]
93. Kunhikrishnan, A.; Bibi, I.; Bolan, N.; Seshadri, B.; Choppala, G.; Niazi, N.K.; Kim, W.I.; Ok, Y.S.; Uchimiya, S.M.; Chang, S.; et al. Biochar for inorganic contaminant management in waste and wastewater. In *Biochar: Production, Characterization and Applications*; CRC Press: Boca Raton, FL, USA, 2015; pp. 167–219.
94. Brown, R.A.; Kercher, A.K.; Nguyen, T.H.; Nagle, D.C.; Ball, W.P. Production and characterization of synthetic wood chars for use as surrogates for natural sorbents. *Org. Geochem.* **2006**, *37*, 321–333. [CrossRef]
95. Rutherford, D.W.; Wershaw, R.L.; Cox, L.G. *Changes in Composition and Porosity Occurring during the Thermal Degradation of Wood and Wood Components*; US Geological Survey, Scientific Investigation Report 2004-5292; USGS: Reston, VA, USA, 2004.
96. Mohan, D.; Pittman, C.U.; Steele, P.H. Pyrolysis of wood/biomass for bio-oil: A critical review. *Energy Fuels* **2006**, *20*, 848–889. [CrossRef]
97. Lehmann, J.; Joseph, S. *Biochar for Environmental Management—Science, Technology and Implementation*, 2nd ed.; Routledge: London, UK; Sterling, VA, USA, 2015.
98. Park, S.H.; Cho, H.J.; Ryu, C.; Park, Y.K. Removal of copper (II) in aqueous solution using pyrolytic biochars derived from red macroalga *Porphyra tenera*. *J. Ind. Eng. Chem.* **2016**, *36*, 314–319. [CrossRef]

99. Zhang, R.H.; Li, Z.G.; Liu, X.D.; Wang, B.C.; Zhou, G.L.; Huang, X.X.; Lin, C.F.; Wang, A.H.; Brooks, M. Immobilization and bioavailability of heavy metals in greenhouse soils amended with rice straw-derived biochar. *Ecol. Eng.* **2017**, *98*, 183–188. [CrossRef]
100. Bogusz, A.; Oleszczuk, P.; Dobrowolski, R. Application of laboratory prepared and commercially available biochars to adsorption of cadmium, copper and zinc ions from water. *Bioresour. Technol.* **2015**, *196*, 540–549. [CrossRef] [PubMed]
101. Ma, Y.; Liu, W.J.; Zhang, N.; Li, Y.S.; Jiang, H.; Sheng, G.P. Polyethylenimine modified biochar adsorbent for hexavalent chromium removal from the aqueous solution. *Bioresour. Technol.* **2014**, *169*, 403–408. [CrossRef]
102. Agrafioti, E.; Kalderis, D.; Diamadopoulos, E. Ca and Fe modified biochars as adsorbents of arsenic and chromium in aqueous solutions. *J. Environ. Manag.* **2014**, *146*, 444–450. [CrossRef] [PubMed]
103. Inyang, M.; Gao, B.; Yao, Y.; Xue, Y.; Zimmerman, A.R.; Pullammanappallil, P.; Cao, X. Removal of heavy metals from aqueous solution by biochars derived from anaerobically digested biomass. *Bioresour. Technol.* **2012**, *110*, 50–56. [CrossRef]
104. Tang, L.; Yu, J.; Pang, Y.; Zeng, G.; Deng, Y.; Wang, J.; Ren, X.; Ye, S.; Peng, B.; Feng, H. Sustainable efficient adsorbent: Alkali-acid modified magnetic biochar derived from sewage sludge for aqueous organic contaminant removal. *Chem. Eng. J.* **2018**, *336*, 160–169. [CrossRef]
105. Ruthiraan, M.; Abdullah, E.C.; Mubarak, N.M.; Noraini, M.N. A promising route of magnetic based materials for removal of cadmium and methylene blue from waste water. *J. Environ. Chem. Eng.* **2017**, *5*, 1447–1455. [CrossRef]
106. Lin, L.; Jiang, W.; Xu, P. Comparative study on pharmaceuticals adsorption in reclaimed water desalination concentrate using biochar: Impact of salts and organic matter. *Sci. Total Environ.* **2017**, *601*, 857–864. [CrossRef]
107. Tan, X.; Liu, Y.; Gu, Y.; Xu, Y.; Zeng, G.M.; Hu, X.J.; Liu, S.B.; Wang, X.; Liu, S.M.; Li, J. Biochar-based nano-composites for the decontamination of wastewater: A review. *Bioresour. Technol.* **2016**, *212*, 318–333. [CrossRef] [PubMed]
108. Rajapaksha, A.U.; Chen, S.S.; Tsang, D.C.; Zhang, M.; Vithanage, M.; Mandal, S.; Gao, B.; Bolan, N.S.; Ok, Y.S. Engineered/designer biochar for contaminant removal/immobilization from soil and water: Potential and implication of biochar modification. *Chemosphere* **2016**, *148*, 276–291. [CrossRef]
109. Igalavithana, A.D.; Lee, S.E.; Lee, Y.H.; Tsang, D.C.W.; Rinklebe, J.; Kwon, E.E.; Ok, Y.S. Heavy metal immobilization and microbial community abundance by vegetable waste and pine cone biochar of agricultural soils. *Chemosphere* **2017**, *174*, 593–603. [CrossRef]
110. Dieguez-Alonso, A.; Anca-Couce, A.; Frišták, V.; Moreno-Jiménez, E.; Bacher, M.; Bucheli, T.D.; Cimò, G.; Conte, P.; Hagemann, N.; Haller, A.; et al. Designing biochar properties through the blending of biomass feedstock with metals: Impact on oxyanions adsorption behavior. *Chemosphere* **2019**, *214*, 743–753. [CrossRef] [PubMed]
111. Zama, E.F.; Zhu, Y.G.; Reid, B.J.; Sun, G.X. The role of biocharproperties in influencing the sorption and desorption of Pb(II), Cd(II) and As(III) in aqueous solution. *J. Clean Prod.* **2017**, *148*, 127–136. [CrossRef]
112. Jiang, J.; Pang, S.Y.; Ma, J. Comment on Adsorption of hydroxyl-and amino-substituted aromatics on carbon nanotubes. *Environ. Sci. Technol.* **2009**, *43*, 2–3. [CrossRef] [PubMed]
113. Kloss, S.; Zehetner, F.; Oburger, E.; Buecker, J.; Kitzler, B.; Wenzel, W.W.; Wimmer, B.; Soja, G. Trace element concentrations in leachates and mustard plant tissue (*Sinapis alba* L.) after biochar application to temperate soils. *Sci. Total Environ.* **2014**, *481*, 498–508. [CrossRef] [PubMed]
114. Frišták, V.; Micháleková-Richveisová, B.; Víglašová, E.; Ďuriška, L.; Galamboš, M.; Moreno-Jimenéz, E.; Pipíška, M.; Soja, G. Sorption separation of Eu and As from single-component systems by Fe-modified biochar: Kinetic and equilibrium study. *J. Iran. Chem. Soc.* **2017**, *14*, 521–530. [CrossRef]
115. Li, X.; Pignatello, J.J.; Wang, Y.; Xing, B. New insight into adsorption mechanism of ionizable compounds on carbon nanotubes. *Environ. Sci. Technol.* **2013**, *47*, 8334–8341. [CrossRef]
116. Kah, M.; Sigmund, G.; Xiao, F.; Hofmann, T. Sorption of ionizable and ionic organic compounds to biochar, activated carbon and other carbonaceous materials. *Water Res.* **2017**, *124*, 673–692. [CrossRef]
117. Sizmur, T.; Fresno, T.; Akgül, G.; Frost, H.; Moreno-Jiménez, E. Biochar modification to enhance sorption of inorganics from water- review. *Bioresour. Technol.* **2018**, *246*, 34–47. [CrossRef]
118. Teixidó, M.; Pignatello, J.J.; Beltrán, J.L.; Granados, M.; Peccia, J. Speciation of the ionizable antibiotic sulfamethazine on black carbon (biochar). *Environ. Sci. Technol.* **2011**, *45*, 10020–10027. [CrossRef]

119. Lin, Y.R.; Teng, H.S. Mesoporous carbons from waste tire char and their application in wastewater discoloration. *Micropor. Mesopor. Mat.* **2002**, *54*, 167–174. [CrossRef]
120. Wu, X.Q.; Xiao, B.D.; Li, R.H.; Wang, C.B.; Huang, J.T.; Wang, Z. Mechanisms and factors affecting sorption of microcystins onto natural sediments. *Environ. Sci. Technol.* **2011**, *45*, 2641–2647. [CrossRef]

Publisher's Note: MDPI stays neutral with regard to jurisdictional claims in published maps and institutional affiliations.

Review

Biochar as an Eco-Friendly and Economical Adsorbent for the Removal of Colorants (Dyes) from Aqueous Environment: A Review

Prithvi Srivatsav [1], Bhaskar Sriharsha Bhargav [1], Vignesh Shanmugasundaram [1], Jayaseelan Arun [2], Kannappan Panchamoorthy Gopinath [1,*] and Amit Bhatnagar [3,*]

1 Department of Chemical Engineering, Sri Sivasubramaniya Nadar College of Engineering, Kalavakkam, Chennai 603110, Tamil Nadu, India; prithvi18043@chemical.ssn.edu.in (P.S.); sriharsha389@gmail.com (B.S.B.); vignesh18059@chemical.ssn.edu.in (V.S.)

2 Centre for Waste Management, International Research Centre, Sathyabama Institute of Science and Technology, Jeppiaar Nagar, Chennai 600119, Tamil Nadu, India; arunjayaseelan93@gmail.com

3 Department of Separation Science, LUT School of Engineering Science, LUT University, Sammonkatu 12, FI-50130 Mikkeli, Finland

* Correspondence: gopinathkp@ssn.edu.in (K.P.G.); Amit.Bhatnagar@lut.fi (A.B.)

Received: 15 November 2020; Accepted: 16 December 2020; Published: 18 December 2020

Abstract: Dyes (colorants) are used in many industrial applications, and effluents of several industries contain toxic dyes. Dyes exhibit toxicity to humans, aquatic organisms, and the environment. Therefore, dyes containing wastewater must be properly treated before discharging to the surrounding water bodies. Among several water treatment technologies, adsorption is the most preferred technique to sequester dyes from water bodies. Many studies have reported the removal of dyes from wastewater using biochar produced from different biomass, e.g., algae and plant biomass, forest, and domestic residues, animal waste, sewage sludge, etc. The aim of this review is to provide an overview of the application of biochar as an eco-friendly and economical adsorbent to remove toxic colorants (dyes) from the aqueous environment. This review highlights the routes of biochar production, such as hydrothermal carbonization, pyrolysis, and hydrothermal liquefaction. Biochar as an adsorbent possesses numerous advantages, such as being eco-friendly, low-cost, and easy to use; various precursors are available in abundance to be converted into biochar, it also has recyclability potential and higher adsorption capacity than other conventional adsorbents. From the literature review, it is clear that biochar is a vital candidate for removal of dyes from wastewater with adsorption capacity of above 80%.

Keywords: adsorption; biochar; dyes removal; wastewater treatment

---

1. Introduction

Water pollution has become a major environmental problem globally. Different types of contaminants, mainly discharged from the industrial and agricultural activities, significantly contribute to water pollution. One of the major contributors to water pollution is the release of untreated dye effluents. These dyes are not only harmful to the plants and aquatic life, but also to human beings. Cancer, allergies, skin diseases are some of the health complications that may arise due to the ingestion and absorption of dye-contaminated water by humans [1]. Biochar has been identified as a potential candidate for wastewater treatment. The use of biochar for wastewater treatment has an added advantage due to the availability of abundance of surface functional groups on biochar surface and having a large surface area. Biochar is produced by the decomposition of biomass by thermochemical methods in the absence or low amount of oxygen. Different biomass feedstocks could be utilized to synthesize biochar, which include algae, crop residues, forests biomass, and animal manures. Torrefaction, hydrothermal carbonization, microwave heating, and gasification are some of the methods available for thermal decomposition of biomass [2,3]. Before biochar can be used, it has to be subjected to pretreatment processes (e.g., sieving, washing, crushing, etc.), followed by pyrolysis.

Biochar is a heterogeneous, carbonaceous char obtained from thermally treating a large amount of organic waste/mixture that has been heated and decomposed completely. Biochar exhibit properties, such as catalytic activity, adsorption efficiency, high porosity, and high surface area, and has been found to enhance various processes, such as anaerobic digestion, soil retention, and adsorption [4]. As other methods of controlling pollution of water/air/soil, such as membrane separation, ion exchange filtration, etc., have exhibited significant economic demerits and low efficiencies, biochar has become an important material to remove the aquatic pollutants [5].

Environmentally friendly, economically feasible, and easily designable adsorbents for the treatment of water have gained popularity in the past few decades. In this regard, use of biochar, obtained from natural sources, such as forest and agricultural biomass, organic waste, and animal waste has become important [6,7]. This review focuses exclusively on how biochar works as an adsorbent for the removal of dyes, present in water ass pollutants. The inherent properties of biochar make it an ideal candidate for the adsorptive removal of dyes from water. Biochar has already been established as a viable non-toxic and environmentally-friendly adsorbent capable of adsorbing a large spectrum of compounds from wastewater [8,9]. The potential of biochar is influenced by a variety of factors starting from the source of biomass from which biochar is prepared along with the method of preparation, reaction conditions, nature of pollutants, and the mechanism of action, all of which have been examined extensively in this review. The use of biochar as being eco-friendly for the removal of dye from wastewater has been examined in the past by several researchers [10–13]. This review aims to summarize and provide a holistic view on the research on biochar as an adsorbent for the removal of dyes from wastewater.

2. Colorants in Water and Toxicity

2.1. Natural and Synthetic Dyes

Dyes can be defined as color-imparting organic compounds that are water or oil soluble; they are distinguished from pigments that are insoluble. Natural dyes are those that are derived or extracted from natural sources, such as animals, flowers, roots, mollusks, minerals, etc. Natural dyes are broadly classified into two types: adjective and substantive [14]. Adjective dyes give permanent color only when used along with a mordant to bind them to the fabric; meanwhile substantive dyes contain a natural mordant, called tannin, and can give fast color without the use of additional mordants [15]. Mordants are compounds that act as a bridge between the fabric molecules and the dye, holding it in place to promote fastness [16]. Weak acids, such as tannic acid and acetic acid, are some natural mordants [17,18], while metal salts, such as copper sulfate and ferrous sulfate, can also be utilized. The use of natural dyes is advantageous as these are relatively non-toxic and renewable in nature, but their use is not cost-effective in the industrial context, and they fail to give an even consistent hue when compared to the synthetic dyes. Natural dyes may still be used in domestic and small-scale operations.

Synthetic dyes are usually unsaturated organic molecules. The first synthetic dye prepared was Mauve, a reddish-purple dye, which quickly degraded under water or direct sunlight to form a pale purple color. The first synthetic dye was derived from coal tar, the product resulting from the carbonization of coal. In the current context, synthetic dyes are more economical and exhibit better color fastness in comparison to natural dyes; thus, they dominate the market. Though they are more economical, synthetic dyes are also decidedly more toxic and polluting than natural dyes, causing environmental pollution and adverse health effects on living organisms [15].

2.2. Classification of Dyes

2.2.1. Chromophores and Color Index (C.I.)

Atoms or groups of atoms are said to be chromogenic if they are able to impart colour to the dye, and different arrangements and numbers of these chromogenic groups in a molecule impart different colors to the dyes. The other atoms or groups of atoms in the molecule, which are bound to the chromophore, and influence or bring a change in the color of the dye, are called auxochromes [19]. Thus, different textile dyes have specific structures that contribute to their characteristic colors. Some examples of chromophores are carbonyl groups, nitro groups, azo groups, or a conjugated pi electron system. Groups such as hydroxyl, aniline, and sulfonic acid act as auxochromes [20]. Dyes are classified by their applications and are given a unique identifying name called the Color Index (CI). As dye structures can be very complex, referring to it by its chemical name can be impractical, and common names can change between regions; thus, the name of the dye is standardized by using its CI. Every dye is issued a generic name and a CI number; the generic name consists of the method of application of the dye (e.g., direct, reactive etc.), the hue, and an identification number. The dye is also given a CI number based on the functional group and configuration of the molecule. For example, the dye commonly known as Remazol Red B is referred to as CI Reactive Red 22 (CI number: 14,824) under CI classification rules [21].

2.2.2. Disperse Dyes

Disperse dyes are mostly non-ionic in nature and are insoluble or very sparingly soluble in water; they are named so because their application involves dispersing the dye into a very fine suspension in solvent [22]. The chromogenic groups in disperse dyes are azo, anthraquinone, or nitro groups. The non-ionic nature of disperse dyes make them an ideal choice for dying synthetic and hydrophobic fibers, such as acetate, polyester, and sometimes nylon and acrylic fibers [23]. This is because these fibers are negatively charged and so reactive or basic dyes cannot be used. Only the non-ionic disperse dyes, which are not affected by surface charge, are able to reliably dye these fabrics. Traditionally, a carrier is added to the dye to improve the dispersion action and to aid dyeing. Disperse dyes can be classified from class A to class D based on their sublimation temperature, with class A having the lowest and class D the highest [24]. Since sublimation temperature depends on the size of the molecule, class A dyes have lower relative molecular size compared to class D dyes. Due to its sparing solubility, non-ionic nature, and non-biodegradability, disperse dyes are particularly harmful when left untreated in wastewater [25].

2.2.3. Direct Dyes

Direct dyes are water-soluble, anionic dyes that, unlike reactive or vat dyes, can be applied directly onto substrates using a neutral or alkaline bath (using sodium chloride or sodium sulfate) [26]. These dyes have an affinity to cellulosic materials, such as paper, cardboard, cotton, etc., but can also be applied on fabrics, such as, rayon and silk, with the use of mordants. The chromogenic functional groups in direct dyes include stilbene, phthalocyanine, oxazine, thiazole, but mainly azo groups [27]. Direct dyes are advantageous compared to other dyes because they have superior lightfastness compared to most reactive dyed hues, even though they cost less and are, thus, more economical. They also require less water use and the salt concentrations of the effluent are much less, compared to most reactive dyes. Some direct dyes, such as, Congo red, are carcinogenic, and have been banned from use [28].

2.2.4. Reactive Dyes

Reactive dyes have become one of the most widely used synthetic dyes in the industrial context due to their excellent wash fastness and bright and varied types of hues [29]. The chromophores in reactive dyes are largely azo groups; blue and green colors are given due to the presence of anthraquinone and phthalocyanine structures [30]. Direct dyes are largely used to dye cellulosic substrates and fibers, although other substrates can be used too. Unlike direct dyes, however, reactive dyes form new covalent bonds with the nucleophilic sites in fabric molecules, thus leading to its remarkable wash fastness [21]. Some of the most commonly utilized groups in reactive dyes are trichloropyrimidine, sulphatoethylsulphone, dichloroquinoxaline, and dichlorotriazin [13]. The major disadvantage regarding reactive dyes is their environmental threat. Effluent from dyeing cotton, using reactive dyes, are extremely polluted, having very high chemical oxygen demand (COD), salt load, and visible color in water [31]. Both, the unfixed dye and its hydrolyzed form, are soluble in water, and thus their removal is particularly challenging [32]. Some reactive dyes are also associated with heavy metals, such as chromium, copper, or nickel, and these can later be released into aquatic ecosystems on degradation of the dye molecule.

2.3. *Toxicity of Dyes*

The global textile industry is estimated to be worth around $1 trillion USD and its contribution towards total world exports is around 7%, employing 35 million people worldwide [33]. Thus, this industry has a high impact on the environment and human health in general, due to the pollution it causes. The most prominent and destructive form of pollution, caused by the textile industry, is the water pollution due to manufacturing of dyes. Textile effluents are both aesthetically polluted, and have high salinity, chemical oxygen demand, and ecotoxicity [34], and due to their increasing ubiquity in surface water, can lead to adverse effects to human and wildlife health and to aquatic ecosystems in general. Most synthetic dyes are highly toxic to humans and aquatic beings, and have acute and chronic effects. For example, reactive dyes are notorious, causing health issues such as, dermatitis, occupational asthma, rhinitis, and other allergic reactions for the workers involved in these dyes manufacturing [32]. Dyes are also mutagenic and carcinogenic in nature [35,36], which leads to chronic effects, such as kidney, urinary bladder, and liver cancer in dye workers. A xanthene dye called erythrosine is carcinogenic, neurotoxic and DNA-damaging for humans and animals alike [30]. Metal complexed dyes, which are widely used for their resistance, have heavy metals, such as copper, nickel, and chromium. When discharged to aquatic environments, these metals can be taken up by fish gills and can be transferred to humans through the food chain [36]. Current treatment methods are inadequate to treat dye effluents effectively, because of their recalcitrant nature in aerobic environments [37], and thus, these substances can linger in soil and lead to bioaccumulation, leading to complications in organisms higher up the food chain [38]. Thus, current effluent treatment techniques are inadequate for the dyeing industry and to prevent the further insemination of surface water with such mutagenic and carcinogenic molecules, we must adopt novel and more effective treatment techniques, such as bioremediation or biochar adsorption.

3. Treatment Technologies for Dyes Removal from Water

3.1. *Coagulation*

Coagulation is one of the most popular wastewater treatment techniques used since the early 20th century. Coagulation is the process of adding chemical compounds to bind particles together until they acquire a large mass to ultimately settle down. These chemicals are known as coagulants and carry a positive charge, which are mixed rapidly in the wastewater for uniform distribution. Most of the dissolved/suspended particles encountered in wastewater carry a negative charge, which are neutralized by these coagulants, thus making these capable of sticking together. Coagulation process is usually employed as a preliminary step in wastewater treatment process. The most frequently used coagulants are iron or aluminum salts. However, wastewater (containing dyes) is rich in color, and have high COD levels. Hence, conventional methods, such as coagulation, prove inefficient and moreover cause the problem of sludge disposal [39]. A combination of the conventional coagulation technique with other treatment methods must be innovated to improve its efficiency. For example, combination of coagulation and adsorption techniques was a feasible way for the removal of reactive dyes from water [40]. The adsorbent used was activated carbon, derived from coconut shells, and the coagulation process was carried out using aluminum chloride as the coagulant. It was found that the removal efficiencies for Orange 16 and Black 5 reactive dyes were 84% and 90%, respectively [40].

3.2. *Advanced Oxidation Processes (AOPs)*

AOPs are chemical treatment methods used to remove the organic/inorganic contaminants present in wastewater by the oxidizing action caused due to the in-situ production of hydroxyl radicals ($^{\bullet}OH$) [41]. These radicals are produced by oxidizing agents (H_2O_2, O_3, $KMnO_4$), catalysts, or UV light. Some of the AOPs include photocatalysis, ozonation etc. [42]. The following sections briefly explain some of the AOPs used in dye removal from wastewater.

3.2.1. Ozonation

Most of the dyes, encountered in wastewater, have polycyclic aromatic structures containing elements, such as nitrogen, metals, and sulfur, which makes it difficult to treat the wastewater by physical, chemical, and biological methods. Conjugated chains present in the dye structures, responsible for imparting color, are actively destroyed by ozonation, which is an AOP method that involves the chemical treatment of wastewater by dissolving ozone in water [43]. Employing ozonation as a treatment technique is advantageous as zero sludge is generated. Degradation of the dye is achieved in a single step and furthermore, ozone decomposes into stable oxygen [44]. However, ozonation as a treatment process very rarely results in complete oxidation, so by-products are usually formed in the effluent stream [45].

3.2.2. Fenton's Reagent and Fenton-Like Processes

Fenton's reagent is a solution of ferrous iron along with hydrogen peroxide, which finds its use as a catalyst in oxidizing various contaminants present in wastewater. A composite of La-Fe-O [46] was used as a photo-Fenton catalyst in the presence of light irradiation and hydrogen peroxide, and it was found that rhodamine B dye was oxidized to 98% within 25 min.

3.3. *Membrane Processes*

Membranes have been in use for wastewater treatment as early as the 1960s. Since membrane processes were too expensive at that time, this process was only chosen for specialized applications. Since the 2000s, membranes have been made cost-effective and are being used with other conventional water treatment processes. A membrane is a thin, semi-permeable material that is attached to a porous support, and is used for the removal of dissolved substances, based on properties, such as, size or charge, when a driving force is applied on it. Membrane processes are used in reverse osmosis (RO), forward osmosis, nanofiltration, and ultrafiltration.

3.3.1. Nanofiltration

The membranes used for nanofiltration have pore size within the range of 0.1–10 nm. Nanofiltration membranes have the advantages of separating dyes with a high molecular weight and also have >90% rejection efficiency for dye removal, making it a promising approach towards dye removal [47]. For example, positively charged polyethylenimine-modified nanofiltration membrane showed semi-xylenol orange, Tropaeolin O, Victoria blue B dyes removal efficiency of 99%, 98.3%, and 99.2%, respectively [48].

3.3.2. Forward Osmosis (FO)

FO is a water treatment process that utilizes the osmotic pressure gradient to separate water from its dissolved solutes. Since FO process is completely devoid for the requirement of a driving force, it is more energy efficient compared to reverse osmosis and other membrane separation processes [49]. A thin-film composite membrane was used under forward osmosis by [50], and it was found that it had a dye rejection rate of ≥96% for commonly used dyes in the textile industries.

3.4. Biological Process

Biological process involves the usage of bacteria, microbes, and other microorganisms to treat wastewater. These biological processes are environmentally friendly, energy saving, generate low sludge, and require zero to minimal amount of chemicals to be used. The efficiency of biological processes could further be increased by varying the environmental conditions to favor the growth of the microorganisms. Algae is widely used by researchers as a potential option for the purpose of biosorption, as algae contains proteins, lipids and functional groups such as amino, carboxylate, sulfate, etc., in its structure [51]. Since algae possesses a wide surface area and excellent binding affinity in its cell structure, it results in high biosorption capabilities [52]. As algae is easily accessible (highly abundant in saltwater oceans and freshwater lakes), the use of algae could subsequently be extended for the dye removal from textile wastewater. For example, chemically (sulfuric acid) modified defatted Laminaria japonica biomass (renewable brown algae) showed methylene blue adsorption capacity of 549.45 mg/g, and quasi-equilibrium was achieved within 60 min under optimal conditions with a biosorbent dose of 0.6 g/L, pH 6 and temperature of 308 K [53].

Bioreactors are also becoming increasingly popular for dyes removal. Usually, a combination of both aerobic and anaerobic processes is used for the treatment of azo dyes. For example, the removal of Alizarin Yellow R dye was studied by the combined process of up-flow bio-electrocatalyzed electrolysis reactor and aerobic bio-contact oxidation reactor in just 6 h of hydraulic retention time [54]. Moreover, the COD removal efficiency and decolorization efficiency was found to be 93.0 ± 0.5% and 93.8 ± 0.7%, respectively, in the process.

3.5. Adsorption Process

As far as dye removal from water is concerned, adsorption has been found as one of the best treatment processes among other conventional water treatment methods due to its low-cost, affordability, greater efficiency, and the fact that it requires minimum maintenance [55]. Adsorption also has the added advantage of producing no detrimental residues and having the capacity to treat a large volumes of water [56]. An adsorbent could also be recycled multiple times for its usage in subsequent treatment processes [57]. Common materials, such as activated carbon, zeolites, activated alumina, silica gel and polymeric adsorbents have been widely used for water treatment. The use of biomaterials for adsorption processes instead of conventional materials is now being the subject of interest by many researchers as the commercial value of biomaterials is low, and also they are available in abundance [58]. Naturally-derived biopolymers, which are hyperreactive, chemically stable, possess good physicochemical properties have garnered significant attention that is worth looking into for employing these biopolymers for the role of green adsorbents [59]. Table 1 highlights the major advantages and disadvantages of treatment techniques, available for dye removal from water.

Table 1. The advantages and disadvantages of various treatment technologies for dyes removal from water.

Treatment Technology for Dyes Removal	Advantages	Disadvantages
Coagulation	• Reduced time for settling of suspended solids • Easy removal of fine particles • Effective in removing bacteria, protozoa, and virus	• High cost to spend for frequent monitoring and accurate dosing • Huge volume of sludge generation
Advanced Oxidation Processes (AOP)	• •OH radicals can treat a wide range of organic material • Zero sludge production	• Fenton's reagent AOP results in iron sludge generation • High capital and maintenance costs are expected

Table 1. *Cont.*

Treatment Technology for Dyes Removal	Advantages	Disadvantages
Membrane Processes	• Minor, valuable products can be recovered from the feed stream • Process can be easily "scaled-up" • No-phase changes involve between the feed and product stream • Eco-friendly as simple and non-toxic materials are used	• High flow rate has the potential to damage the membrane • This process results in membrane fouling effects. Regeneration and extensive cleaning are required. • High equipment cost
Biological Processes	• Almost all biodegradable organic matter is effectively removed • Efficient attenuation of color • Eco-friendly and a common wastewater treatment mechanism	• Slow process • An optimal favorable environment is crucial • Biological sludge generation • Remediation of dye molecules is tough
Adsorption Processes	• Highly efficient process • Applicable for a wide variety of target contaminants • Treatment technology is easy to employ	• Deterioration of adsorbent performance when subjected to multiple operational cycles • Spent adsorbent is likely to be a hazardous waste • Regeneration of adsorbent material is expensive

4. Biochar

4.1. Biochar Synthesis by Various Methods

Biochar is essentially a by-product formed when a large amount of organic matter (biomass) is heated at high temperatures in the absence or low amount of oxygen. Therefore, depending on the source and nature of the raw material, the process to prepare biochar also varies. In the industries however, there are mainly three processes, which dominate when it comes to the preparation of biochar. These are pyrolysis, hydrothermal liquefaction, and hydrothermal carbonization [60,61]. The three processes mentioned are used, depending on the raw material, hence the nature of biochar prepared by these three processes vary from one another.

4.1.1. Hydrothermal Carbonization

Hydrothermal carbonization in simple words is a method used to produce "structured" carbon such as charcoal, biochar, etc., from organic matter [62,63]. The main concept behind hydrothermal carbonization is to replicate natural carbon formation (when bioorganic compounds are exposed to extreme pressures and temperatures over thousands of years), but accomplish the same result in a shorter duration. In essence, hydrothermal carbonization is done by taking a mixture of biomass (animal wastes, plants etc.) and water in a pressure vessel, which is then heated to high temperatures (180–250) °C and extremely high pressures of 10 bars or more [62–66]. At such high temperatures and pressure, the pH of the solution decreases due to the increase in production of oxonium ion. At these specific conditions, more and more organic matter releases into the water due to the low pH conditions, which causes the formation of a sludge-like compound that is predominantly made of carbon. The sludge obtain can be separated and dried. Various compounds such as bio-oil, hydrochar, biochar, carboxylic acids, ketones, aromatics, etc., can then be used for various purposes [64,67]. Depending on the required product, the reaction can be stopped at multiple stages to obtain a variety of products. The biochar produced from regular pyrolysis varies in many aspects from the biochar produced by hydrothermal carbonization; hence, the product obtained from hydrothermal carbonization is called as hydrochar rather than biochar. Hydrochar and biochar vary in properties and conditions in which they are prepared [68], but they are essentially composed of the same chemical constituents and are prepared from the same raw material [64,65].

Hydrochar produced as a result of hydrothermal carbonization has shown to have large amount of polyfuran and N-heterocyclic aromatics, which have a tendency to reduce the mobility of nitrogen thereby, preventing its quick release into the soil [69]. Unlike many types of biochars, it has been seen that hydrochar produced as a result of hydrothermal carbonization has low organic content and reduced mobility of heavy metals, which makes it ideal for soil amendment. Addition of hydrochar to soil has shown to improve water retention capacity, microbial community, nutrient holding capacity, and much more [69,70]. Since the use of hydrochar has grown considerably only during the last decade, research on different aspects of hydrochar is still yet to be studied thoroughly. The extreme conditions of hydrothermal carbonization not only lead to the decomposition and dissolution of organic matter, but also makes heavy metals "bio-available". Therefore, detailed research is needed in order to ascertain the total capacity and effect of hydrochar, especially its use as a soil amendment, and how different hydrochar produced from different raw materials behave.

4.1.2. Hydrothermal Liquefaction

Hydrothermal liquefaction is essentially a decomposition process in which wet biomass is converted into a variety of products including bio-oil (liquid), biochar (solid) by means of thermal depolymerization. Wet biomass is exposed to extreme temperatures (250–550 °C) and pressures of (5–25) MPa. During hydrothermal liquefaction, a variety of chemical processes take place on biomass including hydrolysis, fragmentation, dehydration, aromatization and repolymerization, which results in the formation of biochar along with other compounds [71–74]. As stated before, the nature of the biochar obtained depends on different factors, such as, temperature, pressure, nature of raw material, residence time, etc. [75]. Extensive research into composition of biomass, and hence the properties of biochar, suggests that the main constituents of biomass (lignin, cellulose, and hemi-cellulose) play a key role in influencing the formation, as well as properties of biochar, especially lignin [76–79]. When biomass is heated at 220 °C, lignin gets dissolved very easily into water by hydrolysis leading to the formation of many phenolic compounds. Similarly, cellulose and hemicellulose, which are present in the biomass, also undergo various reaction leading to the formation of char [71]. Rice husks/pinewood was used as feedstock into a 100 mL autoclave along with deionized water. The autoclave was operated for 20 min at 573 K and the products that were formed (bio-oil and biochar) were removed using acetone, and the biochar was subsequently dried in a hot-air oven [71]. At very high temperatures, all of the biomolecules present in biomass, such as lignin, completely degraded and, therefore, the conversion needed to form biochar ceased to occur [71]. Very fast heating and sudden rises in temperatures inhibit biochar formation and slower and steadier processes are preferred [71]. Many studies show that physical properties, such as surface area and volume of biochar, is much lower while producing biochar from hydrothermal liquefaction (HTL) than from pyrolysis, however, recent studies suggest that biochar produced using HTL seems to be effective in adsorbing heavy metals and acting as biological catalysts [80].

Even though water is an abundant resource and is easy to use, it has been seen that use of different types of alcohols (butanol, methanol, etc.) improves the properties of biochar, as well as the amount of biochar that will be produced by HTL [81]. However, it was observed that water does not support polymerization reactions and, hence, production of bio-oil is more preferential as compared to biochar when water is used as a solvent [71]. Chemically prepared biochar as a result of hydrothermal liquefaction, has been found to have very high oxygen content in the form of hydroxyl, carboxyl, cyclic oxygenates, such as phenyls, glycerol, etc., and particularly phenol and lactone hydroxyls in biochar formation from rice husks/pinewood [71,75,81]. The functional groups in each biochar vary significantly and temperature plays a key role in determining the functional group. For example, if biochar is prepared at 280 °C and 360 °C, the functional groups were significantly less in the biochar prepared at higher temperatures [71].

Alcohol as solvent has significant advantages over the use of water, some of which include better reaction conditions, ability to release hydrogen for stabilization, higher reactivity with acid-components to form esters compared to water [81]. Apart from increase in biochar production, bio-oil production is also significantly increased, when alcohol is used as a solvent. Usage of glycerol as co-solvent acts as a high lipid additive and helps in converting low-lipid algae into bio-oil and biochar, which would usually be difficult and would require extreme conditions, which are not economical [81]. In conclusion, hydrothermal liquefaction is a process, which is still improving, and certain characteristics can be helpful under certain specific conditions when necessary.

4.1.3. Pyrolysis

Pyrolysis is mainly used to produce many forms of carbons from biomasses under high temperature and inert conditions. However, as engineering advances, the use of pyrolysis for converting biomass to syngas and biochar has grown significantly. As the other processes described earlier, the products that are formed as a result of pyrolysis significantly depend on the biomass, reaction conditions, and feedstocks used. Bio-oil, which is formed as a pyrolysis result, can easily be upgraded into higher grade fuel and can be used for various applications [82]. Biochar is a valuable product formed from pyrolysis and has seen growth in applications such as improving soil retention, soil amendment, catalyst, carbon sequestration [82] and, hence, learning and gaining knowledge of production of biochar is of great interest. Use of biochar as soil amendment has shown promising results in the agricultural field and, hence, different types of biochars and different type of preparation techniques are studied in order to obtain the most effective biochar, which is capable of increasing soil nutrient availability, porosity, fertility, etc. [82]. Biomass with higher calcium oxide (lime) content when heated at temperature above 450 °C results in biochar with low amounts of oxygen functional groups and increased stability [82]. Further use of additives, such as phosphorus, zinc, iron, etc., during pyrolysis also modify biochar and, hence, expands its applications in various fields [83–85].

In the past few years, there has been extensive research on different methods of pyrolysis, which can affect the conversion of biomass to biochar. Optimization of pyrolysis methods makes it economic and advantageous over other processes. Figure 1 highlights the most preferred techniques for biochar production from various biomasses. The nature of biochar produced, as a result of pyrolysis, depends on many factors, some of which are biomass pre-treatment, reactor type, and dimensions, pressure, residence time in the reactor [86]. Compared to other processes, such as hydrothermal carbonization or hydrothermal liquefaction, pyrolysis has an advantage because of the fact that it can be used for a wide range of biomass and waste materials [87].

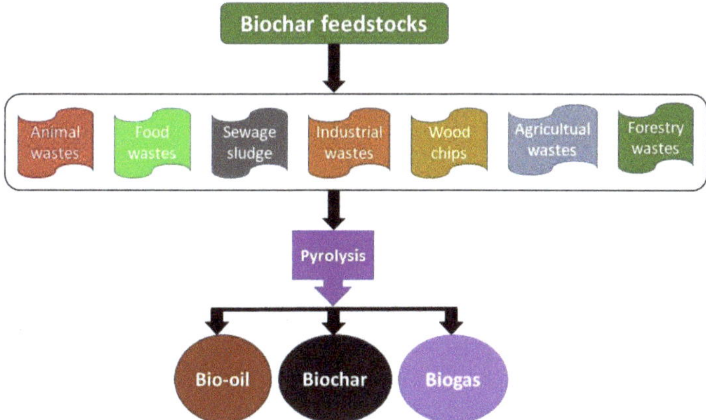

Figure 1. Schematic diagram showing biochar formation routes from different feedstocks.

Reactor designs and reactor types also have an important role to play and information regarding their influence on the type of pyrolysis has been studied [87]. In general, reactors such as auger reactors, rotary and drum Kilns are commonly used for pyrolysis. When the operating temperatures in these reactors are extremely high in the range of 800 °C to 1300 °C, the pyrolysis is referred to as fast pyrolysis [87]. The concept behind fast pyrolysis is that when the temperature is very high and residence time is low, char formation is prevented and the biomass undergoes thermal cracking. Slow pyrolysis, such as the name suggests, involves very slow rise in temperatures and very high residence time, thereby getting maximum exposure of biomass to heat [88]. Contrary to fast pyrolysis, in slow pyrolysis, liquified products are kept to a minimum and solid carbonaceous char is preferentially produced. Finally, pyrolysis is flexible in the sense that it can operate at varying reaction conditions and with different types of feedstocks, hence, resulting in a large variety of products. The biochar obtained as result of pyrolysis of sludge (wastewater sludge) has seen to have many properties, such as nutrient recyclability, heavy metal immobilization, good pore structure, and alkaline nature [89].

Due to its versatility and flexibility and its ability to operate on a wide range of raw feedstock, it is the ideal process to choose. Further, biochar formed because of pyrolysis is extremely valuable, and can be modified in a multitude of ways for various purposes. In recent times, pyrolytic biochar has been used in soil remediation, wastewater treatment, catalytic enhancer, and much more. Since research on this particular subject is still ongoing and different strides are being done on a consistent basis, one can only assume that this process is going to be more efficient and more economical in the future. Supplementary Table S1 depicts different methods of biochar synthesis, based on their sources, methods of preparation, and preparation conditions.

5. Biochar as an Adsorbent

Biochar, being environmentally friendly and extremely versatile, has been seen as a novel material for treating wastewater. Due to biochar adsorption properties, it has been a viable method to adsorb colorants that are potentially toxic and harmful to the environment.

5.1. Effect of Operational Parameters on the Adsorption of Dyes by Biochar

Biochar as an adsorbent works under extremely specific conditions. Concentration of dye/biochar, temperature and solution pH play important roles in determining the efficiency of biochar [90–95]. Nickel modified biochar was capable of adsorbing methylene blue with an adsorption capacity of 479.49 mg/g at 20 °C from wastewater [96]. Initially when the concentration of methylene blue was low, the adsorption by biochar was high due to the large number of active sites still available on the biochar for adsorption. Competitive adsorption was found to hinder the efficiency of adsorption of methylene blue by the biochar [96]. Research on the adsorption and degradation of acid red dye revealed that increase in temperature of the solution increased the rate of decolorization [97]. In this particular study, the temperature was maintained between 30 to 50 degree Celsius and it was observed that increase in temperature increased the rate of molecular collisions and, hence, the rate of the reaction resulting in the rapid increase of adsorption and decolorization of the solution.

One major factor that has constantly been seen to affect the way biochar behaves as an adsorbent is pH. In most studies, we see that the variation in pH causes major effects on the adsorption process using biochar. In study [91], on the investigation of the adsorption of methylene blue from wastewater, authors noted that the pH had a great influence on the degradation process of the dye. They observed that the efficiency of removal of the dye increased from a pH of 2 to 7 and then decreased from a pH of 7 to 11, which was due to the fact that the active sites on the surface of the biochar combined with the abundantly present hydroxyl groups, which caused a decrease in efficiency. A similar study was conducted by other researchers [92] on the adsorption of Congo red dye using biochar, made from organic peel waste, and the effect of different operational parameters on its adsorption was examined. It was observed that when the pH was kept between the range of 2–3, acidic conditions prevailed and the presence of H^+ ions aided the adsorption of Congo red dye onto the active sites present on the biochar. At the same time, however, for a pH of higher than 7, alkali conditions prevailed and the presence of OH^- ions hindered the adsorption of dye onto the biochar by competing for the active sites on the biochar. The results obtained in this study [92] were similar to that of other study [91], as both studies concluded that higher levels of pH are not ideal for the adsorption of dye using biochar due to undesirable electro-static conditions that were developed. The adsorption of organic dyes using food waste biochar was conducted and the effect of different parameters were examined [93]. As far as the effect of pH is concerned [93], authors observed that the removal efficiency of dye decreased by around 5% only, when the pH was altered from 3 to 11. However, when looked at more deeply, it was seen that the decrease in removal efficiency was not uniform overall, but was uniform between pH of 3–9 and a drastic decrease was observed, when pH approached to 11. Authors attributed this effect to the fact that when pH increased, decomposition of hydrogen peroxide decreased along with the conversion of carbonate and bicarbonates to their corresponding acid, which caused low reaction with OH^- radicals, which finally led to the decrease in solubility of iron that was present in the biochar. Therefore, in this particular case observed by [93], we see that the decrease in removal efficiency, due to changes in pH, is not as pronounced as in the other cases. In the experiment pertaining to adsorption of methyl violet dye using biochar conducted by [98], for pH values > 6, it was found that the adsorption of dye was favored, and the trend observed was due to the fact that at very low pH, the concentration of H^+ ions is significant and, hence, they compete along with the cationic dye (methyl violet) for the active adsorption sites on the biochar, thereby decreasing the percentage of removal of the dye. When an anionic dye, such as Congo red, is to be removed, the trend relating to the effect of pH may vary compared to other commonly seen dyes. In another study [99] focusing on the removal of dye using biochar, it was observed that when pH increased from 2 to 11, a significant decrease in removal percentage of dye was observed. Authors of this study [99] inferred from the observations that the decrease in removal efficiency between pH of 2 and 6 was due to the fact that at low pH < 2, the number of H^+ ions were significantly more and, hence, the adsorbent molecules acquired a positive charge and became cationic, increasing the adsorption of an anionic dye, such as Congo red, significantly. We see from the above-mentioned studies that the nature of dyes has played a significant role in determining the optimum pH for adsorption. When a dye is anionic, it has been observed that acidic pH (<2) is favored since the adsorbent surface (biochar) being positively charged due to the increase in presence of H^+ ions and it electrostatically favors attraction of anionic dyes and when a dye is cationic, increase in pH (>2) is favored as higher pH increases the presence of OH^- ions, which in turn makes the surface of biochar negative, which promotes the electrostatic attraction between the positively charged cationic dye and the negatively charged biochar surface [100,101]. From the studies we discussed above, it can be said that the interdependence between pH and adsorption is a key concept needed to understand the process of adsorption and that a large number of factors are responsible for different trends that may be observed. Figure 2 provides the schematic route of adsorption of dye from wastewater by biochar.

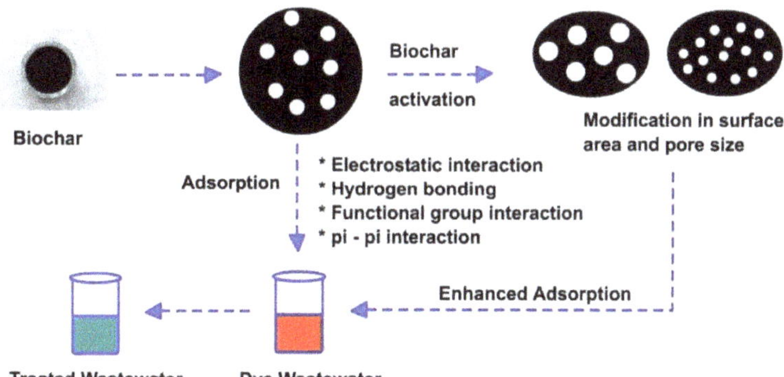

Figure 2. Schematic diagram of dye adsorption mechanism using biochar.

The dosage of adsorbent as well as adsorbate influence the extent of adsorption of dye and are important parameters to be considered. In a study of adsorption of dye using biochar [102], the removal percentage and capacity of adsorption were influenced heavily by the dosage of adsorbent. Even for a small increase in dosage of adsorbent caused the removal percentage to increase significantly (almost an 80 percent increase) and remained constant after a certain maximum dosage was reached. This phenomenon, according to authors [102], was due to the fact that at higher dosages of adsorbent, the number of active sites, as well as surface area of adsorption, will increase drastically, which in turn will increase the removal percentages of the dye present in water. A similar increase in percentage of dye removal with increase in adsorbent dosage was also observed in other studies [99,103,104]. The removal percentage of dye increased significantly for a very small increase in adsorbent (<1 g/100 mL) in all of the above-mentioned studies. This was similar to another study [102], which concluded that the increase in percentage of dye removal was because of the increase in dosage of adsorbent, which in turn increased the active sites for adsorption as well the surface area of adsorbent. However, the percentage of dye removal remained constant after a certain dosage of biochar due to the decrease in amount of dye molecules adsorbed per unit mass of adsorbent and the decrease in specific uptake [99]. Therefore, there will be no change in overall removal efficiency of dye by biochar after an equilibrium point has been reached. A similar conclusion was also achieved by [105], where switchgrass biochar was used as an adsorbent. It was observed that an increase in the dosage of biochar increased the removal percentage of dye and it reached an equilibrium percentage at around 50 mg of adsorbent dosage. Authors of this study [105] related this behavior to the increase in active binding sites and surface area due to the increase in dosage, which is similar to what has been concluded previously by other researchers [97,106]. When we assume the contrary, i.e., if the amount of dye is increased with the constant dosage of the adsorbent, the kinetics will depend on the equilibrium point, where the rate of adsorption of dye is equal to the desorption of already existing dye molecules [99]. The percentage removal will initially increase, and a small decrease may be seen, owing to the fact that the active sites present on the adsorbent (biochar) are completely occupied and the surface area that is free for adsorption decreases, while the number of dye molecules/reactive species adsorbed increases [99,106,107]. However, there are cases where the increase in concentration of dye indirectly increases the adsorption capacity of biochar until an equilibrium point after which no more adsorption is observed as all active sites on the biochar surfaces have been completely occupied by dye molecules, leaving no room for further adsorption [101]. The trends we discussed above, more or less, are similar in many aspects when it comes to the reason behind the increase/decrease in efficiency of dye removal. The operational parameters under which biochar works are of vital importance in order to understand and interpret the process of adsorption or other similar processes that might occur.

5.2. Adsorption Kinetic Studies

In order to understand the adsorption process and the rate of adsorption, kinetic studies are conducted. Kinetic studies have been extensively used by many researchers in order to study the mechanism by which adsorption takes place. Most adsorption kinetic studies apply pre-existing models, which act as guide to illustrate the adsorption mechanism. Pseudo first order, pseudo second order, and particle diffusion models are among the widely used models to describe the adsorption process. A wide range of standard models are available that describe adsorption at different conditions and the most suitable can be chosen for the given experimental conditions [108]. Adsorption of dyes using rice husks biochar was investigated, and from various kinetic models that were tested, pseudo second order model was the most suitable one to describe the experimental observations of the adsorption of malachite green dye using biochar [104], and the Elovich model was used in order to conclude that the mechanism of adsorption was chemisorption. In comparison, a similar study conducted on adsorption of Congo red by manganese composite biochar, it was observed that pseudo first order model was able to explain the kinetics of adsorption for a fixed amount of adsorbent and adsorbate [103]. These kinetic models not only can be used to describe the experimental observations of adsorption capacity, but they also indirectly indicate a way to describe the mechanism of adsorption.

In a study [109], the adsorption of methylene blue was examined using sugarcane bagasse biochar, and it was observed that pseudo second order kinetic model fitted well to the experimental results. It was also inferred that the mechanism of adsorption was chemisorption. Hence, the use of kinetic models is not restricted to only a mathematical result, but can also provide information regarding the mechanism of action. Factors, such as nature of biochar and the nature of the dye to be removed, also play key roles in determining the adsorption kinetics. The adsorption kinetics varied depending on the biochar used and the rate of adsorption varied progressively (fast to slow) [110]. Authors inferred that it was because of inherent properties, such as pore volume and size, that caused the discrepancy in the rate of adsorption and choice of kinetic model, which was in conjunction to similar observations made by other researchers [111–113]. In examination of adsorption of reactive red dye using ball-milled biochar, the adsorption rate was initially fast and then gradually became slower [111]. From calculating the values of coefficient of determination (R^2), it was able to conclude that intra particular diffusion model had the most appropriate fit and hence, the mechanism of adsorption was primarily diffusion, and the removal of dye was influenced mainly by the external surface area [111].

Adsorption of acid chrome blue K onto Cetyl trimethyl ammonium bromide modified magnetic biochar was reported in [112]. The first 600 min showed a rapid increase in adsorption of the dye by biochar until it slowed down and reached an equilibrium stage after 60 min. In most cases, we see that the increase in dosage increases the rate of the adsorption (Section 5.1), i.e., removal efficiency of dye and, hence, the kinetics vary correspondingly. Another study involved the adsorption and degradation of acid red 88 dye from solution using Fe-Mn composite biochar [114]. In this study, authors observed that the change of dosage of biochar increased the removal efficiency and increased the rate of adsorption, which agreed with the pseudo first order kinetic model theoretically. Similar to the method used by [111,115], which inferred using a pseudo second order kinetic model that, in the adsorption of methylene blue using biochar (municipal waste derived biochar), the rate limiting step was the chemisorption step, and that the pseudo second order kinetic model was much more efficient than the other models in describing the particulate and liquid film diffusion and surface adsorption processes.

Thus, we see that the use of kinetic models is not entirely restricted to just correlating the experimental and theoretical values of adsorption efficiency, rate of adsorption and equilibrium adsorption, but can also be used to infer various possible mechanisms of adsorption based on the numerical values that are obtained. Contrary to previous observations by a large number of researchers, it is not always possible to obtain the mechanism of adsorption by using a suitable kinetic model [116]. The reason for this discrepancy stated by [116] was that the mechanisms of dye adsorption by hydrogen bonding (as dyes are present in the solution phase) may have been misrepresented diagrammatically in the previously established standards, which researchers often refer to. Therefore, if we look at what has been observed and concluded by the various researchers of this specific topic (adsorption of dyes using biochar), we see that, even though adsorption kinetic studies have their merits, they may be inherently flawed, if they are not studied extensively and deeply.

Table 2 elaborates the description of different kinetic models that are used in adsorption studies for dyes removal. To summarize, we see that the knowledge regarding kinetics of adsorption is an extremely helpful and intuitive way of understanding how adsorption takes place. Primarily a mathematical concept, one can infer various aspects regarding adsorption from kinetic studies. Considering all the influencing factors and different experimental conditions, one can apply kinetic studies to study the adsorption of dyes or any such pollutant using biochar and gain a much deeper and greater understanding of the concept [108].

Table 2. Adsorption kinetic studies of different dyes onto biochar.

Adsorbent	Adsorbate	Applicable Kinetic Models	Reference
Steam activated spent mushroom substrate (SMS)	Congo red and crystal violet	Pseudo first order Pseudo second order	[100]
Nickel aluminum layered double oxides modified magnetic biochar	Acridine orange	Pseudo first order and Pseudo second order	[101]
Chemically modified lychee seed biochar	Methylene blue	Pseudo second order	[117]
Triethylenetetramine biochar	Sunset yellow dye	Pseudo second order	[94]
Orange peel waste microwave activated biochar	Congo red dye	Pseudo second order	[92]
Fe_2O_3/TiO_2 functionalized biochar	Methylene blue, Methyl orange, Rhodamine B	Pseudo first order	[91]
Switchgrass biochar	Methylene Blue, Orange G, Congo red	Pseudo first and second order model, Intra particular diffusion model	[90]
Biochar derived from *Opuntia ficus-indica* (OFI) cactus	Malachite green	Pseudo second order model, Elovich model	[118]
N-doped biochar	Acid red 18	Intra particular diffusion model	[113]
$Ag\text{-}TiO_2$ biochar	Methyl orange	Pseudo first order kinetic model	[119]
Biochar derived from mixed municipal discarded material	Methylene blue	Pseudo first and second order model Intra particular diffusion model	[115]
Mesoporous nano-zerovalent manganese (nZVMn) and *Phoenix dactylifera* leaves biochar (PBC) composite.	Congo red	Pseudo first and second order model	[103]
Cetyl trimethyl ammonium bromide modified magnetic biochar derived from pine nutshells	Acid chrome blue K	Pseudo first order model Pseudo second order model	[112]
Pristine and ball milled biochar	Reactive red 120	Pseudo-second order kinetic model Intra-particle diffusion model	[111]
Wood based biochar	Acid orange 7	Langmuir-Hinshelwood (L-H) kinetic mode	[107]
Pyrolyzed rice husk biochar	Malachite green	Intraparticle Diffusion Model Elovich models Pseudo second order model Pseudo-first order model	[104]
Animal waste biochar	Basic red 9	Pseudo first order model Pseudo second order model Pseudo Nth order model	[110]

5.3. Adsorption Isotherm Studies

An adsorption isotherm is an empirical equation, which relates the equilibrium concentration of solute, and is adsorbed on the surface of an adsorbent to the concentration of solute present in the liquid solution with which it is in contact with, as a function of pressure under constant temperature. For choosing the optimal isotherm model for the adsorption process, the following criteria must be fulfilled: (1) the isotherm model and the calculated equilibrium data must have a good fit; (2) the function of isotherm must be thermodynamically realistic; and the (3) analytical calculation of concentration from capacity and vice versa must be possible even in ideal scenarios [120].

A biochar prepared from pyrolyzed rice husk showed that the maximum removal of Malachite green dye by the biochar was 99.98% with an adsorbent dosage of 0.2 mg/L, dye concentration of 20 mg/L at pH of 6.2 over a period of 2 h [104]. Freundlich, Dubinin–Radushkevich, Langmuir, and Temkin isotherms were studied in order to determine the best fitting of the models. Langmuir isotherm model was found to have the best fit. This led to the inference that Malachite green dye had uniform binding energy with the active sites on the biochar, which was distributed homogeneously across the adsorbent via the mechanism of physisorption. An eco-friendly biochar was synthesized by the pyrolysis of Ashe Juniper and, thereafter, functionalized by sulfuric acid for the removal of methylene blue, a hazardous dye [121]. The Langmuir model had a higher correlation coefficient (R^2) value of 0.96 compared to the Freundlich model, which only had an R^2 value of 0.94. Since the Langmuir model was determined as the better fit, it implied that the biochar was monolayered, had identical binding sites distributed equally across its surface for dye adsorptive removal.

The fallen leaves from Magnolia Grandiflora L. were converted into biochar by subjecting it to slow pyrolysis for the removal of methylene blue dye [122]. Maximum dye adsorption was observed as 101.27 mg/g at an ionic strength (NaCl) of 0.1 M, pH level of 12 and at 303 K. Langmuir, Freundlich, Temkin, and Harkins–Jura models were tested. The experiments were performed at three temperatures (293 K, 298 K, and 303 K). Langmuir isotherm model was the best-fitted one and the adsorption process of methylene blue onto the biochar was found to be endothermic process. Furthermore, the dimensionless separation factor (R_L) had values of 0.139, 0.273, and 0.139, respectively, which clearly ranged between 0 and 1 for all three temperatures, suggesting favorability for the adsorption process. Freundlich isotherm model also proved favorable as the Freundlich adsorption intensity (1/n) values were 0.12, 0.09, 0.14, respectively, which were also between 0 and 1 for all three different temperatures. Temkin model had shown binding energies of 33.98, 51.36, and 46.78 kJ/mol, respectively, at the above different temperatures, implying that the adsorption process not only consists of an ion-exchange mechanism, as the binding energy should be within the range of 8–16 kJ/mol for the model to be accepted. For the Harkins–Jura model, the correlation coefficient (R^2) values were the lowest compared to other models, implying that multi-layer adsorption process had low correlation. The Langmuir model was concluded to be the best fitted among other models due to a high R^2 value of >0.99. This demonstrated that the biochar surface was homogenous and methylene blue was adsorbed as monolayer, distributed equally throughout its surface.

Empty-fruit bunch fibers were activated by chemo-physical processes to prepare the biochar for the removal of Cibacron Blue 3G-A dye from aqueous solution [123]. The maximum dye removal was 99.05% by the biochar at optimal conditions of an adsorption dosage of 0.10 g/100 mL, dye concentration of 100 mg/L at a pH value of 10, temperature at 343 K over a period of 45 min of contact time. The adsorption isotherm models considered in this study were only Langmuir and Freundlich models. The correlation coefficient (R^2) for both models was greater than 0.9. The dimensionless separation factor (R_L) for Langmuir isotherm was found to be between 0 and 1, suggesting that the adsorption process was feasible. For the Freundlich isotherm, the Freundlich adsorption intensity (1/n) was also between 0 and 1, which was a good indication of favorability of the adsorption process. It was finally concluded that Freundlich isotherm was a better fit compared to Langmuir, due to its high R^2 value compared to the latter. This also led to the inference that the surface of the biochar was heterogeneous, multilayer and the mechanism of adsorption was physisorption. Table 3 summarizes the most common isotherms studies in dye removal studies. Table 4 elaborates the detailed literature on the adsorption capacity of biochar adsorbents on dye removal.

Table 3. Adsorption isotherm studies of different dyes onto biochar.

Adsorbent	Adsorbate	Applicable Isotherm Models	Reference
Mg-Al-layered hydroxide intercalated date-palm biochar composites	Methylene blue	Langmuir, Freundlich	[124]
Rice husk biochar	Congo red	Langmuir, Freundlich	[125]
Cow dung biochar	Congo red	Langmuir, Freundlich	[125]
Magnetic chicken bone biochar	Rhodamine B	Freundlich	[126]
Chitin derived biochar	Methyl violet	Freundlich, Langmuir, Liu	[98]
Switchgrass biochar	Methylene blue	Langmuir, Freundlich	[90]
Switchgrass biochar	Orange G	Langmuir, Freundlich	[90]
Switchgrass biochar	Congo red	Langmuir, Freundlich	[90]
Banana peel extract and $FeSO_4$	Methylene blue	Langmuir, Freundlich, Dubinin-Radushkevich	[127]
Chemically modified lychee seed biochar	Methylene blue	Langmuir, Freundlich, Temkin, Dubinin-Radushkevich	[117]
Bamboo shoot shell biochar	Rhodamine B	Langmuir, Freundlich	[128]
Cetyl trimethyl ammonium bromide modified magnetic biochar from pine nut shells	Acid chrome blue K	Langmuir, Freundlich	[112]
Bovine bone biochar	Basic Red 9	Langmuir, Freundlich	[110]
Fish scale biochar	Basic Red 9	Langmuir, Freundlich	[110]

Table 4. Adsorption capacities of various biochar adsorbents for the removal of different dyes from water.

Adsorbent	Adsorbate	Adsorbent Dose	Contact Time (min)	Adsorption Capacity (mg/g)	Temperature (°C)	pH	References
Chitosan Based Material	Basic Blue 7	0.3 g/L	60	1174 mg/g	25	6	[57]
Wet-torrefied microalgal biochar	Methylene blue	1 g/L	7200	113.00 mg/g	25	6	[129]
Wet-torrefied microalgal biochar	Congo red	2 g/L	240	164.35 mg/g	25	6–8	[129]
Surfactant Modified Chitosan Beads	Crystal violet	0.36 g/L	120	97.09 mg/g	50	6	[130]
Surfactant Modified Chitosan Beads	Tartrazine	0.72 g/L	120	30.03 mg/g	50	3	[130]
Calcium rich biochar from Crab Shell	Malachite green	0.5 g/L	150	12,501.98 mg/g	25	7	[131]
Calcium rich biochar from Crab Shell	Congo red	0.5 g/L	2	20,317 mg/g	25	4	[131]
Ziziphus Lotus stones	Basic Yellow 28	0.5 g/L	180	424 mg/g	20	8	[132]
Ziziphus Lotus stones	Basic Red 46	0.5 g/L	180	307 mg/g	20	8	[132]
Date palm petiole derived biochar	Crystal violet	NA	15	209 mg/g	30	7	[133]
Activated wakame biochar material	Methylene blue	0.06 g/50 mL	NA	841.64 mg/g	20	2–12	[96]
Activated wakame biochar material	Rhodamine B	0.06 g/50 mL	NA	533.77 mg/g	20	2–12	[96]
Activated wakame biochar material	Malachite green	0.06 g/50 mL	NA	4066.96 mg/g	20	2–7	[96]
Activated biochar derived from *Opuntia ficus-indica*	Malachite green	60 mg/100 mL	120	1341.38 mg/g	30	6	[118]
Biochar derived from Caulerpa scalpelliformis	Remazol brilliant blue R	2 g/L	300	129.00 mg/g	30	2	[134]
Biochar derived from Caulerpa scalpelliformis	Remazol brilliant orange 3R	2 g/L	300	130.11 mg/g	30	2	[134]
Biochar derived from Caulerpa scalpelliformis	Remazol brilliant violet 5R	2 g/L	300	139.76 mg/g	30	2	[134]
Biochar derived from Caulerpa scalpelliformis	Remazol black B	2 g/L	300	159.08 mg/g	30	2	[134]

5.4. Mechanisms of Dyes Adsorption onto Biochar

The mechanism of removal of dyes from wastewater using biochar involves many complex interactions (both chemisorption and physisorption) between the adsorbate (dye) and adsorbent (biochar). Current literature leads us to a conclusion that the adsorption involves multiple mechanisms working together, some dominating, others based on prevailing conditions of the system. Depending on the dyes, biochar and solvent, mechanisms, such as pore-filling effect, van der Waals interaction, electrostatic interaction, chemical action, ion exchange, surface complexation, π-π interactions, cation-π interactions could play important roles in the adsorption process [135].

Adsorption of positively charged methylene blue dye (MBD) using municipal sewage sludge and tea waste derived biochar was investigated [11]. The impact of pH on adsorption showed that adsorption was influenced by the pH; for pH lower than 2, the removal efficiency of MBD was only 80%, while it was nearly 100% at a pH of 11 [11]. This strongly suggests that the adsorption mechanism involved electrostatic interaction. At low pH, the abundant H^+ ions occupied the limited bonding sites on biochar and repelled the positively charged MBD molecules, but as pH was increased, the bonded sites were deprotonated and were free for the MBD molecules to interact, thus, greatly increasing the adsorption capacity. The alternate mechanism could involve ion exchange; as the concentration of released metal cations (Ca^{2+}, Na^+, K^+, Mg^{2+}) by the biochar in the equilibrium solution was recorded and it was found that concentration of these ions was greatly increased after MBD adsorption, especially Na^+ and K^+ ions, indicating that these ions were involved in the ion exchange process.

A study of malachite green dye (MGD) on activated biochar, derived from Opuntia ficus-indica, possessing a meso-porous structure showed increased physical adsorption capabilities; which indicated that the pore filling effect was the main mechanism [118]. Hydrogen bonding between the MGD functional groups and the biochar could have supplemented the adsorption process. The activated biochar also possessed amine and hydroxyl groups (according to XRD and FTIR analyses), which are electron rich in nature, while the MGD possessed electron-deficient functional groups as part of the aromatic ring. Thus, depending on the pH of the solution, π-π electron donor–acceptor (EDA) interactions and cation- π interactions would have also been possible.

A study of surface and structural parameters of biochar and their impacts on Q_m (maximum adsorption capacity) of MBD was conducted using biochar, derived from cashew nut shells [136]. It was found that Q_m varied linearly with the volume of pores having a diameter between 1.6 and 2 nm, and adsorption capacity was mainly dependent on these pores. This was because this pore size diameter closely corresponds with the size of the MBD molecule; thus, biochar adsorption can be very selective in the wastewater treatment by closely controlling the volumes and diameter of surface pores. The study also found high correlation between Q_m and presence of weakly acidic phenolic groups on the surface of the biochar. These groups attract MBD molecules by dipole-dipole interactions or hydrogen bonding, both facilitated by the aforementioned phenolic groups.

Removal of MGD, MBD, and Rhodamine B dye (RBD) using biochar was examined and it was found that the intramolecular diffusion process, involved in adsorption, is rate-limiting [96]. The first step is fast, getting progressively slower due to the reduced adsorption intensity and decrease of unoccupied active sorption sites on biochar surface. Fourier Transform Infrared Spectroscopy (FTIR) analysis also showed the presence of hydroxyl and carbonyl groups leading to adsorption by hydrogen bonding and π-π stacking. Biochar derived from crab shells, which were rich in calcium, was studied to adsorb MGD and Congo red Dye (CRD) [131]. Zeta potential analysis showed normal electrostatic interaction activity for both dyes but FTIR and XPS analysis of biochar showed a marked difference before and after adsorption, with typical calcite bands (1440, 873 and 710 cm^{-1}) in biochar missing after CR adsorption, corroborated with the results of XPS spectra (decreased intensity of peaks at 348 and 438 eV) [131]. This suggested that calcite and CaO were involved in CR adsorption by biochar; this theory was further proven by the fact that crab shell biochar was more efficient without acid washing, indicating that the calcium ion interactions in adsorption had a positive effect in dye adsorption.

From the current literature review, it is apparent that there is a lack of studies to assess biochar performance in systems with more than one dye. Studies thus far have only utilized single dyes, adsorbed using batch processes. While studies such as [137,138] used textile effluents, which contained more than one particular dye, they failed to enumerate them and thus, the influence of the interactions of multiple dyes and biochar is unknown. It is essential, in an industrial wastewater treatment setting to determine the performance of biochar while simultaneously adsorbing multiple dyes. As [139] suggests, depending on the type of effluent, multiple treatment technologies could be combined to treat wastewater.

5.5. Potential Drawbacks of Using Biochar as an Adsorbent

In the above-mentioned sections, we extensively reviewed the potential of biochar as an adsorbent for the removal of dyes; however, the use of biochar possesses certain drawbacks. In most of the research that has been citied, we see that biochar is used under extremely specific conditions of dosage, temperature, etc., which determine how the process takes place. Hence, subsequent steps need to be performed to remove the biochar from the wastewater before reusing it for other applications, which can increase the cost of treatment. Few studies [116,126] which investigated the use of biochar as an adsorbent for dyes removal, concluded that in both cases, the efficiency of biochar decreased after prolonged usage. Therefore, we see that biochar as an adsorbent may require additional processes of desorption and that this could cause potentially economic drawbacks. Most commonly used biochars contain a significant amount of heavy metals, thereby increasing the risk of pollution due to excess desorption of these heavy metals [140–142]. Moreover, ref. [142] studied the environmental risk associated with different types of biochar containing heavy metals (derived from sludge of swine wastewater) and examined how they may affect the environment during the application of the biochar. Authors observed that the temperature, at which the biochar was prepared, played a crucial role in immobilization of heavy metal ions and it was found that 600 degrees Celsius was the ideal temperature for immobilization and heavy metals seemed to contaminate the environment at other undesirable temperatures. Moreover, ref. [140] extensively studied the effects of environmentally persistent free radicals on wastewater and soil. The free radicals (produced under certain carbonization and pyrolysis conditions), which may mobilize, may pose a significant threat to the environment if they are not monitored and managed. Furthermore, [140] research led to the conclusion that even though biochar finds a plethora of applications in various fields, it still possesses a significant threat as a pollutant to the environment if not used properly. A similar study was conducted by [141], in which the effect of temperature and raw material on biochar toxicity was studied. Their study suggested that at certain temperatures, heavy metals tend to diffuse and, hence, pollute the surrounding environment (wastewater in this case). It was observed that at exceedingly high temperatures, the heavy metals tend to volatize and temperatures of around 400 degrees proved to increase the volatility of cadmium in comparison to the other heavy metals, hence enabling pollution of the surroundings through heavy metal contamination. Therefore, we see heavy metals diffusion into the surroundings, while biochar is used as an adsorbent, is of crucial environmental concern, and despite its wide application, it does possess certain drawbacks, which need to be taken into account before its usage.

6. Conclusions and Future Perspective

Biochar was found to be an outstanding candidate for the remediation of dyes from wastewater, as evident by the literature reviewed herein. Advanced oxidation processes (AOPs), such as ozonation and Fenton oxidation processes, provide a good removal of dyes and colorants, ranging from 85% to 95% within a time span of 30 min to 1 h. However, these processes have disadvantages, such as being expensive and skill intense. Whereas, the membrane-based processes, such as nanofiltration and forward osmosis, are highly effective, and result in more than 99% removal rates within a few minutes. These membranes are highly vulnerable and need replacement very often. Biochar produced in hydrothermal processing of biomass, upon surface modification, would act as an effective adsorbent that could remove dyes from synthetic and real industrial wastewater. Pyrolysis and hydrothermal carbonization have been the most preferred techniques to processing various types of feedstocks (agricultural waste, algae biomass, sludge, plant residue, etc.) for biochar production. Most of the biochar-based adsorption processes exhibit chemisorption mechanism. Kinetic studies were exclusively carried out to these systems. Pseudo first or second order kinetic models fit well with experimental data. Furthermore, adsorption isotherm studies were also carried out by many researchers. Of all the isotherm models studied, Langmuir or Freundlich isotherm models fit most of the systems with R^2 values higher than 0.95, showing it as a good fit. The adsorption capacities of biochar-based adsorbents sometimes resulted in as low as 30.03 mg/g to as high as 20,317 mg/g. pH values play an important role in deciding the adsorption capacity of the biochar. Many studies were carried out to study the threshold pH values and results showed that in most of the cases, biochar could tolerate lower pH values up to 2 and higher pH values up to 11. Still, the removal efficiencies reached more than 80%.

The future scope in using biochar as adsorbent is huge. Engineered biochar and selective modifications will improve the efficiency of the biochar. Embedding metal frameworks derive catalyst quality biochar, which can be used for many advanced oxidation processes as a cheap catalyst. Moreover, regeneration and reuse of spent biochar adsorbent is a prominent niche in the research of biochar adsorbents. Very few studies have emerged in using spent biochar as soil amendment and additive in various biological degradation processes. Biochar can also be used as biocarriers after using as adsorbents. Studies on biochar–dye interaction will be important, as the problem of dye contamination in water is growing in most of countries worldwide.

Supplementary Materials: The following are available online at http://www.mdpi.com/2073-4441/12/12/3561/s1, Table S1: Summary of most preferred synthesis routes for biochar production from different feedstocks.

Author Contributions: P.S.: Conceptualization, Data curation; B.S.B.: Writing—original draft; V.S.: Writing—original draft; J.A.: Conceptualization; K.P.G.: Project administration, Supervision; A.B.: Review & editing. All authors have read and agreed to the published version of the manuscript.

Funding: This research received no external funding.

Acknowledgments: The authors wish to thank management of SSN College of Engineering and Sathyabama Institute of Science and Technology for the support.

Conflicts of Interest: The authors declare no conflict of interest.

References

1. Lin, Y.T.; Kao, F.Y.; Chen, S.H.; Wey, M.Y.; Tseng, H.H. A facile approach from waste to resource: Reclaimed rubber-derived membrane for dye removal. *J. Taiwan Inst. Chem. Eng.* **2020**, *112*, 286–295. [CrossRef]
2. Kazemi Shariat Panahi, H.; Dehhaghi, M.; Ok, Y.S.; Nizami, A.S.; Khoshnevisan, B.; Mussatto, S.I.; Aghbashlo, M.; Tabatabaei, M.; Lam, S.S. A comprehensive review of engineered biochar: Production, characteristics, and environmental applications. *J. Clean. Prod.* **2020**, *270*, 122462. [CrossRef]
3. Chi, N.T.L.; Anto, S.; Ahamed, T.S.; Kumar, S.S.; Shanmugam, S.; Samuel, M.S.; Mathimani, T.; Brindhadevi, K.; Pugazhendhi, A. A review on biochar production techniques and biochar based catalyst for biofuel production from algae. *Fuel* **2020**, 119411. [CrossRef]
4. Dai, Y.; Zhang, N.; Xing, C.; Cui, Q.; Sun, Q. The adsorption, regeneration and engineering applications of biochar for removal organic pollutants: A review. *Chemosphere* **2019**, *223*, 12–27. [CrossRef] [PubMed]
5. Rodriguez, J.A.; Lustosa Filho, J.F.; Melo, L.C.A.; de Assis, I.R.; de Oliveira, T.S. Influence of pyrolysis temperature and feedstock on the properties of biochars produced from agricultural and industrial wastes. *J. Anal. Appl. Pyrolysis* **2020**, *149*, 104839. [CrossRef]
6. Xu, Y.; Luo, G.; He, S.; Deng, F.; Pang, Q.; Xu, Y.; Yao, H. Efficient removal of elemental mercury by magnetic chlorinated biochars derived from co-pyrolysis of Fe(NO3)3-laden wood and polyvinyl chloride waste. *Fuel* **2019**, *239*, 982–990. [CrossRef]
7. Yang, H.; Ye, S.; Zeng, Z.; Zeng, G.; Tan, X.; Xiao, R.; Wang, J.; Song, B.; Du, L.; Qin, M.; et al. Utilization of biochar for resource recovery from water: A review. *Chem. Eng. J.* **2020**, *397*. [CrossRef]

8. Amen, R.; Yaseen, M.; Mukhtar, A.; Ullah, S.; Al-sehemi, A.G.; Ra, S.; Babar, M.; Lai, C.; Ibrahim, M.; Asif, S.; et al. Lead and cadmium removal from wastewater using eco-friendly biochar adsorbent derived from rice husk, wheat straw, and corncob. *Clean. Eng. Technol.* **2020**. [CrossRef]
9. Santos, G.E.D.S.D.; Lins, P.V.D.S.; Oliveira, L.M.T.D.M.; Da Silva, E.O.; Anastopoulos, I.; Erto, A.; Giannakoudakis, D.A.; De Almeida, A.R.F.; Duarte, J.L.D.S.; Meili, L. Layered double hydroxides/biochar composites as adsorbents for water remediation applications: Recent trends and perspectives. *J. Clean. Prod.* **2020**, 124755. [CrossRef]
10. Leichtweis, J.; Silvestri, S.; Carissimi, E. New composite of pecan nutshells biochar-ZnO for sequential removal of acid red 97 by adsorption and photocatalysis. *Biomass Bioenergy* **2020**, *140*, 105648. [CrossRef]
11. Fan, S.; Tang, J.; Wang, Y.; Li, H.; Zhang, H.; Tang, J.; Wang, Z.; Li, X. Biochar prepared from co-pyrolysis of municipal sewage sludge and tea waste for the adsorption of methylene blue from aqueous solutions: Kinetics, isotherm, thermodynamic and mechanism. *J. Mol. Liq.* **2016**, *220*, 432–441. [CrossRef]
12. Xiang, W.; Zhang, X.; Chen, J.; Zou, W.; He, F.; Hu, X.; Tsang, D.C.W.; Ok, Y.S.; Gao, B. Biochar technology in wastewater treatment: A critical review. *Chemosphere* **2020**, *252*, 126539. [CrossRef] [PubMed]
13. Benkhaya, S.; M' rabet, S.; El Harfi, A. A review on classifications, recent synthesis and applications of textile dyes. *Inorg. Chem. Commun.* **2020**, *115*, 107891. [CrossRef]
14. Bhuiyan, M.A.R.; Islam, A.; Ali, A.; Islam, M.N. Color and chemical constitution of natural dye henna (Lawsonia inermis L) and its application in the coloration of textiles. *J. Clean. Prod.* **2017**, *167*, 14–22. [CrossRef]
15. Zerin, I.; Farzana, N.; Sayem, A.S.M.; Anang, D.M.; Haider, J. *Potentials of Natural Dyes for Textile Applications*; Hashmi, S., Choudhury, I.A., Eds.; Elsevier: Oxford, UK, 2020; pp. 873–883. ISBN 978-0-12-813196-1.
16. Nambela, L.; Haule, L.V.; Mgani, Q. A review on source, chemistry, green synthesis and application of textile colorants. *J. Clean. Prod.* **2020**, *246*, 119036. [CrossRef]
17. Saxena, S.; Raja, A.S.M. *Natural Dyes: Sources, Chemistry, Application and Sustainability Issues BT—Roadmap to Sustainable Textiles and Clothing: Eco-friendly Raw Materials, Technologies, and Processing Methods*; Muthu, S.S., Ed.; Springer Singapore: Singapore, 2014; pp. 37–80. ISBN 978-981-287-065-0.
18. Degano, I.; Mattonai, M.; Sabatini, F.; Colombini, M.P. A mass spectrometric study on tannin degradation within dyed woolen yarns. *Molecules* **2019**, *24*, 2318. [CrossRef]
19. Christie, R. *Colour Chemistry*; Royal Society of Chemistry: Cambridge, UK, 2014; ISBN 1849733287.
20. Raman, C.D.; Kanmani, S. Textile dye degradation using nano zero valent iron: A review. *J. Environ. Manag.* **2016**, *177*, 341–355. [CrossRef]
21. Gürses, A.; Açıkyıldız, M.; Güneş, K.; Gürses, M.S. *Classification of Dye and Pigments BT—Dyes and Pigments*; Gürses, A., Açıkyıldız, M., Güneş, K., Gürses, M.S., Eds.; Springer International Publishing: Cham, Switzerland, 2016; pp. 31–45. ISBN 978-3-319-33892-7.
22. Clark, M. Fundamental Principles of Dyeing. In *Handbook of Textile and Industrial Dyeing*; Woodhead Publishing: Cambridge, UK, 2011; Volume 1, pp. 1–27.
23. Burkinshaw, S.M. *Physico-Chemical Aspects of Textile Coloration*; SDC-Society of Dyers and Colourists; Wiley: Hoboken, NJ, USA, 2015; ISBN 9781118725634.
24. Roy Choudhury, A.K. 2—Dyeing of synthetic fibres. In *Woodhead Publishing Series in Textiles*; Clark, M., Ed.; Woodhead Publishing: Cambridge, UK, 2011; Volume 2, pp. 40–128. ISBN 978-1-84569-696-2.
25. Jamil, A.; Bokhari, T.H.; Javed, T.; Mustafa, R.; Sajid, M.; Noreen, S.; Zuber, M.; Nazir, A.; Iqbal, M.; Jilani, M.I. Photocatalytic degradation of disperse dye Violet-26 using TiO2 and ZnO nanomaterials and process variable optimization. *J. Mater. Res. Technol.* **2020**, *9*, 1119–1128. [CrossRef]
26. Broadbent, A. Basic principles of textile coloration. *Color Res. Appl.* **2003**, *28*, 230–231. [CrossRef]
27. Chattopadhyay, D.P. 4—Chemistry of dyeing. In *Woodhead Publishing Series in Textiles*; Clark, M., Ed.; Woodhead Publishing: Cambridge, UK, 2011; Volume 1, pp. 150–183. ISBN 978-1-84569-695-5.
28. Jalandoni-Buan, A.C.; Decena-Soliven, A.L.A.; Cao, E.P.; Barraquio, V.L.; Barraquio, W.L. Characterization and identification of Congo red decolorizing bacteria from monocultures and consortia. *Philipp. J. Sci.* **2010**, *139*, 71–78.
29. Pereira, L.; Alves, M. *Dyes—Environmental Impact and Remediation BT—Environmental Protection Strategies for Sustainable Development*; Malik, A., Grohmann, E., Eds.; Springer: Dordrecht, The Netherlands, 2012; pp. 111–162, ISBN 978-94-007-1591-2.

30. Pal, P. *Chapter 6—Industry-Specific Water Treatment: Case Studies*; Pal, P., Ed.; Butterworth-Heinemann: Oxford, UK, 2017; pp. 243–511, ISBN 978-0-12-810391-3.
31. Khatri, A.; Peerzada, M.H.; Mohsin, M.; White, M. A review on developments in dyeing cotton fabrics with reactive dyes for reducing effluent pollution. *J. Clean. Prod.* **2015**, *87*, 50–57. [CrossRef]
32. Chavan, R.B. 16—Environmentally friendly dyes. In *Woodhead Publishing Series in Textiles*; Clark, M., Ed.; Woodhead Publishing: Cambridge, UK, 2011; Volume 1, pp. 515–561. ISBN 978-1-84569-695-5.
33. Desore, A.; Narula, S.A. An overview on corporate response towards sustainability issues in textile industry. *Environ. Dev. Sustain.* **2018**, *20*, 1439–1459. [CrossRef]
34. Liang, Z.; Wang, J.; Zhang, Y.; Han, C.; Ma, S.; Chen, J.; Li, G.; An, T. Removal of volatile organic compounds (VOCs) emitted from a textile dyeing wastewater treatment plant and the attenuation of respiratory health risks using a pilot-scale biofilter. *J. Clean. Prod.* **2020**, *253*, 120019. [CrossRef]
35. Khatri, J.; Nidheesh, P.V.; Anantha Singh, T.S.; Suresh Kumar, M. Advanced oxidation processes based on zero-valent aluminium for treating textile wastewater. *Chem. Eng. J.* **2018**, *348*, 67–73. [CrossRef]
36. Lellis, B.; Fávaro-Polonio, C.Z.; Pamphile, J.A.; Polonio, J.C. Effects of textile dyes on health and the environment and bioremediation potential of living organisms. *Biotechnol. Res. Innov.* **2019**, *3*, 275–290. [CrossRef]
37. Vikrant, K.; Giri, B.S.; Raza, N.; Roy, K.; Kim, K.-H.; Rai, B.N.; Singh, R.S. Recent advancements in bioremediation of dye: Current status and challenges. *Bioresour. Technol.* **2018**, *253*, 355–367. [CrossRef]
38. Xiang, X.; Chen, X.; Dai, R.; Luo, Y.; Ma, P.; Ni, S.; Ma, C. Anaerobic digestion of recalcitrant textile dyeing sludge with alternative pretreatment strategies. *Bioresour. Technol.* **2016**, *222*, 252–260. [CrossRef]
39. Issa Hamoud, H.; Finqueneisel, G.; Azambre, B. Removal of binary dyes mixtures with opposite and similar charges by adsorption, coagulation/flocculation and catalytic oxidation in the presence of CeO_2/H_2O_2 Fenton-like system. *J. Environ. Manag.* **2017**, *195*, 195–207. [CrossRef]
40. Furlan, F.R.; de Melo da Silva, L.G.; Morgado, A.F.; de Souza, A.A.U.; Guelli Ulson de Souza, S.M.A. Removal of reactive dyes from aqueous solutions using combined coagulation/flocculation and adsorption on activated carbon. *Resour. Conserv. Recycl.* **2010**, *54*, 283–290. [CrossRef]
41. Babu, D.S.; Srivastava, V.; Nidheesh, P.V.; Kumar, M.S. Detoxification of water and wastewater by advanced oxidation processes. *Sci. Total Environ.* **2019**, *696*, 133961. [CrossRef]
42. Nidheesh, P.V.; Zhou, M.; Oturan, M.A. An overview on the removal of synthetic dyes from water by electrochemical advanced oxidation processes. *Chemosphere* **2018**, *197*, 210–227. [CrossRef] [PubMed]
43. Colindres, P.; Yee-Madeira, H.; Reguera, E. Removal of Reactive Black 5 from aqueous solution by ozone for water reuse in textile dyeing processes. *Desalination* **2010**, *258*, 154–158. [CrossRef]
44. Muniyasamy, A.; Sivaporul, G.; Gopinath, A.; Lakshmanan, R.; Altaee, A.; Achary, A.; Velayudhaperumal Chellam, P. Process development for the degradation of textile azo dyes (mono-, di-, poly-) by advanced oxidation process—Ozonation: Experimental & partial derivative modelling approach. *J. Environ. Manag.* **2020**, *265*. [CrossRef]
45. Mascolo, G.; Lopez, A.; Bozzi, A.; Tiravanti, G. By-products formation during the ozonation of the reactive dye uniblu-A. *Ozone Sci. Eng.* **2002**, *24*, 439–446. [CrossRef]
46. Xu, X.; Geng, A.; Yang, C.; Carabineiro, S.A.C.; Lv, K.; Zhu, J. One-pot synthesis of La—Fe—O @ CN composites as photo-Fenton catalysts for highly e ffi cient removal of organic dyes in wastewater. *Ceram. Int.* **2020**, *46*, 10740–10747. [CrossRef]
47. Jin, J.; Du, X.; Yu, J.; Qin, S.; He, M.; Zhang, K.; Chen, G. High performance nanofiltration membrane based on SMA-PEI cross-linked coating for dye/salt separation. *J. Memb. Sci.* **2020**, *611*, 118307. [CrossRef]
48. Qi, Y.; Zhu, L.; Shen, X.; Sotto, A.; Gao, C.; Shen, J. Polythyleneimine-modified original positive charged nano filtration membrane: Removal of heavy metal ions and dyes. *Sep. Purif. Technol.* **2019**, *222*, 117–124. [CrossRef]
49. Meng, L.; Wu, M.; Chen, H.; Xi, Y.; Huang, M.; Luo, X. Rejection of antimony in dyeing and printing wastewater by forward osmosis. *Sci. Total Environ.* **2020**, *745*. [CrossRef]
50. Lin, C.; Tung, K.; Lin, Y.; Dong, C.; Wu, C. l P re of. *Process Saf. Environ. Prot.* **2020**. [CrossRef]
51. Marzbali, M.H.; Mir, A.A.; Pazoki, M.; Pourjamshidian, R.; Tabeshnia, M. Removal of direct yellow 12 from aqueous solution by adsorption onto spirulina algae as a high-efficiency adsorbent. *J. Environ. Chem. Eng.* **2017**, *5*, 1946–1956. [CrossRef]

52. Ihsanullah, I.; Jamal, A.; Ilyas, M.; Zubair, M.; Khan, G.; Atieh, M.A. Bioremediation of dyes: Current status and prospects. *J. Water Process Eng.* **2020**, *38*, 101680. [CrossRef]
53. Shao, H.; Li, Y.; Zheng, L.; Chen, T.; Liu, J. Removal of methylene blue by chemically modified defatted brown algae Laminaria japonica. *J. Taiwan Inst. Chem. Eng.* **2017**, *80*, 525–532. [CrossRef]
54. Cui, D.; Guo, Y.Q.; Lee, H.S.; Cheng, H.Y.; Liang, B.; Kong, F.Y.; Wang, Y.Z.; Huang, L.P.; Xu, M.Y.; Wang, A.J. Efficient azo dye removal in bioelectrochemical system and post-aerobic bioreactor: Optimization and characterization. *Chem. Eng. J.* **2014**, *243*, 355–363. [CrossRef]
55. Sirajudheen, P.; Karthikeyan, P.; Vigneshwaran, S.; Meenakshi, S. Synthesis and characterization of La(III) supported carboxymethylcellulose-clay composite for toxic dyes removal: Evaluation of adsorption kinetics, isotherms and thermodynamics. *Int. J. Biol. Macromol.* **2020**, *161*, 1117–1126. [CrossRef] [PubMed]
56. Saxena, M.; Sharma, N.; Saxena, R. Highly efficient and rapid removal of a toxic dye: Adsorption kinetics, isotherm, and mechanism studies on functionalized multiwalled carbon nanotubes. *Surf. Interfaces* **2020**, *21*, 100639. [CrossRef]
57. Morais da Silva, P.M.; Camparotto, N.G.; Grego Lira, K.T.; Franco Picone, C.S.; Prediger, P. Adsorptive removal of basic dye onto sustainable chitosan beads: Equilibrium, kinetics, stability, continuous-mode adsorption and mechanism. *Sustain. Chem. Pharm.* **2020**, *18*. [CrossRef]
58. Deniz, F. Adsorption Properties of Low-Cost Biomaterial Derived from *Prunus amygdalus* L. for Dye Removal from Water. *Sci. World J.* **2013**, *2013*, 961671. [CrossRef]
59. Fan, H.; Ma, Y.; Wan, J.; Wang, Y.; Li, Z.; Chen, Y. Adsorption properties and mechanisms of novel biomaterials from banyan aerial roots via simple modification for ciprofloxacin removal. *Sci. Total Environ.* **2020**, *708*, 134630. [CrossRef]
60. Niinipuu, M.; Latham, K.G.; Boily, J.F.; Bergknut, M.; Jansson, S. The impact of hydrothermal carbonization on the surface functionalities of wet waste materials for water treatment applications. *Environ. Sci. Pollut. Res.* **2020**, *27*, 24369–24379. [CrossRef]
61. Sharma, H.B.; Sarmah, A.K.; Dubey, B. Hydrothermal carbonization of renewable waste biomass for solid biofuel production: A discussion on process mechanism, the influence of process parameters, environmental performance and fuel properties of hydrochar. *Renew. Sustain. Energy Rev.* **2020**, *123*, 109761. [CrossRef]
62. Stirling, R.J.; Snape, C.E.; Meredith, W. The impact of hydrothermal carbonisation on the char reactivity of biomass. *Fuel Process. Technol.* **2018**, *177*, 152–158. [CrossRef]
63. Yusuf, I.; Flagiello, F.; Ward, N.I.; Arellano-García, H.; Avignone-Rossa, C.; Felipe-Sotelo, M. Valorisation of banana peels by hydrothermal carbonisation: Potential use of the hydrochar and liquid by-product for water purification and energy conversion. *Bioresour. Technol. Rep.* **2020**, 100582. [CrossRef]
64. Bisinoti, C.; Moreira, A.B.; Laranja, J.; Renata, C.J.; Ferreira, O.P.; Melo, C.A. Semivolatile organic compounds in the products from hydrothermal carbonisation of sugar cane bagasse and vinasse by gas chromatography-mass spectrometry. *Bioresour. Technol. Rep.* **2020**, *12*. [CrossRef]
65. Fregolente, L.G.; Miguel, T.B.A.R.; de Castro Miguel, E.; de Almeida Melo, C.; Moreira, A.B.; Ferreira, O.P.; Bisinoti, M.C. Toxicity evaluation of process water from hydrothermal carbonization of sugarcane industry by-products. *Environ. Sci. Pollut. Res.* **2019**, *26*, 27579–27589. [CrossRef] [PubMed]
66. Ischia, G.; Fiori, L. Hydrothermal Carbonization of Organic Waste and Biomass: A Review on Process, Reactor, and Plant Modeling. *Waste Biomass Valorization* **2020**. [CrossRef]
67. Marzbali, M.H.; Paz-ferreiro, J.; Kundu, S.; Ramezani, M.; Halder, P.; Patel, S.; White, T.; Madapusi, S.; Shah, K. Investigations into distribution and characterisation of products formed during hydrothermal carbonisation of paunch waste. *Biochem. Pharmacol.* **2020**, 104672. [CrossRef]
68. Zhang, Z.; Zhu, Z.; Shen, B.; Liu, L. Insights into biochar and hydrochar production and applications: A review. *Energy* **2019**, *171*, 581–598. [CrossRef]
69. Wang, L.; Chang, Y.; Liu, Q. Fate and distribution of nutrients and heavy metals during hydrothermal carbonization of sewage sludge with implication to land application. *J. Clean. Prod.* **2019**, *225*, 972–983. [CrossRef]
70. Maniscalco, M.P.; Volpe, M.; Messineo, A. Hydrothermal carbonization as a valuable tool for energy and environmental applications: A review. *Energies* **2020**, *13*, 4098. [CrossRef]
71. Ponnusamy, V.K.; Nagappan, S.; Bhosale, R.R.; Lay, C.H.; Duc Nguyen, D.; Pugazhendhi, A.; Chang, S.W.; Kumar, G. Review on sustainable production of biochar through hydrothermal liquefaction: Physico-chemical properties and applications. *Bioresour. Technol.* **2020**, *310*, 123414. [CrossRef]

72. Mathimani, T.; Mallick, N. A review on the hydrothermal processing of microalgal biomass to bio-oil-Knowledge gaps and recent advances. *J. Clean. Prod.* **2019**, *217*, 69–84. [CrossRef]
73. Brindhadevi, K.; Anto, S.; Rene, E.R.; Sekar, M. Effect of reaction temperature on the conversion of algal biomass to bio-oil and biochar through pyrolysis and hydrothermal liquefaction. *Fuel* **2021**, *285*, 119106. [CrossRef]
74. Liu, H.; Ma, M.; Xie, X. New materials from solid residues for investigation the mechanism of biomass hydrothermal liquefaction. *Ind. Crop. Prod.* **2017**, *108*, 63–71. [CrossRef]
75. Sun, Y.; Yang, G.; Zhang, L.; Sun, Z. Preparation of high performance H2S removal biochar by direct fluidized bed carbonization using potato peel waste. *Process Saf. Environ. Prot.* **2017**, *107*, 281–288. [CrossRef]
76. Li, F.; Gui, X.; Ji, W.; Zhou, C. Effect of calcium dihydrogen phosphate addition on carbon retention and stability of biochars derived from cellulose, hemicellulose, and lignin. *Chemosphere* **2020**, *251*, 126335. [CrossRef]
77. Li, J.; Li, Y.; Wu, Y.; Zheng, M. A comparison of biochars from lignin, cellulose and wood as the sorbent to an aromatic pollutant. *J. Hazard. Mater.* **2014**, *280*, 450–457. [CrossRef]
78. Kim, D.; Lee, K.; Park, K.Y. Upgrading the characteristics of biochar from cellulose, lignin, and xylan for solid biofuel production from biomass by hydrothermal carbonization. *J. Ind. Eng. Chem.* **2016**, *42*, 95–100. [CrossRef]
79. Qin, L.; Wu, Y.; Hou, Z.; Jiang, E. Influence of biomass components, temperature and pressure on the pyrolysis behavior and biochar properties of pine nut shells. *Bioresour. Technol.* **2020**, *313*, 123682. [CrossRef]
80. Muppaneni, T.; Reddy, H.K.; Selvaratnam, T.; Dandamudi, K.P.R.; Dungan, B.; Nirmalakhandan, N.; Schaub, T.; Omar Holguin, F.; Voorhies, W.; Lammers, P.; et al. Hydrothermal liquefaction of Cyanidioschyzon merolae and the influence of catalysts on products. *Bioresour. Technol.* **2017**, *223*, 91–97. [CrossRef]
81. Cui, Z.; Cheng, F.; Jarvis, J.M.; Brewer, C.E.; Jena, U. Roles of Co-solvents in hydrothermal liquefaction of low-lipid, high-protein algae. *Bioresour. Technol.* **2020**, *310*, 123454. [CrossRef]
82. Shen, Y.; Yu, S.; Yuan, R.; Wang, P. Biomass pyrolysis with alkaline-earth-metal additive for co-production of bio-oil and biochar-based soil amendment. *Sci. Total Environ.* **2020**, *743*, 140760. [CrossRef] [PubMed]
83. Yuan, S.; Hong, M.; Li, H.; Ye, Z.; Gong, H.; Zhang, J.; Huang, Q.; Tan, Z. Contributions and mechanisms of components in modified biochar to adsorb cadmium in aqueous solution. *Sci. Total Environ.* **2020**, *733*, 139320. [CrossRef] [PubMed]
84. Wu, L.; Zhang, S.; Wang, J.; Ding, X. Phosphorus retention using iron (II/III) modified biochar in saline-alkaline soils: Adsorption, column and field tests. *Environ. Pollut.* **2020**, *261*, 114223. [CrossRef] [PubMed]
85. Nakarmi, A.; Bourdo, S.E.; Ruhl, L.; Kanel, S.; Nadagouda, M.; Kumar Alla, P.; Pavel, I.; Viswanathan, T. Benign zinc oxide betaine-modified biochar nanocomposites for phosphate removal from aqueous solutions. *J. Environ. Manag.* **2020**, *272*, 111048. [CrossRef] [PubMed]
86. Li, Y.; Xing, B.; Ding, Y.; Han, X.; Wang, S. A critical review of the production and advanced utilization of biochar via selective pyrolysis of lignocellulosic biomass. *Bioresour. Technol.* **2020**, *312*, 123614. [CrossRef] [PubMed]
87. Elkhalifa, S.; Al-Ansari, T.; Mackey, H.R.; McKay, G. Food waste to biochars through pyrolysis: A review. *Resour. Conserv. Recycl.* **2019**, *144*, 310–320. [CrossRef]
88. Tripathi, M.; Sahu, J.N.; Ganesan, P. Effect of process parameters on production of biochar from biomass waste through pyrolysis: A review. *Renew. Sustain. Energy Rev.* **2016**, *55*, 467–481. [CrossRef]
89. Wang, X.; Li, C.; Li, Z.; Yu, G.; Wang, Y. Effect of pyrolysis temperature on characteristics, chemical speciation and risk evaluation of heavy metals in biochar derived from textile dyeing sludge. *Ecotoxicol. Environ. Saf.* **2019**, *168*, 45–52. [CrossRef]
90. Park, J.H.; Wang, J.J.; Meng, Y.; Wei, Z.; DeLaune, R.D.; Seo, D.C. Adsorption/desorption behavior of cationic and anionic dyes by biochars prepared at normal and high pyrolysis temperatures. *Colloids Surf. A Physicochem. Eng. Asp.* **2019**, *572*, 274–282. [CrossRef]
91. Chen, X.L.; Li, F.; Chen, H.Y.; Wang, H.J.; Li, G.G. Fe_2O_3/TiO_2 functionalized biochar as a heterogeneous catalyst for dyes degradation in water under Fenton processes. *J. Environ. Chem. Eng.* **2020**, *8*, 103905. [CrossRef]
92. Yek, P.N.Y.; Peng, W.; Wong, C.C.; Liew, R.K.; Ho, Y.L.; Wan Mahari, W.A.; Azwar, E.; Yuan, T.Q.; Tabatabaei, M.; Aghbashlo, M.; et al. Engineered biochar via microwave CO_2 and steam pyrolysis to treat carcinogenic Congo red dye. *J. Hazard. Mater.* **2020**, *395*. [CrossRef] [PubMed]

93. Chu, J.H.; Kang, J.K.; Park, S.J.; Lee, C.G. Application of magnetic biochar derived from food waste in heterogeneous sono-Fenton-like process for removal of organic dyes from aqueous solution. *J. Water Process Eng.* **2020**, *37*, 101455. [CrossRef]
94. Mahmoud, M.E.; Abdelfattah, A.M.; Tharwat, R.M.; Nabil, G.M. Adsorption of negatively charged food tartrazine and sunset yellow dyes onto positively charged triethylenetetramine biochar: Optimization, kinetics and thermodynamic study. *J. Mol. Liq.* **2020**, *318*, 114297. [CrossRef]
95. Zhang, H.; Lu, T.; Wang, M.; Jin, R.; Song, Y.; Zhou, Y.; Qi, Z.; Chen, W. Inhibitory role of citric acid in the adsorption of tetracycline onto biochars: Effects of solution pH and Cu^{2+}. *Colloids Surf. A Physicochem. Eng. Asp.* **2020**, *595*. [CrossRef]
96. Yao, X.; Ji, L.; Guo, J.; Ge, S.; Lu, W.; Chen, Y.; Cai, L.; Wang, Y.; Song, W. An abundant porous biochar material derived from wakame (Undaria pinnatifida) with high adsorption performance for three organic dyes. *Bioresour. Technol.* **2020**, *318*, 124082. [CrossRef] [PubMed]
97. Rubeena, K.K.; Prasad, P.H.; Laiju, A.R.; Nidheesh, P. V Iron impregnated biochars as heterogeneous Fenton catalyst for the degradation of acid red 1 dye. *J. Environ. Manag.* **2018**, *226*, 320–328. [CrossRef] [PubMed]
98. Zazycki, M.A.; Borba, P.A.; Silva, R.N.F.; Peres, E.C.; Perondi, D.; Collazzo, G.C.; Dotto, G.L. Chitin derived biochar as an alternative adsorbent to treat colored effluents containing methyl violet dye. *Adv. Powder Technol.* **2019**, *30*, 1494–1503. [CrossRef]
99. Nautiyal, P.; Subramanian, K.A.; Dastidar, M.G. Adsorptive removal of dye using biochar derived from residual algae after in-situ transesteri fi cation: Alternate use of waste of biodiesel industry. *J. Environ. Manag.* **2016**, *182*, 187–197. [CrossRef]
100. Damertey, D.; Jung, H.; Soo, S.; Sung, D.; Han, S. Decolorization of cationic and anionic dye-laden wastewater by steam- activated biochar produced at an industrial-scale from spent mushroom substrate. *Bioresour. Technol.* **2019**, *277*, 77–86. [CrossRef]
101. Wang, H.; Zhao, W.; Chen, Y.; Li, Y. Nickel aluminum layered double oxides modified magnetic biochar from waste corncob for efficient removal of acridine orange. *Bioresour. Technol.* **2020**, *315*, 123834. [CrossRef]
102. Yu, K.L.; Lee, X.J.; Ong, H.C.; Chen, W.-H.; Chang, J.-S.; Lin, C.-S.; Show, P.L.; Ling, T.C. Adsorptive removal of cationic methylene blue and anionic Congo red dyes using wet-torrefied microalgal biochar: Equilibrium, kinetic and mechanism modeling. *Environ. Pollut.* **2020**, 115986. [CrossRef] [PubMed]
103. Iqbal, J.; Shah, N.S.; Sayed, M.; Niazi, N.K.; Imran, M.; Khan, J.A.; Khan, Z.U.H.; Hussien, A.G.S.; Polychronopoulou, K.; Howari, F. Nano-zerovalent manganese/biochar composite for the adsorptive and oxidative removal of Congo-red dye from aqueous solutions. *J. Hazard. Mater.* **2021**, *403*, 123854. [CrossRef]
104. Ganguly, P.; Sarkhel, R.; Das, P. Synthesis of pyrolyzed biochar and its application for dye removal: Batch, kinetic and isotherm with linear and non-linear mathematical analysis. *Surf. Interfaces* **2020**, *20*, 100616. [CrossRef]
105. Kumar, S.; Abdel-fattah, T.M. Kinetics, isotherm, and thermodynamic studies of the adsorption of reactive red 195 A dye from water by modified Switchgrass Biochar adsorbent. *J. Ind. Eng. Chem.* **2016**. [CrossRef]
106. Sahu, S.; Pahi, S.; Tripathy, S.; Singh, S.K.; Behera, A.; Sahu, U.K.; Patel, R. Adsorption of methylene blue on chemically modified lychee seed biochar: Dynamic, equilibrium, and thermodynamic study. *J. Mol. Liq.* **2020**, *315*, 113743. [CrossRef]
107. Zhu, K.; Wang, X.; Chen, D.; Ren, W.; Lin, H.; Zhang, H. Wood-based biochar as an excellent activator of peroxydisulfate for Acid Orange 7 decolorization. *Chemosphere* **2019**, *231*, 32–40. [CrossRef] [PubMed]
108. Wang, J.; Guo, X. Adsorption kinetic models: Physical meanings, applications, and solving methods. *J. Hazard. Mater.* **2020**, *390*, 122156. [CrossRef] [PubMed]
109. Biswas, S.; Mohapatra, S.S.; Kumari, U.; Meikap, B.C.; Sen, T.K. Batch and continuous closed circuit semi-fluidized bed operation: Removal of MB dye using sugarcane bagasse biochar and alginate composite adsorbents. *J. Environ. Chem. Eng.* **2020**, *8*, 103637. [CrossRef]
110. Côrtes, L.N.; Druzian, S.P.; Streit, A.F.M.; Godinho, M.; Perondi, D.; Collazzo, G.C.; Oliveira, M.L.S.; Cadaval, T.R.S.; Dotto, G.L. Biochars from animal wastes as alternative materials to treat colored effluents containing basic red 9. *J. Environ. Chem. Eng.* **2019**, *7*. [CrossRef]
111. Xu, X.; Xu, Z.; Huang, J.; Gao, B.; Zhao, L.; Qiu, H.; Cao, X. Sorption of reactive red by biochars ball milled in different atmospheres: Co-effect of surface morphology and functional groups. *Chem. Eng. J.* **2020**, 127468. [CrossRef]

112. Wang, H.; Wang, S.; Gao, Y. Cetyl trimethyl ammonium bromide modified magnetic biochar from pine nut shells for efficient removal of acid chrome blue K. *Bioresour. Technol.* **2020**, *312*, 123564. [CrossRef] [PubMed]
113. Wang, L.; Yan, W.; He, C.; Wen, H.; Cai, Z.; Wang, Z.; Chen, Z.; Liu, W. Microwave-assisted preparation of nitrogen-doped biochars by ammonium acetate activation for adsorption of acid red 18. *Appl. Surf. Sci.* **2018**, *433*, 222–231. [CrossRef]
114. Chen, L.; Jiang, X.; Xie, R.; Zhang, Y.; Jin, Y.; Jiang, W. A novel porous biochar-supported Fe-Mn composite as a persulfate activator for the removal of acid red 88. *Sep. Purif. Technol.* **2020**, *250*, 117232. [CrossRef]
115. Hoslett, J.; Ghazal, H.; Mohamad, N.; Jouhara, H. Removal of methylene blue from aqueous solutions by biochar prepared from the pyrolysis of mixed municipal discarded material. *Sci. Total Environ.* **2020**, *714*, 136832. [CrossRef] [PubMed]
116. Tran, H.N. Comments on "High-efficiency removal of dyes from wastewater by fully recycling litchi peel biochar". *Chemosphere* **2020**, *257*, 126444. [CrossRef] [PubMed]
117. Ang, T.N.; Young, B.R.; Taylor, M.; Burrell, R.; Aroua, M.K.; Chen, W.-H.; Baroutian, S. Enrichment of surface oxygen functionalities on activated carbon for adsorptive removal of sevoflurane. *Chemosphere* **2020**, *260*, 127496. [CrossRef] [PubMed]
118. Choudhary, M.; Kumar, R.; Neogi, S. Activated biochar derived from Opuntia ficus-indica for the efficient adsorption of malachite green dye, Cu^{+2} and Ni^{+2} from water. *J. Hazard. Mater.* **2020**, *392*, 122441. [CrossRef]
119. Shan, R.; Lu, L.; Gu, J.; Zhang, Y.; Yuan, H.; Chen, Y.; Luo, B. Photocatalytic degradation of methyl orange by Ag/TiO2/biochar composite catalysts in aqueous solutions. *Mater. Sci. Semicond. Process.* **2020**, *114*, 105088. [CrossRef]
120. Al-Ghouti, M.A.; Da'ana, D.A. Guidelines for the use and interpretation of adsorption isotherm models: A review. *J. Hazard. Mater.* **2020**, *393*, 122383. [CrossRef]
121. Choi, J.; Won, W.; Capareda, S.C. The economical production of functionalized Ashe juniper derived-biochar with high hazardous dye removal efficiency. *Ind. Crop. Prod.* **2019**, *137*, 672–680. [CrossRef]
122. Ji, B.; Wang, J.; Song, H.; Chen, W. Removal of methylene blue from aqueous solutions using biochar derived from a fallen leaf by slow pyrolysis: Behavior and mechanism. *J. Environ. Chem. Eng.* **2019**, *7*, 103036. [CrossRef]
123. Jabar, J.M.; Odusote, Y.A. Removal of cibacron blue 3G-A (CB) dye from aqueous solution using chemo-physically activated biochar from oil palm empty fruit bunch fiber. *Arab. J. Chem.* **2020**, *13*, 5417–5429. [CrossRef]
124. Zubair, M.; Manzar, M.S.; Mu'azu, N.D.; Anil, I.; Blaisi, N.I.; Al-Harthi, M.A. Functionalized MgAl-layered hydroxide intercalated date-palm biochar for Enhanced Uptake of Cationic dye: Kinetics, isotherm and thermodynamic studies. *Appl. Clay Sci.* **2020**, *190*, 105587. [CrossRef]
125. Khan, N.; Chowdhary, P.; Ahmad, A.; Giri, B.S.; Chaturvedi, P. Hydrothermal liquefaction of rice husk and cow dung in Mixed-Bed-Rotating Pyrolyzer and application of biochar for dye removal. *Bioresour. Technol.* **2020**, *309*, 123294. [CrossRef]
126. Oladipo, A.A.; Ifebajo, A.O. Highly efficient magnetic chicken bone biochar for removal of tetracycline and fluorescent dye from wastewater: Two-stage adsorber analysis. *J. Environ. Manag.* **2018**, *209*, 9–16. [CrossRef]
127. Zhang, P.; O'Connor, D.; Wang, Y.; Jiang, L.; Xia, T.; Wang, L.; Tsang, D.C.W.; Ok, Y.S.; Hou, D. A green biochar/iron oxide composite for methylene blue removal. *J. Hazard. Mater.* **2020**, *384*, 121286. [CrossRef]
128. Hou, Y.; Huang, G.; Li, J.; Yang, Q.; Huang, S.; Cai, J. Hydrothermal conversion of bamboo shoot shell to biochar: Preliminary studies of adsorption equilibrium and kinetics for rhodamine B removal. *J. Anal. Appl. Pyrolysis* **2019**, *143*, 104694. [CrossRef]
129. Muhammad, N.Z.; Qamar, D.; Faisal, N.; Muhammad, N.Z.; Munawar, I.; Muhammad, F.N. Effective adsorptive removal of azo dyes over spherical ZnO nanoparticles. *J. Mater. Res. Technol.* **2019**, *8*, 713–725.
130. Pal, P.; Pal, A. Dye removal using waste beads: Efficient utilization of surface-modified chitosan beads generated after lead adsorption process. *J. Water Process Eng.* **2019**, *31*. [CrossRef]
131. Dai, L.; Zhu, W.; He, L.; Tan, F.; Zhu, N.; Zhou, Q.; He, M.; Hu, G. Calcium-rich biochar from crab shell: An unexpected super adsorbent for dye removal. *Bioresour. Technol.* **2018**, *267*, 510–516. [CrossRef]
132. Boudechiche, N.; Fares, M.; Ouyahia, S.; Yazid, H.; Trari, M.; Sadaoui, Z. Comparative study on removal of two basic dyes in aqueous medium by adsorption using activated carbon from Ziziphus lotus stones. *Microchem. J.* **2019**, *146*, 1010–1018. [CrossRef]

133. Chahinez, H.-O.; Abdelkader, O.; Leila, Y.; Tran, H.N. One-stage preparation of palm petiole-derived biochar: Characterization and application for adsorption of crystal violet dye in water. *Environ. Technol. Innov.* **2020**, *19*, 100872. [CrossRef]
134. Gokulan, R.; Avinash, A.; Prabhu, G.G.; Jegan, J. Remediation of remazol dyes by biochar derived from Caulerpa scalpelliformis—An eco-friendly approach. *J. Environ. Chem. Eng.* **2019**, *7*, 103297. [CrossRef]
135. Kah, M.; Sigmund, G.; Xiao, F.; Hofmann, T. Sorption of ionizable and ionic organic compounds to biochar, activated carbon and other carbonaceous materials. *Water Res.* **2017**, *124*, 673–692. [CrossRef] [PubMed]
136. Spagnoli, A.A.; Giannakoudakis, D.A.; Bashkova, S. Adsorption of methylene blue on cashew nut shell based carbons activated with zinc chloride: The role of surface and structural parameters. *J. Mol. Liq.* **2017**, *229*, 465–471. [CrossRef]
137. Del Bubba, M.; Anichini, B.; Bakari, Z.; Bruzzoniti, M.C.; Camisa, R.; Caprini, C.; Checchini, L.; Fibbi, D.; El Ghadraoui, A.; Liguori, F.; et al. Physicochemical properties and sorption capacities of sawdust-based biochars and commercial activated carbons towards ethoxylated alkylphenols and their phenolic metabolites in effluent wastewater from a textile district. *Sci. Total Environ.* **2020**, *708*, 135217. [CrossRef]
138. Santra, B.; Ramrakhiani, L.; Kar, S.; Ghosh, S.; Majumdar, S. Ceramic membrane-based ultrafiltration combined with adsorption by waste derived biochar for textile effluent treatment and management of spent biochar. *J. Environ. Health Sci. Eng.* **2020**. [CrossRef]
139. De Gisi, S.; Notarnicola, M. *Industrial Wastewater Treatment*; Abraham, M., Ed.; Elsevier: Oxford, UK, 2017; pp. 23–42, ISBN 978-0-12-804792-7.
140. Odinga, E.S.; Waigi, M.G.; Gudda, F.O.; Wang, J.; Yang, B.; Hu, X.; Li, S.; Gao, Y. Occurrence, formation, environmental fate and risks of environmentally persistent free radicals in biochars. *Environ. Int.* **2020**, *134*, 105172. [CrossRef]
141. Zhang, Y.; Chen, Z.; Xu, W.; Liao, Q.; Zhang, H.; Hao, S.; Chen, S. Pyrolysis of various phytoremediation residues for biochars: Chemical forms and environmental risk of Cd in biochar. *Bioresour. Technol.* **2020**, *299*, 122581. [CrossRef]
142. Zhang, Q.; Ye, X.; Li, H.; Chen, D.; Xiao, W.; Zhao, S.; Xiong, R.; Li, J. Cumulative effects of pyrolysis temperature and process on properties, chemical speciation, and environmental risks of heavy metals in magnetic biochar derived from coagulation-flocculation sludge of swine wastewater. *J. Environ. Chem. Eng.* **2020**, *8*, 104472. [CrossRef]

Publisher's Note: MDPI stays neutral with regard to jurisdictional claims in published maps and institutional affiliations.

Review

An Overview and Evaluation of Highly Porous Adsorbent Materials for Polycyclic Aromatic Hydrocarbons and Phenols Removal from Wastewater

Zakariyya Uba Zango [1,2,*], Nonni Soraya Sambudi [3,*], Khairulazhar Jumbri [1], Anita Ramli [1], Noor Hana Hanif Abu Bakar [4], Bahruddin Saad [1], Muhammad Nur' Hafiz Rozaini [1], Hamza Ahmad Isiyaka [1], Abubaker Mohammed Osman [5] and Abdelmoneim Sulieman [6]

1 Fundamental and Applied Sciences Department, Universiti Teknologi PETRONAS, Seri Iskandar 32610, Perak, Malaysia; khairulazhar.jumbri@utp.edu.my (K.J.); anita_ramli@utp.edu.my (A.R.); bahrudsaad@gmail.com (B.S.); Muhammad_18000735@utp.edu.my (M.N.H.R.); hamza_18001996@utp.edu.my (H.A.I.)

2 Chemistry Department, Colllege of Natural and Applied Sciences, Al-Qalam University, 2137 Katsina, Nigeria

3 Chemical Engineering Department, Universiti Teknologi PETRONAS, Seri Iskandar 32610, Perak, Malaysia

4 School of Chemical Sciences, Universiti Sains Malaysia, Pinang 11800, Malaysia; hana_hanif@usm.my

5 Chemistry Department, College of Science, King Khalid University, Abha 62529, Saudi Arabia; abubaker123@gmail.com

6 Radiology and Medical Imaging Department, College of Applied Medical Sciences, Prince Sattam Bin Abduaziz University, Alkharj 16278, Saudi Arabia; a.sulieman@psau.edu.sa

* Correspondence: zakariyyazango4@gmail.com (Z.U.Z.); soraya.sambudi@utp.edu.my (N.S.S.); Tel.: +60-174269144 (Z.U.Z.); +60-18957481 (N.S.S.)

Received: 15 September 2020; Accepted: 13 October 2020; Published: 19 October 2020

Abstract: Polycyclic aromatic hydrocarbons (PAHs) and phenolic compounds had been widely recognized as priority organic pollutants in wastewater with toxic effects on both plants and animals. Thus, the remediation of these pollutants has been an active area of research in the field of environmental science and engineering. This review highlighted the advantage of adsorption technology in the removal of PAHs and phenols in wastewater. The literature presented on the applications of various porous carbon materials such as biochar, activated carbon (AC), carbon nanotubes (CNTs), and graphene as potential adsorbents for these pollutants has been critically reviewed and analyzed. Under similar conditions, the use of porous polymers such as Chitosan and molecularly imprinted polymers (MIPs) have been well presented. The high adsorption capacities of advanced porous materials such as mesoporous silica and metal-organic frameworks have been considered and evaluated. The preference of these materials, higher adsorption efficiencies, mechanism of adsorptions, and possible challenges have been discussed. Recommendations have been proposed for commercialization, pilot, and industrial-scale applications of the studied adsorbents towards persistent organic pollutants (POPs) removal from wastewater.

Keywords: adsorption; phenols; polycyclic aromatic hydrocarbons; wastewater

---

1. Introduction

Organic pollutants, such as polycyclic aromatic hydrocarbons (PAHs), phenols, and their derivatives, are one of the major pollutants that are frequently detected in wastewater. They are usually originated from both natural and anthropogenic sources, such as bush-burning, volcanic eruptions, petroleum exploration and refining, mining etc. [1,2]. They are usually run-off into the water bodies where they reside in the ground and surface water, thus resulting into environmental problem for decades. They are highly lipophilic, semi volatile, and accumulative in water and can be transported into various water compartments due to their non-biodegradable and persistent nature [3,4]. They are categorized as persistent organic pollutants (POP's) in the water. When present in the living cells, they are capable of undergoing bioaccumulating and biomagnifying through food webs into different trophic levels of organisms, and subsequently into humans [5].

PAHs are organic compounds consisting of two or more fused benzene ring molecules. They mostly result from petroleum refining, fuel combustion, coal mining and processing, coke production, oil shale pyrolysis, chemicals production, and other industrial processes. Various research conducted in different parts of the world have shown considerable increase in the PAHs concentrations in the receiving water bodies. Thus, PAHs could be ubiquitously transported into the rivers [6,7], fresh and terrestrial water [8], municipal wastewater treatment plants [9], and consequently exposed to the human being and aquatic organism. They are classified based on the number of rings they contained into low molecular weight (LMW-PAHs) and higher molecular weight (HMW-PAHs). The former comprised those PAHs with two to three benzene rings, while the latter are made of four to six rings. In comparisons, the latter are more toxic and persistent to the water due to their low solubility [10,11]. The United States Protection Agency (USEPA) has listed 16 PAHs as potential toxic and carcinogenic and thus they have priority concerned. The structures, molecular weight, and log *p* values of the PAHs are highlighted in Table 1.

Table 1. Chemical structures and log *p* values of polycyclic aromatic hydrocarbons (PAHs).

Compound	Formula	Abbreviation	Structure	Molecular Weight (g/mol)	Partition Coefficient
Naphthalene	$C_{10}H_8$	NAP	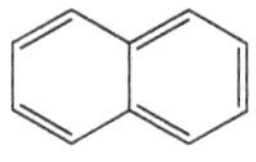	128.17	3.37
Acenaphthylene	$C_{12}H_8$	ACE	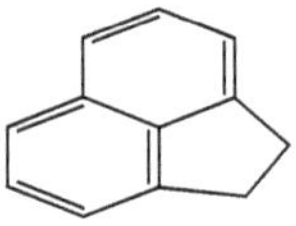	152.1	4.0
Acenaphthene	C_1H_{10}	ACE	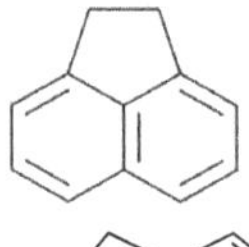	154.21	3.92
Fluorene	$C_{13}H_{10}$	FLU	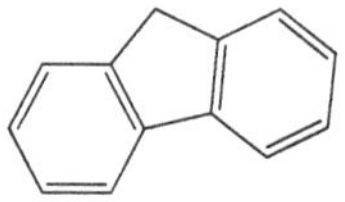	166.22	4.18
Phenanthrene	$C_{14}H_{10}$	PHE	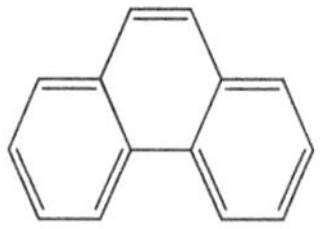	178.23	4.57
Anthracene	$C_{14}H_{10}$	ANT	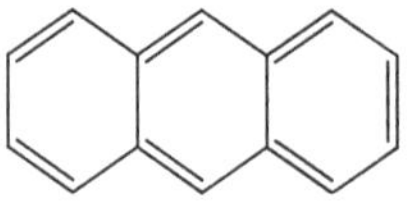	178.23	4.54

Table 1. *Cont.*

Compound	Formula	Abbreviation	Structure	Molecular Weight (g/mol)	Partition Coefficient
Fluoranthene	$C_{16}H_{10}$	FL	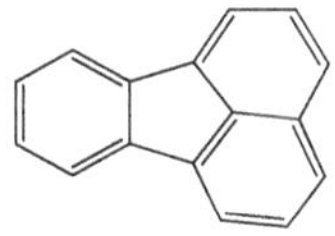	202.25	4.58
Pyrene	$C_{16}H_{10}$	PYR	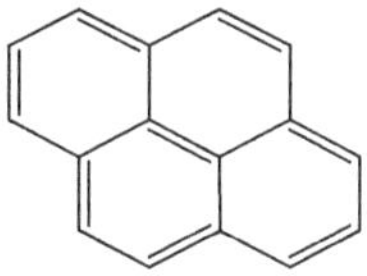	202.25	4.58
Benz[a]anthracene	$C_{18}H_{12}$	B[a]A	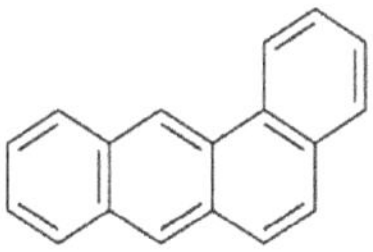	228.3	6.14
Chrysene	$C_{18}H_{12}$	CRY		228.29	5.30
Benzo[b]fluoranthene	$C_{20}H_{12}$	BbF	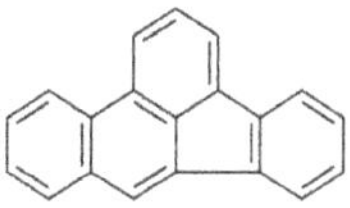	252.3	5.74
Benz[k]fluoranthene	$C_{20}H_{12}$	BkF	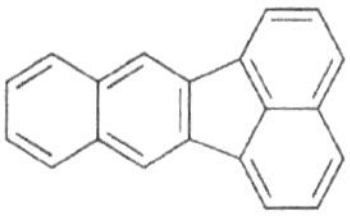	252.3	5.74
Benz[a]pyrene	$C_{20}H_{12}$	BaP	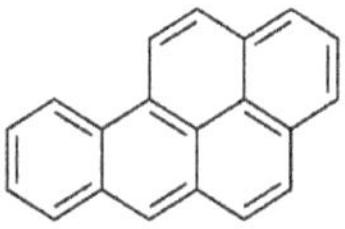	252.3	6.74
Benzo[ghi]perylene	$C_{22}H_{12}$	BhP		276.3	6.52
Dibenz[a,h]anthracene	$C_{22}H_{14}$	DahA	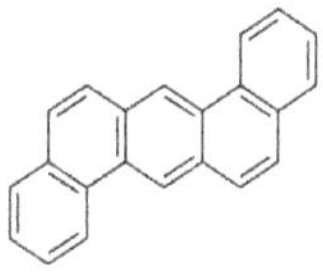	278.4	6.20
Indeno[1,2,3-c,d]pyrene	$C_{22}H_{12}$	IP	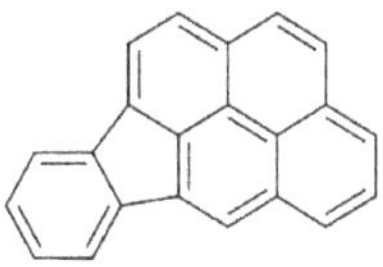	276.3	6.20

Phenols are a class of organic compounds having hydroxyl group (–OH) attached to the carbon atom that is part of the aromatic ring. They have found numerous applications as intermediates for industrial production of petrochemicals, dyes, herbicides, pesticides, pharmaceuticals and cosmetics, etc. [12]. Apart from that, they are also used in household products such as flavorings, cleaners and mouthwash. Thus, they are usually discharged in the wastewater and effluents from those industries [13,14]. Even though small quantity of phenols are produced from the decomposition of plants and animals, but anthropogenic sources has been a major contributor of phenols in the environment wastewater effluents in high concentration up to even several thousands of mg/L [15,16]. Considering their adverse effect to living organisms in the environment, they are also classified as priority pollutants [17,18]. The frequent discharge of phenolics without treatment leads to environmental toxicities to the extent that they are posing serious threats to human health. Some of their toxic effects includes; skin and eyes irritations, respiratory complications, weight loss, diarrhea, vertigo, salivation, and dark coloration of urine etc. [19,20]. Table 2 presents some of the most frequently reported phenols and their dissociation constants (PKa).

Table 2. Chemical structures and log *p* values of polycyclic aromatic hydrocarbons studied.

Compound	Formula	Structure	Molecular Weight (g/mol)	Partition Coefficient
Phenol	C_6H_6O	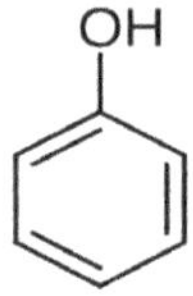 OH	94.11	10.0
Cresol	C_7H_8O	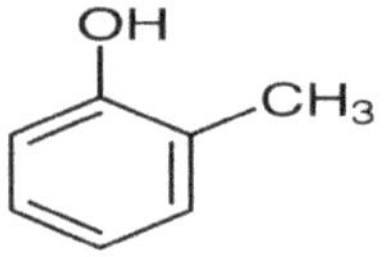 OH, CH_3	108.14	10.28
Resorcinol	$C_6H_6O_2$	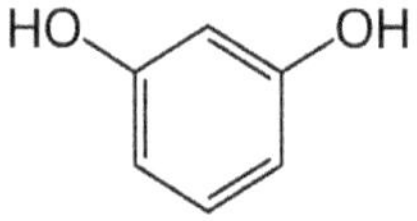 HO, OH	110.1	19.15
Hydroquinone	$C_6H_6O_2$	 HO, OH	110.11	9.9
Pyrogallol	$C_6H_6O_3$	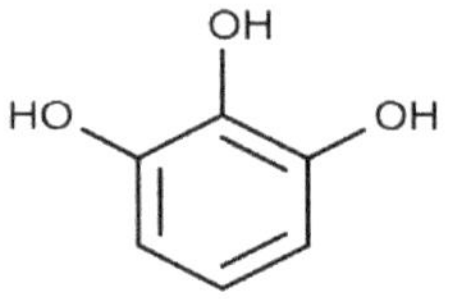 OH, HO, OH	126.11	9.03
Naphthol	C_6H_8O	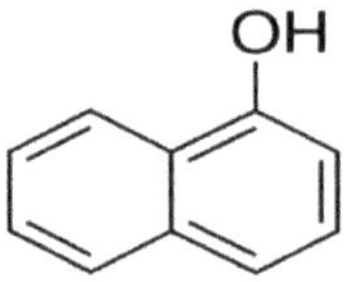 OH	144.17	9.51
Salicylic acid	$C_7H_6O_3$	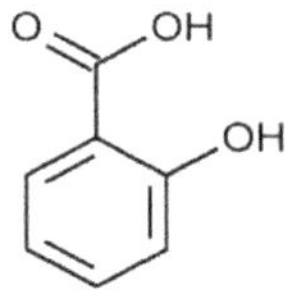 O, OH, OH	178.23	2.97

Table 2. *Cont.*

Compound	Formula	Structure	Molecular Weight (g/mol)	Partition Coefficient
Picric acid	$C_6H_3N_3O_7$	OH, O_2N, NO_2, NO_2	229.1	0.38
Bisphenol A	$C_{15}H_{16}O_2$	HO, OH	228.29	10.29
Chlorophenol	C_6H_5ClO	OH, Cl	128.6	9.12
Nitrophenol	$C_6H_5NO_3$	OH, NO_2	139.10	7.15

In recognition of the mobilities and high toxicities of these pollutants in environmental waters, a prompt response from various environmental monitoring and protection is needed to reduce their concentrations, or to ideally completely eliminate them. Thus, considerable attention has been paid to develop suitable techniques for their effective remediation [21,22]. Over the last three decades, extensive studies have been conducted using different technologies including coagulation and flocculation, phytoremediation, reverse and forward osmosis, chemical oxidation, photocatalytic degradation and adsorption [23–25]. Table 3 highlights the pros and cons of some of the remediation techniques employed. Generally, most of these techniques failed to address the persistent problem, partly because the pollutants are highly lipophilic. In a conventional wastewater treatment plant (WWTPs), they are resistant to bioremediation and photolysis is insignificant for their degradations [26]. Additionally, the excessive chemical remediation might cause secondary pollution because of the unreasonable proportions of the reagents used and formation of oxygenated species that are potentially hazardous [27,28]. Photocatalytic degradation has been limited by the semi conducting property of the materials and the utilization of light energy. Additionally, most of the techniques are non-feasible and non-economical.

Table 3. Pros and Cons of Wastewater treatments process used for PAHs and phenols remediation.

Method	Pros	Cons
Coagulation	– The colloidal particles are easily settled down by the coagulants added to the wastewater	– It is time-consuming – Incomplete removal of the organic pollutants are always the case – It resulted in secondary pollution inform of sludge
Flocculation	– Availability of various flocculants that have been commercialized. – Some of the flocculants have shown a high affinity to the organic pollutants.	– It is highly dependent on the physicochemical parameters of the pollutants; such as pH and ionic strength. – The flocculants often resulted in secondary pollution.
Bioremediation; including the use of microbes such as bacteria, fungi, and algae.	– It was employed for the treatments of PAHs at the contamination site and wastewater treatment plants – Availability and low-cost of various microbes. – It is a simple process	– The method requires nutrients for the microbes to flourish. – Temperature optimization for the condition has been a challenge. – A long time is required to degrade organic pollutants.
Reverse Osmosis	– No chemicals are added to the pollutants; hence no secondary pollutants are generated. – It is available for large scale application using sophisticated reverse osmosis membranes systems	– It is costly, especially for small and medium-scale applications. – It is a time-consuming process.
Chemical oxidations	– The organic pollutants such as PAHs and phenols are easily oxidized to carbon dioxide and water. – No sludge is generated in the process; thus, it is environmentally friendly.	– The high cost of operation and maintenance – Since it involved the use of hydrogen peroxide, it may result in harmful effects to humans if not considerably removed.

Table 3. *Cont.*

Method	Pros	Cons
Solvent extraction and ion exchange process	– It is a rapid process with high efficiency for the removal of PAHs and phenols in wastewater. – The cost of operation is cheap – The extractants are mostly reusable, thus, it is economical.	– Some solvents are toxic; thus, cross-contamination may occur. – The use of volatile solvents may result in fire emission. – Not suitable at the low concentration of the pollutants.
Photocatalytic degradation	– Complete mineralization of the pollutants to carbon dioxide and water – The photocatalysts are upon reusable from the economic point of view. – It is an environmentally friendly technique. – It is mostly operated under room temperature.	– A sufficient light source of energy is required to excite the photocatalyst. – Some of the photocatalysts are expensive, thus not economical for industrial-scale applications. – The photocatalytic reactor is expensive for both purchasing and maintenance.
Adsorption process	– It is simple to design and operate. – Availability of various forms of locally synthetic adsorbents materials. – Low-cost adsorbents and commercial adsorbents are available. – Complete adsorption of the pollutants can be achieved within a short time. – It is an environmentally friendly technique. – Reusability of the adsorbents.	– Adsorbent regeneration is often difficult and costly. – A high temperature is needed to dissolve high molecular weight pollutants. – Spent adsorbent may be considered hazardous but are usually incinerated.

However, adsorption has been recognized as the major technique for the effective removal due to the high affinity the various adsorbents for the organic pollutants. Additionally, the process has been recognized for its simplicity, low-cost, availability of various adsorbents materials, non-environmental toxicity, ease of design, etc. Figure 1 presents the literature survey conducted via web of science and for publications reported and patents for various remediation for PAHs and phenols. The prominence of adsorption is clearly seen among the techniques. It is thus justifiable that this review was conducted for the removal of PAHs and phenols using adsorption techniques. Critical analysis was conducted using various advanced porous adsorbents and the adsorptive performance of the materials have been highlighted. To some extent, the mechanism for the adsorptions have been discussed.

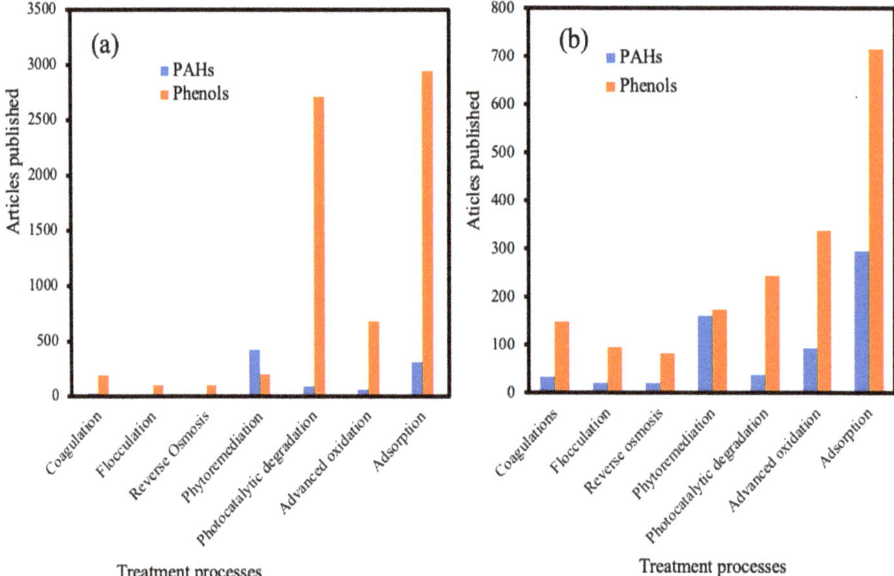

Figure 1. Trends of (**a**) publications and (**b**) patents on various treatment techniques for PAHs and phenols remediation: Data obtained from the web of science using keywords: Adsorptions, Advanced oxidation, Coagulation, Flocculation, PAHs, Phenols, Phytoremediation, Photocatalytic degradation, Reverse osmosis.

1.1. Sources of PAHs and Phenols

Industrial activities driving by energy and power demands have been the major source of PAHs and phenolic compounds in the environment. In particular, the refining of crude oil is usually associated with the production of various by-products such as petrol, diesel, coal tar, etc. [29]. Crude and refined oil leakage have also been widely recognized as one of the major sources of PAHs and phenols in the water [30]. It has been estimated, around 5–30 million tons of crude oil spilt into the ocean every year [31], as a result of either accident or leakage oil tankers. One of the most recent ones is the oil spill which occurred in Ambarnaya river and Mauritius coast in May and July 2020, respectively. In both cases, the spills were reported to occupy large area and consequently washed into the oceans. Petroleum and industrial sludges have also been a major deposit of PAHs and phenols [32,33].

Apart from petroleum and petrochemicals, coal processing to generate heat and power, as well as coke production, has also been identified as another way through which PAHs and phenolic compounds get into the water [34]. The generation of electricity from coal has been widely practiced, taking about one third of the total electricity generation [35]. According to the Environmental Protection Agency, coal fire plants contributed to about a third of the total global water pollution due to the various toxic chemicals such as PAHs, phenolic compound, and heavy metals discharged into the rivers, streams, and other water bodies [36]. Thus, the concentrations of these pollutants in the coal wastewater is high when discharged into the atmosphere and are said to be contributing to the persistent problem of global warming [37].

PAHs and phenols have been identified as one of the major contaminants in wastewater, ans usually exist in complex. The textile and dyeing industry have been a major contributor to PAHs. According to the China environmental statistics yearbook of 2012, discharge of wastewater from dyeing industries was about 2.37 billion. Accordingly, Ning et al., (2014), reported that the total number of 16 different PAHs with concentration range of 1463–16,714 ng/g has been detected in textile dyeing sludge from Guangdong [38]. Natural and anthropogenic sources of phenols have been discussed by Anku et al., (2016), highlighting the contribution of dissolved organic matter in increasing the phenols concentrations in the environment [39]. Floodwater is another source of which PAHs and phenolic compounds get into wastewater. The catastrophe of flooding has been a common phenomenon in different places of the world, particularly in the tropical areas, where rainfall is very high over a long period of the year [40]. For example, Chukwujindu et al. (2017) reported on the prevalent occurrences of flood along the River Niger and River Benue in Nigeria, as a result of excessive rainfall during the wet season [41]. Even though no established finding between the magnitude and frequency of a flood event has been found (especially large floods), the resulting environmental changes has been reported. However, Ciesielczuk et al. (2014) described that PAHs and phenols are usually adsorbed on the humic acids in the floodwaters and thus migrated to the water [42].

1.2. PAHs and Phenols in Wastewater Treatment Plants (WWTPs)

WWTPs has been the secondary source of PAHs and phenols. WWTPs serve as the major reservoir of wastewater collected from a wide range of domestic and industrial sources. Thus, various pollutants from these sources have been found to concentrate along with the solid particles in the wastewater [25]. According to the J. Liu et al. (2011), PAHs were detected in domestic WWTP in Xian, China for a period of one year, originated from domestic usage of petrochemical substances such as oil and grease, cosmetics, etc. Therefore, WWTPs is acting as the secondary source of PAHs from the different sources [43].

1.3. Toxicities of PAHs and Phenols to Humans

Much attention has been given to the study of toxicities of PAHs and phenols to humans. The source of human exposure has been mostly linked to consumption of contaminated water, air inhalation or from infected foods [44,45]. The exposure is more pronounced to communities living at the site of petroleum exploration and refining or near the coastal areas where discharged wastewater is released into rivers, oceans or seas [46]. Industrial water has higher risk of contaminations than the municipal WWTPs as in the former, the wastewater is directly coming from the source such as coke production plants which contains mixture of complex compounds such as PAHs, phenols and nitrogen containing aromatics [47].

The level of bioaccumulation of PAHs in humans is determined by the toxicities which increased with molecular weight. Long term exposure usually resulted in acute toxicities, associated with carcinogenicity and mutagenicity [48]. High exposure increased the risk of cancer of various organs such as lung, breast, prostate, kidney, bladder, stomach, and skin [26]. They are also reported to suppress the immune of the body, thus they are labeled as endocrine disruptors [49]. Chukwujindu et al. (2017), studied the effect of exposure to PAHs due to overflooding of PAH contaminated water with the associated health risk. They employed the technique of incremental lifetime cancer risk (ILCR) to determine the extent of the PAHs exposure among various age groups in the study area. The finding shows that on the average, the total ILCR values obtained in 2014 were 443 and 308 chances in a million among the children and adults, respectively, and that the number of the risk of cancer was 6450 and 4480 chances for children and adults, respectively, about a 15-fold decrease in the average total

ILCR values. Thus, they estimated high risk of exposure among adults than children, attributed to the prolong duration of exposure for the adults [41].

On the other hand, phenols are potential human carcinogens. They are found to exert toxic effects even at lower concentrations. At higher concentrations, phenols has the tendency to coagulate with the proteins in the body, which may result in cellular and cytoplasmic permeability, hence, could cause damage of the sensitive cells, cardiovascular, and central nervous system [50]. Prolonged ingestion of phenols may result into mouth sore, dark urine, and even diarrhea [51].

1.4. Environmental Regulations on PAHs and Phenols

In an effort to eliminate the effect of PAH and other toxic chemicals into the environmental waters, the United Nation has set the regulation to stop the establishment of coal power plants around the globe to tackle the climate change and water pollution challenges [47,52]. The United States Protection Agency (USEPA) has categorized 16 PAHs as priority pollutants of carcinogenic and mutagenic effect in water [25]. To limit the PAHs and other toxic contaminants from the environmental water, E.U has regulations to forbid sludge disposal. The Ministry of Environmental Protection of China has set the level of toxic PAHs and B[a]PYR in wastewater treatment plant as 50 μg/L and 0.03 μg/L [2,9] respectively.

Phenol has been detected in various water samples ranging from hazardous waste sites, surface water, rainwater, ground water, sediments, industrial and urban runoff, as well as the drinking water. The level of toxicity of phenol in human and aquatic animals has been identified as 9–25 mg/L [53]. Considering the toxicity and environmental impact of phenolic compounds in water, the USEPA has included them under the listed of priority pollutants with limited environmental discharge of less than 1 mg/L in the treated effluents and set their maximum content in potable water as 0.5 mg/L [54]. Similarly, the World Health Organization (WHO) has regulated the concentration of phenol in potable drinking water at 0.001 mg/L [55].

2. PAHs and Phenols Remediation in Wastewater

Considering the negative impact of PAHs and phenols with their derivatives in wastewater, various forms of remediation have been proposed. Some of these techniques reported includes conventional methods such as coagulation, flocculation, activated sludge, and bioremediation [42,56]. Other techniques such as membrane technology, photocatalytic degradation, advanced oxidation, and electrochemical catalysis have also been reviewed [57,58]. Some of these techniques have shown considerable promise [26], while some are facing challenges of incomplete removal or degradation of the pollutants, generations of other toxic contaminants [24,59], complexity of the method, as well as the cost of application and maintenance. Hence, the quest for more alternative techniques has seemed endless.

Adsorption techniques using various adsorbent materials from natural and synthetic origins have been well studied due to their prospects to effectively removed the pollutants from water. These technique are simple, economical, and practicable [60]. The ease of operation and non-generation of secondary pollutants are other advantages enjoyed by the method.

2.1. PAHs and Phenols Removal by Adsorption

In the adsorption process, pollutants in the solution are trapped on the surface of suspended particles, known as the adsorbent materials. Adsorption technology is motivated by the availability of a broad spectrum of low-cost adsorbents obtained from abundant naturally occurring and waste substances such as mineral deposits, agricultural waste products, particulate organic matters, and solid industrial wastes [61]. Thus, it provides an alternative for transforming waste materials into useful material for environmental sustainability.

Apart from the naturally occurring materials, synthesized particles with higher surface area and pore volumes such as organic and inorganic polymers, porous carbon materials, graphene, silica-based materials, have been successfully utilized as adsorbents for the PAHs and phenols removal from the wastewater [62,63]. Adsorption process is usually carried out in a batch reactor on a small scale, non-continuous process such as in research laboratories [64], or using a more advanced technique of fixed bed reactors or column for a pilot-scale industrial and water treatment plants application [65]. Villegas et al. (2016) [53] discussed on various methods for phenols removal in wastewater, however, emphasis was laid mainly on activated carbon (AC), those other highly porous adsorbents such as silica and metal-organic frameworks (MOFs) have not been considered.

In this article, we aimed at reviewing and discussing past and present scenario of using various adsorbent materials for the removal of PAHs and phenols from water. The review also provides scientific analysis and propose future research directions on the use of promising adsorbent materials for PAHs and phenols adsorptive removals in the water.

2.2. *Adsorption of PAHs and Phenols onto Porous Carbon Materials*

The use of porous carbon materials as adsorbents for organic pollutants removal from aquatic environment was first mentioned by Walters and Luthy (1984) [66]. They argued that porous carbon adsorbents are superior to soils, sediments, and suspended organic matters in terms of removal of PAHs. This claim has catalyzed discoveries of various porous carbon materials for wastewater remediation. The use of porous carbon materials such as activated carbon (AC) prepared from various agricultural waste materials [67,68], biochar [69,70], carbon nanotubes (CNTs) [71,72], and other derived carbon materials for the adsorption of PAHs have been studied.

2.2.1. Biochar

Biochar is a carbonaceous solid obtained from waste materials such as agricultural wastes, sewage sludge (SSL), and petroleum sludge (PS). For example, sewage sludge is generated as a by-product of sewage treatment. It is composed of many organic and inorganic substances as well as hazardous biological materials [73,74]. Thus, these solid wastes are harmful to the environment and requires proper disposal. Interestingly, pyrolysis is used as an alternative thermochemical technique to convert the waste sludge into useful biochar for effective industrial and environmental applications. It has been identified as one of the major forms of porous derived carbon. Godlewsky et al. (2019) investigated the effect of atmosphere nitrogen (N_2) and carbon dioxide (CO_2) for the production of sludge derived-biochar under pyrolysis temperatures of 500–700 °C. The use of CO_2 atmosphere was able to generate biochar with improved properties such as higher BET surface area, porosity as compared to those obtained under N_2 atmosphere [56].

Biochar that are derived from biomass-derived is by far the most important means to produce the porous carbon materials. It has been estimated that, the world production of dry biomass is about 220 billion tons annually [75], thus pyrolysis treatment to energy and biochar is interesting option [76,77]. Agricultural waste materials such as rice husk [78], wheat straw [79], and palm kernel seeds [80] have been reportedly used for PAHs and phenol adsorption in water.

The review by Bedia et al. (2018) focused mainly on the use of biomass derived biochar for emerging contaminants adsorptions in water streams [81]. In comparison to sludge derived-biochar, biomass derived-biochar has been found to possess higher specific surface area and pore volumes which was attributed to the higher carbon content in the latter [77]. Thus, high removal of PAHs (pyrene and benzo[a]pyrene) has been reported by Qiao et al. (2018) using biomass derived biochar generated from *Enteromorpha prolifera* at 200 °C. The Langmuir monolayer maximum adsorption capacities were estimated to be 187.27 μg/g and 80.00 μg/g, for the pyrene and benzo[a]pyrene, respectively [82]. Recently, Arshad et al. (2019) reported on response surface method (RSM) optimizations of phenol adsorption onto oil–palm bunch derived biochar [80]. Lee et al. (2019) reported on the effective removal of phenol using biochar generated from food waste Factors that affect the adsorption of PAHs and phenols onto biochar include pore size and volume of the adsorbent as well as the concentration of the pollutants [20]. The review by Lamichhane et al., 2016 [25] also highlighted some advantages of biochar in the removal of PAHs. Different biochars for the adsorptions of PAHs and phenols in water streams are summarized in Table 4.

Table 4. Adsorptions of PAHs and phenols using biochars.

Adsorbent	Pollutant	Concentration (mg/L)	Q_e (mg/g)	Equilibrium Time	Ref
RHB Magnetic modified RHB	PHE	5	17 42	1 h	[78]
Biochar graphene/biochar composite	PHE	1	2 3	3 days	[83]
Rice-350-M	PHE	1	43	24 h	[69]
Elephant grass biochar	BaA BkF BaP DahA	0.1	19 19 18 8	250 min	[84]
Poplar Catkins biochar	PHE	30	384	1120 min	[85]
Sewage sludge char	PHE PYR	0.8	- -	24 h	[56]
Wood biochar	PYR				
Wheat straw biochar	PHE FLU PYR	9.1 10.0 11	- - -	48 h	[79]
Chitin biochar	phenol 2-nitrphenol	500	184 206	12 h	
Hizikia fusiformis biochar	phenol	50	30	1440 min	[86]
Black spruce biochar White birch biochar	phenol	200	233 150	18 h 4 h	[87]
Chinese herb biochar	phenol	50	-	360 min	[88]
scots pine bark biochar spruce bark biochar	phenol	500	149 84	24 h	[89]
Bamboo biochar Oak wood biochar	phenol	10	17 13	24 h	[90]
Food waste biochar	phenol	10	3	30 min	[20]
Japanese red pine char Yellow poplar char	Phenol Phenol	200	6 7	30 min	[61]
Pine fruit shell biochar	Phenol	50	27	24 h	[91]

2.2.2. Activated Carbon (AC)

AC are highly porous carbon materials obtained from agricultural wastes, with high carbon contents such as cellulose, hemicellulose, lignin, lipids, proteins, simple sugars, and starch [67,92]. The production of AC involved thermal decomposition of these materials under limited supply of air, followed by the activation to generate highly porous carbon material. Thus, precursors for AC are usually low-cost agricultural waste materials such as husk [93], shells [94], wood [95], sawdust [96], rice husk [66], etc. The use of other materials such as bones, cartilages, coke etc. have also been employed [97]. It has been widely employed for a variety of industrial applications such as gas separations and purifications, liquid and gas storage, super capacitors, electrodes, catalysis, and removal of toxic substances in contaminated wastewater etc. [98]. They are usually characterized by a large surface area and large porosity than most of the conventional adsorbents ever reported. However, the properties are largely dependent on the precursor material, pyrolysis temperature, and the activation conditions. In some cases, the AC produced might contain heteroatoms and mineral matter (ash content) depending on the nature of the raw material used as precursor. A literature has shown that over 26,000 articles have been published by Elsevier on the use of AC for the adsorption of pollutants in aqueous medium. Danish and Ahmad 2018, has reviewed on the production of AC from wood biomass as adsorbent materials. Pore size, pore diameter distribution, of AC as the major contributing factor for the adsorption of the pollutants [92].

Different forms of AC such as AC powdered activated carbons (PAC) and granular activated carbons (GAC) [99,100] from different biomass has been reportedly used for organic pollutants removal from water (Figure 2). Additionally, the use of modified AC to regulate its pore structure and chemical properties for the effective removal has been studied [101]. Wu et al. (2020) recently reported on the preparation of walnut shell AC (WAC) with a surface area of 438.5 m^2/g via microwave-assisted KOH activation process at 900 W for the competitive adsorption of NAP and PHE in aqueous medium. Equilibrium adsorption of 40 min, with the adsorption capacities of 39.58 and 63.37 mg/g for the NAP and PHE respectively were obtained [98]. More porous AC were produced via modified coal-base AC (MCAC) and iron-modified coal-base AC (Fe-MCAC) with BET surface area of 1062 and 1079.67 m^2/g respectively for PAHs adsorption [102]. Similarly, optimizations studies for the phenols adsorption was studied using AC obtained from *Terminalia chebula* (TCAC). Equilibrium time of 24 h with the monolayer adsorption capacity of 36.77 mg/g were observed. Table 5 presented more literature on the application of AC as adsorbent for PAHs and phenols removal.

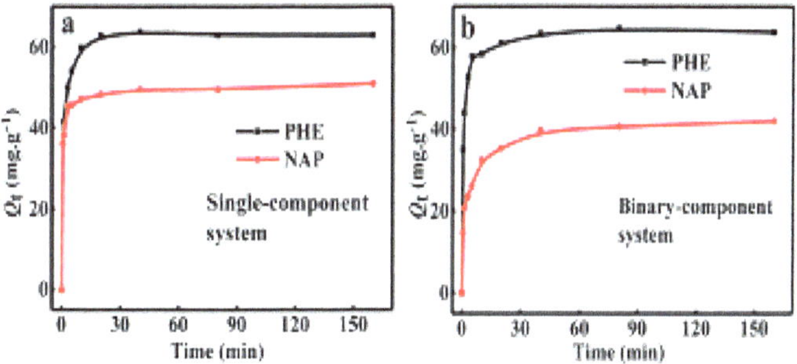

Figure 2. Kinetic studies and adsorption capacities for the removal of NAP and PHE using walnut shell activated carbon (WAC) under (**a**) single and (**b**) binary systems [103]. Copyright: Royal Society of Chemistry (2020).

Table 5. Adsorptions of PAHs and phenols using different forms activated carbon (AC).

Adsorbent	Pollutant	Concentration (mg/g)	Q_e (mg/g)	Equilibrium Time	Ref
Commercial PAC	ANT PYR	-	143 142	4 h, 5 min 4 h, 10 min	[68]
Commercial AC	PHE FLA BaA	120 80 12	46 36 9.7	24 h	[104]
Rice husk AC	NAP PHE PYR	8	14 17 18	24 h	[66]
Flamboyant pod back AC Milk bush kernel AC Rice husk AC	ACE	50	4.7 4 5.6	2 h	[105]
Rice husk AC	NAP PHE PYR	8	15 17.5 18	24 h	[106]
Rice husk AC	PHE FLU	10	- -	60 min 30 min	[107]
Rice husk AC	ACE	50	46	3 h	[93]
Vitis vinifera leaf AC ($Zncl_2$) *Vitis vinifera* leaf AC (H_3PO_4)	PHE	1	94 89	180 min	[108]

Table 5. *Cont.*

Adsorbent	Pollutant	Concentration (mg/g)	Q_e (mg/g)	Equilibrium Time	Ref
Walnut shell AC	NAP PHE	10 10	49.6 63.8	40 min	[103]
flamboyant pod AC milk bush kernel shell AC	NAP	50	294.2 63.3	2 h	[94]
Coal-based AC	NAP PHE PYR	200	78.2 111.4 117.2	20 min	[98]
Fe-modified Coal-based AC	NAP PHE PYR	30	167.8 190.4 20.3	40 min	[101]
Terminalia chebula AC	phenol	100	26.2	24 h	[109]
Commercial AC	phenol	1000	132.3	360 min	[110]
Commercial AC Miswak root AC	phenol phenol	100	142.2	120 min	[111]
Coal-based GAC Coal-based PAC Coconut shell PAC	phenol	300	169.9 176.6 213	180 min	[112]
Cherry stone AC	phenol	25	80	24 h	[113]
Agave Utahensis AC Euphorbia Resinifira AC Banana peel AC	Phenol	14	1.7 1.9 2.0	3h	[114]
Oily sludge AC	phenol	100	434	30 min	[115]
Tea waste AC	phenol	100	132.5	4 h	[67]
Babul sawdust AC	phenol	50	15.3	60 min	[96]
Olive stone AC	phenol	800	44,269	4 h	[116]
Palm oil bunch AC	phenol	40	55.3	2 h	[117]
Wood fiberboard AC	phenol	250	208	120 min	[95]
Avocado kernel AC	phenol	0.7	3.3	100 min	[118]
Date pit AC	Phenol	50	46	1 h	[119]
Sugar cane bagasse AC	Phenol	100	27	30 min	[120]
Lantana camara AC	Phenol	150	112.5	7 h	[121]
Magnetic AC	phenol	80	107.5	15 min	[122]
Granular activated carbon (GAC)	Phenol, 2-chlorophenol 4-chlorophenol	50	4.9 4.3 4.5	60 min	[123]
Red pine AC Yellow poplar AC Commercial AC	Phenol	200	454.5 625 500	30 min	[61]

Generally, AC produced from agricultural wastes displayed better removal capacity and shorter equilibrium time for the adsorption of PAHs and phenols in the aqueous medium as compared to biochar. This is perhaps associated with the higher porosities of the AC. However, in terms of sources, biochar can be generated from different varieties and abundant agricultural waste feedstocks [70]. Additionally, from economic point of view, biochar has cheaper production cost than the AC. Unlike in the latter, the former consumes less energy and activation is necessary in most cases [79]. However, both materials are considered superior to raw biomass in their affinity towards PAHs and phenols [124]. In both cases, the adsorption occurs via hydrogen bonding and π–π electron- donor–acceptor interactions between the active sites on the surface of the AC with the benzene rings of the organic pollutants and the functional groups on the phenol as seen in the Figure 3. For both AC and biochar, better removal capacities are achieved by surface modifications of the porous carbon materials. Additionally, the surface modifications could result in more stable adsorbents [98].

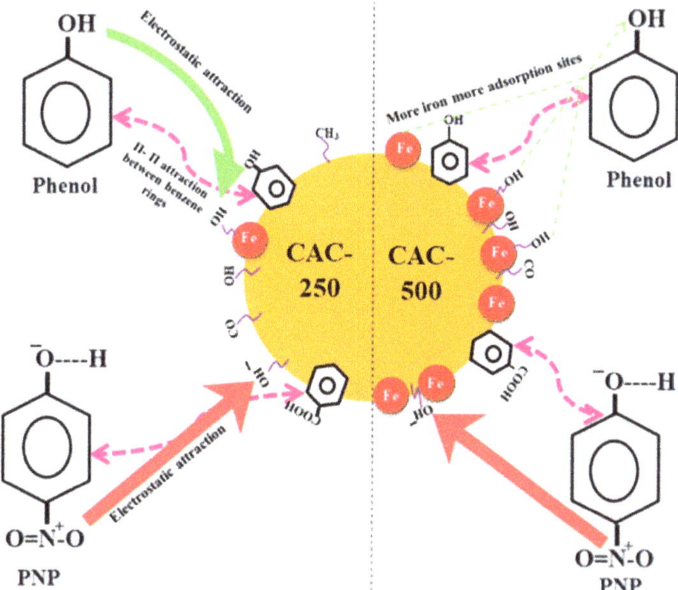

Figure 3. Mechanism for the adsorption of phenols onto activated carbons via hydrogen bonding and π–π interactions [125]. Copyright: Korean Society of Chemical Engineers (2020).

2.3. Other Porous Carbon Materials as Adsorbents for PAHs and Phenols

Apart from biochar and AC, other porous carbon nanomaterials such as graphene, carbon nanotubes (CNTs), nanohorns, nanofibers, fullerenes, soot, molecular carbon sieves, and other carbonaceous-derived materials [126] have been studied. They are chemically heterogenous and high carbon content materials that are usually obtained from either partial or complete combustion of plant materials [127]. Many of the carbon-based nanomaterials have been investigated as adsorption material for water remediation. The most widely employed are graphene and CNTs. They are recognized with extra-ordinary physical and chemical properties that are superior to other carbon-based materials. They possess good chemical and thermal stability with much higher surface area than AC and biochar [62,128]. Their improved surface properties allow them to interact with organic pollutants via covalent and non-covalent bond formation such as hydrogen bonding, electrostatic forces, π-π stacking, van der Waals forces, and hydrophobic interactions [14].

2.3.1. Graphene, Graphene Oxides, and Reduced Graphene Oxides

Graphene is a form of nanomaterial having a honeycomb-like structure of sp^2 hybridized carbon atoms with single atom of graphite layer. The allotrope of graphene is called graphite, having a planar structure [129]. It exhibits outstanding physical and chemical properties, such as good electrical and thermal conductivity, high strength, high specific surface area and pore volumes, flexibility, and negligible thickness [130,131]. Apart from that, graphene chemical mobility, which allow for its modification to form other functionalized advanced carbon materials. Graphene oxides (GOs) and reduced graphene oxide (rGO) are among the most advanced graphene-based engineering materials (Figure 4) [132,133] They are regarded as the fascinating forms of graphene materials and thus have versatile applications in various fields such as biomedicines, sensors, metrology, electronic devices, as well as environmental pollutant's remediation in environmental waters [134].

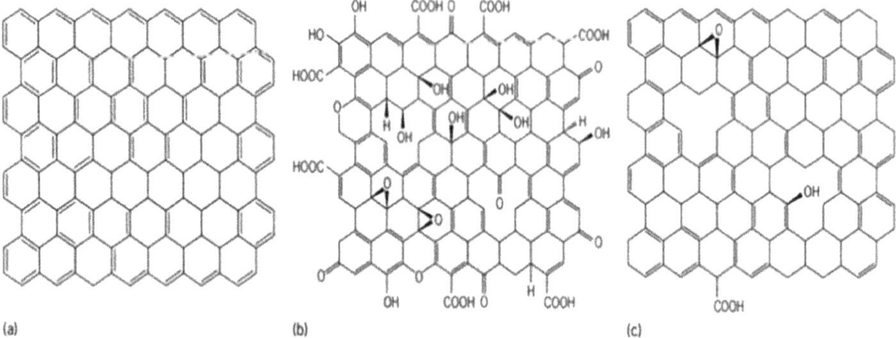

Figure 4. Chrystal structures of (**a**) graphene, (**b**) graphene oxide, and (**c**) reduced graphene oxide [133]. Copyright: Elsevier (2020).

In adsorption studies, their unique chemical properties and high specific surface area has been emphasized. Sun et al. (2013), has reported on the batch adsorptions of NAP, ANT, NAP, and PYR onto graphite, GOs and RGOs. The performance of the materials in decreasing order were rGOs > GOs > graphite, with the former having adsorption capacities of 15.92, 247.17, and 369.06 mg/g for NAP, ANT, and PYR, respectively [135]. Recently, application of graphene wool for the removal of PHE and PYR from aqueous environment has been reported by Adedapo et al. (2019). Effective removal of 99.9 and 99.1% were achieved for PHE and PYR respectively, within 24 h of contact between the PAHs and the adsorbent. The adsorbent material can be repeatedly used with little or no decreased in percentage removals over 8 successive cycles [136].

In the case of phenols removal, graphene and functionalized graphene materials have been reportedly used as adsorbents. Zhou et al. (2017) reported on the fabrication of graphene composite Fe_3O_4@PANI-GO for the batch removal of bisphenol A, t-octyl phenol and α-naphthol in water. The complete removal of the pollutants within shortest time (5 min) for bisphenol A, t-octyl phenol respectively, whereas the α-naphthol experienced slightly longer time, (300 min) to equilibrium. The promising application of the material was demonstrated by its effective reusability (10 successive cycles) without decreased in the removal capacity of the pollutants [137]. The report of Wang et al., (2018) have shown that the adsorption of phenols pollutants; 2-phenylphenol (PPE), bisphenol A (BPA), 4-isopropylphenol (IPE), 4-methylphenol (ME), and phenol (PE) were more effective onto the rGO (Figure 5) as compared to the GO, attributed to the decreased in electron density in the latter, hence the strong π-π interactions between the phenols and the adsorbent [138]. Tian et al. (2019), reported on the rapid adsorption of bisphenol A (3.4 s), onto hybrid material of graphene oxide with amino functionalized polypropylene (PP-g-DMAEMA/GO). The adsorbent possessed dual channel structure which allows bisphenol A to easily access the surface of the GO through the nanochannels of the propylene molecule [139].

In most cases, graphene was shown to have better performance in terms of adsorption capacity as compared to GO and rGO. In comparison to rGO, GO has more oxygen containing functional groups and lesser π electron density [138]. It has been established that the hydrophobic effects of the organic contaminants were the reasons for their strong absorption onto the graphene which occur though π-π interactions [63,140]. In the case of GO, the dense carbonyl groups at the edge of the carbon layers were responsible for the strong adsorption of hydrophobic organic pollutants onto the surface of the material through electrostatic, hydrogen bonding, and anion-π interactions. More details of the adsorptions of PAHs and phenols onto the graphene, GO, rGO, and their derivatives have been highlighted in Table 6.

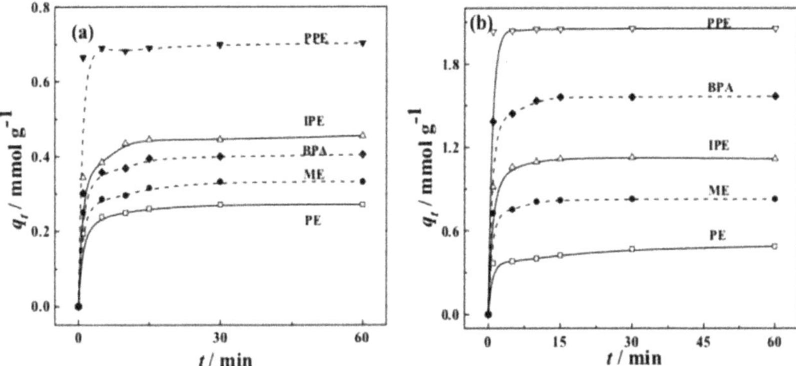

Figure 5. Adsorption capacity (**a**) graphene and (**b**) graphene oxide for phenols removal. Reproduced from ref [138].

Table 6. Adsorptions of PAHs and phenols onto graphene, GO and reduced graphene oxide (rGO) and their composites.

Adsorbent	Pollutant	Concentration (mg/L)	Q_e (mg/g)	Equilibrium Time	Ref
Graphene wool	PHE	0.8	5	24 h	[126]
	PYR	5	200		
	NAP	30	16		
rGO	ANT	0.05	247	7 days	[125]
	PYR	0.1	369		
GO/FeO·Fe_2O_3	NAP	0.1	6	48 h	[64]
Graphene	NAP	4.6	1	4 h	[63]
GO			66		
Graphene	NAP	-	9	10 min	[141]
GO			8		
Sulfonated graphene	NAP	0.39	2	5 days	[142]
rGO@Sand	NAP	0.8	4	120 min	[143]
	ACE		6		
Magnetic graphene nanosheet	PHE	0.5	0.02	1 h	[72]
Magnetic GO			14		
Magnetic chemically GO	PHE	1	30	24 h	[60]
Magnetic rG			26		
	PHE	4.2	1.2		
TiO_2-Graphene	FLAN	4.5	0.9	180 min	[59]
	BaP	2.5	0.5		
GO			21		
rG	NAP	20	52	24 h	[144]
Reduced graphene			145		
	Bisphenol A	5	14	5 min	
Fe_3O_4@PANI-GO	t-octyl phenol	5	31	5 min	[137]
	α-naphthol	10	23	300 min	
	2-phenylphenol	0.470	12	60 min	
	bisphenol A	0.350	1.56		
rGO	4-isopropylphenol	0.587	1.117		[138]
	4-methylphenol	0.550	0.841		
	phenol	0.850	0.483		
CTAB-modified graphite	bisphenol A	300	125	10 min	[129]
Amino-functionalized GO	Bisphenol A	1	2.3	3.4 s	[139]

Table 6. *Cont.*

Adsorbent	Pollutant	Concentration (mg/L)	Q_e (mg/g)	Equilibrium Time	Ref
rGO GO	phenol	50	16 23		
GO	bisphenol A 4-nonylphenol tetrabromobisphenol A 2,4,6-trichlorophenol	20	21 10 11 18	1400 min	[14]
Graphene	4-chloro-2-nitrophenol	10	25	60 min	[145]
GO rGO	Phenol	50	23 16.32	100 min	[146]
MGO-IL	phenol	30	741	12 min	[147]
Graphene	Bisphenol A	10	87	5 h	[62]
Sulfonated graphene	1-naphthol	0.7	2.41	5 days	[142]
Graphene	phenol	50	28	400 min	[148]
GO sheet GO sponge	2,4,6-trichlorophenol	5	10 21	30 min 12 min	[149]
GO	Bisphenol A	-	49	30 min	[150]
GO	tetrabromobisphenol A	1	41	120 min	[151]
GO	p-nitrophenol	200	138	24 h	[51]
GO Chemically GO rG	1-naphthol	50	21 52 145	24 h	[144]
GO Ultrasonic GO Conventional GO Ultrasonic GO	2-chlorophenol	50	32 134 50 209	50 min	[152]

2.3.2. Carbon Nanotubes (CNTs)

CNTs are sheets of pure form of carbon atoms covalently bonded to one another in hexagonal arrays which are rolled into a hollow cylindrical shape, with the outer diameter ranging between 1–100 nm, while the size up to several tens of micrometers [153]. Since its discovery in 1992, CNT have been identified with unique physical and chemical properties such as high strength and specific gravity, good conductivity and thermal stability, high porosity [154,155], making them potential materials have for various applications such as in optical appliances, energy conversion, electrochemical sensors, catalyst and adsorbent, micro analytical devices, biomedical devices, and drug delivery [156,157]. The application of CNTs in environmental remediation have also been investigated [158,159]. CNTs have been employed as potential adsorbents for organic and inorganic pollutants removal from wastewater due to its superior properties and more tailored surface chemistry, presenting higher surface porosities and strong chemical and physical interactions with the organic pollutants than the AC and corresponding carbon-derived adsorbents [160,161]. The surface morphology of the CNTs and the functional groups present in the organic pollutant has been contributing factor for their higher affinity for the pollutants, thus, rapid removal with higher adsorption capacities [162].

The potential of CNTs for environmental remediation of organic pollutants has been well studied. Wu et al. (2016) explored on the affinity of various aromatic compounds onto CNTs. About 22 different organic pollutants including phenols, substituted benzenes, and PAHs, have been considered and the findings confirmed positive correlations between the adsorption capacities of the CNTs based on their mesoporous surfaces to the properties of the pollutants [163]. A review on the prospects and applications of carbon nanostructured materials in water treatment was presented by Selvaraj et al. (2020) [164]. Apur et al. (2015) discussed the adsorption of organic pollutants including PAHs and phenols onto the CNTs [165]. Another report was presented by Ahmad et al. (2019) on the removal of organic pollutants in wastewater by CNTs [166]. The preparations, properties, and modifications of the CNTs have been discussed. The adsorptions of PAHs, pharmaceutical and personal care products, surfactants and pesticides has been explored and the nature of the interactions between the pollutants and the CNTs has been highlighted. However, in most of the review articles written detailed of the optimized adsorption conditions of phenols and PAHs onto the CNTs were not mentioned.

Earlier work on the removal of PAHs from wastewater was conducted by Yang et al. (2006) using commercially obtained single-walled carbon nanotubes (SWCNTs) and multi-walled carbon nanotubes (MWCNTs). The adsorptions were found to be pH independent due to the neutral nature of the PAHs [167]. The adsorption modeling of PAHs onto CNTs has been studied by Kah et al. (2011). Various concentrations of the PAHs were analyzed, achieving higher adsorption capacities at both lower and higher concentrations, attributed to the higher surface area of the CNTs in the range of 200–500 m^2/g [168]. Zhang et al. (2019) reported the use of magnetic CNTs for the removal of PHE from water. The magnetization of the CNTs has been shown to increase the removal efficiency as well as ease of the regeneration and reusability of the adsorbents [72]. The non-linearity of the PAHs and their non-ionic and hydrophobic nature, enable them to interact with CNTs through π-π bonding [169]. Hybrid simulation analysis was employed by Yali et al. (2019) to predict the adsorption coefficient of PAHs onto MWCNT using quantitative structure-property relationship (QSPR) using genetic algorithm-multiple linear regression (GA-MLR). The data was subjected to training and testing. The explanation given was that the electrostatic and steric parameters of PAHs were the major factors responsible for the higher adsorption capacities for the MWCNT. Hence the molecular docking simulation was accurate and reliable to predict the sorption efficiency [170]. Wu et al. (2020), studied the sorption mechanisms for PAHs onto CNTs through density functional theory (DFT) and molecular dynamics simulation. The sorption energies for the individual PAHs (NAP and PHE, −33.48 and −42 kcal/mol, respectively) were calculated from the compass force field. The simulation studies were able to predict that CNTs and PAHs interacted through π-π stacking and the Van der Waals interactions [171].

For the adsorption phenol onto CNTs, researchers have identified multiple interactions via π-π, hydrogen bonds, and electrostatic interactions. Kragulj et al. (2015) reported on the removal of chlorinated phenols (2,4-dichlorophenols, 2,4,6-trichlorophenols, and 2,3,4,5-tetrachlorophenols) onto originally synthesized MWCNT via catalytic chemical vapor deposition (CCVD) and acid functionalization (FMWCNTs). The FMWCNTs has outstanding performance compared to the pristine form as shown by the increased in the BET surface area functionalization (61.3 m^2/g and 600 m^2/g for MWCNT and FMWCNT, respectively). Except for 2,4,6-trichlorophenol, which was slightly lower than 2,4-dichlorophenol, the adsorption was found to be positively correlated with molecular weight of the pollutants [172]. The mechanism for phenol adsorption onto SWCNTs and MWCNTs using quantum mechanics molecular modeling was investigated by Rezakazemi et al. (2018) [136]. The strong affinity of CNTs onto the phenols has been emphasized by the formation of π-π interactions between the delocalized π electrons in the surface of the sp^2 hybridized carbon structure of the CNTs. However, modified CNTs such as oxidized CNT (CNTO) can form hydrogen bonding with the phenol, thus better affinity than the pristine CNT (Figure 6). Other factors affecting the adsorption of phenols onto the CNTs include pH, ionic strength, as well as the dispersion of the CNTs in the solution. Table 7 discussed the adsorptions of phenols and PAHs onto various CNTs.

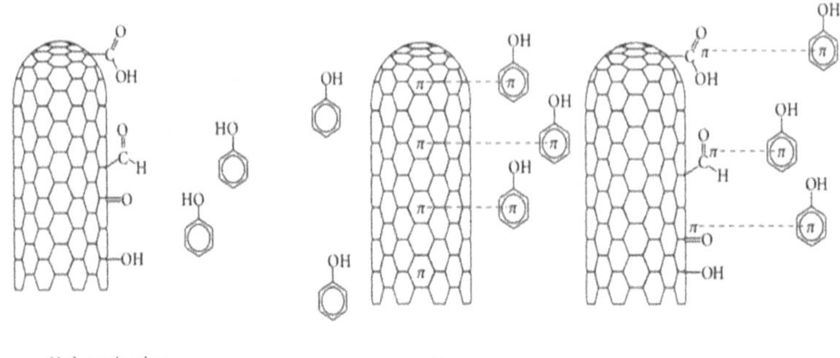

Figure 6. Schematic diagram for the mechanism of phenol adsorption onto carbon nanotubes (CNT). Reproduced from ref [146].

Table 7. Adsorptions of PAHs and phenols onto CNTs and their composites.

Adsorbent	Pollutant	Concentration (mg/L)	Q_e (mg/g)	Equilibrium Time	Ref
MWCNT	NAP	200	69.2	5 days	[173]
Commercial MWCNT	NAP PHE PYR	-	10.9 10.6 10.6	5 days	[167]
Magnetic SWCNT Magnetic MWCNT	PHE	500	24.82 19.37	30 min	[72]
MWCNT	phenol 3-chlorophenol	0.147 1.25	0.13 0.38	3 days	[174]
CNT-polymer bead	p-chlorophenol	100	24	1200 min	[71]
CNT-polyethylene glycol	phenol	20	27	40 min	[54]
CNT-polymer composite	phenol	30	262	50 min	[162]
SWCNT MWCNT	phenol	50	51 65	54 min 36 min	[127]
CNT CNT-Fe_2O_3	phenol	2	1.1 2.8	2 h	[175]
Magnetic-N_2H_4-SH/MWCNT	phenol	40	39	700 min	[176]
Chitosan hydrogel-CNT	phenol	200	404	120 min	[177]
CNT CNT-Al_2O_3	phenol 4-chlorophenol phenol 4-chlorophenol	2	1.1 1.3 2.8 2.8	120 min	[34]
CNT CNT-deep eutectic solvent	phenol	50	62 120	24 h	[13]
Chitosan-MWCNT	phenol	150	88	24 h	[17]
CNT CNT-activated	phenol 2-chlorophenol 4-chlorophenol phenol 2-chlorophenol 4-chlorophenol	150	24 86 52 64 240 105	50 min 50 min 30 min 50 min 50 min 30 min	[178]
MWCNT Nitrogen doped-MWCNT	bisphenol A	0.37	111 440	3 days	[179]
MWCNT	phenol	25	32	300 min	[180]
SWCNT SWCNTO MWCNT MCNTO	phenol	50	20 21 18 16	100 min	[146]
CNT-Mg-Al hydroxide	phenol 4-chlorophenol	50	13 14		[181]
MWCNT	2,4,6-trichlorophenol	120	2.3	48 h	[182]

Thus, the efficiency of the carbon-based materials as adsorbents for PAHs and phenols removal has been well established. AC to a larger extent has been the most widely reported adsorbent. It has been prepared from the various agricultural waste materials under different conditions. Additionally, the commercial AC has been available in the markets. Thus, their removal efficiency towards the PAHs and phenols is superior to that of the corresponding biochar derived from similar agricultural waste. Generally, CNTs have been shown to offer better removal efficiency than their corresponding graphene adsorbents. This is attributed to their remarkable high surface area. The strong affinity of the materials to the pollutants is made possible by the presence of hydrogen and π-π bonding interactions as illustrated in Figure 5.

2.4. Other Porous Materials for PAHs and Phenols Adsorption

Porous polymers (e.g., polymer beads, molecular imprinted polymers (MIPs)) [27], mesoporous chitosan [183], and advanced porous materials, such as mesoporous silica (examples; MCM-41, MCM-48 and SBA-15) [184], metal-organic frameworks (MOFs) [185], and other corresponding composites materials synthesized or are usually synthesized from other known materials. They have been known to possess physical and chemical properties superior to that of precursor components and that of the conventional porous materials. They have ultra-high porosity and surface area, high crystallinity, uniform surface morphology, high chemical and thermal stabilities, and other promising features. Thus, they have found numerous applications in various fields such as catalysis, gas storage, sensors, actuators, drug delivery, microextraction, and wastewater remediation [186,187].

The applications of synthesized polymers and advanced porous materials as adsorbents for PAHs and phenol removals have been investigated by various researchers. In most cases, the efficiency of these materials surpasses that of carbon-based materials such as AC, due to their larger pore size and volumes providing more adsorption sites for the pollutants [188,189]. Additionally, some of these materials have displayed good selectively for the PAHs and other phenolic compounds with the formation of hydrogen bonding and π-π interactions.

2.4.1. Chitosan

The applications of chitosan, a polymer material obtained from the shell of seafoods and natural substances for adsorption of pollutants from wastewater have received considerable attention. The advantages of the material are the low-cost and large content of hydroxyl functional groups on the surface of the materials [190]. The price of commercial chitosan is much lower compared to that of the AC and zeolites and it can be easily synthesized with much less energy consumption [191]. The adsorptions of PAHs and phenols onto the chitosan are predominated by hydrogen bonding, hydrophobic, and π-π interactions [192,193].

Bibi et al. (2015) used membranes synthesized from chitosan and poly vinyl pyrrolidone (CW) and carbon nanotube (CNT). The membranes with a surface area of 246 and 253 m^2/g were used for the adsorption of NAP from aqueous solution. The membranes were applied for the adsorption of NAP from aqueous solution with the removal efficiency of 93 and 97% were achieved within 120 and 150 min, respectively [194]. Crisafully et al. (2008) reported on the comparisons of various natural and commercial adsorbents materials including chitosan for the removal of PAHs [4]. Filho et al. (2018) reported on the synthesis of pectin/chitosan and pectin/chitosan/cyclodextrin polymers for the adsorption of 3 PAHs, PYR, B[b]F, and B[a]P. The adsorption of PAHs and phenols onto chitosan is summarized in Table 8. Efficient removal of the pollutants was achieved with good reusability of the adsorbents [195].

Alves et al. (2019) reported on the modification of chitosan with carbon nanotube composite with enhanced BET surface area of 1130 m^2/g for the adsorption of phenol. Equilibrium was reached within 120 min with adsorption capacity of 404.2 mg/g [177]. A similar finding was reported Bahmani et al. (2019) using chitosan grafted with ZIF-8 for the phenol removal, achieving adsorption capacity of 152.3 mg/g within 30 min [196]. Guo et al. (2019) comprised the of adsorption of phenol using chitosan and MWCNT modified chitosan. The composite material achieved higher adsorption capacity of (86.96 mg/g) than the pristine chitosan with (61.69 mg/g) [17] Karamipour et al. (2020) reported the adsorption of phenol using cellulose acetate/chitosan (CA/Chitosan) composite. The adsorption capacity achieved was 97.43 mg/g within 9 h. A modification of the CA/Chitosan with iron (III) oxide, (Fe_2O_3-CA/Chitosan) improved the physical stability of the adsorbent, achieving good reusability with the adsorption capacity to 163 mg/g [197].

2.4.2. Molecularly Imprinted Polymers (MIPs)

MIPs are obtained from the interactions of organic and or inorganic monomers of polymerizable capabilities which are assembled around a specific template known as the 'imprint', resulting into crosslinking of the monomers to form a solid framework [198]. They are well known for their high porosity resulting from the removal of the template molecules, forming large cavities allows for the easy access of the molecules to the recognition sites [27,199]. The MIPs of inorganic molecularly imprinted materials are more flexible in comparison to those obtained from the organic monomers due to the more availability of the inorganic monomers to be selected from, hence the rigidity of the solid network as well as the adaptable porosity [200].

The adsorptive properties of the MIP towards pollutants removals in wastewater has been emphasized on their large porosity, providing more adsorption sites for the pollutants [201], as highlighted in Table 6. They also offer high selectivity to the pollutants, mainly via hydrogen bonding, ion-pair interactions, hydrophobic interactions, and Van der Waals forces [28]. The adsorptions of PAHs onto the functionalized silica aerogels has been recently reported by Saad et al., (2020). The adsorbent material has proven to be effective in the removal of NAP, ANT, and PYR, with the fluorescence measurements showing effective adsorption and selectivity of the material in the highest order of NAP, ANT, and PYR, respectively [200]. Wei et al. (2015) reported on the selective recognition of PAHs by MIPs. In comparison to the hydrophilic MIP (H-MIP) and co-monomer glycidyl methacrylate (G-MIP), the conventional MIP has higher binding capacity to the PAHs, attributed to the hydrophobic nature of the material [202].

Bhatnagar and Anastopoulos (2017) reported the adsorption of bisphenol A and other phenolic compounds in water [190]. Bayramoglu et al. (2016) reported on the synthesis and selectivity of MIP and non-imprinted polymer (NIP) for the adsorption of bisphenol A, 4-aminophenol, p-toluidine, and 2-napthol in the binary and multiple component system. The adsorption capacity of bisphenol A for the MIP and NIP were 76.7 and 59.9 mg/g, respectively, indicating the selectivity and removal efficiency of the latter [203]. Adsorptive recognition of phenolic compounds bisphenol A, phenol, 4-nitrophenol, 2-amino-4-chlorophenol and 2-napthol has been reported by Lyu et al. (2020), using MIP and ionic liquid mediated MIP (IL-MIP). Higher selective of the bisphenol A has been achieved by the IL-MIP as compared to the conventional MIP, attributed to the complementary of the imprinted cavity of the IL-MIP to the bisphenol structure [198].

Table 8. Adsorption of PAHs and phenols using chitosan and molecularly imprinted polymers (MIPs).

Adsorbent	Pollutant	Concentration (mg/L)	Q_e (mg/g)	Equilibrium Time	Ref
		Chitosan			
Pec-β-CD/Chitosan	PYR B[b]F B[a]P	2	0.1 0.2 0.2	1200 min	[195]
Chitosan Chitosan/CNT	phenol phenol	400	404	120 min	[177]
Chitosan-cyclodextrin	phenol p-chlorophenol p-nitrophenol	150 157 150	34 180 21	-	[193]

Table 8. *Cont.*

Adsorbent	Pollutant	Concentration (mg/L)	Q_e (mg/g)	Equilibrium Time	Ref
chitosan-g-PNVCL/ZIF-8	Phenol	20	152	30 min	[196]
CA/Chitosan Fe_2O_3-CA/Chitosan	phenol	100	97 163	5 h	[197]
Chitosan MWCNT-Chitosan	phenol	150	62 87	24 h	[16]
Fe-EDA/β-CD/Chitosan	phenol p-nitrophenol p-cresol	50	243 274 298	120 min	[204]
Magnetic chitosan	phenol	15	52	50 min	[19]
Resin-chitosan	phenol 4-chlorophenol	50	189 99	200 min 350 min	[205]
EDTA/Chitosan/TiO_2	phenol	50	209	240 min	[206]
Chitosan-Carbon composite	phenol	10	34	1 h	
Magnetic graphene-Chitosan	2-napthol	40	169	60 min	[207]
β-cyclodextrin-chitosan	phenol 2-chlorophenol 4-chlorophenol 2,4-dichlorophenol 2,4,6-trichlorophenol	120	60 71 96 315 376	3 h	[22]
Magnetic-Chitosan	bisphenol A	100	55	120 min	[208]
Glutaraldehyde-Chitosan	phenol	50	21	60 min	[209]
	phenol o-chlorophenol	100	71 51	240 min	[210]
MIPs					
MIP	BaP PYR CHR	1	75.9 7 7	3 h	[211]
MIP	PHE	0.6	16	90 min	[212]
MIP	PYR	1	11	90 min	[213]
MIP	bisphenol A	150	77	120 min	[203]
vermiculite MP magnetic vermiculite MP	bisphenol A	1000	217 274	90 min	[214]
Surface-MIP	phenol p-nitrophenol p-tert-butylphenol	0.75	86 8 7	4 h	[50]
MIP/PES/SiO_2	phenol p-nitrophenol	4000	47 37	8 h	[215]
AMPS-Am-MIP	Phenol	50	97	180 min	[18]
MIP IL-MIP	Bisphenol A	100	87 116	20 min	[198]
Fe_3O_4@SiO_2@PNP-SMIP	4-nitophenol	210	134	60 min	[216]
MIP SMIP	bisphenol-A	20	24 23	240 min	[217]

However, it should be stressed that pristine chitosan, upon long term usage as adsorbents, swells. This has drastically limited its application as potential adsorbent material in wastewater remediation. However, surface modification of the chitosan with other ideal functional materials helps to improve the performance of the material and improve its long-term usage. Some reported chitosan composites used for PAHs and phenols adsorption have been highlighted in Table 6.

2.4.3. Mesoporous Silica

With the advancement in nanotechnology and the attempted reproduction of silica nanoparticles in 1990 in Japan [218], researchers all over the world continue to explore the potential of the materials in various areas. Mesoporous nano silica particles such as Mobil Crystalline Materials (such as MCM-41 and MCM-48) and Santa Barbara Amorphous (such as SBA-15 and SBA-16) have been synthesized from rice husks, characterized and commercialized for various applications such as gas storage, drug delivery, oil–water separations, catalysis, and pollutant remediation [186].

Mesoporous silica has well-defined pore structures (hexagonal mesopores), narrow pore size distributions (4–12 nm), as well as high BET and Langmuir surface-area. Table 6 shows mesoporous silica that have been used for the PAHs and phenols adsorption. The adsorption capacity and equilibrium time have been stated.

Nasreen et al. (2018) synthesized highly porous mesoporous silica SBA-15 and MSU-H with BET surface area of 521 and 580 units, respectively. Both SBA-15 and MSU-H were found to be effective in the removal of NAP and PHE with the complete removal achieved within 20 min [218]. Costa et al. (2020) have reported the adsorptions of PAHs from aqueous solution using mesoporous silica MCM-41-NH_2 that have been hydrothermally synthesized, possessing high retention rate of the pollutants with efficiency in the range of 93–98% [219]. Comparison for the removal efficiency of MCM-48 and SBA-15 for the adsorption of NAP was investigated by Balati et al. (2014) with the latter having better efficiency, attributed to its higher larger pore diameter [220]. Yuan et al. (2018) investigated the mechanism for the PAHs adsorption onto the mesoporous silica. They have identified π-π as the predominant interactions between the pollutants and the adsorbents [221].

The applications of mesoporous silica materials for phenols adsorptions has been investigated. Nasreen et al. (2018) reported that SBA-15 and MSU-H were able to adsorb 4-aminophenol and p-nitrophenol from the aqueous medium. The removal efficiency achieved was above 90% within 20 min. It was suggested that the initial properties of the silica has been retained even after a series of regeneration and reusability, attributed to the high silicon content in the materials [218]. Highly porous mesoporous silica SBA-10 and PA-10 with a BET surface area of 925 and 353 m^2/g were synthesized by Yangui et al. (2017) for the adsorption of phenolic compounds. Adsorption efficiency of 75% and 67% for phenol removal were reported within 2 h when the SBA-15 and PA-10 were modified with amine functional groups [222].

2.4.4. Metal-Organic Frameworks (MOFs)

The application of MOFs in wastewater remediation have recently gained more recognitions due to the great potential of these materials. They are made from the coordination interaction of the metal ion with organic moieties [223]. The metal node is the central building block, while organic moiety is the ligand and together forming a framework of higher tunability [224]. They are known to possess ultra-high porosity (over 90% by volume is empty) with BET surface area higher than that of AC and mesoporous silica (up to 9000 m^2/g) [225]. MOFs have higher crystallinity, thermal and moisture stability, and they can be easily regenerated from the aqueous medium through simple decantation or magnetic separation [226]. The topology of the framework also made it possible to modify the MOFs with other functional groups and materials for versatile and particular applications.

The use of MOFs for PAHs adsorptions recently been published by our groups [227]. The application of Zr-based MOFs; UiO-66(Zr) and NH_2-UiO-66(Zr) have been reported for ANT and CRY adsorptions from the aqueous medium. The BET surface area of the MOFs was 1420 and 985 m^2/g for the UiO-66(Zr) and NH_2-UiO-66(Zr), respectively. Higher removal efficiency has been achieved within short equilibrium time (25–30 min) with percentage removals of 98.6 and 97.9% for the ANT and CRY, respectively, using UiO-66(Zr) MOF. Interestingly, all the reported MOFs have displayed good reusability for the PAHs adsorption. We have also reported on the molecular docking interaction of the Fe-based MOFs, MIL-88(Fe) and NH_2-MIL-88(Fe) with CRY. Figure 7 has illustrated the binding interaction of the MOFs with the CRY molecules with the MIL-88(Fe) having the most stable complex structure with the higher binding energy (ΔG_{bind}) of −3.88 kcal/mol and lower inhibition constant (*Ki*) of 1.58 mM. Thus, the CRY molecules preferred to reside in the internal pores of the MOF. However, for the NH_2-MIL-88(Fe), the binding energy is slightly lower, and the inhibition constant is higher, −3.80 kcal/mol and 1.65 mM respectively, resulting into lower binding stability of the complex. In the NH_2-MIL-88(Fe), the CRY molecules preferred to adsorb on the outer pores of the MOFs. The high selectivity of the MIL-88(Fe) has been emphasized by the larger BET surface area and pore size of 1242 m^2/g and 12.5 nm respectively. The BET surface area and the pore size of the NH_2-MIL-88(Fe) was 941 m^2/g and 8.8 nm respectively [228]. Table 9 highlights the various MOFs used for PAHs and phenols adsorptions in wastewater.

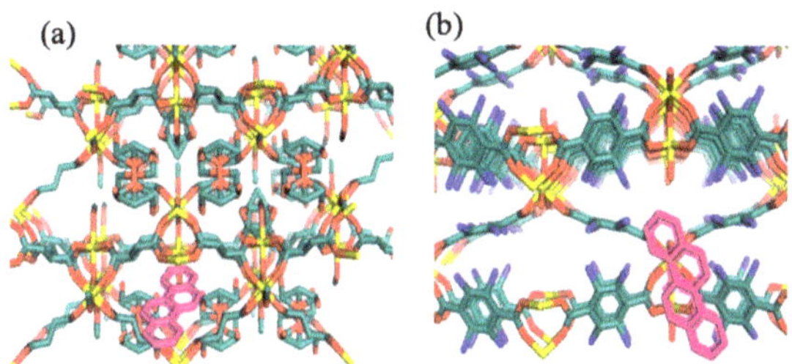

Figure 7. The molecular docking structure of (**a**) MIL-88(Fe) and (**b**) NH_2-MIL-88(Fe) with the CRY molecule, for the binding interactions [229]. Copyright: Elsevier (2020). Metal (Fe), carbon (C), oxygen (O), and nitrogen (N) are colored as yellow, green, red, and blue, respectively. The CRY is indicated in purple.

Adsorptions of phenolic compounds onto MOFs have also been studied. Most of the reported MOFs are promising adsorbents for the removal of the pollutants studied. The famous HKUST-1 has shown good removal efficiency for the p-nitrophenol, with complete removal achieved within 40 min (equilibrium adsorption capacity of 400 mg/g). The high adsorption capacity of the MOF for the p-nitrophenol was attributed to the high affinity of the MOF to the NO_2 side of the pollutant, resulting into hydrogen bonding interactions [229]. The high selectivity and removal efficiency of the crystalline and water stable Zr-based MOF, NH_2-UiO-66(Zr) for the adsorptions of 2,4,6-trinitrophenol, 2,4-dinitrophenol, 2,4,6-tri-nitrotoluene, and 2,4-dinitrotoluene in water was also investigated. Higher equilibrium adsorption capacities was achieved with the hydrogen bonding formation between the MOF and the pollutants [230]. Luo et al. (2019) reported on the enhanced removal of bisphenol A using Al-based composite Al-MOF/SA-CS. The adsorption capacity was 136 mg/g. The adsorption mechanism has been explained by the formation of hydrogen bonding, π-π stacking and cation-π interactions [231].

Table 9. Adsorption of PAHs and phenols onto mesoporous silica and metal-organic frameworks (MOFs).

Adsorbent	Pollutant	Concentration (mg/L)	Q_e (mg/g)	Equilibrium Time	Ref
		Mesoporous Silica			
MCM-41	NAP	15	61	20 min	[184]
Periodic mesoporous organosilica	NAP	10	1.5	24 h	
	ACE		0.9		
	FLU		0.9		
	FLUO		0.7		
	PYR		1.3		
Fe-SBA-15	PYR	0.1	0.034	20 min	[188]
NH_2-SBA-15	NAP	1	1.9	24 h	[232]
	ACE		1.4		
	PHE		0.8		

Table 9. *Cont.*

Adsorbent	Pollutant	Concentration (mg/L)	Q_e (mg/g)	Equilibrium Time	Ref
MCM-41-PABA	NAP B[b]FLUO B[k]FLUO B[b]PYR	200 μg/L	17.83 20 19 19	24 h	[233]
MCM-48 SBA-15	NAP	7	0.7 0.4	24 h	[220]
NH_2-SBA-15	4-chlorophenol 4-bromophenol 4-iodophenol	250	141 115 72	6 h	[234]
MCM-41-	phenol o-chlorophenol	50	12 30	24 h	[222]
MCM-48-Chitosan	Phenol	500	149	24 h	[183]
SBA-15	4-aminophenol 4-nitrophenol	-	10 8	20 min	[235]
Rice husk silica MCM-41	phenol phenol	15 60	4	60 min	[236]
		MOFs			
UiO-66(Zr) NH_2-UiO-66(Zr)	ANT CRY ANT CRY	4	24 24 22 19	30 min 25 min 30 min 25 min	[227]
MIL-88(Fe) NH_2-MIL-88(Fe)	CRY	4	24 22	25 min	[228]
MIL-88(Fe) Mixed-MIL-88(Fe) NH_2-MIL-88(Fe)	ANT	4	24 23 21	30 min	[237]
MIL-88(Fe) NH_2-MIL-88(Fe)	PYR	4	24 23	40 min	[238]
MIL-53(Al) MIL-53(Al)-F127	bisphenol A	250	329 473	90 min 30 min	[239]
MIL-68(Al)/PVDF	p-nitrophenol	10	126	720 min	[240]
HKUST-1(Cu)	p-nitrophenol	200	400	30 min	[229]
SiO_2@MIL-68(Al)	aniline	3000	532	40 s	[241]
[Zn(ATA)(BPD)] MOF-VII	2,4-dichlorophenol	60	- -	90 min	[242]
[Zn(TDC) MOF	2, 4-dichloropheno	60	-	180 min	[243]
MIL-68(Al) CNT@MIL-68(Al)	phenol phenol	1000	118 257	120 min	[244]
NH_2-UiO-66(Zr)	2,4,6-trinitrophenol Styphnic acid 2,4-dinitrophenol 2,4,6-tri- nitrotoluene 2,4-dinitrotoluene	100	23 24 30 0.5 2	36 h	[230]
MIL-53(Al)	dichlorophenoxyacetic acid	150	556	1 h	[245]
MIL–68(Al) MIL–68(Al)/GO	p–nitrophenol	300	271 332	17 h 17 h	[51]
NH_2-MIL-88(Fe)	2,4,6-trinitrophenol	35	164	40 min	[246]
MOF-199(Cu)	Phenol *p*-nitro phenol	500	58 68	300 min 30 min	[247]
Al-MOF/SA-CS	bisphenol A	50	137	18 h	[231]
Cu-BDC MOF Cu-BDC@GrO Cu-BDC@CNT	bisphenol A	200	60 182 164	40 min	[248]
laccase@HKUST-1	bisphenol A	200	-	4 h	[249]

2.5. Pilot and Industrial Scale Adsorption of PAHs and Phenol in Wastewater

Pilot and industrial wastewater treatment plants have been the major repository of PAHs and phenols. In most cases, these pollutants are washed from the soil by heavy rain or directly emitted from petroleum refineries, petrochemical industries, and coal mining and processing areas. Thus, they are deposited into the municipal wastewater and wastewater treatment plants. Various studies have shown the presence of these pollutants at much higher concentrations than the maximum tolerable limit in the water. Thus, the target of wastewater remediation technologies has been the wastewater for both industrial and municipal treatment plants. Application of adsorption technologies at pilot and industrial wastewater has been well explored [250].

Activated carbon has been the adsorbent most widely used for the post-treatment of non-biodegradable pollutants such as PAHs and phenols in the wastewater treatment plants due to its large surface area and uniformity of surface microspores [251]. With the availability of various low-cost and highly porous adsorbents materials, some researchers have endeavored to explore the adsorptive removals of the PAHs and phenols at both pilot and industrial scale treatment plants. Kalmykova et al. (2014) have explored the performance of sand, granular activated carbon (GAC), and peat moss filters for PAHs and phenol adsorption at pilot-plant. The performance of each filter has been assessed with the GAC achieving complete removal of phenols, while the peat filter was able to remove 50–80% of the phenols. For PAHs removal, the percentage achieved by GAC and peat were 50% and 63%, respectively [252]. The application of resin adsorbent, Amberlite XAD16, was reported by Frascari et al. (2016) for the removal of phenols from olive mill wastewater using continuous flow column at pilot-scale. The Amberlite XAD16 was well characterized with BET surface area of 800 m^2/g and average particle size of 0.63 mm. The adsorption capacity of the material was evaluated by the Langmuir model at 40 mg/g with over 70% removal efficiency achieved by the material [253]. El-Nass et al. (2016) have reported on the pilot-scale removal of phenol from refinery wastewater. About 250 kg of GAC with a particle size of 1.5 mm was used as the bed in the packed column of 1.8 m and 0.6 m length and internal diameter respectively. Over 90% removal of the phenols was achieved by the adsorbent, indicating the efficiency of GAC in removal of organic pollutants from the refinery wastewater [251].

3. Conclusions

Ubiquitous transport, distributions, and the fate of PAHs and phenols in wastewater has been a major environmental challenge for decades. These pollutants have been well identified as highly toxic to both fauna and flora and pose carcinogenic and mutagenic effects to humans. Hence, their environmental abatement, particularly from industrial, municipal, and wastewater treatment plants (WWTPs), has been highly desired. Various wastewater remediation technologies have been proposed. This article presented comprehensive review on the use of adsorption technology for PAHs and phenols remediation from water. The application of various porous adsorbents materials has been studied.

Porous carbon materials biochar and AC have discovered various applications in environmental remediation. Those materials have been successfully prepared from various agricultural wastes and been widely used as adsorbents for PAHs and phenols removal. In comparison to biochar, AC has shown more promising and potential removal capacity, attributed to the higher porosity of the AC. Apart from the biochar and AC, other derived porous carbons such as the graphene and CNTs have been widely employed. Their potentially higher removal efficiencies have been commended. This is attributed to their remarkable high surface area and strong affinity to the pollutants through the formation of hydrogen and π-π bonding interactions, as we have illustrated.

Under similar conditions, the use of synthetic polymers, chitosan and MIPs, have been presented. Despite the selectivity of the polymers, the swelling effect of the chitosan limits their applications. However, modified chitosan has been shown to offer better removal capacity than the pristine form. To the larger extent, the potentials of advanced porous materials; mesoporous silica (such as MCM-41,

MCM-48, and SBA-15), and MOFs had been discussed. The superior adsorption capacity of the silica and MOFs adsorbents had been in all way connected to their ultra-porosities and high stability of the materials. Their strong affinity to the PAHs and phenol has been highlighted.

Thus, application of various porous adsorbents has been presented. Findings from previous works from literature have been compiled and analyzed. This review thus highlighted the potential of adsorption technique for the removal of PAHs and phenols from environmental wastewater.

Author Contributions: Conceptualization, Z.U.Z. and N.S.S.; software, K.J.; validation, M.N.H.R.; formal analysis, H.A.I.; investigation, N.H.H.A.B.; resources, N.S.S.; data.; writing—original draft preparation, Z.U.Z.; writing—review and editing, A.R. and A.M.O.; visualization, A.S.; supervision, B.S.; project administration, N.S.S.; funding acquisition, N.S.S., A.R. All authors have read and agreed to the published version of the manuscript.

Funding: This research was funded by YUTP and UTP-UIR, grant number 015LCO-211 and 015MEO-166 and The APC was funded by King Khalid University.

Acknowledgments: We would like to acknowledge the grants received from Universiti Teknologi PETRONAS under YUTP and UTP-UIR with cost centers 015LCO-211 and 015MEO-166 respectively. We also wish to acknowledge King Khalid University, Abha, Kingdom of Saudi Arabia.

Conflicts of Interest: The authors declare no conflict of interest.

References

1. Oh, J.Y.; Choi, S.D.; Kwon, H.O.; Lee, S.E. Leaching of polycyclic aromatic hydrocarbons (PAHs) from industrial wastewater sludge by ultrasonic treatment. *Ultrason. Sonochem.* **2016**, *33*, 61–66. [CrossRef] [PubMed]
2. Zhang, Y.; Zhang, L.; Huang, Z.; Li, Y.; Li, J.; Wu, N.; He, J.; Zhang, Z.; Liu, Y.; Niu, Z. Pollution of polycyclic aromatic hydrocarbons (PAHs) in drinking water of China: Composition, distribution and influencing factors. *Ecotoxicol. Environ. Saf.* **2019**, *177*, 108–116. [CrossRef] [PubMed]
3. Tian, W.; Bai, J.; Liu, K.; Sun, H.; Zhao, Y. Occurrence and removal of polycyclic aromatic hydrocarbons in the wastewater treatment process. *Ecotoxicol. Environ. Saf.* **2012**, *82*, 1–7. [CrossRef] [PubMed]
4. Crisafully, R.; Milhome, M.A.L.; Cavalcante, R.M.; Silveira, E.R.; De Keukeleire, D.; Nascimento, R.F. Removal of some polycyclic aromatic hydrocarbons from petrochemical wastewater using low-cost adsorbents of natural origin. *Bioresour. Technol.* **2008**, *99*, 4515–4519. [CrossRef]
5. Zhang, Y.; Tao, S. Global atmospheric emission inventory of polycyclic aromatic hydrocarbons (PAHs) for 2004. *Atmos. Environ.* **2009**, *43*, 812–819. [CrossRef]
6. Liu, Y.; Zarfl, C.; Basu, N.B.; Cirpka, O.A. Turnover and legacy of sediment-associated PAH in a baseflow-dominated river. *Sci. Total Environ.* **2019**, *671*, 754–764. [CrossRef]
7. Andrade, M.V.F.; Santos, F.R.; Oliveira, A.H.B.; Nascimento, R.F.; Cavalcante, R.M. Influence of sediment parameters on the distribution and fate of PAHs in an estuarine tropical region located in the Brazilian semi-arid (Jaguaribe River, Ceará coast). *Mar. Pollut. Bull.* **2019**, *146*, 703–710. [CrossRef]
8. Zhao, W.; Sui, Q.; Huang, X. Removal and fate of polycyclic aromatic hydrocarbons in a hybrid anaerobic–anoxic–oxic process for highly toxic coke wastewater treatment. *Sci. Total Environ.* **2018**, *635*, 716–724. [CrossRef]
9. Sun, S.; Jia, L.; Li, B.; Yuan, A.; Kong, L.; Qi, H.; Ma, W.; Zhang, A.; Wu, Y. The occurrence and fate of PAHs over multiple years in a wastewater treatment plant of Harbin, Northeast China. *Sci. Total Environ.* **2018**, *624*, 491–498. [CrossRef]
10. De Gisi, S.; Lofrano, G.; Grassi, M.; Notarnicola, M. Characteristics and adsorption capacities of low-cost sorbents for wastewater treatment: A review. *Sustain. Mater. Technol.* **2016**, *9*, 10–40. [CrossRef]
11. Qiao, M.; Qi, W.; Liu, H.; Qu, J. Oxygenated, nitrated, methyl and parent polycyclic aromatic hydrocarbons in rivers of Haihe River System, China: Occurrence, possible formation, and source and fate in a water-shortage area. *Sci. Total Environ.* **2014**, *481*, 178–185. [CrossRef] [PubMed]
12. Zango, Z.U.; Garba, Z.N.; Abu Bakar, N.H.H.; Tan, W.L.; Abu Bakar, M. Adsorption studies of Cu^{2+}–Hal nanocomposites for the removal of 2,4,6-trichlorophenol. *Appl. Clay Sci.* **2016**, *132–133*, 68–78. [CrossRef]
13. Lawal, I.A.; Lawal, M.M.; Azeez, M.A.; Ndungu, P. Theoretical and experimental adsorption studies of phenol and crystal violet dye on carbon nanotube functionalized with deep eutectic solvent. *J. Mol. Liq.* **2019**, *288*. [CrossRef]

14. Catherine, H.N.; Ou, M.H.; Manu, B.; Shih, Y. hsin Adsorption mechanism of emerging and conventional phenolic compounds on graphene oxide nanoflakes in water. *Sci. Total Environ.* **2018**, *635*, 629–638. [CrossRef] [PubMed]
15. Bazrafshan, E.; Mostafapour, F.K.; Mahvi, A.H. Phenol removal from aqueous solutions using pistachio-nut shell ash as a low cost adsorbent. *Fresenius Environ. Bull.* **2012**, *21*, 2962–2968.
16. Soni, U.; Bajpai, J.; Singh, S.K.; Bajpai, A.K. Evaluation of chitosan-carbon based biocomposite for efficient removal of phenols from aqueous solutions. *J. Water Process Eng.* **2017**, *16*, 56–63. [CrossRef]
17. Guo, M.; Wang, J.; Wang, C.; Strong, P.J.; Jiang, P.; Ok, Y.S.; Wang, H. Carbon nanotube-grafted chitosan and its adsorption capacity for phenol in aqueous solution. *Sci. Total Environ.* **2019**, *682*, 340–347. [CrossRef]
18. Zhang, Y.; Qin, L.; Cui, Y.; Liu, W.; Liu, X.; Yang, Y. A hydrophilic surface molecularly imprinted polymer on a spherical porous carbon support for selective phenol removal from coking wastewater. *New Carbon Mater.* **2020**, *35*, 220–231. [CrossRef]
19. Salari, M.; Dehghani, M.H.; Azari, A.; Motevalli, M.D.; Shabanloo, A.; Ali, I. High performance removal of phenol from aqueous solution by magnetic chitosan based on response surface methodology and genetic algorithm. *J. Mol. Liq.* **2019**, *285*, 146–157. [CrossRef]
20. Lee, C.G.; Hong, S.H.; Hong, S.G.; Choi, J.W.; Park, S.J. Production of Biochar from Food Waste and its Application for Phenol Removal from Aqueous Solution. *Water Air Soil Pollut.* **2019**, *230*. [CrossRef]
21. Wang, F.; Haftka, J.J.H.; Sinnige, T.L.; Hermens, J.L.M.; Chen, W. Adsorption of polar, nonpolar, and substituted aromatics to colloidal graphene oxide nanoparticles. *Environ. Pollut.* **2014**, *186*, 226–233. [CrossRef]
22. Zhou, L.C.; Meng, X.G.; Fu, J.W.; Yang, Y.C.; Yang, P.; Mi, C. Highly efficient adsorption of chlorophenols onto chemically modified chitosan. *Appl. Surf. Sci.* **2014**, *292*, 735–741. [CrossRef]
23. Blanchard, M.; Teil, M.J.; Ollivon, D.; Legenti, L.; Chevreuil, M. Polycyclic aromatic hydrocarbons and polychlorobiphenyls in wastewaters and sewage sludges from the Paris area (France). *Environ. Res.* **2004**, *95*, 184–197. [CrossRef] [PubMed]
24. Mojiri, A.; Zhou, J.L.; Ohashi, A.; Ozaki, N.; Kindaichi, T. Comprehensive review of polycyclic aromatic hydrocarbons in water sources, their effects and treatments. *Sci. Total Environ.* **2019**, *696*. [CrossRef] [PubMed]
25. Lamichhane, S.; Bal Krishna, K.C.; Sarukkalige, R. Polycyclic aromatic hydrocarbons (PAHs) removal by sorption: A review. *Chemosphere* **2016**, *148*, 336–353. [CrossRef] [PubMed]
26. Alegbeleye, O.O.; Opeolu, B.O.; Jackson, V.A. Polycyclic Aromatic Hydrocarbons: A Critical Review of Environmental Occurrence and Bioremediation. *Environ. Manag.* **2017**, *60*, 758–783. [CrossRef]
27. Ndunda, E.N.; Mizaikoff, B. Molecularly imprinted polymers for the analysis and removal of polychlorinated aromatic compounds in the environment: A review. *Analyst* **2016**, *141*, 3141–3156. [CrossRef]
28. Kumar, V.; Kim, K.H.; Park, J.W.; Hong, J.; Kumar, S. Graphene and its nanocomposites as a platform for environmental applications. *Chem. Eng. J.* **2017**, *315*, 210–232. [CrossRef]
29. de Souza, C.V.; Corrêa, S.M. Polycyclic aromatic hydrocarbons in diesel emission, diesel fuel and lubricant oil. *Fuel* **2016**, *185*, 925–931. [CrossRef]
30. Akinpelu, A.A.; Ali, M.E.; Johan, M.R.; Saidur, R.; Qurban, M.A.; Saleh, T.A. Polycyclic aromatic hydrocarbons extraction and removal from wastewater by carbon nanotubes: A review of the current technologies, challenges and prospects. *Process Saf. Environ. Prot.* **2019**, *122*, 68–82. [CrossRef]
31. Redondo, J.M.; Platonov, A.K. Self-similar distribution of oil spills in European coastal waters. *Environ. Res. Lett.* **2009**, *4*. [CrossRef]
32. Lai, X.; Ning, X.-A.; He, Y.; Yuan, Y.; Sun, J.; Ke, Y.; Man, X. Treatment of a simulated sludge by ultrasonic zero-valent iron/EDTA/Air process: Interferences of inorganic salts in polyaromatic hydrocarbon removal. *Waste Manag.* **2019**, *85*, 548–556. [CrossRef] [PubMed]
33. Lu, Y.; Zheng, G.; Zhou, W.; Wang, J.; Zhou, L. Bioleaching conditioning increased the bioavailability of polycyclic aromatic hydrocarbons to promote their removal during co-composting of industrial and municipal sewage sludges. *Sci. Total Environ.* **2019**, *665*, 1073–1082. [CrossRef]
34. Ihsanullah; Asmaly, H.A.; Saleh, T.A.; Laoui, T.; Gupta, V.K.; Atieh, M.A. Enhanced adsorption of phenols from liquids by aluminum oxide/carbon nanotubes: Comprehensive study from synthesis to surface properties. *J. Mol. Liq.* **2015**, *206*, 176–182. [CrossRef]
35. Marañón, E.; Vázquez, I.; Rodríguez, J.; Castrillón, L.; Fernández, Y.; López, H. Treatment of coke wastewater in a sequential batch reactor (SBR) at pilot plant scale. *Bioresour. Technol.* **2008**, *99*, 4192–4198. [CrossRef]

36. Bargieł, P.; Zabochnicka-Świątek, M. Technologies of Coke Wastewater Treatment in the Frame of Legislation in Force. *Ochr. Sr. Zasobow Nat.* **2018**, *29*, 11–15. [CrossRef]
37. Pal, P.; Kumar, R. Treatment of coke wastewater: A critical review for developing sustainable management strategies. *Sep. Purif. Rev.* **2014**, *43*, 89–123. [CrossRef]
38. Ning, X.A.; Lin, M.Q.; Shen, L.Z.; Zhang, J.H.; Wang, J.Y.; Wang, Y.J.; Yang, Z.Y.; Liu, J.Y. Levels, composition profiles and risk assessment of polycyclic aromatic hydrocarbons (PAHs) in sludge from ten textile dyeing plants. *Environ. Res.* **2014**, *132*, 112–118. [CrossRef]
39. Anku, W.W.; Mamo, M.A.; Govender, P.P. Phenolic Compounds in Water: Sources, Reactivity, Toxicity and Treatment Methods. *Intech Open* **2016**, 419–443. [CrossRef]
40. Agarwal, T.; Khillare, P.S.; Shridhar, V. PAHs contamination in Bank sediment of the Yamuna River, Delhi, India. *Environ. Monit. Assess.* **2006**, *123*, 151–166. [CrossRef]
41. Iwegbue, C.M.A.; Tesi, G.O.; Overah, L.C.; Emoyan, O.O.; Nwajei, G.E.; Martincigh, B.S. Effects of Flooding on the Sources, Spatiotemporal Characteristics and Human Health Risks of Polycyclic Aromatic Hydrocarbons in Floodplain Soils of the Lower Parts of the River Niger, Nigeria. *Polycycl. Aromat. Compd.* **2017**, *6638*, 1–17. [CrossRef]
42. Ciesielczuk, T.; Kusza, G.; Poluszyńska, J.; Kochanowska, K. Pollution of flooded arable soils with heavy metals and polycyclic aromatic hydrocarbons (PAHs). *Water Air Soil Pollut.* **2014**, *225*. [CrossRef] [PubMed]
43. Liu, J.J.; Wang, X.C.; Fan, B. Characteristics of PAHs adsorption on inorganic particles and activated sludge in domestic wastewater treatment. *Bioresour. Technol.* **2011**, *102*, 5305–5311. [CrossRef] [PubMed]
44. Jin, J.; Wang, S.; Hu, J.; Wu, J.; Li, M.; Wang, Y.; Jin, J. Polychlorinated naphthalenes in human serum samples from an industrial city in Eastern China: Levels, sources, and sex differences. *Ecotoxicol. Environ. Saf.* **2019**, *177*, 86–92. [CrossRef]
45. Liang, X.; Junaid, M.; Wang, Z.; Li, T.; Xu, N. Spatiotemporal distribution, source apportionment and ecological risk assessment of PBDEs and PAHs in the Guanlan River from rapidly urbanizing areas of Shenzhen, China. *Environ. Pollut.* **2019**, *250*, 695–707. [CrossRef] [PubMed]
46. McDonough, C.A.; Khairy, M.A.; Muir, D.C.G.; Lohmann, R. Significance of population centers as sources of gaseous and dissolved PAHs in the lower Great Lakes. *Environ. Sci. Technol.* **2014**, *48*, 7789–7797. [CrossRef] [PubMed]
47. Zhang, W.; Wei, C.; Chai, X.; He, J.; Cai, Y.; Ren, M.; Yan, B.; Peng, P.; Fu, J. The behaviors and fate of polycyclic aromatic hydrocarbons (PAHs) in a coking wastewater treatment plant. *Chemosphere* **2012**, *88*, 174–182. [CrossRef]
48. Li, J.; Dong, H.; Xu, X.; Han, B.; Li, X.; Zhu, C.; Han, C.; Liu, S.; Yang, D.; Xu, Q.; et al. Prediction of the bioaccumulation of PAHs in surface sediments of Bohai Sea, China and quantitative assessment of the related toxicity and health risk to humans. *Mar. Pollut. Bull.* **2016**, *104*, 92–100. [CrossRef]
49. de Jesus, J.H.F.; Cunha, G.d.C.; Cardoso, E.M.C.; Mangrich, A.S.; Romão, L.P.C. Evaluation of waste biomasses and their biochars for removal of polycyclic aromatic hydrocarbons. *J. Environ. Manag.* **2017**, *200*, 186–195. [CrossRef]
50. Qu, Y.; Qin, L.; Liu, X.; Yang, Y. Reasonable design and sifting of microporous carbon nanosphere-based surface molecularly imprinted polymer for selective removal of phenol from wastewater. *Chemosphere* **2020**, *251*. [CrossRef]
51. Wu, Z.; Yuan, X.; Zhong, H.; Wang, H.; Zeng, G.; Chen, X.; Wang, H.; Zhang, L.; Shao, J. Enhanced adsorptive removal of p-nitrophenol from water by aluminum metal-organic framework/reduced graphene oxide composite. *Sci. Rep.* **2016**, *6*, 25638. [CrossRef] [PubMed]
52. Kong, Q.; Wu, H.; Liu, L.; Zhang, F.; Preis, S.; Zhu, S.; Wei, C. Solubilization of polycyclic aromatic hydrocarbons (PAHs) with phenol in coking wastewater treatment system: Interaction and engineering significance. *Sci. Total Environ.* **2018**, *628–629*, 467–473. [CrossRef]
53. Villegas, L.G.C.; Mashhadi, N.; Chen, M.; Mukherjee, D.; Taylor, K.E.; Biswas, N. A Short Review of Techniques for Phenol Removal from Wastewater. *Curr. Pollut. Rep.* **2016**, *2*, 157–167. [CrossRef]
54. Bin-Dahman, O.A.; Saleh, T.A. Synthesis of carbon nanotubes grafted with PEG and its efficiency for the removal of phenol from industrial wastewater. *Environ. Nanotechnol. Monit. Manag.* **2020**, *13*. [CrossRef]
55. Saleh, T.A.; Elsharif, A.M.; Asiri, S.; Mohammed, A.R.I.; Dafalla, H. Synthesis of carbon nanotubes grafted with copolymer of acrylic acid and acrylamide for phenol removal. *Environ. Nanotechnol. Monit. Manag.* **2020**, *14*. [CrossRef]

56. Godlewska, P.; Siatecka, A.; Kończak, M.; Oleszczuk, P. Adsorption capacity of phenanthrene and pyrene to engineered carbon-based adsorbents produced from sewage sludge or sewage sludge-biomass mixture in various gaseous conditions. *Bioresour. Technol.* **2019**, *280*, 421–429. [CrossRef] [PubMed]
57. Giwa, A.; Daer, S.; Ahmed, I.; Marpu, P.R.; Hasan, S.W. Experimental investigation and artificial neural networks ANNs modeling of electrically-enhanced membrane bioreactor for wastewater treatment. *J. Water Process Eng.* **2016**, *11*, 88–97. [CrossRef]
58. Lin, M.; Ning, X.-a.; An, T.; Zhang, J.; Chen, C.; Ke, Y.; Wang, Y.; Zhang, Y.; Sun, J.; Liu, J. Degradation of polycyclic aromatic hydrocarbons (PAHs) in textile dyeing sludge with ultrasound and Fenton processes: Effect of system parameters and synergistic effect study. *J. Hazard. Mater.* **2016**, *307*, 7–16. [CrossRef]
59. Bai, H.; Zhou, J.; Zhang, H.; Tang, G. Enhanced adsorbability and photocatalytic activity of TiO_2-graphene composite for polycyclic aromatic hydrocarbons removal in aqueous phase. *Colloids Surf. B Biointerfaces* **2017**, *150*, 68–77. [CrossRef]
60. Huang, D.; Xu, B.; Wu, J.; Brookes, P.C.; Xu, J. Adsorption and desorption of phenanthrene by magnetic graphene nanomaterials from water: Roles of pH, heavy metal ions and natural organic matter. *Chem. Eng. J.* **2019**, *368*, 390–399. [CrossRef]
61. Hwang, H.; Sahin, O.; Choi, J.W. Manufacturing a super-active carbon using fast pyrolysis char from biomass and correlation study on structural features and phenol adsorption. *RSC Adv.* **2017**, *7*, 42192–42202. [CrossRef]
62. Xu, J.; Wang, L.; Zhu, Y. Decontamination of bisphenol A from aqueous solution by graphene adsorption. *Langmuir* **2012**, *28*, 8418–8425. [CrossRef] [PubMed]
63. Wang, J.; Chen, Z.; Chen, B. Adsorption of polycyclic aromatic hydrocarbons by graphene and graphene oxide nanosheets. *Environ. Sci. Technol.* **2014**, *48*, 4817–4825. [CrossRef] [PubMed]
64. Yang, X.; Li, J.; Wen, T.; Ren, X.; Huang, Y.; Wang, X. Adsorption of naphthalene and its derivatives on magnetic graphene composites and the mechanism investigation. *Colloids Surf. A Physicochem. Eng. Asp.* **2013**, *422*, 118–125. [CrossRef]
65. Mohammad, A.T.; Al-Obaidi, M.A.; Hameed, E.M.; Basheer, B.N.; Mujtaba, I.M. Modelling the chlorophenol removal from wastewater via reverse osmosis process using a multilayer artificial neural network with genetic algorithm. *J. Water Process Eng.* **2020**, *33*. [CrossRef]
66. Yakout, S.M.; Daifullah, A.A.M.; El-Refeey, S.A. Adsorption of Naphthalene, Phenanthrene and Pyrene from Aqueous Solution Using Low-Cost Activated Carbon Derived from Agricultural. *Adsorpt. Sci. Technol.* **2013**, *31*, 293–302. [CrossRef]
67. Gundogdu, A.; Duran, C.; Senturk, H.B.; Soylak, M.; Ozdes, D.; Serencam, H.; Imamoglu, M. Adsorption of phenol from aqueous solution on a low-cost activated carbon produced from tea industry waste: Equilibrium, kinetic, and thermodynamic study. *J. Chem. Eng. Data* **2012**, *57*, 2733–2743. [CrossRef]
68. Rasheed, A.; Farooq, F.; Rafique, U.; Nasreen, S.; Aqeel Ashraf, M. Analysis of sorption efficiency of activated carbon for removal of anthracene and pyrene for wastewater treatment. *Desalin. Water Treat.* **2016**, *57*, 145–150. [CrossRef]
69. Feng, Z.; Zhu, L. Sorption of phenanthrene to biochar modified by base. *Front. Environ. Sci. Eng.* **2018**, *12*, 1–11. [CrossRef]
70. Tong, Y.; McNamara, P.J.; Mayer, B.K. Adsorption of organic micropollutants onto biochar: A review of relevant kinetics, mechanisms and equilibrium. *Environ. Sci. Water Res. Technol.* **2019**, *5*, 821–838. [CrossRef]
71. Xu, L.; Wang, Z.; Ye, S.; Sui, X. Removal of p-chlorophenol from aqueous solutions by carbon nanotube hybrid polymer adsorbents. *Chem. Eng. Res. Des.* **2017**, *123*, 76–83. [CrossRef]
72. Zhang, J.; Li, R.; Ding, G.; Wang, Y.; Wang, C. Sorptive removal of phenanthrene from water by magnetic carbon nanomaterials. *J. Mol. Liq.* **2019**, *293*. [CrossRef]
73. Zielińska, A.; Oleszczuk, P. The conversion of sewage sludge into biochar reduces polycyclic aromatic hydrocarbon content and ecotoxicity but increases trace metal content. *Biomass Bioenergy* **2015**, *75*, 235–244. [CrossRef]
74. Smith, K.M.; Fowler, G.D.; Pullket, S.; Graham, N.J.D. Sewage sludge-based adsorbents: A review of their production, properties and use in water treatment applications. *Water Res.* **2009**, *43*, 2569–2594. [CrossRef]

75. Azargohar, R.; Jacobson, K.L.; Powell, E.E.; Dalai, A.K. Evaluation of properties of fast pyrolysis products obtained, from Canadian waste biomass. *J. Anal. Appl. Pyrolysis* **2013**, *104*, 330–340. [CrossRef]
76. Qiu, M.; Sun, K.; Jin, J.; Gao, B.; Yan, Y.; Han, L.; Wu, F.; Xing, B. Properties of the plant- and manure-derived biochars and their sorption of dibutyl phthalate and phenanthrene. *Sci. Rep.* **2014**, *4*, 5295. [CrossRef]
77. Askeland, M.; Clarke, B.; Paz-Ferreiro, J. Comparative characterization of biochars produced at three selected pyrolysis temperatures from common woody and herbaceous waste streams. *PeerJ* **2019**, *7*, e6784. [CrossRef]
78. Guo, W.; Wang, S.; Wang, Y.; Lu, S.; Gao, Y. Sorptive removal of phenanthrene from aqueous solutions using magnetic and non-magnetic rice husk-derived biochars. *R. Soc. Open Sci.* **2018**, *5*, 1–11. [CrossRef] [PubMed]
79. Li, H.; Qu, R.; Li, C.; Guo, W.; Han, X.; He, F.; Ma, Y.; Xing, B. Selective removal of polycyclic aromatic hydrocarbons (PAHs) from soil washing effluents using biochars produced at different pyrolytic temperatures. *Bioresour. Technol.* **2014**, *163*, 193–198. [CrossRef]
80. Md Arshad, S.H.; Ngadi, N.; Wong, S.; Saidina Amin, N.; Razmi, F.A.; Mohamed, N.B.; Inuwa, I.M.; Abdul Aziz, A. Optimization of phenol adsorption onto biochar from oil palm empty fruit bunch (EFB). *Malays. J. Fundam. Appl. Sci.* **2019**, *15*, 1–5. [CrossRef]
81. Bedia, J.; Peñas-Garzón, M.; Gómez-Avilés, A.; Rodriguez, J.; Belver, C. A Review on the Synthesis and Characterization of Biomass-Derived Carbons for Adsorption of Emerging Contaminants from Water. *J. Carbon Res.* **2018**, *4*, 63. [CrossRef]
82. Qiao, K.; Tian, W.; Bai, J.; Dong, J.; Zhao, J.; Gong, X.; Liu, S. Preparation of biochar from Enteromorpha prolifera and its use for the removal of polycyclic aromatic hydrocarbons (PAHs) from aqueous solution. *Ecotoxicol. Environ. Saf.* **2018**, *149*, 80–87. [CrossRef]
83. Tang, J.; Lv, H.; Gong, Y.; Huang, Y. Preparation and characterization of a novel graphene/biochar composite for aqueous phenanthrene and mercury removal. *Bioresour. Technol.* **2015**, *196*, 355–363. [CrossRef] [PubMed]
84. de Jesus, J.H.F.; Matos, T.T.d.S.; Cunha, G.d.C.; Mangrich, A.S.; Romão, L.P.C. Adsorption of aromatic compounds by biochar: Influence of the type of tropical biomass precursor. *Cellulose* **2019**. [CrossRef]
85. Liu, X.; Sun, J.; Duan, S.; Wang, Y.; Hayat, T.; Alsaedi, A.; Wang, C.; Li, J. A Valuable Biochar from Poplar Catkins with High Adsorption Capacity for Both Organic Pollutants and Inorganic Heavy Metal Ions. *Sci. Rep.* **2017**, *7*, 10033. [CrossRef]
86. Shin, W. seok Adsorption characteristics of phenol and heavy metals on biochar from Hizikia fusiformis. *Environ. Earth Sci.* **2017**, *76*, 782. [CrossRef]
87. Braghiroli, F.L.; Bouafif, H.; Hamza, N.; Neculita, C.M.; Koubaa, A. Production, characterization, and potential of activated biochar as adsorbent for phenolic compounds from leachates in a lumber industry site. *Environ. Sci. Pollut. Res.* **2018**, *25*, 26562–26575. [CrossRef]
88. Zhang, Y.; Tang, Z.; Liu, S.; Xu, H.; Song, Z. Study on adsorption of phenol from aqueous media using biochar of Chinese herb residue. *IOP Conf. Ser. Mater. Sci. Eng.* **2018**, *394*. [CrossRef]
89. Siipola, V.; Pflugmacher, S.; Romar, H.; Wendling, L.; Koukkari, P. Low-Cost Biochar Adsorbents for Water Purification Including Microplastics Removal. *Appl. Sci.* **2020**, *10*, 788. [CrossRef]
90. Jin, D.F.; Xu, Y.Y.; Zhang, M.; Jung, Y.S.; Ok, Y.S. Comparative evaluation for the sorption capacity of four carbonaceous sorbents to phenol. *Chem. Speciat. Bioavailab.* **2016**, *28*, 18–25. [CrossRef]
91. Mohammed, N.A.S.; Abu-Zurayk, R.A.; Hamadneh, I.; Al-Dujaili, A.H. Phenol adsorption on biochar prepared from the pine fruit shells: Equilibrium, kinetic and thermodynamics studies. *J. Environ. Manag.* **2018**, *226*, 377–385. [CrossRef] [PubMed]
92. Danish, M.; Ahmad, T. A review on utilization of wood biomass as a sustainable precursor for activated carbon production and application. *Renew. Sustain. Energy Rev.* **2018**, *87*, 1–21. [CrossRef]
93. Alade, A.O.; Amuda, O.S.; Ibrahim, A.O. Isothermal studies of adsorption of acenaphthene from aqueous solution onto activated carbon produced from rice (Oryza sativa) husk. *Elixir Chem. Eng.* **2012**, *46*, 87–95. [CrossRef]
94. Alade, A.O.; Amuda, O.S.; Afolabi, T.J.; Okoya, A.A. Adsorption of naphthalene onto activated carbons derived from milk bush kernel shell and flamboyant pod. *J. Environ. Chem. Ecotoxicol.* **2012**, *4*, 124–132. [CrossRef]
95. Jin, X.J.; Zhu, Y.M. Absorption of Phenol on Nitrogen-Enriched Activated Carbon from Wood Fiberboard Waste with Chemical Activation by Potassium Carbonate. *J. Chem. Eng. Process Technol.* **2014**, *5*, 4. [CrossRef]
96. Ingole, R.S.; Lataye, D.H. Adsorptive removal of phenol from aqueous solution using activated carbon prepared from Babul sawdust. *J. Hazard. Toxicol. Radioact. Waste* **2015**, *19*, 1–9. [CrossRef]

97. Dąbrowski, A.; Podkościelny, P.; Hubicki, Z.; Barczak, M. Adsorption of phenolic compounds by activated carbon—A critical review. *Chemosphere* **2005**, *58*, 1049–1070. [CrossRef]
98. Xiao, X.; Liu, D.; Yan, Y.; Wu, Z.; Wu, Z.; Cravotto, G. Preparation of activated carbon from Xinjiang region coal by microwave activation and its application in naphthalene, phenanthrene, and pyrene adsorption. *J. Taiwan Inst. Chem. Eng.* **2015**, *53*, 160–167. [CrossRef]
99. Eeshwarasinghe, D.; Loganathan, P.; Vigneswaran, S. Simultaneous removal of polycyclic aromatic hydrocarbons and heavy metals from water using granular activated carbon. *Chemosphere* **2019**, *223*, 616–627. [CrossRef]
100. Rakowska, M.I.; Kupryianchyk, D.; Grotenhuis, T.; Rijnaarts, H.H.M.; Koelmans, A.A. Extraction of sediment-associated polycyclic aromatic hydrocarbons with granular activated carbon. *Environ. Toxicol. Chem.* **2013**, *32*, 304–311. [CrossRef]
101. Ge, X.; Wu, Z.; Wu, Z.; Yan, Y.; Cravotto, G.; Ye, B.C. Enhanced PAHs adsorption using iron-modified coal-based activated carbon via microwave radiation. *J. Taiwan Inst. Chem. Eng.* **2016**, *64*, 235–243. [CrossRef]
102. Ge, X.; Tian, F.; Wu, Z.; Yan, Y.; Cravotto, G.; Wu, Z. Adsorption of naphthalene from aqueous solution on coal-based activated carbon modi fi ed by microwave induction: Microwave power effects. *Chem. Eng. Process.* **2015**, *91*, 67–77. [CrossRef]
103. Wu, Z.; Sun, Z.; Liu, P.; Li, Q.; Yang, R.; Yang, X. Competitive adsorption of naphthalene and phenanthrene on walnut shell based activated carbon and the veri fi cation via theoretical. *RSC Adv.* **2020**, *10*, 10703–10714. [CrossRef]
104. Liu, J.; Chen, J.; Jiang, L.; Yin, X. Adsorption of mixed polycyclic aromatic hydrocarbons in surfactant solutions by activated carbon. *J. Ind. Eng. Chem.* **2014**, *20*, 616–623. [CrossRef]
105. Alade, A.O.; Amuda, O.S.; Afolabi, A.O.; Adelowo, F.E. Adosrption of Acenaphthene unto Ativated Crabon Produced from Agricultural Wastes. *J. Envion. Sci. Technol.* **2012**, *5*, 192–209.
106. Yakout, S.M.; Daifullah, A.A.M.; El-Reefy, S.A. Equilibrium and kinetic studies of sorption of polycyclic aromatic hydrocarbons from water using rice husk activated carbon. *Asian J. Chem.* **2013**, *25*, 10037–10042. [CrossRef]
107. Rinawati; Hidayat, D.; Supriyanto, R.; Permana, D.F. Yunita Adsorption of Polycyclic Aromatic Hydrocarbons using Low-Cost Activated Carbon Derived from Rice Husk. *J. Phys. Conf. Ser.* **2019**, *1338*. [CrossRef]
108. Awe, A.A.; Opeolu, B.O.; Fatoki, O.S.; Ayanda, O.S.; Jackson, V.A.; Snyman, R. Preparation and characterisation of activated carbon from Vitis vinifera leaf litter and its adsorption performance for aqueous phenanthrene. *Appl. Biol. Chem.* **2020**, *63*, 12. [CrossRef]
109. Khare, P.; Kumar, A. Removal of phenol from aqueous solution using carbonized Terminalia chebula-activated carbon: Process parametric optimization using conventional method and Taguchi's experimental design, adsorption kinetic, equilibrium and thermodynamic study. *Appl. Water Sci.* **2012**, *2*, 317–326. [CrossRef]
110. Lin, J.Q.; Yang, S.E.; Duan, J.M.; Wu, J.J.; Jin, L.Y.; Lin, J.M.; Deng, Q.L. The Adsorption Mechanism of Modified Activated Carbon on Phenol. *MATEC Web Conf.* **2016**, *67*. [CrossRef]
111. Biglari, H.; Afsharnia, M.; Javan, N.; Sajadi, S.A. Phenol Removal from Aqueous Solutions by Adsorption on Activated Carbon of Miswak's Root Treated with KMnO4. *Iran. J. Health Sci.* **2016**, *42*, 20–30. [CrossRef]
112. Xie, B.; Qin, J.; Wang, S.; Li, X.; Sun, H.; Chen, W. Adsorption of Phenol on Commercial Activated Carbons: Modelling and Interpretation. *Int. J. Environ. Res. Public Health* **2020**, *17*, 789. [CrossRef]
113. Beker, U.; Ganbold, B.; Dertli, H.; Gülbayir, D.D. Adsorption of phenol by activated carbon: Influence of activation methods and solution pH. *Energy Convers. Manag.* **2010**, *51*, 235–240. [CrossRef]
114. Germain, T.; Lynda, E.; Tchirioua, E. Adsorption of Phenol on Carbon Based on Cactus and Banana Peel. *Aust. J. Basic Appl. Sci.* **2019**, *13*, 64–70. [CrossRef]
115. Mojoudi, N.; Mirghaffari, N.; Soleimani, M.; Shariatmadari, H.; Belver, C.; Bedia, J. Phenol adsorption on high microporous activated carbons prepared from oily sludge: Equilibrium, kinetic and thermodynamic studies. *Sci. Rep.* **2019**, *9*, 19352. [CrossRef]
116. Nouha, S.; Souad, N.S.; Abdelmottalab, O. Enhanced adsorption of phenol using alkaline modified activated carbon prepared from olive stones. *J. Chil. Chem. Soc.* **2019**, *64*, 4352–4359. [CrossRef]
117. Siti Hadjar, M.A.; Astimar, A.A.; Norzita, N.; Noraishah, S.A. Phenol adsorption by activated carbon of different fibre size derived from empty fruit bunches. *J. Oil Palm Res.* **2012**, *24*, 1524–1532.
118. Dejene, K.; Siraj, K.; Kitte, S.A. Kinetic and Thermodynamic Study of Phenol Removal from Water Using Activated Carbon Synthesizes from Avocado Kernel Seed. *Int. Lett. Nat. Sci.* **2016**, *54*, 42–57. [CrossRef]

119. Banat, F.; Al-Asheh, S.; Al-Makhadmeh, L. Utilization of raw and activated date pits for the removal of phenol from aqueous solutions. *Chem. Eng. Technol.* **2004**, *27*, 80–86. [CrossRef]
120. Mohtashami, S.A.; Asasian Kolur, N.; Kaghazchi, T.; Asadi-Kesheh, R.; Soleimani, M. Optimization of sugarcane bagasse activation to achieve adsorbent with high affinity towards phenol. *Turk. J. Chem.* **2018**, *42*, 1720–1735. [CrossRef]
121. Girish, C.R.; Murty, V.R. Adsorption of Phenol from Aqueous Solution Using Lantana camara, Forest Waste: Packed Bed Studies and Prediction of Breakthrough Curves. *Environ. Process.* **2014**, *2*, 773–796. [CrossRef]
122. Mohammadi, S.Z.; Darijani, Z.; Karimi, M.A. Fast and efficient removal of phenol by magnetic activated carbon-cobalt nanoparticles. *J. Alloys Compd.* **2020**, *832*. [CrossRef]
123. Gholizadeh, A.; Kermani, M.; Gholami, M.; Farzadkia, M. Kinetic and isotherm studies of adsorption and biosorption processes in the removal of phenolic compounds from aqueous solutions: Comparative study. *J. Environ. Health Sci. Eng.* **2013**, *11*, 29. [CrossRef] [PubMed]
124. Ioannou, Z.; Simitzis, J. Adsorption kinetics of phenol and 3-nitrophenol from aqueous solutions on conventional and novel carbons. *J. Hazard. Mater.* **2009**, *171*, 954–964. [CrossRef] [PubMed]
125. Yadav, N.; Narayan Maddheshiaya, D.; Rawat, S.; Singh, J. Adsorption and equilibrium studies of phenol and para-nitrophenol by magnetic activated carbon synthesised from cauliflower waste. *Environ. Eng. Res.* **2019**, *25*, 742–752. [CrossRef]
126. Atieh, M.A. Removal of Phenol from Water Different Types of Carbon—A Comparative Analysis. *APCBEE Procedia* **2014**, *10*, 136–141. [CrossRef]
127. Dehghani, M.H.; Mostofi, M.; Alimohammadi, M.; McKay, G.; Yetilmezsoy, K.; Albadarin, A.B.; Heibati, B.; AlGhouti, M.; Mubarak, N.M.; Sahu, J.N. High-performance removal of toxic phenol by single-walled and multi-walled carbon nanotubes: Kinetics, adsorption, mechanism and optimization studies. *J. Ind. Eng. Chem.* **2016**, *35*, 63–74. [CrossRef]
128. Rezakazemi, M.; Kurniawan, T.A.; Albadarin, A.B.; Shirazian, S. Molecular modeling investigation on mechanism of phenol removal from aqueous media by single- and multi-walled carbon nanotubes. *J. Mol. Liq.* **2018**, *271*, 24–30. [CrossRef]
129. Wang, L.C.; Ni, X.J.; Cao, Y.H.; Cao, G.Q. Adsorption behavior of bisphenol A on CTAB-modified graphite. *Appl. Surf. Sci.* **2018**, *428*, 165–170. [CrossRef]
130. Chowdhury, S.; Balasubramanian, R. Recent advances in the use of graphene-family nanoadsorbents for removal of toxic pollutants from wastewater. *Adv. Colloid Interface Sci.* **2014**, *204*, 35–56. [CrossRef]
131. Li, B.; Ou, P.; Wei, Y.; Zhang, X.; Song, J. Polycyclic aromatic hydrocarbons adsorption onto graphene: A DFT and AIMD study. *Materials* **2018**, *11*, 726. [CrossRef] [PubMed]
132. Fraga, T.J.M.; Carvalho, M.N.; Ghislandi, M.G.; Da Motta Sobrinho, M.A. Functionalized graphene-based materials as innovative adsorbents of organic pollutants: A concise overview. *Braz. J. Chem. Eng.* **2019**, *36*, 1–31. [CrossRef]
133. Ali, I.; Basheer, A.A.; Mbianda, X.Y.; Burakov, A.; Galunin, E.; Burakova, I.; Mkrtchyan, E.; Tkachev, A.; Grachev, V. Graphene based adsorbents for remediation of noxious pollutants from wastewater. *Environ. Int.* **2019**, *127*, 160–180. [CrossRef]
134. Bano, Z.; Mazari, S.A.; Saeed, R.M.Y.; Majeed, M.A.; Xia, M.; Memon, A.Q.; Abro, R.; Wang, F. Water decontamination by 3D graphene based materials: A review. *J. Water Process Eng.* **2020**, *36*, 101404. [CrossRef]
135. Sun, Y.; Yang, S.; Zhao, G.; Wang, Q.; Wang, X. Adsorption of polycyclic aromatic hydrocarbons on graphene oxides and reduced graphene oxides. *Chemistry* **2013**, *8*, 2755–2761. [CrossRef]
136. Adeola, A.O.; Forbes, P.B.C. Optimization of the sorption of selected polycyclic aromatic hydrocarbons by regenerable graphene wool. *Water Sci. Technol.* **2020**, *80*, 1931–1943. [CrossRef]
137. Zhou, Q.; Wang, Y.; Xiao, J.; Fan, H. Fabrication and characterisation of magnetic graphene oxide incorporated Fe_3O_4@polyaniline for the removal of bisphenol A, t-octyl-phenol, and α-naphthol from water. *Sci. Rep.* **2017**, *7*, 11316. [CrossRef]
138. Wang, X.; Hu, Y.; Min, J.; Li, S.; Deng, X.; Yuan, S.; Zuo, X. Adsorption characteristics of phenolic compounds on graphene oxide and reduced graphene oxide: A batch experiment combined theory calculation. *Appl. Sci.* **2018**, *8*, 1950. [CrossRef]

139. Tian, J.; Wei, J.; Zhang, H.; Kong, Z.; Zhu, Y.; Qin, Z. Graphene oxide-functionalized dual-scale channels architecture for high-throughput removal of organic pollutants from water. *Chem. Eng. J.* **2019**, *359*, 852–862. [CrossRef]
140. Apul, O.G.; Wang, Q.; Zhou, Y.; Karanfil, T. Adsorption of aromatic organic contaminants by graphene nanosheets: Comparison with carbon nanotubes and activated carbon. *Water Res.* **2013**, *47*, 1648–1654. [CrossRef]
141. Das, P.; Goswami, S.; Maiti, S. Removal of naphthalene present in synthetic waste water using novel Graphene /Graphene Oxide nano sheet synthesized from rice straw: Comparative analysis, isotherm and kinetics. *Front. Nanosci. Nanotechnol.* **2016**, *2*, 38–42. [CrossRef]
142. Zhao, G.; Jiang, L.; He, Y.; Li, J.; Dong, H.; Wang, X.; Hu, W. Sulfonated graphene for persistent aromatic pollutant management. *Adv. Mater.* **2011**, *23*, 3959–3963. [CrossRef]
143. Mortazavi, M.; Baghdadi, M.; Seyed Javadi, N.H.; Torabian, A. The black beads produced by simultaneous thermal reducing and chemical bonding of graphene oxide on the surface of amino-functionalized sand particles: Application for PAHs removal from contaminated waters. *J. Water Process Eng.* **2019**, *31*, 100798. [CrossRef]
144. Wang, J.; Chen, B. Adsorption and coadsorption of organic pollutants and a heavy metal by graphene oxide and reduced graphene materials. *Chem. Eng. J.* **2015**, *281*, 379–388. [CrossRef]
145. Mehrizad, A.; Gharbani, P. Decontamination of 4-chloro-2-nitrophenol from aqueous solution by graphene adsorption: Equilibrium, kinetic, and thermodynamic studies. *Pol. J. Environ. Stud.* **2014**, *23*, 2111–2116. [CrossRef]
146. De La Luz-Asunción, M.; Sánchez-Mendieta, V.; Martínez-Hernández, A.L.; Castaño, V.M.; Velasco-Santos, C. Adsorption of phenol from aqueous solutions by carbon nanomaterials of one and two dimensions: Kinetic and equilibrium studies. *J. Nanomater.* **2015**, *2015*, 405036. [CrossRef]
147. Gholami-Bonabi, L.; Ziaefar, N.; Sheikhloie, H. Removal of phenol from aqueous solutions by magnetic oxide graphene nanoparticles modified with ionic liquids using the Taguchi optimization approach. *Water Sci. Technol.* **2020**, *81*, 228–240. [CrossRef]
148. Li, Y.; Du, Q.; Liu, T.; Sun, J.; Jiao, Y.; Xia, Y.; Xia, L.; Wang, Z.; Zhang, W.; Wang, K.; et al. Equilibrium, kinetic and thermodynamic studies on the adsorption of phenol onto graphene. *Mater. Res. Bull.* **2012**, *47*, 1898–1904. [CrossRef]
149. Wang, J.; Gao, X.; Wang, Y.; Gao, C. Novel Graphene Oxide Sponge synthesized by Freeze-Drying Process for the Removal of 2,4,6-Trichlorophenol. *RSC Adv.* **2014**, *4*, 57476–57482. [CrossRef]
150. Phatthanakittiphong, T.; Seo, G.T. Characteristic evaluation of graphene oxide for bisphenol a adsorption in aqueous solution. *Nanomaterials* **2016**, *6*, 128. [CrossRef]
151. Zhang, Y.; Tang, Y.; Li, S.; Yu, S. Sorption and removal of tetrabromobisphenol A from solution by graphene oxide. *Chem. Eng. J.* **2013**, *222*, 94–100. [CrossRef]
152. Soltani, T.; Lee, B.K. Mechanism of highly efficient adsorption of 2-chlorophenol onto ultrasonic graphene materials: Comparison and equilibrium. *J. Colloid Interface Sci.* **2016**, *481*, 168–180. [CrossRef] [PubMed]
153. Zhou, L.; Zhang, X.; Chen, Y. Modulated synthesis of zirconium metal—Organic framework UiO-66 with enhanced dichloromethane adsorption capacity. *Mater. Lett.* **2017**, *197*, 167–170. [CrossRef]
154. Ong, Y.T.; Ahmad, A.L.; Zein, S.H.S.; Tan, S.H. A review on carbon nanotubes in an environmental protection and green engineering perspective. *Braz. J. Chem. Eng.* **2010**, *27*, 227–242. [CrossRef]
155. Sarkar, B.; Mandal, S.; Tsang, Y.F.; Kumar, P.; Kim, K.H.; Ok, Y.S. Designer carbon nanotubes for contaminant removal in water and wastewater: A critical review. *Sci. Total Environ.* **2018**, *612*, 561–581. [CrossRef]
156. ALOthman, Z.A.; Wabaidur, S.M. Application of carbon nanotubes in extraction and chromatographic analysis: A review. *Arab. J. Chem.* **2019**, *12*, 633–651. [CrossRef]
157. Mallakpour, S.; Khadem, E. Carbon nanotube–metal oxide nanocomposites: Fabrication, properties and applications. *Chem. Eng. J.* **2016**, *302*, 344–367. [CrossRef]
158. Krstić, V.; Urošević, T.; Pešovski, B. A review on adsorbents for treatment of water and wastewaters containing copper ions. *Chem. Eng. Sci.* **2018**, *192*, 273–287. [CrossRef]
159. Zhang, W.; Zeng, Z.; Liu, Z.; Huang, J.; Xiao, R.; Shao, B.; Liu, Y.; Liu, Y.; Tang, W.; Zeng, G.; et al. Effects of carbon nanotubes on biodegradation of pollutants: Positive or negative? *Ecotoxicol. Environ. Saf.* **2020**, *189*. [CrossRef]
160. Yu, J.G.; Zhao, X.H.; Yang, H.; Chen, X.H.; Yang, Q.; Yu, L.Y.; Jiang, J.H.; Chen, X.Q. Aqueous adsorption and removal of organic contaminants by carbon nanotubes. *Sci. Total Environ.* **2014**, *482–483*, 241–251. [CrossRef]

161. Jung, C.; Son, A.; Her, N.; Zoh, K.D.; Cho, J.; Yoon, Y. Removal of endocrine disrupting compounds, pharmaceuticals, and personal care products in water using carbon nanotubes: A review. *J. Ind. Eng. Chem.* **2015**, *27*, 1–11. [CrossRef]
162. Saleh, T.A.; Sarı, A.; Tuzen, M. Carbon nanotubes grafted with poly(trimesoyl, m-phenylenediamine) for enhanced removal of phenol. *J. Environ. Manag.* **2019**, *252*. [CrossRef]
163. Wu, W.; Yang, K.; Chen, W.; Wang, W.; Zhang, J.; Lin, D.; Xing, B. Correlation and prediction of adsorption capacity and affinity of aromatic compounds on carbon nanotubes. *Water Res.* **2016**, *88*, 492–501. [CrossRef]
164. Selvaraj, M.; Hai, A.; Banat, F.; Haija, M.A. Application and prospects of carbon nanostructured materials in water treatment: A review. *J. Water Process Eng.* **2020**, *33*. [CrossRef]
165. Apul, O.G.; Karanfil, T. Adsorption of synthetic organic contaminants by carbon nanotubes: A critical review. *Water Res.* **2015**, *68*, 34–55. [CrossRef] [PubMed]
166. Ahmad, J.; Naeem, S.; Ahmad, M.; Usman, A.R.A.; Al-Wabel, M.I. A critical review on organic micropollutants contamination in wastewater and removal through carbon nanotubes. *J. Environ. Manag.* **2019**, *246*, 214–228. [CrossRef] [PubMed]
167. Yang, K.; Zhu, L.; Xing, B. Adsorption of polycyclic aromatic hydrocarbons by carbon nanomaterials. *Environ. Sci. Technol.* **2006**, *40*, 1855–1861. [CrossRef]
168. Kah, M.; Zhang, X.; Jonker, M.T.O.; Hofmann, T. Measuring and modeling adsorption of PAHs to carbon nanotubes over a six order of magnitude wide concentration range. *Environ. Sci. Technol.* **2011**, *45*, 6011–6017. [CrossRef]
169. Gotovac, S.; Yang, C.M.; Hattori, Y.; Takahashi, K.; Kanoh, H.; Kaneko, K. Adsorption of polyaromatic hydrocarbons on single wall carbon nanotubes of different functionalities and diameters. *J. Colloid Interface Sci.* **2007**, *314*, 18–24. [CrossRef]
170. Pahlavan Yali, Z.; Fatemi, M.H. Prediction of the sorption coefficient for the adsorption of PAHs on MWCNT based on hybrid QSPR-molecular docking approach. *Adsorption* **2019**, *25*, 737–743. [CrossRef]
171. Hu, B.; Gao, Z.; Wang, H.; Wang, J.; Cheng, M. Computational insights into the sorption mechanism of polycyclic aromatic hydrocarbons by carbon nanotube through density functional theory calculation and molecular dynamics simulation. *Comput. Mater. Sci.* **2020**, *179*. [CrossRef]
172. Kragulj, M.; Tričković, J.; Kukovecz, Á.; Jović, B.; Molnar, J.; Rončević, S.; Kónya, Z.; Dalmacija, B. Adsorption of chlorinated phenols on multiwalled carbon nanotubes. *RSC Adv.* **2015**, *5*, 24920–24929. [CrossRef]
173. Yang, K.; Wu, W.; Jing, Q.; Jiang, W.; Xing, B. Competitive adsorption of naphthalene with 2,4-dichlorophenol and 4-chloroaniline on multiwalled carbon nanotubes. *Environ. Sci. Technol.* **2010**, *44*, 3021–3027. [CrossRef] [PubMed]
174. Tóth, A.; Törocsik, A.; Tombácz, E.; László, K. Competitive adsorption of phenol and 3-chlorophenol on purified MWCNTs. *J. Colloid Interface Sci.* **2012**, *387*, 244–249. [CrossRef]
175. Asmaly, H.A.; Abussaud, B.; Ihsanullah; Saleh, T.A.; Gupta, V.K.; Atieh, M.A. Ferric oxide nanoparticles decorated carbon nanotubes and carbon nanofibers: From synthesis to enhanced removal of phenol. *J. Saudi Chem. Soc.* **2015**, *19*, 511–520. [CrossRef]
176. Jiang, L.; Li, S.; Yu, H.; Zou, Z.; Hou, X.; Shen, F.; Li, C.; Yao, X. Amino and thiol modified magnetic multi-walled carbon nanotubes for the simultaneous removal of lead, zinc, and phenol from aqueous solutions. *Appl. Surf. Sci.* **2016**, *369*, 398–413. [CrossRef]
177. Alves, D.C.S.; Gonçalves, J.O.; Coseglio, B.B.; Burgo, T.A.L. Adsorption of phenol onto chitosan hydrogel scaffold modified with carbon nanotubes. *J. Environ. Chem. Eng.* **2019**, *7*, 103460. [CrossRef]
178. Strachowski, P.; Bystrzejewski, M. Comparative studies of sorption of phenolic compounds onto carbon-encapsulated iron nanoparticles, carbon nanotubes and activated carbon. *Colloids Surf. A Physicochem. Eng. Asp.* **2015**, *467*, 113–123. [CrossRef]
179. Yi, L.; Zuo, L.; Wei, C.; Fu, H.; Qu, X.; Zheng, S.; Xu, Z.; Guo, Y.; Li, H.; Zhu, D. Enhanced adsorption of bisphenol A, tylosin, and tetracycline from aqueous solution to nitrogen-doped multiwall carbon nanotubes via cation-π and π-π electron-donor-acceptor (EDA) interactions. *Sci. Total Environ.* **2020**, *719*. [CrossRef]
180. Abdel-Ghani, N.T.; El-Chaghaby, G.A.; Helal, F.S. Individual and competitive adsorption of phenol and nickel onto multiwalled carbon nanotubes. *J. Adv. Res.* **2015**, *6*, 405–415. [CrossRef]
181. Zhang, Z.; Sun, D.; Li, G.; Zhang, B.; Zhang, B.; Qiu, S.; Li, Y.; Wu, T. Calcined products of Mg–Al layered double hydroxides/single-walled carbon nanotubes nanocomposites for expeditious removal of phenol

and 4-chlorophenol from aqueous solutions. *Colloids Surf. A Physicochem. Eng. Asp.* **2019**, *565*, 143–153. [CrossRef]
182. Chen, G.C.; Shan, X.Q.; Wang, Y.S.; Wen, B.; Pei, Z.G.; Xie, Y.N.; Liu, T.; Pignatello, J.J. Adsorption of 2,4,6-trichlorophenol by multi-walled carbon nanotubes as affected by Cu(II). *Water Res.* **2009**, *43*, 2409–2418. [CrossRef] [PubMed]
183. Fathy, M.; Selim, H.; Shahawy, A.E.L. Chitosan/MCM-48 nanocomposite as a potential adsorbent for removing phenol from aqueous solution. *RSC Adv.* **2020**, *10*, 23417–23430. [CrossRef]
184. Albayati, T.M.; Kalash, K.R. Polycyclic aromatic hydrocarbons adsorption from wastewater using different types of prepared mesoporous materials MCM-41in batch and fixed bed column. *Process Saf. Environ. Prot.* **2020**, *133*, 124–136. [CrossRef]
185. Connolly, B.M.; Mehta, J.P.; Moghadam, P.Z.; Wheatley, A.E.H.; Fairen-Jimenez, D. From synthesis to applications: Metal–organic frameworks for an environmentally sustainable future. *Curr. Opin. Green Sustain. Chem.* **2018**, *12*, 47–56. [CrossRef]
186. Jiang, N.; Shang, R.; Heijman, S.G.J.; Rietveld, L.C. High-silica zeolites for adsorption of organic micro-pollutants in water treatment: A review. *Water Res.* **2018**, *144*, 145–161. [CrossRef]
187. Wen, Y.; Zhang, J.; Xu, Q.; Wu, X.; Zhu, Q. Pore surface engineering of metal—Organic frameworks for heterogeneous catalysis. *Coord. Chem. Rev.* **2018**, *376*, 248–276. [CrossRef]
188. Zhang, Z.; Hou, X.; Zhang, X.; Li, H. The synergistic adsorption of pyrene and copper onto Fe(III) functionalized mesoporous silica from aqueous solution. *Colloids Surf. A Physicochem. Eng. Asp.* **2017**, *520*, 39–45. [CrossRef]
189. Nadar, S.S.; Varadan, N.; Suresh, S.; Rao, P.; Ahirrao, D.J.; Adsare, S. Recent progress in nanostructured magnetic framework composites (MFCs): Synthesis and applications. *J. Taiwan Inst. Chem. Eng.* **2018**, *91*, 653–677. [CrossRef]
190. Bhatnagar, A.; Anastopoulos, I. Adsorptive removal of bisphenol A (BPA) from aqueous solution: A review. *Chemosphere* **2017**, *168*, 885–902. [CrossRef]
191. Mahmodi, G.; Zarrintaj, P.; Taghizadeh, A.; Taghizadeh, M.; Manouchehri, S.; Dangwal, S.; Ronte, A.; Ganjali, M.R.; Ramsey, J.D.; Kim, S.J.; et al. From microporous to mesoporous mineral frameworks: An alliance between zeolite and chitosan. *Carbohydr. Res.* **2020**, *489*. [CrossRef] [PubMed]
192. Ahmed, M.J.; Hameed, B.H.; Hummadi, E.H. Review on recent progress in chitosan/chitin-carbonaceous material composites for the adsorption of water pollutants. *Carbohydr. Polym.* **2020**, *247*. [CrossRef] [PubMed]
193. Li, J.M.; Meng, X.G.; Hu, C.W.; Du, J. Adsorption of phenol, p-chlorophenol and p-nitrophenol onto functional chitosan. *Bioresour. Technol.* **2009**, *100*, 1168–1173. [CrossRef]
194. Bibi, S.; Yasin, T.; Hassan, S.; Riaz, M.; Nawaz, M. Chitosan/CNTs green nanocomposite membrane: Synthesis, swelling and polyaromatic hydrocarbons removal. *Mater. Sci. Eng. C* **2015**, *46*, 359–365. [CrossRef] [PubMed]
195. Filho, C.M.C.; Bueno, P.V.A.; Matsushita, A.F.Y.; Rubira, A.F.; Muniz, E.C.; Durães, L.; Murtinho, D.M.B.; Valente, A.J.M. Synthesis, characterization and sorption studies of aromatic compounds by hydrogels of chitosan blended with β-cyclodextrin- and PVA-functionalized pectin. *RSC Adv.* **2018**, *8*, 14609–14622. [CrossRef]
196. Bahmani, E.; Koushkbaghi, S.; Darabi, M.; ZabihiSahebi, A.; Askari, A.; Irani, M. Fabrication of novel chitosan-g-PNVCL/ZIF-8 composite nanofibers for adsorption of Cr(VI), As(V) and phenol in a single and ternary systems. *Carbohydr. Polym.* **2019**, *224*. [CrossRef]
197. Karamipour, A.; Parsi, P.K.; Zahedi, P.; Moosavian, S.M.A. Using Fe_3O_4-coated nanofibers based on cellulose acetate/chitosan for adsorption of Cr(VI), Ni(II) and phenol from aqueous solutions. *Int. J. Biol. Macromol.* **2020**, *154*, 1132–1139. [CrossRef] [PubMed]
198. Lyu, H.; Hu, K.; Chu, Q.; Su, Z.; Xie, Z. Preparation of ionic liquid mediated molecularly imprinted polymer and specific recognition for bisphenol A from aqueous solution. *Microchem. J.* **2020**, *158*. [CrossRef]
199. Ncube, S.; Kunene, P.; Tavengwa, N.T.; Tutu, H.; Richards, H.; Cukrowska, E.; Chimuka, L. Synthesis and characterization of a molecularly imprinted polymer for the isolation of the 16 US-EPA priority polycyclic aromatic hydrocarbons (PAHs) in solution. *J. Environ. Manag.* **2017**, *199*, 192–200. [CrossRef]
200. Saad, N.; Chaaban, M.; Patra, D.; Ghanem, A.; El-rassy, H. Molecularly imprinted phenyl-functionalized silica aerogels: Selective adsorbents for methylxanthines and PAHs. *Microporous Mesoporous Mater.* **2020**, *292*, 109759. [CrossRef]
201. Azizi, A.; Bottaro, C.S. A critical review of molecularly imprinted polymers for the analysis of organic pollutants in environmental water samples. *J. Chromatogr. A* **2020**, *1614*. [CrossRef] [PubMed]

202. Wei, W.; Liang, R.; Wang, Z.; Qin, W. Hydrophilic molecularly imprinted polymers for selective recognition of polycyclic aromatic hydrocarbons in aqueous media. *RSC Adv.* **2015**, *5*, 2659–2662. [CrossRef]
203. Bayramoglu, G., Arica, M.Y.; Liman, G.; Celikbicak, O.; Salih, B. Removal of bisphenol A from aqueous medium using molecularly surface imprinted microbeads. *Chemosphere* **2016**, *150*, 275–284. [CrossRef] [PubMed]
204. Balbino, T.A.C.; Bellato, C.R.; da Silva, A.D.; de Marques Neto, J.O.; de Guimarães, L.M. Magnetic cross-linked chitosan modified with ethylenediamine and β-cyclodextrin for removal of phenolic compounds. *Colloids Surf. A Physicochem. Eng. Asp.* **2020**, *602*. [CrossRef]
205. Heydaripour, J.; Gazi, M.; Oladipo, A.A.; Gulcan, H.O. Porous magnetic resin-g-chitosan beads for adsorptive removal of phenolic compounds. *Int. J. Biol. Macromol.* **2019**, *123*, 1125–1131. [CrossRef] [PubMed]
206. Alizadeh, B.; Delnavaz, M.; Shakeri, A. Removal of Cd(II) and phenol using novel cross-linked magnetic EDTA/chitosan/TiO_2 nanocomposite. *Carbohydr. Polym.* **2018**, *181*, 675–683. [CrossRef] [PubMed]
207. Rebekah, A.; Bharath, G.; Naushad, M.; Viswanathan, C.; Ponpandian, N. Magnetic graphene/chitosan nanocomposite: A promising nano-adsorbent for the removal of 2-naphthol from aqueous solution and their kinetic studies. *Int. J. Biol. Macromol.* **2020**, *159*, 530–538. [CrossRef] [PubMed]
208. Zhou, A.; Chen, W.; Liao, L.; Xie, P.; Zhang, T.C.; Wu, X.; Feng, X. Comparative adsorption of emerging contaminants in water by functional designed magnetic poly(N-isopropylacrylamide)/chitosan hydrogels. *Sci. Total Environ.* **2019**, *671*, 377–387. [CrossRef]
209. Nguyen, M.L.; Huang, C.; Juang, R.S. Synergistic biosorption between phenol and nickel(II) from Binary mixtures on chemically and biologically modified chitosan beads. *Chem. Eng. J.* **2016**, *286*, 68–75. [CrossRef]
210. Nadavala, S.K.; Swayampakula, K.; Boddu, V.M.; Abburi, K. Biosorption of phenol and o-chlorophenol from aqueous solutions on to chitosan-calcium alginate blended beads. *J. Hazard. Mater.* **2009**, *162*, 482–489. [CrossRef]
211. Krupadam, R.J.; Korde, B.A.; Ashokkumar, M.; Kolev, S.D. Novel molecularly imprinted polymeric microspheres for preconcentration and preservation of polycyclic aromatic hydrocarbons from environmental samples. *Anal. Bioanal. Chem.* **2014**, *406*, 5313–5321. [CrossRef] [PubMed]
212. Li, H.; Wang, L. Highly selective detection of polycyclic aromatic hydrocarbons using multifunctional magnetic-luminescent molecularly imprinted polymers. *ACS Appl. Mater. Interfaces* **2013**, *5*, 10502–10509. [CrossRef] [PubMed]
213. Tiu, B.D.B.; Krupadam, R.J.; Advincula, R.C. Pyrene-imprinted polythiophene sensors for detection of polycyclic aromatic hydrocarbons. *Sens. Actuators B Chem.* **2016**, *228*, 693–701. [CrossRef]
214. Saleh, T.A.; Tuzen, M.; Sari, A. Magnetic vermiculite-modified by poly(trimesoyl chloride-melamine) as a sorbent for enhanced removal of bisphenol A. *J. Environ. Chem. Eng.* **2019**, *7*. [CrossRef]
215. An, F.; Gao, B.; Feng, X. Adsorption and recognizing ability of molecular imprinted polymer MIP-PEI/SiO2 towards phenol. *J. Hazard. Mater.* **2008**, *157*, 286–292. [CrossRef]
216. Liang, W.; Lu, Y.; Li, N.; Li, H.; Zhu, F. Microwave-assisted Synthesis of Magnetic Surface Molecular Imprinted Polymer for Adsorption and Solid Phase Extraction of 4-nitrophenol in wastewater. *Microchem. J.* **2020**, 105316. [CrossRef]
217. Zhao, Z.; Fu, D.; Zhang, B. Novel molecularly imprinted polymer prepared by palygorskite as support for selective adsorption of bisphenol A in aqueous solution. *Desalin. Water Treat.* **2016**, *57*, 12433–12442. [CrossRef]
218. Nasreen, S.; Rafique, U.; Ehrman, S.; Ashraf, M.A. Synthesis and characterization of mesoporous silica nanoparticles for environmental remediation of metals, PAHs and phenols. *Ekoloji* **2018**, *27*, 1625–1637.
219. Costa, J.A.S.; Sarmento, V.H.V.; Romão, L.P.C.; Paranhos, C.M. Removal of polycyclic aromatic hydrocarbons from aqueous media with polysulfone/MCM-41 mixed matrix membranes. *J. Membr. Sci.* **2020**, *601*. [CrossRef]
220. Balati, A.; Shahbazi, A.; Amini, M.M.; Hashemi, S.H.; Jadidi, K. Comparison of the efficiency of mesoporous silicas as absorbents for removing naphthalene from contaminated water. *Eur. J. Environ. Sci.* **2014**, *4*, 69–76. [CrossRef]
221. Yuan, P.; Li, X.; Wang, W.; Liu, H.; Yang, H.; Yue, Y.; Bao, X. Tailored design of differently modified mesoporous materials to deeply understand the adsorption mechanism for PAHs. *Langmuir* **2018**. [CrossRef] [PubMed]

222. Yangui, A.; Abderrabba, M.; Sayari, A. Amine-modified mesoporous silica for quantitative adsorption and release of hydroxytyrosol and other phenolic compounds from olive mill wastewater. *J. Taiwan Inst. Chem. Eng.* **2017**, *70*, 111–118. [CrossRef]
223. Zou, D.; Liu, D. Understanding the modifications and applications of highly stable porous frameworks via UiO-66. *Mater. Today Chem.* **2019**, *12*, 139–165. [CrossRef]
224. Elsabawy, K.M.; Fallatah, A.M. New advanced approach of ultra-fast synthesis of ultrahigh-BET-surface area crystalline metal-organic-frameworks MOFs. *Mater. Lett.* **2018**, *224*, 71–74. [CrossRef]
225. Jiang, D.; Chen, M.; Wang, H.; Zeng, G.; Huang, D.; Cheng, M.; Liu, Y.; Xue, W.; Wang, Z.W. The application of different typological and structural MOFs-based materials for the dyes adsorption. *Coord. Chem. Rev.* **2019**, *380*, 471–483. [CrossRef]
226. Mahmoodi, N.M.; Oveisi, M.; Asadi, E. Synthesis of NENU metal-organic framework-graphene oxide nanocomposites and their pollutant removal ability from water using ultrasound. *J. Clean. Prod.* **2019**, *211*, 198–212. [CrossRef]
227. Zango, Z.U.; Sambudi, N.S.; Jumbri, K.; Abu Bakar, N.H.H.; Abdullah, N.A.F.; Negim, E.S.M.; Saad, B. Experimental and molecular docking model studies for the adsorption of polycyclic aromatic hydrocarbons onto UiO-66(Zr) and NH_2-UiO-66(Zr) metal-organic frameworks. *Chem. Eng. Sci.* **2020**, *220*, 115608. [CrossRef]
228. Zango, Z.U.; Abu Bakar, N.H.H.; Sambudi, N.S.; Jumbri, K.; Abdullah, N.A.F.; Abdul Kadir, E.; Saad, B. Adsorption of chrysene in aqueous solution onto MIL-88(Fe) and NH_2-MIL-88(Fe) metal-organic frameworks: Kinetics, isotherms, thermodynamics and docking simulation studies. *J. Environ. Chem. Eng.* **2019**. [CrossRef]
229. Lin, K.A.; Hsieh, Y. Copper-based metal organic framework (MOF), HKUST-1, as an efficient adsorbent to remove p-nitrophenol from water. *J. Taiwan Inst. Chem. Eng.* **2015**, *50*, 223–228.
230. Xu, Z.; Wen, Y.; Tian, L.; Li, G. Efficient and selective adsorption of nitroaromatic explosives by Zr-MOF. *Inorg. Chem. Commun.* **2017**, *77*, 11–13. [CrossRef]
231. Luo, Z.; Chen, H.; Wu, S.; Yang, C.; Cheng, J. Enhanced removal of bisphenol A from aqueous solution by aluminum-based MOF/sodium alginate-chitosan composite beads. *Chemosphere* **2019**, *237*. [CrossRef] [PubMed]
232. Balati, A.; Shahbazi, A.; Amini, M.M.; Hashemi, S.H. Adsorption of polycyclic aromatic hydrocarbons from wastewater by using silica-based organic–inorganic nanohybrid material. *J. Water Reuse Desalin.* **2015**, *5*, 50–63. [CrossRef]
233. Costa, J.A.S.; Sarmento, V.H.V.; Romão, L.P.C.; Paranhos, C.M. Synthesis of functionalized mesoporous material from rice husk ash and its application in the removal of the polycyclic aromatic hydrocarbons. *Environ. Sci. Pollut. Res.* **2019**, *26*, 25476–25490. [CrossRef] [PubMed]
234. Anbia, M.; Amirmahmoodi, S. Adsorption of phenolic compounds from aqueous solutions using functionalized SBA-15 as a nano-sorbent. *Sci. Iran.* **2011**, *18*, 446–452. [CrossRef]
235. Nasreen, S.; Rafique, U.; Ehrman, S.; Ashraf, M.A. Hybrid mesoporous silicates: A distinct aspect to synthesis and application for decontamination of phenols. *Saudi J. Biol. Sci.* **2019**, *26*, 1161–1170. [CrossRef]
236. Asgharnia, H.; Nasehinia, H.; Rostami, R.; Rahmani, M.; Mehdinia, S.M. Phenol removal from aqueous solution using silica and activated carbon derived from rice husk. *Water Pract. Technol.* **2019**, *14*, 897–907. [CrossRef]
237. Zango, Z.U.; Jumbri, K.; Sambudi, N.S.; Abu Bakar, N.H.H.; Abdullah, N.A.F.; Basheer, C.; Saad, B. Removal of anthracene in water by MIL-88(Fe), NH_2-MIL-88(Fe), and mixed-MIL-88(Fe) metal–organic frameworks. *RCS Adv.* **2019**, *9*, 41490–41501. [CrossRef]
238. Zango, Z.U.; Sambudi, N.S.; Jumbri, K.; Abu Bakar, N.H.H.; Saad, B. Removal of Pyrene from Aqueous Solution Using Fe-based Metal-organic Frameworks. *IOP Conf. Ser. Earth Environ. Sci.* **2020**, *549*, 012061. [CrossRef]
239. Zhou, M.; Wu, Y.-n.; Qiao, J.; Zhang, J.; McDonald, A.; Li, G.; Li, F. The removal of bisphenol A from aqueous solutions by MIL-53(Al) and mesostructured MIL-53(Al). *J. Colloid Interface Sci.* **2013**, *405*, 157–163. [CrossRef]
240. Tan, Y.; Sun, Z.; Meng, H.; Han, Y.; Wu, J.; Xu, J.; Xu, Y.; Zhang, X. A new MOFs/polymer hybrid membrane: MIL-68(Al)/PVDF, fabrication and application in high-efficient removal of p-nitrophenol and methylene blue. *Sep. Purif. Technol.* **2019**, *215*, 217–226. [CrossRef]

241. Han, T.; Li, C.; Guo, X.; Huang, H.; Liu, D.; Zhong, C. In-situ synthesis of SiO_2@MOF composites for high-efficiency removal of aniline from aqueous solution. *Appl. Surf. Sci.* **2016**, *390*, 506–512. [CrossRef]
242. Abazari, R.; Salehi, G.; Mahjoub, A.R. Ultrasound-assisted preparation of a nanostructured zinc(II) amine pillar metal-organic framework as a potential sorbent for 2,4-dichlorophenol adsorption from aqueous solution. *Ultrason. Sonochem.* **2018**, *46*, 59–67. [CrossRef] [PubMed]
243. Abazari, R.; Mahjoub, A.R. Ultrasound-assisted synthesis of Zinc(II)-based metal organic framework nanoparticles in the presence of modulator for adsorption enhancement of 2,4-dichlorophenol and amoxicillin. *Ultrason. Sonochem.* **2018**, *42*, 577–584. [CrossRef] [PubMed]
244. Han, T.; Xiao, Y.; Tong, M.; Huang, H.; Liu, D.; Wang, L.; Zhong, C. Synthesis of CNT@MIL-68(Al) composites with improved adsorption capacity for phenol in aqueous solution. *Chem. Eng. J.* **2015**, *275*, 134–141. [CrossRef]
245. Jung, B.K.; Hasan, Z.; Jhung, S.H. Adsorptive removal of 2,4-dichlorophenoxyacetic acid (2,4-D) from water with a metal–organic framework. *Chem. Eng. J.* **2013**, *234*, 99–105. [CrossRef]
246. Guo, H.; Niu, B.; Wu, X.; Zhang, Y.; Ying, S. Effective removal of 2,4,6-trinitrophenol over hexagonal metal–organic framework NH_2-MIL-88B(Fe). *Appl. Organomet. Chem.* **2018**, *33*, e4580. [CrossRef]
247. Giraldo, L.; Bastidas-Barranco, M.; Húmpola, P.; Moreno-Piraján, J.C. Design, synthesis and characterization of MOF-199 and ZIF-8: Applications in the adsorption of phenols derivatives in aqueous solution. *Eur. J. Chem.* **2017**, *8*, 293–304. [CrossRef]
248. Ahsan, M.A.; Jabbari, V.; Islam, M.T.; Turley, R.S.; Dominguez, N.; Kim, H.; Castro, E.; Hernandez-Viezcas, J.A.; Curry, M.L.; Lopez, J.; et al. Sustainable synthesis and remarkable adsorption capacity of MOF/graphene oxide and MOF/CNT based hybrid nanocomposites for the removal of Bisphenol A from water. *Sci. Total Environ.* **2019**, *673*, 306–317. [CrossRef]
249. Zhang, R.; Wang, L.; Han, J.; Wu, J.; Li, C.; Ni, L.; Wang, Y. Improving laccase activity and stability by HKUST-1 with cofactor via one-pot encapsulation and its application for degradation of bisphenol A. *J. Hazard. Mater.* **2020**, *383*. [CrossRef]
250. Kamali, M.; Persson, K.M.; Costa, M.E.; Capela, I. Sustainability criteria for assessing nanotechnology applicability in industrial wastewater treatment: Current status and future outlook. *Environ. Int.* **2019**, *125*, 261–276. [CrossRef]
251. El-Naas, M.H.; Surkatti, R.; Al-Zuhair, S. Petroleum refinery wastewater treatment: A pilot scale study. *J. Water Process Eng.* **2016**, *14*, 71–76. [CrossRef]
252. Kalmykova, Y.; Moona, N.; Strömvall, A.M.; Björklund, K. Sorption and degradation of petroleum hydrocarbons, polycyclic aromatic hydrocarbons, alkylphenols, bisphenol A and phthalates in landfill leachate using sand, activated carbon and peat filters. *Water Res.* **2014**, *56*, 246–257. [CrossRef] [PubMed]
253. Frascari, D.; Bacca, A.E.M.; Zama, F.; Bertin, L.; Fava, F.; Pinelli, D. Olive mill wastewater valorisation through phenolic compounds adsorption in a continuous flow column. *Chem. Eng. J.* **2016**, *283*, 293–303. [CrossRef]

Publisher's Note: MDPI stays neutral with regard to jurisdictional claims in published maps and institutional affiliations.

Review

Chemically Modified Biosorbents and Their Role in the Removal of Emerging Pharmaceutical Waste in the Water System

Adewale Adewuyi

Department of Chemical Sciences, Faculty of Natural Sciences, Redeemer's University, P.M.B 230, Ede, Osun State, Nigeria; walexy62@yahoo.com; Tel.: +234-8035826679

Received: 29 April 2020; Accepted: 26 May 2020; Published: 29 May 2020

Abstract: Presence of pharmaceutically active compounds (PACs) as emerging contaminants in water is a major concern. Recent reports have confirmed the presence of PACs in natural and wastewater systems, which have caused several problems indicating the urgent need for their removal. The current review evaluates the role of chemically modified biosorbents in the removal of PACs in water. Reported biosorbents include plant and animal solid waste, microorganisms and bio-composite. Bio-composites exhibited better prospects when compared with other biosorbents. Types of chemical treatment reported include acid, alkaline, solvent extraction, metal salt impregnation and surface grafting, with alkaline treatment exhibiting better results when compared with other treatments. The biosorption processes mostly obeyed the pseudo-second-order model and the Langmuir isotherm model in a process described mainly by ionic interaction. Desorption and regeneration capacity are very important in selecting an appropriate biosorbent for the biosorption process. Depending on the type of biosorbent, the cost of water treatment per million liters of water was estimated as US $10–US $200, which presents biosorption as a cheap process compared to other known water treatment processes. However, there is a need to conduct large-scale studies on the biosorption process for removing PACs in water.

Keywords: algae; bacteria; biosorbent; biosorption; pharmaceutical waste

---

1. Introduction

Lack of sufficient clean drinking water is a growing challenge in many countries. Most surface and groundwater bodies are contaminated and are not directly safe for consumption without being treated. Surface and ground waters in rural communities that were once clean are now contaminated due to industrialization and abuse in the use and disposal of pharmaceutically active compounds (PACs). Contamination has severely affected the quality of freshwater. This undesired effect harms humans, animals and the ecosystem. Although several measures, such as creating awareness of water quality problems, promoting advanced water cleaning techniques and establishing preventive systems, have been put in place in some countries, the challenge is still persistent with different levels of impact in different countries. Several organic compounds have been reported in water systems; however, the presence of PACs continues to pose a severe problem among these organic compounds. Most PACs are considered as emerging contaminants, since their regulation and limit in domestic and industrial wastewater handling and treatment criteria, in most countries of the world, are not yet specified, most notably in developing countries [1]. They occur in small amounts and are capable of bioaccumulating over a period to cause harm. The danger ascribed to PACs necessitates the development of an environmentally friendly and cost-effective means for their removal in water systems. The consumption of pharmaceutical compounds has increased over the past decade across the world. Their use includes application in disease control and eradication in both human and animals. One major problem of this is the uncontrolled use of antimicrobial agents in veterinary medicine as well as the use of over-the-counter drugs in self-medication [2,3]. The abuse of PACs has led to their presence in the environment. They enter the environment when they are excreted from the body system and sometimes when inappropriately disposed of or applied. Having entered the environment, they are capable of transforming into metabolites under certain environmental conditions, temperature, pressure, light, etc., in a process, which can be described as biodegradation and photodegradation. The metabolites are sometimes more toxic than the parent PACs. This transformation makes it difficult to monitor the presence of PACs in their parent forms in surface and groundwater. The presence of PACs and possibly their metabolites have made pathogenic organisms develop resistance against them and, over time, they lose their potency against these disease-causing microorganisms. These emerging contaminants have gained attention as previous researchers have reported toxic and adverse effect even at low concentration [4]. Apart from the development of drug resistance by pathogenic microorganisms, previous studies have linked feminization of fish species to the presence of natural and synthetic estrogens in water [5–7].

PACs are a wide range of chemical compounds, which may include anti-inflammatories, antibiotics, cytotoxins, birth control pills, synthetic hormones, and statins. They are different from other chemical contaminants found in water because innumerable complex molecules may form them; the molecules are absorbed in the human body and may undergo metabolic reactions that can modify their chemical structure; they may persist in the environment and may be amphiphilic [8]. Several PACs have been detected in surface water, which varies from region to region, as shown in Table 1. Many sources may have contributed to their presence in surface water; primary sources are urban, industrial, domestic and hospital wastewater. Intensive livestock farming, liquid livestock manure, sewage sludge from agricultural practice and effluents from sewage treatment plants may also serve as sources.

Table 1. Pharmaceutically active compounds (PACs) reported in surface water.

PACs	Country	Concentration (ng L^{-1}) *	Reference
Antibiotics			
Erythromycin	Bangladesh	6.46	[9]
Metronidazole	China	5.10	[10]
Sulfamethoxazole	USA	14.73	[11]
Trimethoprim	Mexico	74.00	[12]
Ciprofloxacin	India	6,500,000	[13]
Amoxicillin	Ghana	2.70	[14]
Trimethoprim	Kenya	2650	[15]
Tetracyclin	Ghana	30.00	[14]
Anti-hyperglycemic			
Metformin	Sweden	8.40	[16]
Lipid regulator			
Gemfibrozil	Spain	3735.00	[17]
Analgesics and Anti-inflammatories			
Acetaminophen	UK	9822.00	[18]
Diclofenac	Malaysia	15.49	[19]
Ibuprofen	South-Africa	846.00	[20]
Ketoprofen	Portugal	86.90	[21]
Naproxen	Poland	13.40	[22]
Anti-depressives			
Diazepam	India	305.00	[23]
Fluoxetine	Portugal	25.37	[21]

* = Maximum concentration reported.

The different pathways for the introduction of PACs into the environment can be described as reported by Silva et al. [24] in Figure 1. Apart from their presence in surface water, they have percolated into groundwater, and their presence has also been documented in coastal seawater [25]. The consequences of the presence of PACs in water goes beyond the reduction in water quality to environmental persistence and detrimental effect on aquatic life [26].

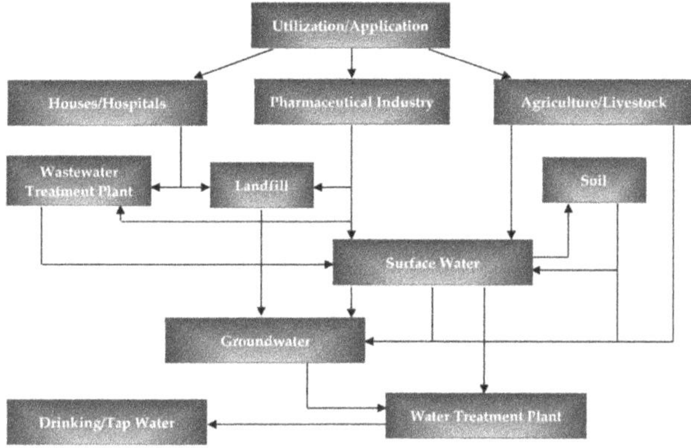

Figure 1. Pathways of PACs' introduction into the environment [24].

There is a need to develop procedures and technology for the complete removal of PACs in water or in wastewater before they are discharged into the environment. Previous techniques used for the removal of PACs from water include chemical, physical and biological approaches, e.g., coagulation, membrane filtration, bioremediation, volatilization, ozonation, flocculation, advance oxidation, photocatalysis, sedimentation, microbial degradation, electrochemical processes and adsorption [27–34]. These approaches differ in effectiveness, sustainability and cost and have different advantages and disadvantages. Unfortunately, sedimentation, coagulation and flocculation processes are not so suitable and effective for the removal of PACs due to the high solubility and mobility of some PACs in water [35]. However, separation processes such as ultra-, nano- and micro-filtrations have been suggested, but their efficiency is mainly dependent on the molecular weight and size of the PACs. Other studies have also shown that the filtration processes might further depend on ionic interaction, membrane functional groups, hydrophobic interaction and age of the membrane [36–38]. Use of advanced electrochemical oxidation has been shown to have the limitation of producing more toxic and recalcitrant by-products than the parent PACs [39,40]. Although ozonation processes have been useful, the toxicity of the intermediate compounds formed during the process is not well understood. It becomes necessary to develop an additional unit operation to remove the intermediate molecules formed, which increases production cost and the need for expertise. The biological process has been useful, but the main challenges are incomplete degradation of PACs and production of intermediate compounds, which either need another procedure to remove them or whose toxicity is not well known.

Several steps have been taken to upgrade these techniques, e.g., attempts at combined advanced oxidation processes, ultrasonic irradiation, electrochemical degradation, photodegradation, adsorption, photo-Fenton, etc. [41–44]; however, the upgrade results from the sophistication of these processes. The biosorption process remains relevant because of its process characteristics such as ease of maintenance, simplicity, ease of modification of biosorbents, inexpensive cost, efficiency of uptake even when the adsorbates (PACs) are in low concentrations, lower selectivity and consistency [26,45]. The outstanding performance of biosorption gives it an edge over the other known processes, most notably the ease of modification, which simplifies the process design. For this reason, this review aims to consider the role of chemically modified biosorbents in the removal of PACs from the water system.

2. Biosorption as a Process for Removing PACs in Water

Biosorption is a mass transfer process that moves substances from the fluid phase to the solid phase. It is a physicochemical and metabolically independent process with mechanisms such as ion exchange, precipitation, absorption, adsorption and surface complexation [46]. It has found application in water treatment and, in this case, removal of PACs from the water system. During this process, substances (adsorbate) in the fluid phase interact with the solid phase and are transferred from the fluid to the solid phase (biosorbent). PACs are the adsorbate in the fluid phase while a suitable material capable of interacting with PACs is the solid phase (biosorbent). The process may be conducted in batch (stirred systems) or continuous mode system (fixed-bed column). The process entails three stages: (I) transport of adsorbate in the fluid phase to the surface of the biosorbent, known as film diffusion, (II) movement of the adsorbate to the pores of the biosorbent, or intraparticle diffusion and (III) fixing of the adsorbate on the surface of the biosorbent, surface bonding. The rate-determining step is at stage I or II, while the III stage is usually very fast [8,47]. Biosorption has been used to remove a wide range of contaminants in water, which has included PACs [29,47,48]. One significant advantage over other processes is that it does not lead to the formation of intermediates [26,27]. The process is reversible, and the adsorbent used can be regenerated by desorption for reuse. However, biosorption may be influenced by certain factors, including pH, temperature, particle size and surface area of the biosorbent. Apart from the surface being large, the surface chemistry must also be appropriate, and the pore size distribution towards PACs must also be correct. Biosorption has received high interest because of the wide range of materials that could be used as biosorbent [49]. The biosorbent used will depend on solubility in water, surface functionality, molecular size and distribution, although it is difficult to predict the performance of any biosorbent at a glance [24,50]. The economic viability of the biosorption process will depend on the adsorbate, biosorbent and the process variable employed, although the biosorbent plays a significant role in this regard. Economic feasibility will depend on the biosorption capacity, regeneration and possible modification of the biosorbent [27,50].

The development of cheap and efficient biosorbent is critical as many of the conventional ones lack satisfactory performance [51], which is the reason for introducing processes such as chemical modification to enhance their performance. Although some materials have been used as adsorbents for the purification of wastewater in the past despite their high purification capacities, the challenge is that a few of them are not biodegradable, which gives natural adsorbents priority over synthetic. Over time, several biosorbents have been modified to improve their capacity for the removal of PACs in the water system. Attention has been placed on biosorbents for water treatment because they are from a natural source, are biodegradable, easy to modify, and most are from renewable sources. Attempts have been made also to make use of waste materials as feedstock for the production of biosorbents. This way, waste from wood, agricultural and food industries are converted to useful and cheap low-cost biosorbents, which has emerged as a cost-effective and efficient alternative for water and wastewater treatment.

2.1. Biosorbents

Biosorbents are biological materials used to remove pollutants passively from a solution. They include biomaterials like agricultural wastes, algae, bacterial, and industrial wastes. They have been receiving encouraging attention because they are from renewable sources, cheap, biodegradable and, after complete usage, do not generate secondary contaminants [52]. Being biological, they contain lots of functional groups, which is the driving force of the hydrophobic interaction they exhibit during the sorption process, which is mostly pH-dependent [53,54]. Many natural materials have been suggested as promising biosorbents for the removal of pollutants in water. These materials can be inactive or dead microbial biomass, as well as living microorganisms. The mechanism by which this is achieved is fully understood for some biosorbents, whereas some require more detailed study. However, the toxicity of some of these biomaterials remains a subject of discussion, which also requires a detailed study.

Biomaterials of various sorption capacities are known. The use of these materials depends on not only their sorption capacity but also reusability over time, which forms the basis for selecting them as biosorbents. One of the major challenges is the selection of the best performing type of biosorbent from a large pool of promising and cheap biomaterials. Although most lignocellulosic materials have a small surface area, effort is focused on how to improve their activity as biosorbents. This does not only include sorption capacity but also improvement in regeneration. Comparison of sorption capacity or efficiency of modified and unmodified biosorbents is essential. The modification might be in the form of pretreatment because the use of untreated biosorbents, most especially plant waste materials, can cause the release of organic compounds into the water being treated. Untreated biosorbents may create high chemical oxygen demand, biological oxygen demand and total organic carbon [55], which leads to depletion of water oxygen content. Therefore, modification of the biosorbent is expected to be encompassing; this cleans the biomaterial by preventing the introduction of unwanted organic compounds into the treated water as well as enhancing the ability of the biomaterial to remove PACs. Different types of chemical modification reported for biosorbents are compared in Table 2.

Table 2. Types of chemical modifications and targeted PACs.

Biosorbent	Type of Chemical Modification	PAC	Qe (mg g^{-1})	pH	Kinetic Model	Time (min)	Reference
Scenedesmus obliquus (Alga)	Alkaline (NaOH)	CEFA PARA IBU TRAM CIP	68.00 58.00 42.00 42.00 39.00	7.80 7.80 7.80 7.80 7.80	PS PS PS PS PS	45 45 45 45 45	[56]
Waste apricot	Metal salt ($ZnCl_2$) impregnation	Naproxen	106.38	5.8	PS	60	[57]
Bone char	Acid (H_2SO_4)	Ibuprofen	56.78	4.00	PS	360	[58]
Spent tea Modified Spent tea	Unmodified Surface grafting (Polyethyleneimine)	Aspirin Aspirin	- -	3.00 3.00	- -	30 30	[59]
cork waste	Alkaline (K_2CO_3)	Ibuprofen	416.70	2–4	PS	360	[60]
Olive waste	Acid impregnation (H_3PO_4)	Ibuprofen Ketoprofen Naproxen Diclofenac	12.60 24.70 39.50 56.20	4.12 4.12 4.12 4.12	PS PS PS PS	1560 1560 1560 1560	[61]
Parthenium hysterophorus weed	Alkaline (NaOH)	Ibuprofen	90.46	4.00	PS	120	[62]
Pine wood chips	Pyrolysis followed by organic solvent	Salicylic acid Ibuprofen	22.70 10.74	2.00	PS PS	960 960	[63]
Sisal waste	Alkaline (K_2CO_3)	Ibuprofen Paracetamol	140.10 124.30	4.00 6.00	PS PS	1440 1440	[64]
Sugar cane bagasse Vegetable sponge	H_2O (Extraction) H_2O (Extraction)	Paracetamol Paracetamol	0.12 0.04	7.00 7.00	- -	- -	[65]

CEFA = Cefadroxil, PARA = Paracetamol, IBU = Ibuprofen, TRAM = Tramadol, CIP = Ciprofloxacin, PS = Pseudo-Second-Order, - = Not determined, Qe = Adsorption capacity.

As shown in Table 2, the use of organic solvent after pyrolysis of pine wood chips removed substances that may have leached from the biosorbent into treated water [63]. Similarly, use of water (H_2O) in the pretreatment of sugar cane bagasse and vegetable sponge before the treatment of water contaminated with paracetamol also removed substances that might have interfered with the use of the biosorbents [65]. Use of alkaline in chemical modification of biosorbents showed better adsorption capacity when compared with other methods. The authors argued that the use of alkaline opens up pores in the biosorbents, which are capable of interacting and trapping the targeted PACs. The kinetic model described the biosorption process for the modified biosorbents as reaching equilibrium in less than 60 min. However, most of the authors allowed the process to keep running for over 60 min to ascertain the consistency of the equilibria process. Moreover, all the sorption processes obeyed pseudo-second-order kinetics at a pH in the range 2–7.8. Most authors claimed that biosorption of PACs is favored at acidic or neutral pH. Unmodified and modified spent teas were evaluated for the removal of aspirin from the water system [58]. The study confirmed that the use of polyethyleneimine surface grafted spent tea performed better at a 65% aspirin removal than unmodified spent tea, which achieved just 1% removal of aspirin. This further shows that chemical modification has the capacity to improve the performance of biosorbents. Many biosorbents are known for the removal of PACs in water. Based on their numerous forms, they may be classified into four categories: natural, industrial wastes, agricultural wastes and forest wastes [66,67].

2.2. Plant and Animal Solid Waste Biosorbent

Plant and animal solid wastes are readily available, cheap and renewable resource materials. They are produced in large amount yearly and usually disposal is a problem. Finding meaningful use of these materials is an important area of research. Making use of these waste materials can help reduce waste load as well as produce economically valuable products [68,69]. The plant-sourced wastes are composed mainly of cellulose with the presence of lignin, proteins, hemicellulose, sugars, lipids and starch, which serve as structural components [70]. Studies have shown the capacity of waste materials to remove PACs from water; the different types of waste reported as biosorbents by the authors of these studies for the removal of PACs in water are presented in Table 3.

The table reveals the importance of pH in the biosorption process. The pH of the solution plays an important role in the sorption process. The pH was related to the pH_{pzc} (point of zero charges) of the studied biosorbents. The pH_{pzc} helps in determining surface properties of the biosorbents. If the pH of the solution is above pH_{pzc}, then the functional groups on the surface of the biosorbents may be protonated by excess H^+ ions whereas if the solution pH is lower than the pHpzc, the surface of the biosorbents may become deprotonated by excess OH^- ions in solution [73]. It was obvious that very high pH did not work well for most PACs; however, most works in the literature reported pH around 5–7. This goes a long way in helping to explain the mechanism of action of the biosorbents towards PACs. Most of the reported mechanisms involved electronic interactions between the surface of the biosorbents and PACs. The concentration of the PACs studied in Table 3 varies. Chen et al. [71] reported a concentration range of 5–20 mg L^{-1} for tetracycline while Wang et al. [75] reported a range of 0.5–32 mg L^{-1}, and a range of 0–100 mg dm^{-3} for diclofenac [72], and Pouretedal et al. [73], for some PACs, reported a range of 20–200 mg L^{-1}. The concentration range studied has an influence on the biosorption capacity exhibited by the biosorbents. The studies further show that biosorption capacity exhibited increases with increase in concentration. However, most PACs are not expected to be present in the environment at very high concentrations; they are expected to occur in the ng L^{-1} but, surprisingly, most authors conducted their studies in ranges (mg L^{-1}), which are higher than reported concentrations of most PACs in surface water as shown in Table 1. The concentration range in the studies may have contributed to the high capacity exhibited by the studied biosorbents. In fact, a range as high as 300 mg L^{-1} was reported in the study of chemically prepared carbon from date palm leaflets towards the sorption of ciprofloxacin, which was higher than the amount expected in an environmental sample. However, attention must be focused on the use of low concentration range to understand further the exact behavior of chemically modified biosorbents at low concentrations, which reflects the precise concentration of these PACs in the environment. PAC solutions used in most studies are synthetic in nature and studies on sorption efficiency of biosorbents in real water samples are lacking. Most reported studies using synthetic PAC solutions did not also consider the effect of salts, organic matters, surfactants, inorganic molecules, etc., which influences the biosorption capacity or process. Therefore, more studies should be conducted to be able to know the exact capacity and prospects of biosorbents.

Table 3. Removal of PACs by some selected biosorbents and their sorption capacity.

Biosorbent	PACs	Adsorption Capacity (mg g^{-1})	Kinetic Model	Mechanism	pH Used	Con (mg L^{-1})	Reference
Rice husk ash	Tetracycline	8.37	PS	Complexation	>7.7	5–20	[71]
Potato peel waste (activated carbon)	Diclofenac	74.00	PS	π-π electron donor-acceptor interaction	5	0–100	[72]
Vine wood	Amoxicillin, Cephalexin, Tetracycline and Penicillin G	1.98–8.41	PS	-	2	20–200	[73]
Bamboo biochar	Sulfamethazine, sulfamethoxazole, and sulfathiazole	25.11–40.11	PS	Lewis acid-base interactions, hydrogen bonding and π-π Electron-donor-acceptor interactions	3–6.5	1–50	[74]
Rice husk	Tetracycline	3.89–13.85	-	π–π electron-donor acceptor	5.5	0.5–32	[75]
Reed straw	Sulfamethoxazole	23.35	-	Hydrogen bonding and π-π electron donor–acceptor	4	5–30	[76]
Date palm	Ciprofloxacin	25.30–53.20	PS	Cation exchange and hydrogen bonding	6	50–300	[77]
Canola biomass	Metronidazole	21.42	PS	Electrostatic interaction	7	0–100	[78]
Groundnut shell	Paracetamol	3.02	-	-	-	10–100	[79]
Brassica nigra	Acetic acid	0.96	-	-	-	0.5 *	[80]

- = Not reported, PS = Pseudo-Second-Order, * = Concentration (N), Con = Concentration.

Table 4 presents the BET (Brunauer-Emmett-Teller) surface area obtained for some biosorbents. The value varies for the different adsorbents depending on the modification process. Plant materials, which are lignocellulose, have low S_{BET} as shown by Silva et al. [81], which demonstrated that plant waste-based biosorbents are feasible alternatives to commercial adsorbents for the removal of PACs in solution. The team reported use of NaOH modified spent coffee grounds, pine bark and cork waste for the removal of fluoxetine from the water with biosorption capacities ranging from 4.74 to 14.31 mg g^{-1}. The sorption capacities exhibited by commercially prepared materials, activated carbon (233.5 mg g^{-1}), zeolite 13 × (32.11 mg g^{-1}) and zeolite 4 A (21.86 mg g^{-1}) were higher than those of spent coffee grounds G (14.31 mg g^{-1}), pine bark (6.53 mg g^{-1}), and cork waste (4.74 mg g^{-1}). However, the NaOH modified biosorbents reflect economic feasibility through cost analysis (0.16–6.85 € g^{-1}) as they are cheaper than the commercial materials, zeolite 4 A (6.85 € g^{-1}), zeolite 13 × (3.13 € g^{-1}) and activated carbon (1.07 € g^{-1}). Apart from economic feasibility, the biosorbents have a positive environmental impact by not introducing solid or liquid waste into the environment. As a form of chemical modification, NaOH-activated carbon was prepared from macadamia nutshell for the removal of tetracycline from the water system [82]. The material showed a microporous structure with a BET surface area of 1524 m^2 g^{-1}. It exhibited an adsorption capacity of 455.33 mg g^{-1} towards tetracycline in a process that can be described by the Elovich kinetic model. The capacity was higher than what was reported for rice husk ash [71] It has become obvious that antibiotics are the most studied classes of PACs with respect to biosorption onto newly sourced biosorbents. This might be because they are the most commonly found in the surface water system and due to the possibility of the emergence of drug-resistant microorganisms. Ciprofloxacin, tetracycline and sulfamethoxazole are among the most reported. The PACs and biosorbents exist in different ionic states (neutral, cationic or anionic) as pH changes. The different states of existence determine their speciation, and thus their π-electron-donating behavior, hydrophobicity and PAC-biosorbent interaction. Biosorption is most favourable when the PAC and biosorbent have different surface charges, which promotes electrostatic interaction. It is important to understand state of existence when designing and planning the removal of PACs in water systems. Study on date palm (*Phoenix dactylifera*) by El-Shafey et al. [77] revealed its capacity to adsorb ciprofloxacin from aqueous media by a sorption mechanism that is mainly related to cation exchange and hydrogen bonding. Previous studies have shown that modified plant-sourced material can exhibit both biosorption and photocatalysis. In this case, the surface of the material breaks down the PACs into smaller molecules and, at the same time, adsorbs. This observation was reported for reed straw supported titanium dioxide, which used synthesis by sol-gel method and was applied for the removal of sulfamethoxazole [76]. The material exhibited both biosorption and photocatalysis properties toward sulfamethoxazole with an adsorption capacity of 23,347 mg g^{-1}, which can be described by Langmuir isotherm. Similarly, Kumar et al. [83] demonstrated the degradation and biosorption of ibuprofen and 2,4-dichlorophenoxyacetic acid using biochar. The degradation was monitored using LC-MS (Liquid chromatography-mass spectrometry) while the toxicity of the degraded products was analyzed by the viability of human peripheral blood cells. This has also shown that biosorbents modified with transition metal salts like Ti, Zn, etc., have the capacity to degrade the PACs, apart from the biosorption of the PACs taking place at the surface of the biosorbent. It is of note that most reported degradation processes did not lead to complete degradation to CO_2 and H_2O. Biosorbents of this class are considered dual-functional because of their ability to exhibit both biosorption and degradation in a single process of water treatment.

Other studies have also shown that biomaterial could be incorporated with nanomaterials to improve on performance. Wang et al. [84] reported this in a study on the removal of ciprofloxacin from aqueous solution by magnetic chitosan grafted graphene oxide. The material exhibited a high sorption capacity of 282.9 mg g^{-1}; the process was pH-dependent. The authors further reported the influence of other ions (Ca and Na), which was significant on the sorption of ciprofloxacin in a process controlled by electrostatic attraction and π-π electron interaction. It has become so obvious that functionalization plays an important role in improving the ability of plant-sourced materials to perform significantly as biosorbents. The authors reported the synthesis of magnetic chitosan grafted graphene oxide nanocomposite as an efficient biosorbent for the removal of ciprofloxacin. The modification led to a reduction in the surface area of graphene oxide from 1685.7m^2 g^{-1} to 388.3 m^2 g^{-1} due to aggregation resulting from chitosan. The average pore diameter was 13.98 nm, which suggested that the biosorbent composite was a mesoporous material. This also revealed that care must be taken when selecting a chemical modification method in order to prevent the selection of methods or steps that may lead to the formation of aggregates, which reduces the surface area. Since both Langmuir and Freundlich isotherms can describe the process, the authors argued that the process of biosorption was likely to be a complex chemisorption process. The bio-composite was regenerated using 100% methanol; the performance was above 80% in the first cycle. However, the biosorption capacity remained at 72% at 4th regeneration. Unfortunately, most authors did not report the regeneration of biosorbents in their studies. Attention should be given to regeneration studies to understand the worth of low-cost biosorbents.

Table 4. Mode of study, S_{BET}, Isotherm, Qe, Isotherm, Desorption and ΔG° for the sorption of PACs on biosorbents.

Biosorbent	PAC	Mode of Study	S_{BET} ($m^2\ g^{-1}$)	Qe ($mg\ g^{-1}$)	Isotherm	Desorption (%)	ΔG° ($kJ\ mol^{-1}$)	Reference
CA-Al-KABs	Ciprofloxacin	Batch	10.595	68.36	Langmuir	-	−0.944	[32]
SCG CW PB	Fluoxetine	Batch	<4 <4 <4	14.31 4.74 6.53	Sips	-	-	[81]
MNS	Tetracycline	Batch	1524	455.33	Temkin	-	-	[82]
MCGO	Ciprofloxacin		388.3	282.9	Freundlich and Langmuir	>80	-	[84]
WPK	Naproxen	Batch	601.9	73.14	Langmuir	-	-	[85]
GS YB CB	Paracetamol	Batch	- - -	1.74 0.77 0.99	Langmuir	-	-	[86]
CTCBW	Acetaminophen	Batch	80.586	-	Freundlich	-		[87]
CBP	Clarithromycin Atenolol		19.26	34.5 39.5	Freundlich Langmuir	-	−838	[88]
ASCK	Paracetamol Amoxicillin	Batch and fixed bed	1908	502.26 282.42	L-F L-F	-	-	[89]
ASK	Paracetamol Amoxicillin	Batch and fixed bed	1635	453.39 228.39	L-F L-F	-	-	[89]
ASP	Paracetamol Amoxicillin	Batch and fixed bed	420	318.84 198.73	L-F L-F	-	-	[89]

CA-Al-KABs = Aluminum-pillared kaolin sodium alginate beads, ΔG^0 = Free Gibbs energy, CW = Cork waste, PB = Pine bark, SCG = Spent coffee grounds, MNS = macadamia nut shells, MCGO = Magnetic chitosan grafted graphene oxide nanocomposite, WPK = Wild plum kernels, GS = Grape stalk, YB = Yohimbe bark, CB = Cork bark, CTCBW = Chemically treated chicken bone waste, CBP = Cuttlefish bone powder, ASCK = NaOH modified Argan waste carbon, ASK = NaOH modified Argan waste, ASP = Phosphoric acid modified Argan waste, L-F = Langmuir-Freundlich isotherm, S_{BET} = Surface area.

Recently, a study by Paunovic et al. [85] documented microwave-functionalized biochar derived from novel lignocellulosic waste biomass for the sorption of ionizable emerging pharmaceuticals in solution. They achieved this by preparing a functionalized biochar from wild plum kernels using simultaneous pyrolysis and microwave potassium hydroxide functionalization. The material removed naproxen from solution at a pH 5–7 having a maximum adsorption capacity of 73.14 mg g^{-1}. The removal was driven by electrostatic attraction as the main mechanism of the process, as shown in Figure 2. Chemical modification by way of first converting biomass to biochar before it is chemically activated or functionalized has received attention for the removal of PACs in a water system. More recently, microwave-assisted techniques have been gaining relevance because of their advantage over the conventional pyrolysis method. Biosorbents prepared through this method have a higher surface area as reported by Paunovic et al. [85] when compared with other direct, simple modifications as reported by Silva et al. [81] in Table 4. The microwave-assisted method is of advantage because the conventional pyrolysis method involves heat transfer to the biomass particles by convection, conduction and radiation mechanisms. During the conventional method, the surface of the biomass is heated first in comparison with the internal zones, resulting in a temperature gradient from the biomass surface to the interior of each particle. The method requires high temperatures, meaning high-energy involvement, which makes it expensive to run. However, the microwave-assisted methods make use of radiation that utilizes volumetric heating within a short time. A relatively low amount of energy is required; the processing time is short with high production yield, which makes it cheaper than the conventional pyrolysis method. The surface area of natural biosorbents has been improved via this means as a way of enhancing sorption capacity for PACs.

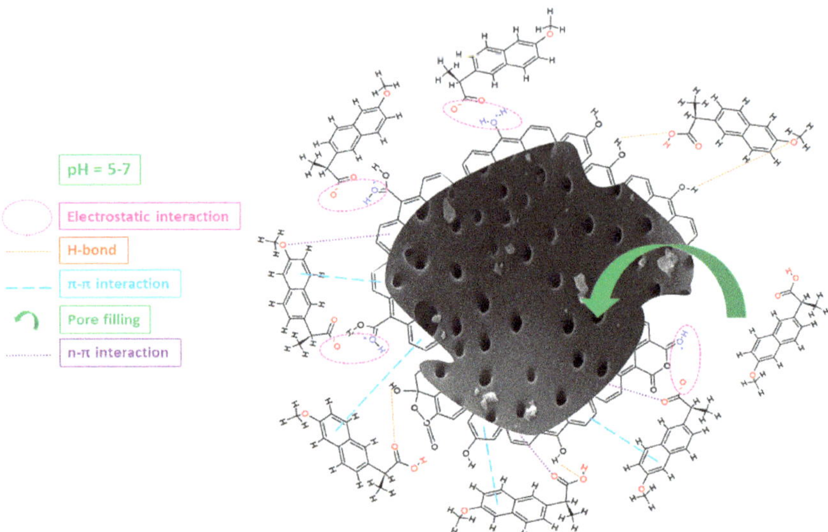

Figure 2. Proposed key sorption mechanisms for the sorption of naproxen on microwave-functionalized biochar [85].

Use of plant waste after simple pretreatment have also been reported. An example of this is the use of grape stalk, yohimbe bark, and cork bark for the removal of paracetamol reported by Villaescusa et al. [86]. The removal process was found not to depend on pH. The grape stalk waste exhibited the best adsorption capacity with a modelling capacity that reflected π-stacking interactions among the lignin, syringyl and guaiacyl moieties in grape stalk and the aromatic ring in paracetamol. Adewuyi et al. [90] also reported the pretreatment of *Adenopus breviflorus* for the removal of 2-chlorophenol in solution. The pretreatment was achieved using different solvent systems. The treatment prevented the introduction of organic molecules from the seed biosorbent into the treated solution in a process which followed the pseudo-second-order model, with isotherm plots fitting well for Freundlich, Langmuir and Temkin models.

Animal waste such as fish bones, cow bones, crab shells, eggshells, etc., have been explored as biosorbents or precursor materials for biosorption. The study by Kizilkaya et al. [91] reported pretreated fish bones obtained from bluefish (*Pomatomus saltotrix*), bogue (*Boop boops*), European anchovy (*Engraulis encrasicolus*) and gilthead seabream (*Sparus aurata*) as low-cost biosorbents. Dahiya et al. [92] also reported the use of pretreated crab and arca shell biomass for biosorption. Ojedokun and Bello [93] published a review on the use of cow dung as biosorbent for the removal of pollutants in water. This has shown an affinity for cations, which suggests that the material may be useful for the removal of positively charged PACs in solution. The review showed cow dung as a bioorganic waste, eco-friendly and inexpensive biosorbent. Previous studies [93,94] showed cow dung to contain calcium sulphate (0.312%), aluminum oxide (20%), calcium oxide (12.48%), magnesium oxide (0.9%), iron oxide (20%) and silica (61%). Another study by El Haddad et al. [95] revealed the use of animal bone meal as biosorbent for the removal of organic compounds in water. The material had an adsorption capacity of 57.15 mg g^{-1} towards reactive yellow 84 dye in a process that is endothermic, and that can be described by Langmuir isotherm. Dyes and PACs are organic molecules; the ability of the material to remove yellow 84 dye in solutions is an indication that the material may be useful for the removal of organic molecules like PACs in solution. Previous work [87] has reported the use of chemically modified chicken bone waste as a biosorbent efficient for the removal of acetaminophen. The process fitted well for both pseudo-second-order kinetic and Freundlich isotherm models. Interestingly, a mixture of activated animal bones from cow, donkey, chicken and horse has also been evaluated as an efficient biosorbent for wastewater treatment [96]. Use of human hair as a biosorbent for the removal of phenol was reported to achieve a removal of 92% in a batch process [97]. The study leveraged on the fact that human hair contains keratin, which is a fibrous proteinaceous material with a large surface area. It reveals the ability of human hair as a possible means for the removal of micro-organic pollutants such as PACs in water. It is a pointer towards finding application for human hair waste produced in large quantities in barbers, serving as a means of converting waste to wealth. Clarithromycin and atenolol, which are common PACs in surface water, were reportedly biosorbed onto cuttlefish bone powder treated with HCl [88]. The biosorbent showed an adsorption capacity of 34.5 mg g^{-1} towards clarithromycin and 39.5 mg g^{-1} towards atenolol in a process that was found to be spontaneous. The $\Delta G°$ for the process was negative (−838 kJ mol^{-1}), which suggests that the process was spontaneous. A similar negative value (−0.944 kJ mol^{-1}) was reported for the sorption of ciprofloxacin on CA-Al-KABs [32]. However, most recent works found in the literature did not consider the effect of temperature on the biosorption of PACs. Most of the published works on biosorption of PACs were carried out focusing on batch operation mode under agitation, which created a gap for limited fixed-bed mode and pilot studies. This makes it difficult to better estimate the potential of biosorbents. It was also evident that most studies did not compare the capacity of the considered low-cost biosorbents with commercially available materials for effective comparison. It is important that future studies consider this.

2.3. *Microorganisms as Biosorbents*

At the early stage of involving microorganisms in the biosorption process, the effort by researchers revealed that inactive/dead microbial biomass could passively bind metal ions. The dead microbial biomass has several advantages over the living microbial biomass such as being cheap, limitation of toxicity, ease of regeneration, exhibiting ion exchange and with a wide range of operational pH and temperature. Subsequently, the effort began to move towards investigating the removal of dyes and other organic pollutants in the water system. Over time, it became evident that biosorption exhibited by microbial biomass does not only depend on the chemical composition of the microbial biomass but also the external physicochemical factors and the matrix chemistry. This mechanism has been reported to be one or a combination of chelation, complexation, adsorption, ion exchange, degradation, electrostatic interaction, microprecipitation, coordination and donor-acceptor interaction [52,98]. Strains of microorganisms have been shown with the capacity to bio-transform toxic compounds into less hazardous forms [99]. The sorption tendency exhibited by a microorganism further depends on the microbial genus, which on the long run defines its cellular composition. They can be used in the form of fine powder or wet cells, which makes resistance to mass transfer negligible. Some biomass has been tested and reported as biosorbent. They include algae, yeast, bacteria and fungi with well-defined structures.

2.3.1. Bacterial

Based on the cell wall and gram staining, bacteria can be divided into Gram-negative and Gram-positive. The cell ranged in diameter between 0.5 to 1.0 μm, and most are unicellular, belonging to the prokaryotes. The cell has four major components: cytoplasm, cell wall, nuclear and cell membrane. The proportions of lipid and protein in the cell membrane vary from one species to the other. They have different shapes such as spiral, rod, cocci and filamentous. The cell wall differs from those of other microorganisms due to the presence of peptidoglycan, which determines its shape and the rigidity of the cell wall. A description of the cell wall of Gram-positive and negative bacterial is shown in Figure 3.

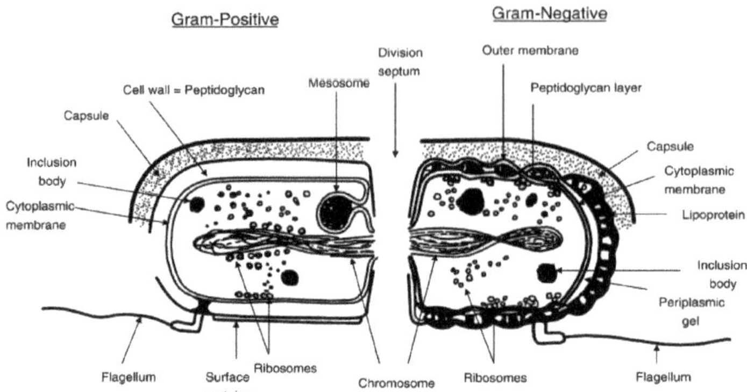

Figure 3. Structure of Gram-positive and negative s [52].

During biosorption by bacteria, the cell wall encounters the PACs first in an interaction where PACs are deposited on its surface. The functional groups such as amine, hydroxyl, carboxyl and phosphonate groups of the cell wall play an important role in the biosorption process since the uptake of the PACs is extracellular [52,100]. Characterization and the biosorption mechanism of bacteria have been studied using different methods such as X-ray diffraction, potentiometric titrations, scanning electron microscopy, Fourier transform infrared spectroscopy, transmission electron microscopy and energy dispersive X-ray microanalysis [101–104]. This has revealed the types, nature and number of binding sites on the surface of the bacteria. Use of Fourier transform infrared spectroscopy has shown band transformations, which enabled the prediction of possible functional groups involved in the biosorption, as well as the acidity and basicity of the bacterial surface involved in the interaction. Energy-dispersive X-ray microanalysis had helped in predicting the chemical and elemental composition of the bacterial, which helps in confirming the involvement of ion exchange mechanism during the biosorption process. X-ray diffraction analysis helps in predicting the chemical nature, while the morphology of the cell surface can be studied using scanning electron microscopy. Both the surface chemistry of the biosorbent and the water chemistry plays an important role in the removal of PACs while certain factors such as solution pH, ionic strength, particle size, concentration and temperature influences the removal rate as well as the adsorption capacity of the bacterial.

Studies have reported different means of enhancing the capacity of bacterial to act as biosorbents. Methods include chemical and genetic modification. Some focused on improving the active binding sites to improve biosorption capacity while less consideration was given to the inhibition sites on the surface of the bacteria. A study by Vijayaraghavan and Yun [105] revealed that the amine group are active towards biosorption of organic compounds via electrostatic interaction. However, on the contrary, the presence of carboxyl functional groups on the same surface repels biosorption from taking place at the surface. The study by Martins et al. [106] reported the use of anaerobic microorganisms to remove ciprofloxacin, 17β-estradiol and sulfamethoxazole from solution as well as elucidate their bio-removal mechanisms. The work provided new insight into the anaerobic bioremediation of ciprofloxacin, 17β-estradiol and sulfamethoxazole. During the process condition used (nitrate- and sulfate-reducing conditions), ciprofloxacin was biodegraded to 80 % under both conditions. Degradation of 17β-estradiol reached 84 % under nitrate-reducing conditions whereas no biodegradation was achieved with respect to sulfamethoxazole. The study further revealed that bioremediation might occur by four mechanisms: (i) extracellular bio-removal by metabolites produced during growth, (ii) biodegradation by co-metabolism, (iii) biodegradation by sole substrate consumption, or (iv) biosorption to the bacterial cells. Sorption steps and biodegradation characteristics of lomefloxacin, ofloxacin, enrofloxacin, ciprofloxacin and norfloxacin have been reported [107]. The study reported sorption of these PACs as the lead process followed by biodegradation. The biodegradation was favored under aerobic condition by temperature increase, which may have been achieved during nitrification through co-metabolism. Co-metabolic activities of heterotrophic denitrifying bacteria and ammonia-oxidizing bacteria in the removal of nitrogen and pharmaceutical and personal care products have been investigated [108]. The study revealed biodegradation as the dominant process for removal of ibuprofen while removal of triclosan was via an equal contribution from adsorption and biodegradation.

Commonly used bacterial strains for remediation of wastewater include *Pseudomonas, Enterobactor, Streptomonas, Aeromonas, Acinetobactor* and *Klebsiella*. Several membrane bioreactors have been developed for large-scale treatment of domestic and industrial wastewater.

Use of membrane reactor has shown several advantages such as flexibility in operation, prolonged microorganism retention, complete removal of suspended solids, low rate of sludge production, treatment of both toxic inorganic and organic pollutants, compact plant size and high rate of degradation [109]. *E-coli* biofilm built on activated carbons was reported [89] for the removal of paracetamol and amoxicillin present in water. The prepared material is microporous with a high surface area. The performance of the material was compared with when the *E-coli* biofilm was not built into the activated carbon. The sorption process without the biofilm showed a rapid kinetic, with sorption capacity being 319 mg g^{-1} in a process that can be described by Langmuir isotherm. Interestingly, the process changed significantly when the biofilm was introduced to the activated carbon. Although the kinetic was slow, the performance changed remarkably with the material exhibiting a biosorption capacity of 465 mg g^{-1} and the process best represented by Langmuir-Freundlich isotherm model having three different stages. This has shown that the use of bacteria in the biosorption process for the removal of PACs in water has a positive impact. As shown in Table 4, the authors focused on the removal of paracetamol and amoxicillin. They argued that the size of PAC plays an important role in its removal; in this case, paracetamol has a lower molecular weight (151.16 g mol^{-1}) compared to amoxicillin (365.4 g mol^{-1}), which may have contributed to the greater sorption capacity exhibited towards paracetamol in the study.

2.3.2. Fungi

This includes yeasts and molds. They are filamentous, with mycelium, which contains a complex mass of filaments. The cell wall is rigid, which gives structural support and shape, mainly consisting of polysaccharide, with polyphosphates, proteins, lipids and inorganic ions. The composition of the cell wall may have contributed significantly to its use as biosorbent for the removal of PACs. Fungi have a promising ability to serve as biosorbents because of its various functional groups, due to the high amount of cell wall materials. It is readily available due to the ease of cultivation in large scale with high yield. Previous work has shown that the use of yeast for environmental research is attractive because it can be genetically modified. It is well characterized, which helps in understanding the biosorption mechanism. The ease of modification shows that it can be manipulated for removing PACs in solution. Four factors that can influence the biosorption of fungi have been reported as: biomass dose in solution, type and nature of biomass, physicochemical factors like temperature, pH, ionic strength, and initial solute concentration [110]. They can be modified chemically or physically to improve capacity to act as biosorbent. The modification can be achieved via improving surface characteristics by exfoliating or masking the functional groups or by making the biosorption sites readily available for sorption [111,112]. They are pretreated chemically to modify the cell wall by creating derivatives with altered sorption abilities and affinities [112]. When fungi are used in water treatment, the free cells have small particle size and low mechanical strength, which creates the need for the application of hydrostatic pressure for a suitable flow rate. However, this leads to disintegration and attrition. Although this material might work well when subjected to the batch process, due to the hydrostatic pressure they become unsuitable for the column packing process, which is applicable in the industrial process [113]. Due to the disintegration and attrition challenge, immobilization techniques have been recommended as means of modification, which may include crosslinking and entrapment.

When this takes place on polymeric matrix, the biosorbent exhibits improvement and advantage in particle size, high biomass loading, minimal clogging, high regeneration and ease of separation [114].

Table 5 shows the different types of fungi reported in the past for the removal of PACs in water. As shown in the table, it is evident that when using fungi for water treatment, biosorption and biodegradation mostly take place together.

Table 5. Selected fungi previously reported for the removal of PACs.

PACs	Fungi	Mechanism of Action	Reference
Estrone, 17 β-estradiol, 17 α-ethinyl-estradiol and estriol	*Myceliophthora thermophile and Trametes versicolor*	Biodegradation Adsorption	[115]
Carbamazepine	*Phanerochaete chrysosporium*	Biodegradation Biosorption	[116]
Sulfapyridine, sulfapyridine, and sulfamethazine	*Trametes versicolor*	Biodegradation Biosorption	[117]
Diclofenac, ibuprofen, naproxen, carbamazepine, and diazepam	*Phanerochaete chrysosporium*	Biodegradation Biosorption	[118]
Carbamazepine, diclofenac, iopromide and venlafaxine	*Trametes versicolor* *Irpex lacteus* *Ganoderma lucidum* *Stropharia rugosoannulata* *Gymnopilus luteofolius* *Agrocybe erebia*	Biosorption	[119]

Sorption of carbamazepine, diclofenac, iopromide and venlafaxine by fungi was reported in a bioreactor treatment [119]. The sorption process was likened to active transport in living cells, which plays an essential role in the sorption process. The removal of the PACs was by both sorption and biodegradation. Recently, *Aspergillus sydowii* and *Aspergillus destruens* were reported for their use under saline conditions for the removal of PACs [120]. Benzo-α-pyrene and phenanthrene were used as a sole carbon source during the study. They achieved a removal of over 90% in a process described to be biodegradation (*Aspergillus sydowii*) and biosorption (*Aspergillus destruens*). Furthermore, a study [121] on the efficiency of *Trametes versicolor* and *Ganoderma lucidum*, to remove thirteen pharmaceutical pollutants with concomitant biodiesel production from the accumulating lipid content after treatment, revealed a 100% removal for diclofenac, gemfibrozil, ibuprofen, progesterone and ranitidine, whereas it showed low removal towards 4-acetamidoantipyrin, clofibric acid, atenolol, caffeine, carbamazepine, hydrochlorothiazide, sulfamethoxazole and sulpiride. However, the combination of *Trametes versicolor* and *Ganoderma lucidum* enhanced efficiency, which was attributed to the interactions developed between both strains. In general, white-rot fungi belonging to the Basidiomycota family has been used in several studies to evaluate potential as a means of removing PACs in water [115,122–125]. They are well known for their ability to degrade lignin. Interestingly, their enzyme system is based on free radicals; they are non-selective and non-specific; they tend to degrade pollutants either by extracellular or intracellular enzymatic route [126]. This capacity and mechanism of degrading lignin is also extended towards PACs. This ability gives them an outstanding edge and attraction in the treatment of wastewater.

2.3.3. Microalgae

There is growing interest in the use of microalgae-based remediation biotechnologies as a means of removing PACs in water. The microalgae-based remediation biotechnology is green, driven by solar energy, eco-friendly, shows fixation and turnover of carbon, and can serve as a source of other useful products [127,128]. Figure 4a shows the different species of microalgae commonly used in bioremediation of wastewater and PACs [129]. The fact that mixotrophic microalgae can maneuver their metabolism process between autotrophic and heterotrophic, making them survive and thrive in extreme environments, gives them an edge over bacteria and fungi that need carbon and other nutrients for growth and degradation of PACs. Just like bacterial and fungi, algae make use of their cell wall for the removal of PACs in solution. The cell wall contains several substances like lignin, pectins, protein, cellulose, hemicellulose and extensin. Due to the main functional groups in the cell wall being phosphoryl, amine and carboxyl groups, the cell wall is negatively charged, which makes them attractive towards cations via electrostatic interaction. Apart from the charge playing an essential role in biosorption, the hydrophobicity, species and structure are also crucial in selecting microalgae for biosorption activities. Studies have shown their ability to remove carbamazepine, ibuprofen, metoprolol, estrone, β-estradiol, etc., from solution. A report by Matamoros et al. [130] showed that mixed microalgae species such as Chlorella and Scenedesmus species could remove tris(2-chloroethyl) phosphate, caffeine, tributyl phosphate, carbamazepine, galaxolide, 4-octylphenol and ibuprofen from solution. The removal of the pollutant was high at a capacity that was not less than 99%. However, the aerated batch reactors inoculated with the mixed microalgae could only efficiently remove 4-octylphenol, galaxolide, and tributyl phosphate as well as ibuprofen and caffeine, but expressed difficulties removing carbamazepine and tris(2-chloroethyl) phosphate. Interestingly, the authors argued that it was the first time that the enhancing effect of microalgae in water treatment had been evaluated. The findings showed that microalgae improved the removal efficiency of ibuprofen by 40%, which further reduced the lag phase of caffeine by three days. It was further shown that microalgae improved the removal efficiency of ibuprofen and caffeine either by releasing exudates, which enhanced the biodegradation processes, or by microalgae biosorption. Studies of *Scenedesmus obliquus* and *Chlorella pyrenoidosa* dead cells have shown their ability to remove progesterone and norgestrel [131]. When PACs are biosorbed by microalgae, they can bioaccumulate with the possibility of causing damage to the cell; however, the concentration at which PACs are present in the environment is safe enough for the microalgae and do not exhibit lethal inhibition [129,132]. *Chlamydomonas Mexicana* and *Scenedesmus obliquus* are examples of species that bioaccumulated carbamazepine [133]. Microalgae have shown biodegradation to be an effective mechanism for the removal of PACs in water. They have biodegraded carbamazepine, ibuprofen, caffeine and tris(2-chloroethyl) phosphate in previous studies [134,135]. Several types of enzymatic reactions have been brought forward to describe the degradation exhibited by microalgae towards PACs, and some of these reactions include oxidation, ring cleavage, demethylation, hydroxylation, hydrogenation, carboxylation, halogenation, decarboxylation and glycosylation [129–135]. The degradation of PACs, as described in Figure 4b may occur in a two phase enzyme system known as phase I and II. In phase I, initial attacks take place using enzyme cytochrome P450, which increases the hydrophilicity of PACs or promote hydrolysis, reduction or oxidation reactions. During phase II reaction, enzymes such as glutathione-S-transferases catalyze the reaction between glutathione and electrophilic molecules; this reaction also protects the system against oxidative damage [129]. Escapa et al. [136] reported the removal of salicylic acid, paracetamol, phosphate and nitrate using *Chlorella sorokiniana*. This study further revealed its use for the biodegradation of metoprolol, diclofenac, paracetamol and ibuprofen.

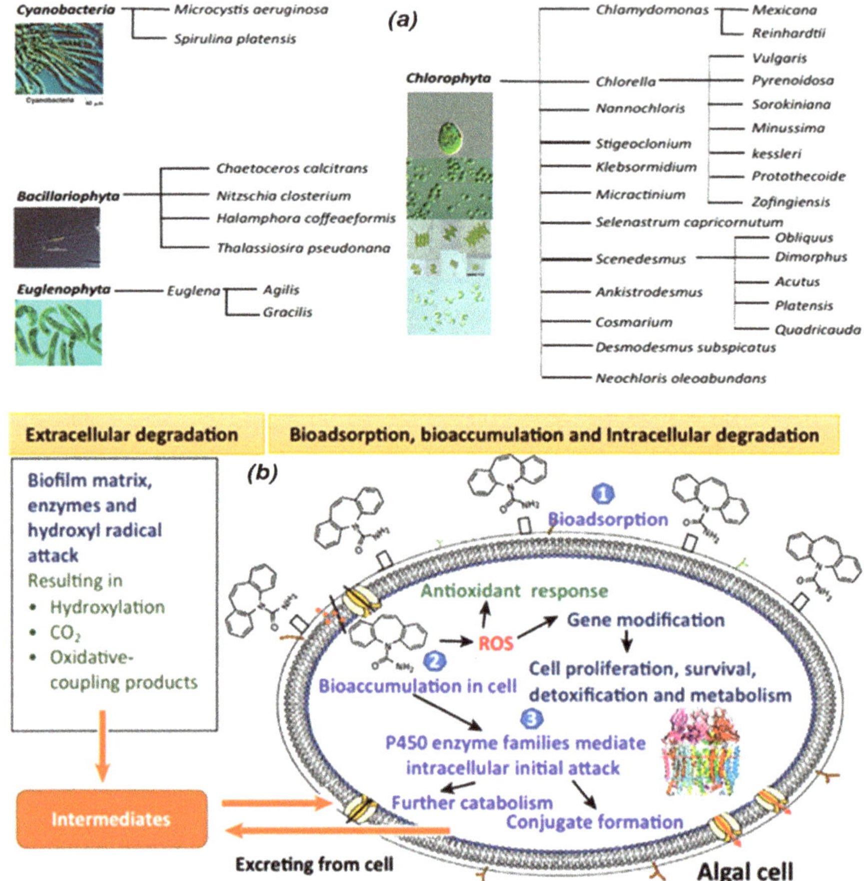

Figure 4. (**a**) Microalgae species commonly used for bioremediation of wastewater, (**b**) Proposed mechanisms of removal of pharmaceutical contaminants and microalgae metabolism [129]. ROS = Reactive oxygen species.

The complete role of these enzymes in the two phase enzyme system is not well understood as research is currently ongoing on how to understand the biotransformation of PACs in the microalgae system. However, metagenomics and meta-transcriptomics analyses are playing a leading role in this regard. Co-metabolism has also been effective in enhancing the ability of microalgae in the removal of PACs. The introduction of organic substrates to the degradation process serves as a means of supporting activities of specific catabolic enzymes in charge of the degradation process. This was demonstrated in the degradation of monooxygenase and N-deethylase with the introduction of acetate [137]. It is important to evaluate the influence of organic substrate before it is introduced into the degradation process because some substrate is capable of having a negative effect on the process by reducing enzyme activity. However, a few factors are understood, such as the role of electron-donating groups, e.g., amine, hydroxyl and alkoxy, in enhancing susceptibility to electrophilic attack by enzymes. It has also become clear that the presence of groups such as nitro, halogen and amide decreases enzymatic biotransformation ability [53,138]. This is also an indication that it is better to understand the structure and functional group composition of PACs before selecting species of microorganisms for its removal. It is also essential to understand the functionality of any organic substrate before inclusion in the degradation process.

2.4. Biocomposite

Bio-composites are materials containing two or more distinct substances, which are brought together to produce a new material with improved performance better than the individual constituent materials. They are biomass-based-materials that can be used for wastewater treatment; they are receiving much attention now because of their biodegradability, high performance and eco-friendliness. Biopolymers remain the major component of bio-composites; these biopolymers are cellulose, chitosan, starch, chitin, alginate, etc. The merit of these biopolymers lies in their non-toxicity, abundance, cost-effectiveness and environmental friendliness [139]. Cellulose, chitin and chitosan are similar in their chemical structure. The only difference is the N-acetyl, amino and hydroxyl functional groups at the C-2 position, as shown in Figure 5.

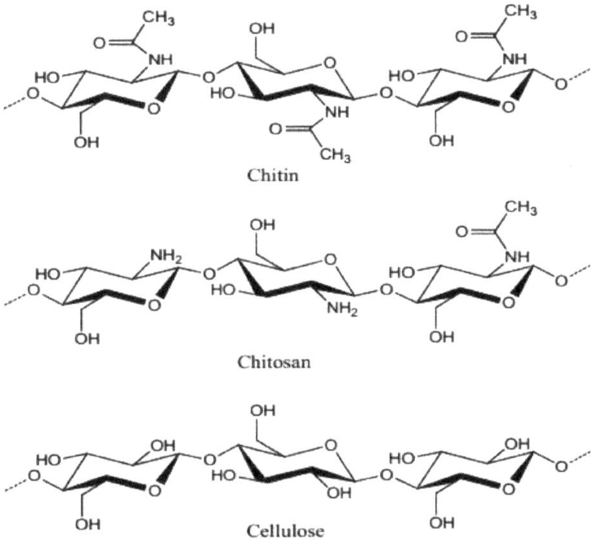

Chitin

Chitosan

Cellulose

Figure 5. Structure of chitosan, chitin and cellulose [140].

These materials have been prepared in combination with other materials such as clay (commonly used clays include kaolinite, vermiculite, montmorillonite and illite), graphene, carbon, etc. Use of biopolymer alone may not be so effective in the removal of PACs from water. However, the combination of these materials (clay, carbon, graphene and natural polymer) plays an important role. Clay and graphene have large surface areas that can enhance removal; moreover, the cation exchange capacity of clays is also essential. An appropriate bio-composite should be inexpensive, abundant, efficient, environmentally friendly, biocompatible and reusable.

Table 6 presents different bio-composites reported in the literature. Karoui et al. [141] reported the preparation of enhanced biosorbent from Tunisian-reed (*Phragmites-Australis) and its use in the removal of* ciprofloxacin antibiotic and methylene blue dye. The composite was designed under response surface methodology. The study revealed a 76.66% removal for ciprofloxacin and 100% removal for methylene blue dye in a process that can be described by the Brouers-Sotolongo-fractal model. A study reported the removal of parabens from wastewater using magnetic waste tyre activated carbon-chitosan composite [142]. The composite has a surface area of 1281 m^2 g^{-1} and pore size of 4.05 nm, which exhibited a high sorption capacity with 100% removal in a process described by Langmuir and Redlich-Peterson isotherm models. Figure 6 shows the adsorption of parabens and the TEM (Transmission electron microscopy) images. The composite contains chitosan, which is the most deacetylated form of chitin. The presence of hydroxyl and amine groups makes it highly reactive with other compounds. The regeneration capacity of the biosorbent was carried out using methanol. The results obtained for the regeneration showed that recovery and adsorption were not affected for up to seven adsorption/desorption cycles. The result of the regeneration demonstrated the outstanding reusability of the bio-composite and its potential use in wastewater treatment.

Table 6. Bio-composites reported for water treatment.

Bio-composite	PAC	S_{BET} ($m^2\ g^{-1}$)	Qe ($mg\ g^{-1}$)	Removal (%)	Desorption (%)	Solvent for Desorption	Reference
CCGC	Metamizol Acetylsalicylic acid Acetaminophen Caffeine	- - - -	6.29 9.92 7.52 8.21	>70 >85 >60 >80	>55 >85 >55 >80	Ethanol and water	[3] [3] [3] [3]
NMVC	4-aminoantipyrine	96	6.53	98.40	76	Water	[30]
MCGO	Ciprofloxacin	388.30	282.90		>80	Methanol	[84]
Phragmites-Australis	Ciprofloxacin	8.90	17.30	76.66	-	-	[141]
MWACC	Methylparaben Propylparaben	1281	85.90 90.00	100 100	>95 >95	Methanol	[142]
APB	Norfloxacin	90.40	5.24	92.70	>84	Methanol	[143]
RGO-M	Ciprofloxacin Norfloxacin	- -	18.22 22.20	- -	- -	- -	[144] [144]
CBM	Diclofenac sodium Tetracycline hydrochloride	27.54	164.00 40.20	>70 >55	>60 >60	Ethanol and water	[145]

CCGC = Chitosan/waste coffee-grounds composite, MWACC = Magnetic waste tyre activated carbon-chitosan, APB = Clay-biochar composite with potato stem and natural attapulgite, - = Not determined, NMVC = Nanocellulose modified vermiculite clay, MCGO = Magnetic chitosan grafted graphene oxide nanocomposite, RGO-M = Graphene oxide/magnetite composites, CBM = Chitosan-based magnetic composite, Qe = Adsorption capacity, S_{BET} = Surface area.

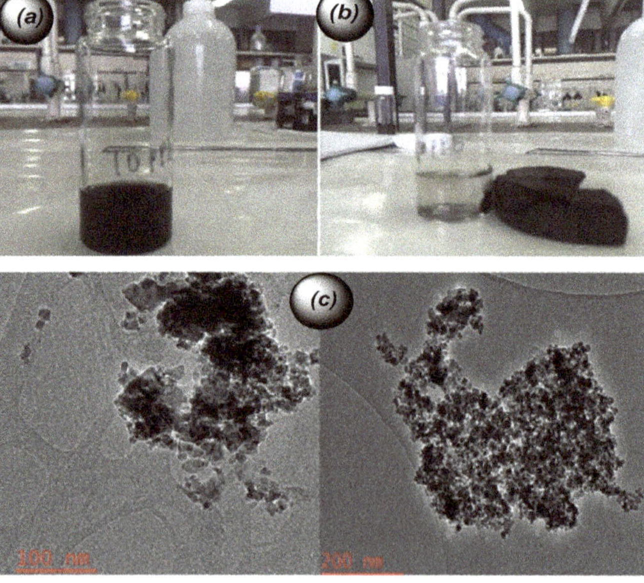

Figure 6. (**a**) Magnetic adsorbent dispersed in sample solution during the adsorption process, (**b**) Separation of sorbent using external magnet and (**c**) TEM of magnetic waste tyre coated activated carbon-chitosan under different magnifications [142].

Clays are hydrous aluminosilicates, which can be categorized mainly into montmorillonite, smectite, kaolinite, ilite, and chlorite. Among these categories, kaolinite, montmorillonite and ilite are mostly used, due to their physicochemical, structural and mechanical properties. The price is low as the cost of clay ranges between \$0.005–0.46 per kg [146,147]. It contains exchangeable cations and anions such as Ca^{2+}, Mg^{2+}, H^+, K^+, Na^+, NH_4^+, Cl^-, PO_4^{3-} and NO^{3-}, which are easily exchanged with other ions without altering the clay structure [147]. Report from the previous study has shown that faces and edges of clay particles can adsorb charged (cations and anions) and neutral pollutants in water [147]. These properties qualify it as a material that can be combined with biopolymer to produce bio-composite. Clay-biochar composites have been produced for the enhancement of sorption capacity of biochar for certain contaminants [148]. Even though several methods have been used for the production of clay-biochar composites, the most frequently used method is pyrolysis [148,149]. Clay-biochar composite prepared by potato stem and natural attapulgite has shown higher sorption capacity for norfloxacin in the pH range of 2.0–11.0 [143]. Major sorption mechanisms involved were electrostatic attraction, hydrogen bonding, hydrophobic interaction, and π-π interactions [150]. The sorption capacity was demonstrated in double distilled tap and river water, and the composite exhibited a removal of 84.14, 82.17 and 87.37%, respectively. The regeneration of the bio-composite with methanol was stable up to the 5th cycle with a capacity greater than 84%. This capacity relates well with values reported for the regeneration of magnetic waste tyre activated carbon-chitosan bio-composite [141], which presents bio-composites as promising materials for wastewater treatment.

Waste coffee-grounds, a poorly explored source of bio-compounds, were combined with chitosan and poly (vinyl alcohol) in order to obtain composite [3]. The composite exhibited a noticeable enhancement for the sorption of metamizol (6.29 mg g^{-1}), acetylsalicylic acid (9.92 mg g^{-1}), acetaminophen (7.52 mg g^{-1}) and caffeine (8.21 mg g^{-1}). The highest removal efficiency was registered at pH 6, which obeyed the pseudo-second-order model and Freundlich isotherm. The composite is cost-effective with remarkable reusability of at least five consecutive biosorption/desorption cycles. Surfactant modified cellulose-montmorillonite composite has also been prepared as a biosorbent [151]. Aluminum-pillared kaolin sodium alginate beads (CA-Al-KABs) were prepared by gelling and solidification processes, which were used for the removal of ciprofloxacin in solution [32]. Sorption of ciprofloxacin was found to be pH-dependent, pseudo-first-order kinetics model controlled and obeyed the Langmuir isotherm model with an adsorption capacity of 68.36 mg g^{-1}. The mechanism for the sorption of ciprofloxacin is electrostatic attraction, as described in Figure 7a. Starch modified smectite clay composite has also been prepared and used for the removal of PACs from wastewater treatment plant effluent [152]. It is evident that most regeneration processes made use of polar solvents or a mixture of organic solvent with water as a means of an aqueous system for the regeneration of bio-composite. Alcohol is the most commonly used organic solvent, which might be due to the polar solubility of most of the PACs.

Graphene is a one-atom-thick, two-dimensional layer of sp^2-hybridized carbon [153]. Its physicochemical, thermal, mechanical and electrical properties are extraordinary, which makes it suitable for different applications. The surface area is large, and its large delocalized π-electron system promotes interaction with other molecules [154,155]. This interaction is a pointer for its use as a material in the removal of PACs in the water system. It can hold oxygen-containing functional groups in its molecule to form reduced graphene (rGO) or graphene oxide (GO). This is an indication that it can be modified by grafting other functional groups onto it or by combination with other materials such as biopolymers. Its exceptional properties such as large surface area, short intra-particle diffusion path-length, high sorption site, low-temperature modification and ease of regeneration present graphene as an excellent material for composite preparation. When such composites are prepared in combination with a biopolymer, they are classified as a new class of fascinating bio-composite. One disadvantage that limits the use of graphene is the possibility of aggregation. However, this challenge is overcome by intercalating particles within the layers of graphene [144]. rGO/magnetite was prepared by Tang et al. [156] and used for the removal of norfloxacin and ciprofloxacin. The process was spontaneous and exothermic with sorption capacity of 22.20 and 18.22 mg g^{-1} for norfloxacin and ciprofloxacin, respectively. The process obeyed Temkin isotherm, Langmuir isotherm and pseudo-second-order kinetic models. It is pH-dependent, and the removal mechanism consisted of electrostatic repulsions and π-π interactions. GO may be incorporated into polymer membrane to boost performance in water treatment. A few reports cover the incorporation of GO into particular; poly (*N*-vinylcarbazole), polyamide; and polysulfone membranes [157]. Inclusion of GO and activated carbon in polymer ultrafiltration membrane has shown improvement in hydrophilicity, electrostatic repulsion and pore size, which remarkably enhance removal of PACs in water [158]. Li et al. [159] reported the degradation of amoxicillin using magnetic TiO_2-GO-Fe_3O_4 composite with a submerged magnetic separation membrane photocatalytic reactor. TiO_2-GO-Fe_3O_4 was prepared with high Photo-Fenton catalytic performance and capability of magnetic recovery. Fe_3O_4 not only enhanced Fenton degradation of amoxicillin but also contributed to the magnetism of the photocatalyst for magnetic separation from treated water. The study revealed an excellent degradation of amoxicillin in a photodegradation process that is described in Figure 7b. It is also obvious in Table 6 that bio-composite prepared with the inclusion of GO exhibited high surface area and high adsorption capacity when compared with other bio-composites which did not include GO. The high surface area exhibited by the bio-composite might be due to the high surface area of GO, although the authors reported that the initial high surface area of GO reduced after inclusion for the preparation of bio-composite, which may be due to the aggregation resulting from the interaction with biomass. Most bio-composites remained steady in their regeneration and adsorption capacity up to 5–8 cycles, which makes them outstanding. Unfortunately, there is limited information on desorption studies or regeneration of plant-based biosorbents, as shown in Table 4, unlike the case for bio-composite. Apart from this, the introduction of magnetic properties into some of the reported bio-composites [142] as described in Figure 6 makes them stand out. The magnetic properties demonstrated by such bio-composites makes it easy to separate the biosorbent from the treated water system, unlike other biosorbents which required filtration and other steps, which adds extra cost to the actual treatment cost.

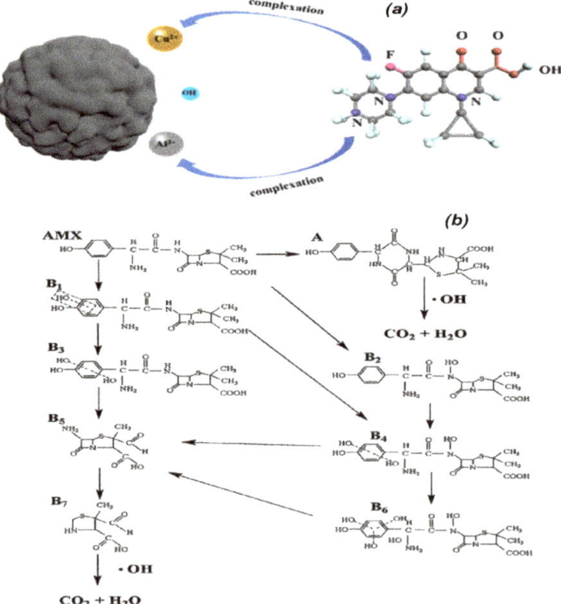

Figure 7. (**a**) The mechanism of ciprofloxacin loaded on CA-Al-KABs [32], (**b**) Mineralization pathways of amoxicillin (AMX) in water by submerged magnetic separation membrane photocatalytic reactor [159].

2.5. Desorption and Regeneration

The reusability of biosorbent is essential. This involves the removal of adsorbate from the surface of the biosorbent after use and the biosorbent should return to almost its original form in morphology and effectiveness. The number of cycles in which biosorbents can be reused goes a long way in determining their economic value and viability. One of the key achievements of biosorbents is their recovery after usage. Developing a suitable system for desorption is very important. It is not sufficient for a biosorbent to exhibit high performance alone, but reusability is equally essential. Therefore, desorption and regeneration are the fundamental processes to check when selecting biosorbents. Many materials have been developed in the past as biosorbents for removal of PACs from the water system. However, some of them are not easily regenerated, which makes their continuous use questionable, as they will have to be discarded after completing a circle or two. Discarding such material may also lead to an environmental pollution problem. It is very important that, after the sorption process, spent biosorbents are separated from the medium, regenerated and recycled. Several methods have been developed for the desorption process. However, the use of eluents is still the most commonly used process. Ability to select the most suitable eluent is very vital, a process, which depends on the type of biosorbent, adsorbate and mechanism of biosorption. An appropriate eluent should not damage or change the structure of the biosorbent, should be environmentally friendly, cheap, have high affinity for the adsorbate, not alter the adsorbate or biosorbent and should easily separate from the adsorbate.

Previously used eluents include mineral acids [30], organic acids and solvents [160], and complexing agents [161]. Desorption may be performed in batch or column, although the desorption process is easier in a packed column arrangement. When the sorption process is in a column, it is better to perform the desorption in the column as well. For example, once the sorption process becomes saturated, it is better to switch to desorption by allowing the eluent to flow through the column. It is important to carefully monitor the process so as not to overstretch the biosorbent; however, this also depends on the strength of the biosorbent. The potential of a desorption process relies on the removal mechanisms and the mechanical stability of the biosorbent. Since most biosorbents exhibit ion-exchange or ionic interaction towards positively charged PACs, use of mild to strong acidic systems should sufficiently achieve the desorption. This becomes beneficial since acidic solutions are common waste, generated in most industries. Therefore, acidic solutions generated during industrial processes can find an application by serving as a means of regenerating spent biosorbents. Chakraborty et al. [162] reported the desorption of ibuprofen from bidirectional activated biochar from sugarcane bagasse. The desorption was carried out using 0.1 N methanol under continuous agitation at 130 rpm for 24 h at 25 °C. It was efficient even after 4 cycles up to values above 65%. Desorption of diclofenac sodium and tetracycline hydrochloride from chitosan-based magnetic composite was achieved using ethanol and water (volume ratio of 1:1) with 1 wt % of NaOH [145]. For effective desorption, the alkaline condition was maintained since Fe_3O_4 is unstable at very low pH. Furthermore, the solution of ethanol-water was used in order to dissolve diclofenac sodium and tetracycline hydrochloride. The process attained a regeneration efficiency of five cycles. Desorption and regeneration of biosorbent are carefully considered when selecting biosorbent for water treatment. Care should be taken to ensure that the integrity of the biosorbent is intact after desorption and regeneration. It has become apparent that selecting the most suitable system for desorption is a challenging task, which requires a detailed understanding of the biosorbent as well as the PACs.

3. Cost Evaluation

Estimating the cost of biosorption and biosorbent is not an easy task, and it is not often reported. Process treatment, transportation, energy consumption, maintenance, process optimization, regeneration, disposal and desorption are factors considered when making an estimate. It depends on the nature of water or wastewater to be treated, as well as volume. However, capital expenditure and running cost will depend on the type and size of the treatment plant. It is better to make use of waste as feedstock for biosorbent production to minimize process cost. Wastes generated from farm produce, domestic and industrial wastes such as bacterial waste from fermentation industries, fungal wastes from food processing industries and sludge from other processing industries are applicable in biosorbent production. Waste disposal is a major problem, therefore making use of these wastes solves the environmental problem and at the same time helps reduce the cost of producing biosorbents for water treatment. One factor that increases production cost is pretreatment given to feedstock. Some of these materials require some pretreatments before processing, and the extra pretreatment may increase production cost. However, it is important to consider an effective but cheap and affordable treatment. It is vital to ensure that the treatment facility is close to the source of waste to minimize cost. Disposal cost should be factored into the cost of biosorption because once the biosorbent is completely utilized in a repeated cycle, it has to be replaced, although the spent biosorbent might find application in other fields like in the production of particleboard, cement, biogas, etc.

Cost evaluation depends on several factors, which makes it difficult to generalize. Therefore, the cost may vary from one biosorbent to the other depending on composition. This will also include whether it is sourced from waste or neat feedstock. Biosorption cost will also depend on the capacity and behavior of the biosorbent used. For example, if the removal process is fast and completed within a short time, less energy is consumed, unlike a biosorbent with slow uptake and long sorption time. The more the energy consumption, the higher the cost of production. The biosorption process is economically feasible compared to other methods such as reverse osmosis, electrodialysis, advance oxidation and electrothermal methods, which cost about US $450 per million liters whereas, depending on the type of biosorbent, the cost of water treatment per million liters is estimated as US $10–US $200 [163]. Use of chemicals is reduced, and when microorganisms are used, they can mineralize and degrade the PACs to forms that are readily adsorbed or wholly converted to CO_2 and H_2O. Different costs have been reported in the literature, for example, bagasse fly ash costs US $0.02 per kg [164], biochar costs US $2.65 per kg [165], chitosan-based biosorbent costs US $8–10 per kg [166] and kaolinite clay costs $0.005–0.46 per kg. The price may change depending on several economic factors.

Future Perspectives for Biosorption

With the current challenge of how to get rid of PACs in the environment, the government needs to develop a policy that will regulate the use of pharmaceutical products. Most countries do not have functional regulations to monitor and control the emergence of PACs in water. Moreover, current emerging PACs in water were not envisaged to occur in the water system; therefore, most of the currently used techniques for water treatment were not developed to cater for them. There is a need to develop efficient technologies that will help remove PACs entirely. However, biosorption has this advantage over other known methods. Nonetheless, most reported works on biosorption were laboratory-based and not on a large scale, which should be an area of focus. For biosorption to gain full operation, there is a need to conduct sufficient research on its use on a large scale for industrial purposes. To apply biosorption in real large-scale situation, there is a need to understand the response of biosorbent to operating parameters such as pH, temperature, particle size and load on a large scale. Most published works were conducted on a small scale in the laboratory. Therefore, it is crucial to know whether the biosorbent will behave differently when subjected to these parameters on a large scale. This information is currently missing.

The fact that it is cheaper to produce biosorbents from waste is an indication that there would be a socio-economic viability for large tons of waste generated from bio-based industrial processes. The waste generated from agriculture is massive. Investing in the conversion of such waste to biosorbents will create jobs and a new avenue for developing biosorption to full industrial scale, although there is the possibility of competition between using agricultural wastes as biosorbent and as feedstock for biogas production. However, the biosorbent can still be used for biogas production after it is completely used up and has repeatedly completed its utilization cycle, which makes it a win-win situation. Even though biosorbents can find use in biogas production, it is crucial to develop good disposal practice for spent biosorbents. It may also be essential to develop other ways of making use of spent biosorbents. Microorganisms as biosorbents have shown exciting results in biosorption, biodegradation and bioaccumulation of PACs. Care must be taken when using living organisms not to create strains that would be resistant by mutating. It is important not to end up creating a new problem while trying to solve environmental challenges. The possibility of using microorganisms creates a new area of research of screening for novel strains of bacteria, fungi and algae that will be useful and efficient in wastewater purification. There is a need to pay more attention to understanding the mechanism by which these microorganisms remove PACs from water. The use of bio-composite has proven very useful for the removal of PACs in water, but the complexity and chemical reagent involved in its production increases cost. There is a need to research more on reducing production cost by reducing production steps and using simple and cheap chemical reagents for bio-composite preparation.

4. Conclusions

This review considers the role of chemically modified biosorbent in removing PACs in water. It identified a few biosorbents and their role. It revealed that modification of biosorbents is vital to improving their capacity. Most biosorbents are prepared from waste, which plays a crucial role in cost reduction. Bio-composites exhibited better prospect as biosorbents when compared with other biosorbents because they contain two or more distinct materials brought together to produce a new material with improved performance better than individual constituent materials. During the sorption process, pH plays an important role; most of the sorption processes were pH-dependent. Use of living microorganisms showed that, apart from biosorption, the organisms are capable of bioaccumulating and biodegrading PACs. The majority of the sorption processes reported followed the pseudo-second-order model and can be described by the Langmuir isotherm model. Production of biosorbents was considered relatively cheap, efficient, and a promising means of removing PACs in the water system. Despite the numerous studies of biosorption, most of the works reported were laboratory-based (small-scale), and there is a need to conduct large-scale studies on the biosorption process for removing PACs in water.

Funding: This research received no external funding

Acknowledgments: Thanks for the support from the Department of Chemical Sciences, Redeemer's University.

Conflicts of Interest: There is no conflict of interest.

References

1. Tran, N.H.; Gin, K.Y.H. Occurrence and removal of pharmaceuticals, hormones, personal care products, and endocrine disrupters in a full-scale water reclamation plant. *Sci. Total Environ.* **2017**, *599–600*, 1503–1516.
2. Kostich, M.S.; Batt, A.L.; Lazorchak, J.M. Concentrations of prioritized pharmaceuticals in effluents from 50 large wastewater treatment plants in the US and implications for risk estimation. *Environ. Pollut.* **2014**, *184*, 354–359. [PubMed]
3. Lessa, E.F.; Nunes, M.L.; Fajardo, A.R. Chitosan/waste coffee-grounds composite: An efficient and eco-friendly adsorbent for removal of pharmaceutical contaminants from water. *Carbohydr. Polym.* **2018**, *189*, 257–266. [PubMed]
4. Larsson, D.G.J. Pollution from drug manufacturing: Review and perspectives. *Philos. Trans. R. Soc. B Biol. Sci.* **2014**, *369*, 1656. [CrossRef] [PubMed]
5. Yan, Z.; Lu, G.; Liu, J.; Jin, S. An integrated assessment of estrogenic contamination and feminization risk in fish in Taihu Lake, China. *Ecotoxicol. Environ. Saf.* **2012**, *84*, 334–340. [CrossRef] [PubMed]
6. Ikhwanuddin, M.; Bahar, H.; Ma, H.; Manan, H. Effect of estrogen hormone, 17β-estradiol on feminization of banana shrimp, *Penaeus merguiensis* (de Man, 1888) postlarvae and the identification of the age of external sex differentiation. *Aquac. Rep.* **2019**, *13*, 100177. [CrossRef]
7. Porseryd, T.; Larsson, J.; Kellner, M.; Bollner, T.; Dinnetz, P.; Hallstrom, I.P. Altered non-reproductive behavior and feminization caused by developmental exposure to 17α-ethinylestradiol persist to adulthood in three-spined stickleback (*Gasterosteus aculeatus*). *Aquatic Toxicol.* **2019**, *207*, 142–152. [CrossRef]

8. Quesada, H.B.; Baptista, A.T.A.; Cusioli, L.F.; Seibert, D.; Bezerra, C.O.; Bergamasco, R. Surface water pollution by pharmaceuticals and an alternative of removal by low-cost adsorbents: A review. *Chemosphere* **2019**, *222*, 766–780. [CrossRef]
9. Hossain, A.; Nakamichi, S.; Habibullah-Al-Mamun, M.; Tani, K.; Masunaga, S.; Matsuda, H. Occurrence and ecological risk of pharmaceuticals in river surface water of Bangladesh. *Environ. Res.* **2018**, *165*, 258–266. [CrossRef]
10. Asghar, M.A.; Zhu, Q.; Sun, S.; Peng, Y.; Shuai, Q. Suspect screening and target quantification of human pharmaceutical residues in the surface water of Wuhan, China, using UHPLC-Q-Orbitrap HRMS. *Sci. Total Environ.* **2018**, *635*, 828–837. [CrossRef]
11. Bean, T.G.; Rattner, B.A.; Lazarus, R.S.; Day, D.D.; Burket, S.R.; Brooks, B.W.; Haddad, S.P.; Bowerman, W.W. Pharmaceuticals in water, fish and osprey nestlings in Delaware River and Bay. *Environ. Pollut.* **2018**, *232*, 533–545. [CrossRef] [PubMed]
12. Rivera-Jaimes, J.A.; Postigo, C.; Melgoza-Aleman, R.M.; Acena, J.; Barcelo, D.; Lopez de Alda, M. Study of pharmaceuticals in surface and wastewater from Cuernavaca, Morelos, Mexico: Occurrence and environmental risk assessment. *Sci. Total Environ.* **2018**, *613–614*, 1263–1274. [CrossRef] [PubMed]
13. Fick, J.; Söderström, H.; Lindberg, R.H.; Phan, C.; Tysklind, M.; Larsson, D.G.J. Contamination of surface, ground, and drinking water from pharmaceutical production. *Environ. Toxicol. Chem.* **2009**, *28*, 2522–2527. [CrossRef] [PubMed]
14. Azanu, D.; Styrishave, B.; Darko, G.; Weisser, J.J.; Abaidoo, R.C. Occurrence and risk assessment of antibiotics in water and lettuce in Ghana. *Sci. Total Environ.* **2018**, *622–623*, 293–305. [CrossRef] [PubMed]
15. Madikizela, L.M.; Tavengwa, N.T.; Chimuka, L. Status of pharmaceuticals in African water bodies: Occurrence, removal and analytical methods. *J. Environ. Manag.* **2017**, *193*, 211–220.
16. Lindim, C.; van Gils, J.; Georgieva, D.; Mekenyan, O.; Cousins, I.T. Evaluation of human pharmaceutical emissions and concentrations in Swedish river basins. *Sci. Total Environ.* **2016**, *572*, 508–519.
17. Carmona, E.; Andreu, V.; Pico, Y. Occurrence of acidic pharmaceuticals and personal care products in Turia River Basin: From waste to drinking water. *Sci. Total Environ.* **2014**, *484*, 53–63.
18. Burns, E.E.; Carter, L.J.; Kolpin, D.W.; Thomas-Oates, J.; Boxall, A.B.A. Temporal and spatial variation in pharmaceutical concentrations in an urban river system. *Water Res.* **2018**, *137*, 72–85. [CrossRef]
19. Praveena, S.M.; Shaifuddin, S.N.M.; Sukiman, S.; Nasir, F.A.M.; Hanafi, Z.; Kamarudin, N.; Ismail, T.H.T.; Aris, A.Z. Pharmaceuticals residues in selected tropical surface water bodies from Selangor (Malaysia): Occurrence and potential risk assessments. *Sci. Total Environ.* **2018**, *642*, 230–240.
20. Matongo, S.; Birungi, G.; Moodley, B.; Ndungu, P. Pharmaceutical residues in water and sediment of msunduzi river, KwaZulu-Natal, South Africa. *Chemosphere* **2015**, *134*, 133–140. [CrossRef]
21. Pereira, A.M.P.T.; Silva, L.J.G.; Laranjeiro, C.S.M.; Meisel, L.M.; Lino, C.M.; Pena, A. Human pharmaceuticals in Portuguese rivers: The impact of water scarcity in the environmental risk. *Sci. Total Environ.* **2017**, *609*, 1182–1191. [PubMed]
22. Caban, M.; Lis, E.; Kumirska, J.; Stepnowski, P. Determination of pharmaceutical residues in drinking water in Poland using a new SPE-GC-MS(SIM) method based on Speedisk extraction disks and DIMETRIS derivatization. *Sci. Total Environ.* **2015**, *538*, 402–411. [CrossRef] [PubMed]
23. Mutiyar, P.K.; Gupta, S.K.; Mittal, A.K. Fate of pharmaceutical active compounds (PhACs) from River Yamuna, India: An ecotoxicological risk assessment approach. *Ecotoxicol. Environ. Saf.* **2018**, *150*, 297–304. [CrossRef]
24. Silva, A.; Delerue-Matos, C.; Figueiredo, S.A.; Freitas, O.M. The use of algae and fungi for removal of pharmaceuticals by bioremediation and biosorption processes: A Review. *Water* **2019**, *11*, 1555.
25. Bottoni, P.; Caroli, S.; Caracciolo, A.B. Pharmaceuticals as priority water contaminants. *Toxicol. Environ. Chem.* **2010**, *92*, 549–565. [CrossRef]
26. Akhtar, J.; Amin, N.A.S.; Shahzad, K. A review on removal of pharmaceuticals from water by adsorption. *Desalin. Water Treat.* **2016**, *57*, 12842–12860. [CrossRef]
27. Hokkanen, S.; Bhatnagar, A.; Sillanpää, M. A review on modification methods to cellulose-based adsorbents to improve adsorption capacity. *Water Res.* **2016**, *91*, 156–173. [CrossRef]
28. Xu, Y.; Liu, T.; Zhang, Y.; Ge, F.; Steel, R.M.; Sun, L. Advances in technologies for pharmaceuticals and personal care products removal. *J. Mater. Chem. A* **2017**, *5*, 12001–12014.

29. Silva, C.P.; Jaria, G.; Otero, M.; Esteves, V.I.; Calisto, V. Waste-based alternative adsorbents for the remediation of pharmaceutical contaminated waters: Has a step forward already been taken? *Bioresour. Technol.* **2018**, *250*, 888–901.
30. Adewuyi, A.; Oderinde, R.A. Chemically modified vermiculite clay: A means to removing emerging contaminant from polluted water system in developing nation. *Polym. Bull.* **2019**, *76*, 4967–4989.
31. Lach, J. Adsorption of chloramphenicol on commercial and modified activated carbons. *Water* **2019**, *11*, 1141. [CrossRef]
32. Hu, Y.; Pan, C.; Zheng, X.; Liu, S.; Hu, F.; Xu, L.; Xu, G.; Peng, X. Removal of ciprofloxacin with aluminum-pillared kaolin sodium alginate beads (CA-Al-KABs): Kinetics, isotherms, and BBD model. *Water* **2020**, *12*, 905. [CrossRef]
33. Rodriguez, A.Z.; Wang, H.; Hu, L.; Zhang, Y.; Xu, P. Treatment of produced water in the permian basin for hydraulic fracturing: Comparison of different coagulation processes and innovative filter media. *Water* **2020**, *12*, 770. [CrossRef]
34. Zhang, G.; Yang, Y.; Lu, Y.; Chen, Y.; Li, W.; Wang, S. Effect of heavy metal Ions on steroid estrogen removal and transport in SAT using DLLME as a detection method of steroid estrogen. *Water* **2020**, *12*, 589. [CrossRef]
35. Jiang, M.; Yang, W.; Zhang, Z.; Yang, Z.; Wang, Y. Adsorption of three pharmaceuticals on two magnetic ion-exchange resins. *J. Environ. Sci.* **2015**, *31*, 226–234.
36. Nghiem, L.D.; Schafer, A.I.; Elimelech, M. Removal of natural hormones by nanofiltration membranes: Measurement, modeling, and mechanisms. *Environ. Sci. Technol.* **2004**, *15*, 1888–1896. [CrossRef] [PubMed]
37. Rana, D.; Scheier, B.; Narbaitz, R.M.; Matsuura, T.; Tabe, S.; Jasim, S.Y.; Khulbe, K.C. Comparison of cellulose acetate (CA) membrane and novel CA membranes containing surface modifying macromolecules to remove pharmaceutical and personal care product micropollutants from drinking water. *J. Membr. Sci.* **2012**, *409*, 346–354.
38. Rajapaksha, A.U.; Premarathna, K.S.D.; Gunarathne, V.; Ahmed, A.; Vithanage, M. Sorptive removal of pharmaceutical and personal care products from water and wastewater. In *Pharmaceuticals and Personal Care Products: Waste Management and Treatment Technology*; Butterworth–Heinemann, Elsevier: Oxford, UK, 2019; pp. 213–238.
39. Feng, Y.; Wang, C.; Liu, J.; Zhang, Z. Electrochemical degradation of 17-alphaethinylestradiol (EE2) and estrogenic activity changes. *J. Environ. Monit.* **2010**, *12*, 404–408. [CrossRef]
40. Sires, I.; Brillas, E. Remediation of water pollution caused by pharmaceutical residues based on electrochemical separation and degradation technologies: A review. *Environ. Int.* **2012**, *40*, 212–229. [CrossRef]
41. Méndez-Arriaga, F.; Torres-Palma, R.; Pétrier, C.; Esplugas, S.; Gimenez, J.; Pulgarin, C. Ultrasonic treatment of water contaminated with ibuprofen. *Water Res.* **2008**, *42*, 4243–4248. [CrossRef]
42. Ciríaco, L.; Anjo, C.; Correia, J.; Pacheco, M.; Lopes, A. Electrochemical degradation of ibuprofen on Ti/Pt/PbO 2 and Si/BDD electrodes. *Electrochim. Acta* **2009**, *54*, 1464–1472. [CrossRef]
43. Madhavan, J.; Grieser, F.; Ashokkumar, M. Combined advanced oxidation processes for the synergistic degradation of ibuprofen in aqueous environments. *J. Hazard. Mater.* **2010**, *178*, 202–208. [CrossRef] [PubMed]
44. Salaeh, S.; Perisic, D.J.; Biosic, M.; Kusic, H.; Babic, S.; Stangar, U.L.; Dionysiou, D.D.; Bozic, A.L. Diclofenac removal by simulated solar assisted photocatalysis using TiO_2-based zeolite catalyst; mechanisms, pathways and environmental aspects. *Chem. Eng. J.* **2016**, *304*, 289–302. [CrossRef]
45. Hasan, Z.; Jhung, S.H. Removal of hazardous organics from water using metalorganic frameworks (MOFs): Plausible mechanisms for selective adsorptions. *J. Hazard. Mater.* **2015**, *283*, 329–339. [CrossRef]
46. Fomina, M.; Gadd, G.M. Biosorption: Current perspectives on concept, definition and application. *Bioresour. Technol.* **2014**, *160*, 3–14. [CrossRef]
47. Worch, E. *Adsorption Technology in Water Treatment*; de Gruyter, W., Ed.; Deutscher Kunstverlag: Berlin, Germany, 2012; p. 332.
48. Rostamian, R.; Behnejad, H. A comprehensive adsorption study and modeling of antibiotics as a pharmaceutical waste by graphene oxide nanosheets. *Ecotoxicol. Environ. Saf.* **2018**, *147*, 117–123. [CrossRef]
49. Basu, S.; Balakrishnan, M. Polyamide thin film composite membranes containing ZIF-8 for the separation of pharmaceutical compounds from aqueous streams. *Sep. Purif. Technol.* **2017**, *179*, 118–125. [CrossRef]

50. Qian, K.; Kumar, A.; Zhang, H.; Bellmer, D.; Huhnke, R. Recent advances in utilization of biochar. *Renew. Sustain. Energy Rev.* **2015**, *42*, 1055–1064. [CrossRef]
51. Šolic, M.; Maletic, S.; Isakovski, M.K.; Nikic, J.; Watson, M.; Kónya, Z.; Trickovic, J. Comparing the adsorption performance of multiwalled carbon nanotubes oxidized by varying degrees for removal of low levels of copper, nickel and chromium (VI) from aqueous solutions. *Water* **2020**, *12*, 723. [CrossRef]
52. Vijayaraghavan, K.; Yun, Y.S. Bacterial biosorbents and biosorption. *Biotechnol. Adv.* **2008**, *26*, 266–291. [CrossRef]
53. Tadkaew, N.; Hai, F.I.; Mcdonald, J.A.; Khan, S.J.; Nghiem, L.D. Removal of trace organics by MBR treatment: The role of molecular properties. *Water Res.* **2011**, *45*, 2439–2451. [CrossRef] [PubMed]
54. Adewuyi, A.; Pereira, F.V. Underutilized *Luffa cylindrica* sponge: A local bio-adsorbent for the removal of Pb (II) pollutant from water system. *Beni-Suef Uni. J. Basic Appl. Sci.* **2017**, *6*, 118–126.
55. Ngah, W.S.W.; Hanafiah, M.A.K.M. Removal of heavy metal ions from wastewater by chemically modified plant wastes as adsorbents: A review. *Bioresour. Technol.* **2008**, *99*, 3935–3948.
56. Ali, M.E.M.; Abd El-Aty, A.M.; Badawy, M.I.; Ali, R.K. Removal of pharmaceutical pollutants from synthetic wastewater using chemically modified biomass of green alga Scenedesmus obliquus. *Ecotoxicol. Environ. Saf.* **2018**, *151*, 144–152. [PubMed]
57. Onal, Y.; Akmil-Basar, C.; Sarıcı-Ozdemir, C. Elucidation of the naproxen sodium adsorption onto activated carbon prepared from waste apricot: Kinetic, equilibrium and thermodynamic characterization. *J. Hazard. Mater.* **2007**, *148*, 727–734. [PubMed]
58. Cazetta, A.L.; Martins, A.C.; Pezoti, O.; Bedin, K.C.; Beltrame, K.K.; Asefa, T.; Almeida, V.C. Synthesis and application of N–S-doped mesoporous carbon obtained from nanocasting method using bone char as heteroatom precursor and template. *Chem. Eng. J.* **2016**, *300*, 54–63.
59. Rosli, N.; Ngadi, N.; Azman, M.A. Synthesis of modified spent tea for aspirin adsorption in aqueous solution. *Inter. J. Rec. Technol. Eng.* **2019**, *8*, 531–534.
60. Mestre, A.S.; Pires, J.; Nogueira, J.M.F.; Carvalho, A.P. Activated carbons for the adsorption of ibuprofen. *Carbon* **2007**, *45*, 1979–1988. [CrossRef]
61. Baccar, R.; Sarrà, M.; Bouzid, J.; Feki, M.; Blánquez, P. Removal of pharmaceutical compounds by activated carbon prepared from agricultural by-product. *Chem. Eng. J.* **2012**, *211–212*, 310–317. [CrossRef]
62. Mondal, S.; Aikat, K.; Halder, G. Biosorptive uptake of ibuprofen by chemically modified Parthenium hysterophorus derived biochar: Equilibrium, kinetics, thermodynamics and modeling. *Ecol. Eng.* **2016**, *92*, 158–172. [CrossRef]
63. Essandoh, M.; Kunwar, B.; Pittman, C.U.; Mohan, D.; Mlsna, T. Sorptive removal of salicylic acid and ibuprofen from aqueous solutions using pine wood fast pyrolysis biochar. *Chem. Eng. J.* **2015**, *265*, 219–227. [CrossRef]
64. Mestre, A.S.; Bexiga, A.S.; Proença, M.; Andrade, M.; Pinto, M.L.; Matos, I.; Fonseca, I.M.; Carvalho, A.P. Activated carbons from sisal waste by chemical activation with K_2CO_3: Kinetics of paracetamol and ibuprofen removal from aqueous solution. *Bioresour. Technol.* **2011**, *102*, 8253–8260. [CrossRef] [PubMed]
65. Ribeiro, A.V.F.N.; Belisário, M.; Galazzi, R.M.; Balthazar, D.C.; Pereira, M.G.; Ribeiro, J.N. Evaluation of two bioadsorbents for removing paracetamol from aqueous media. *Electron. J. Biotechnol.* **2011**, *14*, 1–10.
66. Ornek, A.; Ozacar, M.; Sengil, I.A. Adsorption of Pb(II) onto formaldehyde or sulphuric acid treated acorn waste: Equilibrium and kinetic studies. *Biochem. Eng. J.* **2007**, *37*, 192–200. [CrossRef]
67. Saka, C.; Sahin, O.; Kucuk, M.M. Applications on agricultural and forest waste adsorbentsfor the removal of lead (II) from contaminated waters. *Int. J. Environ. Sci. Technol.* **2012**, *9*, 379–394. [CrossRef]
68. Ramachandran, V.; Pujari, N.; Matey, T.; Kulkarni, S. Enzymatic hydrolysis of cassava using wheat seedlings. *Int. J. Sci. Eng. Technol. Res.* **2014**, *3*, 1216–1219.
69. Kulkarni, S.J. Use of biotechnology for synthesis of various products from different feedstocks-a review. *Int. J. Adv. Res. Bio-Technol.* **2014**, *2*, 1–3.
70. Rajapaksha, A.U.; Vithanage, M.; Ahmad, M.; Seo, D.C.; Cho, J.S.; Lee, S.E.; Lee, S.S.; Ok, Y.S. Enhanced sulfamethazine removal by steam-activated invasive plant-derived biochar. *J. Hazard. Mater.* **2015**, *290*, 43–50.
71. Chen, Y.; Wang, F.; Duan, L.; Yang, H.; Gao, J. Tetracycline adsorption onto rice husk ash, an agricultural waste: Its kinetic and thermodynamic studies. *J. Mol. Liq.* **2016**, *222*, 487–494. [CrossRef]

72. Bernardo, M.; Rodrigues, S.; Lapa, N.; Matos, I.; Lemos, F.; Batista, M.K.S.; Carvalho, A.P.; Fonseca, I. High efficacy on diclofenac removal by activated carbon produced from potato peel waste. *Int. J. Environ. Sci. Technol.* **2016**, *13*, 1989–2000. [CrossRef]
73. Pouretedal, H.R.; Sadegh, N. Effective removal of amoxicillin, cephalexin, tetracycline and penicillin G from aqueous solutions using activated carbon nanoparticles prepared from vine wood. *J. Water Process Eng.* **2014**, *1*, 64–73. [CrossRef]
74. Ahmed, M.B.; Zhou, J.L.; Ngo, H.H.; Guo, W.; Johir, M.A.H.; Sornalingam, K. Single and competitive sorption properties and mechanism of functionalized biochar for removing sulfonamide antibiotics from water. *Chem. Eng. J.* **2017**, *311*, 348–358.
75. Wang, H.; Chu, Y.; Fang, C.; Huang, F.; Song, Y.; Xue, X. Sorption of tetracycline on biochar derived from rice straw under different temperatures. *PLoS ONE* **2017**, *12*, e0182776.
76. Zhang, H.; Wang, Z.; Li, R.; Guo, J.; Li, Y.; Zhu, J.; Xie, X. TiO_2 supported on reed straw biochar as an adsorptive and photocatalytic composite for the efficient degradation of sulfamethoxazole in aqueous matrices. *Chemosphere* **2017**, *185*, 351–360. [CrossRef] [PubMed]
77. El-Shafey, E.S.I.; Al-Lawati, H.; Al-Sumri, A.S. Ciprofloxacin adsorption from aqueous solution onto chemically prepared carbon from date palm leaflets. *J. Environ. Sci.* **2012**, *24*, 1579–1586.
78. Balarak, D.; Mostafapour, F.K. Canola residual as a biosorbent for antibiotic metronidazole removal. *Pharm. Chem. J.* **2016**, *3*, 12–17.
79. N'diaye, A.D.; Bollahi, M.A.; Kankou, M.S.A. Sorption of paracetamol from aqueous solution using groundnut shell as a low cost sorbent. *J. Mater. Environ. Sci.* **2019**, *10*, 553–562.
80. Aziz, S.S.; Mushtaq, S.; Begum, N. Adsorption studies of acetic acid removal from waste water using seeds of *Brassica nigra*. *Int. J. Eng. Res. Appl.* **2017**, *7*, 1–3.
81. Silva, B.; Martins, M.; Rosca, M.; Rocha, V.; Lago, A.; Neves, I.C.; Tavares, T. Waste-based biosorbents as cost-effective alternatives to commercial adsorbents for the retention of fluoxetine from water. *Sep. Purif. Technol.* **2020**, *235*, 116–139.
82. Martins, A.C.; Pezoti, O.; Cazetta, A.L.; Bedin, K.C.; Yamazaki, D.A.S.; Bandoch, G.F.G.; Asefa, T.; Visentainer, J.V.; Almeida, V.C. Removal of tetracycline by NaOH-activated carbon produced from macadamia nut shells: Kinetic and equilibrium studies. *Chem. Eng. J.* **2015**, *260*, 291–299. [CrossRef]
83. Kumar, A.; Sharma, G.; Naushad, M.; Al-Muhtaseb, A.; Hira, I.; Ahamad, T.; Kumar, A.; Ghfar, A.A.; Stadler, F.J. Visible photodegradation of ibuprofen and 2,4-D in simulated waste water using sustainable metal free-hybrids based on carbon nitride and biochar. *J. Environ. Manag.* **2019**, *231*, 1164–1175. [CrossRef]
84. Wang, F.; Yang, B.; Wang, H.; Song, Q.; Tan, F.; Cao, Y. Removal of ciprofloxacin from aqueous solution by a magnetic chitosan grafted graphene oxide composite. *J. Mol. Liq.* **2016**, *222*, 188–194.
85. Paunovic, O.; Pap, S.; Maletic, S.; Taggart, M.A.; Boskovic, N.; Sekulic, M.T. Ionisable emerging pharmaceutical adsorption onto microwave functionalised biochar derived from novel lignocellulosic waste biomass. *J. Colloid Interface Sci.* **2019**, *547*, 350–360. [CrossRef]
86. Villaescusa, I.; Fiol, N.; Poch, J.; Bianchi, A.; Bazzicalupi, C. Mechanism of paracetamol removal by vegetable wastes: The contribution of π-π interactions, hydrogen bonding and hydrophobic effect. *Desalination* **2011**, *270*, 135–142. [CrossRef]
87. Yusoff, N.A.; Ngadi, N.; Alias, H.; Jusoh, M. Chemically treated chicken bone waste as an efficient adsorbent for removal of acetaminophen. *Chem. Eng. Trans.* **2017**, *56*, 925–930.
88. Khazri, H.; Ghorbel-Abid, I.; Kalfat, R.; Trabelsi-Ayadi, M. Extraction of clarithromycin and atenolol by cuttlefish bone powder. *Environ. Technol.* **2018**, *38*, 2662–2668. [CrossRef]
89. Benjedim, S.; Romero-Cano, L.A.; Pérez-Cadenas, A.F.; Bautista-Toledo, M.I.; Lotfi, E.; Carrasco-Marín, F. Removal of emerging pollutants present in water using an E-coli biofilm supported onto activated carbons prepared from argan wastes: Adsorption studies in batch and fixed bed. *Sci. Total Environ.* **2020**, *720*, 137491. [CrossRef]
90. Adewuyi, A.; Göpfert, A.; Adewuyi, O.A.; Wolff, T. Adsorption of 2-chlorophenol onto the surface of underutilized seed of *Adenopus breviflorus*: A potential means of treating waste water. *J. Environ. Chem. Eng.* **2016**, *4*, 664–672. [CrossRef]
91. Kizilkaya, B.; Tekinay, A.A.; Dilgin, Y. Adsorption and removal of Cu (II) ions from aqueous solution using pretreated fish bones. *Desalination* **2010**, *264*, 37–47.

92. Dahiya, S.; Tripathi, R.M.; Hegde, A.G. Biosorption of lead and copper from aqueous solutions by pre-treated crab and arca shell biomass. *Bioresour. Technol.* **2008**, *99*, 179–187. [CrossRef]
93. Ojedokun, A.T.; Bello, O.S. Sequestering heavy metals from wastewater using cow dung. *Water Resour. Ind.* **2016**, *13*, 7–13. [CrossRef]
94. Vasanthakumar, K.; Bhagavanalu, D.V.S. Adsorption of basic dye from its aqueous solution on to bio-organic waste. *J. Ind. Pollut. Control* **2003**, *19*, 20–28.
95. El Haddad, M.; Mamouni, R.; Slimani, R.; Saffaj, N.; Ridaoui, M.; ElAntri, S.; Lazar, S. Adsorptive removal of reactive yellow 84 dye from aqueous solutions onto animal bone meal. *J. Mater. Environ. Sci.* **2012**, *3*, 1019–1026.
96. Nworu, J.S.; Enemose, E.A.; Osideru, O.O.; Emmanuel, O.A. Efficiency of animal (Cow, Donkey, Chicken and Horse) bones, in removal of hexavalent chromium from aqueous solution as a low cost adsorbent. *Am. J. Appl. Chem.* **2019**, *7*, 1–9.
97. Banat, F.A.; Al-Asheh, S. The use of human hair waste as a phenol biosorbent. *Adsorp. Sci. Technol.* **2001**, *19*, 599–608. [CrossRef]
98. Volesky, B.; Schiewer, S. Biosorption of metals. In *Encyclopedia of Bioprocess Technology*; Flickinger, M., Drew, S.W., Eds.; Wiley: New York, NY, USA, 1999; pp. 433–453.
99. Aydin, S. Enhanced biodegradation of antibiotic combinations via the sequential treatment of the sludge resulting from pharmaceutical wastewater treatment using white-rot fungi *Trametes versicolor* and *Bjerkandera adusta*. *Appl. Microbiol. Biotechnol.* **2016**, *100*, 6491–6499. [CrossRef]
100. Doyle, R.J.; Matthews, T.H.; Streips, U.N. Chemical basis for selectivity of metal ions by the Bacillus subtilis cell wall. *J. Bacteriol.* **1980**, *143*, 471–480. [CrossRef]
101. Phoenix, V.R.; Martinez, R.E.; Konhauser, K.O.; Ferris, F.G. Characterization and implications of the cell surface reactivity of Calothrixsp. strain KC97. *Appl. Environ. Microbiol.* **2002**, *68*, 4827–4834. [CrossRef]
102. Kazy, S.K.; Das, S.K.; Sar, P. Lanthanum biosorption by a Pseudomonassp.: Equilibrium studies and chemical characterization. *J. Ind. Microbiol. Biotechnol.* **2006**, *33*, 773–783. [CrossRef]
103. Vannela, R.; Verma, S.K. Cu^{2+} removal and recovery by SpiSORB: Batch stirred and up-flow packed bed columnar reactor systems. *Bioprocess Biosyst. Eng.* **2006**, *29*, 7–17. [CrossRef]
104. Vijayaraghavan, K.; Han, M.H.; Choi, S.B.; Yun, Y.S. Biosorption of reactive black 5 by *Corynebacterium glutamicum* biomass immobilized in alginate and polysulfone matrices. *Chemosphere* **2007**, *68*, 1838–1845. [CrossRef]
105. Vijayaraghavan, K.; Yun, Y.S. Chemical modification and immobilization of *Corynebacterium glutamicum* for biosorption of reactive black 5 from aqueous solution. *Ind. Eng. Chem. Res.* **2007**, *46*, 608–617. [CrossRef]
106. Martins, M.; Sanches, S.; Pereira, I.A.C. Anaerobic biodegradation of pharmaceutical compounds: New insights into the pharmaceutical-degrading bacteria. *J. Hazard. Mater.* **2018**, *357*, 289–297.
107. Wang, L.; Qiang, Z.; Li, Y.; Ben, W. An insight into the removal of fluoroquinolones in activated sludge process: Sorption and biodegradation characteristics. *J. Environ. Sci.* **2017**, *56*, 263–271. [CrossRef]
108. Sun, H.; Wang, T.; Yang, Z.; Yu, C.; Wu, W. Simultaneous removal of nitrogen and pharmaceutical and personal care products from the effluent of waste water treatment plants using aerated solid-phase denitrification system. *Bioresour. Technol.* **2019**, *287*, 121389.
109. Rana, R.S.; Singh, P.; Kandari, V.; Singh, R.; Dobhal, R.; Gupta, S. A review on characterization and bioremediation of pharmaceutical industries' wastewater: An Indian perspective. *Appl. Water Sci.* **2017**, *7*, 1–12. [CrossRef]
110. Yu, J.; Tong, M.; Sun, X.; Li, B. A simple method to prepare poly (amic acid)-modified biomass for enhancement of lead and cadmium adsorption. *Biochem. Eng. J.* **2007**, *33*, 126–133. [CrossRef]
111. Göksungur, Y.; Üren, S.; Güvenc, U. Biosorption of cadmium and lead ion by ethanol treated waste baker's yeast biomass. *Bioresour. Technol.* **2005**, *96*, 103–109. [CrossRef]
112. Dhankhar, R.; Hooda, A. Fungal biosorption—An alternative to meet the challenges of heavy metal pollution in aqueous solutions. *Environ. Technol.* **2011**, *32*, 467–491. [CrossRef]
113. Ridvan, S.; Nalan, Y.; Adil, D. Biosorption of cadmium, lead, mercury and arsenic ions by fungus Penicillium purpurogenum. *Sep. Sci. Technol.* **2003**, *39*, 2039–2053.
114. Gadd, G.M. Accumulation and transformation of metals by microorganisms. In *Biotechnology: A Multi-Volume Comprehensive Treatise, Vol. 10 Special Processes*; Rehm, H.J., Reed, G., Puhler, A., Stadler, P., Eds.; Wiley: Weinheim, Germany, 2001; pp. 225–264.

115. Becker, D.; Rodriguez-Mozaz, S.; Insa, S.; Schoevaart, R.; Barceloó, D.; de Cazes, M.; Belleville, M.P.; Sanchez-Marcano, J.; Misovic, A.; Oehlmann, J.; et al. Removal of endocrine disrupting chemicals in wastewater by enzymatic treatment with fungal laccases. *Org. Process Res. Dev.* **2017**, *21*, 480–491. [CrossRef]
116. Zhang, Y.; Geißen, S.U. Elimination of carbamazepine in a nonsterile fungal bioreactor. *Bioresour. Technol.* **2012**, *112*, 221–227. [CrossRef] [PubMed]
117. Rodriguez-Rodriguez, C.E.; García-Galán, M.J.; Blánquez, P.; Díaz-Cruz, M.S.; Barceló, D.; Caminal, G.; Vicent, T. Continuous degradation of a mixture of sulfonamides by Trametes versicolor and identification of metabolites from sulfapyridine and sulfathiazole. *J. Hazard. Mater.* **2012**, *213–214*, 347–354. [CrossRef] [PubMed]
118. Rodarte-Moralez, A.I.; Feijoo, G.; Moreira, M.T.; Lema, J.M. Operation of stirred tank reactors (STRs) and fixed-bed reactors (FBRs) with free and immobilized Phanerochaete chrysosporium for the continuous removal of pharmaceutical compounds. *Biochem. Eng. J.* **2012**, *66*, 38–45.
119. Lucas, D.; Castellet-Rovira, F.; Villagrasa, M.; Badia-Fabregat, M.; Barceló, D.; Vicent, T.; Caminal, G.; Sarràb, M.; Rodríguez-Mozaz, S. The role of sorption processes in the removal of pharmaceuticals by fungal treatment of wastewater. *Sci. Total Environ.* **2018**, *610–611*, 1147–1153.
120. González-Abradelo, D.; Pérez-Llano, Y.; Peidro-Guzmán, H.; Sánchez-Carbente, M.R.; Folch-Mallol, J.L.; Aranda, E.; Vaidyanathan, V.K.; Cabana, H.; Gunde-Cimerman, N.; Batista-García, R.A. First demonstration that ascomycetous halophilic fungi (*Aspergillus sydowii* and *Aspergillus destruens*) are useful in xenobiotic mycoremediation under high salinity conditions. *Bioresour. Technol.* **2019**, *279*, 287–296. [PubMed]
121. Vasiliadou, I.A.; Sanchez-Vazquez, R.; Molina, R.; Martínez, F.; Melero, J.A.; Bautista, L.F.; Iglesias, J.; Morales, G. Biological removal of pharmaceutical compounds using white-rotfungi with concomitant FAME production of the residual biomass. *J. Environ. Manag.* **2016**, *180*, 228–237. [CrossRef]
122. Badia-Fabregat, M.; Lucas, D.; Pereira, M.A.; Alves, M.A.; Pennanen, T.; Fritze, H.; Rodríguez-Mozaz, S.; Barceló, D.; Vicent, T.; Caminal, G. Continuous fungal treatment of non-sterile veterinary hospital effluent: Pharmaceuticals removal and microbial community assessment. *Appl. Microbiol. Biotechnol.* **2016**, *100*, 2401–2415. [CrossRef]
123. Becker, D.; Varela Della Giustina, S.; Rodriguez-Mozaz, S.; Schoevaart, R.; Barcelo, D.; de Cazes, M.; Belleville, M.P.; Sanchez-Marcano, J.; de Gunzburg, J.; Couillerot, O.; et al. Removal of antibiotics in wastewater by enzymatic treatment with fungal laccase—Degradation of compounds does not always eliminate toxicity. *Bioresour. Technol.* **2016**, *219*, 500–509.
124. Esterhuizen-Londt, M.; Schwartz, K.; Pflugmacher, S. Using aquatic fungi for pharmaceutical bioremediation: Uptake of acetaminophen by Mucor hiemalis does not result in an enzymatic oxidative stress response. *Fungal Biol.* **2016**, *120*, 1249–1257.
125. Palli, L.; Castellet-Rovira, F.; Péerez-Trujillo, M.; Caniani, D.; Sarrá-Adroguer, M.; Gori, R. Preliminary evaluation of *Pleurotus ostreatus* for the removal of selected pharmaceuticals from hospital wastewater. *Biotechnol. Prog.* **2017**, *33*, 1529–1537.
126. Pointing, S.B. Feasibility of bioremediation by white-rot fungi. *Appl. Microbiol. Biotechnol.* **2001**, *57*, 20–33.
127. Wijffels, R.H.; Kruse, O.; Hellingwerf, K.J. Potential of industrial biotechnology with cyanobacteria and eukaryotic microalgae. *Curr. Opin. Biotechnol.* **2013**, *24*, 405–413.
128. Xiong, J.Q.; Govindwar, S.; Kurade, M.B.; Paeng, K.J.; Roh, H.S.; Khan, M.A.; Jeon, B.H. Toxicity of sulfamethazine and sulfamethoxazole and their removal by a green microalga, Scenedesmus obliquus. *Chemosphere* **2019**, *218*, 551–558. [CrossRef]
129. Xiong, J.Q.; Kurade, M.B.; Jeon, B.H. Can microalgae remove pharmaceutical contaminants from water? *Trends Biotechnol.* **2017**, *36*, 30–44. [CrossRef]
130. Matamoros, V.; Uggetti, E.; García, J.; Bayona, J.M. Assessment of the mechanisms involved in the removal of emerging contaminants by microalgae from wastewater: A laboratory scale study. *J. Hazard. Mater.* **2016**, *301*, 197–205. [CrossRef]
131. Peng, F.Q.; Ying, G.G.; Yang, B.; Liu, S.; Lai, H.J.; Liu, Y.S.; Chen, Z.F.; Zhou, G.J. Biotransformation of progesterone and norgestrel by two freshwater microalgae (Scenedesmus obliquus and Chlorella pyrenoidosa): Transformation kinetics and products identification. *Chemosphere* **2014**, *95*, 581–588. [CrossRef]
132. González-Pleiter, M.; Gonzalo, S.; Rodea-Palomares, I.; Leganés, F.; Rosal, R.; Boltes, K.; Marco, E.; Fernández-Piñas, F. Toxicity of five antibiotics and their mixtures towards photosynthetic aquatic organisms: Implications for environmental risk assessment. *Water Res.* **2013**, *47*, 2050–2064. [CrossRef]

133. Xiong, J.Q.; Kurade, M.B.; Abou-Shanab, R.A.I.; Ji, M.K.; Choi, J.; Kim, J.O.; Jeon, B.H. Biodegradation of carbamazepine using freshwater microalgae *Chlamydomonas mexicana* and *Scenedesmus obliquus* and the determination of its metabolic fate. *Bioresour. Technol.* **2016**, *205*, 183–190. [CrossRef]
134. Ding, T.; Yang, M.; Zhang, J.; Lin, K.; Li, J.; Gan, J. Toxicity, degradation and metabolic fate of ibuprofen on freshwater diatom Naviculasp. *J. Hazard. Mater.* **2017**, *330*, 127–134.
135. Hom-Diaz, A.; Jaen-Gil, A.; Bello-Laserna, I.; Rodríguez-Mozaz, S.; Vincent, T.; Barceló, D.; Blánquez, P. Performance of a microalgal photobioreactor treating toilet wastewater: Pharmaceutically active compound removal and biomass harvesting. *Sci. Total Environ.* **2017**, *592*, 1–11. [CrossRef]
136. Escapa, C.; Coimbra, R.; Paniagua, S.; García, A.; Otero, M. Paracetamol and salicylic acid removal from contaminated water by microalgae. *J. Environ. Manag.* **2017**, *203*, 799. [CrossRef] [PubMed]
137. Amorim, C.L.; Moreira, I.S.; Maia, A.S.; Tiritan, M.E.; Castro, P.M.L. Biodegradation of ofloxacin, norfloxacin, and ciprofloxacin as single and mixed substrates by Labrys portucalensis F11. *Appl. Microbiol. Biotechnol.* **2014**, *98*, 3181–3190. [CrossRef] [PubMed]
138. Yang, S.; Hai, F.I.; Nghiem, L.D.; Nguyen, L.N.; Roddick, F.; Price, W.E. Removal of bisphenol a and diclofenac by a novel fungal membrane bioreactor operated under non-sterile conditions. *Int. Biodeterior. Biodegrad.* **2013**, *85*, 483–490. [CrossRef]
139. Zhang, L.; Xia, W.; Teng, B.; Liu, X.; Zhang, W. Zirconium cross-linked chitosan composite: Preparation, characterization and application in adsorption of Cr(VI). *Chem. Eng. J.* **2013**, *229*, 1–8. [CrossRef]
140. Rahim, M.; Haris, M.R.H.M. Application of biopolymer composites in arsenic removal from aqueous medium: A review. *J. Radiat. Res. Appl. Sci.* **2015**, *8*, 255–263. [CrossRef]
141. Karoui, S.; Ben, A.R.; Mougin, K.; Ghorbal, A.; Assadi, A.; Amrane, A. Synthesis of novel biocomposite powder for simultaneous removal of hazardous ciprofloxacin and methylene blue: Central composite design, kinetic and isotherm studies using Brouers-Sotolongo family models. *J. Hazard. Mater.* **2020**, *387*, 121675. [CrossRef]
142. Mashile, G.P.; Mpupa, A.; Nqombolo, A.; Dimpe, K.M.; Nomngongo, P.N. Recyclable magnetic waste tyre activated carbon-chitosan composite as an effective adsorbent rapid and simultaneous removal of methylparaben and propylparaben from aqueous solution and wastewater. *J. Water Process Eng.* **2020**, *33*, 101011. [CrossRef]
143. Li, Y.; Wang, Z.; Xie, X.; Zhu, J.; Li, R.; Qin, T. Removal of norfloxacin from aqueous solution by clay-biochar composite prepared from potato stem and natural attapulgite. *Colloids Surf. A* **2017**, *514*, 126–136. [CrossRef]
144. Wang, H.; Yuan, X.; Wu, Y.; Huang, H.; Peng, X.; Zeng, G.; Zhong, H.; Liang, J.; Ren, M.M. Graphene-based materials, fabrication, characterization and application for the decontamination of wastewater and waste gas and hydrogen storage/generation. *Adv. Colloid Interface Sci.* **2013**, *195–196*, 19–40. [CrossRef]
145. Zhang, S.; Dong, Y.; Yang, Z.; Yang, W.; Wu, J.; Dong, C. Adsorption of pharmaceuticals on chitosan-based magnetic composite particles with core-brush topology. *Chem. Eng. J.* **2016**, *304*, 325–334. [CrossRef]
146. Babel, S.; Kurniawan, T. A Low-cost adsorbents for heavy metals uptake from contaminated water: A review. *J. Hazard. Mater.* **2003**, *97*, 219–243. [CrossRef]
147. Srinivasan, R. Advances in application of natural clay and its composites in removal of biological, organic, and inorganic contaminants from drinking water. *Adv. Mater. Sci. Eng.* **2011**, *2011*, 1–17. [CrossRef]
148. Yao, Y.; Gao, B.; Fang, J.; Zhang, M.; Chen, H.; Zhou, Y.; Creamer, A.E.; Sun, Y.; Yang, L. Characterization and environmental applications of clay-biochar composites. *Chem. Eng. J.* **2014**, *242*, 136–143. [CrossRef]
149. Fosso-Kankeu, E.; Waanders, F.B.; Steyn, F.W. The preparation and characterization of clay-biochar composites for the removal of metal pollutants. In Proceedings of the Seventh International Conference on Latest Trends in Engineering & Technology, Irene, Pretoria, South Africa, 26–27 November 2015.
150. Wang, W.; Zheng, B.; Deng, Z.; Feng, Z.; Fu, L. Kinetics and equilibriums for adsorption of poly(vinyl alcohol) from aqueous solution onto natural bentonite. *Chem. Eng. J.* **2013**, *214*, 343–354. [CrossRef]
151. Kumar, A.S.K.; Kalidhasan, S.; Rajesh, V.; Rajesh, N. Application of cellulose-clay composite biosorbent toward the effective adsorption and removal of chromium from industrial wastewater. *Ind. Eng. Chem. Res.* **2012**, *51*, 58–69. [CrossRef]
152. Amin, M.F.M.; Heijman, S.G.J.; Rietveld, L.C. Clay–starch combination for micropollutants removal from wastewater treatment plant effluent. *Water Sci. Technol.* **2016**, *73*, 1719–1727. [CrossRef] [PubMed]
153. Sophia, A.C.; Lima, E.C.; Allaudeen, N.; Rajan, S. Application of graphene based materials for adsorption of pharmaceutical traces from water and wastewater-a review. *Desalin. Water Treat.* **2016**, *57*, 27573–27586.

154. Stoller, M.D.; Park, S.; Zhu, Y.; An, J.; Ruoff, R.S. Graphene-based ultracapacitors. *Nano Lett.* **2008**, *8*, 3498–3502. [CrossRef]
155. Georgakilas, V.; Otyepka, M.; Bourlinos, A.B.; Chan-dra, V.; Kim, N.; Kemp, K.C.; Hobza, P.; Zboril, R.; Kim, K.S. Functionalization of graphene: Covalent andnon-covalent approaches, derivatives and applications. *Chem. Rev.* **2012**, *112*, 6156–6214. [CrossRef]
156. Tang, Y.; Guo, H.; Xiao, L.; Yu, S.; Gao, N.; Wang, Y. Synthesis of reduced graphene oxide/magnetite composites and investigation of their adsorption performance of fluoroquinolone antibiotics. *Colloids Surf. A Physicochem. Eng. Asp.* **2013**, *424*, 74–80. [CrossRef]
157. Cao, B.; Ansari, A.; Yi, X.; Rodrigues, D.F.; Hu, Y. Gypsum scale formation on graphene oxide modified reverse osmosis membrane. *J. Membr. Sci.* **2018**, *552*, 132–143. [CrossRef]
158. Chu, K.H.; Fathizadeh, M.; Yu, M.; Flora, J.R.V.; Jang, A.; Jang, M.; Park, C.M.; Yoo, S.S.; Her, N.; Yoon, Y. Evaluation of removal mechanisms in a graphene oxide-coated ceramic ultrafiltration membrane for retention of natural organic matter, pharmaceuticals, and inorganic salts. *ACS Appl. Mater. Interfaces* **2017**, *9*, 40369–40377. [CrossRef] [PubMed]
159. Li, Q.; Kong, H.; Li, P.; Shao, J.; He, Y. Photo-Fenton degradation of amoxicillin via magnetic TiO_2-graphene oxide-Fe_3O_4 composite with a submerged magnetic separation membrane photocatalytic reactor (SMSMPR). *J. Hazard. Mater.* **2019**, *373*, 437–446. [CrossRef] [PubMed]
160. Chen, L.; Li, X.; Tanner, E.E.L.; Compton, R.G. Catechol adsorption on graphene nanoplatelets, isotherm, flat to vertical phase transition and desorption kinetics. *Chem. Sci.* **2017**, *8*, 4771–4778. [CrossRef] [PubMed]
161. Oyetibo, G.O.; Ishola, S.T.; Ikeda-Ohtsubo, W.; Miyauchi, K.; Ilori, M.O.; Endo, G. Mercury bioremoval by Yarrowia strains isolated from sediments of mercury polluted estuarine water. *Appl. Microbiol. Biotechnol.* **2015**, *99*, 3651–3657. [CrossRef]
162. Chakraborty, P.; Show, S.; Banerjee, S.; Halder, G. Mechanistic insight into sorptive elimination of ibuprofen employing bidirectional activated biochar from sugarcane bagasse: Performance evaluation and cost estimation. *J. Environ. Chem. Eng.* **2018**, *6*, 5287–5300. [CrossRef]
163. de Andrade, J.l.R.; Oliveira, M.F.; da Silva, M.G.; Vieira, M.G. Adsorption of pharmaceuticals from water and wastewater using nonconventional low-cost materials: A Review. *Ind. Eng. Chem. Res.* **2018**, *57*, 3103–3127. [CrossRef]
164. Gupta, V.K.; Ali, I. Removal of lead and chromium from wastewater using bagasse fly ash a sugar industry waste. *J. Colloid Interface Sci.* **2004**, *271*, 321–328. [CrossRef]
165. Ahmed, M.B.; Zhou, J.L.; Ngo, H.H.; Guo, W. Insight into biochar properties and its cost analysis. *Biomass Bioenergy* **2016**, *84*, 76–86. [CrossRef]
166. Gandhi, M.R.; Meenakshi, S. Recent advancement in heavy metal removal onto silica-based adsorbents and chitosan composites–A review. In *A Book on Ion Exchange, Adsorption and Solvent Extraction*; Naushad, M., Al-Othman, Z.A., Eds.; Nova Science Publishers, Inc.: New York, NY, USA, 2013; pp. 201–230.

 water

Article

Adsorption by Granular Activated Carbon and Nano Zerovalent Iron from Wastewater: A Study on Removal of Selenomethionine and Selenocysteine

Stanley Onyinye Okonji [1], Linlong Yu [2], John Albino Dominic [2], David Pernitsky [3] and Gopal Achari [2,*]

[1] Department of Mechanical & Manufacturing Engineering, University of Calgary, ICT 402, 2500 University Drive NW, Calgary, AB T2N 1N4, Canada; stanley.okonji@ucalgary.ca
[2] Department of Civil Engineering, University of Calgary, ENF 262, 2500 University Drive NW, Calgary, AB T2N 1N4, Canada; linyu@ucalgary.ca (L.Y.); johnalbino.dominic@ucalgary.ca (J.A.D.)
[3] Stantec, 200-325 25 Street SE, Calgary, AB T2A 7H8, Canada; David.Pernitsky@stantec.com
* Correspondence: gachari@ucalgary.ca

Abstract: Selenomethionine (SeMet) and selenocysteine (SeCys) are the most common forms of organic selenium, which is often found in the effluent of industrial wastewater. These organic selenium compounds are toxic, bioavailable and most likely to bioaccumulate in aquatic organisms. This study investigated the use of two adsorbent candidates (granular activated carbon (GAC) and nano zerovalent iron (nZVI)) as treatment technologies for SeMet and SeCys removal. Batch experiments were performed and inductively coupled plasma optical emission spectrometer (ICP-OES) was used for sample analysis. Experimental data showed GAC demonstrated a higher affinity towards the removal of SeMet and SeCys compared to nZVI. The removal efficiency of SeCys and SeMet by GAC was 96.1% and 86.7%, respectively. NZVI adsorption capacity for SeCys was 39.4% and SeMet < 1.1%. Irrespective of the adsorbent, SeMet is more refractory to be adsorbed compared to SeCys. Kinetics data of GAC and nZVI agreed well with the pseudo-second-order model ($R^2 > 0.990$). The experimental data of SeCys was characterized by Langmuir model, indicating monolayer adsorption. The adsorption capacity of nZVI for SeCys increased significantly by about 35%, with a decrease in pH from 9.0 to 4.0, indicating that SeCy removal by nZVI is pH dependent. While electrostatic attraction is considered the driving mechanism for nZVI adsorption, GAC uptake capacity is controlled by weak van der Waal forces. The adsorption of binary adsorbates (SeMet and SeCys) exhibited an inhibitory effect due to the competitive interaction between contaminant molecules.

Keywords: adsorption; nano zerovalent iron; granular activated carbon; organoselenium; water; wastewater; treatment

Citation: Okonji, S.O.; Yu, L.; Dominic, J.A.; Pernitsky, D.; Achari, G. Adsorption by Granular Activated Carbon and Nano Zerovalent Iron from Wastewater: A Study on Removal of Selenomethionine and Selenocysteine. *Water* **2021**, *13*, 23. https://dx.doi.org/10.3390/w13010023

Received: 26 November 2020
Accepted: 22 December 2020
Published: 25 December 2020

1. Introduction

Organoselenium simply refers to compounds that contain selenium (Se) in combination with other elements, such as carbon, oxygen, hydrogen, and nitrogen as part of their structure. The primary form of organic Se is selenoamino acids and selenoproteins [1]. SeMet and SeCys are the most common forms by which Se-amino acids exist in the environment and are uptaken by plants and aquatic organisms [2,3]. Naturally, SeMet and SeCys are organically bound in foods, such as nuts, yeast, eggs, liver, garlic [1,3]. In general, Se is a trace element essential for living organism physiological processes at a concentration range from 63–135 μg/L but exhibits toxicity outside this range of concentration [4].

In aqueous environments, organic selenium can be categorized as an emerging contaminant. It exists in the form of SeMet and SeCys in the effluent of industrial wastewaters [5], emanating from mostly mining, oil and gas refineries and coal-fired power plants. SeMet and SeCys are known to have higher bioavailability than inorganic selenium species as it is readily absorbed [3,5]. As a result, the ecotoxicological effect of organic Se (SeMet and SeCys) in aqueous environments is increased. One of the major concerns of organic selenium is the tendency to accumulate in aquatic organisms for an extended period of time, resulting in the contamination of fish and wildlife diets [6]. Public health is at risk if humans consume selenium-contaminated fish and wildlife [7]. The US EPA recently set environmental Se threshold regulations to be reliant on biotic tissue-based concentration rather than the traditional aqueous concentration limit [8]. This regulation is key to preventing organic Se propensity to bioaccumulate within the food-web. Therefore, it is essential to remove organic selenium contaminants from industrial wastewaters in order to decrease the bioaccumulation of selenium in aquatic life, the adverse impact on the ecosystem and the threat to public health. However, there are limited options available for the industries to meet this regulation.

To date, research on selenium remediation in aqueous media has largely focused on inorganic selenium removal. Various treatment technologies have been explored to remove Se from industrial wastewater. These are broadly classified into physical, chemical and biological treatment processes [9,10]. The following studies investigated these removal technologies: elemental selenium (Se^0) precipitation [11,12], iron coprecipitation methods [13], ion-exchange [14], adsorption using mineral adsorbents [15–21], coagulation [22], electrocoagulation [14] and photocatalytic reduction [23,24]. The biological treatment method is the most commonly used remediation technology for inorganic Se removal [10]. The process relies on the use of microbial mediation for selenium species removal [25–28]. However, the presence of SeMet has been reported in the effluents of an industrial biological treatment system [5]. It stands to reason that biological treatment techniques are among the primary sources of organic Se pollution in wastewaters. The transformation of inorganic selenium species to the organic form is mainly due to the microbial activities [5,29]. As a result, the existence of organic Se (SeMet and SeCys) in wastewaters remains an environmental contaminant of concern and the remediation techniques have rarely been researched.

So far, very few studies had investigated organic selenium removal from wastewaters. Alain (1997) investigated the removal of selenocyanate ($SeCN^-$) from sour crude oil produced wastewater by copper (II) salt precipitation [30]. Meng 2002 [31] studied the removal of selenocyanate ($SeCN^-$) from wastewater using Fe(0) filings through the formation of elemental selenium (Se^0). Sanna (2003) studied seleno-DL-methionine separation from inorganic selenium solution using magnesium-loaded activated charcoal [32]. However, besides our earlier study on SeMet removal [33], no previous research has investigated the removal of SeMet and SeCys from wastewater.

This study was conducted to determine the removal of SeMet and SeCys from wastewaters by (granular activated carbon (GAC) and nano zerovalent iron (nZVI)) adsorption, which have shown promise in treating inorganic selenium [34–36]. In this research, adsorption kinetics and isotherms of the two adsorbents were investigated. The effects of pH, initial adsorbate concentrations on adsorption capacities and the influence of binary adsorption were evaluated. SeMet and SeCys were selected as probe contaminants due to its toxicity and bioavailability in the wastewater, while the choice of adsorbents was driven by inexpensive cost and their being environmental benign [37,38]. This is the first study investigating the mechanism of GAC and nZVI to remove organic selenium from industrial wastewater. The knowledge gain in this study will benefit the development of efficient treatment processes for SeMet and SeCys using GAC and nZVI.

2. Materials and Methods

2.1. Chemicals

All chemicals employed in this study were used as received. Selenomethionine (>98%) and selenocysteine (>98%) were purchased from TCI America. Hydrochloric acid (HCl) (>98%), sodium hydroxide (NaOH) (>98%), nano zerovalent iron (nZVI 60–80 nm >99%) were procured from Sigma-Aldrich Chemical Company. Granular activated carbon was acquired from Evoqua (Pittsburgh).

2.2. Characterization of Adsorbents

The morphological properties of the adsorbents were obtained on a scanning electron microscope (SEM, ThermoFisher, Quanta FEG 250), as shown in Figure 1. BET surface area and total pore volume were obtained by N_2 adsorption at 77 K on a PMI Automated Brunauer–Emmett–Teller (BET), Quantachrome ChemBet (3000 CB-SCL).

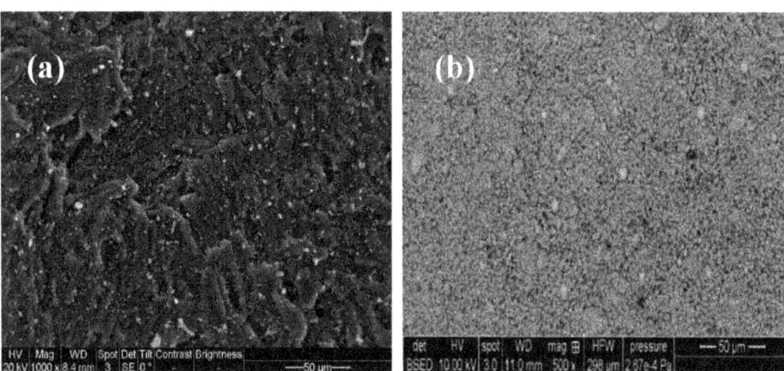

Figure 1. SEM images of granular activated carbon (**a**) and nZVI (**b**).

2.3. Batch Adsorption Studies

Selenocysteine ($C_3H_7NO_2Se$) and Selenomethionine ($C_5H_{11}NO_2Se$) stock solution were prepared separately by dissolving 0.5 g of both compounds in 100 mL of deionized water. The stock solution was diluted appropriately to obtain working solutions of various concentrations, as needed. The concentration of SeMet and SeCys in working solutions, used for adsorbent dose studies, pH studies, initial adsorbate concentrations and adsorption kinetics, was 5 mg/L. Table 1 illustrates the summary of experimental conditions with respect to the adsorbents and their respective concentrations used for evaluating adsorbent dose studies. The effect of pH on the adsorption of SeMet and SeCys was investigated by conducting experiments with different pH conditions (4.0, 7.0 and 9.0). Batch experiments were performed using straight-wall glass jars of 150 mL volume. A magnetic stirrer (VWR 200 model) was used to stir the solutions; the mixtures was agitated at a constant shaking speed of 180 rpm in a temperature-controlled orbital shaker. All experiments were conducted at ambient temperature. The initial pH of the solution was adjusted by adding HCl or NaOH (1 M). At different time intervals, aliquots of 10 mL samples were periodically taken and immediately filtered using 0.22 µm PTFE syringe filter. The collected samples were acidified using 50% nitric acid for a resulting strength of 2% and stored at 4 °C before analysis. For statistical reliability, all the experiments were conducted in duplicate, the samples were analyzed in triplicate and standard deviation was used for error analysis.

Table 1. Summary of experimental conditions for adsorbent dose.

	SeCys		SeMet	
Adsorbent Type	**GAC**	**nZVI**	**GAC**	**nZVI**
Dosage (g/L)	1	-	1	-
	2	2	2	-
	3	-	3	-
	5	-	5	-
	7	7	7	7
	14	14	14	-

2.4. Analysis and Equipment

The total Se concentration in the samples was analyzed using an inductively coupled plasma optical emission spectrometer (ICP-OES) (Thermo Scientific icap 7000 series). The limit of detection and quantitation for selenium was estimated as 0.005 and 0.01 mg/L, respectively. The pH fluctuations in the system with time were measured with a pH digital instrument symphony B20PI VWR. The analysis in this study did not distinguish between the species of organic selenium. The total selenium concentration was measured and analyzed.

2.5. Adsorption Kinetics

Adsorption kinetics were examined for an SeMet and SeCys initial concentration of 5 mg/L using nZVI and GAC adsorbent. An adsorbent loading rate of 7 g/L was used and the adsorption capacity q_e (mg/g) was calculated using the expression (Equation (1)):

$$q_e = \frac{C_o - C_e}{M} V \quad (1)$$

where C_o and C_e represent initial and equilibrium concentrations (mg/L), respectively; M is the mass of the adsorbent (g); V is the volume of the solution (L). A pseudo-second-order kinetic model was applied to the kinetic data; the mathematical expression is shown in Equation (2):

$$\frac{t}{q_t} = \frac{1}{k_2 q_e^2} + \frac{t}{q_e} \quad (2)$$

where q_e and q_t (mg/g) represent the amount of adsorbate adsorbed at equilibrium and time (t), respectively; k_2 is pseudo-second-order kinetic rate constant (g·mg^{-1}min^{-1}) and adsorption time (min). The initial adsorption rate h of the system is equal to $k_2 q_e^2$ (mg·g^{-1} min^{-1}). Equation (3) shows the pseudo-first-order kinetic model:

$$q_t = q_e\left(1 - e^{-kt}\right) \quad (3)$$

where k (min^{-1}) is the rate constant; other parameters in the expression have been defined above.

2.6. Adsorption Isotherm Studies

Adsorption isotherm studies were conducted with various initial concentrations (5 to 47.2 mg/L) for SeMet and SeCys. Freundlich and Langmuir isotherm models were used to evaluate the adsorption isotherm. The Freundlich model describes the relationship between the equilibrium concentration and the adsorption capacity. Equation (4) describes the nonlinear Freundlich adsorption isotherm model [39]:

$$\frac{(C_o - C_e)}{M} V = K_f C_e^{\frac{1}{n}} \quad (4)$$

where C_e is the equilibrium concentration (mg/L), C_o is the initial concentration (mg/L), M is the mass of adsorbent (g), V is the volume of the solution (L), K_f and n are Freundlich constants. The left-hand side of the Equation can be determined from experimental data, and it denotes the mass of adsorbate adsorbed per unit mass of adsorbent. The mathematical expression of Langmuir isotherm model [40] is given in Equation (5):

$$q_e = q_m K_c \frac{C_e}{1 + K_c C_e} \quad (5)$$

where q_e is the adsorption capacity at equilibrium (mgSe/g), q_m (mgSe/g) is the maximum adsorption capacity, K_c (L/mg) defined the equilibrium adsorption constants and C_e is the equilibrium concentration (mg/L). The Langmuir isotherm is based on the assumption that adsorption can occur at a finite number of specific localized sites (monolayer) [40]. Equation (6) can be used to evaluate the Langmuir model further [41].

$$R_L = \frac{1}{1 + K_c C_o} \quad (6)$$

The value of R_L defines the adsorption process as irreversible when ($R_L = 0$), favorable ($1 > R_L > 0$), unfavorable ($R_L > 1$) and linear ($R_L = 1$).

2.7. Parameter Study

The impact of various parameters, including adsorbent dose, pH and initial adsorbate concentration on organic selenium adsorption were evaluated. The experimental conditions were described in Section 2.3 and are summarized in Table 1 (adsorbent dose). The adsorbent amount of 7 g/L was chosen for all experiments based on the optimum dose determined in the dosage studies. For the binary adsorption experiment, a solution containing a mixture of SeMet and SeCys was used to study the effect of organic selenium coexistence. The initial concentration of the organoselenium was 5 mg/L each.

3. Results and Discussion

3.1. Adsorption of Organic Selenium by nZVI and GAC

The batch experiments result for the adsorption of organic selenium species (SeMet and SeCys) using 7 g/L of nZVI and GAC are presented in Figure 2. As shown, the removal of SeCys and SeMet by GAC is more effective than nZVI. After 3 h of adsorption experiments, it was observed that 96% of SeCys and 86.76% of SeMet was removed from water by activated carbon, while 39.44% of SeCys and less than 1.05% of SeMet was transferred to nZVI from the aqueous phase. Physisorption and chemisorption are the two major mechanisms for activated carbon adsorption. The former is caused by relatively weak van der Waals forces formed between the adsorbates and activated carbon's surface. In contrast, the latter is driven by a chemical reaction between the adsorbate molecules and the adsorbent surface [42].

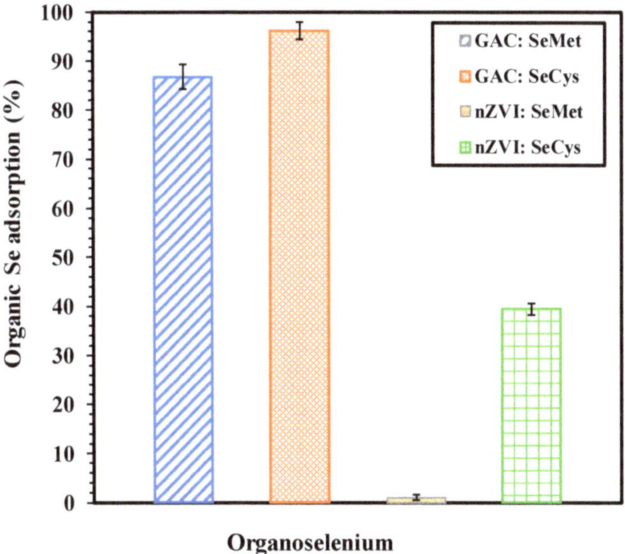

Figure 2. Comparison of organic selenium removal using 7 g/L of GAC and nZVI, pH 7.0, at 25 °C for 3 h period.

Conversely, the elemental iron (nZVI) removal mechanism for organic selenium (selenoyanate) is driven by a corrosion process, which removes the dense oxide layer and activates Fe(0) [31]. Meng [31] demonstrated that, when Fe(0) is mixed in water in the presence of dissolved oxygen (DO), Fe(0) is oxidized to ferrous ions. The oxidation process will give rise to ferric ions, which subsequently form loose ferric hydroxide in water. This phenomenon is referred to as the Fe(0) adsorption process [34], and it describes the nZVI removal mechanism of selenocyanates. In light of selenocyanate being an organic selenium compound and has similar characteristics, such as SeMet and SeCys, it is expected that a similar removal process may be applicable.

In comparison, GAC has better performance and can be attributed to the following reasons—(1) activated carbon possesses a microporous structure, which can lead to large active surface area (~1000 m^2/g) [43], with BET pore volume 0.500 (cm^3/g), and average pore width 2.138 (nm). The second reason is that activated carbon-oxygen surface functional groups (e.g., carboxylic and phenolic groups) can react with SeCys and SeMet to form chemical bonds. SeMet and SeCys are known to contain amino and a carboxylic acid; the functional groups can create hydrogen-bonding with activated carbon surface oxygen [44,45]. Additionally, selenium has an electronegativity similar to sulfur and is capable of forming strong hydrogen bonds identical to sulfur and oxygen [46].

On the other hand, nZVI adsorption was weak; it can be ascribed to the limited number of active sites. The surface area is much smaller than GAC, usually less than 100 m^2/g [47,48]. The adsorption of SeCys and SeMet on metals has not been reported in the literature. However, the mechanism can possibly be understood through the published studies on the interaction between metals, cysteine and methionine. Selenocysteine is an analogue of cysteine with selenium in place of the sulfur, while selenomethionine is analogue of methionine with selenium in place of the sulfur. Therefore, it is expected that the contaminants would have similar behavior and interactions with iron. Generally, the first steps in the adsorption of cysteine or methionine on an iron surface involve the replacement of one or more water molecules adsorbed at the iron surfaces, followed by the formation of iron-cysteine/methionine complexes, as shown in Equations (7)–(9) [49]. It is reported that cysteine can bond to the metal surface through sulfur, two oxygen, and a nitrogen atom in a four-point "quadrangular footprint", while methionine adsorbs on the surface with two oxygen and a nitrogen atom in a "triangular footprint" [50]. The sulfur atom within the methionine molecule does not interact with the metal surface. The reaction between the sulfur atom and the metal substrate can form a strong bond [50,51], indicating that cysteine's adsorption is stronger than methionine. Similar to methionine, the selenium atom in SeMet is not expected to react with iron's surface, hence no strong bond; which explains the weak adsorption of SeMet on the surface of iron compared to selenocysteine.

$$\text{Cystein/methione(sol)} + x\text{H}_2\text{O(ads)} \rightarrow \text{Cystein/methione (ads)} + x\text{H}_2\text{O(sol)} \quad (7)$$

$$\text{Fe} \rightarrow \text{Fe}^{2+} + 2e \quad (8)$$

$$\text{Fe}^{2+} + \text{Cystein/methione (ads)} \rightarrow [\text{Fe} - \text{Cystein/methione}]^{2+}\text{(ads)} \quad (9)$$

3.2. Adsorbent Dosage

Figure 3 presents the result of the adsorbent amount study on the uptake of SeMet and SeCys by GAC and nZVI with an initial concentration of 5 mg/L at pH 7.0. The removal percentage of organoselenium increased with an increase in the dose of GAC and nZVI. As shown in Figure 3, the effect of adsorbent dose on the organic selenium removal was more significant at the lower adsorbent loadings. Approximately 86.3% of SeCys and 57.1% of SeMet were removed by 1.0 g/L of GAC in 3 h. While 12.2% of SeCys was removed by nZVI within the same time frame. About 72.6% and 90% of SeMet was swiftly removed when the GAC dose was increased from 2 to 14 g/L, respectively. While 93% and 97.9% of SeCys were adsorbed from the solution with the same GAC loading. About 39.4% and

56.6% of SeCys were adsorbed by 7 and 14 g/L of nZVI, respectively. The increase in the adsorption capacity of the adsorbent candidates (GAC and nZVI) is attributed to the availability of greater surface area that drives an increase in the number of active adsorption sites; therefore, resulting in a higher removal rate [35,38]. In contrast to nZVI adsorption, an optimum dose of 7 g/L of GAC significantly adsorbed both contaminants.

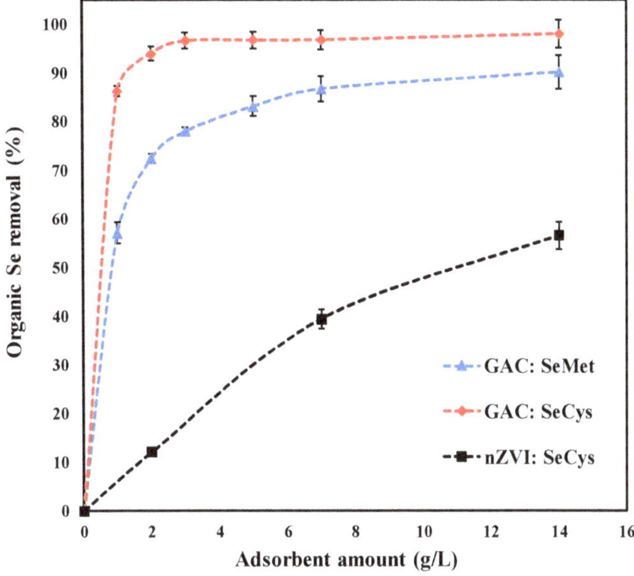

Figure 3. Effect of adsorbent dosage on organic selenium removal, by GAC and nZVI (initial concentration of 5 mg/L, pH 7.0, at 25 °C).

3.3. *Effect of pH*

Figure 4 depicts the results of experiments conducted to determine the pH influence on organic selenium adsorption by GAC and nZVI. The solutions pH level was measured at pH 4.0, 7.0 and 9.0 during mixing for 3 h. As shown in Figure 4, the amount of SeMet and SeCys adsorbed onto GAC for acidic, neutral and alkaline pH was significant. The removal efficiency of SeCys by GAC occurred in the order of 94.7%, 96.1% and 96.8% for pH 4.0, 7.0 and 9.0, respectively. The acidic condition was observed to be slightly less favorable for the adsorption of SeMet by GAC, corresponding to 80.3% removal. The removal of SeMet at pH 7.0 and 9.0 solution was 86.7% and 86.6%, respectively. SeCys and SeMet are zwitterions, containing both amino groups and carboxyl groups. The isoelectric point (Ip) for SeMet and SeCys are 5.75 and 5.54, respectively [51,52]. Increasing the pH from 4.0 to 9.0 would change the net charge of SeCys and SeMet solution from positive to negative and also influence the surface charge of activated carbon, leading to different electrostatic interactions between SeCys and SeMet molecules and activated carbon. In this study, the investigated pHs insignificantly impacted the adsorption of the two organic selenium compounds by GAC, indicating that the adsorption of SeCys and SeMet by GAC was not dominated by electrostatic force. The adsorption of SeCys and SeMet could potentially occur as a result of the hydrogen bonding or the hydrophobic interaction that exists between the hydrophobic part of SeCys and SeMet molecules and the hydrophobic part of the adsorbent, which have been reported as mechanisms for the adsorption of amino acids on the surface of activated carbon [53].

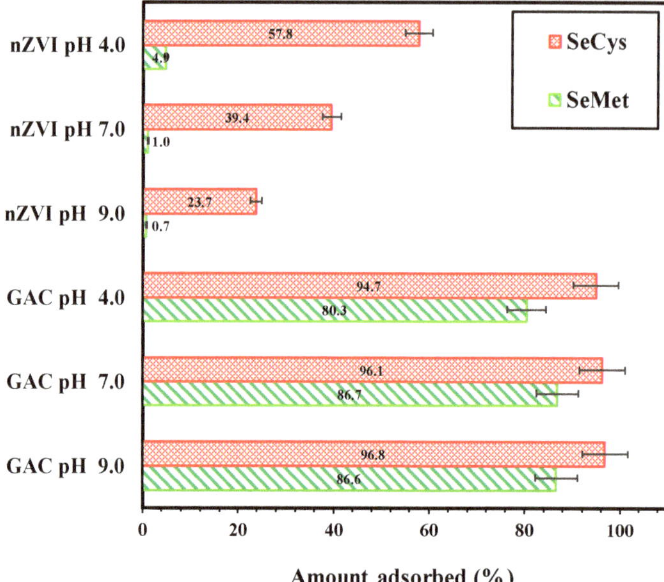

Figure 4. Effect of pH on SeMet and SeCys adsorption by GAC and nZVI, (initial concentration 5 mg/L, 7 g/L, pH 4.0, 7.0 and 9.0, at 25 °C).

On the other hand, the removal of SeCys and SeMet by nZVI decreased remarkably with increasing pH, as shown in Figure 4. The percentages of SeCys adsorbed on nZVI under different pHs are 57.8% for pH 4.0, 39.4% for pH 7.0 and 23.7% for pH 9.0. SeMet removal rate decreased from 4.9% to less than 1% when pH was increased from 4.0 to 9.0, indicating a negative effect of an increase in pH condition. Similar findings have been reported on selenocyanate adsorption (an organic selenium species) by nZVI [31]. Meng et al. [31] demonstrated that selenocyanate's removal rate increased from 50% to 97% when pH was decreased from 8.5 to 5.5.

3.4. Effect of Initial Concentration

GAC performance in treating water containing different concentrations of SeMet and SeCys at pH 7.0 is presented in Figure 5. For SeMet, 4.9 mg/L and 46.8 mg/L of initial concentration were evaluated, while SeCys, 3.8 and 38 mg/L of initial concentration were examined. Both forms of organoselenium were adsorbed continuously as a function of time until equilibrium was reached. However, SeMet appears to have some desorption after 1 h, which was typical in both concentrations (4.9 and 46.8 mg/L). The pseudo-second-order kinetic model adequately describes the data depicted in Table 2. The correlation coefficients (R^2) value was evaluated between 0.999–1.000 (Table 2). As illustrated in Table 2, the rate constant is between 2.50 to 1.80 for SeMet C_o (4.9 and 46.8 mg/L) and 1.30 to 0.25 for SeCys C_o (3.8 and 38 mg/L), respectively. The result demonstrates that the rate constant (K_2) decreased with an increase in SeMet and SeCys concentration. When the initial concentration is increased, the solution takes more time to attain equilibrium. The decrease in K_2, with respect to an increase in the initial concentration, can be attributed to a longer duration that is required for the solution to attain equilibrium; a similar phenomenon has been reported in the literature [54,55].

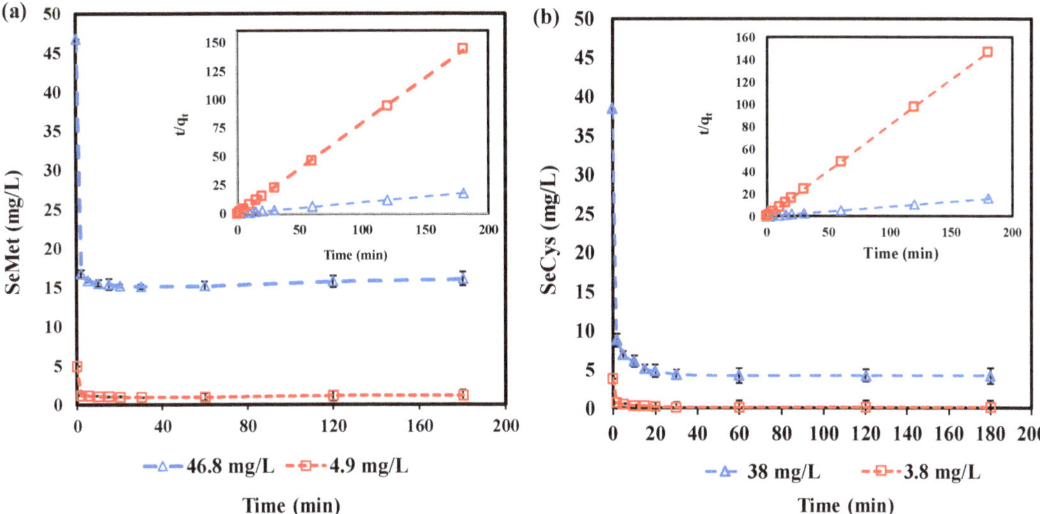

Figure 5. Effect of initial concentration by GAC: (**a**) SeMet (C_o: 46.8 and 4.9 mg/L); (**b**) SeCys (C_o: 38 and 3.8 mg/L); all experiments were conducted at pH 7.0, 7 g/L and 25 °C.

Table 2. Experimental data for adsorbate initial concentration studies with GAC adsorbent, at pH 7.0.

	SeMet		SeCys	
C_o (mg/L)	4.9	46.8	3.8	38
q_e (mg·g^{-1})	1.32	10.54	1.23	11.40
K_2 (g·mg^{-1}·min^{-1})	2.50	1.80	1.30	0.25
R^2	1.000	0.999	1.000	1.000

Furthermore, the experimental data also show that q_e increases with higher C_o, the increment was significant about eight times higher as the initial concentrations of SeMet and SeCys were varied. It can be deduced that the increase in C_o provides the driving force to overcome the mass transfer resistance between the adsorbate and solid phases [56]. It was observed that GAC completely removed 3.8 mg/L of SeCys in 30 min, while the initial concentration of 38 mg/L was reduced to 4.2 mg/L in 3 h. About 30.7 and 3.6 mg/L of SeMet were adsorbed from the initial concentration of 46.8 and 4.9 mg/L, respectively. An upsurge in adsorption capacity as a result of an increase in adsorbate initial concentration has been reported by Aksu [57] and Al-Ghouti [58].

3.5. Adsorption Isotherm

The adsorption data for GAC were evaluated using Langmuir and Freundlich isotherm models. The results of both Langmuir and Freundlich models for SeMet and SeCys adsorption are summarized in Table 3. Freundlich and Langmuir models fitted well with the experimental data ($R^2 > 0.950$). Figure 6a shows that the Freundlich model provided a better fit for SeMet adsorption data compared to the Langmuir model. The correlation coefficient ($R^2 > 0.996$) was higher for the Freundlich model, suggesting that the adsorption of SeMet might occur in multilayers. The parameter K_f of the Freundlich model was calculated to be 1.85, which is related to the adsorption capacity. As reflected in Table 3, the constant n, representing the adsorption intensity, was equal to 1.55, indicating pseudo linear adsorption [59].

Table 3. Adsorption isotherm parameters for organoselenium by GAC.

	Langmuir Model			Freundlich Model		
	q_m (mg/g)	K_c (L/mg)	R^2	K_f	n	R^2
SeMet	18.9	0.10	0.967	1.85	1.55	0.996
SeCys	10.0	0.32	0.999	2.52	2.14	0.950

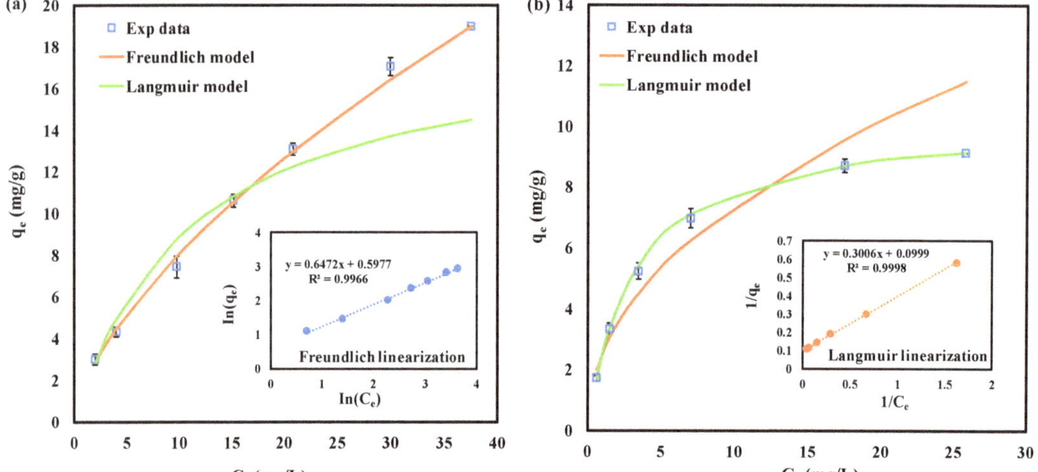

Figure 6. Organic selenium adsorption isotherms by GAC, pH 7.0, at 25 °C. (**a**) SeMet; (**b**) SeCys. The solid lines represent model fits; inset shows a linearization fit.

On the other hand, SeCys adsorption by GAC with corresponding Langmuir plots is presented in Figure 6b. Freundlich isotherm was also used to normalize the adsorption data. As stated in Table 3, the isotherm data fitted better with the Langmuir model (higher coefficients of determination R^2 = 0.990) in contrast to the Freundlich model, suggesting that the adsorption of SeCys is monolayer and it occurred at localized sites [53,57]. The maximum uptake capacity for the Langmuir parameter q_m is 10 mg/g. The dimensionless constant $R_L = 0.082$, an indication that SeCys adsorption process can be considered as favorable [41].

3.6. Adsorption Kinetics

Figure 7a shows the time-dependent data of SeMet and SeCys adsorption by GAC. As shown in the results (Figure 7a), the equilibrium time for SeMet was 20 min, while SeCys took longer than 50 min to achieve equilibrium, indicating that the SeCys adsorption process was progressive. The pseudo-second-order (PSO) kinetics model (inset of Figure 6a) was used to investigate the adsorption kinetics, the rate constant (k_2) values and the maximum adsorption capacity (q_e) for SeMet and SeCys, as presented in Table 4. The correlation coefficients ($R^2 > 0.999$) suggest that the PSO model was a good fit for the adsorbents tested. However, when parameter h, which accounts for the initial adsorption rate, was calculated, the value shows that SeCys adsorption was faster than SeMet adsorption (h value of SeMet is about 33% less than SeCys).

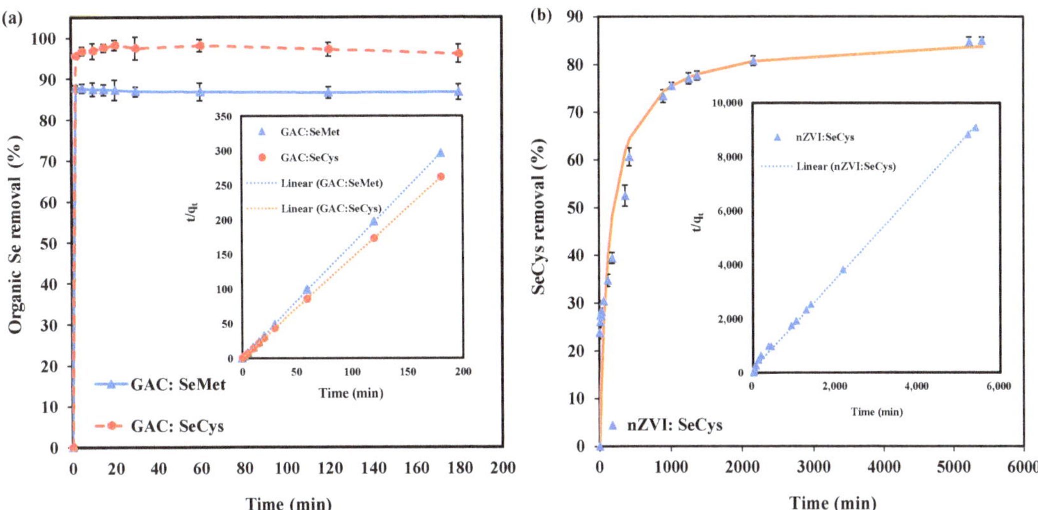

Figure 7. (**a**) Organic selenium adsorption kinetics using 7 g/L of GAC. (**b**) SeCys adsorption kinetics using 7 g/L of nZVI. All experiments were conducted at pH 7.0, 25 °C with an initial concentration of 5 mg/L.

Table 4. Kinetic parameters of pseudo-second-order models for SeMet and SeCys adsorption.

Kinetic Parameter	pH	SeMet	SeCys	
		GAC	nZVI	GAC
q_e (mg·g^{-1})	7	0.61	0.59	0.70
k_2 (g·mg^{-1}·min^{-1})		24	0.012	28
R^2		1.000	0.998	0.999
h (mg·g^{-1}·min^{-1})		8.81	0.004	13.27

The adsorption kinetics of SeCys by nZVI are depicted in Figure 7b. The experiment was performed for 54 h to allow enough time for equilibrium. However, equilibrium was not achieved because the adsorption process was very slow (Table 4). The pseudo-second-order kinetics model showed a good fit for SeCys adsorption data (R^2 > 0.998). Contrary to GAC adsorption, nZVI removal efficiency was found to increase slowly until a final adsorption efficiency of 85% was achieved. Overall, GAC kinetics was very swift compared to nZVI, which can be explained as an outcome of a more significant adsorption site [43].

3.7. Binary Adsorption of SeMet and SeCys

The effect of the coexistence of SeMet and SeCys on wastewaters was investigated using 7 g/L of GAC and nZVI. The initial concentration of organic selenium in the mixture was 10 mg/L (comprising 5 mg/L of SeMet and SeCys each), and the pH of the solution was 7.0. The result presented in Figure 8 shows that GAC and nZVI removed 8.9 mg/L and 1.5 mg/L of Se from the mixture of SeMet and SeCys, respectively, in 3 h. On an individual basis, GAC removed 4.8 mg/L of Se from SeCys solution and 4.3 mg/L of Se from SeMet solution. While nZVI adsorbed 1.9 mg/L of Se from SeCys and 0.1 mg/L of Se from SeMet. The result shows that the total Se removed by GAC from the mixture of SeMet and SeCys solution was lower compared to the sum of Se adsorbed in a single solute system. This phenomenon is an indication of competition between different molecules for available adsorption sites on GAC surface [60,61]. Conversely, the adsorption of SeMet by nZVI in a single solute system was very weak, indicating that the competing effect for active sites on the surface of nZVI was negligible. The inhibition effect of SeMet on the adsorption of

SeCys by GAC and nZVI can possibly occur as a result of the interaction between SeCys and SeMet in the binary system. Organic selenium speciation was not determined in this study; hence, the selenium species removed from the mixture was unknown.

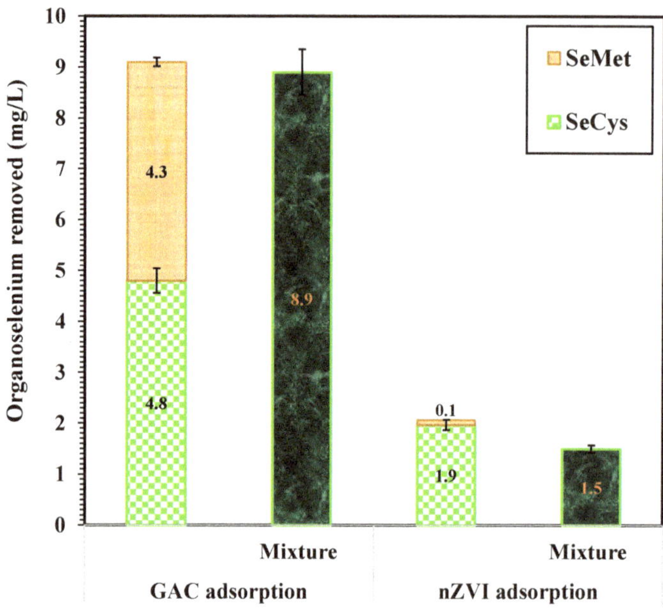

Figure 8. Binary adsorption of SeMet and SeCys by nZVI and GAC, initial concentration of 10 mg/L, 7 g/L at pH 7.0 and 25 °C.

4. Conclusions

The adsorption of SeMet and SeCys by GAC and nZVI under various conditions—different pHs, adsorbate initial concentration, adsorbent dosage, and binary adsorption were investigated. GAC demonstrated a higher affinity towards the removal of SeMet and SeCys, and was, therefore, considered a better adsorbent candidate. An optimum dose of 7 g/L of GAC was found to remove both contaminants effectively. Change in pH had no significant impact on SeCys removal by GAC; nevertheless, more than 93.99% removal was achieved at all pH tested. In the case of nZVI, pH change substantially influenced the adsorption capacity—at pH 4.0, about 57.8% of SeCys was removed. An increase in adsorption capacity with a decrease in pH value was observed for SeMet removal by nZVI. SeCys adsorbed more readily into nZVI compared to SeMet—for all conditions evaluated, SeMet removal by nZVI was less than 5%. The pseudo-second-order kinetics model characterized the adsorption of organoselenium by both adsorbents. The fastest adsorption kinetics was observed with GAC under neutral pH, where an instantaneous removal of organic selenium was observed. Binary adsorption of SeCys and SeMet indicates an inhibitory effect on SeCys removal by SeMet. The adsorption data fitted well with both Langmuir and Freundlich isotherm models.

Author Contributions: Conceptualization, S.O.O., G.A. and D.P.; methodology, S.O.O.; experimental design, S.O.O. and L.Y.; formal analysis, S.O.O. and J.A.D.; investigation, S.O.O.; resources, G.A; writing—original draft preparation, S.O.O.; writing—review and editing, G.A, L.Y., D.P. and J.A.D.; supervision, G.A; project administration, J.A.D.; funding acquisition, G.A. All authors have read and agreed to the published version of the manuscript.

Funding: Natural Sciences and Engineering Research Council of Canada (NSERC).

Data Availability Statement: The data presented in this study are available on request from the corresponding author.

Acknowledgments: The authors gratefully acknowledge the funding support provided by National Sciences and Engineering Research Council of Canada (NSERC) through an Engage grant.

Conflicts of Interest: The authors declare no conflict of interest.

References

1. Maseko, T.; Callahan, D.L.; Dunshea, F.R.; Doronila, A.; Kolev, S.D.; Ng, K. Chemical characterisation and speciation of organic selenium in cultivated selenium-enriched Agaricus bisporus. *Food Chem.* **2013**, *141*, 3681–3687. [CrossRef] [PubMed]
2. Dumont, E.; Vanhaecke, F.; Cornelis, R. Selenium speciation from food source to metabolites: A critical review. *Anal. Bioanal. Chem.* **2006**, *385*, 1304–1323. [CrossRef] [PubMed]
3. Amoako, P.O.; Uden, P.C.; Tyson, J.F. Speciation of selenium dietary supplements; formation of S-(methylseleno) cysteine and other selenium compounds. *Anal. Chim. Acta* **2009**, *652*, 315–323. [CrossRef] [PubMed]
4. Albert, M.; Demesmay, C.; Rocca, J. Analysis of organic and non-organic arsenious or selenious compounds by capillary electrophoresis. *Fresenius' J. Anal. Chem.* **1995**, *351*, 426–432. [CrossRef]
5. LeBlanc, K.L.; Wallschläger, D. Production and release of selenomethionine and related organic selenium species by microorganisms in natural and industrial waters. *Environ. Sci. Technol.* **2016**, *50*, 6164–6171. [CrossRef] [PubMed]
6. Lemly, A.D. Environmental implications of excessive selenium: A review. *Biomed. Environ. Sci.* **1997**, *10*, 415–435.
7. Fan, A.M.; Book, S.A.; Neutra, R.R.; Epstein, D.M. Selenium and human health implications in California's San Joaquin Valley. *J. Toxicol. Environ. Health Part A Curr. Issues* **1988**, *23*, 539–559. [CrossRef]
8. Delos, C. *Draft Aquatic Life Water Quality Criteria for Selenium*; US Environmental Protection Agency: Washington, DC, USA, 2004.
9. CH2M HILL. Review of available technologies for the removal of selenium from water. In *Final Report, Prepared for North American Metals Council (NAMC)*; CH2M HILL: Charlotte, NC, USA; Bellevue, WA, USA, 2010.
10. Khamkhash, A.; Srivastava, V.; Ghosh, T.; Akdogan, G.; Ganguli, R.; Aggarwal, S. Mining-related selenium contamination in Alaska, and the state of current knowledge. *Minerals* **2017**, *7*, 46. [CrossRef]
11. Zhang, Y.; Amrhein, C.; Frankenberger, W.T., Jr. Effect of arsenate and molybdate on removal of selenate from an aqueous solution by zero-valent iron. *Sci. Total Environ.* **2005**, *350*, 1–11. [CrossRef]
12. Mondal, K.; Jegadeesan, G.; Lalvani, S.B. Removal of selenate by Fe and NiFe nanosized particles. *Ind. Eng. Chem. Res.* **2004**, *43*, 4922–4934. [CrossRef]
13. Montgomery, J.M.; Engineers, C. *Water Treatment Principles and Design*; Wiley: New York, NY, USA, 1985.
14. Mavrov, V.; Stamenov, S.; Todorova, E.; Chmiel, H.; Erwe, T. New hybrid electrocoagulation membrane process for removing selenium from industrial wastewater. *Desalination* **2006**, *201*, 290–296. [CrossRef]
15. Balistrieri, L.S.; Chao, T. Selenium Adsorption by Goethite. *Soil Sci. Soc. Am. J.* **1987**, *51*, 1145–1151. [CrossRef]
16. Balistrieri, L.S.; Chao, T. Adsorption of selenium by amorphous iron oxyhydroxide and manganese dioxide. *Geochim. Cosmochim. Acta* **1990**, *54*, 739–751. [CrossRef]
17. Zhang, Y.; Frankenberger, W.T. Factors affecting removal of selenate in agricultural drainage water utilizing rice straw. *Sci. Total Environ.* **2003**, *305*, 207–216. [CrossRef]
18. Peak, D. Adsorption mechanisms of selenium oxyanions at the aluminum oxide/water interface. *J. Colloid Interface Sci.* **2006**, *303*, 337–345. [CrossRef]
19. El-Shafey, E. Sorption of Cd (II) and Se (IV) from aqueous solution using modified rice husk. *J. Hazard. Mater.* **2007**, *147*, 546–555. [CrossRef]
20. Zhang, N.; Gang, D.; Lin, L.-S. Adsorptive removal of parts per million level Selenate using iron-coated GAC Adsorbents. *J. Environ. Eng.* **2010**, *136*, 1089–1095. [CrossRef]
21. Zhang, N.; Lin, L.-S.; Gang, D. Adsorptive selenite removal from water using iron-coated GAC adsorbents. *Water Res.* **2008**, *42*, 3809–3816. [CrossRef]
22. Hu, C.; Chen, Q.; Chen, G.; Liu, H.; Qu, J. Removal of Se (IV) and Se (VI) from drinking water by coagulation. *Sep. Purif. Technol.* **2015**, *142*, 65–70. [CrossRef]
23. Tan, T.; Beydoun, D.; Amal, R. Effects of organic hole scavengers on the photocatalytic reduction of selenium anions. *J. Photochem. Photobiol. A Chem.* **2003**, *159*, 273–280. [CrossRef]
24. Nguyen, V.N.H.; Beydoun, D.; Amal, R. Photocatalytic reduction of selenite and selenate using TiO_2 photocatalyst. *J. Photochem. Photobiol. A Chem.* **2005**, *171*, 113–120. [CrossRef]
25. Zhang, Y.; Moore, J.N. Environmental conditions controlling selenium volatilization from a wetland system. *Environ. Sci. Technol.* **1997**, *31*, 511–517. [CrossRef]
26. Amweg, E.; Stuart, D.; Weston, D. Comparative bioavailability of selenium to aquatic organisms after biological treatment of agricultural drainage water. *Aquat. Toxicol.* **2003**, *63*, 13–25. [CrossRef]
27. Zhang, Y.; Okeke, B.C.; Frankenberger, W.T., Jr. Bacterial reduction of selenate to elemental selenium utilizing molasses as a carbon source. *Bioresour. Technol.* **2008**, *99*, 1267–1273. [CrossRef] [PubMed]

28. Presser, T.S.; Ohlendorf, H.M. Biogeochemical cycling of selenium in the San Joaquin Valley, California, USA. *Environ. Manag.* **1987**, *11*, 805–821. [CrossRef]
29. Ohlendorf, H.M. Bioaccumulation and Effects of Selenium in Wildlife. *Selenium Agric. Environ.* **1989**, *23*, 133–177.
30. Manceau, A.; Gallup, D.L. Removal of Selenocyanate in Water by Precipitation: Characterization of Copper−Selenium Precipitate by X-ray Diffraction, Infrared, and X-ray Absorption Spectroscopy. *Environ. Sci. Technol.* **1997**, *31*, 968–976. [CrossRef]
31. Meng, X.; Bang, S.; Korfiatis, G.P. Removal of selenocyanate from water using elemental iron. *Water Res.* **2002**, *36*, 3867–3873. [CrossRef]
32. Latva, S.; Peräniemi, S.; Ahlgrén, M. Study of metal-loaded activated charcoals for the separation and determination of selenium species by energy dispersive X-ray fluorescence analysis. *Anal. Chim. Acta* **2003**, *478*, 229–235. [CrossRef]
33. Okonji, S.O.; Dominic, J.A.; Pernitsky, D.; Achari, G. Removal and recovery of selenium species from wastewater: Adsorption kinetics and co-precipitation mechanisms. *J. Water Process Eng.* **2020**, *38*, 101666. [CrossRef]
34. Yoon, I.-H.; Kim, K.-W.; Bang, S.; Kim, M.G. Reduction and adsorption mechanisms of selenate by zero-valent iron and related iron corrosion. *Appl. Catal. B Environ.* **2011**, *104*, 185–192. [CrossRef]
35. Wasewar, K.L.; Prasad, B.; Gulipalli, S. Removal of selenium by adsorption onto granular activated carbon (GAC) and powdered activated carbon (PAC). *CLEAN–Soil Air Water* **2009**, *37*, 872–883. [CrossRef]
36. Das, S.; Lindsay, M.B.; Essilfie-Dughan, J.; Hendry, M.J. Dissolved selenium (VI) removal by zero-valent iron under oxic conditions: Influence of sulfate and nitrate. *ACS Omega* **2017**, *24*, 1513–1522. [CrossRef] [PubMed]
37. Zhang, Y.; Wang, J.; Amrhein, C.; Frankenberger, W.T. Removal of selenate from water by zerovalent iron. *J. Environ. Qual.* **2005**, *34*, 487–495. [CrossRef] [PubMed]
38. Liang, L.; Jiang, X.; Yang, W.; Huang, Y.; Guan, X.; Li, L. Kinetics of selenite reduction by zero-valent iron. *Desalination Water Treat.* **2015**, *53*, 2540–2548. [CrossRef]
39. Zelmanov, G.; Semiat, R. Selenium removal from water and its recovery using iron (Fe3+) oxide/hydroxide-based nanoparticles sol (NanoFe) as an adsorbent. *Sep. Purif. Technol.* **2013**, *103*, 167–172. [CrossRef]
40. Langmuir, I. The constitution and fundamental properties of solids and liquids. Part I. Solids. *J. Am. Chem. Soc.* **1916**, *38*, 2221–2295. [CrossRef]
41. Zeng, G.; Liu, Y.; Tang, L.; Yang, G.; Pang, Y.; Zhang, Y.; Zhou, Y.; Li, Z.; Li, M.; Lai, M. Enhancement of Cd (II) adsorption by polyacrylic acid modified magnetic mesoporous carbon. *Chem. Eng. J.* **2015**, *259*, 153–160. [CrossRef]
42. Bansal, R.; Goyal, M. Activated carbon adsorption and environment: Adsorptive removal of organic from water. In *Activated Carbon Adsorption*; Taylor & Francis: Boca Raton, FL, USA, 2005; pp. 297–372.
43. Yang, Y.; Yu, L.; Iranmanesh, S.; Keir, I.; Achari, G. Laboratory and Field Investigation of Sulfolane Removal from Water Using Activated Carbon. *J. Environ. Eng.* **2020**, *146*, 04020022. [CrossRef]
44. Ahnert, F.; Arafat, H.A.; Pinto, N.G. A study of the influence of hydrophobicity of activated carbon on the adsorption equilibrium of aromatics in non-aqueous media. *Adsorption* **2003**, *9*, 311–319. [CrossRef]
45. López-Velandia, C.; Moreno-Barbosa, J.J.; Sierra-Ramirez, R.; Giraldo, L.; Moreno-Piraján, J.C. Adsorption of volatile carboxylic acids on activated carbon synthesized from watermelon shells. *Adsorpt. Sci. Technol.* **2014**, *32*, 227–242. [CrossRef]
46. Mundlapati, V.R.; Sahoo, D.K.; Ghosh, S.; Purame, U.K.; Pandey, S.; Acharya, R.; Pal, N.; Tiwari, P.; Biswal, H.S. Spectroscopic evidence for strong hydrogen bonds with selenomethionine in proteins. *J. Phys. Chem. Lett.* **2017**, *8*, 794–800. [CrossRef] [PubMed]
47. Adusei-Gyamfi, J.; Acha, V. Carriers for nano zerovalent iron (nZVI): Synthesis, application and efficiency. *RSC Adv.* **2016**, *6*, 91025–91044. [CrossRef]
48. Wang, Q.; Kanel, S.R.; Park, H.; Ryu, A.; Choi, H. Controllable synthesis, characterization, and magnetic properties of nanoscale zerovalent iron with specific high Brunauer–Emmett–Teller surface area. *J. Nanoparticle Res.* **2009**, *113*, 749–755. [CrossRef]
49. Oguzie, E.; Li, Y.; Wang, F. Corrosion inhibition and adsorption behavior of methionine on mild steel in sulfuric acid and synergistic effect of iodide ion. *J. Colloid Interface Sci.* **2007**, *310*, 90–98. [CrossRef] [PubMed]
50. Thomsen, L.; Wharmby, M.; Riley, D.; Held, G.; Gladys, M. The adsorption and stability of sulfur containing amino acids on Cu {5 3 1}. *Surf. Sci.* **2009**, *603*, 1253–1261. [CrossRef]
51. Mishra, B.; Priyadarsini, K.; Mohan, H. One-Electron Oxidation of Selenomethionine in Aqueous Solutions. *Barc Newslett.* **2005**, *261*, 115.
52. Bettelheim, F.A.; Brown, W.H.; Campbell, M.K.; Farrell, S.O.; Torres, O. *Introduction to General, Organic and Biochemistry*; Nelson Education: Toronto, ON, Canada, 2012.
53. Cermakova, L.; Kopecka, I.; Pivokonsky, M.; Pivokonska, L.; Janda, V. Removal of cyanobacterial amino acids in water treatment by activated carbon adsorption. *Sep. Purif. Technol.* **2017**, *173*, 330–338. [CrossRef]
54. Gupta, S.S.; Bhattacharyya, K.G. Kinetics of adsorption of metal ions on inorganic materials: A review. *Adv. Colloid Interface Sci.* **2011**, *162*, 39–58. [CrossRef]
55. Allen, S.J.; Gan, Q.; Matthews, R.; Johnson, P.A. Kinetic modeling of the adsorption of basic dyes by kudzu. *J. Colloid Interface Sci.* **2005**, *286*, 101–109. [CrossRef]
56. Wang, W.; Wang, J.; Guo, Y.; Zhu, C.; Pan, F.; Wu, R.; Wang, C. Removal of multiple nitrosamines from aqueous solution by nanoscale zero-valent iron supported on granular activated carbon: Influencing factors and reaction mechanism. *Sci. Total Environ.* **2018**, *639*, 934–943. [CrossRef] [PubMed]

57. Aksu, Z.; Dönmez, G. A comparative study on the biosorption characteristics of some yeasts for Remazol Blue reactive dye. *Chemosphere* **2003**, *50*, 1075–1083. [CrossRef]
58. Al-Ghouti, M.A.; Khraisheh, M.A.; Ahmad, M.N.; Allen, S. Adsorption behaviour of methylene blue onto Jordanian diatomite: A kinetic study. *J. Hazard. Mater.* **2009**, *165*, 589–598. [CrossRef] [PubMed]
59. Tseng, R.-L.; Wu, F.-C. Inferring the favorable adsorption level and the concurrent multi-stage process with the Freundlich constant. *J. Hazard. Mater.* **2008**, *155*, 277–287. [CrossRef]
60. Jain, J.S.; Snoeyink, V.L. Adsorption from bisolute systems on active carbon. *J. Water Pollut. Control Fed.* **1973**, *45*, 2463–2479.
61. Noroozi, B.; Sorial, G.; Bahrami, H.; Arami, M. Adsorption of binary mixtures of cationic dyes. *Dyes Pigment.* **2008**, *76*, 784–791. [CrossRef]

Article

Removal of Arsenic(III) Ion from Aqueous Media Using Complex Nickel-Aluminum and Nickel-Aluminum-Zirconium Hydroxides

Fumihiko Ogata [1], Noriaki Nagai [1], Megumu Toda [2], Masashi Otani [2], Chalermpong Saenjum [3,4], Takehiro Nakamura [1] and Naohito Kawasaki [1,5,*]

1 Faculty of Pharmacy, Kindai University, 3-4-1 Kowakae, Higashi-Osaka, Osaka 577-8502, Japan; ogata@phar.kindai.ac.jp (F.O.); nagai_n@phar.kindai.ac.jp (N.N.); nakamura@phar.kindai.ac.jp (T.N.)
2 Kansai Catalyst Co. Ltd., 1-3-13, Kashiwagi-cho, Sakai-ku, Sakai, Osaka 590-0837, Japan; megumu.toda@kansyoku.co.jp (M.T.); masashi.ootani@kansyoku.co.jp (M.O.)
3 Faculty of Pharmacy, Chiang Mai University, Suthep Road, Muang District, Chiang Mai 50200, Thailand; chalermpong.saenjum@gmail.com
4 Cluster of Excellence on Biodiversity-based Economics and Society (B.BES-CMU), Chiang Mai University, Suthep Road, Muang District, Chiang Mai 50200, Thailand
5 Antiaging Center, Kindai University, 3-4-1 Kowakae, Higashi-Osaka, Osaka 577-8502, Japan
* Correspondence: kawasaki@phar.kindai.ac.jp; Tel.: +81-6-4307-4012

Received: 12 May 2020; Accepted: 12 June 2020; Published: 14 June 2020

Abstract: The technology of wastewater treatment involving removal of heavy metals using complex metal hydroxides is reported. In this study, complex nickel-aluminum (NA11 and NA12) and nickel-aluminum-zirconium (NAZ1 and NAZ2) hydroxides were prepared for the removal of arsenite ions, As(III), from aqueous solution. The characteristics of each adsorbent were evaluated, and the adsorption capacity and adsorption mechanism were determined. The adsorption capacity of As(III) on NAZ1 (15.3 mg g^{-1}) was greater than that on NA11 (9.3 mg g^{-1}). Coverage is directly related to the specific surface area with a correlation coefficient of 0.921. Ion exchange involving sulfate ions in the interlayer of the adsorbent also plays a role in the mechanism of As(III) adsorption as demonstrated by correlation coefficients of 0.797 and 0.944 for the NA11 and NAZ1, respectively. The results demonstrate the usefulness of NAZ1 in removing As(III) from aqueous media.

Keywords: complex nickel-aluminum-zirconium hydroxide; arsenic ion; adsorption

1. Introduction

The United Nations has adopted 17 goals as part of its 2030 agenda to develop a sustainable future for all. Issues of clean water and sanitation (Goal No. 6) and undersea water (Goal No. 14) are particularly relevant to the development of an acceptable global water environment [1]. Heavy metals and their compounds are useful in many areas of human endeavor [2]. However, their excessive use leads to an increasing presence in aqueous environments [3]. Arsenic (As) pollution in aquatic systems is a global issue owing to its high toxicity and chronic effects on human health [4,5]. Arsenic pollution in drinking water is a major concern in developing countries. Previous studies have identified a relationship between arsenic exposure and diseases such as asthma and hyperglycemia [6,7]. Arsenic and inorganic arsenic compounds are classified as Group 1 human carcinogens by the International Agency for Research on Cancer [8]. The World Health Organization has established 0.01 mg L^{-1} as the maximum permissible arsenic content in potable water to protect human health [9]. The predominant forms of inorganic arsenic in aquatic systems are As(V) (arsenate) and As(III) (arsenite). Arsenic is a ubiquitous element, occurring from natural sources (such as wearing of arsenic bearing minerals and volcanic eruptions) and anthropogenic activities (such as mining, glassware production, dyes, drugs, and pesticide industries) [10–12]. As(III) is highly toxic, soluble, and mobile compared to As(V) [13,14]. Therefore, there is a need for simple, efficient, and inexpensive techniques to remove As(III) from aquatic systems.

Adsorption techniques using a variety of materials have been shown to be superior to conventional techniques such as precipitation, reverse osmosis, filtration, and biological treatment for As(III) removal [2,15]. Complex metal hydroxides are useful adsorbents for the removal of heavy metals from aqueous systems [16,17]. A complex metal hydroxide, or layered double hydroxide, contains divalent and trivalent cations at the octahedral positions of brucite-like layers. Anions are incorporated in the interlayer region for charge neutrality [18,19]. Multi-metal hydroxides exhibit substantially different physicochemical characteristics than their single-metal counter parts. Their properties depend on the identities and molar ratios of the constituent metals [20]. Many researchers have reported As(III) removal using adsorbents such as a Mn-oxide-doped Al oxide [15], an iron-zirconium binary oxide [13], a calcined Mg-Fe-La hydrotalcite-like compound [21], a nickel-aluminum hydroxide [16], and an Fe-Mg type hydrotalcite [17].

Incorporation of zirconium (Zr^{4+}) into a complex metal hydroxide such as a Mg-Al-layered double hydroxide has recently been examined [22,23]. Incorporation increases the charge in the metal layer and promises to be an attractive means of generating unique physicochemical properties. Positively charged brucite-like sheets have been prepared containing Zr^{4+}, which randomly occupies the octahedral holes in a close-packed configuration of hydroxide ions [22]. Adsorption of phosphate ion from aqueous media using the Zr^{4+}-containing complex metal hydroxide has been reported [22,23]. However, there have been no reports regarding As(III) adsorption from aqueous media. Arsenic as As(V) (arsenate) or As(III) (arsenite) adopts chemical forms similar to those of phosphorus as P(V) (phosphate) or P(III) (phosphite). We previously reported that nickel-aluminum-zirconium hydroxide is useful in the removal of phosphate ions from an aqueous phase [24]. This paper focuses on the relationship between As(III) adsorption and the complex nickel-aluminum and nickel-aluminum-zirconium hydroxides. The objectives of this work are to determine the physicochemical characteristics of complex nickel-aluminum and nickel-aluminum-zirconium hydroxides as a function of composition, to evaluate their ability to adsorb As(III) from aqueous media, and to investigate the mechanism of As(III) adsorption.

2. Materials and Methods

2.1. Materials and Chemicals

Nickel-aluminum and nickel-aluminum-zirconium hydroxides of different compositions were obtained from Kansai Catalyst Co., Ltd., Japan. Nickel-to-aluminum molar ratios of 1 and 0.5 are designated as NA11 and NA12, respectively. Nickel:aluminum:zirconium molar ratios of 0.9:1.0:0.09 and 0.9:0.2:0.09 are denoted as NAZ1 and NAZ2, respectively. Sulfate was included as an exchangeable anion in the interlayer of NA and NAZ. These materials were synthesized by the following method. Nickel(II) sulfate hexahydrate, aluminum sulfate octahydrate, and zirconium(IV) sulfate tetrahydrate were mixed with distilled water (230 g) at heating. The reaction mixture was added to the distilled water (400 g) at pH 9.0 for 800 rpm at 25 °C. The solution pH was adjusted using a 25% sodium hydroxide solution. After mixing for 2 h, the suspension was filtered, washed, and dried at 110 °C for 12 h. Arsenic(III) standard solution (As(III), As_2O_3 and NaOH in water pH 5.0 with HCl), hydrochloric acid, and sodium hydroxide were purchased from FUJIFILM Wako Pure Chemical Co., Japan. The purity grade of all reagents was special grade and the quality of water (resistivity) was 18.2 MΩ·cm at 25 °C.

2.2. Physicochemical Properties of Adsorbents

The NA11, NA12, NAZ1, and NAZ2 adsorbents were characterized as follows. The morphologies and crystallinities of the adsorbents were determined by scanning electron microscopy with an SU1510 instrument (Hitachi High-technologies Co., Tokyo, Japan) and a Mini Flex II X-ray diffraction analyzer (Rigaku Co., Tokyo, Japan), respectively. The surface area was determined by a specific surface analyzer N42-25E (Quantachrome Instruments Japan G.K., Kanagawa, Japan). Surface hydroxyl groups were assayed using the fluoride-ion adsorption method [25]. Briefly, the adsorbent (0.125 g) was mixed with a 0.01 mol L^{-1} NaF solution (50 mL) at pH 4.6. The solution pH was adjusted by 0.2 mol L^{-1} acetic acid solution and 0.2 mol L^{-1} acetate buffer solution. The reaction mixture was shaken for 24 h at 100 rpm and 25 °C and filtered through a 0.45-μm membrane filter. The concentration of fluoride ion before and after adsorption was measured using absorption spectrophotometry (DR890, HACH, Loveland, CO, USA). The amount of fluoride ion adsorbed was calculated (the ratio of the concentration of fluoride ion and the concentration of hydroxyl group is 1:1). The surface pH of the adsorbent was measured as follows. A 0.1-g quantity of adsorbent was added to 50 mL distilled water at pH 7.0. The suspension was shaken for 2 h at 100 rpm and 25 °C and filtered through a 0.45-μm membrane filter. The solution pH was measured with a digital pH meter (F-73, Horiba, Ltd., Kyoto, Japan) [26].

2.3. Adsorption Experiments

NA11, NA12, NAZ1, and NAZ2 in 0.1-g quantities were mixed with 50 mL of 100 mg L^{-1} As(III) solution to determine the extent of As(III) adsorption. Mixtures were shaken for 24 h at 100 rpm and filtered through a 0.45-μm membrane filter. The As(III) concentration of the filtrate was measured by inductively coupled plasma optical emission spectrometry using an iCAP 7600 Duo instrument (ICP-OES, Thermo Fisher Scientific Inc., Kanagawa, Japan). The quantity of As(III) adsorbed was calculated from the difference in As(III) concentration before and after adsorption. To examine the effect of contact time, 0.1-g quantities of NA11 and NAZ1 were mixed with 50 mL of a 100 mg L^{-1} As(III) solution and shaken for 0.5, 1, 3, 6, 9, 12, 16, 20, and 24 h at 100 rpm and 25 °C. The effect of pH was studied by mixing the same adsorbents with 50 mL of 100 mg L^{-1} As(III) and adjusting the solution pH from 2 to 10 using hydrochloric acid or sodium hydroxide. The same adsorbents were mixed with 50 mL of 1, 10, 30, 50, 70, and 100 mg L^{-1} As(III) solutions to study the effect of concentration. These solutions also were shaken for 24 h at 100 rpm and 25 °C. The quantity of sulfate ion released from adsorbent was measured to elucidate the adsorption mechanism of As(III).

Sulfate ion concentration was determined by ion chromatography using a DIONEX ICS-900 instrument (Thermo Fisher Scientific Inc., Tokyo, Japan), an IonPac AS12A system (4 × 200 mm, Thermo Fisher Scientific Inc., Tokyo, Japan), and an AMMS 300 (4 mm, Thermo Fisher Scientific Inc., Tokyo, Japan) micro-membrane filter suppressor. The regenerant and mobile phase were 12.5 mmol L^{-1} H_2SO_4, and 2.7 mmol L^{-1} Na_2CO_3 + 0.3 mmol L^{-1} $NaHCO_3$, respectively. The flow rate and sample volume were 1.0 mL min^{-1} and 10 μL under ambient conditions. The quantity of sulfate ions released from adsorbent were calculated using the levels before and after adsorption in Equation (1):

$$q = (C_e - C_0)V/W, \quad (1)$$

where q is the quantity of sulfate ion released (mg g^{-1}); C_0 is the concentration before adsorption (mg L^{-1}); C_e is the concentration after adsorption (mg L^{-1}); V is the solvent volume (L); and W is the weight of the adsorbent (g). Data are presented as the mean ± standard deviation from 2–3 experiments.

3. Results and Discussion

3.1. Physicochemical Properties of Adsorbents

The SEM images of each adsorbent are shown in Figure 1. The adsorbents are not perfectly spherical in shape, but no significant differences are observed in the appearance of the adsorbents in this study.

Figure 1. SEM images of adsorbents.

Figure 2 contains the X-ray diffraction patterns. Peaks at (003), (006), (015), and (113) are observed for all adsorbents indicating that they can be indexed to the crystal structure of the complex metal hydroxide. Peaks (003) and (006) are attributed to basal reflections that correspond to the stacking of the brucite-like sheets [27]. Some NAZ peaks are comparable to those in NA, which indicates that Zr^{4+} has been successfully incorporated into the octahedral layer of the complex nickel-aluminum hydroxide. Zirconium(IV) insertion alters the distances between metals and between the octahedral layers. Similar trends have been reported in previous investigations [22,23].

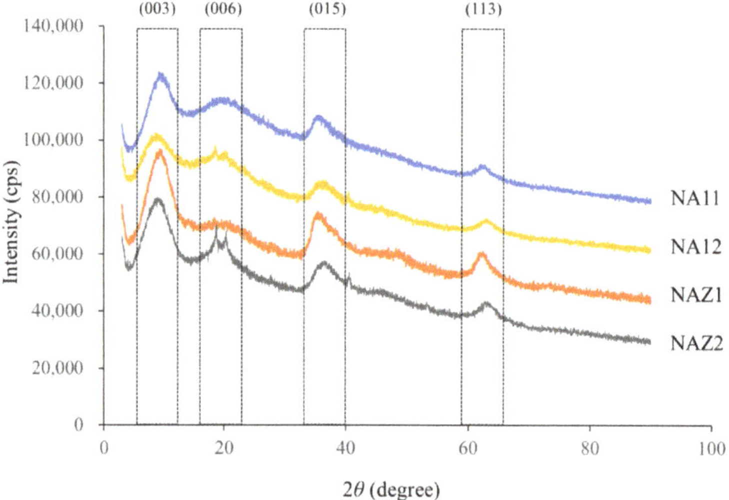

Figure 2. XRD patterns of adsorbents.

Some characteristics of the adsorbents are shown in Table 1. Surface pH and the number of hydroxyl groups of NA are higher than those of NAZ. However, the specific surface area of NAZ is greater than that of NA. The specific surface area of NAZ1 (51.9 m^2 g^{-1}) is greater than that of previously reported (Mg(AlZr)-LDHs(CO_3 or Cl) = 2.2–39 $m^2\ g^{-1}$ and Zr-MgAl = 49 $m^2\ g^{-1}$) adsorbents, but not of (Zr-MgAl-HT = 154 $m^2\ g^{-1}$) [17,18]. These results indicate that incorporation of Zr^{4+} into the octahedral layer of the complex nickel-aluminum hydroxide results in an increase of the specific surface area.

Table 1. Characteristics of adsorbents.

Adsorbents	Surface pH	Number of Hydroxyl Groups (mmol g^{-1})	Specific Surface Area ($m^2\ g^{-1}$)
NA11	7.98	1.92	22.8
NA12	7.63	1.62	26.4
NAZ1	6.18	1.08	51.9
NAZ2	6.21	1.51	27.8

3.2. Adsorption Capability of As(III)

The quantities of As(III) adsorbed onto each adsorbent are shown in Figure 3. The amount adsorbed increases as NA11 < NA12 < NAZ2 < NAZ1, which indicates that NAZ1 should be the most useful for removal of As(III) from aqueous media. Relationships between the quantity of As(III) adsorbed and the three parameters in Table 1 were statistically assessed. The correlation coefficients between the amount adsorbed and the surface pH, amount of hydroxyl groups, and specific surface area were negative 0.570, negative 0.738, and positive 0.921, respectively. Therefore, specific surface area most strongly influences As(III) adsorption capability in solution. The surface properties of adsorbents are very important in the interactions with As(III) in the aqueous phase. Additionally, the effect of zirconium incorporation on the characteristic of adsorbent and adsorption capability of As(III) was evaluated in detail. Thus, NA11 and NAZ1 were selected to evaluate As(III) adsorption in the following experiments.

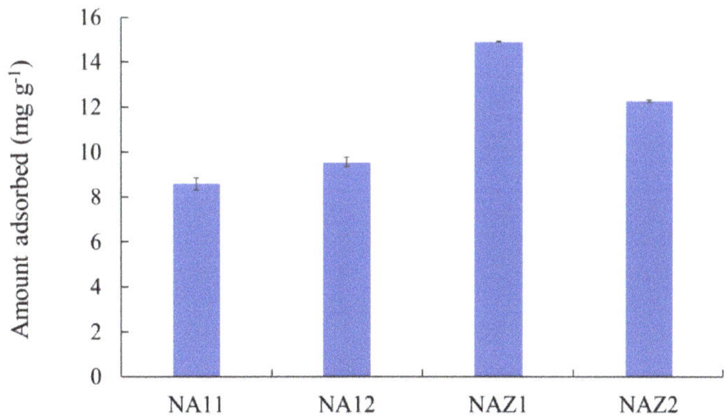

Figure 3. Amount of As(III) adsorbed. Initial concentration: 100 mg L^{-1}, solvent volume: 50 mL, adsorbent: 0.1 g, contact time: 24 h, temperature: 25 °C, agitation speed: 100 rpm.

3.3. Adsorption Kinetics

The kinetics of As(III) adsorption onto NA11 and NAZ1 were conducted at an initial concentration of 100 mg L^{-1}. Adsorption equilibrium was reached within 24 h under the experimental conditions (Figure 4). It is clear that adsorption of As(III) onto NA11 and NAZ1 is complicated, because the time to reach equilibrium would be short if the process were controlled by a single factor [21]. Thus, the adsorption mechanism was investigated by comparing results to both pseudo-first-order (2) and pseudo-second-order models (3) [28,29]:

$$\ln(q_e - q_t) = \ln q_e - k_1 t, \tag{2}$$

$$\frac{t}{q_t} = \frac{t}{q_e} + \frac{1}{k_2 \times q_e^{2'}} \tag{3}$$

where q_e and q_t are the amounts of As(III) adsorbed (mg g^{-1}) at equilibrium and given time t, and k_1 and k_2 are the pseudo-first-order (h^{-1}) and pseudo-second-order (g mg^{-1} h^{-1}) rate constants. Table 2 shows the results of fitting the experimental data to the pseudo-first-order (coefficient of determination = 0.817–0.961) and pseudo-second-order (coefficient of determination = 0.978–0.992) models. The adsorption of As(III) onto NA11 and NAZ1 appears to occur by a process of chemisorption. The value of k_2 with NAZ1 is less than that with NA11, which indicates that As(III) adsorption is favored at NAZ1. This conclusion is supported by the result in Figure 3, which shows that a greater amount of As(III) is adsorbed by NAZ1. Adsorption equilibrium also tends to be established more rapidly using NA11 than NAZ1. Similar trends have been reported in previous studies [13,30]. Additionally, a previous study pointed out the fact that more caution should be exercised in the analysis of kinetic data [31]. Thus, the chi-square analysis (χ^2) was conducted for the evaluation of kinetic models to avoid including errors. The low value of χ^2 indicates that the kinetic model fits to the experimental data [32]. The value of χ^2 in the pseudo-first-order model was higher than that in the pseudo-second-order model (Table 2), which indicates that the pseudo-second-order model is considered to be better for the obtained data in this study.

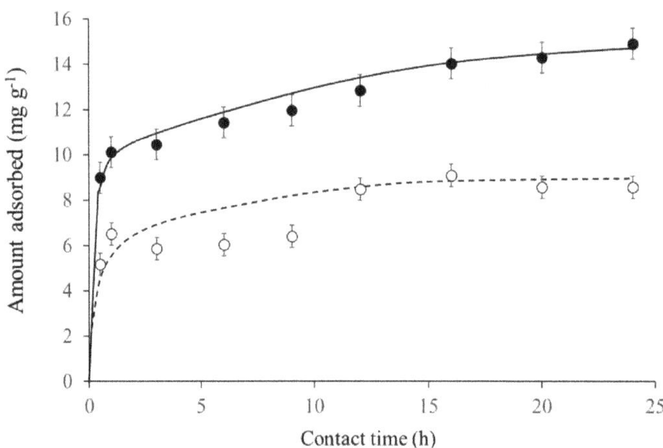

Figure 4. Effect of contact time on the adsorption of As(III) using NA11 and NAZ1. Initial concentration 100 mg L^{-1}, solvent volume 50 mL, adsorbent 0.1 g, contact time 0.5, 1, 3, 6, 9, 12, 16, 20, and 24 h, temperature 25 °C, agitation speed 100 rpm, •: NAZ1, ○: NA11.

Table 2. Kinetic parameters for the adsorption of As(III).

Adsorbents	q_e	Pseudo-First-Order Model				Pseudo-Second-Order Model			
		k_1 (h^{-1})	q_e (mg g^{-1})	r^2	χ^2	k_2 (g mg^{-1} h^{-1})	q_e (mg g^{-1})	r^2	χ^2
NA11	8.6	0.093	3.9	0.817	1.8×10^2	0.110	9.1	0.978	1.8
NAZ1	14.9	6.4	6.4	0.961	2.9×10	0.066	15.2	0.992	1.4

3.4. Effect of pH on the Adsorption of As(III)

Solution pH is a critical factor in the removal of As(III) from aqueous solution. Figure 5 shows the effect of pH on As(III) adsorption using NA11 and NAZ1. The quantity of As(III) adsorbed tends to increase with increasing pH. A similar trend using zirconium oxide-ethanolamine has been reported in a previous study [12]. Here, solution pH tends to increase or decrease slightly after adsorption indicating that the surface of NA11 and NAZ1 may be useful as a neutralizing agent in the wastewater treatment of highly acidic or basic As(III) samples [12,33]. The $pK_{a,1}$ of H_3AsO_3 is 9.2, which indicates that As(III) is a neutral species at pH 9.2. At pH > 9.2, anionic forms of As(III) are produced that undergo ion exchange with sulfate ion from NA11 and NAZ11 under the experimental conditions (see Section 3.5).

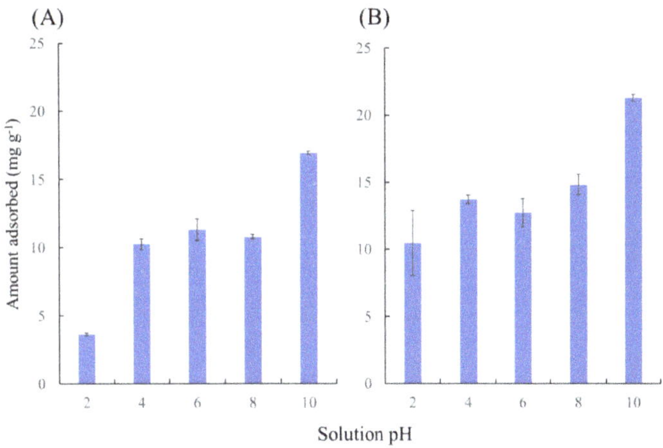

Figure 5. Effect of pH on the adsorption of As(III) using NA11(**A**) and NAZ1(**B**). Initial concentration 100 mg L^{-1}, solvent volume 50 mL, adsorbent 0.1 g, pH 2, 4, 6, 8, and 10, contact time 24 h, temperature 25 °C, agitation speed 100 rpm.

3.5. As(III) Adsorption Isotherms

Figure 6 displays adsorption isotherms of As(III) at different temperatures. The data provide considerable insight to the adsorptive behavior of NA11 and NAZ1 towards As(III) in aqueous solution. The quantities of As(III) adsorbed increase with increasing temperatures (10 < 25 < 45 °C), which indicates that chemisorption is involved in the adsorption of As(III) on NA11 and NAZ1. The solution pH was 5.5–6.4 and 5.9–6.2 for NA11 and NAZ1 in all experimental temperatures. Thus, the predominant form of As(III) was not affected by the temperature under our experimental conditions.

Langmuir and Freundlich models were applied to provide a more quantitative description of the relationship between the quantity of As(III) adsorbed and the solute concentration. The Langmuir model is described by Equation (4) [34]:

$$1/q = 1/(W_s a C_e) + 1/W_s, \tag{4}$$

where q is the quantity of As(III) adsorbed (mg g^{-1}), W_s is the maximum amount of As(III) adsorbed (mg g^{-1}), a is the Langmuir isotherm constant (binding energy) (L mg^{-1}), and C_e is the equilibrium concentration (mg L^{-1}). The Langmuir isotherm is applicable to monolayer coverage by the adsorbate. The Freundlich model is described by Equation (5) [35]:

$$\log q = \frac{1}{n} \log C + \log k, \tag{5}$$

where k and $1/n$ are the adsorption capacity and strength of adsorption, respectively. The Freundlich model describes systems where the surface of the adsorbent is heterogeneous and has sites of different energy [13]. Table 3 shows that the Freundlich model (≥0.970) provides better correlation than the Langmuir model (≥0.965) except for NAZ1 at 25 °C. Previous studies have reported similar behavior using Fe-based backwashing sludge and polymer adsorbents [33,36]. The maximum adsorption capacity (W_s) of As(III) increases from 10 °C to 45 °C. It is worth noting that in the experimental conditions used, equilibrium data (Figure 6) suggest that the plateau for maximum adsorption was not reached. The maximum adsorbed amounts calculated in the Langmuir fitting, however, are lower than some values measured experimentally, which may be due to the use of a linearized equation (Equation (4)) to mathematically model the results. This indicates that the adsorption of As(III) on NA11 and NAZ1 is an endothermic process, which was supported by the results in Figure 6. In the Freundlich model, n is the heterogeneity factor. Therefore, when $1/n = 0.1–0.5$, adsorption occurs easily, and when $1/n > 2$, adsorption is difficult [37]. The values of $1/n = 0.13–0.97$ obtained in this study indicate that As(III) adsorption onto NA11 and NAZ1 is favored in aqueous solution.

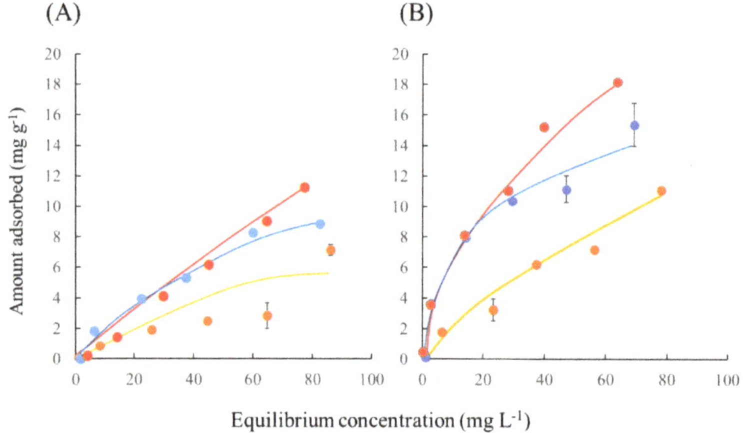

Figure 6. Effect of temperature on the adsorption of As(III) using NA11 (**A**) and NAZ1 (**B**). Initial concentration 1, 10, 30, 50, 70, and100 mg L^{-1}, solvent volume 50 mL, adsorbent 0.1 g, contact time 24 h, pH 5.5-6.4 and 5.9-6.2 for NA11 and NAZ1, temperature 10 (•), 25 (•), and 45 °C (•), agitation speed 100 rpm.

Table 3. Langmuir and Freundlich constants for the adsorption of As(III).

Adsorbents	Temperature (°C)	Langmuir Constants			Freundlich Constants		
		W_s (mg g^{-1})	a (L mg^{-1})	r^2	log$k$	$1/n$	r^2
NA11	10	6.1	0.02	0.965	−0.1	0.91	0.970
	25	9.3	0.04	0.982	−0.1	0.56	0.991
	45	15.4	0.006	0.989	−1.4	0.13	0.996
NAZ1	10	6.3	0.09	0.996	−0.3	0.70	0.989
	25	15.3	0.01	0.935	−0.4	0.97	0.768
	45	16.6	0.10	0.977	0.29	0.54	0.999

In this study, sulfate ions are present in the interlayer of the NA11 and NAZ1 absorbents. Figure 7 shows the relationship between the amount of As(III) adsorbed and the amount of sulfate ion released from NA11 and NAZ1. The correlation coefficients for release from NA11 and NAZ1 are 0.797 and 0.944, respectively. The results indicate that the release of sulfate from NA11 and NAZ1 by ion exchange has a role in the adsorption mechanism of As(III).

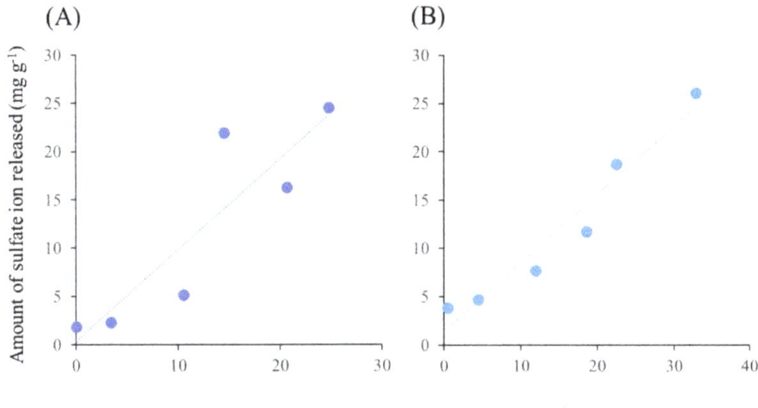

Figure 7. Relationship between amount of As(III) adsorbed and amount of sulfate ion released using NA11 (**A**) and NAZ1 (**B**). Initial concentration 1, 10, 30, 50, 70, and100 mg L^{-1}, solvent volume 50 mL, adsorbent 0.1 g, pH 10, contact time 24 h, temperature 25 °C, agitation speed 100 rpm.

3.6. Comparison of Adsorption Capability of As(III)

Table 4 summarizes comparisons of As(III) adsorption capability of NA11 and NAZ1 with other reported adsorbents [38–41]. The NAZ1 exhibited a good potential to be used for removal of As(III) from aqueous media compared to other reported adsorbents (except for drinking-water treatment residuals (WTRs)). From these comparisons, NAZ1 is expected to be employed in a commercial process in the future for As(III) adsorption.

Table 4. Comparison of As(III) adsorption capacity of NA11 and NAZ1 with other reported adsorbents.

Adsorbents	Adsorption Capability (mg/g)	pH	Temp. (°C)	Initial Concentration (mg/L)	Contact Time (h)	Adsorbent (g/L)	Ref.
ZrPACM-43	0.8	<5	50	100	2	13	[3]
Al-HDTMA-sericite	0.433	4.5	-	2	Approximately 6.7	1	[38]
Iron hydroxide-coated alumina	9.0	6.6	20 × 0.5	7.5–135	48	1–25	[39]
Silica gel impregnated with ferric hydroxide	4.5	7.0	20–23	1	15	-	[40]
WTRs	15	6.0–6.5	23	375–3000	48	25–200	[41]
NA11	9.3	5.5–6.4	25	100	24	2	This study
NAZ1	15.3	5.9–6.2	25	100	24	2	This study

4. Conclusions

The complex metal hydroxides NA11 and NAZ1 were prepared to study As(III) removal from aqueous media. The specific surface area of NAZ1 (51.9 m^2 g^{-1}) was greater than that of NA11 (22.8 m^2 g^{-1}). The quantity of As(III) adsorbed onto NAZ1 (15.3 mg g^{-1}) was also greater than that adsorbed onto NA11 (9.3 mg g^{-1}). The correlation coefficient between the quantity adsorbed and the specific surface area was 0.921. The adsorption kinetics followed the pseudo-second-order model (coefficient of determination using NA11 and NAZ1 was 0.978 and 0.992, respectively). Adsorption isotherms data established that As(III) adsorption was an endothermic process. Ion exchange with sulfate ions in the interlayer of the adsorbent was also involved in the As(III) adsorption mechanism.

Author Contributions: Conceptualization, N.K. and F.O.; investigation, N.N., C.S., and T.N.; resources, M.T. and M.O.; writing—original draft preparation, F.O.; writing—review and editing, F.O. and C.S.; project administration, N.K. All authors have read and agreed to the published version of the manuscript.

Funding: This research was funded by The Research Foundation for Pharmaceutical Sciences.

Conflicts of Interest: The authors declare no conflict of interest.

References

1. Transforming Our World: The 2030 Agenda for Sustainable Development. Available online: https://sustainabledevelopment.un.org/post2015/transformingourworld (accessed on 8 May 2020).
2. Mohan, D.; Charles, P. Arsenic removal from water/wastewater using adsorbents-a critical review. *J. Hazard. Mater.* **2007**, *142*, 1–53. [CrossRef]
3. Mandal, S.; Sahu, M.K.; Patel, R.K. Adsorption studies of arsenic(III) removal from water by zirconium polyacrylamide hybrid material (ZrPACM-34). *Water Resour. Ind.* **2013**, *4*, 51–67. [CrossRef]
4. Hughes, M.F. Arsenic toxicity and potential mechanisms of action. *Toxicol. Lett.* **2002**, *133*, 1–16. [CrossRef]
5. Mandal, B.K.; Suzuki, K.T. Arsenic round the world: A review. *Talanta* **2002**, *58*, 201–235. [CrossRef]
6. Siddique, A.E.; Rahman, M.; Hossain, M.I.; Karim, Y.; Hasibuzzaman, M.M.; Biswas, S.; Islam, M.S.; Rahman, A.; Hossen, F.; Mondal, V.; et al. Association between chronic arsenic exposure and the characteristic features of asthma. *Chemosphere* **2020**, *246*, 129750. [CrossRef]
7. Paul, S.K.; Islam, M.S.; Hasibuzzaman, M.M.; Hossain, F.; Anjum, A.; Saud, Z.A.; Haque, M.M.; Sultana, P.; Haque, A.; Andric, K.B.; et al. Higher risk of hyperglycemia with greater susceptibility in females in chronic arsenic-exposed individuals in Bangladesh. *Sci. Total Environ.* **2019**, *668*, 1004–1012. [CrossRef]
8. IARC Monographs on the Identification of Carcinogenic Hazards to Human. Available online: https://monographs.iarc.fr/agents-classified-by-the-iarc/ (accessed on 31 May 2020).
9. Arsenic in Drinking Water. Available online: https://www.who.int/water_sanitation_health/water-quality/guidelines/arsenic-information/en/ (accessed on 31 May 2020).
10. Vaishya, R.C.; Gupta, S.K. Modeling arsenic(III) adsorption from water by sulfate-modified iron oxide-coated sand (SMIOCS). *J. Chem. Technol. Biotechnol.* **2002**, *89*, 73–80. [CrossRef]
11. Martinson, C.A.; Reddy, K.J. Adsorption of arsenic(III) and arsenic(V) by cupric oxide nanoparticles. *J. Colloid Interf. Sci.* **2009**, *336*, 406–411. [CrossRef]
12. Mandal, S.; Padhi, T.; Patel, R.K. Studies on the removal of arsenic(III) from water by a novel hybrid material. *J. Hazard. Mater.* **2011**, *192*, 899–908. [CrossRef]
13. Ren, Z.; Zhang, G.; Paul Chen, J. Adsorptive removal of arsenic from water by an iron-zirconium binary oxide adsorbent. *J. Colloid Interf. Sci.* **2011**, *358*, 230–237. [CrossRef]
14. Ferguson, J.F.; Gavis, J. A review of arsenic cycle in natural waters. *Water Res.* **1972**, *6*, 1259–1274. [CrossRef]
15. Wu, K.; Liu, T.; Xue, W.; Wang, X. Arsenic(III) oxidation/adsorption behaviors on a new bimetal adsorbent of Mn-oxide-doped Al oxide. *Chem. Eng. J.* **2012**, *192*, 343–349. [CrossRef]
16. Ogata, F.; Nakamura, T.; Toda, M.; Otani, M.; Kawasaki, N. Evaluation of nickel-aluminum complex hydroxide for adsorption of chromium(VI) ion. *Chem. Pharm. Bull.* **2020**, *68*, 70–76. [CrossRef]
17. Ogata, F.; Ueta, E.; Kawasaki, N. Characteristics of a novel adsorbent Fe-Mg-type hydrotalcite and its adsorption capability of As(III) and Cr(VI) from aqueous solution. *J. Ind. Eng. Chem.* **2018**, *59*, 56–63. [CrossRef]
18. Zhang, X.; Guo, L.; Huang, H.; Jiang, Y.; Li, M.; Leng, Y. Removal of phosphorus by the core-shell bio-ceramic/Zn-layered double hydroxides (LDHs) composites for municipal wastewater treatment in constructed rapid infiltration system. *Water Res.* **2016**, *96*, 280–291. [CrossRef]
19. Radha, A.V.; Kamath, P.V. Aging of trivalent metal hydroxide/oxide gels in divalent metal salt solutions: Mechanism of formation of layered double hydroxides (LDHs). *Bull. Mater. Sci.* **2003**, *26*, 661–666. [CrossRef]
20. Ogata, F.; Nagai, N.; Toda, M.; Otani, M.; Nakamura, T.; Kawasaki, N. Evaluation of the interaction between borate ions and nickel-aluminum complex hydroxide for purification of wastewater. *Chem. Pharm. Bull.* **2019**, *67*, 487–492. [CrossRef]

21. Wu, P.; Xia, L.; Liu, Y.; Wu, J.; Chen, Q.; Song, S. Simultaneous sorption of arsenate and fluoride on calcined Mg-Fe-La hydrotalcite-like compound from water. *ACS Sustain. Chem. Eng.* **2018**, *6*, 16287–16297. [CrossRef]
22. Chitrakar, R.; Tezuka, S.; Sonoda, A.; Sakane, K.; Ooi, K.; Hirotsu, T. Synthesis and phosphate uptake behavior of Zr4+ incorporated MgAl-layered double hydroxides. *J. Colloid Interf. Sci.* **2007**, *313*, 53–63. [CrossRef]
23. Miyauchi, H.; Yamamoto, T.; Chitrakar, R.; Makita, Y.; Wang, Z.; Kawai, J.; Hirotsu, T. Phosphate adsorption site on zirconium ion modified MgAl-layered double hydroxides. *Top Catal.* **2009**, *52*, 714–723. [CrossRef]
24. Ogata, F.; Iijima, S.; Toda, M.; Otani, M.; Nakamura, T.; Kawasaki, N. Characterization and phosphate adsorption capability of novel nickel-aluminum-zirconium complex hydroxide. *Chem. Pharm. Bull.* **2020**, *68*, 292–297. [CrossRef]
25. Ymashita, T.; Ozawa, Y.; Nakajima, N.; Murata, T. Collection of uranium from sea water with hydrous oxide adsorbents. 1. Ion exchange properties and uranium adsorption of hydrous titanium(IV) oxide. *Nippon Kagaku Kaishi* **1978**, *8*, 1057–1061. [CrossRef]
26. Ogata, F.; Ueta, E.; Toda, M.; Otani, M.; Kawasaki, N. Adsorption of phosphate ions from an aqueous solution by calcined nickel-cobalt binary hydroxide. *Water Sci. Technol.* **2017**, *75*, 94–105. [CrossRef]
27. Türk, T.; Alp, İ. Arsenic removal from aqueous solutions with Fe-hydrotalcite supported magnetite nanoparticle. *J. Ind. Eng. Chem.* **2014**, *20*, 732–738. [CrossRef]
28. Lagergren, S. Zur theorie der sogenannten adsorption geloster stoffie, *Kunglia Svenska Vetenskapsakademiens. Handlingar* **1898**, *24*, 1–39.
29. Ho, Y.S.; McKay, G. Pseudo-second order model for sorption process. *Pro Biochem.* **1999**, *34*, 451–465. [CrossRef]
30. Raven, K.P.; Jain, A.; Loeppert, R.H. Arsenite and arsenate adsorption on ferrihydrite: Kinetics, equilibrium, and adsorption evelopes. *Environ. Sci. Technol.* **1998**, *32*, 344–349. [CrossRef]
31. Simonin, J.P. On the comparison of pseudo-first order and pseudo-second order rate laws in the modeling of adsorption kinetics. *Chem. Eng. J.* **2016**, *300*, 254–263. [CrossRef]
32. Ho, Y.S. Selection of optimum sorption isotherm. *Carbon* **2004**, *42*, 2115–2116. [CrossRef]
33. Wu, K.; Liu, R.; Li, T.; Liu, H.; Peng, J.; Qu, J. Removal of arsenic(III) from aqueous solution using a low-cost by-product in Fe-removal plants-Fe-based backwashing sludge. *Chem. Eng. J.* **2013**, *226*, 292–401. [CrossRef]
34. Langmuir, I. The constitution and fundamental properties of solids and liquids. *J. Am. Chem. Soc.* **1916**, *38*, 2221–2295. [CrossRef]
35. Freundlich, H.M.T. Over the adsorption in solution. *J. Phys. Chem.* **1906**, *57*, 385–471.
36. Li, H.T.; Xu, M.C.; Shi, Q.; He, B.L. Isotherm analysis of phenol adsorption on polymeric adsorbents from nanoaqueous solution. *J. Colloid. Interf. Sci.* **2004**, *271*, 47–54. [CrossRef]
37. Abe, I.; Hayashi, K.; Kitagawa, M. Studies on the adsorption of surfactants on activated carbons. I. Adsorption of nonionic surfactants. *Yukagaku* **1976**, *25*, 145–150.
38. Tiwari, D.; Lee, S.M. Novel hybrid materials in the remediation of ground waters contaminated with As(III) and As(V). *Chem. Eng. J.* **2012**, *204-206*, 23–31. [CrossRef]
39. Hlavay, J.; Polyak, K. Determination of surface properties of iron hydroxide-coated alumina adsorbent prepared for removal of arsenic from drinking water. *J. Colloid Interf. Sci.* **2005**, *284*, 71–77. [CrossRef]
40. Yoshida, I.; Kobayashi, H.; Ueno, K. Selective adsorption of arsenic ions on silica gel impregnated with ferric hydroxide. *Anal. Lett.* **1976**, *9*, 1125–1133. [CrossRef]
41. Makris, K.C.; Sarkar, D.; Datta, R. Evaluating a drinking-water waste by-product as a novel sorbent for arsenic. *Chemosphere* **2006**, *64*, 730–741. [CrossRef]

Article

Uptake and Recovery of Gold from Simulated Hydrometallurgical Liquors by Adsorption on Pine Bark Tannin Resin

Maria Beatriz Q. L. F. Torrinha, Hugo A. M. Bacelo, Sílvia C. R. Santos *, Rui A. R. Boaventura and Cidália M. S. Botelho

Laboratory of Separation and Reaction Engineering—Laboratory of Catalysis and Materials (LSRE-LCM), Chemical Engineering Department, Faculdade de Engenharia da Universidade do Porto, Rua Dr. Roberto Frias, 4200-465 Porto, Portugal; up201305818@fe.up.pt (M.B.Q.L.F.T.); up201204895@fe.up.pt (H.A.M.B.); bventura@fe.up.pt (R.A.R.B.); cbotelho@fe.up.pt (C.M.S.B.)

* Correspondence: scrs@fe.up.pt

Received: 6 November 2020; Accepted: 7 December 2020; Published: 9 December 2020

Abstract: The recovery of critical and precious metals from waste electrical and electronic equipment (WEEE) is an environmental and economic imperative. Biosorption has been considered a key technology for the selective extraction of gold from hydrometallurgical liquors obtained in the chemical leaching of e-waste. In this work, the potential of tannin resins prepared from *Pinus pinaster* bark to sequester and recover gold(III) from hydrochloric acid and aqua regia solutions was assessed. Equilibrium isotherms were experimentally determined and maximum adsorption capacities of 343 ± 38 and 270 ± 19 mg g^{-1} were found for Au uptake from HCl and HCl/HNO_3 (3:1 *v/v*) solutions containing 1.0 mol L^{-1} H^+. Higher levels of acidity (and chloride ligands) significantly impaired the adsorption of gold from both kinds of leaching solutions, especially in the aqua regia system, in which the adsorbent underperformed. Pseudo-first and pseudo-second order models successfully described the kinetic data. The adsorbent presented high selectivity towards gold. Actually, in simulated aqua regia WEEE liquors, Au(III) was extensively adsorbed, compared to Cu(II), Fe(III), Ni(II), Pd(II), and Zn(II). In three adsorption–desorption cycles, the adsorption capacity of the regenerated adsorbent moderately decreased (19%), although the gold elution in acidic thiourea solution had been quite limited. Future research is needed to examine more closely the elution of gold from the exhausted adsorbents. The results obtained in this work show good perspectives as regards the application of pine bark tannin resins for the selective extraction of Au from electronic waste leach liquors.

Keywords: precious metals; hydrometallurgical processing; biosorption; selectivity; pine bark

---

1. Introduction

Gold is a precious metal commonly used in jewellery and electronics [1], but also in medicine and as a catalyst in various chemical processes [2]. Waste electrical and electronic equipment (WEEE) is a fast-growing waste stream. In 2019, 53.6 Mt of e-waste were generated in the world, corresponding to an average of 7.3 kg per capita [3]. Globally, however, only 17.4% was documented to be formally collected and recycled [3]. In Europe, the management of WEEE is regulated by the Directive 2012/19/EU [4], which states that, from 2019, member states should achieve an annual collection rate of WEEE of 65% of the average weight of EEE placed on the market in the three preceding years. Although Europe was the continent presenting the highest rate (42.5%), collection and recycling must increase even further to meet the target [3]. WEEE contains bulky metals, along with plastics and toxic elements, but also technology metals which, in the frame of a circular economy, should be recovered. The content of precious/critical metals in WEEE vary considerably, but in certain types of WEEE can even exceed those found in primary sources [5–7]. For instance, levels of 250–2050 ppm of gold have been reported in waste printed circuit boards [8]. However, greener and more economically competitive routes are necessary to extract metals from WEEE and achieve sustainable processes.

The recovery of precious metals from e-waste usually involves dismantling, comminution, and separation [8–10], chemical pre-treatments to dissolve base metals [10,11], followed by pyrometallurgical and/or hydrometallurgical processes. Despite the challenges related to the use of strong acids in leaching processes, and the need to minimise the loss of chemicals, hydrometallurgical processes have been considered a cleaner option to be integrated at the back-end of a recycling scheme for the recovery and production of high purity metals [6,7,12]. They also have lower capital costs and benefit from easier implementation at the small scale. Cyanide has been the dominating leaching reagent of Au from primary resources [13] and possibly the most reliable and economic for WEEE [10]. However, the toxicity of cyanide makes its use prohibitive in modern processes. Alternative leaching reagents have been proposed to dissolve Au from WEEE, including chloride, aqua regia, thiourea and thiosulfate [13]. The subsequent stage of a typical flowsheet is the extraction of gold from leaching liquors. Among the various techniques that can be implemented, adsorption has been the dominant one, especially in cyanide liquors, and due to a favourable cost-effectiveness [14], in which activated carbon has been for decades the adsorbent of choice [15,16]. The current interest in moving forward a more benign recovery has been motivating the search for alternative adsorbents, at more competitive prices and able to sequester gold from non-cyanide leach liquors. The use of bio-derived adsorbents has been regarded as a key technology [17], offering advantages related to the ready availability of the biomass, to the simple synthesis and chemical functionality, which usually make them good sequestrants of metals. Important criteria to be met by biosorbents for gold recovery include: (i) high uptake capacity of the metal from actual matrices; (ii) selectivity, i.e., adsorbents should be able to uptake gold to the detriment of other base metals and noble metals coexisting in solution; (iii) fast adsorption kinetics; and (iv) feasible subsequent recovery of the adsorbed metal. In addition, the adsorbents should have particle sizes suitable for use in fixed bed systems, avoiding pressure drop and column clogging, and should have also good chemical stability in strong acidic solutions, considering this is the condition of typical gold-bearing liquors [18]. Various adsorbents have been investigated for gold recovery, including commercial resins [19,20], crosslinked polyethyleneimine resin [19], and a wide range of natural-derived adsorbents such as polyethylenimine modified Ca^{2+}-alginate fibres [18], polyethylenimine modified *Lagerstroemia speciosa* leaves [21], banana peel derivatives [22], raw date pits [23], and tannin-derived materials [6,24–26]. The successful uptake of gold has been explained by electrostatic interaction between the gold complexes in solution and the adsorbent surface, followed by a reduction mechanism, involving amine, hydroxyl, and aldehyde groups of the adsorbents [18,21,27,28].

Tannins are natural polyphenolic macromolecules present in leaves, wood, bark, seeds and fruits of various plants. Their abundance in nature, easy extraction with water, richness in hydroxyl groups and easy tailoring have made tannin materials interesting biosorbents for the removal of contaminants from water and for the uptake and recovery of precious metals [26]. Tannin resins may be prepared through a crosslinking reaction (gelification) with aldehydes [24,25]. Further chemical modifications, such as anchoring of amine groups [24,29] are possible to improve the adsorptive performance of the materials. Immobilization in support materials [28,30] and magnetization [24] have been also investigated to enhance solid-liquid separation. Outstanding uptake capacities of gold have been reported for different tannin-derived biosorbents [24,27,29], explained by the oxidation of polyhydroxyl groups and the reduction of the metal to its elemental form [24,28]. Literature on the recovery of gold by tannin adsorbents has been however limited to few sources of the precursor, namely tannin acid, commercial tannins, valonea, and persimmon tannins [26,28–30]. Despite the fact that pine bark tannin resins have been investigated to remove pollutants from water, in a remediation perspective [31–33], no studies were found in the literature evaluating the uptake of precious metals by this material and subsequent recovery. More research is also needed to obtain results in aqua regia solutions and stronger chloride and acidic media.

This work comes in response to the need of economical and environmentally competitive processes for the recovery of precious metals from WEEE. The adsorption of gold(III) by a tannin resin prepared from bark of maritime pine (*Pinus pinaster*) is studied. The bark of this native Mediterranean tree is a forest residue and a by-product of wood conversion industry, which can give a contribute to bioeconomy. Chloride and aqua regia solutions were used, and the uptake of gold was investigated from single and multi-metal systems, simulating actual hydrometallurgical liquors. The final recovery of the metal was achieved by elution.

2. Materials and Methods

2.1. Adsorbent

A tannin resin, prepared from maritime pine (*P. pinaster*) bark, was used in this work as adsorbent. *P. pinaster* bark was collected (direct sampling) from the coastal North region of Portugal, milled using a regular coffee grinder resulting a final particle size distribution of (percentages by mass, % *w/w*): 40% in the fraction <0.15 mm, 24% in the range 0.15–0.50 mm, 29% of 0.50–1.0 mm, and 7% of 1.0–2.0 mm. Tannins were extracted from the pine bark in batch mode, using an alkaline aqueous solution of 7.5 % NaOH (% *w/w*, in respect to the bark), and a bark-to-liquid ratio of 1:6 (*w/w*). The extraction was carried out under magnetic stirring at a temperature of 90 °C and for a contact time of 60 min. These extraction conditions were the best found in a previous work [33]. After solid-liquid separation (suction filtration, using a vacuum pump), the resulting solution was neutralized with HCl 2.0 mol L^{-1} and subjected to freeze-drying (FreeZone 2.5 Plus, Labconco, Kansas City, MO, USA). NaOH and HCl solutions were prepared from analytical grade salt and acid, respectively. The obtained solid extract was converted into an insoluble material by a cross-linking reaction (gelification). For this purpose, 3.0 g of tannin extract were solubilized in 12.0 mL of a 0.25 mol L^{-1} NaOH solution and mixed with 0.6 mL of formaldehyde (36% *w/w*, analytical grade). The reaction was carried out at 80 °C in digestion vessels for 8 h. The precipitate was dried, milled, washed with HCl 0.05 mol L^{-1} (prepared from analytical grade 37% *w/w* acid) and distilled water, and dried at 65 °C. These operating conditions were also selected according to the optimization studies previously conducted [33]. Particle size analysis was performed by laser diffraction (Coulter LS230, Beckman Coulter, Pasadena, CA, USA): 32%, 43%, 14%, and 11%, respectively found for particles sizes < 0.15 mm, 0.15–0.50 mm, 0.50–1.0 mm, and 1.0–2.0 mm. The fraction 0.15–0.50 mm was selected and used as adsorbent in the present work. The point of zero charge of the pine bark tannin resin was 6.2 ± 0.1 [33]. The high chemical stability of the adsorbent at acidic conditions was demonstrated by low values of dissolved organic carbon concentration at pH 2.0 (4.98 ± 0.07 mg-C L^{-1}), for a tannin resin dosage of 2.0 g L^{-1} and a contact time of 24 h [33].

Scanning electron microscopy (SEM) and energy dispersion spectroscopy (EDS) were used to observe the morphology and detect the chemical elemental composition in the tannin adsorbent surface, before and after adsorption of gold. The SEM/EDS exam was performed at CEMUP-LMEV (Materials Centre of the University of Porto—Laboratory for Scanning Electron Microscopy and X-ray Microanalysis) using a high resolution (Schottky) environmental scanning electron microscope with X-ray microanalysis and electron backscattered diffraction analysis (Quanta 400 FEG ESEM/EDAX Genesis X4M). Samples were coated with a gold/palladium thin film by sputtering, using the SPI Module Sputter Coater equipment.

2.2. Gold Solutions

In this work, gold(III) was used as adsorbate. Aqueous solutions of the metal, with concentrations ranging from 10 to 550 mg-Au L^{-1}, were prepared by dilution of a commercial standard 1000 mg-Au L^{-1}, containing 2 mol L^{-1} HCl. Considering the aim of the study, to evaluate the uptake of Au(III) from chloride and aqua regia liquors, two types of strong acidic solutions containing different levels of H^+ were used as aqueous matrices: hydrochloric acid (HCl) and a mixture of hydrochloric acid and nitric acid (HCl/HNO_3 in a ratio of 3:1 (*v/v*), molar ratio of 2.6:1.0). The desired HCl and HNO_3 concentrations were obtained by the addition of analytical grade commercial acid solutions (HCl 37% *w/w*, density 1.19 g mL^{-1}; HNO_3 65% *w/w*, density 1.39 g mL^{-1}) and taking into account the pre-existing HCl concentration in the gold standard solution.

2.3. Analytic Methods

The analysis of dissolved metals (Au and Cu, Fe, Ni, Pd, Zn—multi-metal experiments) in aqueous solutions was performed by AAS (Atomic Absorption Spectroscopy), with flame atomization (GBC Scientific 932 Plus, Australia). The calibration plots were taken daily and accepted for coefficients of determination (R^2) higher than 0.995. In particular, the dissolved gold was measured by air-acetylene flame, using a hollow-cathode lamp operating at 242.8 nm, intensity of 4.0 mA. Potassium nitrate (analytical grade) was added to the standards and samples at final concentrations of 2000 mg-K L^{-1}, in order to suppress ionization.

2.4. Adsorption Studies

Adsorption studies were conducted in batch mode. Accurately measured volumes of Au solutions (15.0 mL) were put in contact with the required amount of adsorbent (depending on the adsorbent dosages defined in the Sections 2.4.1–2.4.4) in Erlenmeyer closed flasks. Suspensions were stirred in an orbital shaker, operating at 280 rpm. Samples taken from the suspensions were filtered using cellulose acetate membranes (0.45 μm porosity), the liquid phase was diluted when necessary and analysed for metal concentration. All the experiments were made in duplicate and results presented as the average values with corresponding uncertainties (absolute deviations or propagated errors).

2.4.1. Effect of the Leaching Reagent

Aqueous solutions of 100 mg-Au L^{-1} were prepared in HCl, using acid concentrations varying from 0.2 to 3.7 mol L^{-1}, and in aqua regia (HCl/HNO_3) dilutions with total H^+ varying from 0.3 to 4.2 mol L^{-1}. These solutions were mixed with the tannin resin (adsorbent dosage of 2.0 g L^{-1}) and stirred for two days. The uptake percentages of gold from the initial solution was calculated by Equation (1), where C_0 and C_e denote gold concentrations (mg L^{-1}) in the initial and the final solutions, respectively.

$$\text{Uptake } (\%) = \frac{C_0 - C_e}{C_0} \times 100 \tag{1}$$

2.4.2. Adsorption Equilibrium Isotherms

Adsorption equilibrium data allow for the design and optimization of the adsorption systems and provide information on the capacity of an adsorbent to accumulate the adsorbate on its surface. Equilibrium isotherms were obtained for the adsorption of Au(III) by the tannin resin at 20 °C and for different aqueous matrices: HCl 1.0 mol L^{-1} and 2.0 mol L^{-1}, and aqua regia (3:1 *v/v* HCl/HNO_3) dilutions corresponding to 1.0, 1.4 and 2.0 mol L^{-1} total H^+ levels. The experimental data was obtained using solutions with initial concentrations of Au in the range 10–550 mg L^{-1} and a constant solid-to-liquid ratio (*S/L* = 1.0 g L^{-1}). Adsorbent dosages in the range 0.3–3.0 g L^{-1} have been reported in literature [34–36] for gold uptake. Usually, a higher adsorbent dosage provides a higher uptake efficiency, but a lower usage capacity of the solid [31]. Considering the results obtained with 2.0 g L^{-1} (effect of leaching reagent, previous subsection), it was decided to use a lower solid-to-liquid ratio (1.0 g L^{-1}) to ensure that final measurable gold concentrations (significantly different from zero) would be obtained to plot the isotherm. Suspensions were stirred for 72 h (enough to attain the steady state). The amount of metal adsorbed per gram of adsorbent in the equilibrium (q_e, in mg

Au per g of tannin resin) was calculated as a function of the final Au concentration in equilibrium (C_e, mg L^{-1}) by the mass balance expressed by Equation (2).

$$q_e = \frac{C_0 - C_e}{S/L} \tag{2}$$

Langmuir [37] and Freundlich [38] models, expressed by Equations (3) and (4), respectively, were fitted to experimental data by non-linear regression. In Equation (3), Q_m (mg g^{-1}) symbolizes the maximum uptake capacity, corresponding to the monolayer coverage assumed by the Langmuir model, and K_L denotes a constant related to the energy of adsorption. In Equation (4), K_F and n are constants related to the adsorption capacity and adsorption intensity, respectively.

$$q_e = \frac{Q_m K_L C_e}{1 + K_L C_e} \tag{3}$$

$$q_e = K_F {C_e}^{1/n} \tag{4}$$

2.4.3. Kinetic Study

The contact time effect on the amount of adsorbed Au was studied using 1.0 mol L^{-1} HCl solution and aqua regia at the same HCl level (total H^+ concentration of 1.4 mol L^{-1}), adsorbent dosages of 0.5, 1.0 and 2.0 g L^{-1}, and different initial Au concentrations (100 mg L^{-1} and 300 mg L^{-1}). The liquid solutions were stirred with the tannin resin for different periods of time, at room temperature (20 ± 2 °C). Control assays (with no adsorbent) were conducted in parallel. After designated times, suspensions were immediately filtered, and the liquid phase analysed for Au concentration (C). The amount of Au adsorbed per gram of tannin resin (q, mg g^{-1}) at each contact time (t) was calculated by Equation (5).

$$q = \frac{C_0 - C}{S/L} \tag{5}$$

Lagergren's pseudo-first order and pseudo-second order reaction-based models have been widely applied in numerous adsorption systems, due to their simplicity and the commonly good fittings generated. Despite their questionable significance and fairly empirical nature [39], these models were also employed in the present work. Lagergren's pseudo-first order [40] and pseudo-second order models [41,42] are represented by Equations (6) and (7). In these expressions, q and q_e (mg g^{-1}) symbolize the amount of gold adsorbed per mass unit of adsorbent, at time t (h) and at equilibrium, respectively, and k_1 (h^{-1}) and k_2 (g mg^{-1} h^{-1}) are the kinetic constants.

$$q = q_e[1 - exp(-k_1 t)] \tag{6}$$

$$q = q_e \frac{q_e k_2 t}{1 + q_e k_2 t} \tag{7}$$

In addition, an intraparticle diffusion model was also used. Linear Driving Force (LDF) approximation (Equation (8)) [43] assumes that the uptake rate of an adsorbate is proportional to the difference between the adsorbed phase concentration at the solid/fluid interface (q^*, given by Langmuir model) and the average adsorbed phase concentration in the particle (q). The calculation of the LDF constant (k_{LDF}) was done using the explicit equation (Equation (12)) that results from Equation (8), combined with the use of the variables expressed in Equations (9)–(11) [44]. The solver tool in Excel was used to obtain k_{LDF} constants, by minimizing the sum of squared residuals. The calculation of D_h, based on Equation (13) [45], assumes spherical adsorbent particles (r_p denotes the average particle radius) and a parabolic profile of metal concentration inside the particle.

$$\frac{\partial q}{\partial t} = k_{LDF} \times (q^* - q) \tag{8}$$

$$\xi = \frac{Q_m}{C_0} \times S/L \tag{9}$$

$$y = \frac{C}{C_0} \tag{10}$$

$$a = \xi - 1 + \frac{1}{K_L \cdot C_0};\ b = \frac{1}{K_L \cdot C_0};\ \alpha = \frac{-a + \sqrt{a^2 + 4b}}{2};\ \beta = \frac{-a - \sqrt{a^2 + 4b}}{2} \tag{11}$$

$$t = -\frac{1}{k_{LDF}} \left\{ \frac{1}{2b} \times ln\left[\frac{y^2 + ay - b}{a - b + 1} \right] + \left(1 - \frac{a}{2b}\right)\left(\frac{1}{\alpha - \beta}\right) \times ln\left[\frac{(1-\beta)(y-\alpha)}{(1-\alpha)(y-\beta)} \right] \right\} \tag{12}$$

$$k_{LDF} = \frac{15D_h}{{r_p}^2} \tag{13}$$

2.4.4. Competitive Adsorption and Selectivity

The dissolution of precious metals from WEEE is usually preceded by chemical pre-treatments to dissolve base metals (e.g., copper, lead) [10]. However, certain amounts of these metals remain in the residue, together with the precious metals, and are leached out with gold in the subsequent steps. The presence of these metals in solution may affect the adsorption capacity of the adsorbent, due to possible competition. In addition, if an adsorbent sequesters significant amounts of other metals, besides gold, the generation of a final high purity recovered product may be compromised. Assessing the performance of an adsorbent in a multi-metal solution simulating an actual hydrometallurgical liquor is therefore important. Multi-metal solutions containing Au(III) (200 mg L^{-1}), Cu(II) (200 mg L^{-1}), Fe(III) (150 mg L^{-1}), Ni(II) (80 mg L^{-1}), Zn(II) (10 mg L^{-1}) and Pd(II) (40 mg L^{-1}) were prepared in aqua regia (0.75, 1.3 and 2.0 mol L^{-1} H^+ solutions). The desired concentration of each metal was obtained by the dilution of commercial metal standards (1000 mg L^{-1}) or, in the case of Cu(II), by the dissolution of $CuCl_2 \cdot 2H_2O$ salt. The metal composition of the simulated solutions was based on the levels reported in literature for aqua regia liquors of WEEE [25,29]. Simulated leaching solutions were then mixed with the tannin resin at an adsorbent dosage of 1.0 g L^{-1}. After 72 h of stirring, the concentration of each metal was analysed in the liquid and uptake percentages and adsorbed amounts calculated (Equations (1) and (2)). Control assays were conducted in parallel, in similar conditions, but using single-metal solutions (200 mg-Au L^{-1}).

2.5. Desorption and Regeneration

Desorption and regeneration studies were conducted in order to evaluate the recovery of gold by elution from exhausted adsorbents and the possibility to reuse the regenerated material. A sample of Au-loaded adsorbent was prepared by stirring the tannin resin (1.0 g L^{-1}) with 500 mg L^{-1} Au solution, prepared in HCl 1.0 mol L^{-1} H^+. After the saturation with gold, the adsorbent was separated from the remaining liquid by vacuum filtration, washed with distilled water and dried at 50 °C overnight. The first desorption stage was conducted using a solid-to-liquid ratio of 2.5 g L^{-1} and two different eluents: acid thiourea solution (0.5 mol L^{-1} thiourea and 0.5 mol L^{-1} HCl) and HCl 50% (*v/v*). After 3 days stirring, the liquid phase was analysed for Au and the desorbed percentage calculated. Adsorption and desorption steps were repeated for 3 cycles, using the thiourea solution as eluent.

3. Results and Discussion

3.1. Effect of the Leaching Solution

Figure 1 presents the results on gold(III) uptake from aqueous matrices containing different levels of HCl and HCl/HNO_3. In both aqueous systems, the acidity of the medium affects the amount of gold adsorbed by tannin resin.

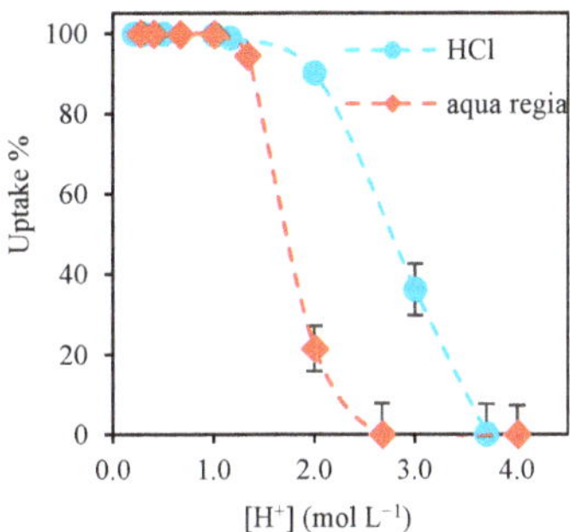

Figure 1. Effect of hydrogen ion concentration ($[H^+]$) on the uptake of gold by the pine bark tannin resin from HCl and HCl/HNO_3 (aqua regia) aqueous solutions.

Total removal of gold from HCl solutions is achieved up to 1.2 mol L^{-1} H^+. Above 2.0 mol L^{-1} HCl, the uptake of the metal is seriously affected, and for 3.7 mol L^{-1} HCl is practically suppressed. Results can be explained considering the effect of H^+, but also the presence of Cl- ligands, and their effect on the adsorbate species (gold complexes) and on the adsorbent surface. The adsorption mechanism that has been proposed for the uptake of gold by tannin materials is based on an electrostatic attraction followed by the reduction of Au(III) and the oxidation of hydroxyl groups [19,24]. The redox reaction is expressed by Equation (14), where the reduction of the chlorogold complex is accompanied by the oxidation of hydroxyl to carbonyl groups [46].

$$AuCl_4^- + 3\,R\text{-}OH \rightarrow Au^0 + 3\,R = O + 3\,H^+ + 4\,Cl^- \quad (14)$$

Under the HCl concentrations used in this study, Au should be present in solution as tetrachloroaurate(III) complex ($AuCl_4^-$) (speciation diagram presented in [47]), and the resin surface should present a high positive charge. These conditions favour the uptake of Au complexes by electrostatic attraction, which explains the good results observed up to 1.2 mol L^{-1} HCl. However, an excessive increase in HCl concentration (and consequently, in H^+ and Cl^- levels) impairs the reduction of gold (Equation (14)), and chloride ions, abundantly present in solution, may also compete with $AuCl_4^-$ to the active sites of the adsorbent [19,24]. This explains the marked reduction in the gold uptake observed for HCl levels above 2.0 mol L^{-1}.

With regards to the aqua regia, Figure 1 shows that 100% of gold in HCl/HNO_3 solutions is sequestered by the tannin resin for acidic levels below 1.0 mol-H^+ L^{-1}. The adsorptive ability sharply decreased for higher acid concentrations and ceases at 2.7 mol L^{-1} H^+ (corresponding to 2.0 and 0.7 mol L^{-1} of HCl and HNO_3, respectively), results that can be explained in the same way as for the HCl solutions. As it can be seen, for acidity levels higher than 1.2 mol-H^+ L^{-1}, the tannin resin exhibits a higher performance for the uptake of Au(III) from HCl solutions than from aqua regia, and it is much more affected by the acid concentration in the latter system than in the former. This means that the presence of nitric acid in solution impairs the uptake of Au by the adsorbent, being a result of the strong oxidizing capacity of aqua regia which hinders the reduction of Au(III) [25].

The results here obtained are generally in line with those reported in literature. Fan et al. [25] studied the adsorption of Au(III) by a persimmon tannin in HCl and HNO_3 media. Under the studied conditions, the adsorbent performance was only significantly reduced for HCl and HNO_3 concentrations higher than 3.74 and 3 mol L^{-1}, respectively. The adsorbent performed better in HCl than in HNO_3 aqueous medium. Yi et al. [29] studied the effect of HCl concentration on the adsorption of gold(III) by ethylenediamine modified persimmon tannin adsorbent and reported a gradual decrease in the Au uptake when the H^+ concentration rose from 0.1 to 5.82 mol L^{-1}. Al-Saidi [23] reported an increase in Au(III) removal by raw date pits for HCl concentrations up to 0.5 mol L^{-1}, but a decrease in the uptake efficiency above this level. Most of literature studies were conducted using HCl aqueous matrices. In the present study we have investigated also the use of aqua regia because it is also employed as leaching solution for WEEE, superior to chloride leaching in some criteria [10,13]. Several works have also evaluated the uptake of gold under HCl levels equal and lower than 0.1 mol L^{-1}, corresponding to pH values between 1 and 5–6 [18,21,22], which are weaker acidic conditions than those used in the present work. In fact, our experiments were carried out by using 1.0–2.0 mol H^+ L^{-1} (HCl and HCl/HNO_3 solutions), closer to the H^+ concentration in actual leach liquors of WEEE, even though that acid levels below 1.0 mol L^{-1} would assure a greater adsorbent performance.

3.2. Adsorption Equilibrium Isotherms

Figure 2 shows adsorption isotherms obtained for the uptake of gold by the tannin resin, in different acidic systems (HCl and aqua regia).

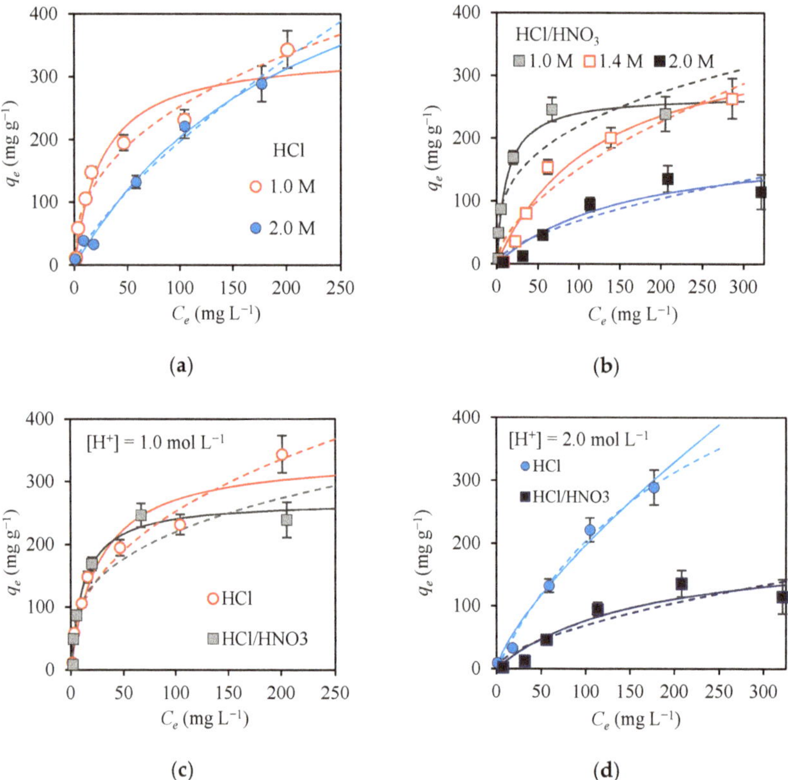

Figure 2. Equilibrium isotherms for the adsorption of gold by the pine bark tannin resin from different acidic matrices (20 °C, S/L = 1.0 g L^{-1}): effect of H^+ concentration in (**a**) HCl and (**b**) aqua regia solutions; effect of the leaching solution for a total H^+ concentration of (**c**) 1.0 and (**d**) 2.0 mol L^{-1}. C_e denotes the gold concentration in the liquid phase, in equilibrium with the metal adsorbed amount, q_e. Langmuir and Freundlich models are represented by solid and dashed lines, respectively.

In line with the previous results, Figure 2a,b shows a negative effect of increasing the solution acidity. The slope of the isotherms for low equilibrium concentrations gives indication about the affinity adsorbent/adsorbate, and in both charts (Figure 2a,b), the decrease of adsorption is evident for higher H^+ concentrations. However, the results show that the negative influence of acid concentrations tends to cease as the Au concentration increases. The adsorbed amounts of Au from 1.0 mol L^{-1} HCl solution are significantly higher than the ones recorded at 2.0 mol L^{-1} only for equilibrium gold concentrations below 100 mg L^{-1}, but are practically the same for higher adsorbate levels (Figure 2a). The same happens when it comes to aqua regia (Figure 2b), as the isotherms measured at 1.0 and 1.4 mol L^{-1} H^+ get closer for Au concentrations of ca. 200 mg L^{-1}. This is reflected by the similarity of the maximum adsorbed amounts obtained experimentally ($q_{e,m}$, Table 1) for 1.0 and 2.0 mol L^{-1} HCl, and for 1.0 and 1.4 mol L^{-1} H^+ aqua regia. The use of 2.0 mol L^{-1} of H^+, however, impaired more dramatically the uptake ability, as even for higher Au levels the adsorbed amounts remained significantly below the values obtained at 1.0 and 1.4 mol L^{-1}.

Table 1. Parameters of Langmuir and Freundlich isotherms for the adsorption of gold from HCl and aqua regia solutions using pine bark tannin resin, at 20 °C, and comparison between the maximum adsorbed amounts obtained experimentally ($q_{e,m}$) and from Langmuir modelling (Q_m).

	exp.	Langmuir				Freundlich			
	$q_{e,m}$ (mg g^{-1})	Q_m (mg g^{-1})	$K_L \times 10^3$ (L mg^{-1})	R	SE (mg g^{-1})	K_F ($mg^{1-1/n}g^{-1}L^{1/n}$)	n	R	SE (mg g^{-1})
HCl Solution									
1.0 mol L^{-1} H^+	344 ± 30	343 ± 38	38 ± 13	0.97	31.0	38 ± 7	2.4 ± 0.3	0.98	23.7
2.0 mol L^{-1} H^+	289 ± 28	675 ± 152	4 ± 2	1.00	12.6	7 ± 2	1.4 ± 0.1	0.99	15.7
Aqua Regia Solution									
1.0 mol L^{-1} H^+	246 ± 19	270 ± 19	81 ± 21	0.99	19.3	51 ± 21	3 ± 2	0.91	47.2
1.4 mol L^{-1} H^+	264 ± 32	386 ± 62	8 ± 3	0.98	20.9	10 ± 6	1.7 ± 0.3	0.96	31.5
2.0 mol L^{-1} H^+	136 ± 21	200 ± 65	6 ± 4	0.95	19.3	5 ± 4	1.7 ± 0.5	0.92	24.8

Figure 2d compares isotherms measured from HCl and aqua regia matrices, under the same total H^+ concentrations. For 1.0 mol L^{-1} H^+, the two isotherms are quite similar, mainly for low gold concentrations. For 2.0 mol L^{-1} H^+, the performance of the tannin resin is unquestionably worst under aqua regia solutions (Figure 2d). In this case, the maximum adsorbed amounts, recorded within the experimental conditions were 289 mg g^{-1} in HCl solution and 136 mg g^{-1} in aqua regia solution (Table 1). Such behaviour should be related to the strong oxidizing power of aqua regia that inhibits the reduction of Au and limits the sequestering ability of the adsorbent. Therefore, the use of HCl as a leaching agent for gold in hydrometallurgical processes has proven to create more favourable conditions than aqua regia, considering its subsequent uptake and recovery.

The parameters of Langmuir and Freundlich models are shown in Table 1. The coefficients of correlation (R) and the standard error of the regressions (SE) are also presented, as measures of the quality of the fittings. Modelled curves are illustrated in Figure 2. As it can be seen, both equilibrium models described quite well the experimental data. Considering the higher correlation coefficients and the lower standard errors, the Langmuir model provided better fittings (exception is the isotherm measured at 1.0 mol L^{-1} HCl). Despite this, in the experimental range of concentrations studied and in certain conditions (e.g., 2.0 mol L^{-1} HCl; 1.4 and 2.0 mol L^{-1} HNO_3/HCl), the maximum experimental adsorbed amounts (denoted as $q_{e,m}$, in Table 1) are lower than the Q_m values predicted by Langmuir model. This is an indication that the monolayer capacity was not attained, which means that the adsorbent still has capacity to sequester more gold ions from solution. Indeed, other studies present similar observations. A continuous rise of the isotherm curve, even for Au equilibrium concentrations of 1500–2000 mg L^{-1}, has been reported for the adsorption of gold by crosslinked polyethylenimine/calcium-alginate fibres, and attributed to a continuous reduction process of Au(III) [18].

In a general way, and despite the negative effect of H^+ and aqua regia, pine bark tannin resin showed an excellent ability to sequester and accumulate gold on its surface in all the conditions tested. Table 2 presents maximum adsorption capacities of gold(III) reported in the literature for different adsorbents. In order to allow a suitable comparison, maximum adsorbed amounts obtained experimentally ($q_{e,m}$) and from Langmuir modelling (Q_m) are presented, together with the most relevant operating conditions. The pine bark tannin resin presents adsorbed amounts in the range of the reported values for other biosorbents. Looking only at the uptake capacities of gold, its performance seems to be comparable to the banana peel derived adsorbent and PEI-modified *L. speciosa* leaves, superior to some other biosorbents or synthetic polymers (e.g., commercial resin Lewatit TP214, raw date pits, activated rice husk), but worse than other tannin adsorbents. However, it is worth noting that the adsorption capacity of the pine bark tannin resin was evaluated here under more harsh acidic conditions (1.0 mol L^{-1} acid solutions) than most of the data collected from literature and referred in Table 2. Although strong acidic conditions are more representative of the actual conditions found industrially, it is known that they impair the uptake of gold (Figure 1; [25,29]). On this basis, it can be said that pine bark tannin resin proved to be a promising material for gold uptake.

Table 2. Maximum adsorption capacities reported in literature for the adsorption of gold from chloride media by different adsorbents: maximum adsorbed amounts, obtained experimentally ($q_{e,m}$) and calculated monolayer capacities (Q_m) by Langmuir modelling.

Adsorbent	T (K)	pH or [H⁺]	C_e (mg L^{-1})	$q_{e,m}$ (mg g^{-1})	Q_m (mg g^{-1})	Ref.
Commercial resin IRA400		pH 2	0–215		902.3	[19]
Commercial resin Lewatit TP214	298	pH 6.1	35–225		108.7	[20]
Polyaniline modified by TMP	298	pH 4	0–300	881	883	[48]
Crosslinked PEI resins		pH 2	0–210		943.5	[19]
NIPA gel	323	1 mol L^{-1}	0–790		125.5	[49]
Activated rice husk	298	pH 6.1	50–260		93.46	[20]
PEI-alginate fibers	298	0.1 mol L^{-1}	0–2000	1240	1404	[18]
GA-PEI-alginate fibers	298	0.1 mol/L	0–1500	2325	2182	[18]
Cellulose acetate fibers	298	2 mol L^{-1}	0–800	110		[50]
Raw date pits	298	0.5 mol L^{-1}	0–35	78	61	[23]
Banana peel	298	pH 1	0–1200	370.18	377.2	[22]
Banana peel (lipid extraction)	298	pH 1	0–1000	475.48	448.4	[22]
Oil palm trunk (dewaxed)	303	pH 2	0–120	91.47	95.16	[35]
PEI-modified *L. speciosa* leaves	298	pH 1	0–200	282	286	[21]
Persimmon resin	303	pH 2	150–351	≈965	1905	[36]
Persimmon peel gel	303	0.1 mol L^{-1}	0–11 × 10^3	1.8 × 10^3		[51]
Crosslinked persimmon tannin gel	303	0.1 mol L^{-1}	0–2.4 × 10^3	1517		[34]
TEPA-persimmon tannin gel	303	0.1 mol L^{-1}	0–1.2 × 10^3		1168	[6]
EDA-modified persimmon tannin	303	0.1 mol L^{-1}	0–1.5 × 10^3		1550.4	[29]
persimmon tannin onto Fe3O4@SiO2 microspheres	298	pH 5	(0.2–1.8) × 10^3	860	917.4	[24]
persimmon tannin functionalized viscose fiber	298	pH 2	0–120	528	536	[28]
Sericin and alginate particles chemically crosslinked by proanthocyanidins	298	pH 2.5–3	0–140	196.1	188.4	[27]
Pine bark tannin resin	298	1 mol L^{-1}	0–200	344	343	This work
Pine bark tannin resin	298	1 mol L^{-1} (HCl/HNO_3)	0–200	246	270	This work

EDA—ethylenediamine; GA–glutaraldehyde; NIPA–N-isopropylacrylamide; PEI–polyethylenimine; TEPA–tetraethylenepentamine; TMP–trimethyl phosphate.

3.3. Kinetic Study

The effect of contact time on the uptake of gold was studied at 20 °C, for different adsorbate concentrations, solid-to-liquid ratios and acidic aqueous matrices. The kinetic profiles are presented in Figure 3 and show a relatively slow adsorptive process, with equilibrium times of two or three days. Similar contact times were found by Gurung et al. [34], with persimmon tannin resins, although other authors report considerably lower values (8 h) [36]. The slow kinetics on the gold uptake appears to be a shortcoming of many biosorbents, but this has been explained by the reductive process of gold, which is essential for high uptake capacities.

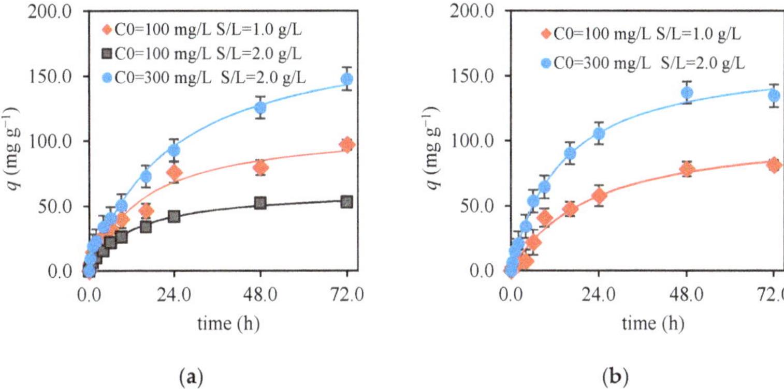

Figure 3. Effect of contact time on the uptake of gold by the pine bark tannin resin at 20 °C, different initial gold concentrations, adsorbent dosages and using aqueous solutions of (**a**) 1.0 mol L^{-1} HCl (**a**) and (**b**) aqua regia 1.0 mol L^{-1} H^+: experimental data and pseudo-second order modelling.

The adsorption dynamics is a result of a multi-step process involving [52,53]: (i) transport through the external film up to the surface of the adsorbent, (ii) intraparticle diffusion (in pores and/or over the surface), and (iii) final adsorption reaction on the active sites. The last step is usually very fast and do not determine the overall adsorption rate [52]. In well-stirred batch systems, the external resistance of mass transfer can be also considered negligible. Despite the fact that the adsorption reaction is also usually fast, Lagergren's pseudo-first order and pseudo-second order reaction-based models have been widely applied in numerous adsorption systems, due to their simplicity and the commonly good fittings generated. Table 3 presents kinetic parameters for Lagergren's pseudo-first and pseudo-second order models. As it can be seen, both models describe very well the experimental data, with all coefficients of correlation equal or higher than 0.98. Despite the fact that the pseudo-second order model (Figure 3) has mostly provided the lowest SE, the pseudo-first order predicted equilibrium adsorbed amounts (q_e) closest to the experimental values.

Table 3. Kinetic parameters of reaction-based models (k, kinetic constants; q_e, equilibrium adsorbed amounts) for the adsorption of gold from HCl and aqua regia solutions by the pine bark tannin resin, at 20 °C and for different initial Au(III) concentrations (C_0) and adsorbent dosages (S/L).

		Pseudo-First Order Model				**Pseudo-Second Order Model**			
C_0 (mg L^{-1})	S/L (g L^{-1})	q_e (mg g^{-1})	$k_1 \times 10^2$ (h^{-1})	R	SE (mg g^{-1})	q_e (mg g^{-1})	$k_2 \times 10^4$ (g mg^{-1} h^{-1})	R	SE (mg g^{-1})
				1.0 mol L^{-1} HCl					
100	1.0	91 ± 6	6 ± 1	0.98	7.2	112 ± 9	6 ± 2	0.98	6.2
100	2.0	52 ± 2	7.7 ± 0.6	0.99	2.2	63 ± 2	1.3 ± 0.1	1.00	1.4
300	2.0	147 ± 8	4.7 ± 0.6	0.99	7.5	190 ± 12	2.3 ± 0.5	0.99	6.0
				1.4 mol L^{-1} H^+ Aqua Regia Solution					
100	1.0	84 ± 4	5.3 ± 0.7	0.99	4.5	109 ± 9	4 ± 1	0.99	4.8
300	2.0	136 ± 3	7.2 ± 0.5	1.0	4.2	167 ± 5	4.4 ± 0.5	1.0	3.8

The kinetics of gold uptake by the tannin resin was also analysed considering an intraparticle diffusion-controlled process. Assuming a negligible mass transfer resistance in the outer layer and considering that intraparticle resistance is governed by surface diffusion (as this adsorbent is essentially a non-porous material), homogeneous solid diffusion model (HSDM) combined with Linear Driving Force (LDF) approximation were used to model kinetic data and to calculate the average values of the homogeneous diffusion coefficients (D_h) (equations in Section 2.4.3). Table 4 presents k_{LDF} and D_h estimated values. The good agreement between the LDF fitting and the experimental data (coefficients of correlation equal or higher than 0.98) shows that this model can also successfully describe the adsorption of gold by the tannin resin. The homogeneous solid diffusivity coefficients calculated were in the range 10^{-14}–10^{-13} $m^2\ s^{-1}$. Saman et al. [35] reported values in the order of magnitude of 10^{-13} $m^2\ s^{-1}$ for the adsorption of Au(III) by oil palm trunk biosorbents. For the same adsorbent dosage, the results show an increase in D_h with the initial adsorbate concentration, which has been reported in various adsorptive systems [44,54], including on the uptake of gold cyanide by activated carbon [16]. For the same initial adsorbate concentrations and adsorbent dosages, lower k_{LDF} and D_h values were obtained for the aqua regia solutions, in comparison to HCl, which indicates slower adsorption kinetics.

Table 4. Kinetic constant (k_{LDF}) of LDF approximation and solid diffusivity coefficients (D_h) calculated by HSDM model for the adsorption of gold from HCl and aqua regia solutions by the pine bark tannin resin.

C_0 (mg L^{-1})	S/L (g L^{-1})	R	k_{LDF} (h^{-1})	D_h (m^2 s^{-1})
1.0 mol L^{-1} HCl				
100	1.0	0.98	0.076	3.6×10^{-14}
100	2.0	0.98	0.051	2.5×10^{-14}
300	2.0	0.99	0.215	1.1×10^{-13}
1.4 mol L^{-1} H$^+$ Aqua Regia Solution				
100	1.0	1.00	0.021	1.0×10^{-14}
300	2.0	0.99	0.087	4.3×10^{-14}

3.4. Competitive Adsorption and Selectivity

The tannin resin was tested in multi-metal solutions containing Cu(II), Fe(III), Ni(II), Zn(II), Pd(II), and Au(III). The results are depicted in Figure 4.

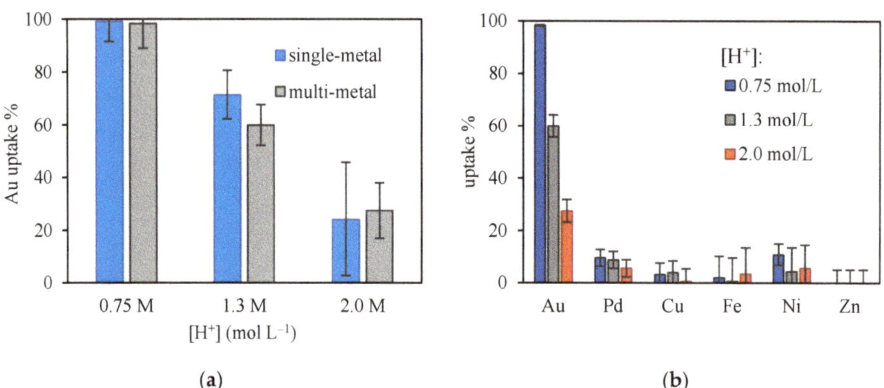

Figure 4. Percentages of metals extracted from simulated liquors containing aqua regia in different H^+ concentrations by the pine bark tannin resin from (20 °C, S/L 1.0 g L^{-1}, initial Au concentration 200 mg L^{-1}): (**a**) comparison of Au uptake from single and multi-metal solutions; (**b**) uptake of metals from the multi-metal solution.

As it can be seen (Figure 4a), the ability of the tannin resin to uptake gold from the simulated liquors is practically the same as that found in single-metal solutions. The uptake percentages of gold obtained in multi-metal systems (Figure 4b) were noticeably higher than those of the other metals (which has never surpassed 11%), particularly for total acidities of 1.0 and 1.4 mol L^{-1}. For instance, at acidic level of 0.75 mol L^{-1} H^+, almost 100% of the gold in solution was adsorbed by the tannin resin, whereas Ni and Pd were very weakly extracted (11 ± 4% and 10 ± 3%, respectively), and other metals remained practically solubilized (uptake percentages < 3%). In the presence of high chloride concentrations and low pH, gold and platinum group metals are present as anionic chloro-complexes, which favour their uptake by the positively charged adsorbent surface. Palladium has similar properties and behaviour as Au, namely a low reduction potential. It has been reported that $PdCl_4^-$ can be also adsorbed by tannin-adsorbents through the oxidation of the functional groups of tannins and reduction to Pd(0) [24], which justify its co-adsorption. Base metals, on the other hand, are present in solution as cationic species and are electrostatically repelled. Indeed, in biosorption processes, anions are typically adsorbed at low pH (1.0–3.0), whereas metal cations are typically adsorbed under optimum conditions of pH 5.0–5.5 [55]. This explains the reduced competition observed between gold and base metals. The present results are generally in line with results published in literature, which have shown very good selectivity for gold over other metals for different HCl and HNO_3 media [25].

3.5. SEM and EDS Analysis

Figure 5 presents the SEM images of the tannin resin adsorbent, before and after gold uptake. As it can be observed, the adsorbent particles present an irregular shape and a faceted non-porous surface (Figure 5a). After adsorbing gold from HCl (Figure 5b) and aqua regia solutions (Figure 5d,e), the adsorbent surface presented an accumulation of metal particles, identified as gold through the EDS spectra performed in the micro areas denoted as Z1 and Z2 (Figure 5f). These images are in line with the quantitative adsorption results previously obtained, considering that Figure 5d (Au-loaded adsorbents in HCl and aqua regia solutions 1.0 mol L^{-1} H^+, respectively) show high amounts of gold particles, whereas Figure 5e shows a much scarce distribution. The EDS spectra obtained in areas not covered by gold particles (not illustrated) showed an abundance of carbon and oxygen (also observed before adsorption), but the additional presence of chlorine. This confirms the competition between chloride and gold to the adsorbent surface.

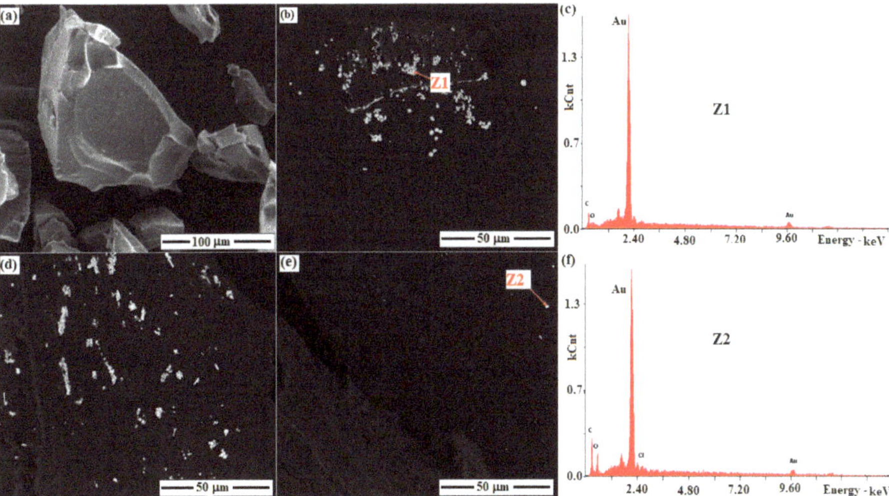

Figure 5. SEM images obtained for the pine bark tannin resin (**a**) before contact with gold, and after adsorption of gold (**b**) from 1.0 mol L^{-1} HCl, and from aqua regia solutions (**d**) 1.0 mol L^{-1} H^+ and (**e**) 2.0 mol L^{-1} H^+. EDS spectra obtained in the areas denoted as (**c**) Z1 and (**f**) Z2.

3.6. Desorption and Regeneration

The final recovery of gold from the loaded adsorbent and the regeneration of the adsorbent are important for industrial applications. A first desorption test of gold(III) was conducted using HCl 50%

(v/v) and 0.5 mol L^{-1} acidic thiourea as eluents. Negligible desorption percentage (<1%) was obtained in HCl eluent, whereas 57% was registered using the acidic thiourea solution. In fact, formulations of thiourea with HCl have been identified in literature as the most effective desorbing solutions for gold [18,22,27,56]. The desorption mechanism of the process is based on the displacement of the electrostatically adsorbed Au(III), that complexes with chloride in the HCl eluent, and on the oxidation of the elemental gold forming the gold(I)-thiourea $Au(CS[NH_2]_2)_2^+$ [18,56]. Details about the reactions of gold dissolution through the use of acid thiourea solution can be found in literature [57,58].

Adsorption–desorption assays were then conducted for three cycles, using the acidic thiourea solution. The results are presented in Figure 6.

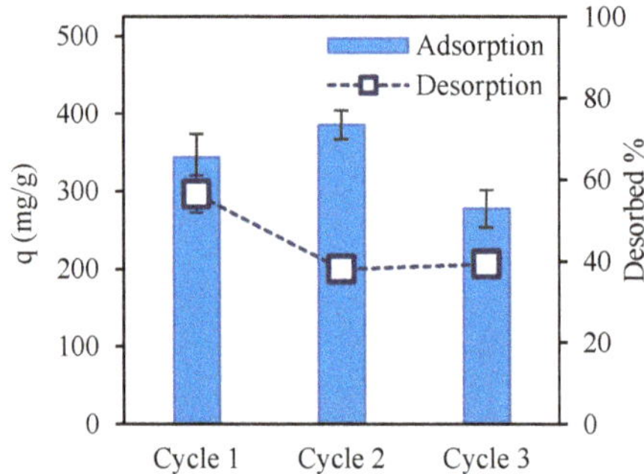

Figure 6. Results obtained in the adsorption (C_0 = 500 mg L^{-1}, S/L = 1.0 g L^{-1}, 20 °C) and desorption cycles (eluent: 0.5 mol L^{-1} thiourea and 0.5 mol L^{-1} HCl solution, S/L = 2.5 g L^{-1}, 20 °C).

As it can be seen, the desorption efficiency observed in the first cycle (57%) decreased to 38% and 39% in the two subsequent desorption steps. In spite of that, the regeneration ability of the tannin resin was quite good, as the adsorbed amount remained practically constant in two cycles and reduced moderately (19%) in the third one, indicating that the regenerated tannin resin still has capacity to accumulate gold in its surface. These results are promising, but the necessity of more research is evident to find an improved solution (e.g., increasing HCl concentration) to desorb gold more efficiently from the pine bark tannin resins. Although from environmental and cost-effectiveness points of view, the desorption and regeneration seem to be the most interesting options to recover gold, incineration could be considered in future as it is a direct way to obtain metallic gold.

4. Conclusions

Tannin resins prepared from maritime pine (*Pinus pinaster*) bark were evaluated for the uptake and recovery of gold from simulated e-waste hydrometallurgical liquors. Higher extraction efficiencies of gold were obtained from HCl solutions, in comparison to aqua regia, but excessive levels of HCl caused a severe impact in the metal uptake from both aqueous systems. Even though, at strong acidic levels and in both acidic systems, the pine bark tannin resin presented a good performance, with maximum adsorption capacities ranging from 200 and 343 mg g^{-1} (2.0 and 1.0 mol L^{-1} H^+, respectively). The adsorption process was found to be a rather slow process (2–3 days to be completed) and successfully described by the homogeneous solid diffusion model and LDF approximation. The selectivity of the adsorbent towards gold was evaluated using simulated hydrometallurgical liquors. The amount of gold extracted by the tannin resin from multi-metal solutions was similar to the values registered in the single-metal solutions and the co-adsorption of Pd(II), Cu(II), Fe(II), Ni(II) and Zn(II) was in general low. Desorption and regeneration studies conducted through three adsorption–desorption cycles, using a solution of 0.5 mol L^{-1} thiourea and 0.5 mol L^{-1} HCl as eluent, indicate desorption percentages of 38–57%, although with only a mild loss of the adsorption capacity of the regenerated adsorbent.

Considering the harsher conditions at which the pine bark tannin resin was evaluated, it can be said that it presents analogous or even better performance when compared to other biosorbents or synthetic resins. A deeper analysis on the final recovery of gold from the exhausted adsorbents will be considered in forthcoming work. More research is needed in this field to improve the circularity in WEEE, particularly by the promotion of benign practices.

Author Contributions: Investigation, M.B.Q.L.F.T., H.A.M.B., and S.C.R.S.; conceptualization, S.C.R.S.; formal analysis, S.C.R.S., M.B.T. and H.A.M.B.; funding acquisition and resources, C.M.S.B. and R.A.R.B.; project administration, C.M.S.B.; methodology, S.C.R.S.; supervision, S.C.R.S., R.A.R.B., and C.M.S.B.; validation, S.C.R.S., C.M.S.B., and R.A.R.B.; visualization, M.B.Q.L.F.T., H.A.M.B., and S.C.R.S.; writing—original draft—S.C.R.S.; writing—review and editing—M.B.Q.L.F.T., H.A.M.B., R.A.R.B., and C.M.S.B. All authors have read and agreed to the published version of the manuscript.

Funding: This work was financially supported by: Base Funding—UIDB/50020/2020 of the Associate Laboratory LSRE-LCM—funded by national funds through FCT/MCTES (PIDDAC). S. Santos acknowledges postdoctoral scholarship (SFRH/BPD/117387/2016), awarded by the Portuguese Science and Technology Foundation (FCT) and financed by National Funds and by the European Social Fund (ESF) through the Human Capital Operational Programme (POCH). H. Bacelo acknowledges PhD scholarship (PD/BD/135062/2017) funded by FCT.

Conflicts of Interest: The authors declare no conflict of interest.

References

1. Perez, J.P.H.; Folens, K.; Leus, K.; Vanhaecke, F.; Van Der Voort, P.; Du Laing, G. Progress in hydrometallurgical technologies to recover critical raw materials and precious metals from low-concentrated streams. *Resour. Conserv. Recycl.* **2019**, *142*, 177–188. [CrossRef]
2. Falahati, M.; Attar, F.; Sharifi, M.; Saboury, A.A.; Salihi, A.; Aziz, F.M.; Kostova, I.; Burda, C.; Priecel, P.; Lopez-Sanchez, J.A.; et al. Gold nanomaterials as key suppliers in biological and chemical sensing, catalysis, and medicine. *Biochim. Biophys. Acta BBA Gen. Subj.* **2020**, *1864*, 129435. [CrossRef] [PubMed]
3. Forti, V.; Baldé, C.P.; Kuehr, R.; Bel, G. *The Global E-waste Monitor 2020: Quantities, Flows and the Circular Economy Potential*; United Nations University (UNU)/United Nations Institute for Training and Research (UNITAR)–co-hosted SCYCLE Programme, International Telecommunication Union (ITU) & International Solid Waste Association (ISWA): Bonn, Germany; Geneva, Switzerland; Rotterdam, The Netherlands, 2020.
4. European Parliament and Council. *Directive 2012/19/EU of the European Parliament and of the Council of 4 July 2012 on Waste Electrical and Electronic Equipment (WEEE)*; European Parliament and Council: Brussels, Belgium, 2012.
5. Akcil, A.; Erust, C.; Gahan, C.S.; Ozgun, M.; Sahin, M.; Tuncuk, A. Precious metal recovery from waste printed circuit boards using cyanide and non-cyanide lixiviants-A review. *Waste Manag.* **2015**, *45*, 258–271. [CrossRef] [PubMed]
6. Gurung, M.; Adhikari, B.B.; Kawakita, H.; Ohto, K.; Inoue, K.; Alam, S. Recovery of gold and silver from spent mobile phones by means of acidothiourea leaching followed by adsorption using biosorbent prepared from persimmon tannin. *Hydrometallurgy* **2013**, *133*, 84–93. [CrossRef]
7. Rao, M.D.; Singh, K.K.; Morrison, C.A.; Love, J.B. Challenges and opportunities in the recovery of gold from electronic waste. *RSC Adv.* **2020**, *10*, 4300–4309. [CrossRef]
8. Kaya, M. Recovery of metals and nonmetals from electronic waste by physical and chemical recycling processes. *Waste Manag.* **2016**, *57*, 64–90. [CrossRef]
9. Ventura, E.; Futuro, A.; Pinho, S.C.; Almeida, M.; Dias, J.M. Physical and thermal processing of Waste Printed Circuit Boards aiming for the recovery of gold and copper. *J. Environ. Manag.* **2018**, *223*, 297–305. [CrossRef]
10. Zhang, Y.; Liu, S.; Xie, H.; Zeng, X.; Li, J. Current Status on Leaching Precious Metals from Waste Printed Circuit Boards. *Procedia Environ. Sci.* **2012**, *16*, 560–568. [CrossRef]
11. Camelino, S.; Rao, J.; Padilla, R.L.; Lucci, R. Initial Studies about Gold Leaching from Printed Circuit Boards (PCB's) of Waste Cell Phones. *Procedia Mater. Sci.* **2015**, *9*, 105–112. [CrossRef]
12. Diaz, L.A.; Lister, T.E. Economic evaluation of an electrochemical process for the recovery of metals from electronic waste. *Waste Manag.* **2018**, *74*, 384–392. [CrossRef]
13. Birich, A.; Mohamed, S.R.; Friedrich, B. Screening of Non-cyanide Leaching Reagents for Gold Recovery from Waste Electric and Electronic Equipment. *J. Sustain. Met.* **2018**, *4*, 265–275. [CrossRef]

14. Korolev, I.; Kolehmainen, E.; Haapalainen, M.; Yliniemi, K.; Lundström, M. Gold Recovery from Chloride Leaching Solutions by Electrodeposition-Redox Replacement Method. In Proceedings of the 10th European Metallurgical Conference (EMC 2019), Düsseldorf, Germany, 23–26 June 2019.
15. Sayiner, B.; Acarkan, N. Effect of Silver, Nickel and Copper Cyanides on Gold Adsorption on Activated Carbon in Cyanide Leach Solutions. *Physicochem. Probl. Miner. Process.* **2014**, *50*, 277–287. [CrossRef]
16. Leroux, J.D.; Bryson, A.W.; Young, B.D. A Comparison of Several Kinetic-Models for the Adsorption of Gold Cyanide onto Activated Carbon. *J. S. Afr. Inst. Min. Metall.* **1991**, *91*, 95–103.
17. Dodson, J.R.; Parker, H.L.; García, A.M.; Hicken, A.; Asemave, K.; Farmer, T.J.; He, H.; Clark, J.H.; Hunt, A.J. Bio-derived materials as a green route for precious & critical metal recovery and re-use. *Green Chem.* **2015**, *17*, 1951–1965. [CrossRef]
18. Bediako, J.K.; Lin, S.; Sarkar, A.K.; Zhao, Y.; Choi, J.-W.; Song, M.-H.; Wei, W.; Reddy, D.H.K.; Cho, C.-W.; Yun, Y.-S. Benignly-fabricated crosslinked polyethylenimine/calcium-alginate fibers as high-performance adsorbents for effective recovery of gold. *J. Clean. Prod.* **2020**, *252*, 119389. [CrossRef]
19. Liu, F.; Zhou, L.; Wang, W.; Yu, G.; Huang, J. Adsorptive recovery of Au(III) from aqueous solution using crosslinked polyethyleneimine resins. *Chemosphere* **2019**, *241*, 125122. [CrossRef]
20. Morcali, M.H.; Zeytuncu, B.; Ozlem, E.; Aktas, S. Studies of Gold Adsorption from Chloride Media. *Mater. Res.* **2015**, *18*, 660–667. [CrossRef]
21. Choudhary, B.C.; Choudhary, B.C.; Borse, A.U.; Garole, D.J. Surface functionalized biomass for adsorption and recovery of gold from electronic scrap and refinery wastewater. *Sep. Purif. Technol.* **2018**, *195*, 260–270. [CrossRef]
22. Bediako, J.K.; Choi, J.-W.; Song, M.-H.; Zhao, Y.; Lin, S.; Sarkar, A.K.; Cho, C.-W.; Yun, Y.-S. Recovery of gold via adsorption-incineration techniques using banana peel and its derivatives: Selectivity and mechanisms. *Waste Manag.* **2020**, *113*, 225–235. [CrossRef]
23. Al-Saidi, H. The fast recovery of gold(III) ions from aqueous solutions using raw date pits: Kinetic, thermodynamic and equilibrium studies. *J. Saudi Chem. Soc.* **2016**, *20*, 615–624. [CrossRef]
24. Fan, R.; Min, H.; Hong, X.; Yi, Q.; Liu, W.; Zhang, Q.; Luo, Z. Plant tannin immobilized Fe3O4@SiO2 microspheres: A novel and green magnetic bio-sorbent with superior adsorption capacities for gold and palladium. *J. Hazard. Mater.* **2019**, *364*, 780–790. [CrossRef] [PubMed]
25. Fan, R.; Xie, F.; Guan, X.; Zhang, Q.; Luo, Z. Selective adsorption and recovery of Au(III) from three kinds of acidic systems by persimmon residual based bio-sorbent: A method for gold recycling from e-wastes. *Bioresour. Technol.* **2014**, *163*, 167–171. [CrossRef] [PubMed]
26. Santos, S.C.; Bacelo, H.A.M.; Boaventura, R.A.R.; Botelho, C.M.S. Tannin-Adsorbents for Water Decontamination and for the Recovery of Critical Metals: Current State and Future Perspectives. *Biotechnol. J.* **2019**, *14*, e1900060. [CrossRef] [PubMed]
27. Santos, N.T.D.G.; Moraes, L.F.; Da Silva, M.G.C.; Vieira, M.G.A. Recovery of gold through adsorption onto sericin and alginate particles chemically crosslinked by proanthocyanidins. *J. Clean. Prod.* **2020**, *253*, 119925. [CrossRef]
28. Liu, F.; Wang, S.; Chen, S. Adsorption behavior of Au(III) and Pd(II) on persimmon tannin functionalized viscose fiber and the mechanism. *Int. J. Biol. Macromol.* **2020**, *152*, 1242–1251. [CrossRef]
29. Yi, Q.; Fan, R.; Xie, F.; Min, H.; Zhang, Q.; Luo, Z. Selective Recovery of Au(III) and Pd(II) from Waste PCBs Using Ethylenediamine Modified Persimmon Tannin Adsorbent. *Procedia Environ. Sci.* **2016**, *31*, 185–194. [CrossRef]
30. Wang, Z.; Li, X.; Liang, H.; Ning, J.; Zhou, Z.; Li, G. Equilibrium, kinetics and mechanism of Au^{3+}, Pd^{2+} and Ag^{+} ions adsorption from aqueous solutions by graphene oxide functionalized persimmon tannin. *Mater. Sci. Eng. C* **2017**, *79*, 227–236. [CrossRef]
31. Bacelo, H.A.; Vieira, B.R.; Santos, S.C.; Boaventura, R.A.; Botelho, C.M. Recovery and valorization of tannins from a forest waste as an adsorbent for antimony uptake. *J. Clean. Prod.* **2018**, *198*, 1324–1335. [CrossRef]
32. Sánchez-Martín, J.; Beltrán-Heredia, J.; Gibello-Pérez, P. Adsorbent biopolymers from tannin extracts for water treatment. *Chem. Eng. J.* **2011**, *168*, 1241–1247. [CrossRef]
33. Bacelo, H.A.; Santos, S.C.; Botelho, C.M. Removal of arsenic from water by an iron-loaded resin prepared from Pinus pinaster bark tannins. *Euro-Mediterr. J. Environ. Integr.* **2020**, *5*, 1–17. [CrossRef]
34. Gurung, M.; Adhikari, B.B.; Kawakita, H.; Ohto, K.; Inoue, K.; Alam, S. Recovery of Au(III) by using low cost adsorbent prepared from persimmon tannin extract. *Chem. Eng. J.* **2011**, *174*, 556–563. [CrossRef]

35. Saman, N.; Tan, J.-W.; Mohtar, S.S.; Kong, H.; Lye, J.W.P.; Johari, K.; Hassan, H.; Mat, H. Selective biosorption of aurum(III) from aqueous solution using oil palm trunk (OPT) biosorbents: Equilibrium, kinetic and mechanism analyses. *Biochem. Eng. J.* **2018**, *136*, 78–87. [CrossRef]
36. Xie, F.; Fan, Z.; Zhang, Q.; Luo, Z. Selective adsorption of Au3+from aqueous solutions using persimmon powder-formaldehyde resin. *J. Appl. Polym. Sci.* **2013**, *130*, 3937–3946. [CrossRef]
37. Langmuir, I. The Adsorption of Gases on Plane Surfaces of Glass, Mica and Platinum. *J. Am. Chem. Soc.* **1918**, *40*, 1361–1403. [CrossRef]
38. Freundlich, H.M.F. Over the adsorption in solution. *J. Phys. Chem.* **1906**, *57*, 385–471.
39. Tien, C. Batch Adsorption Models and Model Applications. In *Introduction to Adsorption*; Tien, C., Ed.; Elsevier: Amsterdam, The Netherlands, 2019; Chapter 5; pp. 119–153. [CrossRef]
40. Zur Theorie der sogenannten Adsorption gelöster Stoffe. *Z. Chem. Ind. Kolloide* **1907**, *2*, 15. [CrossRef]
41. Blanchard, G.; Maunaye, M.; Martin, G. Removal of heavy metals from waters by means of natural zeolites. *Water Res.* **1984**, *18*, 1501–1507. [CrossRef]
42. Ho, Y.S. Adsorption of Heavy Metals from Waste Streams by Peat. Ph.D. Thesis, The University of Birmingham, Birmingham, UK, 1995.
43. Glueckauf, E. Theory of chromatography. Part 10-Formulae for diffusion into spheres and their application to chromatography. *Trans. Faraday Soc.* **1955**, *51*, 1540–1551. [CrossRef]
44. Santos, S.C.; Boaventura, R. Adsorption of cationic and anionic azo dyes on sepiolite clay: Equilibrium and kinetic studies in batch mode. *J. Environ. Chem. Eng.* **2016**, *4*, 1473–1483. [CrossRef]
45. Tien, C. Adsorbate Uptake and Equations Describing Adsorption Processes. In *Introduction to Adsorption*; Tien, C., Ed.; Elsevier: Amsterdam, The Netherlands, 2019; Chapter 4; pp. 87–118.
46. Ogata, T.; Nakano, Y. Mechanisms of gold recovery from aqueous solutions using a novel tannin gel adsorbent synthesized from natural condensed tannin. *Water Res.* **2005**, *39*, 4281–4286. [CrossRef]
47. Bergamini, M.F.; Santos, D.P.; Zanoni, M.V.B. Screen-printed carbon electrode modified with poly-L-histidine applied to gold(III) determination. *J. Braz. Chem. Soc.* **2009**, *20*, 100–106. [CrossRef]
48. Wang, C.; Zhao, J.; Zhang, L.; Wang, C.; Wang, S. Efficient and Selective Adsorption of Gold Ions from Wastewater with Polyaniline Modified by Trimethyl Phosphate: Adsorption Mechanism and Application. *Polymers* **2019**, *11*, 652. [CrossRef] [PubMed]
49. Tokuyama, H.; Kanehara, A. Temperature swing adsorption of gold(III) ions on poly(N-isopropylacrylamide) gel. *React. Funct. Polym.* **2007**, *67*, 136–143. [CrossRef]
50. Yang, J.; Kubota, F.; Baba, Y.; Kamiya, N.; Goto, M. Application of cellulose acetate to the selective adsorption and recovery of Au(III). *Carbohydr. Polym.* **2014**, *111*, 768–774. [CrossRef] [PubMed]
51. Parajuli, D.; Kawakita, H.; Inoue, K.; Ohto, K.; Kajiyama, K. Persimmon peel gel for the selective recovery of gold. *Hydrometallurgy* **2007**, *87*, 133–139. [CrossRef]
52. Malash, G.F.; El-Khaiary, M.I. Piecewise linear regression: A statistical method for the analysis of experimental adsorption data by the intraparticle-diffusion models. *Chem. Eng. J.* **2010**, *163*, 256–263. [CrossRef]
53. Yao, C.; Chen, T. A new simplified method for estimating film mass transfer and surface diffusion coefficients from batch adsorption kinetic data. *Chem. Eng. J.* **2015**, *265*, 93–99. [CrossRef]
54. Da Silva, J.S.; Da Rosa, M.P.; Beck, P.H.; Peres, E.C.; Dotto, G.L.; Kessler, F.; Grasel, F.D.S. Preparation of an alternative adsorbent from Acacia Mearnsii wastes through acetosolv method and its application for dye removal. *J. Clean. Prod.* **2018**, *180*, 386–394. [CrossRef]
55. Mack, C.; Wilhelmi, B.; Duncan, J.; Burgess, J. Biosorption of precious metals. *Biotechnol. Adv.* **2007**, *25*, 264–271. [CrossRef]
56. Wei, W.; Reddy, D.H.K.; Bediako, J.K.; Yun, Y.-S. Aliquat-336-impregnated alginate capsule as a green sorbent for selective recovery of gold from metal mixtures. *Chem. Eng. J.* **2016**, *289*, 413–422. [CrossRef]
57. Li, J.; Miller, J. Reaction kinetics for gold dissolution in acid thiourea solution using formamidine disulfide as oxidant. *Hydrometallurgy* **2002**, *63*, 215–223. [CrossRef]

58. Đurović, M.; Puchta, R.; Bugarčić, Ž.D.; Van Eldik, R. Studies on the reactions of [AuCl4]− with different nucleophiles in aqueous solution. *Dalton Trans.* **2014**, *43*, 8620–8632. [CrossRef]

Article

Cd(II) and Pb(II) Adsorption Using a Composite Obtained from *Moringa oleifera* Lam. Cellulose Nanofibrils Impregnated with Iron Nanoparticles

Adriana Vázquez-Guerrero [1], Raúl Cortés-Martínez [2,*], Ruth Alfaro-Cuevas-Villanueva [3], Eric M. Rivera-Muñoz [4] and Rafael Huirache-Acuña [1,*]

1 División de Estudios de Posgrado, Facultad Ingeniería Química, Universidad Michoacana de San Nicolás de Hidalgo Edificio M, Ciudad Universitaria, C.P. 58030 Morelia, Mexico; adris.mokona@gmail.com
2 Facultad de Químico Farmacobiología, Universidad Michoacana de San Nicolás de Hidalgo, Tzintzuntzan 173, Colonia Matamoros, C.P. 58240 Morelia, Mexico
3 Instituto de Investigaciones en Ciencias de la Tierra, Universidad Michoacana de San Nicolás de Hidalgo, Edificio U5, Ciudad Universitaria, C.P. 58030 Morelia, Mexico; racv11@gmail.com
4 Centro de Física Aplicada y Tecnología Avanzada, Universidad Nacional Autónoma de México, Campus Juriquilla, C.P. 76230 Querétaro, Mexico; emrivera@fata.unam.mx
* Correspondence: raulcortesmtz@gmail.com (R.C.-M.); rafael_huirache@yahoo.it (R.H.-A.)

Abstract: This work informs on the green synthesis of a novel adsorbent and its adsorption capacity. The adsorbent was synthesized by the combination of iron nanoparticles and cellulose nanofibers (FeNPs/NFCs). Cellulose nanofibers (NFCs) were obtained from Moringa (*Moringa oleifera* Lam.) by a pulping Kraft process, acid hydrolysis, and ultrasonic methods. The adsorption method has advantages such as high heavy metal removal in water treatment. Therefore, cadmium (Cd) and lead (Pb) adsorption with FeNP/NFC from aqueous solutions in batch systems was investigated. The kinetic, isotherm, and thermodynamic parameters, as well as the adsorption capacities of FeNP/NFC in each system at different temperatures, were evaluated. The adsorption kinetic data were fitted to mathematical models, so the pseudo-second-order kinetic model described both Cd and Pb. The kinetic rate constant (K_2), was higher for Cd than for Pb, indicating that the metal adsorption was very fast. The adsorption isotherm data were best described by the Langmuir–Freundlich model for Pb multilayer adsorption. The Langmuir model described Cd monolayer sorption. However, experimental maximum adsorption capacities ($q_{e\,exp}$) for Cd (>12 mg/g) were lower than those for Pb (>80 mg/g). In conclusion, iron nanoparticles on the FeNP/NFC composite improved Cd and Pb selectivity during adsorption processes, indicating the process' spontaneous and exothermic nature.

Keywords: cadmium; lead; adsorption; cellulose nanofibrils; iron nanoparticles

Citation: Vázquez-Guerrero, A.; Cortés-Martínez, R.; Alfaro-Cuevas-Villanueva, R.; Rivera-Muñoz, E.M.; Huirache-Acuña, R. Cd(II) and Pb(II) Adsorption Using a Composite Obtained from *Moringa oleifera* Lam. Cellulose Nanofibrils Impregnated with Iron Nanoparticles. *Water* **2021**, *13*, 89. https://doi.org/10.3390/w13010089

Received: 26 November 2020
Accepted: 28 December 2020
Published: 3 January 2021

1. Introduction

Water pollution by heavy metals is one of the most severe environmental problems, with industrial processes being their main anthropogenic sources. Heavy metals can be distinguished from other pollutants because they are not biodegradable, and they can be bioaccumulated in different food chains. Furthermore, these metals have been linked to carcinogenic, embryotoxic, teratogenic, and mutagenic effects [1,2]. Consequently, the permissible limits for cadmium (Cd) and lead (Pb) in drinking water established in 2017 by the World Health Organization's (WHO) standards are 0.003 and 0.01 mg/L, respectively. The United States Environmental Protection Agency (USEPA) has set the maximum contaminant levels (MCLs) in drinking water at 0.015 mg/L for lead and 0.005 mg/L for cadmium [3–5]. Cadmium and lead are some of the heavy metals introduced as anthropogenic pollutants worldwide by various processes. Cadmium contamination may occur due to the manufacture of alloys, batteries, pigments, plastics, mining, and refining processes [6]. Cadmium causes severe damage to kidneys and bones. Its accumulation in the human body leads to nausea, headaches, fatigue, salivation, diarrhea, muscular cramps, renal degradation, chronic pulmonary problems, skeletal deformity, low levels of iron in the blood, and breathlessness (dyspnea) [7,8].

On the other hand, common anthropogenic causes of lead contamination in groundwater are mining, smelting, fossil fuel combustion, solid waste incineration, batteries, paints, cables, ceramics, and glass manufacturing [9]. In children, overexposure to lead causes swelling of the optic nerve, ataxia, brain damage (encephalopathy), convulsions, seizures, and impaired consciousness. In adults, lead exposure causes high blood pressure, damage to the reproductive organs, fever, headaches, fatigue, vomiting, anorexia, abdominal pain, constipation, joint pain, incoordination, insomnia, irritability, and various symptoms related to the nervous systems [7].

Many techniques have been employed to reduce the adverse effects of cadmium and lead in water bodies, including chemical oxidation/reduction, chemical precipitation, electrochemical processes, ion exchange, membrane filtration, and others [10,11]. These processes can be difficult to operate, and they show low selectivity. However, among these methods, adsorption is flexible, low cost, and a high-efficiency technique [12]. Adsorption is a mass transfer process by which a substance is transferred from the liquid phase to the solid's surface by physical or chemical interactions.

Other studies have demonstrated the use of biomaterials as efficient, low-cost, abundant, biodegradable, and highly stable adsorbents [13]. Cellulose is an abundant natural polymer and is commonly found in wood, seeds, algae, and agro-industrial waste. The Moringa tree (*Moringa oleifera* Lam), a member of the Moringaceae family, can grow around the tropics. It is native to sub-Himalayan areas of northwestern India. It can be easily cultivated and adapts to semiarid climates, as it is a polyvalent tropical tree. In Michoacan, Mexico, the generation of residual biomass product from its plantation along the Pacific coast has been reported, with an average of 300 to 500 kg per hectare. This waste from leaves, stems, and branches concentrates wasted cellulose in large amounts [14,15]. Among the various functional polymers exist cellulose nanofibers (NFCs). Such fibers are large flexible fiber structures consisting of alternating crystalline and amorphous domains in a net-like structure. NFCs can be extracted from a cellulosic material, moderately degraded by chemical methods, thermal processes, and mechanical processes. Additionally, NFCs have been obtained from combined pulping Kraft process, acid hydrolysis, and ultrasonic methods [16,17].

The adsorption of metals by this type of material can be attributed to their structural compounds with various functional groups, such as carboxyl, carbonyl, sulfide, and hydroxyl. Many of these hydroxyl groups allow chemical modification and improve adsorption performance [18]. Cellulose tends to form an aggregate structure due to the hydrogen bonds being present, limiting adsorption; hence, modification of the cellulose structure and surface can help to avoid adsorption restraints. One of these modifications involves impregnation with metallic nanoparticles. Such a change can enhance adsorption properties in the biopolymer [19]. It has been reported that many metal oxide nanoparticles and other inorganic nanoparticles (i.e., zeolite nanoparticles) effectively and quickly remove metal contaminants in wastewater treatment processes [11,20].

The appearance of iron nanoparticles in remediation is attributed to their low toxicity and redox capacity when reacting with water. They have a large surface areas and high reactivity [21]. The use of iron nanoparticles has been reported in different carbon-based materials, such as starch and biopolymers, as well as in mineral adsorbents, such as clays. These materials act as supports for obtaining nanocomposites [22–26]. Moreover, functionalization with organic solvents helps to mitigate nanoparticles' agglomeration on the composite and improves dispersion. Different modifiers are used to abate nanoparticles' mobility in water, such as carboxymethyl cellulose or cellulose, since they are negatively charged molecules. These nanoparticles (NPs) are generally considered to be the most mobile in aqueous solutions due to electrostatic repulsion [27].

Several processes have been utilized to prepare nanoadsorbents. These are often considered as time-consuming processes, have high reagent requirements, and release toxic substances into the environment [28]. Otherwise, nanoparticles synthesized by plant products or extracts are more stable. These green syntheses are more manageable than conventional techniques. Such ecological approaches are cost-effective, simple, and sustainable and do not involve any toxic agents [29]. Different green synthesis routes use agro-industrial wastes, biopolymers, solvent replacement, plants, microorganisms, etc. [30]. Nanoparticles can be incorporated into the biopolymer through different mechanisms, such as chemical coprecipitation, to obtain the final composite material. Still, the use of plant extracts (organic compounds) can influence the reduction process. They can also stabilize the nanoparticles, promoting bonding to the biopolymer [31].

Fruits, seeds, stem, and leaf extracts from different plants contain useful polyphenols for metal reduction. These substances have been researched as bioreducing agents to synthesize nanoparticles, such as *Vitex negundo* L., *E. condylocarpa*, and *Amaranthus dubius* [32–35]; including *Rumex acetosa* [36], *Eucalyptus globulus* [37], *Mukia maderaspatana* [38], *Camellia sinensis* [39], *Rosemarinus officinalis* [40], *Citrus maxima* [41], *Azadirachta Indica* [30,42], green tea [43], eucalyptus [44], and *Canna indica* [21]. Additionally, waste materials such as banana peel and citrus juice waste have also been used.

Green plant leaves have recently been studied for the eco-friendly synthesis of iron nanoparticles, using Fe(II) or Fe(III) chloride solutions as precursors [30]. This synthesis is possible because these plants contain bioactive components, mostly polyphenols, that act as reducing and capping agents. Moreover, several applications and syntheses of metal oxide nanoparticles, particularly for iron, have been reported [45–49]. Jitendra et al. [50] reported that *A. indica* leaf extract assisted in the green synthesis of highly crystalline α-Fe_2O_3 nanomaterials with well-defined unique morphologies. Similarly, the green-synthesized mesoporous α-Fe_2O_3 had a surface area four times larger than the commercial α-Fe_2O_3 nanoparticles [39]. Moringa leaf contains various photochemical type-like flavonoids, such as polyphenols, rutin, kaempferol, chlorogenic acid, quercetin-malonyl-glucoside, glucosinolate, and isothiocyanates, with high antioxidant capacities. These are an alternative reducing agent for synthesizing iron oxide NPs [51,52].

Therefore, the present study aims to evaluate the iron nanoparticles and cellulose nanofibers (FeNPs/NFCs) performance for the removal of Cd(II) and Pb(II) from aqueous solutions and to examine the structure of the composite obtained from the cellulose nanofiber's (from branch wood residues) contact with iron nanoparticles (from green synthesis with $FeCl_3$ and leaf tea extract). Furthermore, the influence of contact time, temperature, and pH on adsorption experiments was determined, as were the adsorption mechanisms and thermodynamic, kinetic, and isothermal parameters of the studied systems.

2. Materials and Methods

2.1. Preparation of FeNP/NFC Composite

Large amounts of wood residues obtained from the pruning of the Moringa tree were collected from an orchard located in the municipality of Nueva Italia in the State of Michoacán, Mexico. These wood residues were used for cellulose extraction by Kraft pulping, followed by acid hydrolysis.

The cellulose nanofibrils (NFCs) were obtained using a high-intensity ultrasonication technique combined with cryo-crushing with liquid nitrogen. Wood chips from *Moringa oleifera* Lam. were cooked by Kraft pulping in a Kamir digester (METROTEC, model LB-16) with a capacity of 15 L and a speed of 5.16 r.p.m. The Kraft Pulping process was carried out with the following operating parameters: the temperature ranged from 160 to 170 °C with a boiling time of 20–30 min, a white liquor with sulfidity levels between 20 and 27% was used, using 200 g of wood chips (dry basis).

Chemical post-treatment to wood fibers by acid hydrolysis was performed with an H_2SO_4 solution 40% (*v/v*) at 80 °C for 2 h. The separation of fibers was carried out using a high-intensity ultrasonication technique combined with cryo-crushing with liquid nitrogen.

The mechanical cryo-crushing with ice crystal formation slashed the vegetal fiber wall and released wall fragments. The ultrasonic treatment was carried out for 40 min at 60 kHz, and then the material was dried in an oven. Mechanical fibrillation was performed using liquid nitrogen and applying high shear forces.

The synthesis of the composite material (FeNP/NFC) was carried out by putting in contact 4 g of cellulose (Kraft pulp) and 40 mL of leaf extract. Fresh *Moringa oleifera* Lam. leaves were collected and carefully washed with deionized water, and they were dried at room temperature with constant ventilation. Then, 20 g of dried leaves sample were mixed with 150 mL distilled water at 60 °C for 45 min. The supernatant was filtered with a Whatman filter paper to produce the leaf extract, and was stored at 4 °C in the refrigerator. A 0.5 M solution of $FeCl_3.6H_2O$ was subsequently adjusted at pH 8, using 0.1 N HCl and 0.1 N NaOH solutions. The leaf extract (pH 5.6) was added dropwise to the $FeCl_3$ solution with continuous stirring at 60 °C for 180 min. The modified material FeNP/NFC was filtered and washed several times with distilled water to eliminate any excess of reactive components from the material; the modified material was allowed to dry and it was stored in a desiccator.

2.2. *Characterization of FeNP/NFC Composite*

Different methods were used to perform the characterization of Kraft pulp and composite material in order to determine their morphologies, semiquantitative elemental properties, and surface properties. Scanning Electron Microscopy (SEM) images were obtained using a JEOL JSM IT300LV microscope (Tokyo, Japan).

Fourier Transform Infrared Spectroscopy (FTIR) was used to examine the changes in the functional groups induced by the adsorbent synthesis and the adsorption process. The samples were analyzed in a Bruker® Tensor 27 FTIR spectrophotometer (Bruker Optics Inc., Billerica, MA, USA).

The zeta potential (ZP) was used to measure the electrical potential on the adsorbent's surface. From the values of the zeta potential at different values 3 to 10 pH, the surface acidity or basicity and the isoelectric point were determined [16]. The FeNP/NFC material's zeta potential was measured with a Nano Brook 90 Plus Zeta instrument.

X-ray Diffraction (XRD) measurements were conducted to determine the presence of the cellulose crystalline phase in the samples, using a Rigaku Ultima IV (Rikagu, Tokio, Japan). The diffracted intensity of the Cu-Kα (λ = 1.54 Å) was measured in a 2θ range. Data analysis to identify crystalline and amorphous phases was carried out using Match V. 1.10[a] 2003–2010 Software (Crystal Impact, Bonn, Germany).

The FeNP/NFC composite oxides were analyzed on an energy dispersive X-Ray fluorescence spectrometer model Bruker S2PUMA with an Ag tube. Samples were analyzed to define the form of oxides.

2.3. *Adsorption Experiments*

2.3.1. Kinetics

Batch-type experiments with FeNP/NC were performed to determine Cd and Pb removal. Samples with 10 mL aliquots of a 2 mg/L metal solution were put in contact with 0.05 g of composite material in plastic flasks. The pHs of the initial solution for Pb and Cd were 2 and 3, respectively. The solutions' pHs were adjusted to and maintained at 5 (by adding 0.01 M HCl or 0.05 M NaOH to the solutions) due to the cationic chemical species (II) present in the solution. The flasks were shaken, separately, at 100 rpm and 25 °C. The solutions' concentrations were 2 mg/L for both $Pb(NO_3)_2$ and $Cd(NO_3)_2{\cdot}4H_2O$. The contact times ranged from 5 to 150 min to acquire the equilibrium time. After the contact time was achieved, the supernatants were separated by filtration, and the residual solution was analyzed for each metal by Flame Atomic Absorption Spectrometry (FAAS) (model Perkin Elmer® AANALYST 200, Waltham, MA, USA). The solutions were refrigerated at 4 °C and stored in amber bottles (plastic and glass) to avoid precipitation and oxidation during analysis. The experiments were performed in triplicate to determine reproducibility.

The Cd and Pb adsorption concentrations at equilibrium time during adsorption were calculated by:

$$q_t = \frac{(C_o - C_t)V}{m} \tag{1}$$

where q_t is the amount of ion adsorbed, at a time (t), per weight of adsorbent (mg/g); C_o is the initial concentration of heavy metal (mg/L); C_t is the concentration of heavy metal at that equilibrium time (mg/L); V is the volume of solution (L), and m is the mass of the adsorbent (g).

A batch removal system requires information about the optimum conditions that can only be acquired from kinetic models. Moreover, adsorption kinetics studies and their adjustment to empirical models allow the evaluation of removal mechanisms and are fundamental for selecting optimal design parameters for future effluent treatments. The data obtained from the kinetics tests were fitted to the Elovich, Lagergren (pseudo-first-order), and pseudo-second-order kinetic models by nonlinear regression analysis using Statistica 7.0 software. Such kinetic models were used in this research due to their wide use and acceptance in heavy metals removal studies.

The general expression of the pseudo-first-order rate, also known as the Lagergren model, is represented by the following equation [53].

$$q_t = q_e\left(1 - e^{-K_l t}\right) \tag{2}$$

where K_L is the Lagergren rate constant (1/min), q_t is the amount of ions adsorbed, at a time (t), per weight of adsorbent (mg/g), and q_e is the amount of ions adsorbed at the equilibrium (mg/g).

The pseudo-second-order model is based on the assumption that the rate-limiting step may be chemisorption [54]. Alternatively, the pseudo-second-order model is expressed in its linear form as follows:

$$q_t = \frac{K_2 q_e^2 t}{1 + 2K_2 q_e t} \tag{3}$$

where K_2 is the pseudo-second-order rate constant for adsorption (g/mg min), q_t is the amount of ions adsorbed, at a time (t), per weight of adsorbent (mg/g), and q_e is the amount of ions adsorbed at the equilibrium (mg/g).

The Elovich equation is represented by the following [55], commonly used in the kinetics of the chemisorption of gases on solids:

$$q_t = \frac{1}{b}\ln(1 + abt) \tag{4}$$

where q_t is the amount of ion adsorbed, at a time (t), per weight of adsorbent (mg/g), q_e is the concentration of solute removed at the equilibrium per weight of adsorbent (mg/g), K_L is the pseudo-first-order kinetic constant (1/min), K_2 is the pseudo-second-order rate constant of sorption (g/mg.min), and a and b are the Elovich constants related to the initial adsorption rate (mg/g min) and desorption rate (g/mg), respectively.

2.3.2. Isotherms

Adsorption isotherms were performed with Cd and Pb aqueous solutions at different concentrations to determine the systems' equilibrium parameters. Plastic vials were used to put 0.05 g of FeNP/NC in contact with 10 mL of Pb and Cd solutions, separately, at different concentrations, ranging from 5 to 500 mg/L. Then, the vials were placed in a thermostat-adjusted water bath agitator. They were shaken at 100 rpm at different temperatures (25, 30, and 40 °C). All experiments were performed in triplicate. According to the chemical equilibrium diagram for cation ions in an aqueous solution, the pH values were selected. Cd ions predominate at pH values between 2 and 8 and Pb ions with pH values between 3 and 5.5. The pH 5 value was then set and adjusted with 0.01 M HCl and 0.05 M NaOH solutions, as required, before the adsorption experiments.

The experimental data from sorption isotherms of Pb and Cd were fitted to the following isotherm models by nonlinear regression analysis. The equation that expresses the Freundlich model is:

$$Q_e = K_f C_e^{\frac{1}{n}} \tag{5}$$

where Q_e is the adsorption capacity of Pb(II) and Cd(II) of the adsorbents (mg/g), C_e is the equilibrium concentration of the metal ion in solution (mg/L), K_f is the equilibrium constant indicative of adsorption capacity, and n is the adsorption equilibrium constant whose reciprocal is indicative of adsorption intensity. The Freundlich model assumes that the adsorption process happened on heterogeneous surfaces [56].

Langmuir model could be represented by Equation (6) [57]:

$$Q_e = \frac{Q_0 K_L C_e}{1 + K_L C_e} \tag{6}$$

where Q_0 represents the maximum monolayer adsorption capacity of Pb and Cd, and K_L is the Langmuir constant (L/g). The Langmuir model is established based on forming a complete monolayer on the surface (mg/g).

Langmuir–Freundlich model is represented by Equation (7) [58]:

$$Q_e = \frac{K_{LF} C_e^{\frac{1}{n}}}{1 + a_{LF} C_e^{\frac{1}{n}}} \tag{7}$$

where K_{LF} and a_{LF} are empirical constants.

2.4. Thermodynamic Studies

The thermodynamic results were calculated from isotherm data of Pb and Cd at 25, 30, and 40 °C. The thermodynamic parameters were estimated to analyze the influence of temperature on metal ion adsorption, such as Gibb's free energy change ΔG (kJ/mol), enthalpy change ΔH (kJ/mol), and entropy change ΔS (J/mol K) [4,59], using the following thermodynamic equations:

$$K_c = \frac{C_{ad}}{C_e} \tag{8}$$

$$C_e = C_i\,(1 - F_e) \tag{9}$$

$$C_{ad} = C_i\, F_e \tag{10}$$

$$K_c = \frac{F_e}{1 - F_e} \tag{11}$$

$$\Delta G^\circ = -RTlnK_c \tag{12}$$

$$\Delta H^\circ = R\left(\frac{T_1 T_2}{T_1 - T_2}\right)\ln\frac{K_{c1}}{K_{c2}} \tag{13}$$

$$\Delta S^\circ = \frac{\Delta H^\circ - \Delta G^\circ}{T} \tag{14}$$

where R = 8.314 J/mol K is the universal gas constant; K_c (mL/g) is the equilibrium constant; C_e is the concentration of the metal in the solution at equilibrium (mg/L); C_i is the initial concentration of the metal in the solution at equilibrium (mg/L); F_e is the fractional conversion of the sorption at equilibrium, and T is the thermodynamic temperature (K).

Van't Hoff's equation [59] allowed us to obtain the ΔH° and ΔS° values graphically:

$$lnK_C = \frac{-\Delta H^\circ}{R\,x\,T} + \frac{\Delta S^\circ}{R} \tag{15}$$

A graph of lnK_c on the abscissa axis and $1/T$ on the ordinate axis should be linear, and the intercept would be equivalent to $\Delta S°/R$, while the slope would be numerically equal to $\Delta H°/R$.

3. Results and Discussion

3.1. Characterization of FeNP/NFC Composite

As shown in Figure 1, the surface morphology is significantly different before and after NFC impregnation with iron nanoparticles. After the hydrolysis and cryo-crushing process, NFC exhibited a cavernous network structure that has a smooth surface. Figure 1a,b, show the formation of several clefts or folds, which will help distribute the nanoparticles on the support material. The spaces between the microfibers collapse for wood and plant cellulose sources due to water molecules' outflow. When the microfibers dry, all the hydrogen bonds between the fibrils are displaced, forming weak links. It is essential to indicate the effect of pH on the material because hydrolysis was generated in an acidic medium. The concentration of hydronium ions incorporated into the material increased and, therefore, they are attributed to agglomeration behavior [60,61]. Cellulose nanofibers tend to aggregate due to hydrogen bonding between the OH and COOH- groups [62,63].

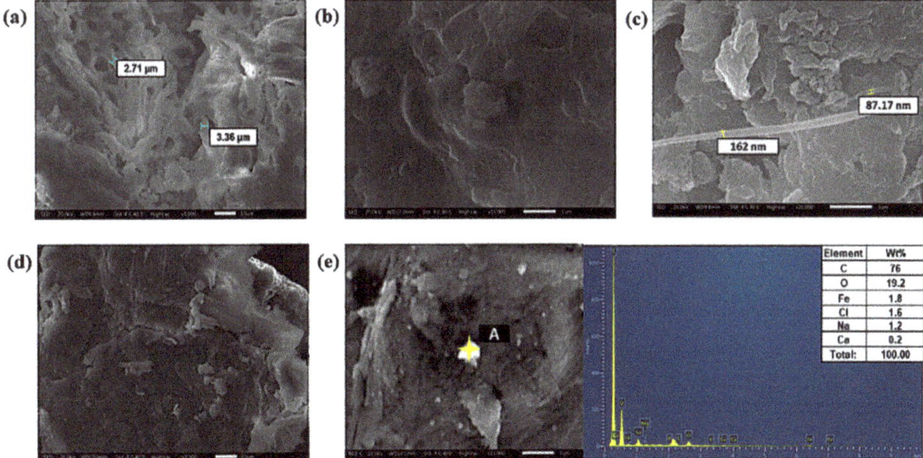

Figure 1. Scanning electron microscopy (SEM) micrographs of Cellulose nanofibers (NFCs): (**a**) X1000, (**b**) X15,000, (**c**) X20,000; iron nanoparticles and cellulose nanofibers (FeNPs/NFCs): (**d**) X1000, (**e**) X1500.

The cryo-crushing process formed ice crystals that helped to slash vegetal fiber and expose more surfaces, which allowed access to chemical activation of various functional groups. Because of nanofibers' exposure (Figure 1c), it can be established that the mechanical cryo-crushing and ultrasonic treatment were suitable for defibrillation of Moringa's Kraft pulp. The average diameter of nanofibers reported in other studies ranged from 50 to 100 nm, depending on the cellulose source [63–66]. In this case, the nanofibers' diameters ranged from 80 to 160 nm, as shown in Figure 1c, indicating nanofibers' presence. These nanofibers are not clearly seen in Figure 1a,b due to the micrographs' magnification. However, a nonhomogeneous accumulation of fibers can be observed in these micrographs (Figure 1a,b), which can also be denominated as microfibrillated cellulose (depending on the size and production process), consisting of cellulose nanofibers and nanofiber bundles [67,68].

Moreover, an increase in cellulose crystallinity has also been reported in NFC production from various sources [63,67]. In this case, such an increase was observed after the Moringa's NFC synthesis process, compared to the Kraft pulp cellulose crystallinity, due to the breaking down of the hierarchical structure of the cellulose into individualized nanofibers of high crystallinity, with a reduction in amorphous parts [67]. The crystallinity of the Moringa's NFC will be discussed further below. Furthermore, FTIR analyses of several NFC samples from different sources proved that chemical treatment also led to partial removal of hemicelluloses and lignin from the fibers' structures. Still, the molecular structure of cellulose remained unchanged after chemical and mechanical treatments [64,67]. This behavior was also observed in the NFC from Moringa and will also be discussed further. Based on the above, the presence of cellulose nanofibrils in the Moringa composite can be confirmed.

In Figure 1d,e, micrographs of FeNP/NFC are observed. The surface showed a distribution of small particles which are related to NPs. Results for punctual Energy Dispersive X-Ray Spectroscopy (EDS) analysis (point A in Figure 1e) indicate high concentrations of carbon (C), oxygen (O), iron (Fe), chlorine (Cl), sodium (Na), and calcium (Ca), Figure 1e. Chlorine comes from the reactive solution used in FeNP/NCF preparation. The results indicate a certain percentage of Fe adhered to the surface, which is related to the synthesis of NPs. The alkaline pH system adjusted in the synthesis of nanoparticles helps incorporate hydroxyl groups (–OH) and NPs to the matrix. Hemicelluloses that still partially persist in the structure can inhibit nanofibrils' coalescence during the synthesis of iron nanoparticles. The formation of several slits is observed, which helped distribute the nanoparticles on the support material and capture these ions [61].

The elemental mapping was performed to the FeNP/NFC samples after the adsorption process. In Figure 2a,b, the main elements present on the materials' surfaces are carbon and oxygen in a higher proportion in weight, and low concentrations of dispersed forms of Cd and Pb. The sulfate (S), Ca, and Fe compounds contribute to Cd aggregates' adsorption on the surface of the material observed at various sites. Sodium comes from cooking liquors of the Kraft pulping process: NaOH, NaS, and Na_2CO_3. Calcium and potassium are the main elements present in the wood. The sample's composition with adhered particles of Cd is shown in Figure 3b at 5000×, where the spot analysis indicates solid Cd particles. The results demonstrate the presence of Cd at around 11% (*w/w*), forming various phases on the material. The sample analysis shows a decrease in Fe (with values <0.15%) compared to the samples before Cd adsorption, indicating a possible ion-exchange sorption mechanism. The spot analysis of the sample that adsorbed Pb (Figure 3a) reached a concentration near 1.05% (*w/w*). The carboxylic acids, phenols, and alcohols on the surface of the adsorbent were protonated, and the surface was negatively charged, allowing adsorption.

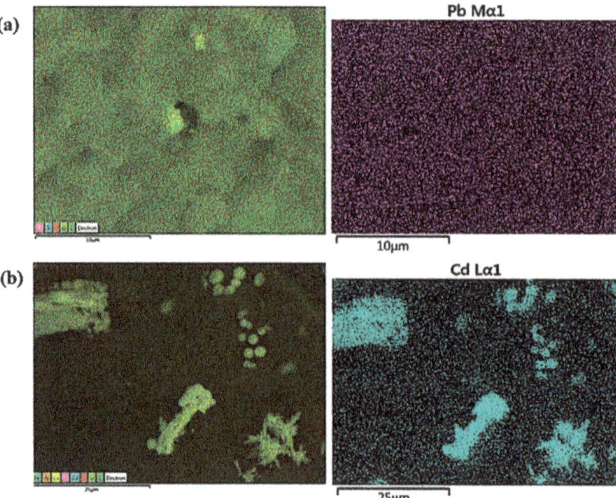

Figure 2. Elemental mapping of FeNP/NFC samples after adsorption of (**a**) lead (Pb) and (**b**) cadmium (Cd).

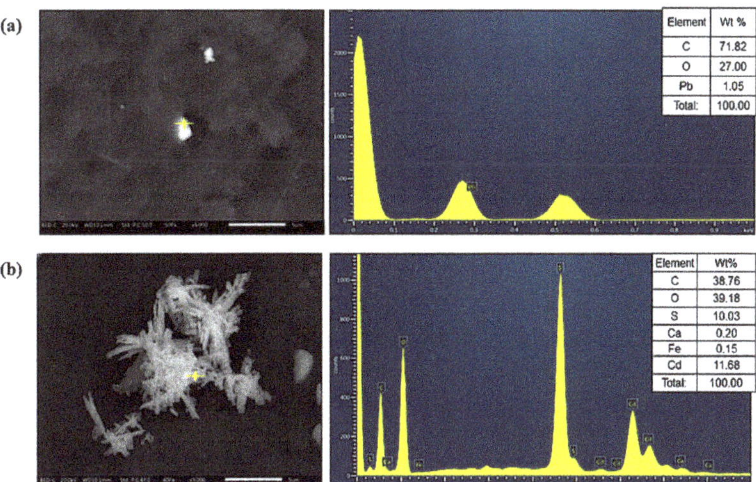

Figure 3. Spot analysis of composite samples after adsorption of (**a**) Pb and (**b**) Cd.

Figure 4 shows the FTIR spectra of NFC and FeNP/NFC. For both spectra, the broad peaks at 3406 and 3374 cm^{-1} are associated with -OH stretching vibration. The bands at 1636 and 1611 cm^{-1} are attributed to the water bending vibration. Displacement of these peaks is observed due to their intervention during the synthesis of the new material. In the case of the FeNP/NFC sample, the smaller peak at 2909 cm^{-1} corresponds to the -CH stretching of the aromatic rings and methylene (-CH_2), which are associated with functional groups involved in the green synthesis of nanoparticles. The appearance of a band at 1456 cm^{-1} can be assigned to a -CH_2 bending vibration and -OH deformation vibration, for the phenolic or alcoholic group.

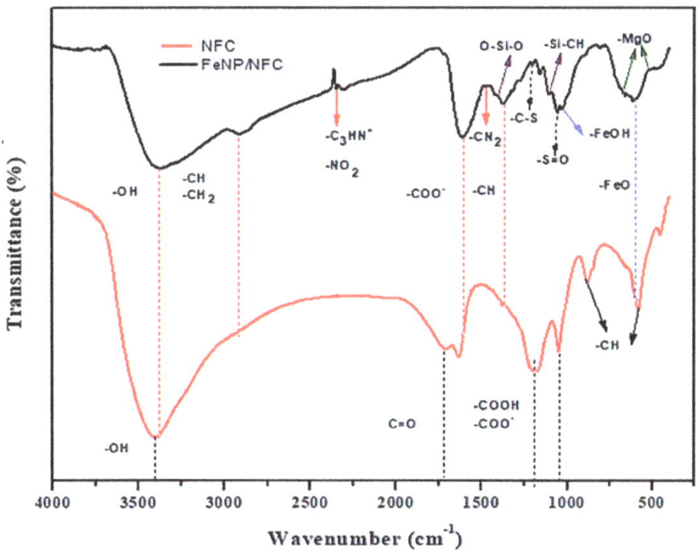

Figure 4. Fourier Transform Infrared Spectroscopy (FTIR) spectra of NFC and FeNP/NFC samples.

Significant changes were observed in the bands from 1500 to 1000 cm^{-1}. The bands from 1502 to 1418 cm^{-1} correspond to the aromatic elongation of C=C bonds and aromatic deformation vibration of C-C bonds. The stretching of carboxylic acids and esters peak is present at 1271 cm^{-1}. Ether and free alcohol C-O stretching can also be observed at the bands from 1000 to 1280 cm^{-1}. The adsorption peaks at 1161 and 661 cm^{-1} are attributed to asymmetric C-O-C and deformation of out-of-plane hydrogen of the C-H bond in the aromatic ring [69].

After the Fe nanoparticles' impregnation on NFC, some peaks' intensities increased in the spectrum of the FeNP/NFC composite (Figure 4)—mainly the bands that appeared in the region between 600 and 400 cm^{-1}, which correspond to Fe oxides. The FTIR spectrum of FeNP/NFC exhibited peaks at 1611, 1551, 1463, 1376, 1031, and 600 cm^{-1}. The broad peak at 601 cm^{-1} was attributed to stretching vibrations of Fe-O.

On the other hand, the O-Si-O and Si-CH functional groups were observed in the cellulosic material; thus, main peaks with stretching and bending vibrations from 1430 to 1100 cm^{-1} were noticed. The Si can also react with the noncellulosic components, where it can contribute to the Fe-O bond from 580 to 603 cm^{-1} due to the formation of the Fe-O-Si bond [70]. Furthermore, the peaks with lower intensities, at 511 and 675 cm^{-1}, such as -MgO, are related to other oxides present in the material. These sites intervened in the adsorption and were largely attributed to the wood's inorganic portion, and a smaller portion to the final material [71]

The peaks around 1376 and 1463 cm^{-1} increased their intensity; these are attributed to the -CH symmetric vibration of the polymeric matrix of NFC, mainly due to a strong hydrogen bond between the NFC groups and iron nanoparticles. The broad peak at 1031 cm^{-1} was associated with Fe-OH vibration; it indicates the incorporation of iron onto the surface [69]. The stretching vibration of O=S=O was observed at 1384 cm^{-1}; the C-S group was found at 1200 cm^{-1} and the S=O group was identified at 1053 cm^{-1}. These bands confirmed the presence of the sulfonic group. The effect of alkaline and acid treatments on cellulosic materials leads to surface activation with various functional groups such as -SH, -SOH_2, -SO, -$(CH_3O)_2SO_2$, and C-S, which can help in the adsorption process [4,56]. Bands at 2340, 2346, and 1478 cm^{-1} are typical of N vibrations, such as -NO_2, C_3HN^+, and CN_2; these groups were formed by the cellulose synthesis and extraction [72].

The Fourier Transform Infrared Spectroscopy (FTIR) analysis reveals (Figure 4) that the FeNP/NFC was esterified during the synthesis process at 1721 cm^{-1}; this band is usually attributed to C=O stretching vibration, and a decrease in this band was observed for the FeNP/NFC sample. A new weak absorption peak appeared at 1427 cm^{-1}, which could be attributed to the bending vibration of OH in carboxyl groups. On the other hand, a large amount of -COOH groups had been introduced to cellulose after the modification; these groups possess strong adsorption properties for heavy metal ions [73].

Figure 5 shows the FTIR spectra of FeNP/NFC adsorbent after Pb adsorption (PbRM) and after Cd adsorption (CdRM). It was observed that metal adsorption generated slight displacements in some bands and significant changes in the band frequencies of groups such as -OH, -COO-, C=O, C-O, and C-O-C; this suggests that such ionizable functional groups on the adsorbent surface can bond with the metal ion. For CdRM and PbRM, the stretching vibration peak of C-H observed at bands from 2800 to 3000 cm^{-1} disappeared, indicating that C-H may participate in the cross-linking reaction. Moreover, at the 1630 cm^{-1} wavenumber, band intensity changes can be attributed to the $Pb(NO_2)$ stretching vibration. Therefore, it can be established that amine groups in FeNP/NFC play an essential role in lead adsorption on this material by sharing one electron pair of the nitrogen atom with the Pb ions. The decrease in free carboxylic acid, iron oxide, iron hydroxides, amide, nitrogen dioxide, magnesium oxide, sulfur groups, and other possible sites indicates the same type changes upon sorption of Pb and Cd ions.

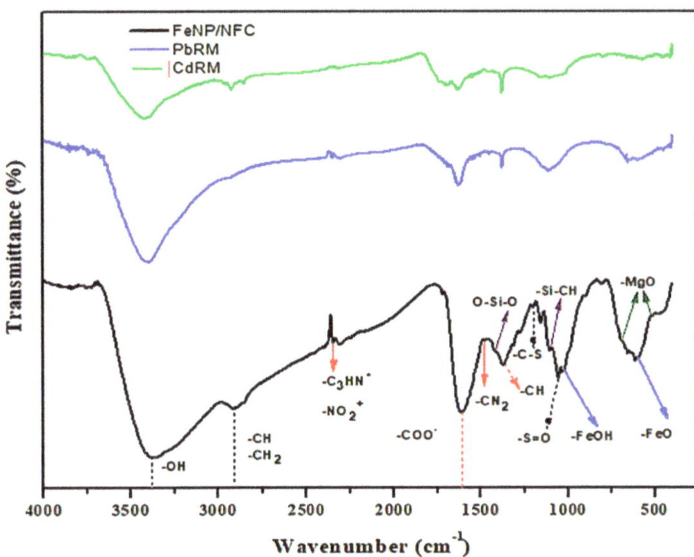

Figure 5. FTIR spectra of FeNP/NFC, adsorbent after Pb adsorption (PbRM) and after Cd adsorption (CdRM).

The zeta potential (ZP) value of FeNP/NFC increased from −11 to −23 mV; this lower negative value is due both to the polyphenolic components present in the extract during the synthesis of the material and also to the cellulose structure of the support material for the Fe nanoparticles (Figure 6) [34]. Different modification strategies create ionic groups on the surfaces of NFCs. Carboxymethylation, oxidation, and sulfonation are three routes that can add ionic groups on the surface of the cellulose. Carboxymethylation causes surfaces to be negatively charged, promoting stable suspension from carboxymethylated fibers. The oxidation process includes hydroxyl groups, which oxidize to aldehyde groups, followed by carboxyl groups. Sulfonation is considered as a way to induce an anionic charge on the surface with different functional groups such as -SO, -SH, and -HSO_3^- [74,75].

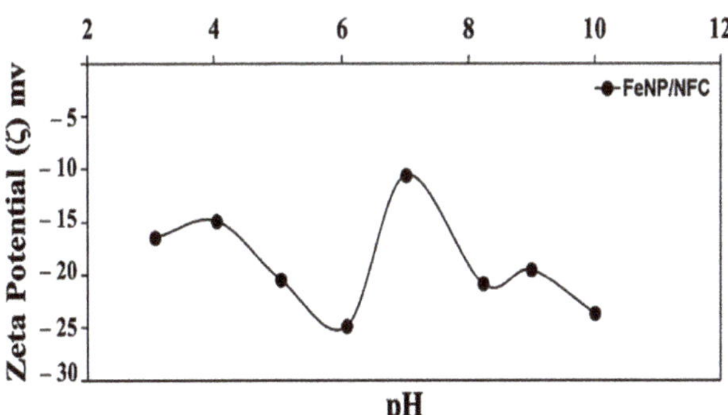

Figure 6. Zeta potential (ZP) of FeNP/NFC composite.

There is no evidence of an isoelectric point in the sample, which confirms that charge dissociation is virtually independent of pH. Under acidic conditions, as the pH decreases, the -COO- groups are converted to -COOH resulting in a lower density of surface charge, and therefore a reduction in mobility. Additionally, an increase in pH did not change the particle mobility significantly. In all of the pH ranges, the Zeta potential values were negative; therefore, it can be considered that the pH does not affect the surface loading properties of the material. Particles with low zeta potential have tended to agglomerate, while particles with high zeta potentials have been electrically stable.

On the other hand, the crystalline structure of cellulose is one of the most studied structural problems in polymer science, and its accurate description of the crystal structure remains a topic of discussion [76,77]. Therefore, an analysis of the crystal structure by XRD was carried out to determine the composite material's crystalline state.

As shown in XRD patterns (Figure 7), there is strong diffraction at 14.63°, 16.7°, and 22.42° of FeNP/NFC, corresponding to the (101), (10 ī), and (002) crystalline planes for cellulose II, into the sample. The acid hydrolysis and cryo-crushing methods destroy the stable hydrogen bonds, generate free hydroxyl groups, and increase cellulose reactivity [73]. They were indexed in the structure of cellulose (JCPDS No. 50-2241). The diffraction intensity was weak for the (101) and (10 ī) crystal planes. These two peaks overlap with each other because they have a lower amount of lignin and hemicellulose, resulting from the adsorbent's synthesis process.

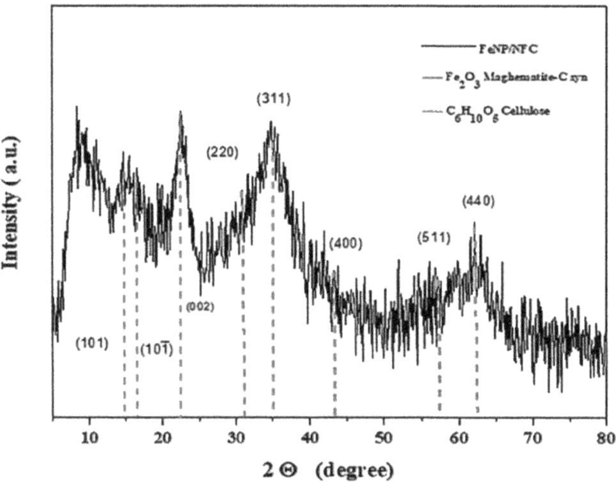

Figure 7. X-ray Diffraction (XRD) patterns of the FeNP/NFC, prepared with Moringa leaf extract.

It is difficult for chemical reactions to take place in the crystalline region because of its stable structure. Hence, it becomes an easily accessible and penetrable amorphous structure, increasing iron nanoparticles' introduction [4,73]. XRD patterns show peaks related to amorphous and crystalline compounds (Figure 7). The nanoparticles in the material presented peaks at around 30° to 62°, which could be due to crystalline materials on organic matter.

The iron oxide phase in FeNP/NFC shows some intense XRD peaks—identified as maghematite (Fe_2O_3). These peaks and indices were 30.79° (220), 35.63° (311), 43.51° (400), 57.22° (511), and 62.79° (440), as can be observed in Figure 7. The diffraction peaks belong to iron oxide maghematite-C syn (No. 00-024-0081) with a rhombohedral structure. Different amorphous materials impregnated with nanoparticles have been reported, finding that the iron nanoparticles prepared with green tea extract were efficient in different aspects [34]. The polyphenolic compounds from tea do not possess a reduction potential negative enough to reduce from Fe^{3+} to Fe^0 but can reduce Fe^{3+} to Fe^{2+}, forming oxide/hydroxide nanoparticles.

The Energy Dispersive X-ray Fluorescence (EDXRF) analysis is shown in the spectrum of Figure 8. The peaks around 0.8, 6.4, and 7.05 keV are related to the Fe binding energies. Therefore, the EDXRF spectrum for Fe_2O_3 from the FeNP/NFC material, synthesized with aqueous extract of Moringa leaves, confirmed nanoparticles' presence. The number of radicals present in the hybrid sample revealed the photocatalytic activity of the nanoparticles. The other intense peaks shown in the EDXRF spectrum are marked as salts' structures derived from the extraction of the cellulose and its initial composition from the raw material derived from Moringa wood (Figure 8). The signal peak at 2.98 keV is assigned to silver (Ag), derived from its presence on the sample holder and tube in the instrument employed for the analysis; hence, the peak of iron looks smaller than silver. Potassium is essential for the growth of the tree, which can be seen in the graph's 3.16 keV intensity. The EDXRF spectrum analysis indicates that the sample has a composition of approximately 32.81% Fe_2O_3 and is only one of the existing oxide groups in the material (Table 1). According to this result, the weight percent of metal oxides on cellulose increased, confirming that the iron nanoparticles were successfully incorporated. Other than iron, different compounds were identified, such as MgO, SiO_2, P_2O_5, SO_3, and Cl (Table 1). Inorganic material in wood is found as carbonates, oxalates, silicates, phosphates, and sulfates; or organized as carboxyl groups in the cellulosic materials [78].

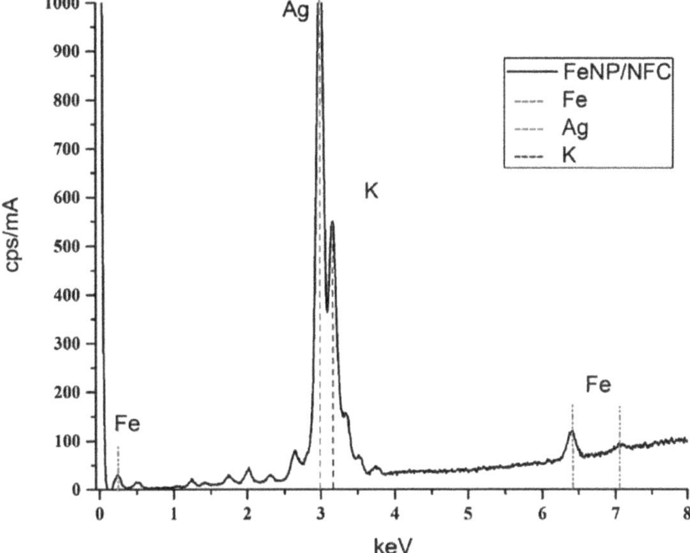

Figure 8. Energy Dispersive X-ray Fluorescence (EDXRF) spectrum of the FeNP/NFC composite.

Table 1. Composition of oxide groups in iron nanoparticles and cellulose nanofibers (FeNPs/NFCs).

Formula	
Concentration (%)	
Formula	**Concentration (%)**
MgO	13.55
SiO_2	20.65
P_2O_5	24.04
SO_3	3.68
Cl	5.26
Fe_2O_3	32.81

Generally, tropical woods contain inorganic salts of calcium, potassium, and magnesium [78,79]. Mineral transport into the plants can occur through natural routes from the soil, where minerals are in abundance. Silica, phosphorus, magnesium, and potassium components of biomass contribute to the plant´s rigidness, growth, and posture [78,80,81]. Sulfur oxide is attributed to cooking Kraft pulp in the composition of the reagents in the white liquor. Chlorine can be attributed to the reagent used ($FeCl_3$) in the nanoparticle synthesis process [82–84]. The inorganic substances found contributed to the various chemical reaction processes that took place during the synthesis of the composite, such as carboxymethylation, oxidation, hydrolysis, and sulfonation.

3.2. Adsorption Experiments

3.2.1. Kinetics

The adsorption kinetics results are shown in Figure 9. The adsorption kinetics of Pb(II) and Cd(II) ions in FeNP/NFC were significantly different. Cd's adsorption in the initial stage was very fast in the first 5 min, where 22% of the total cadmium elimination was obtained. Still, it gradually slowed down and finally reached equilibrium with a maximum efficiency of 32%. In the case of Pb, the adsorption capacity was slow in the first 5 min, rapidly increased between 5 and 20 min, and then slowed down—being more constant up to 75 min (with adsorption of 40%). This adsorption system had an efficiency above 58%. The equilibrium time was selected after 150 min for both cations, and it was ensured that the adsorption equilibrium was reached in later experiments.

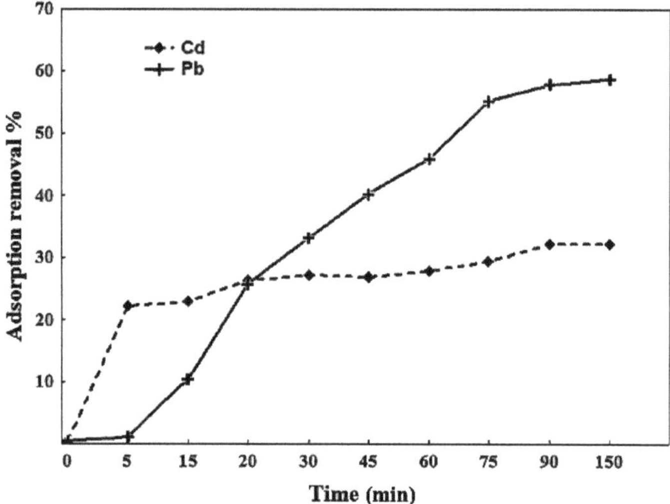

Figure 9. Cd and Pb adsorption onto FeNP/NFC composite as a function of time.

The adsorption phenomenon occurs at the surface, where the transport process and the diffusion of the adsorption are carried out between the film and the adsorbent. Moreover, the second phase is characterized by a slow adsorption rate due to the ions' slow diffusion and surface phenomena, which indicates the rate-limiting step and the type of adsorption governing the system. The effective adsorption area on the adsorbent's surface has many vacant bonding sites for the metal ions, and all resulted in easy adsorption. As adsorption proceeded, the concentration of metal ions and active sites decreased, resulting in a lower adsorption rate when the systems reached equilibrium.

The adsorption kinetic data were fitted to Pseudo-first-order, pseudo-second-order, and Elovich models. The correspondent parameters for each model are shown in Table 2. The high correlation coefficient values ($R = 0.9858$) obtained by the pseudo-first-order model for Pb adsorption indicates that sorption occurs exclusively at one site per ion and involves physisorption. This model suggests that the adsorption is superficial, and one solute molecule is adsorbed at one defined specific site. These sites were made up of the same type of functional groups. For both adsorption kinetics, the calculated values (qe_{calc}) and experimental values ($q_{e\ exp}$) were similar. The theoretical or estimated values are considered the adsorption capacity calculated by kinetic modeling. Thus, $q_{e\ calc}$ from the pseudo-first-order model resulted in 0.1348 mg/g for Cd and 0.3401 mg/g for Pb. Adsorption capacity experimental values ($q_{e\ exp}$) were obtained from the experimental data graphs (Figure 9). Consequently, $q_{e\ exp}$ for Cd sorption was 0.1513 mg/g and 0.3109 mg/g for Pb.

Table 2. Parameters of kinetic models for cadmium (Cd) and lead (Pb) adsorption onto FeNP/NFC.

Model	Parameters	Metal Ions	
		Cd	Pb
	$q_{e\ exp.}$ [1] (mg/g)	0.1513	0.3109
Pseudo-first-order	K_L (1/min)	0.2445	0.0220
	$q_{e\ calc.}$ (mg/g)	0.1348	0.3401
	R	0.9197	0.9858
Pseudo-second-order	K_2 (1/min)	0.4486	0.0170
	$q_{e\ calc.}$ (mg/g)	0.2937	0.7781
	R	0.9611	0.9734
Elovich	a (1/min)	0.8112	0.00478
	b (mg/g)	59.077	2.4590
	R	0.9895	0.9204

[1] Obtained from experimental data.

The pseudo-first-order model can only be applied to describe the initial stage of adsorption. However, the pseudo-second-order model can predict the entire period of the adsorption behaviors. Therefore, the pseudo-second-order model can describe the adsorption behavior of the ions studied. In this model, the individual molecules join in two adsorption sites of the surface. They indicate that the sorption processes are entirely controlled by chemical processes and highlight their particularity in the electrical charge during equilibrium, such as the exchange of electrons or covalent forces.

According to the pseudo-second-order model constant (K_2), it can be observed that the adsorption rate is higher for Cd (K_2 = 0.4486 1/min) than for Pb (K_2 = 0.0170 1/min). This fact corroborates the differences found for both cations' adsorption kinetics, Cd, which implies a different equilibrium reaction for Cd compared to Pb. Cd's kinetic reaction rate indicates that it is only involved with the functional groups on the external surface. Pb's kinetics consist of multiple slow reactions between the outer surface and the material's internal pores. Additionally, the adsorption of Cd ions by the adsorbent was described with the Elovich equation (R = 0.9895). However, when the Elovich model fitted Pb data, low correlation coefficients were observed. This model indicates that the adsorbent's active sites are heterogeneous and exhibit different activation energies, based on a second-order reaction mechanism for a heterogeneous reaction process. Consequently, this model suggests that Cd removal occurs by a chemisorption process concerning the transfer or exchange of electrons, taking into account the metal's oxidation state and the possible formation of complexes involved in it [85].

3.2.2. Isotherms

Figure 10 shows the adsorption isotherms of Cd in FeNP/NFC at different temperatures. Considering the Langmuir, Freundlich, and Langmuir–Freundlich models presented, the three models had a good fit for the three different temperatures, with R values > 0.9700 (Table 3). However, in solid–liquid systems, many factors come into play, such as hydration forces, mass transfer, etc., which makes it more complicated to define equilibrium behavior in a sorption system, and obeying a single isotherm does not fully reflect the entire adsorption process. Moreover, the isotherm's suitability may be affected by experimental conditions, mainly the solute's concentration range. Both Langmuir and Freundlich isotherms may adequately describe the same set of liquid–solid adsorption data at specific concentration ranges. The Langmuir–Freundlich isotherm model gives information about the adsorption on homogeneous and heterogeneous surfaces with the possibility of multilayer adsorption.

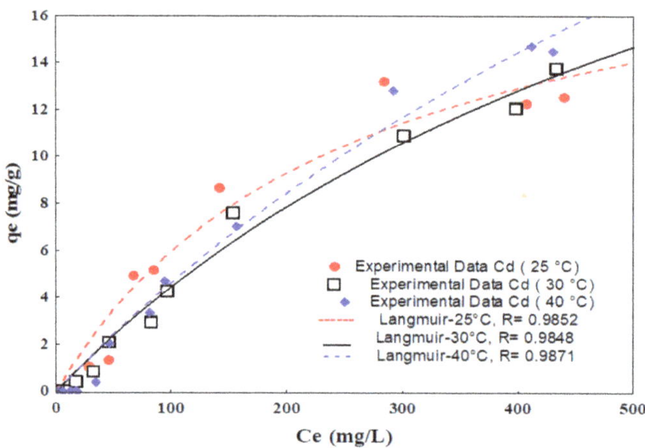

Figure 10. Cd adsorption isotherms onto FeNP/NFC composite at 25, 30, and 40 °C.

Table 3. Parameters of kinetic models for Cd and Pb adsorption onto FeNP/NFC.

Model	Parameters	Metal Ions					
		Cd			Pb		
		25 °C	30 °C	40 °C	25 °C	30 °C	40 °C
	$q_{e\ exp.}$ [1] (mg/g)	12.538	13.7377	14.4896	81.4064	89.2212	89.0882
Langmuir	$Q_{0\ calc.}$ (mg/g)	21.233	34.999	49.697	467.307	956.02	357.549
	b	0.0040	0.0014	0.0010	0.0131	0.0084	0.0263
	R	0.9789	0.9929	0.9924	0.9149	0.7274	0.7294
Freundlich	K_F (mg/g) (L/mg)$^{1/n}$	0.2601	0.0922	0.0775	4.1092	0.7209	1.4545
	n	1.5275	1.2117	1.1443	0.8916	0.4163	0.5055
	R	0.9617	0.9895	0.9891	0.9262	0.9180	0.8707
Langmuir–Freundlich	K_{LF} (mg/g)(L/mg)$^{1/nLF}$	0.8453	1.0635	0.2825	0.17961	0.3134	0.2995
	a_{LF} (mg/L)	1×10^{-4}	2×10^{-6}	3×10^{-6}	0.1394	0.1223	0.1511
	n_{LF}	0.4763	0.4052	0.6554	3.1486	3.0864	3.2310
	R	0.9234	0.8864	0.9731	0.9934	0.9390	0.9040

[1] Obtained from experimental data.

The Langmuir model was the one that best described all the isotherms of Cd at 25, 30, and 40 °C, with adsorption capacities of 21.233, 34.999, and 46.697 mg/g for each temperature, respectively. The Langmuir model was used to describe the monolayer adsorption at the adsorbent's surface, considering the similar types of vacancies and similar forces binding Cd molecules to the adsorbent's surface. The value of b in the Langmuir model suggests that the isotherm was favorable ($0 < b < 1$). In this context, the low b values reflect that adsorption is favorable: b > 1 is unfavorable, $b = 1$ is linear, $0 < b < 1$

is favorable or $b = 0$ is irreversible. In this case, the respective b values for each evaluated temperature indicate that the adsorption process was favorable [12,86]. The Langmuir equation is valid for monolayer sorption onto a completely homogeneous surface with a finite number of identical sites and little interaction between adsorbed molecules. This fact means that once a metal ion occupies a site, it cannot have another place for adsorption, so the adsorption behavior is gradually reflected in the different tested concentrations of Cd.

Lead (Pb) isotherm data were also fitted to the same models mentioned above. The results are shown in Figure 11. It was noticed that the Langmuir–Freundlich model best described such data at 25, 30, and 40 °C, according to their correlation coefficients (R), shown in Table 3. This model describes the equilibrium relationship between adsorbate binding and heterogeneous or homogeneous surfaces [72]. These surfaces have spaces with different forces, where the adsorbate molecules bind to the surface. The isotherm at high solute concentrations indicated a duality of both monolayer and multilayer adsorption [72]. This fact reinforces the idea that the governing mechanism is more complicated for Pb (II) than for the weak bond of Cd (II). As a result, a $Q_{0\ \mathrm{cal}}$ value of 467.307 mg/g was obtained for the Pb system at 25 °C, and the $Q_{0\ \mathrm{cal}}$ values at 30 and 40 °C were 956.02 and 357.549 mg/g, respectively, as shown in Table 3. Some discrepancies regarding Q_0 can be observed at 30 and 40 °C because of the relatively low R coefficients obtained in the Langmuir model's fittings in such cases. These results complement together with the Freundlich model, so the adsorption intensity (n) was always <1 for Pb adsorption cases (Table 3). These values ($0 < n < 1$) indicated favorable adsorption. These facts support that the lead adsorption was superficial and multilayered, considered as a heterogeneous adsorption type. Additionally, the adsorption mechanisms varied depending on the solution's concentrations. In this context, it is essential to emphasize that the FeNP/NFC composite can be considered highly efficient in removing Pb ions from aqueous solutions under these experimental conditions.

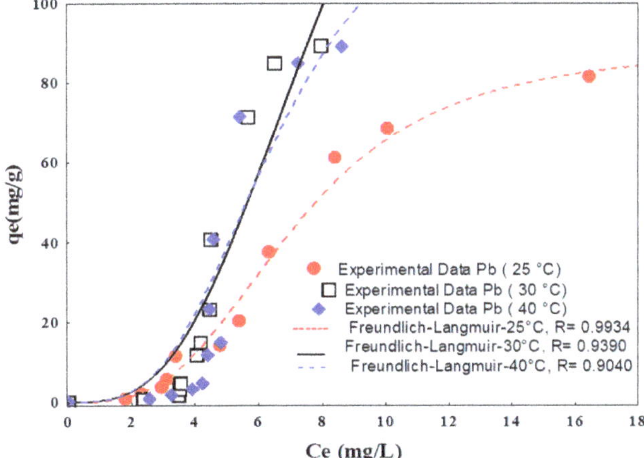

Figure 11. Pb adsorption isotherms onto FeNP/NFC composite at 25, 30, and 40 °C.

Polymers are heterogeneous materials containing sorption sites with various binding affinities and selectivities [45]. Hence, it can be suggested that the FeNP/NFC adsorbent has a heterogeneous surface, and Pb and Cd sorption onto this composite may present a multi-layer behavior. The adsorption capacity of FeNP/NFC increased as the solute concentration rose until it reached the highest adsorption capacity. An increase in experimental adsorption capacity ($q_{e\ \mathrm{exp}}$) was observed in high-temperature conditions (Table 3) for both ions.

Previous studies indicated three explanations in the absorption behavior of specific metallic ions towards the adsorbent: radio, electronegativity, and charge [87]. It can be inferred from the results that the order of heavy metal ion removal efficiency under the experimental conditions was as follows: Pb(II) > Cd(II). The sizes of the ionic radius of the elements that are Pb (0.119 nm) > Cd (0.095 nm) should be taken into account. However, some studies conclude that the order of adsorption efficiency is due to the hydrated ionic radius of Pb (0.45 nm) < Cd (0.5 nm), highlighting a relatively smaller lead ion compared to cadmium. In aqueous solutions, ions form hydrate complexes with water, thus increasing the ionic size. Therefore, hydrated ions are transported to the surface of the material and diffuse into the pores [87–89].

The electronegativity is higher for lead ions (1.87–2.34) compared to cadmium (1.69). Phenomena, such as electrostatic interactions, covalent bonds, and ion exchange, could occur at the cellulose wall where cation bonding takes place. Adsorption or chelation involving hydroxyl functions, close to carboxylates or iron nanoparticles (-FeO, -FeOH), can increase the cation binding level [87].

The pH is an important parameter influencing the adsorption process at the water–sorbent interfaces. However, Pb and Cd's adsorption on this adsorbent is independent of the pH according to the zeta potential results of the material. As mentioned above, the solution's pH is reduced after the sorption process; this occurs because the deprotonation of the acidic functional groups of the adsorbent releases H_3O^+, allowing the metal cations to be adsorbed. The final pH value, obtained after the sorption experiments, was 3.6. Cd and Pb's removal by adsorption, reported in the literature, ranges from 3 to 8 for this type of sample [73,74]. Therefore, all further kinetic adsorption experiments were performed at a pH of 5.

Some polymers have shown that various surfaces are nonionogenic and acquire a negative charge in aqueous solutions. It has been shown that the electrokinetic charges of polymer surfaces become more negative as surface hydrophobicity increases. FeNP/NFC contains carboxyl (-COOH) and hydroxyl (-OH) functional groups as part of their polymer structures. Therefore, the negative electrokinetic charge acquired in the material can be explained by the dissociation of the mentioned functional groups [90]. The zeta potential was negative in all the pH values investigated; thus, this material can help remove Cd and Pb in some chemical species present in the solution. Both metals have the same charge (+2). It makes adsorption favorable in the system under study, though it depends on the chemical species present. Cadmium exists as a different form of Cd^{2+}, $Cd(OH)^+$, $Cd(OH)_2$, and $Cd(OH)_3^-$ at different pH values. At pH < 8.0, the predominant Cd species are Cd^{2+} [91]. Furthermore, Pb has a different form in aqueous solution at different pH values, such as Pb^{2+}, $Pb(OH)^+$, $Pb(OH)_2$, $Pb(OH)_3^-$, and $Pb(OH)_4^{2-}$. Pb^{2+} is the dominant species under acidic conditions because the Pb^{2+} and Cd^{2+} ions begin to hydrolyze and then form a minimal amount of other chemical species. Some of these species could be unfavorable for adsorption on negatively charged materials at pH < 3 in solution, so a value of 5 as the initial pH was selected for all adsorption experiments. The intense competition between hydronium ions and cations (Cd or Pb ions) decreases the metal ion interaction with the sorbent's binding sites by greater repulsive forces.

According to the Giles classification [92], sorption isotherms of Cd at 25, 30 and 40 °C can be included in S-type subgroup 2 (S-2). This fact indicates cooperative or complex adsorption with molecules that tend to be adsorbed in rows or groups. In celluloses, where free radical ions were involved in the adsorption, such as those present in this adsorbent, the solutes formed bonds with aromatic nuclei or their surroundings. Similarly, for the Pb isotherms at 25, 30, and 40 °C, the S-2 type can be described. The shapes of isotherms suggested one orientation of the adsorbed molecules stacked on the adsorbent's surface. The polymeric hydrocarbons in this material clearly cannot have simple surfaces because of their complex three-dimensional structures. Nevertheless, they produce S-curves. The cause is not, in this case, competition with water in the solvent, but it is possibly promoted by some form or interlocking of the adsorbed molecules [92,93]. Being classified in subgroup 2 indicates that monolayer capacity has reached its maximum. The plateau's height determines the surface cover for each solute molecule; additionally, electrostatic forces were involved in the adsorption. As seen in Figure 11, the length of the plateau showed an exposed surface, which easily attracted Cd and Pb ions. In nearly all cases, the isotherms have the S-shape. The adsorption sites appear to be isolated and negatively charged atoms.

A comparison of the maximum experimental adsorption capacities with similar adsorbents are shown in Table 4 for Cd and Pb. Differences in experimental conditions have been considered when comparing with other adsorbents. The advantage of using FeNP/NFC as an adsorbent is significant, and it can be observed that, compared with the other materials, it is highly efficient for removing Cd and, especially, Pb ions.

The adsorption capacities of synthesized nanocomposites in this study can be compared to other available adsorbents. Prasher et al. [94] indicated that the Pb(II) adsorption capacity obtained for the red algae biosorbent was superior compared to Cd(II) under similar experimental conditions. In some studies, this sorption behavior was also explained by the concepts of ionizing radio, electronegativity, and ion charge.

Table 4. Cd and Pb adsorption capacities of different adsorbents.

Adsorbent	$q_{e\ exp.}$ (mg/g)		Temperature (°C)	pH	Adsorbent Mass (g)	Solution Volume (mL)	Reference
	Cd	Pb					
FeNP/NFC	12.538 13.737 14.489	81.406 89.221 89.088	25 30 40	5 5 5	0.05 0.05 0.05	10 10 10	This study
Magnetic Bauhinia purpurea (Kaniar) powders	0.83 0.86 2.20	14.14 1.02 1.35	25 30 40	5	2	50	[4]
Biochar (waste agro-industry) FeSO4·7H2O	38.3 to 75.3	179-311	25	5	0.04	20	[95]
Banana peels	5.71	2.18	25	3	1.5	50	[96]
Polymer-modified magnetic nanoparticles	29.6	3.103	25	1 to 8	0.05	50	[97]
Chitosan/iron oxide nanocomposite	4.106	3.795	25	3	0.05	20,000	[98]
Sugar beet pulp	0.98	1.00	25	5	0.4	20	[99]
Sawdust ($Fe3O_4$/SC)	63 22 25	– – –	27	6.5	0.4 1.2 2	50	[100]
Fe3O4 nanoparticles onto cellulose acetate nanofibers	–	44	27	6	0.1	50	[101]

Biochar from agro-industrial wastes (nut shield, wheat straw, grape stalk, grape husk, and plum stone) was evaluated by Trakal et al. [95]. The results indicated that this material was strengthened after Fe oxide impregnation, increasing Cd and Pb sorption capacities. Such higher capacities were attributed to the surface complexation of metals with carboxyl and hydroxyl functional groups. In this case, Pb's adsorption capacity was higher than that of Cd, with different experimental conditions.

3.3. Thermodynamic Parameters

The analysis of these parameters allows the estimation of the feasibility of the adsorption process and the effect of temperature. The results of ΔH and ΔS were obtained from the slope and intercept of the graph lnK_c versus $1/T$ (Figure 12), using isotherm data. Parameters $\Delta G°$, $\Delta H°$, and $\Delta S°$ for the metal ions adsorption with FeNP/NFC are listed in Table 5.

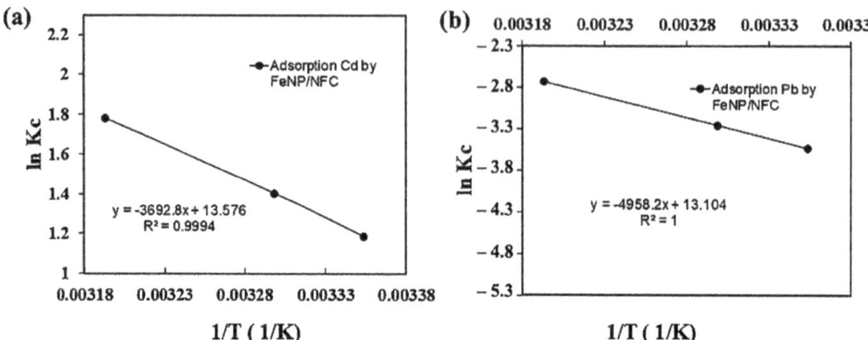

Figure 12. Van't Hoff's equation model for (**a**) Cd, and (**b**) Pb.

Table 5. Thermodynamic parameters for Pb and Cd adsorption on FeNP/NFC.

Metal Ions	$\Delta G°$ (kJ/mol)			$\Delta H°$ (kJ/mol)	$\Delta S°$ (KJ/mol K)
	25 °C	30 °C	40 °C		
Cd	−64.339	−64.904	−66.031	−30.701	0.1128
Pb	−73.704	−74.250	−75.339	−41.222	0.1089

The negative values of ΔG indicate a spontaneous reaction of the adsorption process for the FeNP/NFC. However, only a slight decrease in Pb and Cd maximum adsorption capacities was observed as temperature increased. Therefore, a large amount of heat was consumed for the Pb and Cd ions to transfer from water into the surface of FeNP/NFC [26]. The $\Delta H°$ results are negative for both cations, indicating that the adsorption process is exothermic; the values of $\Delta H° < 0$ and $\Delta S > 0$ imply that it is spontaneous at all temperatures. The positive value of ΔS suggests an increased disorder in the system during the process of adsorption. A redistribution of energy between the heavy metals and FeNP/NFC occurred, which suggests increasing randomness with structural or solvation changes occurring at the solid/liquid interface [4]. Acidic functional groups in the adsorbent, such as carboxylic and carboxylate, decreased. The solvent–sorbent interactions were similar; therefore, the change in entropy decreased [59].

3.4. Adsorption Mechanism

A possible mechanism of nanoparticle formation is illustrated in Figure 13. Ferric chloride hexahydrate hydrolyzes to form the ferric hydroxide, releasing H^+. The leaves extract partially reduces the ferric hydroxide to form NPs. Small Fe_2O_3 crystals were formed and placed at the coordination sites in a monodisperse form on the carrier. Physical or thermodynamic principles can also contribute to nanocrystals' self-assembly [3]. They can provide heterogeneous coordination or nucleation sites needed to form complexes with the ion. The aldehyde groups of the extract oxidize the cellulose material's corresponding structures [102,103]. The predominant flavonoids and polyphenols identified in the Moringa leaf extracts include rutin, kaempferol or isorhamnetin, isothiocyanates, chlorogenic acid, and quercetin-malonyl-glucoside. All these processes contribute to metal ion reduction [39,52]. The free radical scavenging activity of *Moringa oleifera* Lam leaf has been reported to be higher than synthetic antioxidant counterparts such as butylated hydroxytoluene (BHT), rutin, and ascorbic acid [51].

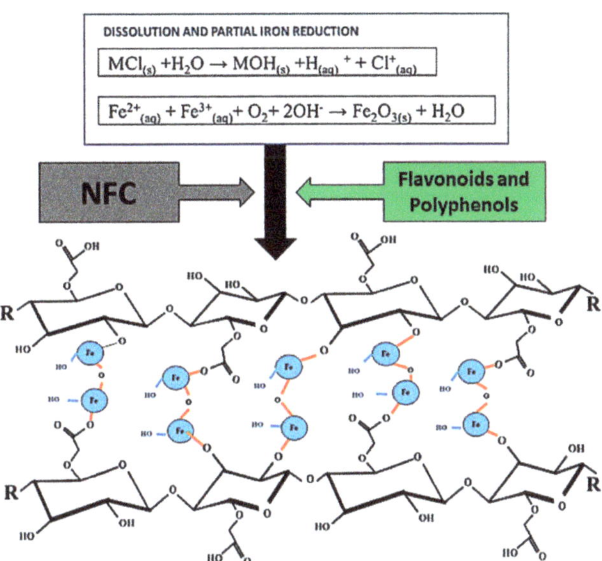

Figure 13. Schematic diagram of iron nanoparticle formation.

Cellulose nanofibers (NFCs) contain cellulose, simple sugars, and carbonaceous compounds that can bind to heavy metals. In nanoparticle synthesis, various compounds such as secondary metabolites and phytochemicals (phenolics, flavonoids, and fatty acids) have been introduced; these contribute to the metals' adsorptions. The significant importance of the polymeric matrix support and the nanoparticles' green synthesis processes can be attributed to carbonyl's complexation capacity with the metal ions. Therefore, it is observed that carbonyls are partially converted into carboxylic anions [97]. The FTIR confirmed this conversion in NFC and FeNP/NFC samples where functional groups such as carbonyl (-C=O), carboxylate (-COO), carboxyl (-COOH) in the range from 1000 to 1640 cm^{-1} are shown. These functional groups contribute to the adsorption mechanisms in the CdRM and PbRM samples of the FTIR spectra. They showed a significant reduction in the 1270, 1268, and 1610 cm^{-1} peaks of the carboxyl group (-COOH). Other groups also participated, such as the sulfonic group (-SO) with a band at 1059 cm^{-1}, iron hydroxide group (-FeOH) at 1023 cm^{-1}, and iron oxide group (-FeO) at 601 cm^{-1}.

The overall mechanism is a solid–liquid interfacial reaction, where Cd^{2+} and Pb^{2+} are represented as the metal ions (M^{2+}), Figure 14:

The mechanism for removing Pb^{2+} and Cd^{2+} ions was adsorption onto metallic oxides and nonmetallic sites through ion exchange, electrostatic bonds, and hydrogen bridge bonds. Some oxides (MgO, SiO_2, SO_3, Fe_2O_3) cause ligand chelation due to oxygen in the range from 4.5 to 5 pH used in these experiments (Figure 14b). The negative charge of the adsorbent, independently of the solution's pH, helps in the adsorption of these ions. These factors contribute to the formation of internal and external complexes by the various hydroxyl (-OH) groups present in the material, as shown in Figure 14c,d [56]. Electrostatic interactions would only occur with the weakest bound cations (Figure 14a). Hydroxyl (-OH) groups close to carboxyl (-COOH) groups would participate in the complexation of the highly bound cations and allow at least two functional carboxyl groups to contain a divalent cation.

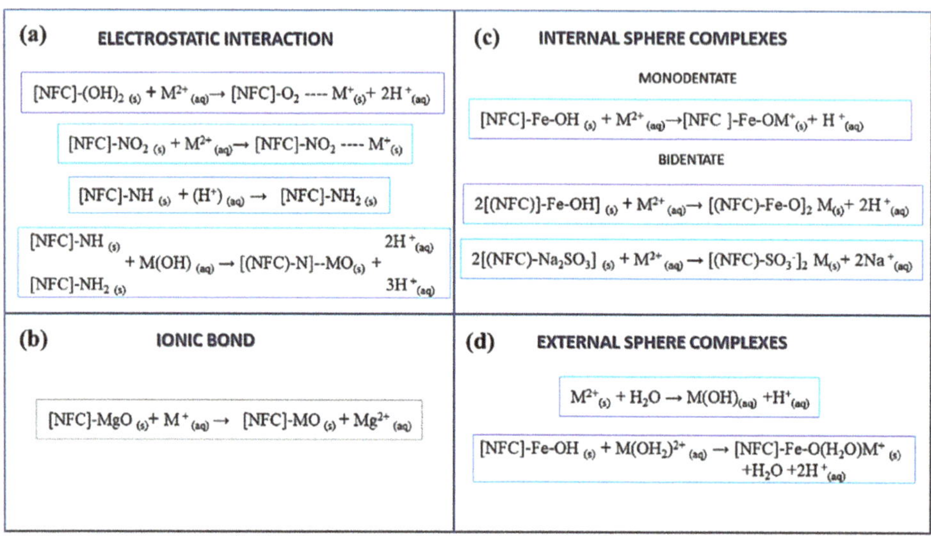

Figure 14. Metal ion (M^{2+}) adsorption mechanisms on composite adsorbent: (**a**) electrostatic interactions, (**b**) ionic bond formation, (**c**) internal sphere complexes formation, (**d**) external sphere complexes formation. [3,56,85,102,103].

The inner-sphere complex is formed by a covalent bond between the oxide's oxygen on the adsorbent's surface (-FeO) or the polymer's oxygen with the metal ion (M^{2+}). These possible mechanisms cause the loss of protons to the solutions, reducing the pH. Monodentate and bidentate complexes are formed depending on the presence of hydroxyl groups (Figure 15). The outer-sphere complex is formed when water is present between the adsorbed metal ion and the hydroxyl group on the iron (Fe-OH) surface. This complex can be distinguished because the bond is weaker than the inner-sphere complexes that form an electrostatic bond between the metal ions and the functional groups on the surface [3].

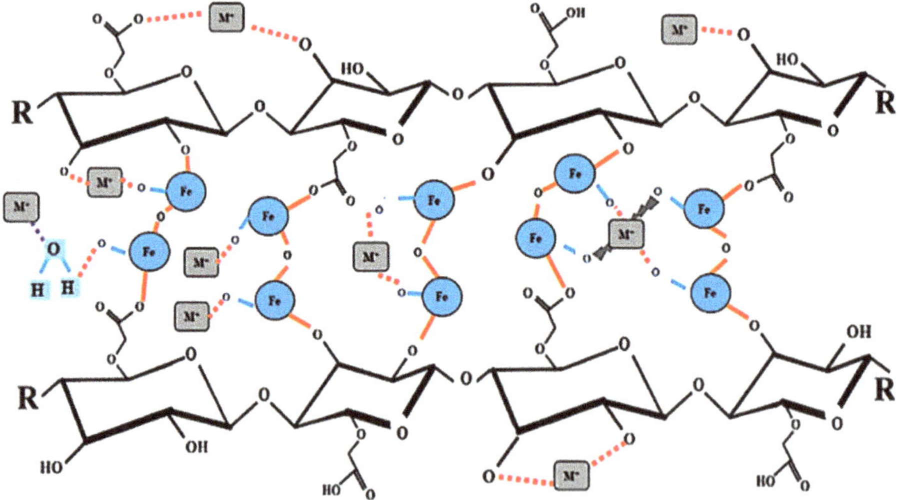

Figure 15. Scheme of adsorption mechanisms on FeNP/NFC.

The Kraft pulping and acid hydrolysis processes formed the anionic sulfonated cellulose derivatives, where the oxidation of the cellulose and sulfonation with the reagents used in these processes occurs. Material modifications, such as mercerization or sulfurization, increase the adsorption capacity of lignocellulosic materials. Sulfonated nanocelluloses have a large number of binding sites due to the different surfaces observed in the material. The participation in the adsorption of the sulfonic groups was important, as well as the carboxyl or hydroxyl groups. Sulfur forms complexes with ligands with a more covalent character instead of ionic, where metal ions (M^{2+}) bind with two oxygenates to form complexes, as illustrated in Figure 16a:

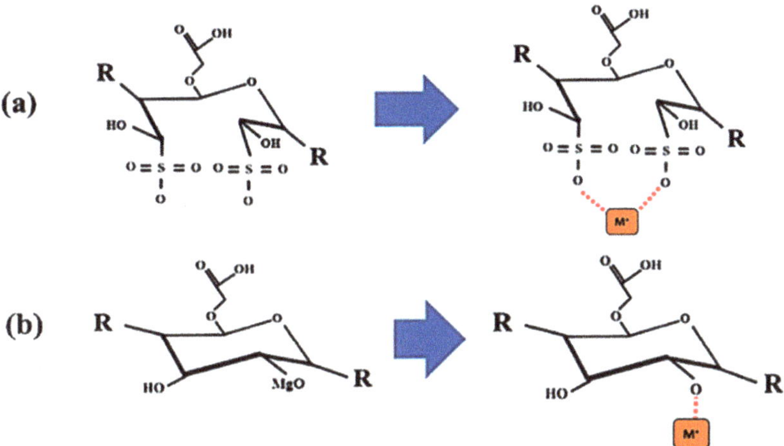

Figure 16. Adsorption mechanism on FeNP/NFC: (**a**) sulfur forms complexes and (**b**) ionic bonds.

The exchange of metal ions (M^2) with H^+ ions of hydroxyl groups on the surface occurs because the cations in the aqueous solutions are hydrated; therefore, they are attracted to the surface's functional group of interest. Mg^{2+} can participate in the adsorption. This result indicates that the magnesium ions, which were part of the MgO crystals in the FeNP/NFC composite, exchange with the M^{2+} during adsorption—Figure 16b [67]. As with many other adsorbent materials, there is the possibility that two or more mechanisms are involved in adsorption, as there are several active groups in FeNP/NFC.

The amino group on NFC, as mentioned above, influences the mechanism of lead removal. Metal ions can interact and bind with two binding sites on the surface [104]. There seems to be a significant impact on the ionization ability of –NH by pH, which affects the adsorption capacity of target ions. H^+ concentration is high in solution when pH is low; thus, –NH of material is protonated in this environment. This behavior will enhance the electrostatic attraction of anions, further increasing target ions' adsorption capacity [105].

4. Conclusions

Having Fe nanoparticles on the FeNP/NFC composite improved Cd and Pb selectivity during removal processes. The various chemical reaction processes (carboxymethylation, oxidation, hydrolysis, and sulfonation) occurred during the Kraft process, and acid hydrolysis increased the composite's heterogeneous surface. The chemical structure of the FeNP/NFC surface (adsorption sites and hydroxyl coordination) depends on the oxides' morphologies and the crystal structure. The high content of minerals or inorganic substances does not interfere with the adsorption process; otherwise, they helped to functionalize the material and diversify the adsorption mechanisms. Cd and Pb ions' speciation in an aqueous solution at pH 5 affects and determines the charge that will allow adsorption. As seen in the zeta potential results, the composite is negatively charged, so cations were easily adsorbed. Cd and Pb sorption mechanisms are related to electrostatic interactions, hydrogen bonding, and ion exchange with biopolymer structures. The strong electrostatic interactions between the -OH ions and the carboxyl groups were essential to form inner-sphere and outer-sphere complexes in the material. The thermodynamic results indicate that both processes have exothermic reactions which are irreversible and stable. The pseudo-first-order model describes Pb sorption kinetics well. Still, the Elovich equation was also a good model to describe Cd sorption kinetics. Based on the adsorption capacities obtained in this study, FeNP/NFC can be considered as a highly efficient adsorbent to remove these ions from aqueous solutions, particularly for Pb.

Author Contributions: Conceptualization, A.V.-G., R.C.-M., and R.H.-A.; methodology, A.V.-G. and R.C.-M.; software, A.V.-G.; validation, A.V.-G, R.A.-C.-V., and R.H.-A.; formal analysis, E.M.R.-M. and R.H.-A; investigation, A.V.-G. and E.M.R.-M.; resources, R.A.-C.-V., R.H.-A. and R.C.-M.; data curation, A.V.-G. and E.M.R.-M.; writing—original draft preparation, A.V.-G.; writing—review and editing, R.C.-M., R.H.-A., and R.A.-C.-V.; supervision, R.C.-M., R.H.-A., and R.A.-C.-V.; funding acquisition, R.A.-C.-V. All authors have read and agreed to the published version of the manuscript.

Funding: This research was funded by Coordinación de la Investigación Científica-UMSNH, grant number CIC-UMSNH-2020, and the APC was funded by PRODEP.

Institutional Review Board Statement: Not applicable.

Informed Consent Statement: Not applicable.

Acknowledgments: We appreciate the technical support provided by B. Millán-Malo from CFATA-UNAM.

Conflicts of Interest: The authors declare no conflict of interest. The funders had no role in the design of the study; in the collection, analyses, or interpretation of data; in the writing of the manuscript, or in the decision to publish the results.

References

1. Riani, E.; Cordova, M.R.; Arifin, Z. Heavy Metal Pollution and Its Relation to the Malformation of Green Mussels Cultured in Muara Kamal Waters, Jakarta Bay, Indonesia. *Mar. Pollut. Bull.* **2018**, *133*, 664–670. [CrossRef] [PubMed]
2. Fijałkowska, G.; Szewczuk-Karpisz, K.; Wiśniewska, M. Chromium(VI) and Lead(II) Accumulation at the Montmorillonite/Aqueous Solution Interface in the Presence of Polyacrylamide Containing Quaternary Amine Groups. *J. Mol. Liq.* **2019**, *293*, 111514. [CrossRef]
3. Kumari, M.; Pittman, C.U.; Mohan, D. Heavy Metals [Chromium (VI) and Lead (II)] Removal from Water Using Mesoporous Magnetite (Fe3O4) Nanospheres. *J. Colloid Interface Sci.* **2015**, *442*, 120–132. [CrossRef] [PubMed]
4. Sharma, R.; Sarswat, A.; Pittman, C.U.; Mohan, D. Cadmium and Lead Remediation Using Magnetic and Non-Magnetic Sustainable Biosorbents Derived from Bauhinia Purpurea Pods. *RSC Adv.* **2017**, *7*, 8606–8624. [CrossRef]
5. Koju, N.K.; Song, X.; Wang, Q.; Hu, Z.; Colombo, C. Cadmium Removal from Simulated Groundwater Using Alumina Nanoparticles: Behaviors and Mechanisms. *Environ. Pollut.* **2018**, *240*, 255–266. [CrossRef]
6. Chen, Z.; Zhang, J.; Huang, L.; Yuan, Z.; Li, Z.; Liu, M. Removal of Cd and Pb with Biochar Made from Dairy Manure at Low Temperature. *J. Integr. Agric.* **2019**, *18*, 201–210. [CrossRef]
7. Jaishankar, M.; Tseten, T.; Anbalagan, N.; Mathew, B.B.; Beeregowda, K.N. Toxicity, Mechanism and Health Effects of Some Heavy Metals. *Interdiscip. Toxicol.* **2014**, *7*, 60–72. [CrossRef]
8. Mohan, D.; Pittman, C.U.; Bricka, M.; Smith, F.; Yancey, B.; Mohammad, J.; Steele, P.H.; Alexandre-Franco, M.F.; Gómez-Serrano, V.; Gong, H. Sorption of Arsenic, Cadmium, and Lead by Chars Produced from Fast Pyrolysis of Wood and Bark during Bio-Oil Production. *J. Colloid Interface Sci.* **2007**, *310*, 57–73. [CrossRef]
9. Mager, E.M.; Brix, K.V.; Gerdes, R.M.; Ryan, A.C.; Grosell, M. Effects of Water Chemistry on the Chronic Toxicity of Lead to the Cladoceran, Ceriodaphnia Dubia. *Ecotoxicol. Environ. Saf.* **2011**, *74*, 238–243. [CrossRef]
10. Akinyeye, O.J.; Ibigbami, T.B.; Odeja, O.O.; Sosanolu, O.M. Evaluation of Kinetics and Equilibrium Studies of Biosorption Potentials of Bamboo Stem Biomass for Removal of Lead (II) and Cadmium (II) Ions from Aqueous Solution. *Afr. J. Pure Appl. Chem.* **2020**, *14*, 24–41. [CrossRef]
11. Isawi, H. Using Zeolite/Polyvinyl Alcohol/Sodium Alginate Nanocomposite Beads for Removal of Some Heavy Metals from Wastewater. *Arab. J. Chem.* **2020**, *13*, 5691–5716. [CrossRef]
12. Ahmad, R.; Mirza, A. Facile One Pot Green Synthesis of Chitosan-Iron Oxide (CS-Fe2O3) Nanocomposite: Removal of Pb(II) and Cd(II) from Synthetic and Industrial Wastewater. *J. Clean. Prod.* **2018**, *186*, 342–352. [CrossRef]
13. O'Connell, D.W.; Birkinshaw, C.; O'Dwyer, T.F. Heavy Metal Adsorbents Prepared from the Modification of Cellulose: A Review. *Bioresour. Technol.* **2008**, *99*, 6709–6724. [CrossRef] [PubMed]

14. Valdés Rodríguez, O.A.; Palacios Wassenaar, O.M.; Ruíz Hernández, R.; Pérez Vásquez, A. Moringa and Ricinus Association Potential in the Sub-Tropics of Veracruz. *Rev. Mex. Cienc. Agríc.* **2014**, *5*, 1673–1686.
15. Olson, M.E.; Fahey, J.W. Moringa oleifera: Un árbol multiusos para las zonas tropicales secas. *Rev. Mex. Biodivers.* **2011**, *82*, 1071–1082. [CrossRef]
16. Kargarzadeh, H.; Ahmad, I.; Abdullah, I.; Dufresne, A.; Zainudin, S.Y.; Sheltami, R.M. Effects of Hydrolysis Conditions on the Morphology, Crystallinity, and Thermal Stability of Cellulose Nanocrystals Extracted from Kenaf Bast Fibers. *Cellulose* **2012**, *19*, 855–866. [CrossRef]
17. Ramos-Vargas, S.; Huirache-Acuña, R.; Guadalupe Rutiaga-Quiñones, J.; Cortés-Martínez, R. Effective Lead Removal from Aqueous Solutions Using Cellulose Nanofibers Obtained from Water Hyacinth. *Water Supply* **2020**, *20*, 2715–2736. [CrossRef]
18. Mahfoudhi, N.; Boufi, S. Nanocellulose as a Novel Nanostructured Adsorbent for Environmental Remediation: A Review. *Cellulose* **2017**, *24*, 1171–1197. [CrossRef]
19. Rahmani, A.; Mousavi, H.Z.; Fazli, M. Effect of Nanostructure Alumina on Adsorption of Heavy Metals. *Desalination* **2010**, *253*, 94–100. [CrossRef]
20. Mondal, P.; Purkait, M.K. Green Synthesized Iron Nanoparticles Supported on PH Responsive Polymeric Membrane for Nitrobenzene Reduction and Fluoride Rejection Study: Optimization Approach. *J. Clean. Prod.* **2018**, *170*, 1111–1123. [CrossRef]
21. Bolade, O.P.; Williams, A.B.; Benson, N.U. Green Synthesis of Iron-Based Nanomaterials for Environmental Remediation: A Review. *Environ. Nanotechnol. Monit. Manag.* **2020**, *13*, 100279. [CrossRef]
22. Abujaber, F.; Zougagh, M.; Jodeh, S.; Ríos, Á.; Guzmán Bernardo, F.J.; Rodríguez Martín-Doimeadios, R.C. Magnetic Cellulose Nanoparticles Coated with Ionic Liquid as a New Material for the Simple and Fast Monitoring of Emerging Pollutants in Waters by Magnetic Solid Phase Extraction. *Microchem. J.* **2018**, *137*, 490–495. [CrossRef]
23. Krishna, R.; Dias, C.; Ventura, J.; Titus, E. Green and Facile Decoration of Fe3O4 Nanoparticles on Reduced Graphene Oxide. *Mater. Today Proc.* **2016**, *3*, 2807–2813. [CrossRef]
24. Ostovan, A.; Ghaedi, M.; Arabi, M. Fabrication of Water-Compatible Superparamagnetic Molecularly Imprinted Biopolymer for Clean Separation of Baclofen from Bio-Fluid Samples: A Mild and Green Approach. *Talanta* **2018**, *179*, 760–768. [CrossRef]
25. Soliemanzadeh, A.; Fekri, M. The Application of Green Tea Extract to Prepare Bentonite-Supported Nanoscale Zero-Valent Iron and Its Performance on Removal of Cr(VI): Effect of Relative Parameters and Soil Experiments. *Microporous Mesoporous Mater.* **2017**, *239*, 60–69. [CrossRef]
26. Wang, X.; Le, L.; Alvarez, P.J.J.; Li, F.; Liu, K. Synthesis and Characterization of Green Agents Coated Pd/Fe Bimetallic Nanoparticles. *J. Taiwan Inst. Chem. Eng.* **2015**, *50*, 297–305. [CrossRef]
27. Zhang, Y.; Chen, Y.; Westerhoff, P.; Hristovski, K.; Crittenden, J.C. Stability of Commercial Metal Oxide Nanoparticles in Water. *Water Res.* **2008**, *42*, 2204–2212. [CrossRef]
28. Liu, L.; Fan, S. Removal of Cadmium in Aqueous Solution Using Wheat Straw Biochar: Effect of Minerals and Mechanism. *Environ. Sci. Pollut. Res.* **2018**, *25*, 8688–8700. [CrossRef]
29. Bibi, I.; Nazar, N.; Ata, S.; Sultan, M.; Ali, A.; Abbas, A.; Jilani, K.; Kamal, S.; Sarim, F.M.; Khan, M.I.; et al. Green Synthesis of Iron Oxide Nanoparticles Using Pomegranate Seeds Extract and Photocatalytic Activity Evaluation for the Degradation of Textile Dye. *J. Mater. Res. Technol.* **2019**, *8*, 6115–6124. [CrossRef]
30. Bolade, O.P.; Akinsiku, A.A.; Adeyemi, A.O.; Williams, A.B.; Benson, N.U. Dataset on Phytochemical Screening, FTIR and GC–MS Characterisation of Azadirachta Indica and Cymbopogon Citratus as Reducing and Stabilising Agents for Nanoparticles Synthesis. *Data Brief* **2018**, *20*, 917–926. [CrossRef]
31. Akhtar, M.S.; Panwar, J.; Yun, Y.-S. Biogenic Synthesis of Metallic Nanoparticles by Plant Extracts. *ACS Sustain. Chem. Eng.* **2013**, *1*, 591–602. [CrossRef]
32. Kharissova, O.V.; Dias, H.V.R.; Kharisov, B.I.; Pérez, B.O.; Pérez, V.M.J. The Greener Synthesis of Nanoparticles. *Trends Biotechnol.* **2013**, *31*, 240–248. [CrossRef] [PubMed]
33. Nasrollahzadeh, M.; Mohammad Sajadi, S.; Rostami-Vartooni, A.; Khalaj, M. Green Synthesis of Pd/Fe3O4 Nanoparticles Using Euphorbia Condylocarpa M. Bieb Root Extract and Their Catalytic Applications as Magnetically Recoverable and Stable Recyclable Catalysts for the Phosphine-Free Sonogashira and Suzuki Coupling Reactions. *J. Mol. Catal. Chem.* **2015**, *396*, 31–39. [CrossRef]
34. Harshiny, M.; Iswarya, C.N.; Matheswaran, M. Biogenic Synthesis of Iron Nanoparticles Using Amaranthus Dubius Leaf Extract as a Reducing Agent. *Powder Technol.* **2015**, *286*, 744–749. [CrossRef]
35. Cai, Y.; Shen, Y.; Xie, A.; Li, S.; Wang, X. Green Synthesis of Soya Bean Sprouts-Mediated Superparamagnetic Fe3O4 Nanoparticles. *J. Magn. Magn. Mater.* **2010**, *322*, 2938–2943. [CrossRef]
36. Makarov, V.V.; Makarova, S.S.; Love, A.J.; Sinitsyna, O.V.; Dudnik, A.O.; Yaminsky, I.V.; Taliansky, M.E.; Kalinina, N.O. Biosynthesis of Stable Iron Oxide Nanoparticles in Aqueous Extracts of Hordeum Vulgare and Rumex Acetosa Plants. *Langmuir* **2014**, *30*, 5982–5988. [CrossRef]
37. Balamurugan, M.; Saravanan, S.; Soga, T. Synthesis of Iron Oxide Nanoparticles by Using *Eucalyptus Globulus* Plant Extract. *E-J. Surf. Sci. Nanotechnol.* **2014**, *12*, 363–367. [CrossRef]
38. Harshiny, M.; Matheswaran, M.; Arthanareeswaran, G.; Kumaran, S.; Rajasree, S. Enhancement of Antibacterial Properties of Silver Nanoparticles–Ceftriaxone Conjugate through Mukia Maderaspatana Leaf Extract Mediated Synthesis. *Ecotoxicol. Environ. Saf.* **2015**, *121*, 135–141. [CrossRef]

39. Ahmmad, B.; Leonard, K.; Shariful Islam, M.; Kurawaki, J.; Muruganandham, M.; Ohkubo, T.; Kuroda, Y. Green Synthesis of Mesoporous Hematite (α-Fe2O3) Nanoparticles and Their Photocatalytic Activity. *Adv. Powder Technol.* **2013**, *24*, 160–167. [CrossRef]
40. Wang, Z.; Fang, C.; Megharaj, M. Characterization of Iron–Polyphenol Nanoparticles Synthesized by Three Plant Extracts and Their Fenton Oxidation of Azo Dye. *ACS Sustain. Chem. Eng.* **2014**, *2*, 1022–1025. [CrossRef]
41. Li, J.; Hu, J.; Xiao, L.; Wang, Y.; Wang, X. Interaction Mechanisms between α-Fe2O3, γ-Fe2O3 and Fe3O4 Nanoparticles and Citrus Maxima Seedlings. *Sci. Total Environ.* **2018**, *625*, 677–685. [CrossRef] [PubMed]
42. Devatha, C.P.; K, J.; Patil, M. Effect of Green Synthesized Iron Nanoparticles by Azardirachta Indica in Different Proportions on Antibacterial Activity. *Environ. Nanotechnol. Monit. Manag.* **2018**, *9*, 85–94. [CrossRef]
43. Zhu, F.; Ma, S.; Liu, T.; Deng, X. Green Synthesis of Nano Zero-Valent Iron/Cu by Green Tea to Remove Hexavalent Chromium from Groundwater. *J. Clean. Prod.* **2018**, *174*, 184–190. [CrossRef]
44. Weng, X.; Guo, M.; Luo, F.; Chen, Z. One-Step Green Synthesis of Bimetallic Fe/Ni Nanoparticles by Eucalyptus Leaf Extract: Biomolecules Identification, Characterization and Catalytic Activity. *Chem. Eng. J.* **2017**, *308*, 904–911. [CrossRef]
45. Abdelghany, T.M.; Al-Rajhi, A.M.H.; Al Abboud, M.A.; Alawlaqi, M.M.; Ganash Magdah, A.; Helmy, E.A.M.; Mabrouk, A.S. Recent Advances in Green Synthesis of Silver Nanoparticles and Their Applications: About Future Directions. A Review. *BioNanoScience* **2018**, *8*, 5–16. [CrossRef]
46. Ahmed, S.; Ahmad, M.; Swami, B.L.; Ikram, S. A Review on Plants Extract Mediated Synthesis of Silver Nanoparticles for Antimicrobial Applications: A Green Expertise. *J. Adv. Res.* **2016**, *7*, 17–28. [CrossRef] [PubMed]
47. Iqbal, A.; Iqbal, K.; Li, B.; Gong, D.; Qin, W. Recent Advances in Iron Nanoparticles: Preparation, Properties, Biological and Environmental Application. *J. Nanosci. Nanotechnol.* **2017**, *17*, 4386–4409. [CrossRef]
48. Saif, S.; Tahir, A.; Chen, Y. Green Synthesis of Iron Nanoparticles and Their Environmental Applications and Implications. *Nanomaterials* **2016**, *6*, 209. [CrossRef]
49. Rauwel, P.; Küünal, S.; Ferdov, S.; Rauwel, E. A Review on the Green Synthesis of Silver Nanoparticles and Their Morphologies Studied via TEM. Available online: https://www.hindawi.com/journals/amse/2015/682749/ (accessed on 4 October 2020).
50. Sharma, J.K.; Srivastava, P.; Akhtar, M.S.; Singh, G.; Ameen, S. α-Fe2O3 Hexagonal Cones Synthesized from the Leaf Extract of Azadirachta Indica and Its Thermal Catalytic Activity. *New J. Chem.* **2015**, *39*, 7105–7111. [CrossRef]
51. Falowo, A.B.; Mukumbo, F.E.; Idamokoro, E.M.; Lorenzo, J.M.; Afolayan, A.J.; Muchenje, V. Multi-Functional Application of Moringa Oleifera Lam. in Nutrition and Animal Food Products: A Review. *Food Res. Int.* **2018**, *106*, 317–334. [CrossRef]
52. Waterman, C.; Cheng, D.M.; Rojas-Silva, P.; Poulev, A.; Dreifus, J.; Lila, M.A.; Raskin, I. Stable, Water Extractable Isothiocyanates from Moringa Oleifera Leaves Attenuate Inflammation in Vitro. *Phytochemistry* **2014**, *103*, 114–122. [CrossRef] [PubMed]
53. Lagergren, S. Zur theorie der sogenannten adsorption geloster stoffe. Kungliga Svenska Vetenskapsakademiens. *Handlingar* **1898**, *24*, 1–39.
54. Ho, Y.S.; McKay, G.; Wase, D.A.J.; Forster, C.F. Study of the Sorption of Divalent Metal Ions on to Peat. *Adsorpt. Sci. Technol.* **2000**, *18*, 639–650. [CrossRef]
55. Low, M.J.D. Kinetics of Chemisorption of Gases on Solids. Available online: https://pubs.acs.org/doi/pdf/10.1021/cr60205a003 (accessed on 3 October 2020).
56. Chen, K.; He, J.; Li, Y.; Cai, X.; Zhang, K.; Liu, T.; Hu, Y.; Lin, D.; Kong, L.; Liu, J. Removal of Cadmium and Lead Ions from Water by Sulfonated Magnetic Nanoparticle Adsorbents. *J. Colloid Interface Sci.* **2017**, *494*, 307–316. [CrossRef] [PubMed]
57. Langmuir, I. The Adsorption of Gases on Plane Surfaces of Glass, Mica and Platinum. *J. Am. Chem. Soc.* **1918**, *40*, 1361–1403. [CrossRef]
58. Papageorgiou, S.K.; Kouvelos, E.P.; Katsaros, F.K. Calcium Alginate Beads from Laminaria Digitata for the Removal of Cu^{+2} and Cd^{+2} from Dilute Aqueous Metal Solutions. *Desalination* **2008**, *224*, 293–306. [CrossRef]
59. Acemioğlu, B. Removal of Fe(II) Ions from Aqueous Solution by Calabrian Pine Bark Wastes. *Bioresour. Technol.* **2004**, *93*, 99–102. [CrossRef]
60. Sivarathnakumar, S.; Baskar, G.; Kumar, R.P.; Bharathiraja, B. Bioethanol Production by the Utilisation of Moringa Oleifera Stem with Sono-Assisted Acid/Alkali Hydrolysis Approach. *Int. J. Environ. Sustain. Dev.* **2016**, *15*, 392–403. [CrossRef]
61. Zhao, H.; Kwak, J.H.; Conrad Zhang, Z.; Brown, H.M.; Arey, B.W.; Holladay, J.E. Studying Cellulose Fiber Structure by SEM, XRD, NMR and Acid Hydrolysis. *Carbohydr. Polym.* **2007**, *68*, 235–241. [CrossRef]
62. Lavoine, N.; Desloges, I.; Dufresne, A.; Bras, J. Microfibrillated Cellulose—Its Barrier Properties and Applications in Cellulosic Materials: A Review. *Carbohydr. Polym.* **2012**, *90*, 735–764. [CrossRef]
63. Kumar, A.; Negi, Y.S.; Choudhary, V.; Bhardwaj, N.K. Characterization of Cellulose Nanocrystals Produced by Acid-Hydrolysis from Sugarcane Bagasse as Agro-Waste. *J. Mater. Phys. Chem.* **2014**, *2*, 1–8. [CrossRef]
64. Zhao, Y.; Xu, C.; Xing, C.; Shi, X.; Matuana, L.M.; Zhou, H.; Ma, X. Fabrication and Characteristics of Cellulose Nanofibril Films from Coconut Palm Petiole Prepared by Different Mechanical Processing. *Ind. Crops Prod.* **2015**, *65*, 96–101. [CrossRef]
65. Abdul Khalil, H.P.S.; Davoudpour, Y.; Islam, M.N.; Mustapha, A.; Sudesh, K.; Dungani, R.; Jawaid, M. Production and Modification of Nanofibrillated Cellulose Using Various Mechanical Processes: A Review. *Carbohydr. Polym.* **2014**, *99*, 649–665. [CrossRef] [PubMed]

66. Tonoli, G.H.D.; Teixeira, E.M.; Corrêa, A.C.; Marconcini, J.M.; Caixeta, L.A.; Pereira-da-Silva, M.A.; Mattoso, L.H.C. Cellulose Micro/Nanofibres from Eucalyptus Kraft Pulp: Preparation and Properties. *Carbohydr. Polym.* **2012**, *89*, 80–88. [CrossRef] [PubMed]
67. Kalia, S.; Dufresne, A.; Cherian, B.M.; Kaith, B.S.; Avérous, L.; Njuguna, J.; Nassiopoulos, E. Cellulose-Based Bio- and Nanocomposites: A Review. Available online: https://www.hindawi.com/journals/ijps/2011/837875/ (accessed on 27 December 2020).
68. Dong, H.; Snyder, J.F.; Tran, D.T.; Leadore, J.L. Hydrogel, Aerogel and Film of Cellulose Nanofibrils Functionalized with Silver Nanoparticles. *Carbohydr. Polym.* **2013**, *95*, 760–767. [CrossRef]
69. Sahu, U.K.; Mahapatra, S.S.; Patel, R.K. Synthesis and Characterization of an Eco-Friendly Composite of Jute Fiber and Fe_2O_3 Nanoparticles and Its Application as an Adsorbent for Removal of As(V) from Water. *J. Mol. Liq.* **2017**, *237*, 313–321. [CrossRef]
70. Nagappan, S.; Ha, H.M.; Park, S.S.; Jo, N.-J.; Ha, C.-S. One-Pot Synthesis of Multi-Functional Magnetite–Polysilsesquioxane Hybrid Nanoparticles for the Selective Fe^{3+} and Some Heavy Metal Ions Adsorption. *RSC Adv.* **2017**, *7*, 19106–19116. [CrossRef]
71. Cao, C.-Y.; Qu, J.; Wei, F.; Liu, H.; Song, W.-G. Superb Adsorption Capacity and Mechanism of Flowerlike Magnesium Oxide Nanostructures for Lead and Cadmium Ions. *ACS Appl. Mater. Interfaces* **2012**, *4*, 4283–4287. [CrossRef]
72. Xu, Q.; Jiang, L. Infrared Spectra of the M(NO)n (M = Sn, Pb; n = 1, 2) and PbNO- Molecules. *Inorg. Chem.* **2006**, *45*, 8648–8654. [CrossRef]
73. Qin, X.; Zhou, J.; Huang, A.; Guan, J.; Zhang, Q.; Huang, Z.; Hu, H.; Zhang, Y.; Yang, M.; Wu, J.; et al. A Green Technology for the Synthesis of Cellulose Succinate for Efficient Adsorption of Cd(II) and Pb(II) Ions. *RSC Adv.* **2016**, *6*, 26817–26825. [CrossRef]
74. Asmaly, H.A.; Abussaud, B.; Ihsanullah; Saleh, T.A.; Bukhari, A.A.; Laoui, T.; Shemsi, A.M.; Gupta, V.K.; Atieh, M.A. Evaluation of Micro- and Nano-Carbon-Based Adsorbents for the Removal of Phenol from Aqueous Solutions. *Toxicol. Environ. Chem.* **2015**, *97*, 1164–1179. [CrossRef]
75. Taipale, T.; Österberg, M.; Nykänen, A.; Ruokolainen, J.; Laine, J. Effect of Microfibrillated Cellulose and Fines on the Drainage of Kraft Pulp Suspension and Paper Strength. *Cellulose* **2010**, *17*, 1005–1020. [CrossRef]
76. Capron, I.; Rojas, O.J.; Bordes, R. Behavior of Nanocelluloses at Interfaces. *Curr. Opin. Colloid Interface Sci.* **2017**, *29*, 83–95. [CrossRef]
77. Visakh, P.M.; Mathew, A.P.; Oksman, K.; Thomas, S. Starch-Based Bionanocomposites: Processing and Properties. In *Polysaccharide Building Blocks*; John Wiley & Sons, Ltd.: Hoboken, NJ, USA, 2012; pp. 287–306. ISBN 978-1-118-22948-4.
78. Fengel, D.; Wegener, G. *Wood: Chemistry, Ultrastructure, Reactions*; Walter de Gruyter: Berlin, Germany, 2011; ISBN 978-3-11-083965-4.
79. Martínez-Pérez, R.; Pedraza-Bucio, F.E.; Orihuela-Equihua, R.; López-Albarrán, P.; Rutiaga-Quiñones, J.G. Calorific Value and Inorganic Material of Ten Mexican Wood Species. *Wood Res.* **2015**, *60*, 12.
80. Argyropoulos, D.S. Wood and Cellulosic Chemistry. Second Edition, Revised and Expanded Edited by David N.-S. Hon (Clemson University) and Nubuo Shiraishi (Kyoto University). Marcel Dekker: New York and Basel. 2001. Vii + 914 Pp. $250.00. ISBN 0-8247-0024-4. *J. Am. Chem. Soc.* **2001**, *123*, 8880–8881. [CrossRef]
81. Ngangyo-Heya, M.; Foroughbahchk-Pournavab, R.; Carrillo-Parra, A.; Rutiaga-Quiñones, J.G.; Zelinski, V.; Pintor-Ibarra, L.F. Calorific Value and Chemical Composition of Five Semi-Arid Mexican Tree Species. *Forests* **2016**, *7*, 58. [CrossRef]
82. Muthukumar, H.; Matheswaran, M. Amaranthus Spinosus Leaf Extract Mediated FeO Nanoparticles: Physicochemical Traits, Photocatalytic and Antioxidant Activity. *ACS Sustain. Chem. Eng.* **2015**, *3*, 3149–3156. [CrossRef]
83. Zhu, X.; Song, T.; Lv, Z.; Ji, G. High-Efficiency and Low-Cost α-Fe2O3 Nanoparticles-Coated Volcanic Rock for Cd(II) Removal from Wastewater. *Process Saf. Environ. Prot.* **2016**, *104*, 373–381. [CrossRef]
84. Tajik, E.; Naeimi, A.; Amiri, A. Fabrication of Iron Oxide Nanoparticles, and Green Catalytic Application of an Immobilized Novel Iron Schiff on Wood Cellulose. *Cellulose* **2018**, *25*, 915–923. [CrossRef]
85. Gupta, V.K.; Nayak, A. Cadmium Removal and Recovery from Aqueous Solutions by Novel Adsorbents Prepared from Orange Peel and Fe2O3 Nanoparticles. *Chem. Eng. J.* **2012**, *180*, 81–90. [CrossRef]
86. Foo, K.Y.; Hameed, B.H. Insights into the Modeling of Adsorption Isotherm Systems. *Chem. Eng. J.* **2010**, *156*, 2–10. [CrossRef]
87. Papageorgiou, S.K.; Katsaros, F.K.; Kouvelos, E.P.; Nolan, J.W.; Le Deit, H.; Kanellopoulos, N.K. Heavy Metal Sorption by Calcium Alginate Beads from Laminaria Digitata. *J. Hazard. Mater.* **2006**, *137*, 1765–1772. [CrossRef] [PubMed]
88. Wang, F.; Lu, X.; Li, X. Selective Removals of Heavy Metals (Pb2+, Cu2+, and Cd2+) from Wastewater by Gelation with Alginate for Effective Metal Recovery. *J. Hazard. Mater.* **2016**, *308*, 75–83. [CrossRef] [PubMed]
89. Pawar, R.R.; Lalhmunsiama; Kim, M.; Kim, J.-G.; Hong, S.-M.; Sawant, S.Y.; Lee, S.M. Efficient Removal of Hazardous Lead, Cadmium, and Arsenic from Aqueous Environment by Iron Oxide Modified Clay-Activated Carbon Composite Beads. *Appl. Clay Sci.* **2018**, *162*, 339–350. [CrossRef]
90. Elimelech, M.; Chen, W.H.; Waypa, J.J. Measuring the Zeta (Electrokinetic) Potential of Reverse Osmosis Membranes by a Streaming Potential Analyzer. *Desalination* **1994**, *95*, 269–286. [CrossRef]
91. Zirino, A.; Yamamoto, S. A pH-Dependent Model for The Chemical Speciation of Copper, Zinc, Cadmium, and Lead in Seawater. *Limnol. Oceanogr.* **1972**, *17*, 661–671. [CrossRef]
92. Giles, C.H.; D'Silva, A.P.; Easton, I.A. A General Treatment and Classification of the Solute Adsorption Isotherm Part. II. Experimental Interpretation. *J. Colloid Interface Sci.* **1974**, *47*, 766–778. [CrossRef]
93. Giles, C.H.; Smith, D.; Huitson, A. A General Treatment and Classification of the Solute Adsorption Isotherm. I. Theoretical. *J. Colloid Interface Sci.* **1974**, *47*, 755–765. [CrossRef]

94. Prasher, S.O.; Beaugeard, M.; Hawari, J.; Bera, P.; Patel, R.M.; Kim, S.H. Biosorption of Heavy Metals by Red Algae (Palmaria Palmata). *Environ. Technol.* **2004**, *25*, 1097–1106. [CrossRef]
95. Trakal, L.; Veselská, V.; Šafařík, I.; Vítková, M.; Číhalová, S.; Komárek, M. Lead and Cadmium Sorption Mechanisms on Magnetically Modified Biochars. *Bioresour. Technol.* **2016**, *203*, 318–324. [CrossRef]
96. Anwar, J.; Shafique, U.; Waheed-uz-Zaman; Salman, M.; Dar, A.; Anwar, S. Removal of Pb(II) and Cd(II) from Water by Adsorption on Peels of Banana. *Bioresour. Technol.* **2010**, *101*, 1752–1755. [CrossRef] [PubMed]
97. Ge, F.; Li, M.-M.; Ye, H.; Zhao, B.-X. Effective Removal of Heavy Metal Ions Cd2+, Zn2+, Pb2+, Cu2+ from Aqueous Solution by Polymer-Modified Magnetic Nanoparticles. *J. Hazard. Mater.* **2012**, *211–212*, 366–372. [CrossRef] [PubMed]
98. Keshvardoostchokami, M.; Babaei, L.; Zamani, A.A.; Parizanganeh, A.H.; Piri, F. Synthesized chitosan/iron oxide nanocomposite and shrimp shell in removal of nickel, cadmium and lead from aqueous solution. *Global J. Environ. Sci. Manag.* **2017**, *3*, 267–278.
99. Pehlivan, E.; Yanık, B.H.; Ahmetli, G.; Pehlivan, M. Equilibrium Isotherm Studies for the Uptake of Cadmium and Lead Ions onto Sugar Beet Pulp. *Bioresour. Technol.* **2008**, *99*, 3520–3527. [CrossRef] [PubMed]
100. Kataria, N.; Garg, V.K. Green Synthesis of Fe3O4 Nanoparticles Loaded Sawdust Carbon for Cadmium (II) Removal from Water: Regeneration and Mechanism. *Chemosphere* **2018**, *208*, 818–828. [CrossRef]
101. Shalaby, T.I.; El-Kady, M.F.; Zaki, A.E.H.M.; El-Kholy, S.M. Preparation and Application of Magnetite Nanoparticles Immobilized on Cellulose Acetate Nanofibers for Lead Removal from Polluted Water. *Water Supply* **2017**, *17*, 176–187. [CrossRef]
102. Mahdavi, M.; Namvar, F.; Ahmad, M.B.; Mohamad, R. Green Biosynthesis and Characterization of Magnetic Iron Oxide (Fe3O4) Nanoparticles Using Seaweed (Sargassum Muticum) Aqueous Extract. *Molecules* **2013**, *18*, 5954–5964. [CrossRef]
103. Gao, S.; Shi, Y.; Zhang, S.; Jiang, K.; Yang, S.; Li, Z.; Takayama-Muromachi, E. Biopolymer-Assisted Green Synthesis of Iron Oxide Nanoparticles and Their Magnetic Properties. *J. Phys. Chem. C* **2008**, *112*, 10398–10401. [CrossRef]
104. Hokkanen, S.; Repo, E.; Suopajärvi, T.; Liimatainen, H.; Niinimaa, J.; Sillanpää, M. Adsorption of Ni(II), Cu(II) and Cd(II) from Aqueous Solutions by Amino Modified Nanostructured Microfibrillated Cellulose. *Cellulose* **2014**, *21*, 1471–1487. [CrossRef]
105. Zhao, T.; Feng, T. Application of Modified Chitosan Microspheres for Nitrate and Phosphate Adsorption from Aqueous Solution. *RSC Adv.* **2016**, *6*, 90878–90886. [CrossRef]

Article

Removal of Pb^{2+} from Aqueous Solutions Using K-Type Zeolite Synthesized from Coal Fly Ash

Yuhei Kobayashi [1], Fumihiko Ogata [1], Chalermpong Saenjum [2,3], Takehiro Nakamura [1] and Naohito Kawasaki [1,4,*]

1 Faculty of Pharmacy, Kindai University, 3-4-1 Kowakae, Higashi-Osaka, Osaka 577-8502, Japan; 1944420001t@kindai.ac.jp (Y.K.); ogata@phar.kindai.ac.jp (F.O.); nakamura@phar.kindai.ac.jp (T.N.)
2 Faculty of Pharmacy, Chiang Mai University, Suthep Road, Muang District, Chiang Mai 50200, Thailand; chalermpong.saenjum@gmail.com
3 Cluster of Excellence on Biodiversity-Based Economics and Society (B.BES-CMU), Chiang Mai University, Suthep Road, Muang District, Chiang Mai 50200, Thailand
4 Antiaging Center, Kindai University, 3-4-1 Kowakae, Higashi-Osaka, Osaka 577-8502, Japan
* Correspondence: kawasaki@phar.kindai.ac.jp; Tel.: +81-6-4307-4012

Received: 29 July 2020; Accepted: 22 August 2020; Published: 24 August 2020

Abstract: In this study, a novel zeolite (K-type zeolite) was synthesized from coal fly ash (FA), and adsorption capacity on Pb^{2+} was assessed. Six types of zeolite (FA1, FA3, FA6, FA12, FA24, and FA48) were prepared, and their physicochemical properties, such as surface functional groups, cation exchange capacity, pH_{pzc}, specific surface area, and pore volume, were evaluated. The quantity of Pb^{2+} adsorbed by the prepared zeolites followed the order FA $<$ FA1 $<$ FA3 $<$ FA6 $<$ FA12 $<$ FA24 $<$ FA48. Current results indicate that the level of Pb^{2+} adsorbed was strongly related to the surface characteristics of the adsorbent. Additionally, the correlation coefficient between the amounts of Pb^{2+} adsorbed and K^+ released from FA48 was 0.958. Thus, ion exchange with K^+ in the interlayer of FA48 is critical for the removal of Pb^{2+} from aqueous media. The new binding energies of Pb(4f) at 135 and 140 eV were detected after adsorption. Moreover, FA48 showed selectivity for Pb^{2+} adsorption in binary solution systems containing cations. The results revealed that FA48 could be useful for removing Pb^{2+} from aqueous media.

Keywords: K-type zeolite; fly ash; lead; adsorption

1. Introduction

In recent years, issues faced by the aquatic environment, such as water wastage (Goal 6) and the presence of plastic bags in the ocean (Goal 14), have become major concerns for sustainable societal development [1]. Several heavy metals have been contaminating the aquatic environment through human activities, which are non-biodegradable and accumulate in the ecosystem via the food chain, causing various health problems and diseases [2]. Among these heavy metals, lead (Pb^{2+}), mercury (Hg^{2+}), and cadmium (Cd^{2+}) have been defined as the "big three" harmful heavy metals posing the greatest threat to humans, animals, and the environment [3,4]. Numerous studies have previously identified a relationship between Pb^{2+} exposure and human health impacts, such as neurotoxicity, cardiovascular problems, kidney damage, hormonal imbalances, and decreased musculoskeletal function [5,6]. Pb^{2+} has been classified by the International Agency for Research on Cancer as Group 2B which is possibly carcinogenic to humans. Additionally, drinking water problems involving Pb^{2+} have occurred in several countries [7–10]. Therefore, the World Health Organization and U.S. Environmental Protection Agency have established maximum permissible Pb^{2+} contents for drinking water of 0.01 and 0.015 mg L^{-1}, respectively. Thus, the removal of Pb^{2+} from aqueous media is an important global issue.

In Japan, the demand for coal-fired power plants increased after the Fukushima Daiichi Nuclear Power Station accident in 2001. Coal fly ash, a byproduct of coal combustion, is produced from coal-fired power plants (approximately 11.5 million tons in Japan in 2017) [11]. Approximately 750 million tons of coal fly ash are produced globally, and only 25% of this ash is recycled in cement, concrete, soil conditioner, and fertilizer materials [12–14]. Therefore, a recycling technology for coal fly ash must be developed to utilize its unused value [14]. SiO_2, Al_2O_3, and Fe_2O_3 were identified as the major chemical components of coal fly ash. Various studies have previously reported the adsorption capability of heavy metals using coal fly ash; however, they suggested a lower adsorption capacity than that of conventional adsorbents [15,16].

Different activation methods have been modified and developed to improve the adsorption capacity of coal fly ash [17]. Such methods include the conversion of coal fly ash to zeolite named hydrothermal activation, which is useful for decreasing the amount of coal fly ash and preparing an adsorbent with a high adsorption capacity for heavy metals. Zeolite is a general term for crystalline hydrous aluminosilicate, which consists of a three-dimensional network structure of SiO_4–AlO_4 tetrahedrons with Si and Al as basic constituents. Their structures contain channels and cavities of different sizes with unique physicochemical properties, such as their adsorption, ion exchange, molecular sieving, and catalytic abilities [18–21]. Additionally, the physicochemical properties of zeolite prepared from coal fly ash following the activation using hydrothermal technique are directly affected by certain parameters, such as the coal fly ash composition, alkaline solution concentration, reaction temperature, time, and pressure, volume ratio of the alkaline solution, and amount of coal fly ash [14,22–24]. Therefore, suitable conditions for preparing zeolite from coal fly ash generated by coal-fired power plants in Japan must be identified. Our previous studies elucidated the composition of coal fly ash generated from the Tachibana-Wan Thermal Power Station in Japan [25]. Moreover, Na-type zeolite could be produced from coal fly ash, and adsorption capability on heavy metals was studied and evaluated [26]. Additionally, various previous studies have also investigated the adsorption of Pb^{2+} from aqueous media using industrial/agriculture wastes such as wool, sawdust, sugar beet pulp, and *Azadirachta indiea* (Neem) leaf [27–30]. However, the physicochemical properties of other types of zeolite produced from coal fly ash generated by the Tachibana-wan Thermal Power Station and its heavy metals adsorption capacity have not yet been reported.

Sodium hydroxide is often used in the conversion of FA into zeolite by hydrothermal activation, and the ion in the prepared zeolite can then be exchanged with sodium ions in aqueous media. Some studies have already reported the synthesis and characterization of zeolite prepared from FA using a sodium hydroxide solution [12,18,19,23,24,26]. However, few have been conducted on the synthesis and physicochemical properties of zeolite prepared from fly ash using potassium hydroxide solution. The preparation of K-type zeolite from coal ash for use as a fertilizer in the agricultural field has been reported [14]. However, those studies did not sufficiently evaluate the adsorption capacity for heavy metals. Additionally, the cation exchange capacity (ion exchange capacity) in zeolite is one of the most important adsorption mechanisms. This indicates that sodium ion (or potassium ion) is exchanged with cation (Pb^{2+} in this study) using Na-type zeolite (or K-type zeolite). The ionic radii of sodium ion and potassium ion are 1.80 and 2.20 Å, respectively. In addition, the (hydrate) radii of lead ion is over 1.8 Å [17,31]. Therefore, it was expected that K-type zeolite would be useful for removal of Pb^{2+} from aqueous solution compared to Na-type zeolite.

Therefore, the aim of this study was to synthesize K-type zeolite from coal fly ash following the activation using hydrothermal technique and evaluate adsorption capability on Pb^{2+} from aqueous media. The influences of various parameters including initial concentration, temperature, pH, contact time, and selectivity were additionally explored, and the Pb^{2+} adsorption mechanism of K-type zeolite was elucidated.

2. Materials and Methods

2.1. Materials and Chemicals

Standard solutions of Pb^{2+} ($Pb(NO_3)_2$ in 0.1 mol L^{-1} HNO_3), Na^+ (NaCl in water), Mg^{2+} ($Mg(NO_3)_2$ in 0.1 mol L^{-1} HNO_3), K^+ (KCl in water), Ca^{2+} ($CaCO_3$ in 0.1 mol L^{-1} HNO_3), Ni^{2+} ($Ni(NO_3)_2$ in 0.1 mol L^{-1} HNO_3), Cu^{2+} ($Cu(NO_3)_2$ in 0.1 mol L^{-1} HNO_3), Zn^{2+} ($Zn(NO_3)_2$ in 0.1 mol L^{-1} HNO_3), Sr^{2+} ($SrCO_3$ in 0.1 mol L^{-1} HNO_3), and Cd^{2+} ($Cd(NO_3)_2$ in 0.1 mol L^{-1} HNO_3), potassium hydroxide, nitric acid, and sodium hydroxide were purchased from FUJIFILM Wako Pure Chemical Co., Osaka, Japan. Coal fly ash (FA, JIS Type-II) was obtained from the Tachibana-wan Thermal Power Station (Shikoku Electric Power, Inc., Takamatsu, Japan). K-type zeolites were produced following a previously reported method [32]. Briefly, FA (3.0 g) was mixed with 3-mol L^{-1} potassium hydroxide solution in the volume of 240 mL. The reaction solution was then shaken and heated at 93 °C for 1, 3, 6, 12, 24, and 48 h. Afterwards, the suspensions were filtered through a 0.45-μm membrane filter (Advantec MFS, Inc., Tokyo, Japan), and obtained samples were washed with distilled water and then dried at 50 °C for 1 day. The samples heated for different durations were denoted as FA1, FA3, FA6, FA12, FA24, and FA48, respectively.

2.2. Physicochemical Properties of the Adsorbents

The physicochemical properties of each adsorbent were analyzed as follows. The crystal structure and morphology were analyzed using a MiniFlex II (Rigaku, Tokyo, Japan) and SU1510 (Hitachi Ltd., Tokyo, Japan), respectively. The specific surface area and pore volume were measured using a NOVA4200*e* instrument (Yuasa Ionic, Kyoto, Japan). The cation exchange capacity (CEC) and pH_{pzc} were determined following the Japanese Industrial Standard Method (JIS K 1478: 2009) and the method reported by Faria et al. [33]. The concentrations of acidic or basic functional groups in the adsorbents were measured following the Boehm titration method [34]. Additionally, the binding energy was analyzed using an AXIS-NOVA instrument (Shimadzu Co., Ltd., Kyoto, Japan). Finally, the solution pH was measured using an F-73S digital pH meter (HORIBA, Ltd., Kyoto, Japan).

2.3. Adsorption Capacity of Pb^{2+}

Each adsorbent (0.01 g) was mixed with 50 mL of the Pb^{2+} solution at 50 mg L^{-1} (pH 3.0). The reaction solution was shaken at 100 rpm and 25 °C for 24 h, and then filtered through a 0.45-μm membrane filter. The concentration of Pb^{2+} was measured using an iCAP-7600 Duo (Thermo Fisher Scientific Inc., Tokyo, Japan). The quantity of Pb^{2+} adsorbed was calculated as the difference between the Pb^{2+} concentrations before and after adsorption.

2.4. Effect of Initial Concentration, Temperature, pH, Contact Time, and Coexisting Ions on Pb^{2+} Adsorption

First, 0.01 g of FA48 was mixed with a 50-mL Pb^{2+} solution at concentrations of 10, 20, 30, 40, and 50 mg L^{-1} (the concentration range was determined by ecological risk assessment of sediment and total concentrations of heavy metals in sewage sludge from wastewater discharging area) [35–37]. The reaction solution was shaken at 100 rpm and 7, 25, and 45 °C for 24 h using water bath shaker MM-10 (TAITEC Co., Nagoya, Japan), and then filtered through a 0.45-μm membrane filter. Second, to evaluate the pH, 0.01 g of FA48 was mixed with 50 mL of a Pb^{2+} solution at concentrations of 10, 30, and 50 mg L^{-1}. The pH of the solution was adjusted between 2, 3, 5, 7, and 9 using nitric acid or sodium hydroxide solutions. The reaction solution was then shaken at 100 rpm and 25 °C for 24 h. To evaluate the contact time, 0.01 g of FA48 was mixed with 50 mL of a Pb^{2+} at a concentration of 50 mg L^{-1}. The reaction solution was then shaken at 100 rpm and 25 °C for 0.5, 1, 3, 6, 18, 21, 24, 30, 42, and 48 h. The quantity of Pb^{2+} adsorbed in each case was calculated as described above, and the data are presented as the mean ± standard deviation. Additionally, to evaluate the adsorption mechanism, the quantity of K^+ released from FA48 in the adsorption isotherm experiment was also measured using an iCAP-7600 Duo device. Finally, to evaluate the Pb^{2+} adsorption selectivity of the zeolite, 0.01 g of FA48 was mixed with 50 mL of a binary solution system at a concentration of 10 mg L^{-1} (Pb^{2+} and Na^{+}, Mg^{2+}, K^{+}, Ca^{2+}, Ni^{2+}, Cu^{2+}, Zn^{2+}, Sr^{2+}, or Cd^{2+}). The reaction solution was shaken at 100 rpm and 25 °C for 24 h. The concentration of each metal was measured using an iCAP-7600 Duo device. The quantity of each metal adsorbed was also calculated based on the differences in concentration before and after adsorption.

3. Results and Discussion

3.1. Physicochemical Properties

Zeolite is a highly porous aluminosilicate containing various channels and cavities. Ion-exchangeable cations, such as sodium, potassium, and calcium were used to balance the negative charge [21,38]. The scanning electron microscopy (SEM) images of each zeolite are shown in Figure 1. Spherical particles with different diameters were observed in FA. The spheres and agglomerates of FA1, FA3, FA6, and FA12 could be maintained, and some changes on the surfaces of each type of FA were observed under the tested experimental conditions. The spheres and agglomerates of FA24 and FA48 significantly changed to different crystal shapes and their particle sizes decreased. The synthesis of zeolite from fly ash involves three steps, i.e., dissolution, condensation, and crystallization [39]. Similar trends were observed in previous studies [12,17].

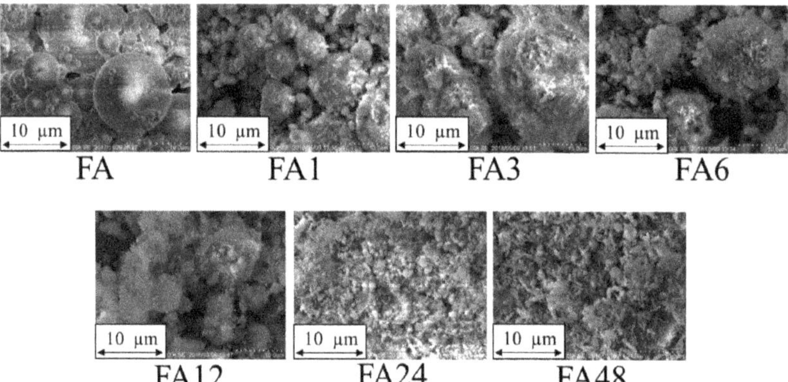

Figure 1. SEM images of adsorbents. Magnification is 3000 diameters.

The X-ray diffraction (XRD) patterns of the samples are shown in Figure 2. The FA was composed of quartz (SiO_2) and mullite ($3Al_2O_3{\cdot}2SiO_2$), and no significant changes in the XRD patterns of FA, FA1, FA3, FA6, and FA12 under our experimental conditions were observed. However, the structure of Zeolite F ($(K_{13.5}Si_{10}Al_{10}O_{40})(OH)_3{\cdot}13H_2O$) appeared in FA24 and FA48, indicating that FA was converted into zeolite. Less time was required to convert FA into zeolite using sodium hydroxide (approximately 6 h) than that required when using potassium hydroxide (approximately 24 h). Trends similarly were observed in a previous study [14]. The physicochemical properties of the adsorbents are shown in Table 1, and the concentrations of basic functional groups of FA24 and FA48 were higher than those of the other adsorbents. However, the concentrations of acidic functional groups decreased with raising the time of alkaline activation. The pH_{pzc} were not significantly different between the different types of FAs in this study. The CEC and pore volume ($d \leq 20$ (Å)), which greatly influence the adsorption capacity, of FA24 and FA48 were 12.4–26.4 times and 100 times higher than those of other FAs, indicating that the hydrothermal activation method significantly affected the CEC and pore volume ($d \leq 20$ (Å)) under our experimental conditions. Additionally, the specific surface area and pore volume (such as $20 < d \leq 500$ (Å), total, and mean pore diameter) of FA6 and FA12 were higher than those of the other FAs. Our previous study reported the characteristics of Na-type zeolite [26], and the CEC and pore volume ($d \leq 20$ (Å)) of the K-type zeolite prepared in this study exceeded those of the Na-type zeolite. Thus, hydrothermal activation using FA and potassium hydroxide was useful for producing novel zeolites to remove metals from aqueous media.

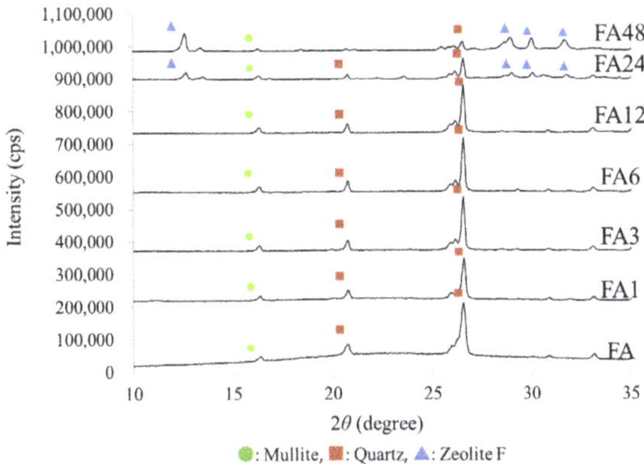

Figure 2. XRD patterns of adsorbents.

Table 1. Physicochemical properties of adsorbents.

Adsorbents		FA	FA1	FA3	FA6	FA12	FA24	FA48
Concentration of basic functional groups (mmol g^{-1})		0	0.41	0.77	1.38	1.36	1.48	1.43
Concentration of acidic functional groups (mmol g^{-1})		0.10	0.24	0.30	0.31	0.31	0.24	0.24
CEC (mmol g^{-1})	pH 5	0.34	1.98	1.17	1.63	2.27	7.90	8.98
	pH 10	0.19	0.65	1.55	2.09	3.46	11.17	11.17
pH_{pzc}		9.8	9.3	9.3	9.5	9.7	10.4	10.4
Specific surface area (m^2 g^{-1})		1.4	15.1	31.5	53.3	54.5	50.3	47.3
Pore volume (μL g^{-1})	$d \leq 20$ (Å)	0.1	0.9	0	0.5	0.2	10.0	10.0
	$20 < d \leq 500$ (Å)	2.0	41.9	97.4	161.5	185.0	105.0	99.0
	Total	2.2	63.0	139.0	221.0	220.0	151.0	131.0
Mean pore diameter (Å)		57.0	167.2	176.7	165.9	161.6	120.1	110.7

3.2. Adsorption Capacity of Pb^{2+}

FA, FA1, FA3, FA6, FA12, FA24, and FA48 adsorbed 1.04, 1.45, 4.87, 8.11, 13.54, 54.29, and 55.53 mg g^{-1} of Pb^{2+}, respectively. The relationships between the quantity of Pb^{2+} adsorbed and the parameters in Table 1 were statistically assessed (Table 2), and the correlation coefficients value between the level of Pb^{2+} adsorbed and CEC, pH_{pzc}, and pore volume ($d \leq 20$ (Å)) were positive, at 0.986–0.999, 0.921, and 0.980, respectively. Therefore, these properties have the greatest influence on the Pb^{2+} adsorption capacity in solution. Additionally, FA24 and FA48 adsorbed more Pb^{2+} than Na-type zeolite (approximately 30 mg g^{-1}) under the same experimental conditions [26]. This is because K-type zeolite has a high CEC and pore volume ($d \leq 20$ (Å)) than Na-type zeolite. In this study, FA48 was selected to study and evaluate the adsorption capability on Pb^{2+} in the following experiments.

3.3. Pb^{2+} Adsorption Isotherms

The Pb^{2+} adsorption isotherms are shown in Figure 3. The quantity of Pb^{2+} adsorbed increased with raising initial concentration. Amount adsorbed was 2.0 times (7 °C), 2.36 times (25 °C), and 2.44 times (45 °C) raised from 10 to 50 mg L^{-1} of initial concentration. Additionally, the quantity of Pb^{2+} adsorbed also increased with raising temperature, thereby indicating that Pb^{2+} adsorption using FA48 was endothermic. In this study, the distribution of Pb^{2+} between the aqueous media and FA48 was described using adsorption isotherms based on a set of assumptions related to the heterogeneity or homogeneity of the FA48 [24]. Thus, two useful isotherm models, i.e., Freundlich and Langmuir, were selected [40,41]. The Freundlich isotherm model is applicable to adsorption onto heterogeneous surfaces, whereas the Langmuir isotherm is applicable to monolayer and homogeneous adsorption [17]. The Freundlich and Langmuir models are described by Equations (1) and (2), respectively:

$$\log q = \frac{1}{n} \log C_e + \log k, \quad (1)$$

$$1/q = 1/(W_s a C_e) + 1/W_s, \quad (2)$$

where q is the quantity of Pb^{2+} adsorbed (mg g^{-1}), C_e is the equilibrium concentration (mg L^{-1}), and k and n are the adsorption capacity and intensity, respectively. Moreover, W_s is the maximum quantity of adsorbed Pb^{2+} (mg g^{-1}), and a is the coefficient reflecting the relative sorption and desorption rates at equilibrium (L mg^{-1}). The Freundlich and Langmuir constants for the adsorption of Pb^{2+} are summarized in Table 3. In this study, the isotherm data were fitted to both the Freundlich (correlation coefficient: ≥0.991) and Langmuir (correlation coefficient: ≥0.960) models. The maximum adsorption capacity (W_s) of Pb^{2+} increased with raising temperature, thereby indicating that the adsorption of Pb^{2+} by FA48 is an endothermic process (Figure 3). Additionally, when the value of $1/n$ is 0.1–0.5, adsorption readily occurs. However, when the value of $1/n$ exceeds 2, adsorption is difficult [42]. Herein, the value of $1/n$ was 0.30–0.32, and the adsorption of Pb^{2+} using FA48 in aqueous solutions were more favorable.

Table 2. Correlation coefficients between quantity adsorbed and physicochemical properties.

Adsorbents		FAs
Concentration of basic functional groups (mmol g^{-1})		0.677
Concentration of acidic functional groups (mmol g^{-1})		0.044
CEC (mmol g^{-1})	pH 5	0.986
	pH 10	0.999
pH_{pzc}		0.921
Specific surface area (m^2 g^{-1})		0.549
Pore volume (μL g^{-1})	$d \leq 20$ (Å)	0.980
	$20 < d \leq 500$ (Å)	0.201
	Total	0.227
Mean pore diameter (Å)		0.262

To evaluate the adsorption mechanism of Pb^{2+} using FA48, the relationships between the quantities of Pb^{2+} and K^+ adsorbed by and released from FA48 were explored, respectively (Figure 4A). The correlation coefficient was positive at 0.958 under our experimental conditions, suggesting that ion exchange was involved in the adsorption of Pb^{2+}. As mentioned above, the correlation coefficient of the relationship between the quantity of Pb^{2+} adsorbed and the CEC value was 0.986–0.999, demonstrating that ion exchange was involved in the removal capacity on Pb^{2+} from aqueous media. A similar result was achieved using Na-type zeolite in a previous study [26]. Additionally, the binding energies of lead before and after adsorption were measured (Figure 4B). New peaks of Pb(4f) were detected at 135 and 140 eV after adsorption, indicating that Pb^{2+} was adsorbed onto the surface of FA48 in this study. Thus, the surface properties of FA48 are vital in the removal ability on Pb^{2+} from aqueous media, and these results agree with those in Table 2.

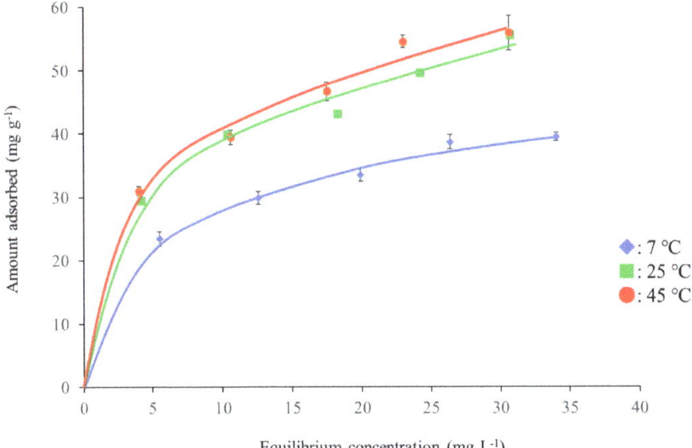

Figure 3. Adsorption isotherms of Pb^{2+} using FA48 at different temperatures. Initial concentration 10, 20, 30, 40, and 50 mg L^{-1}, solvent volume 50 mL, adsorbent 0.01 g, contact time 24 h.

Table 3. Langmuir and Freundlich constants for the adsorption of Pb^{2+}.

Adsorbent	Temperatures (°C)	Langmuir Constants			Freundlich Constants		
		W_s (mg g^{-1})	a (L mg^{-1})	r	log$k$	1/n	r
FA48	7	74.1	0.20	0.982	1.38	0.30	0.994
	25	98.0	0.24	0.987	1.51	0.30	0.997
	45	102.0	0.25	0.960	1.52	0.25	0.960

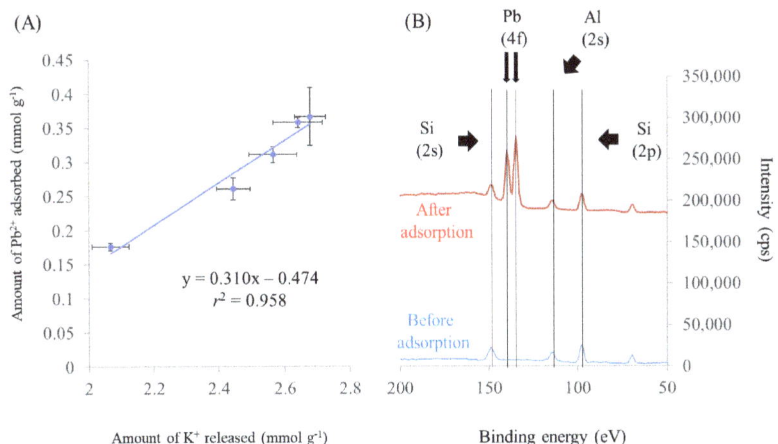

Figure 4. Correlation coefficient between quantity of Pb^{2+} adsorbed and quantity of K^+ released from FA48 (**A**) and binding energy of lead onto FA48 surface before and after adsorption (**B**).

3.4. Effect of pH on Pb^{2+} Adsorption

The solution pH is critical for determining the capacity of Pb^{2+} adsorption from aqueous media. Figure 5 exerts the effect of pH on the adsorption of Pb^{2+}. The quantity adsorbed increased as the pH raised from 2 to 5 under our experimental conditions. Above pH 5, the adsorption capacity decreased sharply. Under acidic conditions, the FA48's surface will be covered with protons (H^+), with which Pb^{2+} competes for adsorption sites. A previous study reported that zeolite could preferentially adsorb H^+ from aqueous media also containing heavy metal ions [43]. The edge groups with a positive charge (Al-OH_2^+ and Si-OH_2^+) on the FA48 surface are explained below [44].

$$\begin{array}{c} \text{-AlOH} + H_2O\ (H^+ + OH^-) \rightarrow \text{-AlOH}_2^+ + OH^- \\ \text{-SiOH} + H_2O\ (H^+ + OH^-) \rightarrow \text{-SiOH}_2^+ + OH^- \end{array} \quad (3)$$

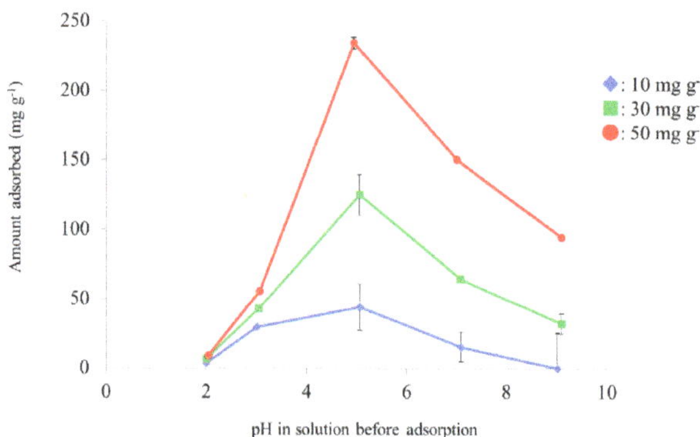

Figure 5. Effect of pH on the adsorption of Pb^{2+} onto FA48. Initial concentration 10, 30, and 50 mg L^{-1}, solvent volume 50 mL, adsorbent 0.01 g, pH 2, 3, 5, 7, and 9, contact time 24 h, temperature 25 °C, agitation speed 100 rpm.

Thus, electrostatic repulsion easily occurred between them, and it was also difficult for ion exchange to occur between Pb^{2+} and K^+ in aqueous media. As the pH increased, the deprotonation of hydroxyl groups increased, indicating that the number of anionic sites (adsorption sites) on FA48 increased, thereby contributing to the adsorption of Pb^{2+} [17]. However, this did not increase the removal percentage of Pb^{2+} beyond pH 6.0 in our experimental condition. The forms of Pb^{2+} in the solution are Pb^{2+}, $Pb(OH)^+$, $Pb(OH)_2$, $Pb(OH)_3^-$, $Pb(OH)_4^{2-}$, $Pb_2(OH)_3^+$, and $Pb_3(OH)_4^{2+}$ [45], and the predominant species of Pb^{2+} at pH < 7 are Pb^{2+} and $Pb(OH)^+$, whereas $Pb(OH)_2$, $Pb(OH)_3^-$, and $Pb(OH)_4^{2-}$ are the main forms of Pb^{2+} at pH > 7–8. Thereby the complexion between free Pb^{2+} available and FA48 surface was not occurred in our experimental condition. Similar trend was reported by a previous study [46]. Additionally, there is deprotonation as basicity increases beyond pH 6.0 [46], suggesting that electron repulsion occurred between the dissolved negatively charged Pb species and the negatively charged surface of FA48. Thus, the Pb^{2+} adsorption capacity decreased under basic conditions. However, further studies are necessary for elucidate the effect of pH on the adsorption of Pb^{2+} using FA48 from aqueous media in detail.

3.5. *Effect of Contact Time on* Pb^{2+} *Adsorption*

The effect of contact time on the adsorption capability of Pb^{2+} is shown in Figure 6. The adsorption equilibrium was attained within approximately 6 h under our experimental conditions. Kinetic experiments were conducted to indicate the rate of adsorption and the potential of the rate-limiting step [2]. The pseudo-first-order (Equation (4)) and pseudo-second-order (Equation (5)) models were employed to evaluate these factors and can be described as follows [47,48].

$$\ln(q_{e,exp} - q_t) = \ln q_{e,cal} - k_1 t, \quad (4)$$

$$\frac{t}{q_t} = \frac{t}{q_{e,cal}} + \frac{1}{k_2 \times q_{e,cal}^2}, \tag{5}$$

where $q_{e,exp}$ and $q_{e,cal}$ are the quantities of Pb^{2+} adsorbed in the experiment and calculation (mg g^{-1}), and k_1 and k_2 are the pseudo-first-order (h^{-1}) and pseudo-second-order (g mg^{-1} h^{-1}) rate constants, respectively. The rapid adsorption during the initial stage was due to the availability of adsorption sites on FA48. Afterwards, the adsorption equilibrium may have been reached due to the limited mass transfer of Pb^{2+} from the bulk liquid phase to the external surface of FA48 [24]. Similar trends were reported in previous studies [49,50]. Table 4 summarizes the kinetic constants of the pseudo-first-order and second-order models. The kinetic data fitted to the pseudo-second-order model (correlation coefficient: 0.998) better than the pseudo-first-order model (correlation coefficient: 0.602). Therefore, the adsorption mechanism of Pb^{2+} using FA48 in this study may be chemisorption. Additionally, the value of $q_{e,exp}$ (99.5 mg g^{-1}) was similar to that of $q_{e,cal}$ (98.0 mg g^{-1}) in the pseudo-second-order model. Thus, chemical adsorption, one of the adsorption mechanisms, could be the rate-limiting step in the adsorption of Pb^{2+} using FA48 [3].

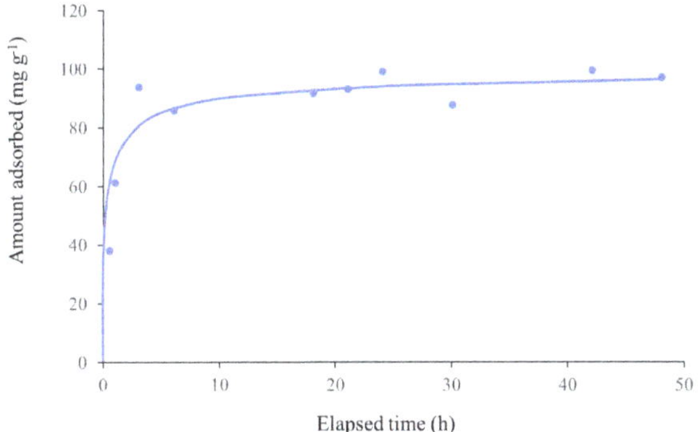

Figure 6. Effect of contact time on the adsorption of Pb^{2+} using FA48. Initial concentration 50 mg L^{-1}, solvent volume 50 mL, adsorbent 0.01 g, contact time 0.5, 1, 3, 6, 18, 21, 24, 30, 42, and 48 h, temperature 25 °C, agitation speed 100 rpm.

Table 4. Kinetic parameters for the adsorption of Pb^{2+}.

Adsorbent	$q_{e,exp}$	Pseudo-First-Order Model			Pseudo-Second-Order Model		
		k_1 (h^{-1})	$q_{e,cal}$ (mg g^{-1})	r	k_2 (g mg^{-1} h^{-1})	$q_{e,cal}$ (mg g^{-1})	r
FA48	99.5	0.057	19.7	0.602	0.010	98.0	0.998

3.6. *Effect of Coexisting Ions on Pb^{2+} Adsorption*

In this study, the effect of coexisting ions on the adsorption capability of Pb^{2+} using FA48 was demonstrated in Table 5. Na^{+}, Mg^{2+}, K^{+}, Ca^{2+}, Ni^{+}, Cu^{2+}, Zn^{2+}, Sr^{2+}, and Cd^{2+} were selected as the components of the binary solution system, which exist in aquatic environments or wastewater from factories (human activities) [38,51]. The adsorption percentage of Pb^{2+} using FA48 in a single solution system was 75.7%. The Pb^{2+} adsorption selectivity of FA48 in a binary solution system (adsorption percentage was approximately >60.7%) exceeded that of other cations under our experimental conditions. These results indicate that the adsorption capacity of FA48 was influenced by cationic factors, such as charge density and hydrate ion diameter, and the accessibility of the active sites of FA48 [51]. The radius of hydrated ions and electronegativity of Pb^{2+} exceed those of other cations, such as Ni^{2+}, Cu^{2+}, Cd^{2+}, and Zn^{2+}. Similar trends were reported in previous studies [38,51].

Additionally, the adsorption capacity of cations excluding Pb^{2+} was significantly lower than that of Pb^{2+} in this study. Thus, FA48 is useful for the selective removal of Pb^{2+} from aqueous media.

Table 5. Adsorption capacity of Pb^{2+} in binary solution system.

Components in Binary Solution	Removal Percentage of Pb^{2+} (%)	Removal of Other Cations (%)
Pb^{2+} + Na^{+}	60.7	0.2
Pb^{2+} + Mg^{2+}	69.4	0
Pb^{2+} + K^{+}	75.7	0
Pb^{2+} + Ca^{2+}	62.4	0
Pb^{2+} + Ni^{+}	73.2	0
Pb^{2+} + Cu^{2+}	62.7	3.7
Pb^{2+} + Zn^{2+}	63.7	0.9
Pb^{2+} + Sr^{2+}	70.0	0.7
Pb^{2+} + Cd^{2+}	64.4	4.1

4. Conclusions

In this study, K type-zeolite (FA48) was synthesized from coal fly ash. The CEC, pH_{pzc}, and pore volume ($d \leq 20$ (Å)) of FA48 were higher than those of other zeolites. Additionally, the quantity of Pb^{2+} adsorbed using FA48 (55.53 mg g^{-1}) exceeded that of other zeolites. The adsorption of Pb^{2+} was related to physicochemical properties, such as the CEC (0.986–0.999), pH_{pzc} (0.921), and pore volume ($d \leq 20$ (Å)) (0.980). The relationship between the amounts of Pb^{2+} and K^{+} adsorbed by and released from FA48 was 0.958, respectively. Thus, ion exchange with K^{+} in the interlayer of FA48 is strongly related to the adsorption capacity. Furthermore, the binding energy of lead at 135 and 140 eV could be detected after adsorption, indicating that the adsorbent's surface characteristics were critical for the removal of Pb^{2+} from aqueous media. The adsorption isotherms and kinetics data demonstrated that the adsorption of Pb^{2+} using FA48 was an endothermic process. Finally, FA48 exhibited selectivity for Pb^{2+} adsorption from a binary solution system containing cations. These results provide useful information for the recycling of coal fly ash and the removal of Pb^{2+} from aqueous media.

Author Contributions: Conceptualization, N.K. and F.O.; investigation, Y.K., C.S. and T.N.; writing—original draft preparation, Y.K. and F.O.; writing—review and editing, Y.K., F.O. and C.S.; project administration, N.K. All authors have read and agreed to the published version of the manuscript.

Funding: This research received no external funding.

Conflicts of Interest: The authors declare no conflict of interest.

References

1. About the Sustainable Development Goals. Available online: https://www.un.org/sustainabledevelopment/sustainable-development-goals/ (accessed on 19 May 2020).
2. Venkata Ramana, D.K.; Kumar Reddy, D.H.; Kumar, B.N.; Seshaiah, K.; Chandra Rao, G.P.; Lu, C. Adsorption of Pb(II) from aqueous solutions by chemically modified zeolite supported carbon nanotubes: Equilibrium, kinetic, and thermodynamic studies. *Sep. Sci. Technol.* **2013**, *48*, 403–412. [CrossRef]
3. Kragovic, M.; Dakovic, A.; Markovic, M.; Krstic, J.; Gatta, G.D.; Rotiroti, N. Characterization of lead sorption by the natural and Fe(III)-modified zeolite. *Appl. Surf. Sci.* **2013**, *283*, 764–774. [CrossRef]
4. Hamidpour, M.; Kalbasib, M.; Afyunib, M.; Shariatmadarib, H.; Holmic, P.E.; Hansenc, H.C.B. Sorption hysteresis of Cd(II) and Pb(II) on natural zeolite and bentonite. *J. Hazard. Mater.* **2010**, *181*, 686–691. [CrossRef] [PubMed]
5. Lead Poisoning and Health. Available online: https://www.who.int/news-room/fact-sheets/detail/lead-poisoning-and-health (accessed on 19 May 2020).
6. Sarma, P.R.; Chattopadhyay, A.; Zhan, C.; Sharma, S.K.; Geng, L.; Hsiao, B.S. Lead removal from water using carboxycellulose nanofibers prepared by nitro-oxidation method. *Nature* **2018**, *25*, 1961–1973.
7. Torrice, M. How lead ended up in flint's tap water. *Chem. Eng. News* **2016**, *94*, 26–29.

8. Edwards, M.; Triantafyllidou, S.; Best, D. Elevated blood lead in young children due to lead-contaminated drinking water: Washington, DC, 2001–2004. *Environ. Sci. Technol.* **2009**, *43*, 1618–1623. [CrossRef] [PubMed]
9. Mankikar, D.; Campbell, C.; Greenberg, R. Evaluation of a home-based environmental and educational intervention to improve health in vulnerable households: Southeastern Pennsylvania lead and healthy homes program. *Int. J. Environ. Res. Public Health* **2016**, *13*, 900. [CrossRef] [PubMed]
10. Pieper, K.J.; Krometis, L.A.; Gallagher, D.; Benham, B.; Edwards, M. Profiling private water system to identify patterns of waterborne lead exposure. *Environ. Sci. Technol.* **2015**, *49*, 12697–12704. [CrossRef] [PubMed]
11. Japan Coal Energy Center. Available online: http://www.jcoal.or.jp/ashdb/ashstatistics/upload/H30_ashstatistics.pdf (accessed on 19 May 2020).
12. Franus, W.; Wdowin, M.; Franus, M. Synthesis and characterization of zeolites prepared from industrial fly ash. *Environ. Monit. Assess.* **2014**, *186*, 5721–5729. [CrossRef]
13. Blissett, R.S.; Rowson, N.A. A review of the multi-component utilization of coal fly ash. *Fuel* **2012**, *97*, 1–23. [CrossRef]
14. Flores, C.G.; Schneider, H.; Marcillio, N.R.; Ferret, L.; Oliveira, J.C.P. Potassic zeolites from Brazilian coal ash for use as a fertilizer in agriculture. *Waste Manag.* **2017**, *70*, 263–271. [CrossRef] [PubMed]
15. Visa, M.; Isac, L.; Dute, A. Fly ash adsorbents for multi-cation wastewater treatment. *App. Surf. Sci.* **2012**, *258*, 6345–6352. [CrossRef]
16. Ahmaruzzaman, M. A review on the utilization of fly ash. *Prog. Energy Combust. Sci.* **2010**, *36*, 327–363. [CrossRef]
17. He, K.; Chen, Y.; Tang, Z.; Hu, Y. Removal of heavy metal ions from aqueous solution by zeolite synthesized from fly ash. *Environ. Sci. Pollut. Res.* **2016**, *23*, 2778–2788. [CrossRef]
18. Wdowin, M.; Franus, M.; Panek, R.; Badura, L.; Rranus, W. The conversation technology of fly ash into zeolites. *Clean Technol. Environ. Policy* **2014**, *16*, 1217–1223. [CrossRef]
19. Derkowski, A.; Franus, W.; Beran, E.; Czimerova, A. Properties and potential applications of zeolitic materials produced from fly ash using simple method of synthesis. *Powder Technol.* **2006**, *166*, 47–54. [CrossRef]
20. Derkowski, A.; Franus, W.; Waniak-Nowicka, H.; Czimerova, A. Textural properties vs. CEC and EGME retention of Na-X zeolite prepared from fly ash at room temperature. *Int. J. Miner. Process.* **2007**, *82*, 57–68. [CrossRef]
21. Shaheen, S.M.; Derbalah, A.S.; Moghanm, F.S. Removal of heavy metals from aqueous solution by zeolite in competitive sorption system. *Int. J. Environ. Sci. Dev.* **2012**, *3*, 362–367. [CrossRef]
22. Siddique, R. Performance characteristics of high-volume class F fly ash concrete. *Cem. Concr. Res.* **2004**, *34*, 487–493. [CrossRef]
23. Izidoro, J.C.; Fungaro, D.A.; Abbott, J.E.; Wang, S. Synthesis of zeolite X and A from fly ashes for cadmium and zinc removal from aqueous solutions in single and binary ion systems. *Fuel* **2013**, *103*, 827–834. [CrossRef]
24. Golbad, S.; Khoshnoud, P.; Abu-Zahra, N. Hydrothermal synthesis of hydroxyl sodalite from fly ash for the removal of lead ions from water. *Int. J. Environ. Sci. Technol.* **2017**, *14*, 135–142. [CrossRef]
25. Ogata, F.; Tominaga, H.; Yabutani, H.; Taga, A.; Kawasaki, N. Recovery of molybdenum from fly ash by gibbsite. *Toxicol. Environ. Chem.* **2011**, *93*, 635–642. [CrossRef]
26. Kobayashi, Y.; Ogata, F.; Nakamura, T.; Kawasaki, N. Synthesis of novel zeolites produced from fly ash hydrothermal treatment in alkaline solution and its evaluation as an adsorbent for heavy metal. *J. Environ. Chem. Eng.* **2020**, *8*, 103687. [CrossRef]
27. Ogata, F.; Tominaga, H.; Kangawa, M.; Inoue, K.; Kawasaki, N. Adsorption capacity of Cu(II) and Pb(II) onto carbon fiber produced from wool. *J. Oleo Sci.* **2012**, *3*, 149–154. [CrossRef] [PubMed]
28. Yu, B.; Zhang, Y.; Shukla, A.; Shukla, S.S.; Dorris, K.L. The removal of heavy metals from aqueous solutions by sawdust adsorption removal of lead and comparison of its adsorption with copper. *J. Hazard. Mater.* **2001**, *B84*, 83–94. [CrossRef]
29. Pehlivan, E.; Yamk, B.H.; Ahmetli, G.; Pehlivan, M. Equilibrium isotherm studies for the uptake of cadmium and lead ions onto sugar beet pulp. *Bioresour. Technol.* **2008**, *99*, 3520–3527. [CrossRef] [PubMed]
30. Battacharyya, K.G.; Sharma, A. Adsorption of Pb(II) from aqueous solution by Azadirachta indiea (Neem) leaf powder. *J. Hazard. Mater.* **2004**, *113*, 97–109. [CrossRef]
31. Slater, J.C. Atomic radii in crystals. *J. Chem. Phys.* **1964**, *41*, 3199–3204. [CrossRef]

32. Ogata, F.; Iwata, Y.; Kawasaki, N. Properties of novel adsorbent produced by hydrothermal treatment of waste fly ash in alkaline solution and its capability for adsorption of tungsten from aqueous solution. *J. Environ. Chem. Eng.* **2015**, *3*, 333–338. [CrossRef]
33. Faria, P.C.C.; Orfao, J.J.M.; Pereira, M.F.R. Adsorption of anionic and cationic dyes on activated carbons with different surface chemistries. *Water Res.* **2004**, *38*, 2043–2052. [CrossRef]
34. Boehm, H.P. Chemical identification of surface groups. *J. Adv. Catal. Sci. Technol.* **1966**, *16*, 179–274.
35. Nassar, N.N. Rapid removal and recovery of Pb(II) from wastewater by magnetic nanoadsorbents. *J. Hazard. Mater.* **2010**, *184*, 538–546. [CrossRef] [PubMed]
36. Kwon, Y.T.; Lee, C.W. Ecological risk assessment of sediment in wastewater discharging area by means of metal speciation. *Microchem. J.* **2001**, *70*, 255–264. [CrossRef]
37. Wang, C.; Hu, X.; Chen, M.L.; Wu, Y.H. Total concentrations and fractions of Cd, Cr, Pb, Cu, Ni and Zn in sewage sludge municipal and industrial wastewater treatment plants. *J. Hazard. Mater.* **2005**, *B119*, 245–249. [CrossRef] [PubMed]
38. Hernandez-Montoya, V.; Perez-Cruz, M.; Mendoza-Castillo, D.I.; Moreno-Virgen, M.R.; Bonilla-Petriciolet, A. Competitive adsorption of syes and heavy metals on zeolitic structures. *J. Environ. Manag.* **2013**, *116*, 213–221. [CrossRef]
39. Murayama, N.; Ymamamoto, H.; Shibata, J. Mechansim of zeolite synthesis from coal fly ash by alkali hydrothermal reaction. *Int. J. Miner. Process.* **2002**, *64*, 1–17. [CrossRef]
40. Freundlich, H.M.T. Over the adsorption in solution. *J. Phy. Chem.* **1906**, *57*, 385–471.
41. Langmuir, I. The constitution and fundamental properties of solids and liquids. *J. Am. Chem. Soc.* **1916**, *38*, 2221–2295. [CrossRef]
42. Abe, I.; Hayashi, K.; Kitagawa, M. Studies on the adsorption of surfactants on activated carbons. I. Adsorption of nonionic surfactants. *Yukagaku* **1976**, *25*, 145–150.
43. Myroslav, S.; Boguslaw, B.; Artur, T.P.; Jacek, N. Study of the selection mechanism of heavy metal (Pb^{2+}, Cu^{2+}, Ni^{2+}, and Cd^{2+}) adsorption on clinoptilolite. *J. Colloid Interface Sci.* **2006**, *304*, 21–28.
44. Sari, A.; Tuzen, M.; Citak, D.; Soylak, M. Equilibrium, kinetic and thermodynamic studies of adsorption of Pb(II) from aqueous solution onto Turkish kaolinite clay. *J. Hazard. Mater.* **2007**, *149*, 283–291. [CrossRef] [PubMed]
45. Yang, X.; Yang, S.; Yang, S.; Hu, J.; Tan, X.; Wang, X. Effect of pH, ionic strength and temperature on sorption of Pb(II) on NKF-6 zeolite studied by batch technique. *Chem. Eng. J.* **2011**, *168*, 86–93. [CrossRef]
46. Alswat, A.A.; Ahmad, M.B.; Saleh, T.A. Zeolite modified with copper oxide and iron oxide for lead and arsenic adsorption from aqueous solutions. *J. Water Supply Res. Technol. AQUA* **2016**, *65*, 465–479. [CrossRef]
47. Lagergren, S. Zur theorie der sogenannten adsorption geloster stoffie, *Kunglia Svenska Vetenskapsakademiens. Handlingar* **1898**, *24*, 1–39.
48. Ho, Y.S.; McKay, G. Pseudo-second order model for sorption process. *Process Biochem.* **1999**, *34*, 451–465. [CrossRef]
49. Xie, J.; Wang, Z.; Wu, D.; Zhang, Z.; Kong, H. Synthesis of zeolite/aluminum oxide hydrate from coal ash: A new type of adsorbent for simultaneous removal of cationic and anionic pollutants. *J. Ind. Eng. Chem.* **2013**, *52*, 14890–14897. [CrossRef]
50. Yadanaparthi, S.K.P.; Graybill, D.; Wandruszka, R.V. Adsorbents for the removal of arsenic, cadmium, and lead from contaminated waters. *J. Hazard. Mater.* **2009**, *171*, 1–15. [CrossRef]
51. Merrikhpour, H.; Jalali, M. Comparative and competitive adsorption of cadmium, copper, nickel, and lead ions by Iranian natural zeolite. *Clean Technol. Environ. Policy* **2013**, *15*, 303–316. [CrossRef]

Article

Mechanistic Study of Pb^{2+} Removal from Aqueous Solutions Using Eggshells

Mohamed A. Hamouda [1,2,*], Haliemeh Sweidan [1], Munjed A. Maraqa [1,2] and Hilal El-Hassan [1]

1 Department of Civil and Environmental Engineering, United Arab Emirates University, P.O. Box 15551, Al Ain, UAE; hsweidan@uaeu.ac.ae (H.S.); m.maraqa@uaeu.ac.ae (M.A.M.); helhassan@uaeu.ac.ae (H.E.-H.)

2 National Water Center, United Arab Emirates University, P.O. Box 15551, Al Ain, UAE

* Correspondence: m.hamouda@uaeu.ac.ae; Tel.: +971-3-713-5155

Received: 27 July 2020; Accepted: 4 September 2020; Published: 9 September 2020

Abstract: This study investigates the impact of eggshell particle size and solid-to-water (s/w) ratio on lead (Pb^{2+}) removal from aqueous solution. Collected raw eggshells were washed, crushed, and sieved into two particle sizes (<150 and 150–500 μm). Batch Pb^{2+} removal experiments were conducted at different s/w ratios with initial Pb^{2+} concentrations of up to 70 mg/L. The contribution of precipitation to Pb^{2+} removal was simulated by quantifying removal using eggshell water, whereas sorbed Pb^{2+} was quantified by acid digestion. Results indicated that eggshell particle sizes did not affect Pb^{2+} removal. High removal (up to 99%) of Pb^{2+} was achieved for low initial Pb^{2+} concentrations (<30 mg/L) across all s/w ratios studied. However, higher removal capacity was observed at lower s/w ratios. In addition, results confirmed that precipitation played a major role in the removal of Pb^{2+} by eggshells. Yet, this role decreased as the s/w ratio and initial concentration of Pb^{2+} increased. A predictive relationship that relates the normalized removal capacity of eggshells to the s/w ratio was developed to potentially facilitate the design of the reactor.

Keywords: Pb^{2+} removal; sorption; precipitation; waste recycling; removal mechanisms; solid-to-water ratio

1. Introduction

The presence of lead (Pb^{2+}) in industrial wastewater has become a noteworthy source of water pollution and poses a significant health hazard to humans [1]. Physiological damage to human kidneys, liver, brain, and nervous system can happen as a result of ingesting levels of Pb^{2+} higher than the body's tolerance [2]. In addition, elevated levels of Pb^{2+} in the human body have been shown to have negative health effects including fatigue, a decreased reproductive ability, problems in the digestive tract, and anemia [1,3]. In fact, sterility, stillbirths, and neonatal deaths are associated with constant exposure to Pb^{2+} [1]. Pb^{2+} poisoning is especially dangerous for children, as high levels of Pb^{2+} in the blood stream are associated with depression of brain development and cognitive skills [4–6]. Pb^{2+} poisoning can be contracted by humans through polluted drinking water, ingestion from food, inhalation of contaminated dust, or occupational exposure [1,2,4,6–9]. It can infiltrate and contaminate the water sources, and from there, contaminate every level of the food chain [1,2,10]. It has been difficult to scientifically specify a level under which Pb^{2+} is no longer associated with negative health effects, therefore many jurisdictions focus on the ability of treatment technologies to completely remove Pb^{2+} from water. The maximum acceptable concentration of Pb^{2+} in drinking water is becoming more stringent, with some jurisdictions assigning a level as low as 5 μg/L [11], while the US EPA has set the maximum contaminant level goal for Pb^{2+} in drinking water at zero but an action level at 15 μg/L [12].

Sources of Pb^{2+} in water originate mainly from industrial activities such as mining, smelting, printing, and metal plating. In addition, manufacturing of batteries, paints, alloys, ceramic glass, and plastics contribute to increased Pb^{2+} pollution in water bodies [1]. Wastewater from battery manufacturing contains, on average, a Pb^{2+} content of 0.5–25 mg/L [13–16]. Regulations for industrial effluent discharge or reuse vary considerably between different jurisdictions. Certain jurisdictions specify a maximum Pb^{2+} level for discharge of effluent wastewater into the environment; this level can be as low as 0.1 mg/L (Saudi Arabia and Oman) or as high as 1 mg/L (Tunisia) [17].

Several methods were proposed and investigated for the removal of Pb^{2+} from industrial wastewater including ion exchange, filtration and membrane processes, electro-dialysis, chemical precipitation, solvent extraction, and chemical coagulation [1,2]. In addition, electrocoagulation, membrane electrodialysis, and biosorption were also employed for Pb^{2+} removal [14]. A widely used method for Pb^{2+} removal from industrial wastewater is to precipitate Pb^{2+} with caustic soda at a pH 8.5–9.2 in the presence of Fe (III) salts, followed by the addition of a polyelectrolyte, which facilitates the flocculation of the Pb^{2+} precipitate. The effluent is further treated with sedimentation and filtration, after which the Pb^{2+} concentration is reduced to the legal allowable limit [16]. Sorption by activated carbon is also a convenient and efficient treatment method. However, the use of commercial activated carbon for sorption can be costly [18]. Therefore, there is an increasing need to re-use biowaste in wastewater treatment [19]. Many studies have focused their efforts on deriving substitute sorbents from waste materials, particularly agricultural waste [20]. This not only helps to reduce the cost of sorbents, but it also helps recycle agricultural waste. Waste materials that have been used for Pb^{2+} removal include: *Chlorella vulgaris*, chicken feathers reinforced with chitosan, cow bone, leather, coconut shell, peach and apricot stones, *Alocacée* shell, *Mimosaceae* husk, and *Burseraceae* sawdust [1,21–25].

Eggshells are an example of an agricultural waste that could be used for Pb^{2+} removal. In fact, the per capita consumption of eggs in the United Arab Emirates (UAE) was 8.3 kg in 2013 [26]. Previous research reported that eggshells consist of (weight %) calcium carbonate (94%), magnesium carbonate (1%), calcium phosphate (1%), and organic matter (4%) [27]. Eggshells were studied for removal of various pollutants from water, including heavy metals such as copper, iron, cadmium, and Pb^{2+} [28–33]. Other work investigated the removal of dyes such as C.I. Reactive Yellow 205, methylene blue, and malachite green using eggshells [34–36]. However, few of these studies provided a detailed investigation of the mechanisms responsible for contaminant removal. It is plausible that, in addition to sorption, other removal mechanisms could be responsible for contaminant removal when using a naturally occurring material as a sorbent. Such mechanisms include ion exchange, as reported by Andersson et al. [37] for the removal of heavy metals from water onto the calcite surface, precipitation [38–40], or surface reactions [41]. However, only few studies investigated the impact of operation conditions (eggshell particle size and dose) on the mechanisms of Pb^{2+} removal. Nonetheless, existing literature on the use of eggshells for Pb^{2+} removal acknowledged that precipitation and sorption occurred [13,28,30,42]. However, the reviewed studies offer no quantification of the contribution of each mechanism, i.e., precipitation and sorption, which is an important aspect of process design.

Therefore, the objectives of this study were twofold: first, to identify the mechanisms responsible for the removal of Pb^{2+} from aqueous solutions using eggshells and to quantify the contribution of each of these mechanisms towards total removal. The authors were specifically interested in isolating the role of precipitation from surface attachment. While surface attachment may take different forms (i.e., sorption, ion exchange, complexation, etc.), for simplicity, it is referred to as sorption from here on. The second objective was to investigate the impact of the applied eggshells mass-to-solution volume ratio on Pb^{2+} removal. Results of this study provide novel and beneficial information for the design of eggshell reactors for optimal removal of Pb^{2+}, while also contributing to sustainable waste management by recycling eggshell material to treat wastewater.

2. Materials and Methods

A general framework for investigating and characterizing the use of biowaste materials for treatment and contaminant removal was previously published [19]. In this study, the framework is applied to carry out a mechanistic study of the removal of Pb^{2+} from aqueous solutions using eggshells (Figure 1).

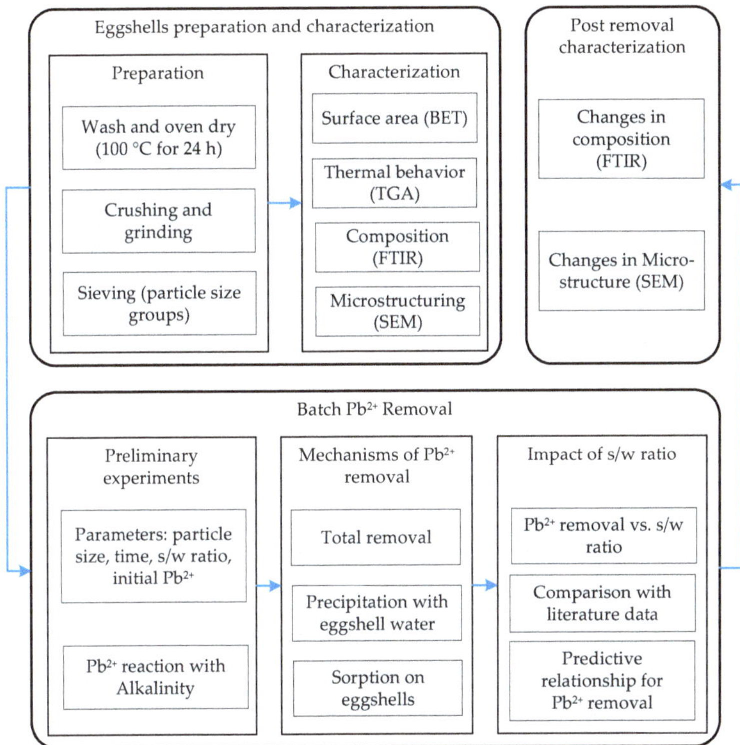

Figure 1. Summary of applied methodology.

2.1. Materials

Hanna Instruments 0.1M standard Pb^{2+} solution (HI4012-01) was used to prepare all Pb^{2+} solutions throughout this study. The required amount of standard Pb^{2+} solution was diluted in ultrapure water (Milli-Q® IQ 7000 Ultrapure, Merck, Darmstadt, Germany) as required to prepare the needed Pb^{2+} concentrations. In addition, H_2SO_4 (95–97% assay), ethanol (95% purity), phenolphthalein (Riedel-de Haën™, indicator grade, Honeywell International Inc., Charlotte, NC, US), and methyl orange powder (Fluka™, indicator grade, Honeywell International Inc., Charlotte, NC, US) were used for alkalinity measurements.

2.2. Preparation of Eggshells

Eggshells were collected from a local restaurant in the city of Al Ain, UAE. Two batches were collected from the restaurant across a period of two weeks; each batch contained the eggshells collected by the restaurant for the week. All batches of collected eggshells were then mixed and rinsed several times with deionized water to wash away any impurities. They were then dried at 100 °C for 24 h in a conventional drying oven (ULE 400, Memmert, Schwabach, Germany). The eggshells were then ground with a pulverizer (LC-53, Gilson Company Inc., Lewis Center, OH, USA). The distance between the blades was set at about 0.7 mm to obtain particles in the range of 0.8–1.0 mm. This particle size was required to help physically remove the inner membranes. The eggshell inner membranes were then removed by soaking the eggshells in deionized water and mixing them, after which the inner membranes floated to the top and were physically removed. The eggshells were then rinsed several times to remove any residual membranes, dried again in the oven for 24 h at 100 °C, and further ground in the pulverizer. Subsequently, the obtained particles were sieved using stainless steel ASTM test sieves (Gilson Company Inc., Lewis Center, OH, USA). The resulting eggshells were classified into 4 groups depending on size, namely <150 μm, 150–500 μm, 500–800 μm, and >800 μm. The ground eggshells were stored in plastic sealable bags at room temperature until use.

2.3. Eggshells Characterization

The powdered eggshells were characterized to evaluate their potential performance for Pb^{2+} removal. Eggshell particles underwent nitrogen gas adsorption for characterization of their surface area and porosity. This was carried out with a Quantachrome Autosorb-1-C volumetric gas sorption instrument at 77K. Before measurements, samples were degassed at 150 °C for one hour. Further, Brunauer–Emmett–Teller (BET) theory was used to calculate surface area, and pore size distributions were determined by the Barett–Joyner–Halenda (BJH) model based on the desorption branch of the N_2 isotherms. Fourier-transform infrared (FTIR) spectroscopy and scanning electron microscopy (SEM) analyses were used to investigate the changes to the microstructure of eggshells as a result of Pb^{2+} removal, as will be discussed in detail in Section 2.8. Based on the results of eggshell particles characterization, it was decided to use only two particle sizes (<150 μm of d_{50} = 75 μm and 150–500 μm of d_{50} = 300 μm) in the Pb^{2+} removal experiments, as they had the highest surface area. In addition, the study focused on particle size 150–500 μm, which was thought to be more practical and feasible to scale to column studies.

2.4. Pb^{2+} Removal Experiments

For all Pb^{2+} removal batch experiments, Pb^{2+} solutions at the desired concentrations were prepared as stock solutions from which sample volumes of 50 or 100 mL were obtained. Specific amounts of crushed eggshell powder were added to the samples to achieve the desired solid-to-water (s/w) ratio. Samples were then subject to shaking in a water bath shaker (MaXturdy 18, Daihan Scientific Co., GANG-WON -DO, South Korea) at a temperature of 25 °C and rotational speed of 150 rpm. If required, they were filtered using syringe filters (Thermo Scientific™, PTFE, 0.45 μm, 25 mm). Measurements of Pb^{2+}, Mg^{2+}, and Ca^{2+} concentration in solution were taken by Varian ICP-OES 720 ES with standards by Chem-Lab™, Belgium. A Thermo Scientific™ Orion™ 3-star pH meter was used to carry out pH measurements. Alkalinity was measured by titration, according to method 2320 [43].

For all the experiments explained in the sections below, samples were analyzed in triplicate, and the coefficient of variation for Pb^{2+} concentration ranged from 0.3% to 8.4%. Average readings for the initial and final Pb^{2+} concentrations were used to calculate the capacity of removal (m in mg/g), regardless of the mechanism of removal, as shown in Equation (1): where V is the volume of solution used (L), M is the mass of eggshells used (g), C_0 is the initial Pb^{2+} concentration (mg/L), and C_e is the final (equilibrium) Pb^{2+} concentration (mg/L). The removal percentage of Pb^{2+} was also calculated using Equation (2).

$$m = \frac{(C_0 - C_e)V}{M} \tag{1}$$

$$\text{Pb2+ removal (\%)} = \frac{(C_0 - C_e)}{C_0} \times 100 \tag{2}$$

2.5. Preliminary Investigation of Pb^{2+} Removal

A screening experiment was conducted to determine the adequate contact time (until equilibrium) and to demonstrate the potential of eggshells for Pb^{2+} removal. At an s/w ratio of 1:200, eggshells were added to Pb^{2+} solutions of concentrations 0 to 70 mg/L for a contact time of 2 h. To help theorize the removal mechanisms involved, other parameters besides Pb^{2+} were monitored including pH, alkalinity, Ca^{2+}, and Mg^{2+}.

To investigate the potential precipitation of Pb^{2+} due to the presence of dissolved carbonates and bicarbonates, stock solutions of 1000 mg/L of Na_2CO_3 and $NaHCO_3$ were prepared and then left to react with Pb^{2+} in solution. One milliliter of the 1000 mg/L Na_2CO_3 stock solution was added to the 20 mg/L (0.19 meq/L) of Pb^{2+} solution and filled to the 100 mL mark in a volumetric flask. The resulting solution had a concentration of 10.07 mg/L (0.19 meq/L) Na_2CO_3. Furthermore, 1.62 mL of the 1000 mg/L Na_2CO_3 solution was added to the 20 mg/L Pb^{2+} solution and filled to the 100 mL mark in a volumetric flask, leading to a solution of 16.2 mg/L (0.19 meq/L) $NaHCO_3$. These solutions were then left for 2 h to observe any precipitation of Pb^{2+} by $CO_3{}^{2-}$ or $HCO_3{}^{-}$. Resulting solutions were then filtered and the Pb^{2+} concentration, pH, and total alkalinity were measured before and after the addition of Na_2CO_3 and $NaHCO_3$.

2.6. Effect of Particle Size and s/w Ratio on Removal of Pb^{2+}

In order to investigate the effect of eggshell particle size and s/w ratio on the removal of Pb^{2+}, 500 mL of Pb^{2+} solutions of 0 to 70 mg/L were prepared and their initial pH, Ca^{2+}, Mg^{2+}, and Pb^{2+} concentrations were measured. Then, 100 mL of each solution was added to 0.5 g of eggshells of particle size 150–500 µm, giving an s/w ratio of 1:200. The samples were shaken for 2 h then filtered. The filtrate was then analyzed for pH, alkalinity, Pb^{2+}, Ca^{2+}, and Mg^{2+} concentration. This process was also repeated for an s/w ratio of 1:100, 1:1000, and 1:2000 for eggshells of particle sizes 150–500 µm and <150 µm. This process is detailed in Figure 2.

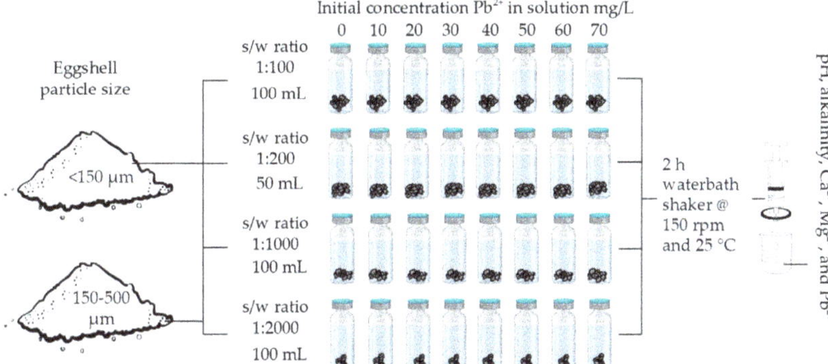

Figure 2. Experimental design for investigating the impact of solid-to-water (s/w) ratio and particle size on Pb^{2+} removal.

2.7. Differentiation between Removal of Pb^{2+} by Sorption and Precipitation

To elucidate the role of sorption versus precipitation in Pb^{2+} removal by eggshells, these mechanisms have to be isolated. To do so, eggshells used in the previous experiment (Figure 2) with an s/w ratio of 1:2000 and Pb^{2+} concentrations 0 to 60 mg/L were collected and separated from the solution with a sieve (number 120 mesh size 125 µm). The solution passing through the sieve was filtered through a 0.45 µm membrane, and a portion of the filtrate was used to rinse the eggshell particles to remove any Pb^{2+} precipitate on the surface. The eggshells were then dried in an oven at 60 °C for 16 h, and then subjected to ICP analysis after undergoing acid digestion according to method 3050B [44]. This was carried out to directly measure the amount of Pb^{2+} sorbed on the surface of the eggshells. The exact amount of Pb^{2+} sorbed was calculated using mass balance after accounting for the amount present in the water film originally surrounding the wet eggshells.

Another experiment was conducted to quantify the removal of Pb^{2+} by precipitation using eggshell extracts at different s/w ratios. A volume of 600 mL of eggshell water (ESW) with an s/w ratio of 1:200 was prepared using 150–500 μm eggshells. The mixture was mixed for 2 h in a water bath shaker at 150 rpm and 25 °C. The mixture was then decanted and filtered. The filtrate gave an ESW sample, which was used in the preparation of Pb^{2+} solutions. The pH, alkalinity, Ca^{2+}, and Mg^{2+} concentration of the ESW were measured. The ESW was then used to prepare 50 mL solutions of concentrations 0 to 70 mg/L Pb^{2+}. The solutions were mixed again for 2 h. Subsequently, they were filtered and analyzed for alkalinity, pH, Ca^{2+}, Mg^{2+}, and Pb^{2+} concentration. This process was repeated for the same eggshell size using an s/w ratio of 1:2000. This process is detailed in Figure 3.

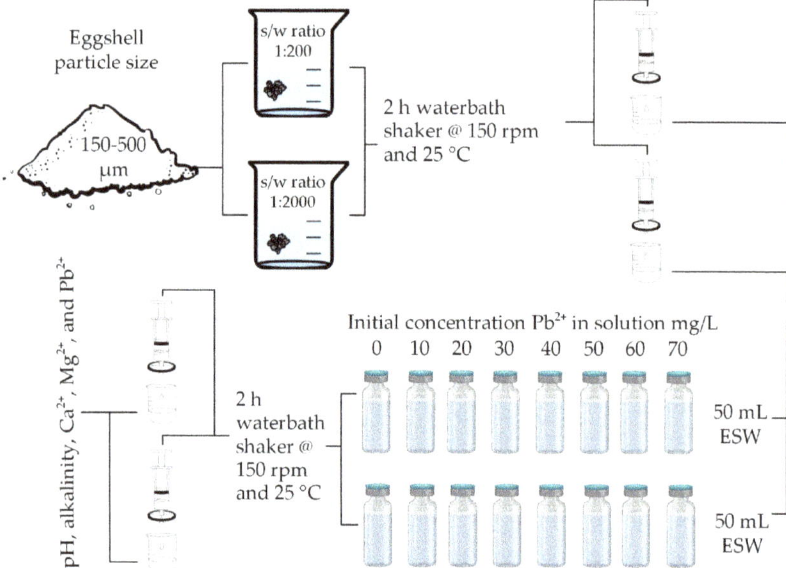

Figure 3. Experimental design for investigation of removal of Pb^{2+} in eggshell water (ESW).

2.8. Microstructure Characterization of Pb^{2+} Removal

FTIR and SEM analyses were used to investigate the changes in functional groups and morphology of selected eggshells post-exposure to Pb^{2+} and examine the removal mechanisms. FTIR spectroscopy was carried out on raw eggshells using a Shimadzu, IR-Prestige-21 spectrometer in transmittance mode at a resolution of 4 cm^{-1} from 500 to 4000 cm^{-1}. Similarly, it was performed on typical used eggshells after Pb^{2+} removal in a solution of 40 mg/L Pb^{2+} and an s/w ratio of 1:200. Samples were formulated by pulverizing eggshells into a powder, which was then sieved through an 80-micron sieve to remove coarse particles. Obtained powder samples were mixed with potassium bromide at a 3:1 ratio, by mass, and made into pellets for FTIR testing.

A JEOL-JSM 6390A scanning electron microscope was used to examine the morphology of eggshells with a particle size of 150–500 μm before and after Pb^{2+} removal. Pb^{2+} solutions of 20 mg/L at s/w ratios of 1:200 and 1:2000 were utilized. To separate the eggshells from the solution, the mixture was passed through a 125-micron sieve, rinsed with deionized water to remove any precipitated solids on the eggshell surface, and dried in the oven at 100 °C for 24 h. Samples were sputter-coated with a thin gold layer and examined in high-vacuum mode at 15 kV and magnification ranging from 100–1000.

3. Results and Discussion

3.1. Raw Eggshells Characterization

Removal of contaminants from solution by a biomass material depends on the material's surface area and accessibility to active sites [45]. Nitrogen adsorption curves were used to evaluate the surface area according to BET theory, while N_2 desorption curves were used to evaluate the pore size distributions based on the BJH model. The results are presented in Table 1 and Figure 4. As expected, the surface area was found to decrease with increasing eggshell particle size. It is worth noting that the BET surface area for eggshells was found to be comparable to the surface areas of other types of biomass used for sorption, as olive stones, tomato husks, and Bacillus badius (Table 1). In addition, the surface area of eggshells found in this study for particle size <150 µm, 0.977 m^2/g, was similar to that reported by Tsai et al. [35] for eggshells of particle size 77 µm, 1.023 m^2/g. Figure 4 shows a representative curve for the nitrogen adsorption of eggshells of particle size <150 µm; the eggshells follow a type II isotherm with negligible hysteresis. This indicates a lack of affinity between eggshells and N_2 molecules [35]. The N_2 adsorption isotherm signifies poor pore properties, owing possibly to a non-porous surface. Tsai et al. [35] also reached a similar conclusion when analyzing eggshells.

Table 1. BET and Langmuir surface area calculations of biowaste material at various particle size ranges.

Material	Size (µm)	Surface Area (m^2/g)	Pore Volume (10^3 cm^3/g)	Pore Size (nm)	Reference
Eggshells	<150	0.977	3.544	14.8	This work
	150–500	0.056	-	-	This work
Eggshells	77.9	1.023	6.5	-	[35]
Olive stones	<1000	0.163	1.84	45.302	[46]
Tomato husks	75–150	0.68	0.0015	-	[47]

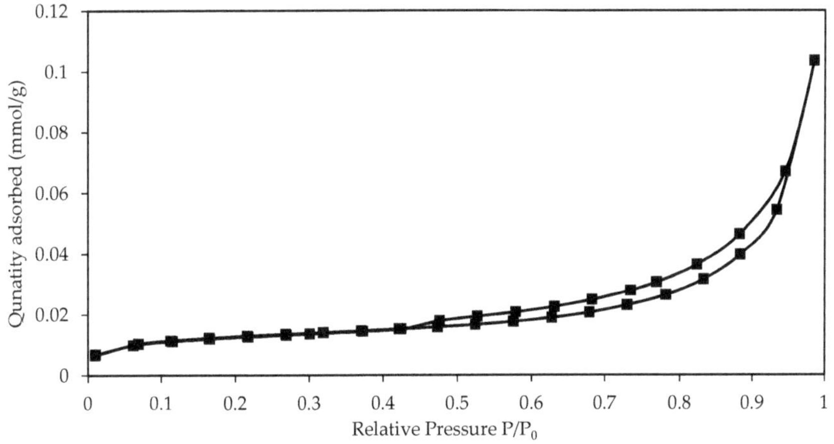

Figure 4. Nitrogen adsorption isotherm curves for eggshell powder of particle size <150 µm.

The thermogravimetric analysis of raw eggshells has been examined in previous studies [48–50]. Such analysis highlights the change in eggshell powder mass with temperature and gives an indication of its composition. There is general agreement that the decomposition of raw eggshells with temperature has two regions of mass loss. In the first region, a minor mass loss (around 3%) indicates the decomposition of organic matter and occurs from 100 to 550 °C. Meanwhile, in the second region, a major mass loss (around 40%) takes place at higher temperatures up to 850 °C, signifying the decomposition of calcium carbonate [48–50].

3.2. Preliminary Pb^{2+} Removal Experiments

A screening experiment of removal with time indicated that Pb^{2+} removal happened over a short time period. The experiment was done using an initial Pb^{2+} concentration of 100 mg/L and s/w ratio of 1:100; measurements of Pb^{2+} concentration were taken at 30 min and every hour for 6 h. Little differences in removal were observed after 30 min of contact time. Therefore, to ensure equilibrium status is achieved, 2 h was chosen as the contact time for all further experiments. Another screening experiment was conducted to investigate the capacity of eggshells for Pb^{2+} removal at an s/w ratio of 1:200. The results indicated that eggshells showed a good potential for Pb^{2+} removal compared to previously investigated biowaste materials [1,23,46]. A removal capacity (m) of 8.18 mg/g (pH = 5.9) was found for an initial Pb^{2+} concentration of 70 mg/L, at which an equilibrium aqueous concentration of 29.5 mg/L was reached. During the analysis of Pb^{2+} solution after contact with eggshells, it was noticed that Ca^{2+} and Mg^{2+} were released into the solution by eggshells. Therefore, it was hypothesized that there could be other removal mechanisms, in addition to sorption, which could be responsible for the removal of Pb^{2+}. Previous studies have noted that Pb^{2+} precipitation due to reaction with carbonates could play a role in Pb^{2+} removal [51,52].

To confirm the potential precipitation of Pb^{2+} due to reaction with carbonates, an investigation was conducted to determine the extent of Pb^{2+} precipitation by reaction with CO_3^{2-} and HCO_3^-. $NaHCO_3$ and Na_2CO_3 were added to 20 mg/L Pb^{2+} solution. As expected, the pH and alkalinity of Na_2CO_3 and $NaHCO_3$ solutions decreased when added to a Pb^{2+} solution. Indeed, immediately after the addition of Na_2CO_3, the pH was 8.29 and then decreased to 6.09, and alkalinity dropped from 14 to 11 mg/L as $CaCO_3$. Similarly, after the addition of $NaHCO_3$, the pH decreased from 7.31 to 5.94 and the alkalinity dropped from 16 to 5.6 mg/L as $CaCO_3$. The drop in alkalinity was accompanied by a drop in Pb^{2+} concentration from 23.66 to 0.75 and 2.53 mg/L for Na_2CO_3 and $NaHCO_3$ solutions, respectively. This indicates that alkalinity was consumed and Pb^{2+} was precipitated upon contact with CO_3^{2-} and HCO_3^-. These findings provide evidence that Pb^{2+} is precipitated by carbonates that are released by eggshells in solution.

3.3. Effect of Particle Size and s/w Ratio on Removal of Pb^{2+}

The method outlined in Section 2.6 was used to investigate the effect of s/w ratio and particle size on the eggshell capacity for Pb^{2+} removal. The results, presented in Figure 5, show that eggshell particle size generally did not have a measurable effect on the removal capacity, but a higher removal capacity was achieved with the smaller particle size (<150 μm) at high initial Pb^{2+} concentrations (>50 mg/L). This effect at a higher initial concentration of Pb^{2+} was observed in previous studies where m decreased with the increase in eggshell particle size [42], or with the increase in the particle size of calcite and aragonite [51]. Despite the large difference in the surface area of the two eggshell sizes used in this study (Table 1), a rather insignificant difference in the eggshell capacity for Pb^{2+} removal was noticed between the two sizes for initial Pb^{2+} concentrations below 50 mg/L (Figure 5). However, eggshell particle size could result in a significantly lower removal capacity for larger particles and higher initial Pb^{2+} concentration.

Furthermore, the removal capacity of eggshells decreases as the s/w ratio increases; at s/w ratios 1:1000 and 1:2000, the removal capacity is higher than that for ratios 1:100 and 1:200 (Figure 6). If the removal of Pb^{2+} is due to sorption, then a decrease in the eggshell removal capacity with the increase in the s/w ratio may be due to less effective mixing at a higher density of eggshells in the solution causing reduced accessibility to sorption sites. On the other hand, if the removal of Pb^{2+} is due to precipitation, then a decrease in the eggshell capacity with the increase in the s/w ratio could be due to a decreased dissolution of $CaCO_3$ per unit mass of eggshells at higher s/w ratios. In addition, if precipitation occurs on the eggshell surface (microprecipitation) then less sites become available for sorption. In this batch experiment mode, the highest removal capacity was found to be about 70 mg/g which occurred for particle size <150 μm at an s/w ratio of 1:2000 and an initial Pb^{2+} concentration of about 60 mg/L. Furthermore, high percentage removals (up to 99%) were achieved for low Pb^{2+} concentrations (<30 mg/L) across all s/w ratios studied (not shown here). The low concentration of solute ensured high removal.

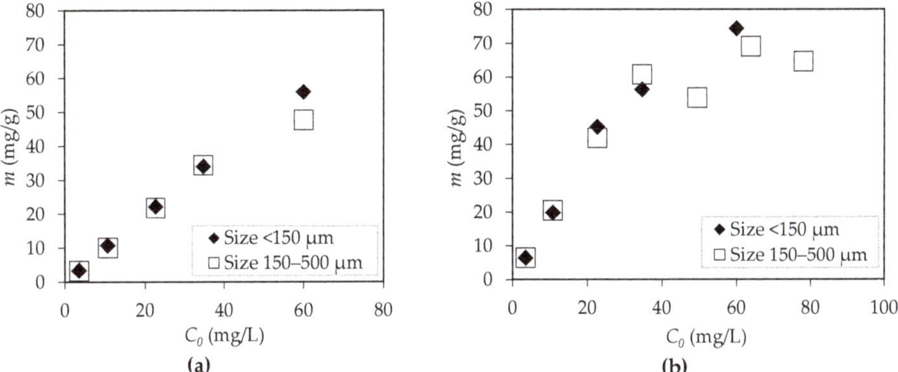

Figure 5. Removal of Pb^{2+} versus initial Pb^{2+} concentration for particle size 150–500 μm and <150 μm at (**a**) s/w ratio 1:1000 (pH increased from 5.49 (SD 0.25) to 8.64 (SD 0.64) for particle size <150 μm and from 5.49 (SD 0.25) to 6.92 (SD 0.67) for particle size 150–500 μm) and (**b**) s/w ratio 1:2000 (pH increased from 5.49 (SD 0.25) to 6.89 (SD 0.72) for particle size <150 μm and from 5.49 (SD 0.25) to 6.86 (SD 0.72) for particle size 150–500 μm).

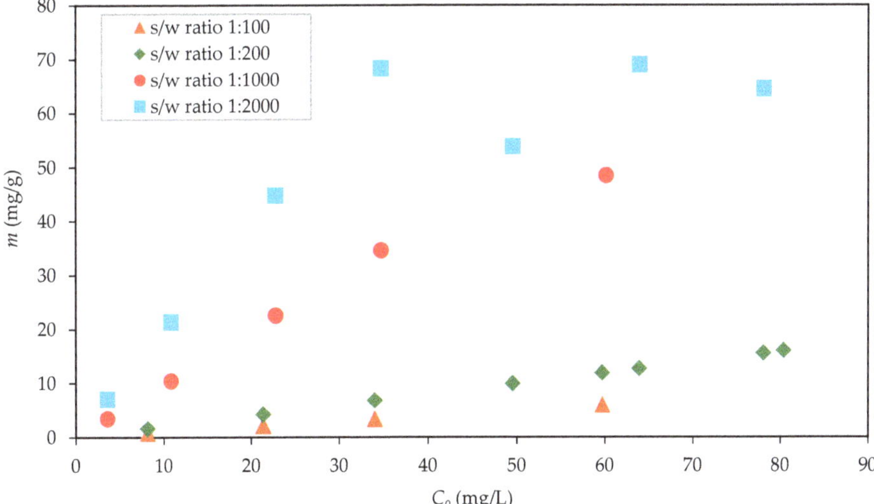

Figure 6. Removal capacity of eggshells for Pb^{2+} removal as a function of initial Pb^{2+} concentration and s/w ratio for particle size 150–500 μm (pH increased from 5.76 (SD 0.48) to 8.35 (SD 0.27) for s/w ratio 1:100; from 5.71 (SD 0.37) to 7.94 (SD 1.06) for s/w ratio 1:200; from 5.49 (SD 0.25) to 6.92 (SD 0.67) for s/w ratio 1:1000; and from 5.49 (SD 0.25) to 6.86 (SD 0.72) for s/w ratio 1:2000).

Previous studies (Table 2) showed variations in the capacity of eggshells employed for Pb^{2+} removal, with values as low as 0.12 mg/g [29] to as high as 156 mg/g [42]. Such variations could be due to differences in the experimental conditions employed, including differences in the initial concentration, particle size, and s/w ratio. However, these studies only included one s/w ratio in their investigation. As demonstrated in this study, the s/w ratio has a significant impact on the removal capacity (m). To gain insight into the effect of s/w ratio, normalized removal capacity (m/C_0) is plotted versus s/w ratio as shown in Figure 7. Normalized capacity values for each s/w ratio used in this study were obtained from the slope of the best fit line with zero intercept of the corresponding data set presented in Figure 6. Data from previous studies [28–30,42] were superimposed for comparison purposes. Since these previous studies only included one s/w ratio in their investigation, an average value of the slope (m/C_0) was taken for each study. One study by Soares et al. [30] showed a decreasing trend of (m/C_0) with the increase in C_0 and therefore the selected data points (Figure 7) represent those before a plateau for removal capacity (m) was reached. The data of Figure 8 show a linear trend on a log–log scale.

Table 2. Eggshell capacity for removal of Pb^{2+} as reported by other studies.

Study	Eggshell Size (μm)	s/w Ratio	pH	C_0 (mg/L)	m (mg/g)
[42] [a]	750	1:500	5.0	523, 1045	156
[28]	<1000	1:40	5.5	10–150	2.24
[30] [b]	<500	1:100	5.0	100–1435	12.9
[29]	210–400	1:40	NA	3	0.12

[a] Only the results for the 750 μm particle size are considered here. [b] Considered values represent those before a plateau for removal capacity was reached.

From Figure 7, it is clear that the impact of the s/w ratio is significant. The best fit relationship for the data in Figure 7 is given in Equation (3). The relation shows that the slope (m/C_0) is almost inversely proportional to the s/w ratio. Equation (3) is valid only in the range of the s/w ratio presented in Figure 8 and before the removal capacity of eggshells (m) reaches a plateau, and for particle size <1 mm.

$$\log(m/C_0) = -0.933\ \log(\text{s/w ratio}) - 2.9 \quad R^2 = 0.99 \tag{3}$$

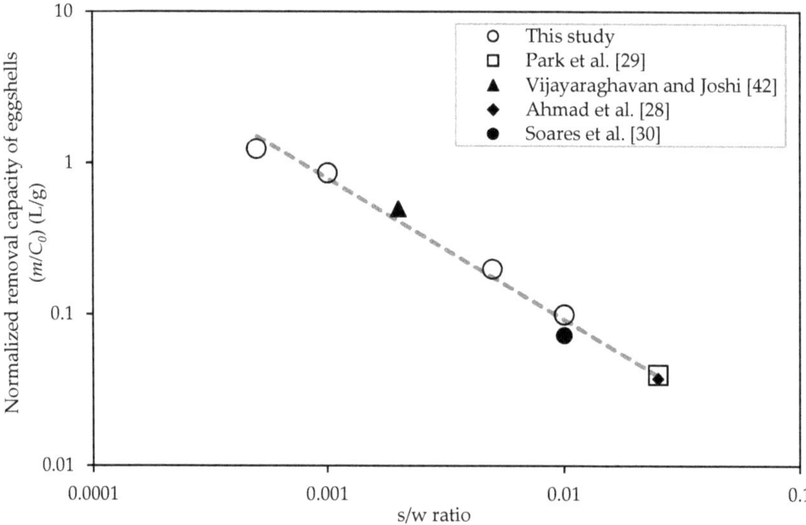

Figure 7. Comparison between normalized capacity of eggshells for Pb^{2+} removal (m/C_0) versus s/w ratio used in this study and previous studies.

The predictive relationship (Equation (3)) could be used in designing batch reactors utilizing eggshells for Pb^{2+} removal from wastewater. Specifically, this relationship could be used to optimize the design conditions and configuration of the batch reactors (single versus sequential batch reactors).

For example, based on Equation (3), 250 kg of eggshells would be required to reduce Pb^{2+} concentration from 55 to 1 mg/L with a reaction time of 2 h for 10 m^3 of contaminated water if one batch reactor is used. However, if two or three batch reactors in series were to be used, the mass of eggshells would be reduced to 80 and 12 kg, respectively, with a reaction time of 2 h for each reactor. Thus, using multiple reactors in series reduces the required mass of eggshells and consequently reduces the mass of generated waste. However, the use of multiple reactors as compared to a single reactor would entail higher construction and operation costs. As such, Equation (3) could serve as a guide to reach an optimal system design subject to economic and environmental constraints.

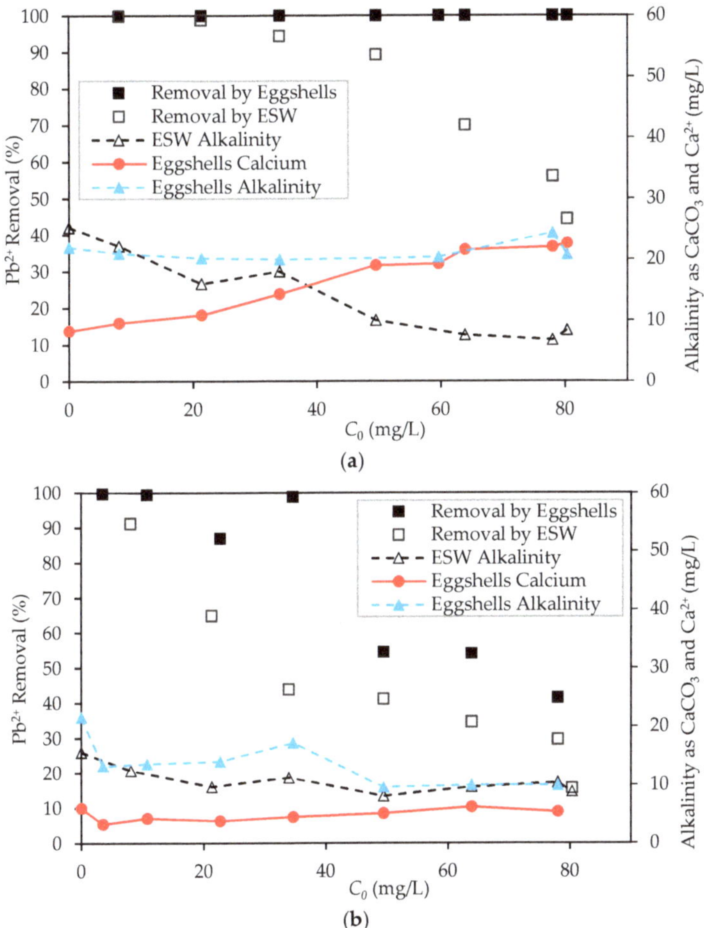

Figure 8. Comparison of percentage Pb^{2+} removal using eggshells of size 150–500 µm and ESW at (**a**) s/w ratio 1:200 (pH increased from 5.71 (SD 0.37) to 7.94 (SD 1.06) with eggshells, whereas with ESW the pH instantly drops from 9.36 (ESW only) to 6.18 (SD 0.75) right after mixing with lead solution) and (**b**) s/w ratio 1:2000 (pH increased from 5.49 (SD 0.25) to 6.86 (SD 0.72) with eggshells, whereas with ESW the pH instantly drops from 9.36 (ESW only) to 6.06 (SD 0.51) right after mixing with lead solution).

3.4. Differentiation between Removal of Pb^{2+} by Sorption and Precipitation

To study the removal of Pb^{2+} by precipitation, the method outlined in Section 2.7 was followed. The percentage of Pb^{2+} removal as a function of the initial Pb^{2+} in solution (C_0) is presented in Figure 8.

The figure compares the total Pb^{2+} removal obtained when eggshells were physically in solution and Pb^{2+} removal due to precipitation with eggshell water (ESW). Clearly, precipitation plays a major role in the removal of Pb^{2+} by eggshells. At a low initial concentration, Pb^{2+} removal could entirely be attributed to precipitation due to the presence of carbonates in the solution. However, the role of precipitation is reduced as the s/w ratio and initial concentration of Pb^{2+} become higher (Figure 8a). The difference between the removal in the two cases (eggshells and ESW) could be attributed to sorption. However, it was observed that aqueous Ca^{2+} in equilibrium with eggshells increased with the increase in initial Pb^{2+} concentration, while alkalinity remained almost constant (Figure 8a). The level of Ca^{2+}, in the case of ESW, remained the same at 8.3 and 7.5 mg/L for s/w ratios 1:200 and 1:2000, respectively, at all initial Pb^{2+} concentrations. Thus, comparison between Pb^{2+} removal with ESW versus that of eggshells may provide an underestimation of the extent of precipitation. Further, the increase in Ca^{2+} in the presence of eggshells was more pronounced at the high s/w ratio (Figure 8a) compared to that for the low s/w ratio (Figure 8b). Such a finding was accompanied by an almost constant alkalinity due to the continuous release of carbonates by eggshells. This could explain the higher difference between the removal of Pb^{2+} by eggshells and ESW at the high s/w ratio (Figure 8a).

To provide a better elucidation of the roles of different removal mechanisms. The authors resorted to direct quantification of the amount of sorbed Pb^{2+} by acid digestion. Precipitation was then calculated as the difference between total removal and percentage of Pb^{2+} sorbed on the eggshells. This quantification is presented in Figure 9, which distinguishes Pb^{2+} removed by sorption from that removed by precipitation at an s/w ratio of 1:2000. Though the percentage of Pb^{2+} removal declined with increasing initial Pb^{2+} concentration, precipitation continued to be more prominent than sorption up to an initial Pb^{2+} concentration of 60 mg/L, after which removal by precipitation significantly dropped (Figure 9a) and became almost similar to removal by sorption. On the other hand, the contribution of sorption to the percent removal was the same at different initial concentrations but slightly decreased after 60 mg/L. These findings are consistent with those of Ahmad et al. [28], who concluded that precipitation may become important for the removal of Pb^{2+} and Cu^{2+} by waste eggshells. It should be noted that some studies suggest that microprecipitation of Pb^{2+} by carbonates present in eggshells was followed by adsorption of the metal carbonates on the eggshell surface [30,42]. Thus, it is difficult to confirm that all the sorbed Pb was in the form of Pb^{2+}, as there could be some $PbCO_3$ remaining on the surface, which could result in an underestimation of the role of precipitation in Pb removal.

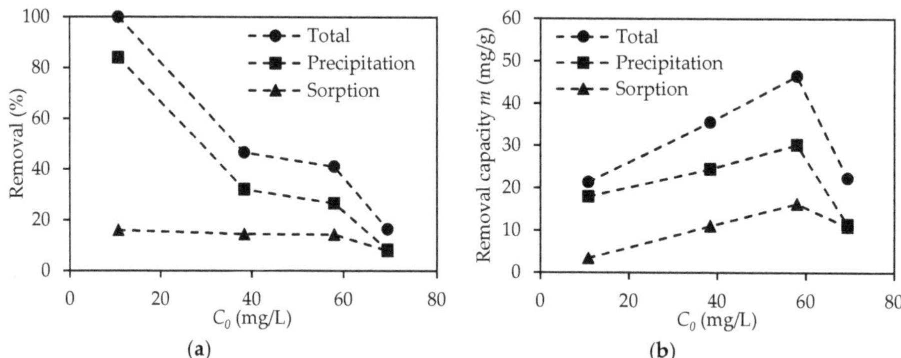

Figure 9. Removal of Pb^{2+} by sorption and precipitation versus initial Pb^{2+} concentration (size 150–500 μm and s/w ratio of 1:2000) expressed in (**a**) percent removal and (**b**) removal capacity.

Figure 9b shows the changes in removal capacity of eggshells as a function of initial Pb^{2+} concentration. The contribution of both sorption and precipitation to the removal capacity increased with the increase in the initial Pb^{2+}, but then both dropped after 60 mg/L. This behavior was also observed for the total removal capacity at this s/w ratio (1:2000) and particle size (150–500 μm) as shown in Figures 5b and 6. The expectation is for the removal capacity (*m*) to reach a plateau value after a certain initial concentration, as shown for Pb^{2+} removal by eggshells in the work of Soares et al. (2016) and Cu^{2+} removal by eggshells in the work of Ahmad et al. (2012), in which the solution pH was maintained at a fixed value (at 5 and 5.5, respectively). However, in this study, the observed decline, rather than a plateau, in the removal capacity (*m*) could be attributed to a decrease in the available carbonates needed for the precipitation reaction, coupled with a decrease in the solution pH. The latter could be the reason why the observed removal capacity by sorption dropped as well.

3.5. Thermodynamic and Stiochiometric Analysis

Several Pb^{2+} solid phases could occur in the Pb-$CaCO_3$-H_2O system including PbO, $Pb(OH)_2$, $PbCO_3$ (cerussite), $Pb_3(CO_3)_2(OH)_2$ (hydrocerussite), and $Pb_{10}(OH)_6(CO_3)_6$ (plumbonacrite). PbO and plumbonacrite are thermodynamically stable at very high pH values (pH > 12) [53]. Pure $Pb(OH)_2$(s) is believed not to exist, but the basic salt, used to prepare the Pb solution, combined with aqueous $Pb(OH)_2$ could exist as a solid form at high pH [54]. In this study, the solution pH started at around pH 5.5 (only Pb^{2+} in the water) and increased to less than 9.0 after two hours from the addition of eggshells, making lead carbonate, in the form of cerussite or/and hydrocerussite, the more probable solid that could be formed. The chemical reactions describing the formation of cerussite and hydrocerussite along with their solubility product (K_{sp}) values [52] are

$$Pb^{2+}(aq) + CO_3{}^{2-}(aq) = PbCO_3(s) \qquad K_{sp} = 7.9 \times 10^{-14} \tag{4}$$

$$3Pb^{2+}(aq) + 2CO_3{}^{2-}(aq) + 2OH^-(aq) = Pb_3(CO_3)_2(OH)_2(s) \qquad K_{sp} = 3.16 \times 10^{-46} \tag{5}$$

Thermodynamic analysis was conducted to investigate the possible formation of cerussite or hydrocerussite using the range of values of Pb^{2+}, $CO_3{}^{2-}$, and pH encountered in this study. Pb^{2+} ranged from 10 to 70 mg/L (4.8×10^{-5} to 3.3×10^{-4} mol/L) and alkalinity ranged from 10 to 40 mg/L as $CaCO_3$. If all alkalinity is due to the release of $CO_3{}^{2-}$ from the eggshells, then the molar concentration of $CO_3{}^{2-}$ would range from 1×10^{-4} to 4.2×10^{-4} mol/L. The activity coefficients of Pb^{2+} and $CO_3{}^{2-}$ were determined using the Debye–Huckel law as described by Snoeyink and Jenkins [55] and fell in the range of 0.77 to 0.89 (an average value of 0.83 was considered here for Pb^{2+} and $CO_3{}^{2-}$). Based on that, the values of the ion activity product for cerussite ($\{Pb\}\{CO_3\}$, where { } refers to ion activity = activity coefficient × molar concentration) range from 3.3×10^{-9} to 9.7×10^{-8}. These values are higher than the K_{sp} of cerussite (7.9×10^{-14}), indicating favorable conditions for the formation of this solid. Analogous calculations of the ion activity product for hydrocerussite ($\{Pb\}^3\{CO_3\}^2\{OH\}^2$) resulted in values that range from 4.4×10^{-39} to 2.6×10^{-28}, which are higher than its K_{sp} (3.16×10^{-46}), and hence the conditions also favor the formation of hydrocerussite.

The above analysis indicates that the precipitate formed in the Pb-eggshell system of this study could be cerussite or/and hydrocerussite. There is no agreement in the literature about the relative stability of cerussite and hydrocerussite. For aqueous Pb^{2+} in contact with $CaCO_3$ in open systems, Bilinski and Schindler [56] concluded that hydrocerussite is the most stable solid phase, whereas Taylor and Lopata [53] indicated that cerussite is thermodynamically more stable than hydrocerussite at all pH values. On the other hand, for a closed aqueous system (i.e., with no exchange of atmospheric CO_2), hydrocerussite becomes more stable at pH > 7 [51].

Stoichiometric analysis was carried out to investigate the extent to which the experimental data of Pb^{2+} precipitation with eggshells are aligned with the formation of cerussite and hydrocerussite. Figure 10 plots the concentration of reacted Pb^{2+} (expressed as the difference between the initial and final Pb^{2+} concentration in solution) versus the drop in the carbonate level estimated from the alkalinity measurement of the control solution (with eggshells but without Pb^{2+}) and of the Pb^{2+} solution for two s/w ratios. The dotted line in Figure 10 represents the stoichiometry of the cerussite formation. This line has a slope of 1.0 since 1 mol of Pb^{2+} reacts with 1 mol carbonate to form cerussite. The solid line represents the stoichiometry for hydrocerussite formation and has a slope of 1.5 (3 mol of Pb^{2+} reacts with 2 mol of carbonate to form hydrocerussite). Most of the experimental data lie within the two lines, which is consistent with the notion that the removal of Pb^{2+} from the eggshell solution is due to the formation of lead carbonate. As shown in Figure 10, some of the experimental data are above the lines of cerussite and hydrocerussite formation. This is expected and could be attributed to two reasons: (1) not all Pb^{2+} precipitated but some was sorbed on the eggshells which resulted in a higher reacted Pb^{2+} than what could be predicted by stoichiometry, and (2) the eggshells could release additional carbonate during the reaction to reach a new equilibrium state and this could result in an apparent lower CO_3^{2-} change than that estimated based on the reaction stoichiometry. Nonetheless, the observed drop in Pb^{2+} versus the drop in alkalinity is justified based on the stoichiometry for the formation of lead carbonate.

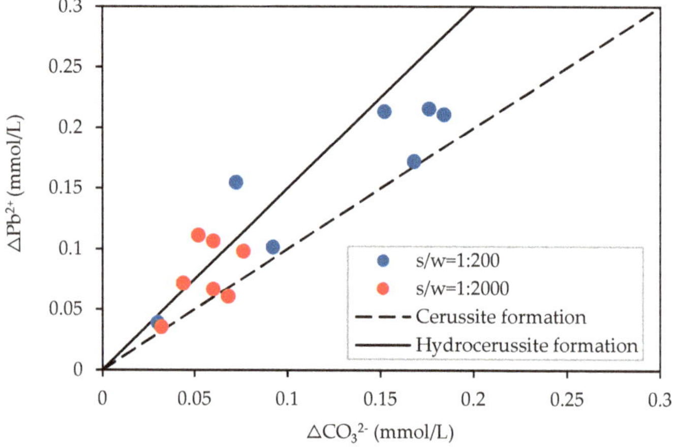

Figure 10. Alignment of experimental results of eggshells (size 150–500 μm) with the stoichiometry for the formation of cerussite and hydrocerussite. ΔPb^{2+} refers to the difference in the initial and final Pb^{2+} concentration in solution, while ΔCO_3^{2-} refers to the drop in the carbonate level estimated as the difference in the alkalinity of the control solution and of the Pb^{2+} solution for each bottle.

3.6. FTIR Analysis

The change in functional groups of eggshells before and after Pb^{2+} removal was examined using FTIR spectroscopy. Figure 11 shows the typical FTIR spectrum of raw and used eggshells with 40 mg/L Pb^{2+} solution. The infrared bands at 712 and 872 cm^{-1} are associated with the respective in-plane and out-of-plane bending vibrations of calcium carbonate ($CaCO_3$) in its polymorph form of calcite [35,57]. A broad band in the range of 1300–1500 cm^{-1}, centered at 1422 cm^{-1}, is recognized as a C–O stretching mode vibration of $CaCO_3$ [57]. Such bands have also been identified as v_4, v_2, and v_3 vibration modes of carbonate ion, respectively [58]. These bands are observed in both samples, providing evidence to the stability and inertness of calcite when exposed to the Pb^{2+} solution. Furthermore, raw eggshells were characterized by a major absorption peak at 2350 cm^{-1}, corresponding to a C=O stretching vibration [59]. However, such a peak was not detected in the used eggshells sample, indicating that this carbonate polymorph was less stable and was released into the Pb^{2+} solution to cause removal by precipitation.

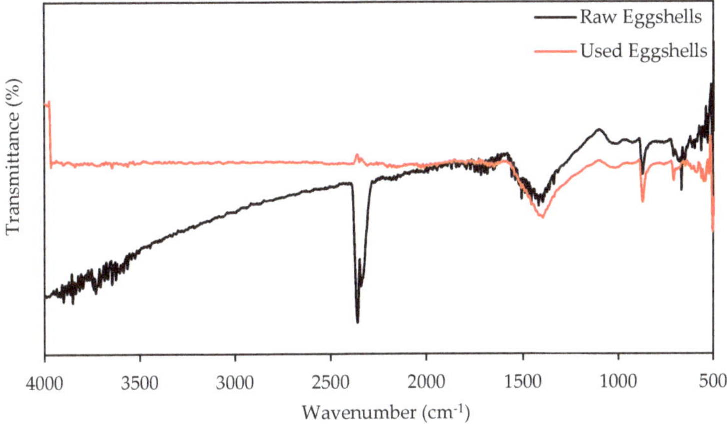

Figure 11. FTIR analysis of eggshells before and after Pb^{2+} removal (40 mg/L Pb^{2+}, s/w ratio 1:200, and particle size 150–500 μm).

3.7. SEM Analysis

SEM micrographs of Figures 12–14 illustrate the eggshells with a particle size of 150–500 μm before and after Pb^{2+} removal. Figure 12 of raw eggshells (processed based on the procedure in Section 2.2) shows the agglomeration of small particles, with the absence of connecting channels. This supports the BET results, suggesting eggshells to be a non-porous material. This conclusion is further supported by the N_2 isotherm type II curve (Figure 4), corresponding to non-porous materials. Similar findings have been reported for raw eggshells in other work [29,35]. In addition, the micrographs show that the surface texture of raw eggshells is rough and uneven.

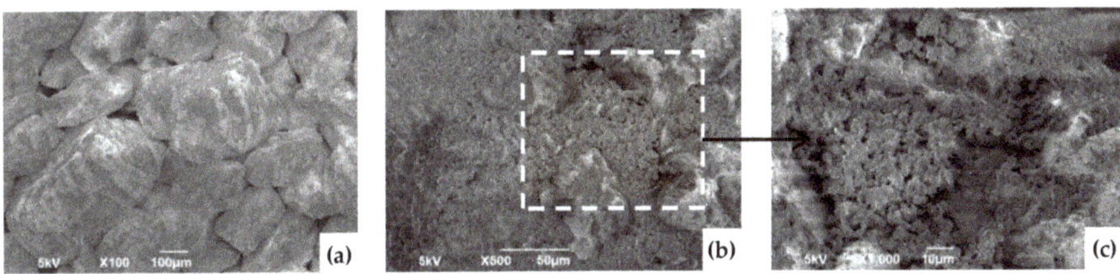

Figure 12. Raw eggshells, particle size 150–500 μm at magnification (**a**) × 100, (**b**) × 500, and (**c**) × 1000.

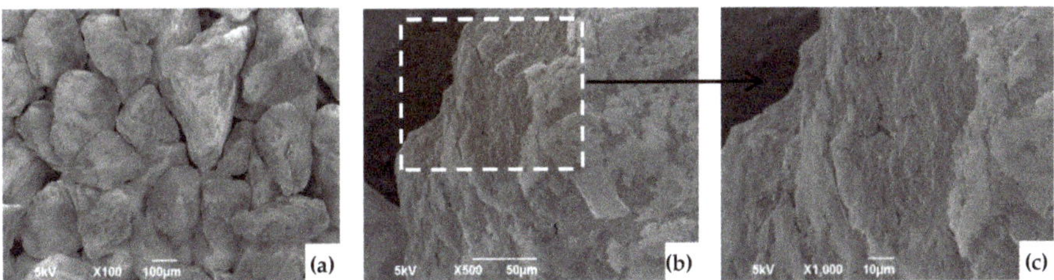

Figure 13. Eggshells, particle size 150–500 μm, after Pb^{2+} removal process with s/w ratio of 1:200 and 20 mg/L Pb^{2+} solution at magnification (**a**) ×100, (**b**) ×500, and (**c**) ×1000.

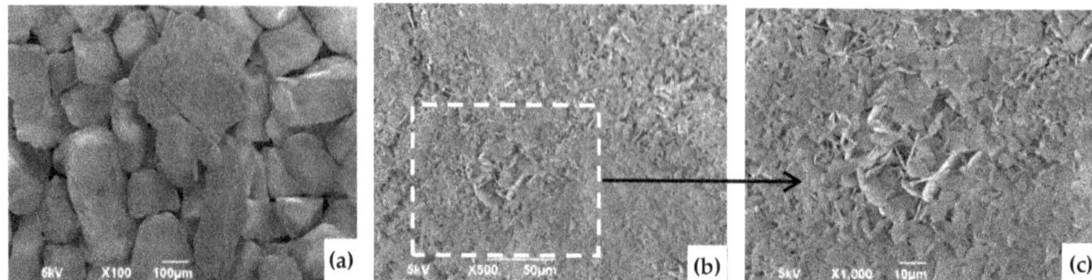

Figure 14. Eggshells, particle size 150–500 μm, after Pb^{2+} removal process with s/w ratio of 1:2000 and 20 mg/L Pb^{2+} solution at magnification (**a**) ×100, (**b**) ×500, and (**c**) ×1000.

The removal process induced changes to the morphology of the eggshells. Figure 13 presents the microstructure of eggshells post-exposure to Pb^{2+} with an s/w ratio of 1:200. It is clear that the surface of the eggshells is different from that of the raw eggshells. In fact, the surface is denser and less porous, owing to the deposition of Pb^{2+} and formation of lead carbonate ($PbCO_3$). At a lower s/w ratio of 1:2000 (Figure 14), more Pb^{2+} has been deposited on the eggshells surface, leading to the growth of needle-like $PbCO_3$ crystals. The morphology of $PbCO_3$ illustrated herein is consistent with that reported in previous studies [60,61]. This is also consistent with the trend of a higher removal capacity of eggshells (m) at lower s/w ratios (Figure 6). Nevertheless, SEM micrographs presented herein cannot be used to differentiate the mechanism of Pb^{2+} deposition onto the eggshells, i.e., microprecipitation or sorption.

4. Conclusions

Eggshells are a common biowaste that could be recycled for the removal of Pb^{2+} from wastewater, thus promoting the use of nature-based sustainable solutions for both waste management and wastewater treatment. In this study, the influence of the particle size of ground eggshells and s/w ratio on removal of Pb^{2+} from aqueous solutions was investigated. Results show that the percent Pb^{2+} removal was not significantly different for particle sizes 150–500 μm and <150 μm despite the large difference in the surface area of the two eggshell sizes. A high percentage removal (up to 99%) of Pb^{2+} by eggshells was achieved for low initial Pb^{2+} concentrations (<30 mg/L) across all s/w ratios studied. SEM images confirmed that the surface of eggshells used for Pb^{2+} removal became denser and less porous due to Pb^{2+} deposition and lead carbonate formation. Needle-like $PbCO_3$ crystals were observed in samples with a lower s/w ratio, which coincides with the trend where eggshells exhibited higher removal capacity at lower s/w ratios. A developed predictive relationship suggests that the removal capacity of eggshells normalized to the initial Pb^{2+} concentration is almost inversely proportional to the s/w ratio. The relationship explains differences in the Pb^{2+} removal capacities of eggshells reported by others. The developed relationship could be utilized to design batch reactors for Pb^{2+} removal by eggshells subject to economic and environmental constraints.

The study also attempted to quantify the role of different removal mechanisms in the total removal of Pb^{2+}. Results confirmed that precipitation played a major role in the removal of Pb^{2+} by eggshells. However, this role was reduced as the s/w ratio and initial concentration of Pb^{2+} became higher. The difference between the extent of Pb^{2+} removal by eggshells and eggshell water suggested that sorption also played a role in Pb^{2+} removal. This was confirmed by the direct quantification of the concentration of Pb^{2+} sorbed on the eggshells and by the results from FTIR spectroscopy. Indeed, FTIR spectroscopy of eggshells highlighted the presence of high- and low-stability calcium carbonate polymorphs that existed in raw eggshells. While the former polymorph was detected in the used eggshells sample, the latter was not, signifying its release into the Pb^{2+} solution to cause removal by precipitation. It was also observed that the contribution of both sorption and precipitation dropped at the high concentration. It was speculated that such decline was due to a decrease in the available carbonates needed for precipitation coupled with a decrease in the solution pH.

Author Contributions: Conceptualization, M.A.H., M.A.M. and H.E.-H.; data curation, M.A.H., H.S. and M.A.M.; formal analysis, M.A.H., H.S. and M.A.M.; funding acquisition, M.A.H. and H.E.-H.; investigation, H.S.; methodology, M.A.H., M.A.M. and H.E.-H.; project administration, M.A.H. and H.E.-H.; resources, H.E.-H.; supervision, M.A.H., M.A.M. and H.E.-H.; validation, M.A.H. and M.A.M.; visualization, M.A.H., M.A.M. and H.E.-H.; writing—original draft, H.S.; writing—review and editing, M.A.H., M.A.M. and H.E.-H. All authors have read and agreed to the published version of the manuscript.

Funding: This research was funded by UAE University, grant number 31N322; the National Water Center at UAE University grant number 31R150; and The APC was funded by the National Water Center at UAE University grant number 31R150.

Conflicts of Interest: The authors declare no conflict of interest. The funders had no role in the design of the study; in the collection, analyses, or interpretation of data; in the writing of the manuscript, or in the decision to publish the results.

References

1. Anantha, R.K.; Kota, S. Removal of lead by adsorption with the renewable biopolymer composite of feather (*Dromaius novaehollandiae*) and chitosan (*Agaricus bisporus*). *Environ. Technol. Innov.* **2016**, *6*, 11–26. [CrossRef]
2. Gupta, V.K.; Agarwal, S.; Saleh, T.A. Synthesis and characterization of alumina-coated carbon nanotubes and their application for lead removal. *J. Hazard. Mater.* **2011**, *185*, 17–23. [CrossRef] [PubMed]
3. Ritchey, A.K.; O'Brien, S.H.; Keller, F.G. Chapter 152—Hematologic manifestations of childhood illness. In *Hematology*, 7th ed.; Hoffman, R., Benz, E.J., Silberstein, L.E., Heslop, H.E., Weitz, J.I., Anastasi, J., Salama, M.E., Abutalib, S.A., Eds.; Elsevier: Amsterdam, The Netherlands, 2018; pp. 2215–2237.e9, ISBN 978-0-323-35762-3.
4. Burke, D.M.; Morris, M.A.; Holmes, J.D. Chemical oxidation of mesoporous carbon foams for lead ion adsorption. *Sep. Purif. Technol.* **2013**, *104*, 150–159. [CrossRef]
5. Canfield, R.L.; Henderson, C.R.; Cory-Slechta, D.A.; Cox, C.; Jusko, T.A.; Lanphear, B.P. Intellectual impairment in children with blood lead concentrations below 10 µg per deciliter. *N. Engl. J. Med.* **2003**, *348*, 1517–1526. [CrossRef] [PubMed]
6. Lanphear, B.P.; Matte, T.D.; Rogers, J.; Clickner, R.P.; Dietz, B.; Bornschein, R.L.; Succop, P.; Mahaffey, K.R.; Dixon, S.; Galke, W.; et al. The contribution of lead-contaminated house dust and residential soil to children's blood lead levels: A pooled analysis of 12 epidemiologic studies. *Environ. Res.* **1998**, *79*, 51–68. [CrossRef]
7. Cooper, W.C.; Wong, O.; Kheifets, L. Mortality among employees of lead battery plants and lead-producing plants, 1947–1980. *Scand. J. Work Environ. Health* **1985**, *11*, 331–345. [CrossRef]
8. Lancranjan, I.; Popescu, H.I.; GAvănescu, O.; Klepsch, I.; Serbănescu, M. Reproductive ability of workmen occupationally exposed to lead. *Arch. Environ. Health* **1975**, *30*, 396–401. [CrossRef]
9. Lanphear, B.P.; Roghmann, K.J. Pathways of lead exposure in urban children. *Environ. Res.* **1997**, *74*, 67–73. [CrossRef]
10. Mahmoud, M.T.; Hamouda, M.A.; Al Kendi, R.R.; Mohamed, M.M. Health risk assessment of household drinking water in a district in the UAE. *Water* **2018**, *10*, 1726. [CrossRef]
11. Health Canada. *Guidelines for Canadian Drinking Water Quality—Summary Table*; Water and Air Quality Bureau; Healthy Environments and Consumer Safety Branch, Health Canada: Ottawa, ON, Canada, 2019.
12. US EPA. *National Primary Drinking Water Regulations*; EPA 816-F-09-004; Office of Water United States Environmental Protection Agency: Washington, DC, USA, 2009.
13. Arunlertaree, C.; Kaewsomboon, W.; Kumsopa, A.; Pokethitiyook, P.; Panyawathanakit, P. Removal of lead from battery manufacturing wastewater by egg shell. *Songklanakarin J. Sci. Technol.* **2007**, *29*, 13.
14. Dermentzis, K.; Valsamidou, E.; Marmanis, D. Simultaneous removal of acidity and lead from acid lead battery wastewater by aluminum and iron electrocoagulation. *J. Eng. Sci. Technol. Rev.* **2012**, *5*, 1–5. [CrossRef]
15. Khaoya, S.; Pancharoen, U. Removal of lead (II) from battery industry wastewater by HFSLM. *Int. J. Chem. Eng. Appl.* **2012**, 98–103. [CrossRef]
16. Macchi, G.; Pagano, M.; Santori, M.; Tiravanti, G. Battery industry wastewater: Pb removal and produced sludge. *Water Res.* **1993**, *27*, 1511–1518. [CrossRef]
17. WHO. *A Compendium of Standards for Wastewater Reuse in the Eastern Mediterranean Region*; World Health Organization, Regional Office for the Eastern Mediterranean: Cairo, Egypt, 2006.
18. Babel, S.; Kurniawan, T.A. Low-cost adsorbents for heavy metals uptake from contaminated water: A review. *J. Hazard. Mater.* **2003**, *97*, 219–243. [CrossRef]

19. Sweidan, H.; Hamouda, M.; El-Hassan, H.; Maraqa, M. A framework for the investigation of biowaste materials as potential adsorbents for water treatment. In Proceedings of the 4th World Congress on Civil, Structural, and Environmental Engineering, Rome, Italy, 7–9 April 2019.
20. Podstawczyk, D.; Witek-Krowiak, A.; Chojnacka, K.; Sadowski, Z. Biosorption of malachite green by eggshells: Mechanism identification and process optimization. *Bioresour. Technol.* **2014**, *160*, 161–165. [CrossRef]
21. Eba, F.; Biboutou, R.K.; Nlo, J.N.; Bibalou, Y.G.; Oyo, M. Lead removal in aqueous solution by activated carbons prepared from *Cola edulis* shell (Alocacée), *Pentaclethra macrophylla* husk (Mimosaceae) and *Aucoumea klaineana* sawdust (Burseraceae). *Afr. J. Environ. Sci. Technol.* **2011**, *5*, 197–204. [CrossRef]
22. El-Naas, M.H.; Al-Rub, F.A.; Ashour, I.; Al Marzouqi, M. Effect of competitive interference on the biosorption of lead(II) by *Chlorella vulgaris*. *Chem. Eng. Process. Process Intensif.* **2007**, *46*, 1391–1399. [CrossRef]
23. Rashed, M.N. Fruit stones from industrial waste for the removal of lead ions from polluted water. *Environ. Monit. Assess.* **2006**, *119*, 31–41. [CrossRef]
24. Sekar, M.; Sakthi, V.; Rengaraj, S. Kinetics and equilibrium adsorption study of lead(II) onto activated carbon prepared from coconut shell. *J. Colloid Interface Sci.* **2004**, *279*, 307–313. [CrossRef]
25. Yan, T.; Luo, X.; Lin, X.; Yang, J. Preparation, characterization and adsorption properties for lead (II) of alkali-activated porous leather particles. *Colloids Surf. A Physicochem. Eng. Asp.* **2017**, *512*, 7–16. [CrossRef]
26. Ritchie, H.; Roser, M. Meat and Dairy Production. Available online: https://ourworldindata.org/meat-production (accessed on 7 September 2020).
27. Stadelman, W.J. EGGS|Structure and Composition. In *Encyclopedia of Food Sciences and Nutrition*, 2nd ed.; Caballero, B., Ed.; Academic Press: Oxford, UK, 2003; pp. 2005–2009. ISBN 978-0-12-227055-0.
28. Ahmad, M.; Usman, A.R.A.; Lee, S.S.; Kim, S.-C.; Joo, J.-H.; Yang, J.E.; Ok, Y.S. Eggshell and coral wastes as low cost sorbents for the removal of Pb^{2+}, Cd^{2+} and Cu^{2+} from aqueous solutions. *J. Ind. Eng. Chem.* **2012**, *18*, 198–204. [CrossRef]
29. Park, H.J.; Jeong, S.W.; Yang, J.K.; Kim, B.G.; Lee, S.M. Removal of heavy metals using waste eggshell. *J. Environ. Sci.* **2007**, *19*, 1436–1441. [CrossRef]
30. Soares, M.; Marto, S.; Quina, M.; Gando-Ferreira, L.; Quinta-Ferreira, R. Evaluation of eggshell-rich compost as biosorbent for removal of Pb(II) from aqueous solutions. *Water Air Soil Pollut.* **2016**, *227*, 1–16. [CrossRef]
31. Vijayaraghavan, K.; Jegan, J.; Palanivelu, K.; Velan, M. Removal and recovery of copper from aqueous solution by eggshell in a packed column. *Miner. Eng.* **2005**, *18*, 545–547. [CrossRef]
32. Yeddou, N.; Bensmaili, A. Equilibrium and kinetic modelling of iron adsorption by eggshells in a batch system: Effect of temperature. *Desalination* **2007**, *206*, 127–134. [CrossRef]
33. Zhang, T.; Tu, Z.; Lu, G.; Duan, X.; Yi, X.; Guo, C.; Dang, Z. Removal of heavy metals from acid mine drainage using chicken eggshells in column mode. *J. Environ. Manag.* **2017**, *188*, 1–8. [CrossRef]
34. Pramanpol, N.; Nitayapat, N. Adsorption of reactive dye by eggshell and its membrane. *Agric. Nat. Resour.* **2006**, *40*, 192–197.
35. Tsai, W.T.; Yang, J.M.; Lai, C.W.; Cheng, Y.H.; Lin, C.C.; Yeh, C.W. Characterization and adsorption properties of eggshells and eggshell membrane. *Bioresour. Technol.* **2006**, *97*, 488–493. [CrossRef]
36. Witek-Krowiak, A.; Chojnacka, K.; Podstawczyk, D.; Dawiec, A.; Pokomeda, K. Application of response surface methodology and artificial neural network methods in modelling and optimization of biosorption process. *Bioresour. Technol.* **2014**, *160*, 150–160. [CrossRef]
37. Andersson, M.P.; Sakuma, H.; Stipp, S.L.S. Strontium, nickel, cadmium, and lead substitution into calcite, studied by density functional theory. *Langmuir* **2014**, *30*, 6129–6133. [CrossRef]
38. Davis, A.D.; Webb, C.J.; Sorensen, J.L.; Dixon, D.J.; Hudson, R. Geochemical thermodynamics of cadmium removal from water with limestone. *Environ. Earth Sci.* **2018**, *77*, 37. [CrossRef]
39. Lee, Y.-C.; Kim, E.J.; Yang, J.-W.; Shin, H.-J. Removal of malachite green by adsorption and precipitation using aminopropyl functionalized magnesium phyllosilicate. *J. Hazard. Mater.* **2011**, *192*, 62–70. [CrossRef] [PubMed]
40. Xie, Y.; Giammar, D.E. Effects of flow and water chemistry on lead release rates from pipe scales. *Water Res.* **2011**, *45*, 6525–6534. [CrossRef] [PubMed]
41. Xie, L.; Giammar, D.E. Chapter 13 Influence of phosphate on adsorption and surface precipitation of lead on iron oxide surfaces. In *Developments in Earth and Environmental Sciences*; Barnett, M.O., Kent, D.B., Eds.; Adsorption of Metals by Geomedia II: Variables, Mechanisms, and Model Applications; Elsevier: Amsterdam, The Netherlands, 2007; Volume 7, pp. 349–373.

42. Vijayaraghavan, K.; Joshi, U.M. Chicken eggshells remove Pb(II) ions from synthetic wastewater. *Environ. Eng. Sci.* **2013**, *30*, 67–73. [CrossRef]
43. APHA; AWWA; WEF. *Standard Methods for the Examination of Water and Wastewater*, 23rd ed.; Rice, E.W., Baird, R.B., Eaton, A.D., Eds.; American Public Health Association: Washington, DC, USA, 2017.
44. U.S. EPA. *Method 3050B: Acid Digestion of Sediments, Sludges, and Soils, Revision 2*; U.S. EPA: Washington, DC, USA, 1996.
45. Spagnoli, A.A.; Giannakoudakis, D.A.; Bashkova, S. Adsorption of methylene blue on cashew nut shell based carbons activated with zinc chloride: The role of surface and structural parameters. *J. Mol. Liq.* **2017**, *229*, 465–471. [CrossRef]
46. Martín-Lara, M.A.; Blázquez, G.; Ronda, A.; Pérez, A.; Calero, M. Development and characterization of biosorbents to remove heavy metals from aqueous solutions by chemical treatment of olive stone. *Ind. Eng. Chem. Res.* **2013**, *52*, 10809–10819. [CrossRef]
47. García-Mendieta, A.; Olguín, M.T.; Solache-Ríos, M. Biosorption properties of green tomato husk (Physalis philadelphica Lam) for iron, manganese and iron–manganese from aqueous systems. *Desalination* **2012**, *284*, 167–174. [CrossRef]
48. Queirós, M.V.A.; Bezerra, M.N.; Feitosa, J.P.A. Composite superabsorbent hydrogel of acrylic copolymer and eggshell: Effect of biofiller addition. *J. Braz. Chem. Soc.* **2017**, *28*, 2004–2012. [CrossRef]
49. da Silva Castro, L.; Barañano, A.G.; Pinheiro, C.J.G.; Menini, L.; Pinheiro, P.F. Biodiesel production from cotton oil using heterogeneous CaO catalysts from eggshells prepared at different calcination temperatures. *Green Process. Synth.* **2019**, *8*, 235–244. [CrossRef]
50. Habte, L.; Shiferaw, N.; Mulatu, D.; Thenepalli, T.; Chilakala, R.; Ahn, J.W. Synthesis of nano-calcium oxide from waste eggshell by sol-gel method. *Sustainability* **2019**, *11*, 3196. [CrossRef]
51. Godelitsas, A.; Astilleros, J.M.; Hallam, K.; Harissopoulos, S.; Putnis, A. Interaction of calcium carbonates with lead in aqueous solutions. *Environ. Sci. Technol.* **2003**, *37*, 3351–3360. [CrossRef] [PubMed]
52. Macchi, G.; Marani, D.; Pagano, M.; Bagnuolo, G. A bench study on lead removal from battery manufacturing wastewater by carbonate precipitation. *Water Res.* **1996**, *30*, 3032–3036. [CrossRef]
53. Taylor, P.; Lopata, V.J. Stability and solubility relationships between some solids in the system $PbO–CO_2–H_2O$. *Can. J. Chem.* **1984**, *62*, 395–402. [CrossRef]
54. Powell, K.J.; Brown, P.L.; Byrne, R.H.; Gajda, T.; Hefter, G.; Leuz, A.-K.; Sjöberg, S.; Wanner, H. Chemical speciation of environmentally significant metals with inorganic ligands. Part 3: The Pb^{2+} + OH^-, Cl^-, $CO_3{}^{2-}$, $SO_4{}^{2-}$, and $PO_4{}^{3-}$ systems (IUPAC Technical Report). *Pure Appl. Chem.* **2009**, *81*, 2425–2476. [CrossRef]
55. Snoeyink, V.L.; Jenkins, D. *Water Chemistry*; Wiley: New York, NY, USA, 1980; ISBN 978-0-471-05196-1.
56. Bilinski, H.; Schindler, P. Solubility and equilibrium constants of lead in carbonate solutions (25 °C, I = 0.3 mol dm^{-3}). *Geochim. Cosmochim. Acta* **1982**, *46*, 921–928. [CrossRef]
57. Li, Z.; Yang, D.-P.; Chen, Y.; Du, Z.; Guo, Y.; Huang, J.; Li, Q. Waste eggshells to valuable $Co_3O_4/CaCO_3$ materials as efficient catalysts for VOCs oxidation. *Mol. Catal.* **2020**, *483*, 110766. [CrossRef]
58. Abeywardena, M.R.; Elkaduwe, R.K.W.H.M.K.; Karunarathne, D.G.G.P.; Pitawala, H.M.T.G.A.; Rajapakse, R.M.G.; Manipura, A.; Mantilaka, M.M.M.G.P.G. Surfactant assisted synthesis of precipitated calcium carbonate nanoparticles using dolomite: Effect of pH on morphology and particle size. *Adv. Powder Technol.* **2020**, *31*, 269–278. [CrossRef]
59. Ismail, S.; Ahmed, A.S.; Anr, R.; Hamdan, S. Biodiesel production from castor oil by using calcium oxide derived from mud clam shell. *J. Renew. Energy* **2016**, *2016*, 8. [CrossRef]
60. Seignez, N.; Gauthier, A.; Bulteel, D.; Buatier, M.; Recourt, P.; Damidot, D.; Potdevin, J.L. Effect of Pb-rich and Fe-rich entities during alteration of a partially vitrified metallurgical waste. *J. Hazard. Mater.* **2007**, *149*, 418–431. [CrossRef]
61. Ng, D.-Q.; Lin, Y.-P. Effects of pH value, chloride and sulfate concentrations on galvanic corrosion between lead and copper in drinking water. *Environ. Chem.* **2015**, *13*, 602–610. [CrossRef]

Article

Adsorption Mechanisms and Characteristics of Hg^{2+} Removal by Different Fractions of Biochar

Xiaoli Guo, Menghong Li *, Aijv Liu, Man Jiang, Xiaoyin Niu and Xinpeng Liu

Department of Resource and Environmental Engineering, Shandong University of Technology, Zibo 255000, China; gwmy1027@163.com (X.G.); liuajsdut@gmail.com (A.L.); jiangman@sdut.edu.cn (M.J.); zbnxy@sdut.edu.cn (X.N.); wangbaohua1996@163.com (X.L.)

* Correspondence: zblmh76180@sdut.edu.cn; Tel.: +86-139-5336-9334

Received: 27 June 2020; Accepted: 22 July 2020; Published: 24 July 2020

Abstract: The adsorption mechanisms of mercury ion (Hg^{2+}) by different fractions of biochar were studied, providing a theoretical basis and practical value for the use of biochar to remediate mercury contamination in water. Biochar (RC) was prepared using corn straw as the raw material. It was then fractionated, resulting in inorganic carbon (IC), organic carbon (OC), hydroxyl-blocked carbon (BHC), and carboxyl-blocked carbon (BCC). Before and after Hg^{2+} adsorption, the biochar fractions were characterized by several techniques, such as energy-dispersive X-ray spectroscopy (EDS), Fourier-transform infrared spectroscopy (FTIR), and X-ray photoelectron spectroscopy (XPS). Obtained results indicate that the reaction mechanisms of RC for Hg^{2+} removal mainly include electrostatic adsorption, ion exchange, reduction, precipitation, and complexation. The equilibrium adsorption capacity of RC for Hg^{2+} is 75.56 mg/g, and the adsorption contribution rates of IC and OC are approximately 22.4% and 77.6%, respectively. Despite the lower rate, IC shows the largest adsorption capacity, of 92.63 mg/g. This is attributed to all the mechanisms involved in Hg^{2+} adsorption by IC, with ion exchange being the main reaction mechanism (accounting for 39.8%). The main adsorption mechanism of OC is the complexation of carboxyl and hydroxyl groups with Hg^{2+}, accounting for 71.6% of the total OC contribution. BHC and BCC adsorb mercury mainly via the reduction–adsorption mechanism, accounting for 54.6% and 54.5%, respectively. Among all the adsorption mechanisms, the complexation reaction of carboxyl and hydroxyl groups with Hg^{2+} is the dominant effect.

Keywords: biochar; different fractions; Hg^{2+}; characterization; adsorption mechanism

---

1. Introduction

Mercury is a highly toxic, non-essential heavy metal. Even at low concentrations, it poses potential threat to human health [1]. Inorganic mercury is the most common form of mercury found in aquatic ecology. It stably accumulates in aqueous solutions and does not degrade easily. Under natural conditions, it easily reacts with chemicals and microorganisms via a methylation reaction, producing mercury forms with increased toxicity, such as methylmercury or dimethylmercury [2]. According to United States Environmental Protection Agency (EPA) regulations, mercury concentrations in treated wastewater and drinking water must be less than 10 and 2 μg/L, respectively [3]. Therefore, it is of great significance to study mercury removal from water. Treatment methods for mercury removal mainly include chemical precipitation, ion exchange, reduction, coagulation, solvent extraction, and adsorption methods. Among them, adsorption is widely applied due to its advantages of fast action, superior removal efficiency, inexpensive nature of adsorbents, and easy access to raw materials that can be used as adsorbents [4–6]. Biochar is the most commonly used adsorbent material, showing high adsorption efficiency [7]. In previous studies, commonly used raw materials for biochar preparation included walnut shells [8], flax shive [9], guava skin [10], fruit shell of Terminalia catappa [11], rice husk [12], sugarcane bagasse [13], coconut husk [14], and peanut husk [15], all showing good adsorption results. In general, biochar adsorbs heavy metals mainly via combining metal ions with aromatic alcohols or acids, and carbonates [16]. The main adsorption pathways are through electrostatic effect, reduction, complexation, and cation exchange. Kong et al. [17] showed that the phenolic hydroxyl groups contained in soybean stalk-based biochar had a strong reduction effect, reducing Hg^{2+} to Hg^0 during Hg^{2+} adsorption. Das et al. [18] studied Hg^{2+} adsorption by Aspergillus versicolor biomass and suggested complexation as the main mechanism between oxygen-containing functional groups and Hg^{2+}. Kílıç et al. [19] observed the release of a large concentration of basic metal ions (Ca^{2+}, Na^+, and K^+) during Hg^{2+} adsorption by activated sludge biomass, indicating that an ion exchange mechanism was involved in the adsorption process. Dong et al. [20] used XPS to analyze the adsorption mechanisms of Hg^{2+} by Brazilian pepper biochars at different pyrolysis temperatures. It was shown that biochar prepared at relatively low temperatures (300 and 450 °C) adsorbs Hg^{2+} mainly through complexation with hydroxyl and carboxyl functional groups, while biomass carbon prepared at 600 °C has a graphite-like structure and Hg–Cπ is formed between Hg^{2+} and the C=C structure. As a heterogeneous material, biochar is mainly composed of inorganic mineral components (ash) and organic carbonaceous components. Current studies on adsorption of pollutants by biochar mainly focus on the overall removal effect of biochar, with research on the adsorption mechanisms of pollutants by different fractions of biochar being very limited. The contribution rate of each fraction of the adsorbent during the adsorption mechanism for a certain pollutant needs to be further studied.

Herein, biochar was prepared using corn stover as the raw material, which was then fractionated to prepare inorganic, organic, hydroxyl-blocked, and carboxyl-blocked carbonaceous fragments. Through physicochemical analysis (element composition, pH value, and isoelectric point pH_{pzc}), Boehm titration and surface analysis (FTIR, XPS, and EDS) measurements before and after Hg^{2+} adsorption, the mercury adsorption mechanisms, and structure–property relationships of biochar and its fractions were studied. This research has important practical significance for mercury removal from water by biochar.

2. Materials and Methods

2.1. Preparation of Biochar Fractions

Raw carbon was synthesized from cornstalk (sourced from Da Xu Village, Zibo, Shandong, China) and air-dried outdoors. It was ground when the moisture content dropped below 10% and was put in a ceramic crucible for pyrolysis and carbonization at 500 °C with a rate of 20 °C/min for 6 h under N_2 atmosphere. The biochar removed from the furnace was cooled in a desiccator, weighed, and stored in airtight plastic bags. Raw carbon was referred to as RC. RC was placed in a tube furnace, the temperature of which was raised to 600 °C at a rate of 20 °C/min and kept for 4 h to obtain inorganic carbon (IC). RC was thoroughly mixed for 24 h with an acid mixture (0.3 M HF + 0.1 M HCl), at a solid/liquid ratio of 1:250 (g/mL). The mixture was then washed and dried to obtain organic carbon (OC). Meanwhile, a quantity of 9.0 g of RC was thoroughly mixed with 633 mL of anhydrous methanol and 5.4 mL of 0.1 M HCl to block carboxylic acid groups (–COOH) by methyl esterification of carboxyl groups. After being stirred for 6 h, the mixture was washed and dried to obtained carboxyl-blocked carbon (BCC). Similarly, hydroxyl-blocked carbon (BHC) was prepared by blocking hydroxyl groups (–OH) via thoroughly mixing and stirring 5.0 g of biochar and 100 mL of HCHO for 6 h [21–23]. The yield of each fraction of biochar was calculated.

2.2. Physicochemical Properties of Biochar Fractions

Typically, 1.0 g of biochar was weighed and placed in a 100 mL covered conical flask. Meanwhile, 20 mL of deionized water was boiled and let to cool before being added into the flask. Subsequently, the mixture was shaken at 120 r/min and 25 °C for 24 h. It was then taken out and let to sit still for

5 min. The pH of the solution was determined using a Mettler Delta 320 pH Meter (Mettler Toledo, Guangzhou, China) [24].

For a typical test, 50 mL of 0.01 M KNO_3 solution was pipetted into a 100 mL covered conical flask. The initial pH value was adjusted to a value within the range of 2–10, using 0.10 M HCl or NaOH solution, and denoted as pH_i. Then, 0.1 g biochar was added and the mixture was shaken at 120 r/min and 25 °C, for 48 h. Subsequently, the mixture was filtered and the pH of the filtrate was marked as pH_f. In the case where ΔpH ($\Delta pH = pH_f - pH_i$) is 0, the value of pH_i (or pH_f) is the pH_{pzc} of the biochar [25].

A Vario EL cube elemental analyzer (Elementar, Karlsruhe, Germany) was used to determine the C, H, and N content of biochar and its fractions, and the H/C value was calculated.

2.3. Determination of Oxygen-Containing Functional Groups on Biochar Surface

Boehm titration [26] was applied to quantitatively determine the oxygen-containing functional groups on the surface of biochar. Specifically, 1.0 g of biochar was placed in a 100 mL covered conical flask, and 25 mL of 0.05 M alkaline solution ($NaHCO_3$, Na_2CO_3, and NaOH) and 25 mL of 0.05 M acidic solution (HCl) were added, respectively. The mixture was shaken at 120 r/min and 25 °C for 24 h before being subjected to filtration. Subsequently, back titration was carried out with 0.05 M NaOH and HCl [27] to determine the content of acidic and basic groups on the surface of biochar.

2.4. Adsorption Experiment and Characterization before and after Hg^{2+} Adsorption

A $HgCl_2$ 3 mM solution was prepared. Then, 1.0 g of biochar (passed through a 200-mesh sieve) and 200 mL $HgCl_2$ solution were placed into a 250 mL covered conical flask. The mixture was shaken at 120 r/min and 25 °C for 24 h and filtrated upon reaching adsorption equilibrium. The pH of the filtrate was measured, and the concentration of Hg^{2+} was determined via an inductively coupled plasma mass spectrometer (ICP-MS, Santa Clara, CA, USA). On this basis, the Hg^{2+} amount adsorbed was calculated. Meanwhile, the leaching experiments of all biochars were conducted under the same conditions, with samples withdrawn at certain time intervals (3, 5, 10, 20, 30, 60, 90, 120, 150, 180, 240, 300, 420, 540, and 720 min). After filtration, the changes in the concentrations of K^+, Na^+, and Hg^{2+} in the leaching solution were measured.

The surface functional groups of the samples before and after mercury adsorption were identified by Fourier-transform infrared spectroscopy (FTIR), using a Nicolet 5700 machine (Thermo Nicolet, Waltham, MA, USA). The change in the valence state of Hg^{2+} before and after mercury adsorption was determined by X-ray photoelectron spectroscopy (XPS), using an Axis Ultr DLD machine (Thermo Fisher Scientific, Waltham, MA, USA) and analyzing the experimental data using the software Peak 4.1. (Hampton, MA, USA) The main elemental composition of biochar was detected by an energy-dispersive spectrometer (EDS), using a Quanta 250 (FEI, Hillsboro, OR, USA).

3. Results and Discussion

3.1. Physicochemical Properties of Biochar and Analysis of Surface Functional Groups

The different biochar fractions examined displayed varied physicochemical properties, as well as difference in the concentration of surface functional groups. According to Lehmann [28] and Shaaban et al. [29], when the pH value of the biochar is greater than pH_{pzc}, the biochar surface is negatively charged, which combined with positively charged ions present in the solution, leads to weakened competitiveness of H^+. As shown in Table 1, the pH of the biochar was greater than pH_{pzc}, so the surfaces of the five biochar fractions are all negatively charged, posing an electrostatic adsorption effect on Hg^{2+}. OC has the lowest pH with the highest contents of acidic functional groups such as –OH and –COOH groups. This may be caused due to OC not being cleaned thoroughly after acid washing or due to a replacement of the cations in the functional groups, such as –COOM (M is a metal cation) by H^+ during the acid washing process, resulting in the acidity of OC [30,31]. The pH value of OC did not change significantly after Hg^{2+} adsorption, which may be due to the release of H^+ by the complexation reaction between carboxyl/hydroxyl functional groups and Hg^{2+} [32]. Hg^{2+} mainly exists in the anion forms of $HgCl^-$ and $HgClO^-$ in acidic solutions [33]. The carboxyl and hydroxyl functional groups present on OC surface consume H^+ by protonation, forming positively charged $-OH^{2+}$ and $-COOH^{2+}$, which combine with $HgCl^-$ and $HgClO^-$ through electrostatic effect to remove Hg^{2+}. IC has the highest concentration of strongly basic functional groups, possibly attributed to it being the product of RC pyrolysis and carbonization at 600 °C. Thy et al. [34] showed that after high-temperature pyrolysis, graphitization of biochar occurs via polycondensation of low-temperature aromatization, which removes organic components, resulting in increased alkali metal ions. The pH of IC dropped significantly after Hg^{2+} adsorption, attributed to the fact that Hg^{2+} easily reacts with OH^- to form $Hg_2(OH)_2$ precipitation under alkaline conditions, resulting in decreased concentration of OH^- [35]. Although the concentration of carboxyl and hydroxyl oxygen-containing functional groups in BHC and BCC decreased significantly, the H/C value of BHC and BCC decreased, and the number of lactone groups increased significantly. Dougherty et al. [36] showed that a smaller H/C value represents a lower degree of carbonation and a higher degree of aromatization. They showed that the strength of the cation $-\pi$ effect is mainly determined by the aromaticity degree of biochar surface, i.e., the more abundant the $-\pi$ conjugated aromatic structure, the stronger the electron donating ability of the biochar, and thereby, the more significant the reduction effect. Therefore, BHC and BCC show a strong reduction effect.

Table 1. Physicochemical properties and the number of surface functional groups of biochar (mmol/g).

Biochar	pH_{pzc}	pH	pH after Adsorption	H/C	Carboxyl	Lactone Group	Phenolic Hydroxyl	Acid Functional Groups	Basic Functional Groups
RC	9.3	9.5	7.6	0.036	0.370	0.050	0.125	0.545	0.970
IC	10.4	10.9	8.6	0.314	0.105	0.210	0.080	0.395	1.390
OC	3.3	3.6	3.3	0.043	0.495	0.090	0.355	0.940	0.285
BHC	8.2	8.4	6.4	0.017	0.400	0.190	0.015	0.605	0.905
BCC	8.1	8.3	6.5	0.018	0.180	0.195	0.175	0.550	0.895

RC: raw carbon; IC: inorganic carbon; OC: organic carbon; BHC: hydroxyl-blocked carbon; BCC: carboxyl-blocked carbon.

3.2. FTIR Analysis of Biochar Fractions before and after Hg^{2+} Adsorption

The FTIR spectra of different biochar fractions before and after Hg^{2+} adsorption are shown in Figure 1. A significant characteristic peak at 3431 cm^{-1} was observed for all the five biochar fractions. This was assigned to –OH stretching vibration of alcohol hydroxyl groups and phenolic hydroxyl groups formed via intermolecular hydrogen bonding, indicating the large concentration of –OH contained in the biochar fractions [37]. The broad peak around 1578 cm^{-1} was assigned to the stretching vibration of C=O in the lactone group and C=C in mononuclear aromatic hydrocarbons [38], while the stretching vibration of –COOH resulted in the broad peak observed around 1425 cm^{-1} [39]. Therefore, it can be concluded that biochar contains oxygen-containing functional groups, such as –COOH, –OH, and lactone groups, observations consistent with the results in Table 1.

The absorption peaks weakened after Hg^{2+} adsorption for all the five biochar fractions, with the most significant changes observed for absorption bands at 3431 and 1425 cm^{-1}, indicating that the concentration of –OH and –COOH groups were significantly affected by Hg^{2+} adsorption. The absorbance of OC, RC, and BCC at 3431 cm^{-1} decreased by 0.474, 0.327, and 0.185, respectively, after Hg^{2+} adsorption, indicating that the –OH groups possibly reacted with Hg^{2+}. This is consistent with the perspective proposed by Herrero et al. [40] that the phenolic hydroxyl groups on the adsorbent surface undergo a complexation reaction with Hg^{2+}. At 1425 cm^{-1}, the absorbance of OC, RC, and BHC decreased by 0.178, 0.103, and 0.083, respectively, after Hg^{2+} adsorption. Zhang [41] and Goyal [42] et al. also found that the oxygen-containing functional groups (e.g., –COOH) on the biochar surface reacted with Hg^{2+}, which weakened the vibration signal of –COOH in FTIR.

Before Hg^{2+} adsorption, the concentration of lactone functional groups in BHC and BCC was big (Table 1). After Hg^{2+} adsorption, the absorbance of BHC and BCC at 1578 cm^{-1} decreased by 0.166 and 0.182, respectively, indicating that the C=O in the lactone group and C=C in aromatic hydrocarbons engaged in a reaction with Hg^{2+}. A smaller H/C value represents a more significant reduction effect of cation $-\pi$, as the $-\pi$ electron of the benzene ring can reduce Hg^{2+} to Hg^{+} [36,43]. This also proves that BHC and BCC have a strong reduction effect, in accordance with Table 1.

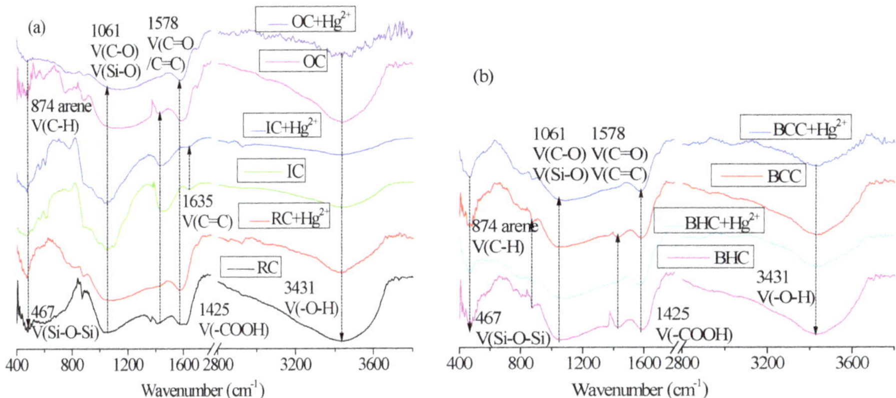

Figure 1. FTIR spectra before and after Hg^{2+} adsorption by biochar (**a**): RC, IC, and OC; (**b**): BHC and BCC.

3.3. XPS Analysis of Biochar Fractions before and after Hg^{2+} Adsorption

XPS characterization was carried out for biochar fractions before and after mercury adsorption, to analyze the valence changes of C, O, and Hg and to further clarify the adsorption mechanisms of mercury by biochar. Table 2 shows the peak area changes of different forms of C and O in the five biochar fractions before and after mercury adsorption. After mercury adsorption by RC and OC, the contents of carboxyl C (288 eV) decreased by 6.4% and 9.5%, and the contents of carboxyl O (534.6 eV) decreased by 9.6% and 13.0%, respectively. Meanwhile, the contents of hydroxyl O (532.4 eV) decreased by 4.7% and 5.6% in RC and OC, respectively. These results are consistent with the FTIR analysis, indicating that the –COOH and –OH functional groups on the biochar surface complexed with Hg^{2+}. The variation of carboxyl and hydroxyl groups on the OC surface is more significant than that on RC surface, attributed to OC being weakly acidic whereas RC is alkaline. Hg^{2+} is more likely to form $Hg(OH)_2$ under alkaline conditions [44], but OC contains more –COOH and –OH functional groups and, therefore, engages more Hg^{2+}. The content of hydroxyl O on BCC surface decreased by 7.1%, while the contents of carboxyl C and O on BHC surface dropped by 5.3% and 6.8%, respectively. This indicates that the –COOH groups of BHC and –OH groups of BCC also contribute greatly to mercury removal. After Hg^{2+} adsorption by IC, BHC, and BCC, the areas of C peaks at 284.7 (C=C) and 531.7 eV (C=O) decreased significantly, and the O peak at 531.7 eV decreased by 3.1%, 13.0%, and 15.9%, respectively. These changes indicate that C=C and C=O are also involved in Hg^{2+} removal, by combining with mercury to form Hg–Cπ and remove Hg^{2+} [45].

Table 2. Elemental binding energy and mass content of different biochar components before and after Hg^{2+} adsorption.

Element	BE(eV) [a]	PA(%) [b] of RC		PA(%) [b] of IC		PA(%) [b] of OC		PA(%) [b] of BHC		PA(%) [b] of BCC	
		BS [c]	AS [d]	BS	AS	BS	AS	BS	AS	BS	AS
C_{1s}(C=C)	284.7	54.0%	49.7%	55.8%	51.2%	50.6%	52.1%	61.4%	55.5%	60.7%	52.8%
C_{1s}(C–O)	285.6	25.7%	36.7%	14.0%	27.1%	12.9%	24.1%	5.7%	25.4%	7.5%	23.6%
C_{1s}(C=O)	287.1	8.0%	7.7%	22.2%	18.6%	16.4%	15.1%	22.1%	13.6%	27.4%	19.7%
C_{1s}(–COO)	288.8	12.3%	5.9%	8.0%	3.1%	18.1%	8.6%	10.8%	5.5%	4.5%	4.0%
O_{1s}(C=O)	531.7	25.5%	33.2%	41.2%	38.1%	27.4%	37.7%	51.5%	38.5%	53.4%	37.5%
O_{1s}(hydroxyl)	532.4	23.5%	18.8%	15.5%	11.7%	25.5%	16.9%	7.4%	7.2%	21.7%	14.6%
O_{1s}(C–O)	533.6	25.2%	31.7%	21.3%	36.9%	17.0%	28.3%	15.1%	35.1%	19.8%	40.9%
O_{1s}(COO)	534.6	25.8%	16.3%	21.0%	13.3%	30.2%	17.2%	26.1%	19.3%	5.1%	5.0%

[a] BE: binding energy, [b] PA: peak area, [c] BS: before sorption of Hg^{2+}, [d] AS: after sorption of Hg^{2+}.

The XPS pattern analysis of Hg_{4f} (Figure 2) confirms that a certain amount of Hg^{2+} was adsorbed on biochar. Obvious $Hg_{4f\,7/2}$ and $Hg_{4f\,5/2}$ peaks were observed for RC, IC, and OC, at the binding energies of 101.50 and 101.80 eV, which are two states of spin orbits with a split value of 4.1 eV (distance between two Hg_{4f} peaks), representing the complexes of $(–COO)_2Hg$ and $(–O)_2Hg$ [46,47]. According to the areas of fitted peaks assigned to $(–COO)_2Hg$ and $(–O)_2Hg$, the complexations by IC, RC, and OC account for 39.8%, 68.9%, and 71.6%, respectively, with the corresponding $(–COO)_2Hg$ taking up 18.0%, 51.0%, and 64.6%, respectively. Therefore, it can be concluded that the main adsorption mechanism of RC and OC is a complexation, dominated by –COOH functional groups. Manivannan [48], Wang [49], and Hyland [50] et al. have discovered that the peaks at the binding energies of 102.5 ($Hg_{4f\,7/2}$) and 106.4 eV ($Hg_{4f\,5/2}$) correspond to Hg^{2+}; the fitted peak at 103.6 eV is attributed to Hg_2Cl_2; and the peak at 99.8 eV is assigned to Hg^0. The fitted peak of Hg^+ at 103.6 eV is present in the XPS spectra of all the five biochar factions, with the fitted peak of Hg^0 at 99.8 eV being observed only for IC (Figure 2). According to FTIR analysis and the significant decrease of the concentration of phenolic hydroxyl groups on biochar (except BHC) after Hg^{2+} adsorption (Figure 1, Table 2), there may be a reduction effect of phenolic hydroxyl groups and $-\pi$ electrons during mercury adsorption, causing the reduction of Hg^{2+} to Hg^+ and Hg^0. Such reduction reactions account for 25.2%, 36.4%, and 13.0% in the Hg^{2+} adsorption by RC, OC, and IC, respectively.

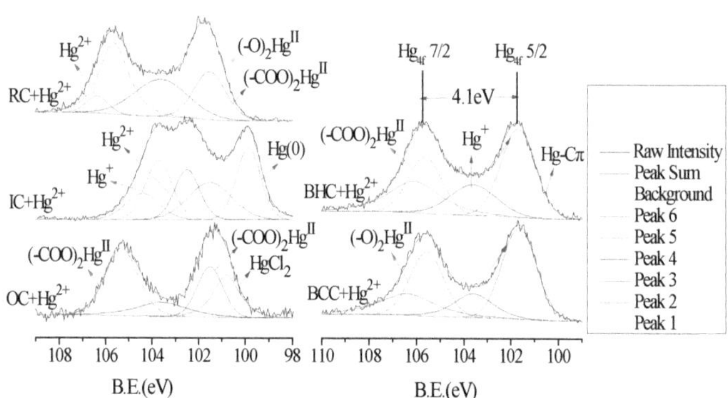

Figure 2. XPS spectra of Hg_{4f} after adsorption of Hg^{2+} by biochars.

According to the above analysis, the fitted peaks at 105.6 and 105.9 eV represent $(–COO)_2Hg$ and $(–O)_2Hg$, respectively, indicating that the complexation reaction accounts for 23.3% and 28.3% in the adsorptions by BHC and BCC, respectively. Xu et al. [51] assigned the fitted peak around 101.7 eV to Hg–Cπ, formed from the reduction of Hg^{2+} to Hg^+ by the $-\pi$ bond and the further combination of Hg^+ with C=C and C=O. Its peak area accounts for 37.1% and 41.4% in the XPS spectra of BHC and BCC, respectively. According to the areas of fitted peaks at 103.6 eV (Hg_2Cl_2) in the XPS spectra of BHC and

BCC, reduction accounts for 17.5% and 13.1%, respectively. Thus, the overall reduction effect of BHC and BCC contribute 54.6% and 54.5%, respectively. Therefore, the main adsorption mechanism of BHC and BCC is by reduction.

3.4. EDS Analysis of Biochar Fractions before and after Hg^{2+} Adsorption and Leaching Experiments

According to the EDS spectra of the five biochar fractions before and after Hg^{2+} adsorption and leaching experiments of all biochar, IC contains the most basic cations. Therefore, only IC analysis results were displayed. As shown in Figure 3a, after Hg^{2+} adsorption by IC, the concentration of metal cations, such as K^+, Na^+, Mg^{2+}, and Al^{2+}, on the biochar decreased, accompanied by the appearance of a relatively obvious mercury peak, indicating that there may be an ion exchange mechanism during the adsorption process. Among the various cations, the most significant reduction was observed for K^+, followed by Na^+, implying that Hg^{2+} is more prone to ion exchange with K^+ and Na^+. As suggested by Figure 3b, the K^+ and Na^+ concentrations gradually increased with time, and the Hg^{2+} concentration simultaneously decreased, until the adsorption equilibrium. The adsorption contribution rates of K^+ and Na^+ (30.6% and 10.3%, respectively) were calculated, and based on the concentration of K^+, Na^+, and Hg^{2+} at the adsorption equilibrium, the adsorption contribution rate of ion exchange was 40.9%. According to studies carried out by Kílıç [19] and Carro et al. [52], K^+ and Na^+ are more prone to ion exchange with Hg^+. The contents of Hg^+ and Hg^{2+} determined by XPS analysis are 17.3% and 23.9%, respectively. Hence, in the total ion exchange of 40.9%, Hg^+ and Hg^{2+} take up 17.3% and 23.6%, respectively. Associating with XPS analysis, the complexation and reduction reactions during Hg^{2+} adsorption by IC account for 39.8% and 36.4%, respectively. Therefore, the main mechanism for Hg^{2+} adsorption by IC is ion exchange. In addition, the adsorption contribution rates of ion exchange in the removal of Hg^{2+} (40.9%) by IC is higher in comparison with other adsorbents such as 31.7% by activated sludge biomass [19] and 35.0% by algal biomass [52].

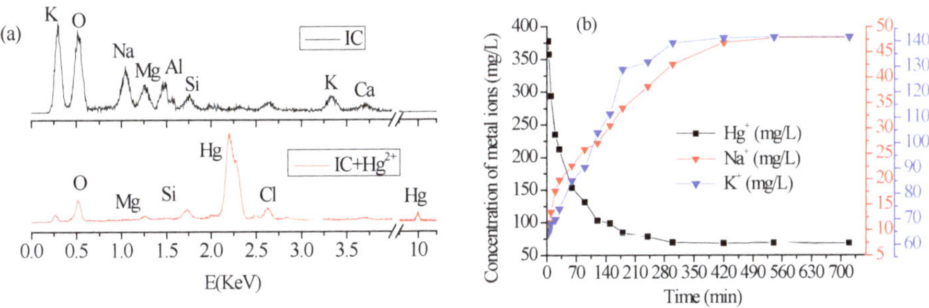

Figure 3. (**a**) EDS analysis before and after adsorption of Hg^{2+} by IC; (**b**) The concentration change of K^+, Na^+, and Hg^{2+} with oscillation time after adsorption of Hg^{2+} by IC.

3.5. Adsorption Mechanisms

All five biochar fractions have negatively charged surfaces and can adsorb positively charged Hg^{2+} via, for example, electrostatic effects (Table 1). According to the above analysis, the main adsorption mechanisms and contribution rates of Hg^{2+} adsorption by RC and its fractions were determined. Based on Table 3, the equilibrium adsorption amount of Hg^{2+} by RC is 75.56 mg/g. As calculated based on the mass fractions of IC and OC, the theoretical adsorption capacity of RC is 75.52 mg/g, which is very close to the measured value of 75.56 mg/g, indicating that the adsorption contribution rates of inorganic and organic components in RC account for approximately 22.4% and 77.6%, respectively. The main adsorption mechanism of RC and OC in the Hg^{2+} adsorption process is based on a complexation reaction, where oxygen-containing functional groups, such as –COOH and –OH groups, react with Hg^{2+} to form $(–COO)_2Hg$ and $(–O)_2Hg$ complexes. The complexation effects of RC and OC contribute

68.9% and 71.6%, respectively. In the complexation reaction of RC, the contribution rates of the adsorption by –COOH and –OH groups are 51.0% and 17.9%, respectively. The reaction formulae are as follows [17]:

$$2\,(\text{-COOH}) + Hg^{2+} \rightarrow (\text{-COO})_2\text{-Hg} + 2H^+, \quad (1)$$

$$2\,(\text{-OH}) + Hg^{2+} \rightarrow (\text{-O})_2\text{-Hg} + 2H^+. \quad (2)$$

IC has the maximum equilibrium capacity for Hg^{2+} adsorption of 92.63 mg/g. This may be due to the reduction of Hg^{2+} to Hg^+ and Hg^0 by $-\pi$ electrons and phenolic hydroxyl groups (36.4%), the complexation reaction between carboxyl/hydroxyl functional groups and Hg^{2+} (39.8%), and the ion exchange of Hg^{2+} and Hg^+ with K^+ and Na^+ (40.9%) occurring simultaneously. The main adsorption mechanism of IC is ion exchange, during which the ion exchange of Hg^+ and Hg^{2+} contributes 17.3% and 23.6%, respectively.

As shown in Table 3, the equilibrium capacities for Hg^{2+} adsorption by BHC and BCC are significantly lower compared to IC, being 66.30 and 61.13 mg/g, respectively. In addition, the calculated theoretical adsorption capacities of BHC and BCC, based on their mass fractions, are 65.96 and 61.43 mg/g, respectively, which are close to the measured values. The $-\pi$ electrons and phenolic hydroxyl groups on the surfaces of BHC and BCC reduce Hg^{2+} to Hg^+, and total reduction ratios take up 54.6% and 54.5%, respectively. According to the adsorption capacities of BHC and BCC, the adsorption contribution rates of the –COOH groups on BHC surface and the –OH functional groups on BCC surface are 20.4% and 12.3%, respectively. As for RC, –COOH and –OH groups contribute 51.0% and 17.9% respectively, to the total adsorption. The reason for this difference may be attributed to the presence of some –COOH and –OH functional groups remaining after the blocking of the biochar functional groups (Table 1).

Table 3. Contribution rate (%) and adsorption quantity of Hg^{2+} (mg/g) at equilibrium by biochars.

Biochars	Mass Percentage (%)	Adsorption Quantity of Hg^{2+} (mg/g)	Adsorption Mechanism	Adsorption Contribution Rate (%)
RC	100.0	75.56	Complexation with –COOH and –OH	68.9
			Reduction reaction	25.2
IC	22.4	92.63	Complexation with –COOH and –OH	39.8
			Reduction reaction	36.4
			Cation exchange	41.2
OC	77.6	70.57	Complexation with –COOH and –OH	71.6
			Reduction reaction	13.0
BHC	87.3	66.30	Complexation with –COOH	23.3
			Reduction reaction	54.6
BCC	81.3	61.13	Complexation with –OH	28.3
			Reduction reaction	54.5

The adsorption mechanisms of Hg^{2+} removal by RC include electrostatic adsorption, ion exchange, reduction, precipitation, and complexation. Among all adsorption mechanisms, the complexation reaction of –COOH/–OH groups with Hg^{2+} plays the dominating roles, and the adsorption contribution rate of –COOH functional groups is greater than that of –OH functional groups.

The adsorption capacity of Hg^{2+} by corn-straw-based biochar (75.56 mg/g) was found to be higher than some values reported in literature, such as 35.71 mg/g by sugarcane bagasse [13], 22.82 mg/g by peanut husk biochar [15], 19.30 mg/g by activated sludge biomass [19], and 24.20 mg/g by Brazilian pepper biochars [20]. It is, however, similar to the value determined for Aspergillus versicolor biomass (75.60 mg/g) [18], and slightly lower than the adsorption capacities reported for flax shive sorbent (89.50 mg/g) [9] and fruit shell of Terminalia catappa (82.93 mg/g) [11]. Moreover, the irreversible complexation reaction of –COOH/–OH groups with Hg^{2+} is the main adsorption mechanism, which reduces the secondary pollution to aqueous solution even depositing it to the water bottom. Therefore, it is helpful to provide theoretical basis and data support for the special adsorption of functional adsorbents and has practical significance for the treatment of mercury pollution in water. In addition, by studying the adsorption mechanisms of biochar different fractions for Hg^{2+}, the contribution rate of each fraction of the adsorbent to Hg^{2+} was calculated, providing a certain theoretical basis for the research on the removal of Hg from biochar.

4. Conclusions

Except for OC, which is weakly acidic, the other four biochar fractions are basic (IC has the strongest basicity), and their pH values are significantly lower after Hg^{2+} adsorption. In addition, the pH values of all five biochar fractions are greater than their corresponding pH_{pzc} values. The surfaces of the biochar fractions are negatively charged and adsorb positively charged Hg^{2+}.

The equilibrium adsorption capacity of RC for Hg^{2+} is 75.56 mg/g, with IC and OC contributing 22.4% and 77.6%, respectively. IC has the largest equilibrium adsorption capacity of 92.63 mg/g, which may be due to all the adsorption mechanisms involved in the adsorption process. Moreover, the equilibrium adsorption capacities of BHC and BCC are significantly lower, being 66.30 and 61.13 mg/g, respectively.

There are five reaction mechanisms identified for Hg^{2+} removal by biochar: electrostatic adsorption, cation exchange, precipitation, reduction effect of $-\pi$ and phenolic hydroxyl groups, and complexation by –OH and –COOH groups. The main adsorption mechanism of RC and OC is the complexation of oxygen-containing functional groups with Hg^{2+}, accounting for 68.9% and 71.6%, respectively. The contribution rates of –COOH and –OH groups in RC complexation reaction are 51.0% and 17.9%, respectively, while those in OC complexation reaction are 64.6% and 7.0%, respectively. Among the five examined fractions, IC plays a major role in mercury adsorption by RC. All adsorption mechanisms identified are involved in the adsorption process by IC, with ion exchange being the main one, accounting for 39.8%. The reduction effect of phenolic hydroxyl groups and $-\pi$ electrons is the main mechanism for Hg^{2+} removal by BHC and BCC, taking up 54.6% and 54.5% of the total adsorption for each fragment, respectively. In addition, the adsorption contribution rates of –COOH groups on BHC surface and –OH functional groups on BCC surface are 20.4% and 12.3%, respectively. Among all the adsorption mechanisms, the complexation reaction of –COOH/–OH groups with Hg^{2+} plays the dominating roles, and the adsorption contribution rate of –COOH functional groups is greater than that of –OH functional groups. These conclusions play a guiding role in the remediation mechanism of heavy metals by biochar in water. The prepared RC could be a promising low-cost adsorbent for the treatment of Hg^{2+}-contaminated water.

Author Contributions: Methodology, M.L.; data curation and writing—original draft preparation, X.G.; supervision, A.L.; funding acquisition, M.J. and X.N.; formal analysis, X.L. All authors have read and agreed to the published version of the manuscript.

Funding: This research was funded by the National Natural Science Foundation of China (41771348, 41877122, 41671322, 41703099); Doctor of Natural Science Foundation of Shandong province, China (ZR2018BEE015, ZR2019BB045); Youth Science Foundation of the National Natural Science Foundation of China (51804188).

Acknowledgments: The authors are also thankful to the analytical testing center of Shandong University of Technology.

Conflicts of Interest: The authors declare no conflict of interest.

References

1. Lyu, H.H.; Xia, S.Y.; Tang, J.C.; Zhang, Y.R.; Cao, B.; Shen, B.X. Thiol-Modified biochar synthesized by a facile ball-milling method for enhanced sorption of inorganic Hg^{2+} and organic CH_3Hg^+. *J. Haz. Mater.* **2019**, *384*, 121357. [CrossRef] [PubMed]
2. Inbaraj, B.S.; Wang, J.; Lu, J.; Siao, F.; Chen, B. Adsorption of toxic mercury(II) by an extracellular biopolymer poly(γ-glutamic acid). *Bioresour. Technol.* **2009**, *100*, 200–207. [CrossRef]

3. Zhu, Y.C.; Peng, S.C.; Lu, P.P.; Chen, T.H.; Yang, Y. Mercury Removal from Aqueous Solutions Using Modified Pyrite: A Column Experiment. *Minerals* **2019**, *10*, 43. [CrossRef]
4. Mahmudov, R.; Huang, C.P. Selective adsorption of oxyanions on activated carbon exemplified by Filtrasorb 400 (F400). *Sep. Purif. Technol.* **2011**, *77*, 294–300. [CrossRef]
5. Wang, T.; Wu, J.W.; Zhang, Y.S.; Liu, J.; Sui, Z.F.; Zhang, H.C.; Chen, W.Y.; Norris, P.; Pan, W.P. Increasing the chlorine active sites in the micropores of biochar for improved mercury adsorption. *Fuel* **2018**, *229*, 60–67. [CrossRef]
6. Liu, Z.Y.; Yang, W.; Xu, W.; Liu, Y.X. Removal of elemental mercury by bio-chars derived from seaweed impregnated with potassium iodine. *Environ. Sci. Health* **2018**, *339*, 468–478. [CrossRef]
7. Xu, Y.L. *Thermodynamic Properties of Biochar Preparation and Sorption Characteristics and Mechanisms of Cadmium onto Biochars*; Zhejiang University: Hangzhou, China, 2013.
8. Zabihi, M.; Ahmadpour, A.; Haghighi, A.A. Removal of mercury from water by carbonaceous sorbents derived from walnut shell. *J. Hazard. Mater.* **2008**, *167*, 230–236. [CrossRef]
9. Cox, M.; El-Shafey, E.I.; Pichugin, A.A.; Appleton, Q. Removal of mercury (II) from aqueous solution on a carbonaceous sorbent prepared from flax shive. *J. Chem. Technol. Biotechnol.* **2000**, *75*, 427–435. [CrossRef]
10. Minaxi, B.; Lohani Amarika, S.; Rupainwar, D.C.; Dhar, D.N. Studies on efficiency of guava (Psidium guajava) bark as bioadsorbent for removal of Hg(II) from aqueous solutions. *J. Hazard. Mater.* **2008**, *159*, 626–629. [CrossRef]
11. Stephen Inbaraj, B.; Sulochana, N. Mercury adsorption on a carbon sorbent derived from fruit shell of Terminalia catappa. *J. Hazard. Mater.* **2005**, *133*, 283–290. [CrossRef]
12. El-Shafey, E.I. Removal of Zn(II) and Hg(II) from aqueous solution on a carbonaceous sorbent chemically prepared from rice husk. *J. Hazard. Mater.* **2009**, *175*, 319–327. [CrossRef] [PubMed]
13. Khoramzadeh, E.; Nasernejad, B.; Halladj, R. Mercury biosorption from aqueous solutions by Sugarcane Bagasse. *J. Taiwan Inst. Chem. Eng.* **2013**, *44*, 266–269. [CrossRef]
14. Johari, K.; Saman, N.; Song, S.T.; Chin, C.S.; Kong, H.; Mat, H. Adsorption enhancement of elemental mercury by various surface modified coconut husk as eco-friendly low-cost adsorbents. *Int. Biodeter. Biodegr.* **2016**, *109*, 45–52. [CrossRef]
15. Xue, Y.W.; Gao, B.; Yao, Y.; Inyang, M.D.; Zhang, M.; Zimmerman, A.R.; Ro, K.S. Hydrogen peroxide modification enhances the ability of biochar (hydrochar) produced from hydrothermal carbonization of peanut hull to remove aqueous heavy metals: Batch and column tests. *Chem. Eng. J.* **2012**, *200*, 673–680. [CrossRef]
16. Li, H.B.; Dong, X.L.; da Silva, E.B.; de Oliveira, L.M.; Chen, Y.; Ma, L.Q. Mechanisms of metal sorption by biochars: Biochar characteristics and modifications. *Chemosphere* **2017**, *178*, 466–478. [CrossRef]
17. Kong, H.; He, J.; Gao, Y.; Wu, H.; Zhu, X. Cosorption of phenanthrene and mercury(II) from aqueous solution by soybean stalk-based biochar. *J. Agric. Food. Chem.* **2011**, *59*, 12116–12123. [CrossRef]
18. Das, S.K.; Das, A.R.; Guha, A.K. A study on the adsorption mechanism of mercury on Aspergillus versicolor biomass. *Environ. Sci. Technol.* **2007**, *41*, 8281–8287. [CrossRef]
19. Kílıç, M.; Keskin, M.E.; Mazlum, S.; Mazlum, N. Hg(II) and Pb(II) adsorption on activated sludge biomass: Effective biosorption mechanism. *Int. J. Miner. Process.* **2008**, *87*, 1–8. [CrossRef]
20. Dong, X.L.; Ma, L.Q.; Zhu, Y.J.; Li, Y.C.; Gu, B. Mechanistic investigation of mercury sorption by Brazilian pepper biochars of different pyrolytic temperatures based on X-ray photoelectron spectroscopy and flow calorimetry. *Environ. Sci. Technol.* **2013**, *47*, 12156–12164. [CrossRef]
21. Gardea-Torresdey, J.L.; Becker-Hapak, M.K.; Hosea, J.M.; Darnall, D.W. Effect of chemical modification of algal carboxyl groups on metal ion binding. *Environ. Sci. Technol.* **1990**, *24*, 1372–1378. [CrossRef]
22. Chen, J.P.; Yang, L. Study of a heavy metal biosorption onto raw and chemically modified Sargassum sp. via spectroscopic and modeling analysis. *Langmuir* **2006**, *22*, 8906–8914. [CrossRef] [PubMed]
23. Lu, H.L.; Zhang, W.H.; Yang, Y.X.; Huang, X.; Wang, S.; Qiu, R. Relative distribution of Pb^{2+} sorption mechanisms by sludge-derived biochar. *Water Res.* **2012**, *46*, 854–862. [CrossRef] [PubMed]
24. Wang, Z.Y.; Liu, G.C.; Zheng, H.; Li, F.M.; Ngo, H.H.; Guo, W.S.; Liu, C.; Chen, L.; Xing, B.S. Investigating the mechanisms of biochar's removal of lead from solution. *Bioresour. Technol.* **2015**, *177*, 308–317. [CrossRef] [PubMed]
25. Ofomaja, A.E.; Naidoo, E.B.; Modise, S.J. Removal of copper(II) from aqueous solution by pine and base modified pine cone powder as biosorbent. *J. Hazard. Mater.* **2009**, *168*, 909–917. [CrossRef]

26. Boehm, H.P. Some aspects of the surface chemistry of carbon blacks and other carbons. *Carbon* **1994**, *32*, 759–769. [CrossRef]
27. Goertzen, S.L.; Thériault, K.D.; Oickle, A.M.; Tarasuk, A.C.; Andreas, H.A. Standardization of the Boehm titration. Part, I. CO_2 expulsion and endpoint determination. *Carbon* **2010**, *48*, 1252–1261. [CrossRef]
28. Lehmann, J. A handful of carbon. *Nature* **2007**, *447*, 143–144. [CrossRef] [PubMed]
29. Shaaban, A.; Se, S.M.; Dimin, M.F.; Juoi, J.M.; Mohd Husin, M.H.; Mitan, N.M.M. Influence of heating temperature and holding time on biochars derived from rubber wood sawdust via slow pyrolysis. *J. Anal. Appl. Pyrol.* **2014**, *107*, 31–39. [CrossRef]
30. Sari, A.; Tuzen, M. Removal of mercury (II) from aqueous solution using moss (Drepanocladus revolvens) biomass: Equilibrium, thermodynamic and kinetic studies. *J. Hazard. Mater.* **2009**, *171*, 500–507. [CrossRef]
31. Jia, M.Y.; Wang, F.; Bian, Y.R.; Jin, X.; Song, Y.; Kengara, F.O.; Xu, R.K.; Jiang, X. Effects of pH and metal ions on oxytetracycline sorption to maize-straw-derived biochar. *Bioresour. Technol.* **2013**, *136*, 87–93. [CrossRef]
32. Tüzün, İ.; Bayramoğlu, G.; Yalçın, E.; Başaran, G.; Celik, G.; Arica, M.Y. Equilibrium and kinetic studies on biosorption of Hg(II), Cd(II) and Pb(II) ions onto microalgae Chlamydomonas reinhardtii. *J. Environ. Manag.* **2005**, *77*, 85–92. [CrossRef]
33. Walcarius, A.; Delacôte, C. Mercury(II) binding to thiol-functionalized mesoporoussilicas: Critical effect of pH and sorbent properties on capacity and selectivity. *Anal. Chim. Acta* **2005**, *547*, 3–13. [CrossRef]
34. Thy, P.; Jenkins, B.M.; Grundvig, S.; Shiraki, R.; Lesher, C. High temperature elemental losses and mineralogical changes in common biomass ashes. *Fuel* **2006**, *85*, 783–795. [CrossRef]
35. Yardim, M.F.; Budinova, T.; Ekinci, E.; Petrov, N.; Razvigorova, M.; Minkova, V. Removal of mercury(II) from aqueous solution by activated carbon obtained from furfural. *Chemosphere* **2003**, *52*, 835–841. [CrossRef]
36. Dougherty, D.A. The cation -π interaction. *Acc. Chem. Res.* **2012**, *46*, 885–893. [CrossRef] [PubMed]
37. Tang, L.; Yang, G.D.; Zeng, G.M.; Cai, Y.; Li, S.S.; Zhou, Y.Y.; Pang, Y.; Liu, Y.Y.; Zhang, Y.; Luna, B. Synergistic effect of iron doped ordered mesoporous carbon on adsorption-coupled reduction of hexavalent chromium and the relative mechanism study. *Chem. Eng. J.* **2014**, *239*, 114–122. [CrossRef]
38. Chen, T.; Zhang, Y.; Wang, H.; Lu, W.; Zhou, Z.; Zhang, Y.; Ren, L. Influence of pyrolysis temperature on characteristics and heavy metal adsorptive performance of biochar derived from municipal sewage sludge. *Bioresour. Technol.* **2014**, *164*, 47–54. [CrossRef] [PubMed]
39. Lammers, K.; Arbuckle-Keil, G.; Dighton, J. FT-IR study of the changes in carbohydrate chemistry of three New Jersey pine barrens leaf litters during simulated control burning. *Soil Biol. Biochem.* **2009**, *41*, 340–347. [CrossRef]
40. Herrero, R.; Lodeiro, P.; Rey-Castro, C.; Vilariño, T.; Sastre de Vicente, M.E. Removal of inorganic mercury from aqueous solutions by biomass of the marine macroalga Cystoseira baccata. *Water Res.* **2005**, *39*, 3199–3210. [CrossRef]
41. Zhang, S.; Yang, X.; Ju, M.T.; Liu, L.; Zheng, K. Mercury adsorption to aged biochar and its management in China. *Environ. Sci. Pollut. Res.* **2019**, *26*, 4867–4877. [CrossRef]
42. Goyal, M.; Bhagat, M.; Dhawan, R. Removal of mercury from water by fixed bed activated carbon columns. *J. Hazard. Mater.* **2009**, *171*, 1009–1015. [CrossRef] [PubMed]
43. Carro, L.; Anagnostopoulos, V.; Lodeiro, P.; Barriada, J.L.; Herrero, R.; Sastre de Vicente, M.E. A dynamic proof of mercury elimination from solution through a combined sorption-reduction process. *Bioresour. Technol.* **2010**, *101*, 8969–8974. [CrossRef] [PubMed]
44. Pulido-Novicio, L.; Kurimoto, Y.; Aoyama, M.; Seki, K.; Doi, S.; Hata, T.; Ishihara, S.; Imamura, Y. Adsorption of mercury by sugi wood carbonized at 1000 °C. *J. Wood Sci.* **2001**, *47*, 159–162. [CrossRef]
45. Park, J.H.; Wang, J.J.; Zhou, B.; Mikhael, J.E.R.; DeLaune, R.D. Removing mercury from aqueous solution using sulfurized biochar and associated mechanisms. *Environ. Pollut.* **2018**, *244*, 627–635. [CrossRef]
46. Bond, A.M.; Miao, W.; Raston, C.L. Mercury(II) immobilized on carbon nanotubes: Synthesis, characterization, and redox properties. *Langmuir* **2000**, *16*, 6004–6012. [CrossRef]
47. Behra, P.; Bonnissel-Gissinger, P.; Alnot, M.; Revel, R.; Ehrhardt, J.J. XPS and XAS study of the sorption of Hg(II) onto pyrite. *Langmuir* **2001**, *17*, 3970–3979. [CrossRef]
48. Manivannan, S.; Park, S.; Jeong, J.; Kim, K. Aggregation-Free optical and colorimetric detection of Hg(II) with M13 bacteriophage-templated Au nanowires. *Biosens. Bioelectron.* **2020**, *161*, 112237. [CrossRef] [PubMed]
49. Wang, J.; Deng, B.L.; Chen, H.; Wang, X.R.; Zheng, J.Z. Removal of aqueous Hg(II) by polyaniline: Sorption characteristics and mechanisms. *Environ. Sci. Technol.* **2009**, *43*, 5223–5228. [CrossRef]

50. Hyland, M.M.; Jean, G.E.; Bancroft, G.M. XPS and AES studies of Hg(II) sorption and desorption reactions on sulphide minerals. *Pergamon* **1990**, *54*, 1957–1967. [CrossRef]
51. Xu, X.Y.; Ariette, S.; Xu, N.; Cao, X.D. Comparison of the characteristics and mechanisms of Hg(II) sorption by biochars and activated carbon. *J. Colloid Interf. Sci.* **2016**, *463*, 55–60. [CrossRef]
52. Carro, L.; Barriada, J.L.; Herrero, R.; Sastre de Vicente, M.E. Adsorptive behavior of mercury on algal biomass: Competition with divalent cations and organic compounds. *J. Hazard. Mater.* **2011**, *192*, 284–291. [CrossRef] [PubMed]

Article

Removal Efficiencies of Manganese and Iron Using Pristine and Phosphoric Acid Pre-Treated Biochars Made from Banana Peels

Hyunjoon Kim [1,†], Ryun-Ah Ko [1,†], Sungyun Lee [2,*] and Kangmin Chon [1,*]

1 Department of Environmental Engineering, College of Engineering, Kangwon National University, Kangwondaehak-gil, 1, Chuncheon-si, Gangwon-do 24341, Korea; guswns863@naver.com (H.K.); rh6468@naver.com (R.-A.K.)

2 Department of Civil Environmental Engineering, School of Disaster Prevention and Environmental Engineering, Kyungpook National University, 2559, Gyeongsang-daero, Sangju-si, Gyeongsangbuk-do 37224, Korea

* Correspondence: sungyunlee@knu.ac.kr (S.L.); kmchon@kangwon.ac.kr (K.C.); Tel.: +82-54-530-1446 (S.L.); +82-33-250-6352 (K.C.)

† These authors contributed equally to this work and should be considered as co-first authors.

Received: 2 March 2020; Accepted: 15 April 2020; Published: 20 April 2020

Abstract: The purpose of this study was to compare the removal efficiencies of manganese (Mn) and iron (Fe) using pristine banana peel biochar (BPB) and phosphoric acid pre-treated biochars (PBPB) derived from banana peels. The removal efficiencies of Mn and Fe were investigated under different adsorbent dosages (0.4–2 g L^{-1}), temperatures (15–45 °C), and ionic strengths (0–0.1 M), and were directly correlated to the differences in physicochemical properties of BPB and PBPB, to identify the removal mechanisms of heavy metals by adsorption processes. The removal of Mn by PBPB obeyed the Freundlich isotherm model while the removal of Mn and Fe by BPB followed the Langmuir isotherm model. However, the removal of Fe by PBPB followed both Freundlich and Langmuir isotherm models. The removal efficiencies of Mn and Fe by BPB and PBPB increased with increasing temperatures and decreased with increasing ionic strengths. PBPB more effectively removed Mn and Fe compared to BPB due to its higher content of oxygen-containing functional groups (O/C ratio of PBPB = 0.45; O/C ratio of BPB = 0.01), higher surface area (PBPB = 27.41 m^2 g^{-1}; BPB = 11.32 m^2 g^{-1}), and slightly greater pore volume (PBPB = 0.03 cm^3 g^{-1}; BPB = 0.027 cm^3 g^{-1}). These observations clearly show that phosphoric acid pre-treatment can improve the physicochemical properties of biochar prepared from banana peels, which is closely related to the removal of heavy metals by adsorption processes.

Keywords: biochar; manganese; iron; banana peel; adsorption; pre-treatment; phosphoric acid

---

1. Introduction

Water pollution by heavy metals released from various industrial activities such as metal plating and cleaning, mining, refineries, coatings, batteries, and automobile radiators, is an emerging environmental issue in water treatment engineering since heavy metals may pose adverse effects on human health and aquatic ecosystems due to their high toxicity, carcinogenicity, and non-biodegradability [1,2]. Among various heavy metals used in industrial activities, manganese (Mn) and iron (Fe) are known to be major inorganic pollutants affecting water quality [3]. Although Mn is essential to activate enzymes in the human system, high Mn concentrations can generate respiratory diseases, and continuous administration can cause neurotoxicity risk in humans [3,4]. In the case of Fe, it can cause undesirable aesthetic concerns (i.e., metallic tastes) and lead to the growth of ferrobacteria related to odor problems [3]. The World Health Organization (WHO) have set a safe drinking water concentration of 0.05 and 0.3 mg L^{-1} for Mn and Fe, respectively [5]. Based on these reasons, there is great need to develop an economic and efficient method for removing heavy metals from wastewater.

Several treatment techniques, including membrane filtration (e.g., reverse osmosis and nanofiltration), chemical precipitation, and oxidation/reduction, are available for the removal of heavy metals, including Mn and Fe, from industrial wastewater [6–10]. A commonly practiced Mn and Fe treatment approach is to chemically oxidize dissolved Mn(II) to particulate Mn(IV) or dissolved Fe(II) to particulate Fe(III), followed by physical separation of the insoluble precipitates from water using clarification and filtration processes [9,11]. However, most of those are not applicable for wastewater treatment due to their low removal efficiencies when the heavy metal concentrations are lower than 100 mg L^{-1} [12]. In contrast, activated carbon may effectively remove heavy metals from wastewater even at low concentrations [13]. Despite this advantage, the use of the activated carbon adsorption process for the removal of heavy metals has been limited as it requires high maintenance and operational costs [14]. In recent years, biochars, which can be produced at low cost, have attracted great attention as an alternative to activated carbon [15–18]. Biochar is an ecofriendly adsorbent produced using by-products of the agricultural industries and wastes from various crops, and is effective for removing heavy metals from wastewater [18–21].

Over the wide range of crops, bananas cultivated in more than 130 countries are regarded to be one of the most widely grown tropical fruits in the world [22]. The world production of bananas was approximately 117.9 million tons in 2015 [23], and about 7 million tons of banana peel wastes are produced annually (the proportion of banana peels in total dry weight = 25–30%) [24]. Currently, most banana peel wastes are used as natural fertilizers on soils in agricultural fields, and some of them are fed to animals [25]. Banana peels contain a large amount of pectins which are complex polysaccharides consisting of galacturonic acids, arabinoses, galactoses, and rhamnoses. Among them, galacturonic acids have a strong binding capacity to the metal cations in the aqueous phases due to the presence of carboxyl groups [22,26]. Therefore, biochar derived from banana peels is considered to be a promising option for removing heavy metals effectively from wastewater.

Raw biochars have showed feasibility for adsorbent material to remove contaminants including heavy metals and organic pollutants [17,18,21,27,28]. However, the sorption capacities can be enhanced by treatment with acids, nanocomposites, and activation agents [17,29,30]. For example, Chu et al. showed that phosphoric acid treatment improved the porosity of biochars from pine sawdust, cellulose, and lignin [31]. The modified biochars provided better sorption of carbamazepine and bisphenol A. The sorption for 15 different pesticides by biochars from rice straw and corn stover was also increased by phosphoric acid treatment due to increased functional groups and aromatization of the biochars [32]. Considering the effects of phosphoric acid treatment on porosity and the modification of functional groups of biochars, the treatment method can be applied to improve removal efficiencies of heavy metals by biochars from other sources. In addition, phosphoric acid has advantages in low pyrolysis temperature, low cost, and low corrosivity to the equipment [31]. Nevertheless, to the best of our knowledge, phosphoric acid pre-treatment has not been used to enhance the physicochemical properties of biochars from banana peels in association with the adsorption of heavy metals.

The main purpose of this study was to evaluate the effects of phosphate pre-treatment on the adsorption of heavy metals (i.e., Mn and Fe) using biochars prepared from banana peels. First, the physicochemical characteristics of pristine and phosphoric acid pre-treated biochars derived from banana peels were rigorously characterized. Then, various adsorption experiments were conducted to investigate optimum adsorbent dosages, adsorption kinetics, adsorption isotherms, and effects of temperature and ionic strengths on adsorption. The improved adsorption efficiencies by phosphoric acid treatment were analyzed based on the physicochemical characteristics of the biochar, and the adsorption mechanisms of Mn and Fe were discussed. This study improves our understanding of the effect of phosphoric acid treatment on modification of the surface structure and functional groups of biochars for heavy metal removal.

2. Materials and Methods

2.1. Materials

Banana peel wastes were collected in Chuncheon-si (Gangwon-do, Korea). Mn ($KMnO_4$, oxidation state = +7, concentration = 1000 mg L^{-1}) and Fe ($Fe(NH_4)_2{\cdot}(SO_4)_2$, oxidation state = +2, concentration = 1000 mg L^{-1}) AA standard solutions and phosphoric acid (H_3PO_4), were purchased from Daejung Chemicals & Metals (Siheung-Si, Gyeonggi-Do, Korea). Mn(VII) and Fe(II) solutions for adsorption experiments were prepared by diluting the concentrated standard solutions in deionized water.

2.2. Production of Biochar

The banana peel wastes were dried in an oven at 105 °C for 24 h, ground using a mortar, washed several times with deionized (DI) water to remove impurities on their surfaces, and then dried in the oven at 105 °C for 12 h. From the resulting banana peel powder, 50 g was immersed in a 500 mL phosphoric acid solution (20 wt. %) for 2 h to activate adsorption sites, and dried at 105 °C for 12 h. Pristine and phosphoric acid pre-treated banana peel wastes (weight of each banana peel waste = 20 g) were pyrolyzed in the tubular furnace (PyroTech, Namyangju, Gyeonggi-do, Korea) at 600 °C (heating rates = 0.2 °C min^{-1}) under N_2 conditions (N_2 flow rate = 0.25 L min^{-1}) for 2 h and then cooled to room temperature. Biochars prepared from banana peel wastes were washed several times with DI water, filtered with a 0.7 μm GF/F filter (Whatman, Maidstone, UK), and then dried in the oven at 105 °C for 12 h. Biochars produced from pristine and phosphoric acid pre-treated banana peel wastes are defined as BPB (banana peel biochar without pre-treatment) and PBPB (pre-treated banana peel biochar), respectively.

2.3. Characterization of Biochar

Element composition of BPB and PBPB was analyzed using an elemental analyzer (EuroEA3000 CHNS-O, Euro Vector S.p.A, Via Tortona, Milan, Italy). The ash content was calculated by subtracting the quantities of carbon (C), hydrogen (H), nitrogen (N), and oxygen (O) from the total mass fraction of the adsorbents. The specific surface area of biochar was determined with a Bronauer–Emmett–Teller (BET) analyzer (BELSORP-mini II, MicrotracBEL, Japan). The functional group composition of the adsorbents was identified using attenuated total reflectance-Fourier transform infrared spectroscopy (ATR-FTIR) (Frontier Optica, Perkin Elmer, Waltham, MA, USA).

2.4. Adsorption Experiments

2.4.1. Optimal Adsorbent Dosages

Prior to the adsorption kinetics experiments, the optimum dosage of BPB and PBPB for each heavy metal was determined. The adsorbents (dosage = 0.02–4 g L^{-1}) were added to 25 mL of heavy metal solution (each metal concentration = 10 mg L^{-1}, pH = 7.0) and then mixed at 150 rpm and 25 °C using a shaking incubator (VS-8480, Vision Scientific, Daejeon-Si, Korea) for 3 h. All the adsorption tests were repeated three times to minimize experimental errors.

2.4.2. Adsorption Kinetics Analysis

For adsorption kinetics experiments, the optimum dosage of each adsorbent was added to 25 mL of sample solutions (each heavy metal concentration = 10 mg L^{-1}) and mixed at 150 rpm using the shaking incubator for 0–24 h (temperature = 25 °C, pH = 7.0). The concentrations of Mn and Fe at the initial and equilibrium states were measured using colorimetric methods with a UV-Vis spectrophotometer (UV-1280, Shimadzu, Kyoto, Japan) at UV absorbances of 525 and 510 nm, respectively. All the adsorption tests were repeated three times to minimize experimental errors. The amount of adsorbed heavy metals at time t (q_t (mg g^{-1})) was calculated as follows [14]:

$$q_t = \frac{(C_0 - C)V}{m} \tag{1}$$

where C_0 and C are the initial and final concentrations (mg L^{-1}) of heavy metals in the solutions, V is the volume (L) of the solution, and m is the mass (g) of the used adsorbents.

The removal efficiencies of heavy metals were calculated using Equation (2) [14]:

$$\text{Removal efficiency of heavy metal } (\%) = \frac{(C_0 - C_e)}{C_0} \times 100 \tag{2}$$

where C_e represents the concentrations of each heavy metal (mg L^{-1}) at equilibrium of the solutions.

The adsorption kinetics of Mn and Fe were investigated using Equations (3) and (4):

$$\text{Pseudo-first-order model}: \quad \log(q_e - q_t) = \log(q_e) - \frac{k_1 t}{2.303} \tag{3}$$

$$\text{Pseudo-second-order model}: \quad \frac{t}{q_t} = \frac{1}{k_2 q_e^2} + \frac{t}{q_e} \tag{4}$$

where k_1 (min^{-1}) is the constant of the pseudo-first-order equation, and k_2 (g mg^{-1} min^{-1}) is the constant of the pseudo-second-order equation. q_e (mg g^{-1}) is the adsorption capacity at equilibrium.

2.4.3. Adsorption Isotherm Analysis

The adsorption isotherms of Mn and Fe by BPB and PBPB were identified with 6 different initial concentrations (each heavy metal concentration = 1–10 mg L^{-1}) under controlled conditions (agitation time = 24 h, mixing speed = 150 rpm, temperature = 25 °C, pH = 7.0). The adsorption results were analyzed using the Langmuir and Freundlich isotherm models.

$$\text{Langmuir isotherm}: \quad q_e = \frac{q_{\max} K_L c_e}{1 + K_L c_e} \tag{5}$$

where q_e (mg g^{-1}) is the maximum monolayer adsorption capacity of heavy metals, and K_L (L mg^{-1}) is the equilibrium constant of the Langmuir equation.

$$\text{Freundlich isotherm}: \quad q_e = K_F c_e^{1/n} \tag{6}$$

where K_F ($mg^{1-(1/n)}$ $L^{1/n}$ g^{-1}) is the Freundlich adsorption constant, and n is the dimensionless empirical coefficient related to adsorption strength, which depends on the surface heterogeneity.

2.4.4. Effects of Temperatures and Ionic Strengths

The effects of temperature and ionic strength on the adsorption of heavy metals by BPB and PBPB were investigated at various temperatures (15–45 °C) and ionic strength conditions (0.005–0.1 M) (each heavy metal concentration = 10 mg L^{-1}, agitation time = 6 h, mixing speed = 150 rpm, temperature = 25 °C, pH = 7.0). The removal efficiencies of Mn and Fe using BPB and PBPB were calculated using Equations (2) and (3) as described in the previous section.

3. Results and Discussion

3.1. Physical Properties of Biochar

Figure 1 illustrates the functional group composition of BPB and PBPB, measured using ATR-FTIR. The functional group composition of BPB and PBPB was similar, but the intensities of the IR peaks of the functional group composition affecting the adsorption of heavy metals were different. The IR peaks related to O-H stretching and C-O stretching of alcohols appeared at 3500–3000 cm^{-1} (BPB = 3443 cm^{-1}, PBPB = 3431 cm^{-1}) and 1210–1100 cm^{-1} (BPB = 1103 cm^{-1}, PBPB = 1180 cm^{-1}) due to the presence of alcohol functional groups (-CH_2OH-) derived from the cellulose component of the banana peels [33]. The IR peaks related to N-H stretching of amides were found in the range of 1650–1550 cm^{-1} (BPB = 1565 cm^{-1}, PBPB = 1575 cm^{-1}). Furthermore, C-H stretching of alkanes and C-H stretching of aromatics exhibited relatively strong IR peaks in the range of 1500–1300 cm^{-1} and 900–670 cm^{-1}, respectively [34]. The IR peak derived from C-O stretching of alcohols exhibited a relatively high IR intensity in PBPB compared to BPB in the range of 1210–1100 cm^{-1} because the phosphoric acid pre-treatment promoted the formation of oxygen-containing functional groups closely associated with the adsorption of heavy metals [34,35].

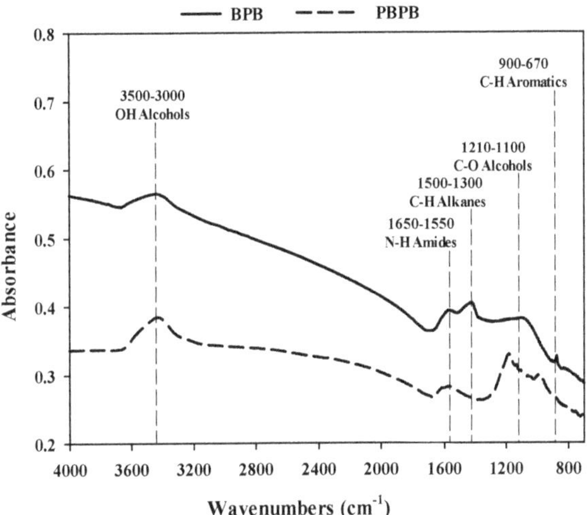

Figure 1. Attenuated total reflectance (ATR)-FTIR spectra of banana peel biochar without pre-treatment (BPB) and pre-treated banana peel biochar (PBPB).

Table 1 shows the elemental composition and surface properties of BPB and PBPB. Although the hydrogen, nitrogen, and ash content of BPB and PBPB were similar (H content of BPB = 1.6%, H content of PBPB = 1.9%; N content of BPB = 2.9%, N content of PBPB = 2.2%; ash content of BPB = 19.0%, ash content of PBPB = 16.4%), PBPB had lower carbon content (54.7%) and higher oxygen content (24.8%) compared to BPB (C content = 75.4%, O content = 1.1%). Therefore, the H/C, O/C, and N/C ratios of PBPB were much higher than those of BPB. These observations indicate that increases in hydroxyl and carboxyl functional groups in PBPB after phosphoric acid pre-treatment may enhance electrostatic attractions between heavy metals and the adsorbent surfaces intimately related to the adsorption of heavy metals [36]. Furthermore, the specific surface area and total pore volume of PBPP (specific surface area = 27.41 m^2 g^{-1}; total pore volume = 0.032 cm^3 g^{-1}) were considerably greater compared to BPB (BPB = 11.32 m^2 g^{-1}; total pore volume = 0.027 cm^3 g^{-1}). These results suggest that heavy metals can be adsorbed more readily by PBPB than BPB due to its abundance of adsorption sites [12].

Table 1. Physicochemical properties of BPB and PBPB.

	Elements Composition (%)					Atomic Ratio			S_{BET} (m^2 g^{-1})	Pore Volume (cm^3 g^{-1})
	C	H	O	N	Ash	H/C	O/C	N/C		
BPB	75.4	1.6	1.1	2.9	19.0	0.022	0.014	0.038	11.32	0.027
PBPB	54.7	1.9	24.8	2.2	16.4	0.035	0.453	0.039	27.41	0.032

3.2. Effects of Adsorbent Dosages

Figure 2 presents the change in the removal efficiencies of Mn and Fe as a function of adsorbent dosages. The removal efficiencies of Mn and Fe using BPB and PBPB were increased with increasing adsorbent dosages because of the increased availability of adsorption sites on the adsorbent surfaces [14]. The lower removal efficiencies of Mn by BPB and PBPB compared to those of Fe might be attributed to differences in the electronegativity and ion radius of Mn and Fe. Since Fe has a higher electronegativity than Mn (Fe = 1.8, Mn = 1.5) and a small ion radius, it can more easily diffused into the pores of PBPB and BPB [37,38]. In the equilibrium state, PBPB exhibited higher removal efficiencies of Mn and Fe than BPB (removal efficiency of Mn by BPB = 32%, removal efficiency of Mn by PBPB = 46%, removal efficiency of Fe by PBPB = 96%, removal efficiency of Fe by BPB = 85%). These results mean that phosphate pre-treatment promoted the formation of functional groups on the biochar surfaces capable of adsorbing heavy metals [27,39]. The removal efficiencies of Mn by BPB and PBPB reached the steady state at an adsorbent dosage of 3 g L^{-1} while the steady states of Fe adsorption occurred at 0.1 g L^{-1} for BPB and 0.3 g L^{-1} for PBPB, respectively. Hence, the PBPB and BPB dosages obtained at their steady states for Mn and Fe were applied for further adsorption experiments.

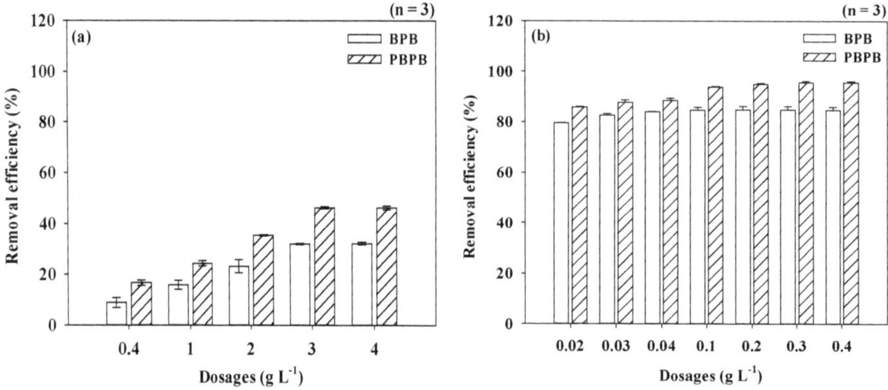

Figure 2. Effects of biochar dosages on the adsorption of heavy metals using BPB and PBPB: (**a**) Mn and (**b**) Fe (agitation time = 3 h, agitation speed = 150 rpm, initial concentration of Mn and Fe = 10 mg L^{-1}, pH = 7, and temperature = 25 °C).

3.3. Adsorption Kinetics of Mn and Fe

Figure 3 depicts the adsorption kinetics of Mn and Fe by BPB and PBPB. The adsorption proceeded rapidly in the beginning for both heavy metals (Mn ≤360 min; Fe ≤180 min) and almost reached equilibrium at 10 h. A possible explanation for these results is that the availability of adsorption sites plays key roles in the adsorption of Fe and Mn by BPB and PBPB [19]. Table 2 presents the kinetic model parameters for the adsorption of Mn and Fe by BPB and PBPB. Based on the correlation coefficient values (R^2), the pseudo-second-order model better described the adsorption of Mn and Fe by BPB and PBPB than the pseudo-first-order model. These observations indicate that the adsorption of Mn and Fe by BPB and PBPB is predominantly governed by chemical adsorption (i.e., covalent bonding or ion/electron exchange) [27]. As shown in Figure 3, the adsorption capacities at equilibrium (q_e) of Mn and Fe by BPB were 1.14 and 31.61 mg g^{-1}, respectively. Meanwhile, the q_e of Mn and Fe by PBPB were 2.03 and 32.99 mg g^{-1}, respectively. The higher q_e values by PBPB support the assumption that PBPB is more effective for the adsorption of heavy metals than BPB due to its higher content of oxygen-containing functional groups [34,35].

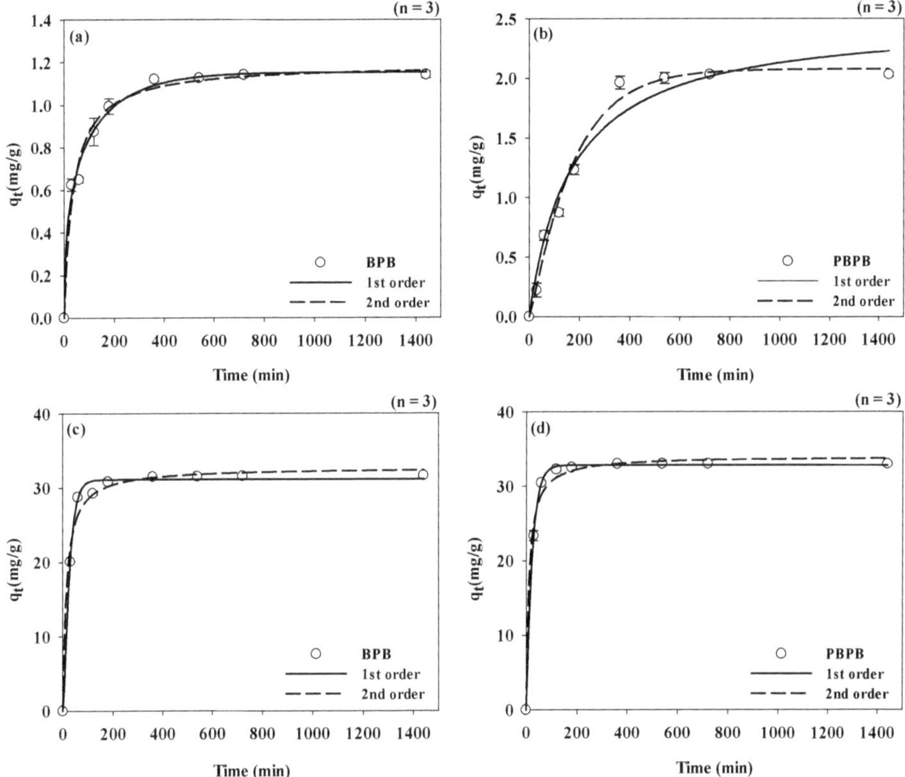

Figure 3. Adsorption kinetics of heavy metals onto BPB and PBPB: (**a**,**b**) Mn and (**c**,**d**) Fe (agitation time = 24 h, agitation speed = 150 rpm, adsorbent dosage of BPB and PBPB for Mn = 3 g L^{-1}, adsorbent dosage of BPB for Fe = 0.1 g L^{-1}, adsorbent dosage of PBPB for Fe = 0.3 g L^{-1}, initial concentration of Mn and Fe = 10 mg L^{-1}, pH = 7, and temperature = 25 °C).

Table 2. Kinetic model parameters for the adsorption of Mn and Fe by BPB and PBPB (n = 3).

			k	R^2
BPB	Mn	First-order (min^{-1})	0.008	0.944
		Second-order (g mg^{-1} min^{-1})	0.029	0.999
	Fe	First-order (min^{-1})	0.007	0.867
		Second-order (g mg^{-1} min^{-1})	0.004	0.999
PBPB	Mn	First-order (min^{-1})	0.009	0.965
		Second-order (g mg^{-1} min^{-1})	0.003	0.977
	Fe	First-order (min^{-1})	0.019	0.892
		Second-order (g mg^{-1} min^{-1})	0.006	0.999

3.4. Adsorption Isotherms of Mn and Fe

The adsorption mechanisms of Mn and Fe by BPB and PBPB were analyzed using the Langmuir and Freundlich adsorption isotherm models (Figure 4 and Table 3). The Langmuir isotherm model was well-fitted to the adsorption of Mn and Fe by BPB (R^2 of Mn = 0.972, R^2 of Fe = 0.869). These results imply that monolayer adsorption is responsible for the adsorption of Mn and Fe by BPB [7]. The adsorption of Mn by PBPB followed the Freundlich isotherm model (R^2 = 0.993) more closely than the Langmuir isotherm model (R^2 = 0.898). It is evident that multilayer adsorption strongly contributes to the adsorption of Mn by PBPB [13]. However, the adsorption of Fe by PBPB followed both Freundlich (R^2 = 0.933) and Langmuir isotherms (R^2 = 0.949). The n value of the Freundlich isotherm model was used to examine the adsorption affinity of Mn and Fe onto BPB and PBPB (Table 3): (i) $n > 1$ (favorable), (ii) $n = 1$ (linear), and (iii) $n < 1$ (unfavorable) [30]. The adsorption of Mn and Fe by BPB (n value of Mn = 7.267, n value of Fe = 1.069) and the adsorption of Mn by PBPB (n value = 2.471) were favorable whereas the adsorption of Fe by PBPB (n value = 0.977) was not favorable. The separation parameter R_L value, based on $R_L = 1/(1 + K_L C_0)$ of the Langmuir isotherm model, was also calculated to assess the adsorption preference of Mn and Fe toward BPB and PBPB: (i) $R_L > 1$ (unfavorable), (ii) $R_L = 1$ (linear), (iii) $1 > R_L > 0$ (favorable), and (iv) $R_L = 0$ (irreversible) [40]. Since the R_L values of Mn and Fe by BPB and PBPB were in the range of 0–1 (R_L of Mn by BPB = 0.003; R_L of Fe by BPB = 0.146; R_L of Mn by PBPB = 0.012; R_L of Fe by PBPB = 0.156), the adsorption of Mn and Fe by BPB and PBPB seemed to be favorable.

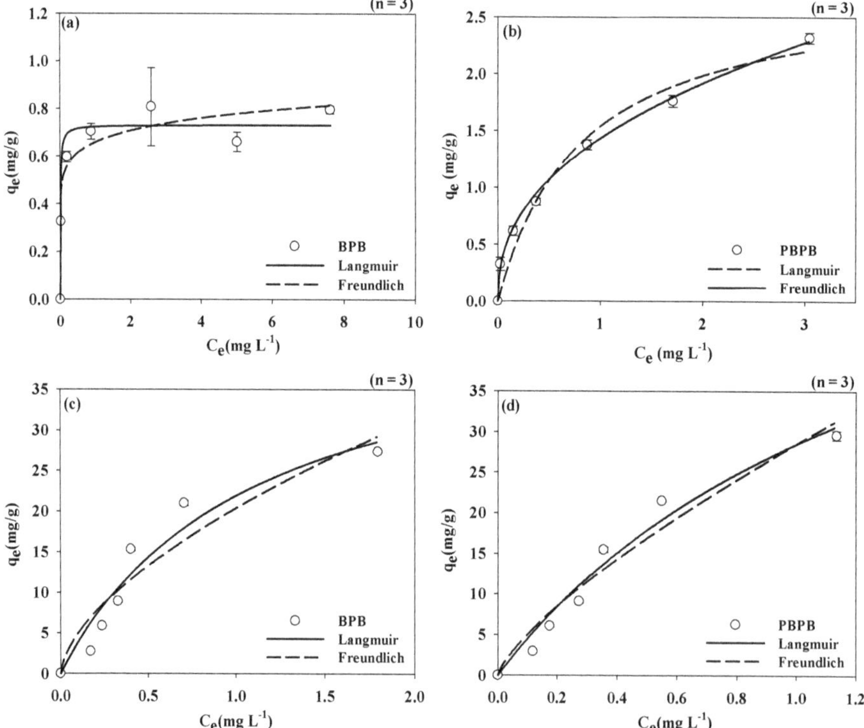

Figure 4. Adsorption isotherms of heavy metals onto BPB and PBPB: (**a**,**b**) Mn and (**c**,**d**) Fe (agitation time = 24 h, agitation speed = 150 rpm, adsorbent dosage of BPB and PBPB for Mn = 3 g L^{-1}, adsorbent dosage of BPB for Fe = 0.1 g L^{-1}, adsorbent dosage of PBPB for Fe = 0.3 g L^{-1}, initial concentration of Mn and Fe = 10 mg L^{-1}, pH = 7, and temperature = 25 °C).

Table 3. Isotherm model parameters for adsorption of Mn and Fe by BPB and PBPB (n = 3).

			Mn	Fe
BPB	Langmuir	K_L (L mg^{-1})	32.204	0.583
		q_e (mg g^{-1})	0.796	27.355
		R_L	0.003	0.146
		R^2	0.972	0.869
	Freundlich	K_F ($mg^{1-(1/n)}$ $L^{1/n}$ g^{-1})	0.821	3.864
		n	7.267	1.069
		R^2	0.81	0.826
PBPB	Langmuir	K_L (dm^3 mg^{-1})	7.943	0.540
		q_e (mg g^{-1})	2.319	29.55
		R_L	0.012	0.156
		R^2	0.898	0.949
	Freundlich	K_F ($mg^{1-(1/n)}$ $L^{1/n}$ g^{-1})	1.161	4.611
		n	2.471	0.977
		R^2	0.993	0.933

3.5. Effects of Temperature on Adsorption of Mn and Fe

The effects of temperature on the adsorption of Mn and Fe by BPB and PBPB are compared in Figure 5. The removal efficiencies of Mn and Fe by both BPB and PBPB were increased with increasing temperatures from 15 (removal efficiency of Mn by BPB = 17%, removal efficiency of Mn by PBPB = 46%, removal efficiency of Fe by BPB = 88%, removal efficiency of Fe by PBPB = 97%) to 45 °C (removal efficiency of Mn by BPB = 32%, removal efficiency of Mn by PBPB = 65%, removal efficiency of Fe by BPB = 97%, removal efficiency of Fe by PBPB = 99%). These results suggest that high temperatures may provide sufficient energy for the adsorption of heavy metals on the surficial and interior layers of biochars [41]. From the higher removal efficiencies of Mn and Fe by PBPB than BPB, it can be concluded that the abundance of oxygen-containing functional groups facilitates the adsorption of heavy metals by carbonaceous adsorbents [36].

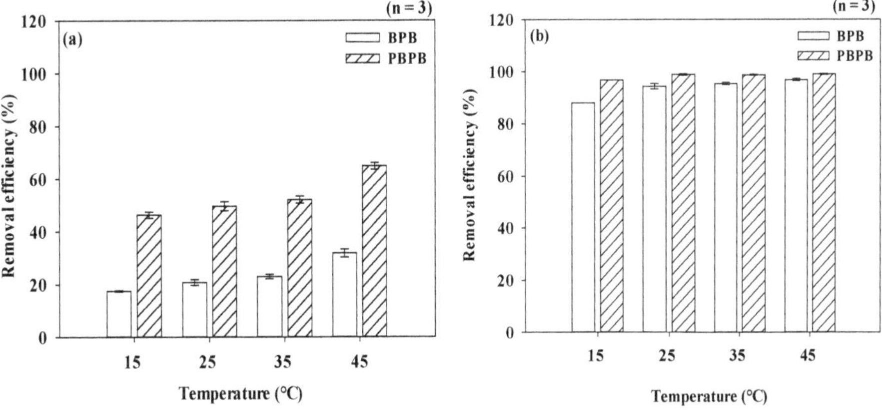

Figure 5. Effects of temperature on the adsorption of heavy metals using BPB and PBPB: (**a**) Mn and (**b**) Fe (agitation time = 6 h, agitation speed = 150 rpm, adsorbent dosage of BPB and PBPB for Mn = 3 g L^{-1}, adsorbent dosage of BPB for Fe = 0.1 g L^{-1}, adsorbent dosage of PBPB for Fe = 0.3 g L^{-1}, initial concentration of Mn and Fe = 10 mg L^{-1}, pH = 7, and temperature = 15–45 °C).

3.6. Effects of Ionic Strength on Adsorption of Mn and Fe

Figure 6 illustrates the effects of ionic strength on the removal of heavy metals by BPB and PBPB. The removal efficiencies of Mn and Fe by BPB and PBPB were gradually decreased with increasing ionic strengths. For example, when ionic strength was increased from 0 to 0.1 M, the removal efficiencies of Mn by BPB and PBPB were decreased from 35% and 54% to 21% and 32%, respectively. Meanwhile, the removal efficiencies of Fe by BPB and PBPB were decreased from 95% and 99% to 84% and 96%, respectively. These observations imply that increases of ionic strength may reinforce the electrostatic repulsion between heavy metals and the biochar surfaces and reduce the availability of adsorption sites on the biochar surfaces through the aggregation of biochars [42]. The removal efficiencies of Fe by BPB and PBPB were less affected by the changes in ionic strength compared to the removal efficiencies of Mn by BPB and PBPB because Fe has a smaller ion radius and higher electronegativity than Mn [37,38]. These inherent natures allowed Fe to exhibit higher attractive charges in the nucleus on the electron orbital [37,38]. Therefore, Fe more easily penetrated into the pores of biochars compared to Mn [37,38]. In addition, the higher removal efficiencies of Mn and Fe by PBPB (removal efficiency of Mn = 32–54%; removal efficiency of Fe = 96–99%) than BPB (removal efficiency of Mn = 21–35%; removal efficiency of Fe = 84–95%) at all the tested ionic strengths provide evidence that the surface structural features and oxygen-containing functional group abundance govern the adsorption of heavy metals by carbonaceous adsorbents [12,21,22].

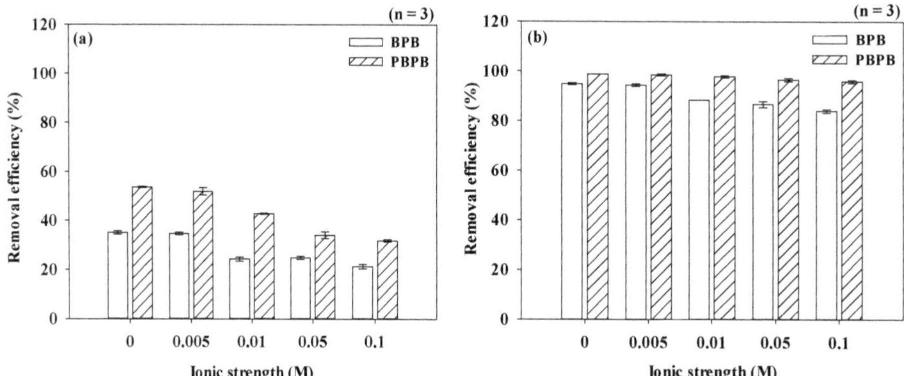

Figure 6. Effects of ionic strength on the adsorption of heavy metals using BPB and PBPB: (**a**) Mn and (**b**) Fe (agitation time = 6 h, agitation speed = 150 rpm, adsorbent dosage of BPB and PBPB for Mn = 3 g L^{-1}, adsorbent dosage of BPB for Fe = 0.1 g L^{-1}, adsorbent dosage of PBPB for Fe = 0.3 g L^{-1}, initial concentration = 10 mg L^{-1}, pH = 7, and temperature = 25 °C).

4. Conclusions

In this study, phosphoric acid pre-treatment on banana peel biochar was investigated for enhancement of Mn and Fe removal efficiencies. The physicochemical characteristics of pristine and phosphoric acid pre-treated biochars were characterized using the elemental analyzer, BET, and ATR-FTIR. These characterizations revealed that phosphoric acid pre-treatment facilitated the formation of oxygen-containing functional groups (i.e., hydroxyl and carboxyl functional groups), which could enhance the adsorption of heavy metals on the biochar surfaces. In addition, phosphoric acid pre-treatment improved specific surface area and pore volume for more adsorption sites for heavy metals. The adsorption experiments showed that phosphoric acid pre-treatment can improve the removal of Mn and Fe significantly. The results of the adsorption kinetics of Mn and Fe by BPB and PBPB were well-matched to the pseudo-second-order, indicating that adsorption was predominantly governed by chemical adsorption. The adsorption of Mn and Fe by BPB obeyed the Langmuir isotherm model, whereas the Freundlich isotherm model described the adsorption of Mn by PBPB well. The adsorption of Fe by PBPB followed both Freundlich and Langmuir isotherm models. Furthermore, the adsorption of Mn and Fe by both BPB and PBPB increased with increasing temperature. However, the removal efficiencies were decreased with increasing ionic strength. From the excellent adsorption performance of Mn and Fe by BPB compared to BPB, it can be concluded that phosphoric acid pre-treatment is a promising method to enhance Mn and Fe removal by banana peel biochar.

Author Contributions: H.K.: methodology and conceptualization, R.-A.K.: formal analysis and data curation, S.L.: writing—original draft preparation, K.C.: writing—review, editing, and funding acquisition. All authors have read and agree to the published version of the manuscript.

Funding: This study was supported by the Basic Research Program through the National Research Foundation of Korea (NRF) funded by the Ministry of Science, ICT and Future Planning (NRF-2018R1D1A1B07044718) and partially supported by a 2017 Research Grant from Kangwon National University (Republic of Korea).

Conflicts of Interest: The authors declare no conflict of interest.

References

1. Ihsanullah; Abbas, A.; Al-Amer, A.M.; Laoui, T.; Al-Marri, M.J.; Nasser, M.S.; Khraisheh, M.; Atieh, M.A. Heavy metal removal from aqueous solution by advanced carbon nanotubes: Critical review of adsorption applications. *Sep. Purif. Technol.* **2016**, *157*, 141–161. [CrossRef]
2. Wang, F.; Yuanfeng, P.; Cai, P.; Guo, T.; Xiao, H. Single and binary adsorption of heavy metal ions from aqueous solutions using sugarcane cellulose-based adsorbent. *Bioresour. Technol.* **2017**, *241*, 482–490. [CrossRef]
3. Elsehly, E.M.; Chechenin, N.G.; Bukunov, K.; Makunin, A.V.; Priselkova, A.B.; Vorobyeva, E.A.; Motaweh, H.A. Removal of iron and manganese from aqueous solutions using carbon nanotube filters. *Water Supply* **2015**, *16*, 347–353. [CrossRef]
4. Idrees, M.; Batool, S.; Ullah, H.; Hussain, Q.; Al-Wabel, M.I.; Ahmad, M.; Hussain, A.; Riaz, M.; Ok, Y.S.; Kong, J. Adsorption and thermodynamic mechanisms of manganese removal from aqueous media by biowaste-derived biochars. *J. Mol. Liq.* **2018**, *266*, 373–380. [CrossRef]
5. Kwakye-Awuah, B.; Sefa-Ntiri, B.; Von-Kiti, E.; Nkrumah, I.; Williams, C. Adsorptive Removal of Iron and Manganese from Groundwater Samples in Ghana by Zeolite Y Synthesized from Bauxite and Kaolin. *Water* **2019**, *11*, 1912. [CrossRef]
6. Al-Rashdi, B.; Johnson, D.; Hilal, N. Removal of heavy metal ions by nanofiltration. *Desalination* **2013**, *315*, 2–17. [CrossRef]
7. Chen, Q.; Yao, Y.; Li, X.; Lu, J.; Zhou, J.; Huang, Z. Comparison of heavy metal removals from aqueous solutions by chemical precipitation and characteristics of precipitates. *J. Water Process. Eng.* **2018**, *26*, 289–300. [CrossRef]
8. Bazrafshan, E.; Mohamadi, L.; Ansari-Moghaddam, A.; Mahvi, A.H. Heavy metals removal from aqueous environments by electrocoagulation process- a systematic review. *J. Environ. Health Sci. Eng.* **2015**, *13*, 74. [CrossRef]
9. Khatri, N.; Tyagi, S.; Rawtani, D. Recent strategies for the removal of iron from water: A review. *J. Water Process. Eng.* **2017**, *19*, 291–304. [CrossRef]
10. Patil, D.S.; Chavan, S.M.; Oubagaranadin, J.U.K. A review of technologies for manganese removal from wastewaters. *J. Environ. Chem. Eng.* **2016**, *4*, 468–487. [CrossRef]
11. Tobiason, J.E.; Bazilio, A.; Goodwill, J.; Mai, X.; Nguyen, C. Manganese Removal from Drinking Water Sources. *Curr. Pollut. Rep.* **2016**, *2*, 168–177. [CrossRef]
12. Pérez-Marín, A.B.; Zapata, V.M.; Ortuño, J.F.; Aguilar, M.I.; Saez, J.; Lloréns, M. Removal of cadmium from aqueous solutions by adsorption onto orange waste. *J. Hazard. Mater.* **2007**, *139*, 122–131. [CrossRef]
13. Demirbas, A. Heavy metal adsorption onto agro-based waste materials: A review. *J. Hazard. Mater.* **2008**, *157*, 220–229. [CrossRef]
14. Pellera, F.-M.; Giannis, A.; Kalderis, D.; Anastasiadou, K.; Stegmann, R.; Wang, J.-Y.; Gidarakos, E. Adsorption of Cu(II) ions from aqueous solutions on biochars prepared from agricultural by-products. *J. Environ. Manag.* **2012**, *96*, 35–42. [CrossRef]

15. Liu, Z.; Zhang, F.-S. Removal of lead from water using biochars prepared from hydrothermal liquefaction of biomass. *J. Hazard. Mater.* **2009**, *167*, 933–939. [CrossRef]
16. Liu, Z.; Zhang, F.-S.; Wu, J. Characterization and application of chars produced from pinewood pyrolysis and hydrothermal treatment. *Fuel* **2010**, *89*, 510–514. [CrossRef]
17. Jun, B.-M.; Kim, Y.; Han, J.; Yoona, Y.; Kim, J.; Park, C.M. Preparation of Activated Biochar-Supported Magnetite Composite for Adsorption of Polychlorinated Phenols from Aqueous Solutions. *Water* **2019**, *11*, 1899. [CrossRef]
18. Lucaci, A.R.; Bulgariu, D.; Ahmad, I.; Lisa, G.; Mocanu, A.M.; Bulgariu, L. Potential Use of Biochar from Various Waste Biomass as Biosorbent in Co(II) Removal Processes. *Water* **2019**, *11*, 1565. [CrossRef]
19. Gupta, P.; Ann, T.-W.; Lee, S.-M. Use of biochar to enhance constructed wetland performance in wastewater reclamation. *Environ. Eng. Res.* **2016**, *21*, 36–44. [CrossRef]
20. Zhao, J.; Shen, X.-J.; Domene, X.; Alcañiz, J.-M.; Liao, X.; Palet, C. Comparison of biochars derived from different types of feedstock and their potential for heavy metal removal in multiple-metal solutions. *Sci. Rep.* **2019**, *9*, 9869. [CrossRef]
21. Wang, X.; Li, X.; Liu, G.; He, Y.; Chen, C.; Liu, X.; Li, G.; Gu, Y.; Zhao, Y.; He, Y. Mixed heavy metal removal from wastewater by using discarded mushroom-stick biochar: Adsorption properties and mechanisms. *Environ. Sci. Process. Impacts* **2019**, *21*, 584–592. [CrossRef] [PubMed]
22. Vilardi, G.; Di Palma, L.; Ochando-Pulido, J.M. Heavy metals adsorption by banana peels micro-powder: Equilibrium modeling by non-linear models. *Chin. J. Chem. Eng.* **2018**, *26*, 455–464. [CrossRef]
23. Martínez-Ruano, J.A.; Caballero-Galván, A.S.; Restrepo-Serna, D.L.; Alzate, C.A.C. Techno-economic and environmental assessment of biogas production from banana peel (Musa paradisiaca) in a biorefinery concept. *Environ. Sci. Pollut. Res.* **2018**, *25*, 35971–35980. [CrossRef] [PubMed]
24. Sagar, N.A.; Pareek, S.; Sharma, S.; Yahia, E.M.; Lobo, M.G. Fruit and Vegetable Waste: Bioactive Compounds, Their Extraction, and Possible Utilization. *Compr. Rev. Food Sci. Food Saf.* **2018**, *17*, 512–531. [CrossRef]
25. Omulo, G.; Banadda, N.; Kabenge, I.; Seay, J.R. Optimizing slow pyrolysis of banana peels wastes using response surface methodology. *Environ. Eng. Res.* **2018**, *24*, 354–361. [CrossRef]
26. Ibarra-Rodríguez, D.; Lizardi-Mendoza, J.; López-Maldonado, E.A.; Oropeza-Guzman, M.T.; Diana, I.-R.; Jaime, L.-M. Capacity of 'nopal' pectin as a dual coagulant-flocculant agent for heavy metals removal. *Chem. Eng. J.* **2017**, *323*, 19–28. [CrossRef]
27. Zhou, N.; Chen, H.; Xi, J.; Yao, D.; Zhou, Z.; Tian, Y.; Lu, X. Biochars with excellent Pb(II) adsorption property produced from fresh and dehydrated banana peels via hydrothermal carbonization. *Bioresour. Technol.* **2017**, *232*, 204–210. [CrossRef]
28. Amin, M.T.; Alazba, A.A.; Shafiq, M. Removal of Copper and Lead using Banana Biochar in Batch Adsorption Systems: Isotherms and Kinetic Studies. *Arab. J. Sci. Eng.* **2017**, *43*, 5711–5722. [CrossRef]
29. Tan, X.; Liu, S.; Liu, Y.; Gu, Y.-L.; Zeng, G.-M.; Hu, X.; Wang, X.; Liu, S.-H.; Jiang, L. Biochar as potential sustainable precursors for activated carbon production: Multiple applications in environmental protection and energy storage. *Bioresour. Technol.* **2017**, *227*, 359–372. [CrossRef]
30. Wang, Z.; Wu, J.; He, T.; Wu, J. Corn stalks char from fast pyrolysis as precursor material for preparation of activated carbon in fluidized bed reactor. *Bioresour. Technol.* **2014**, *167*, 551–554. [CrossRef]
31. Chu, G.; Zhao, J.; Huang, Y.; Zhou, D.; Liu, Y.; Wu, M.; Peng, H.; Zhao, Q.; Pan, B.; Steinberg, C.E. Phosphoric acid pretreatment enhances the specific surface areas of biochars by generation of micropores. *Environ. Pollut.* **2018**, *240*, 1–9. [CrossRef]
32. Taha, S.M.; Amer, M.E.; Elmarsafy, A.E.; Elkady, M.Y. Adsorption of 15 different pesticides on untreated and phosphoric acid treated biochar and charcoal from water. *J. Environ. Chem. Eng.* **2014**, *2*, 2013–2025. [CrossRef]
33. Wu, C.; Wang, Z.; Huang, J.; Williams, P.T. Pyrolysis/gasification of cellulose, hemicellulose and lignin for hydrogen production in the presence of various nickel-based catalysts. *Fuel* **2013**, *106*, 697–706. [CrossRef]
34. Coates, J. Interpretation of infrared spectra, a practical approach. In *Encyclopedia of Analytical Chemistry: Applications, Theory and Instrumentation*; John Wiey & Sons, Inc.: Hoboken, NJ, USA, 2006; pp. 10815–10837.
35. O'Connell, D.W.; Birkinshaw, C.; O'Dwyer, T.F. Heavy metal adsorbents prepared from the modification of cellulose: A review. *Bioresour. Technol.* **2008**, *99*, 6709–6724. [CrossRef] [PubMed]

36. Wang, Z.; Liu, G.; Zheng, H.; Li, F.; Ngo, H.-H.; Guo, W.; Liu, C.; Chen, L.; Xing, B. Investigating the mechanisms of biochar's removal of lead from solution. *Bioresour. Technol.* **2015**, *177*, 308–317. [CrossRef] [PubMed]
37. Dastgheib, S.A.; Rockstraw, D.A. A model for the adsorption of single metal ion solutes in aqueous solution onto activated carbon produced from pecan shells. *Carbon* **2002**, *40*, 1843–1851. [CrossRef]
38. Bin Jusoh, A.; Cheng, W.; Low, W.; Nora'Aini, A.; Noor, M.M.M. Study on the removal of iron and manganese in groundwater by granular activated carbon. *Desalination* **2005**, *182*, 347–353. [CrossRef]
39. Moreno-Castilla, C.; Álvarez-Merino, M.A.; López-Ramón, M.; Rivera-Utrilla, J. Cadmium Ion Adsorption on Different Carbon Adsorbents from Aqueous Solutions. Effect of Surface Chemistry, Pore Texture, Ionic Strength, and Dissolved Natural Organic Matter. *Langmuir* **2004**, *20*, 8142–8148. [CrossRef]
40. Gorgievski, M.; Božić, D.; Stanković, V.; Strbac, N.; Šerbula, S. Kinetics, equilibrium and mechanism of Cu2+, Ni2+ and Zn2+ ions biosorption using wheat straw. *Ecol. Eng.* **2013**, *58*, 113–122. [CrossRef]
41. Abdelhafez, A.A.; Li, J. Removal of Pb(II) from aqueous solution by using biochars derived from sugar cane bagasse and orange peel. *J. Taiwan Inst. Chem. Eng.* **2016**, *61*, 367–375. [CrossRef]
42. Park, C.M.; Han, J.; Chu, K.H.; Al-Hamadani, Y.; Her, N.; Heo, J.; Yoona, Y. Influence of solution pH, ionic strength, and humic acid on cadmium adsorption onto activated biochar: Experiment and modeling. *J. Ind. Eng. Chem.* **2017**, *48*, 186–193. [CrossRef]

Article

A Biosorption-Pyrolysis Process for Removal of Pb from Aqueous Solution and Subsequent Immobilization of Pb in the Char

Yue Wang, Jinhong Lü *, Dongqing Feng, Sen Guo and Jianfa Li

College of Chemistry and Chemical Engineering, Shaoxing University, Zhejiang 312000, China; wangyue835@163.com (Y.W.); fengdongqing@yeah.net (D.F.); gsen1041@gmail.com (S.G.); ljf@usx.edu.cn (J.L.)
* Correspondence: lvjinhong@usx.edu.cn; Tel.: +86-575-8834-1524

Received: 19 July 2020; Accepted: 18 August 2020; Published: 25 August 2020

Abstract: The application of biosorption in the removal of heavy metals from water faces a challenge of safe disposal of contaminated biomass. In this study, a potential solution for this problem was proposed by using a biosorption-pyrolysis process featured by pretreatment of biomass with phosphoric acid (PA). The PA pretreatment of biomass increased the removal efficiency of heavy metal Pb from water by sorption, and subsequent pyrolysis helped immobilize Pb in the residual char. The results indicate that most (>95%) of the Pb adsorbed by the PA-pretreated biomass was retained in the char, and that the lower pyrolysis temperature (350 °C) is more favorable for Pb immobilization. In this way, the bioavailable Pb in the char was hardly detected, while the Pb leachable in acidic solution decreased to <3% of total Pb in the char. However, higher pyrolysis temperature (450 °C) is unfavorable for Pb immobilization, as both the leachable and bioavailable Pb increased to >28%. The reason should be related to the formation of elemental Pb and unstable Pb compounds during pyrolysis at 450 °C, according to the X-ray diffraction study.

Keywords: heavy metal; biosorption; biochar; immobilization; phosphoric acid

---

1. Introduction

Water pollution by heavy metals, such as Pb, is one of the most serious environmental issues in the world. In comparison with other technologies for removing heavy metals from water, biosorption has the advantages of low operation cost, abundant biomass availability, and easy acceptance by the public [1–5]. Many kinds of plant-derived biomass wastes have been used for sorption of heavy metals, such as Pb, Cd, and Cu [6–8]. However, the practical applications of biosorption in the removal of heavy metals face two challenges. The first is that the sorption capacity of many indigenous biosorbents needs to be enhanced. Except for some biosorbents derived from algae and wheat bran, the maximum sorption of Pb by many lignocellulosic materials (seed husks, hulls and leaves, etc.) has shown to be <50 mg/g [6,9]. For seeking the solution of this problem, chemical modifications of biomass have been used for enhancing the sorption capacity [10–12]. The chemical modifying agents, including inorganic bases (e.g., NaOH), mineral acids (e.g., HNO_3), and organic compounds (e.g., urea) can enhance the ion exchange, and increase the functional groups and metal holding capacity of biosorbents. Among these modifying agents, phosphoric acid (PA) has shown to be a good candidate. PA has been used to modify olive pomace [13], corncob [14], rice husk [15], bagasse [16], wheat straw biochar [17] and so forth, so that their sorption to heavy metals was greatly enhanced. The second and more significant challenge is related to the safe disposal of the contaminated biomass after sorption of heavy metals. As the biomass is readily decomposed, the adsorbed heavy metals would be released or leached from the contaminated biomass if it was not properly disposed.

The proposed methods for disposal of heavy metal-contaminated biomass include incineration [18], composting [19], and pyrolysis [20], among others. Both incineration and composting have the problem of secondary pollution because the heavy metals will be dispersed into surroundings. In comparison, pyrolysis is operated in a closed system, and dispersion of heavy metals is controllable. Thus, pyrolysis has been used more frequently for post-treatment of heavy metal-contaminated biomass [21–26]. Pyrolysis is useful for valorization of contaminated biomass by transforming it into bio-fuel [27,28], and has also been used for preparing biochars that are good adsorbents to many pollutants [29–31]. The distribution of heavy metals in various fractions of pyrolysis product (e.g., oil or char) depends on the specific pyrolysis technique and heavy metal species [32–34]. For example, higher pyrolysis temperatures resulted in more emission of heavy metals into volatile fractions (e.g., oil) [27,35]. In contrast, slow pyrolysis at relatively low temperatures is more efficient for retention of heavy metals in the char [36,37]. Furthermore, at the same pyrolysis conditions, the heavy metals Cu, Zn, and Pb are more readily retained in the char than Cd [38]. As the char is more resistant to degradation than the biomass, pyrolysis is a feasible solution for disposal of heavy metal-contaminated biomass if most of the heavy metals are retained in the char and their stability in the char is guaranteed. In our previous studies, we investigated the stability of Pb and Cd in the char, and found that addition of phosphates in the contaminated biomass before pyrolysis enhanced the retention and stability of heavy metals [39–41].

In this work, a novel biosorption-pyrolysis process using PA-pretreated biomass was used to attempt to remove Pb from aqueous solution and subsequently immobilize Pb in the char. The specific objectives were to test two hypotheses: the first is that the PA pretreatment of biomass will enhance its sorption to Pb, and the second and more significant is that the subsequent pyrolysis of the contaminated biomass will increase the stability of Pb in the char. Different from our previous studies using phosphates as the additive for pyrolysis of the contaminated biomass [39], the biomass derived from an aquatic plant was firstly treated with PA, and then used for sorption of Pb in aqueous solution. The Pb-contaminated biomass was pyrolyzed, and the retention and stability of Pb in the char was evaluated. The comparison with the biosorption-pyrolysis process using un-pretreated biomass was also performed, and the immobilization mechanism of Pb in the char was investigated by using X-ray diffraction (XRD) and infrared spectra (IR).

2. Materials and Methods

2.1. Biomass and PA Pretreatment

The plant biomass was derived from the *Hydrocotyle verticillata* species, an aquatic plant that is widely distributed in tropical and subtropical zones. Its great adaptability and vitality makes it a good choice for phytoremediation of polluted or eutrophic water. The floating part (leaves and stems) of the *Hydrocotyle* plant was collected from a local river at Shaoxing, Zhejiang Province of China. The biomass was dried in an oven at 100 °C to a constant weight, and pulverized to a size of <0.15 mm. Elemental analysis using the EA3000 elemental analyser (Euro Vector, Italy) indicates that the pristine biomass (Mass) is composed of C 40.7%, O 29.7%, H 5.9%, N 1.5%, and S 0.6% by weight. The ash content of the biomass measured at 750 °C is 15.7% by weight. Phosphoric acid (H_3PO_4, PA) of analytical grade (Sinopharm Chemical Reagent Co. Ltd., China) was used for pre-treating the biomass. The biomass powder was soaked in 30 wt% PA solution for 24 h, then washed by water until the pH of the effluent reached 6.0. Two PA dosages with the mass ratio of PA/biomass = 0.5 or 1.0 were chosen according to the preliminary experiments. The dry solid residue obtained by PA pretreatment is labeled hereafter as $PA_{0.5}$Mass or $PA_{1.0}$Mass, respectively, according to the PA dosage (0.5 or 1.0) used for pretreatment.

2.2. Pb Removal and Sorption Experiments

$Pb(NO_3)_2$ of analytical grade (Sinopharm Chemical Reagent Co. Ltd., Shanghai, China) was used to prepare Pb solutions for removal and sorption experiments. Of the biosorbent (un-pretreated biomass (Mass) or PA-pretreated biomass ($PA_{0.5}$Mass or $PA_{1.0}$Mass)), 50 mg was added to 20 mL of aqueous Pb solution with initial concentrations (C_0, mg/L) ranging from 5 to 50 mg/L. The slurry was adjusted to an initial pH of 5.0 with diluted HNO_3 solution, and placed in a thermostatic shaker bath at 25.0 ± 0.2 °C and shaken for 6 h when the sorption reached equilibrium according to our preliminary experiments. The supernatant solution was sampled and the Pb concentration (C_e) in solution was analyzed by the flame atomic absorption spectrometry (AAS) (AA-7000, Shimadzu, Japan). The removal efficiency (%) was calculated by Equation (1).

$$\text{Removal efficiency}(\%) = \frac{C_0 - C_e}{C_0} \times 100\% \tag{1}$$

The isothermal sorption experiments were conducted in a way similar to the removal experiments, except that a higher C_0 range (50–300 mg/L) was used, so as to leave a certain equilibrium Pb concentration (C_e) in the solution. The quantity adsorbed (Q_e, mg/g) was calculated by Equation (2):

$$Q_e = (C_0 - C_e) \times \frac{V}{m} \tag{2}$$

where V is the volume of aqueous Pb solution (0.02 L) and m is the mass of biomass used for sorption experiments (0.05 g). The isothermal sorption data were further analyzed with the Langmuir and Freundlich sorption isotherm models. The Langmuir model has been expressed in Equation (3):

$$Q_e = \frac{Q_m b C_e}{1 + b C_e} \tag{3}$$

where Q_m gives the maximum sorption capacity (mg/g) based on the monolayer adsorption model, and b (L/mg) is the Langmuir constant. The empirical Freundlich model has been expressed in Equation (4):

$$Q_e = k_F {C_e}^{1/n} \tag{4}$$

where k_F is the Freundlich sorption constant, and n gives the sorption intensity of biosorbents.

2.3. Pyrolysis of Pb-Contaminated Biomass

To obtain an adequate amount of contaminated biomass for subsequent pyrolysis, the biomass sample (1.5 g) was mixed with 500 mL of Pb solution with C_0 = 100 mg/L at initial pH = 5.0, and the mixture was stirred magnetically at room temperature for 6 h. Then the solid residue was filtered out and dried at 100 °C. The Pb-contaminated biomass powder was marked as Pb@Mass, Pb@$PA_{0.5}$Mass, or Pb@$PA_{1.0}$Mass according to the biomass sample used for biosorption. Pyrolysis of the Pb-contaminated biomass was carried out at 350 or 450 °C in a quartz tube furnace filled with N_2 gas. The relatively lower pyrolysis temperature was used because a higher temperature will lead to less retention of Pb in the char, according to a previous report [39]. The resulting Pb-concentrated chars are referred to hereafter as Pb@Char350, Pb@$PA_{0.5}$Char350, Pb@$PA_{1.0}$Char350, Pb@Char450, Pb@$PA_{0.5}$Char450, and Pb@$PA_{1.0}$Char450, respectively, depending on the temperature and feedstock biomass used for pyrolysis. The amount of Pb loaded in the biomass (Q_{mass}, mg/g) and then concentrated in the chars (Q_{char}, mg/g) was determined using a microwave digestion method [40], and the Pb concentration in the dilute digestion solution was determined using AAS. Retention efficiency of Pb (%) in the char after pyrolysis was calculated by Equation (5).

$$\text{Retention efficiency in the char}(\%) = \frac{Q_{char}}{Q_{mass}} \times \text{Yield of char}(\%) \tag{5}$$

2.4. Evaluation of the Stability of Pb in the Solid Samples

The bioavailable Pb, namely that available for utilization by living organisms, was evaluated with the DTPA extraction method using a buffer DTPA (diethylene triamine pentacetate acid) solution according to the ISO standard (ISO 14870-2001). The leachable Pb in the samples was evaluated based on the amount of extractable Pb with the acid solution used in the Toxicity Characteristic Leaching Procedure (TCLP, EPA Test Method 1311). The details for the DTPA extraction and the TCLP test follow the procedures reported previously [40,42].

2.5. Sample Analysis

Surface composition of the biomass samples was determined with a JSM-6360LV scanning electron microscope (SEM) (JEOL, Tokyo, Japan) equipped with an X-act energy dispersive X-ray spectrometer (EDS) (Oxford, UK). Infrared spectra (IR) were recorded in the 4000–400 cm^{-1} region on a Nexus FT-IR spectrophotometer (Nicolet, Glendale, WI, USA) using a KBr pellet. The crystalline species in the biomass and chars were identified by X-ray diffraction (XRD) using D/MAX3A (Rigaku, Tokyo, Japan) equipment with CuKα radiation and a goniometer rate of 4°/min.

3. Results and Discussion

3.1. Influence of PA Pretreatment on the Removal of Pb by Biomass

The results in Figure 1a indicate that the removal efficiency of Pb by the *Hydrocotyle* biomass (Mass) is enhanced after PA treatment, despite that the sample ($PA_{1.0}$Mass) pretreated with more PA performed worse than the counterpart using less PA ($PA_{0.5}$Mass). At the dosage of biomass used in this study (2.5 g/L), Pb is almost completely removed by $PA_{0.5}$Mass when the initial concentration (C_0) of Pb (II) ranged from 5~50 mg/L. In contrast, the removal efficiency of Pb by the un-pretreated biomass dropped to <85% with the C_0 increase to 50 mg/L. The enhanced removal efficiency corresponds well to the improved sorption of Pb by biomass after PA pretreatment (Figure 1b). The maximum sorption (Q_m) calculated with the Langmuir model (Equation (3)) increased to 78.7 mg/g for $PA_{0.5}$Mass from 69.6 mg/g for the indigenous biomass (Mass) (Table 1). The Q_m values of the biosorbents in this study are close to those lignocellulosic materials with high sorption capacity to Pb reported in literature [6,7]. In addition, the n values calculated with the Freundlich model are much larger than 1 (Table 1), indicating that all the biosorbents favored the sorption of Pb [17]. The kinetic analysis indicates that the sorption of Pb by all three biosorbents (Mass, $PA_{0.5}$Mass, and $PA_{1.0}$Mass) fitted well to the pseudo-second-order model ($R^2 > 0.99$), and the sorption rate by the biomass was increased by PA-pretreatment according to the calculated rate constants (Table S1 in the Supplementary Materials). The quantity of Pb adsorbed by all three biosorbents (Mass, $PA_{0.5}$Mass and $PA_{1.0}$Mass) increased with the initial pH and the initial Pb concentrations in solution (Figures S2 and S3 in the Supplementary Materials). Therefore, the *Hydrocotyle* biomass and its PA-pretreated derivatives are potential sorbents for removal of heavy metals from water.

The relatively high sorption of Pb by the *Hydrocotyle* biomass and its PA-pretreated derivatives is related to their surface properties. Figure 2 shows the SEM images of the un-pretreated biomass (Figure 2a) and the PA-pretreated biomass (Figure 2b). The rough surface of $PA_{0.5}$Mass favors the sorption by providing more reactive sites. The specific surface area measured by the BET method is 7.24 m^2/g for $PA_{0.5}$Mass, which is higher than that of the un-pretreated biomass (1.42 m^2/g). Thus, PA-pretreatment increases the surface area of biomass and makes more reactive sites available for the sorption of Pb. The EDS analysis indicates that there are some inorganic minerals composed of Ca and K on the biomass surface, which are useful for sorption of Pb through cation exchange. For example, Pb can replace K in carbonate and sulfate to form precipitates, and this is a major pathway for sorption of Pb by those biomasses and biochars rich in inorganic minerals [43–45]. The sorption of Pb by biomass is also related to the surface functional groups. As can be seen in Figure 3, the bands at 1735, 1440, 1160–1060 cm^{-1} indicate oxygen-containing functional groups (C=O, C–OH, and C–O) on the biomass surface, which facilitate the sorption of Pb through surface complexation that has also been reported to contribute much to biosorption of heavy metals [46,47]. Some previous studies have shown that PA treatment can strengthen the acid functional groups of biomass [13,16,48], due to the dissociation of these groups. In addition, PA pretreatment of the biomass introduces a new PO_4^{3-} group (at 1023 cm^{-1}) that can bind Pb through formation of phosphates, according to the IR spectra (Figure 3). However, pretreatment with a higher dosage of PA (1.0 g/g dry biomass) may result in over-acidification of biomass, which may inhibit the precipitation of Pb, so that the decreased removal efficiency and sorption capacity was observed on $PA_{1.0}$Mass in comparison with $PA_{0.5}$Mass (Figure 1). The over-acidification of biomass is confirmed by the change of pH before and after sorption of Pb. For example, at the same initial Pb concentration (e.g., C_0 = 250 mg/L) and same initial pH = 5.0, pH in the solution after sorption of Pb dropped to 4.8, 4.7, and 4.4, respectively for the sorption systems using Mass, $PA_{0.5}$Mass, and $PA_{1.0}$Mass. Namely, the pH in the $PA_{1.0}$Mass system is much lower than the other two systems, and lower pH favors dissolution of Pb and impacts its removal by biomass.

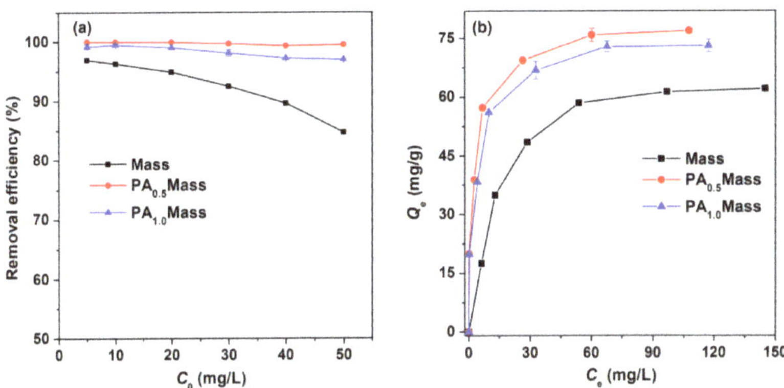

Figure 1. (**a**) Removal of Pb and (**b**) sorption isotherms (at 25 °C) of Pb by the *Hydrocotyle* biomass (Mass) and the biomass samples after PA pretreatment ($PA_{0.5}$Mass and $PA_{1.0}$Mass).

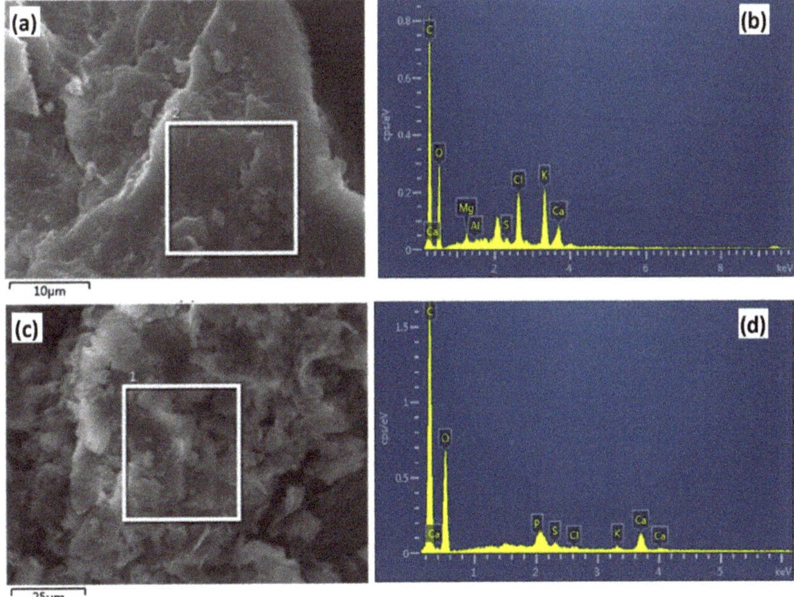

Figure 2. SEM image and EDS mapping results of (**a**,**b**) the *Hydrocotyle* biomass, and (**c**,**d**) PA-pretreated biomass ($PA_{0.5}$Mass).

Table 1. Fitting results of the sorption data with the Langmuir and the Freundlich models.

Models	Langmuir			Freundlich		
	Q_m (mg/g)	b (L/mg)	R^2	k_F	n	R^2
Mass	69.6	0.0788	0.992	20.6	4.21	0.855
$PA_{0.5}$Mass	78.7	0.373	0.996	36.7	5.72	0.853
$PA_{1.0}$Mass	76.9	0.242	0.993	32.8	5.39	0.839

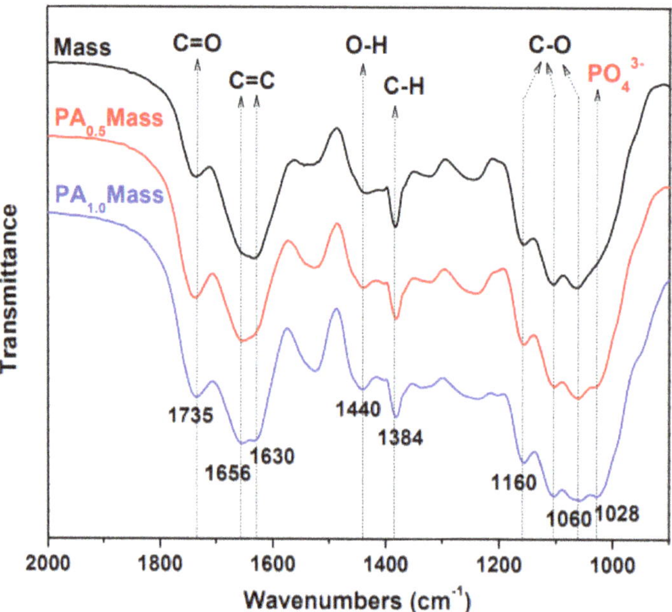

Figure 3. IR spectra of the *Hydrocotyle* biomass and its PA-pretreated derivatives.

3.2. *Influence of Pyrolysis on Pb Retention in the Char*

Pyrolysis of the contaminated biomass at high temperatures will increase the risk of emission of Pb with volatile matters, and decrease the retention of Pb in the char, according to the previous studies [40–42]. Thus, two relatively low temperatures (350 and 450 °C) were used for pyrolysis of the Pb-contaminated biomass in this work, and the retention efficiency of Pb in various chars is shown in Figure 4. As can be seen, the Pb retentions in the two chars derived from the un-pretreated biomass were both less than 85%, and higher pyrolysis temperature (450 °C) led to lower retention efficiency due to the emission of Pb with volatile matters [27,35]. In comparison, the Pb retentions in the four chars ($PA_{0.5}$char350, $PA_{1.0}$char350, $PA_{0.5}$char450, and $PA_{1.0}$char450) derived from the PA-pretreated biomass were all >95%, and a higher dosage of PA used for pretreatment favors retention of more Pb in the char, indicating that PA pretreatment helped inhibit the emission of Pb during pyrolysis. PA is often used for activation of carbonization of biomass [49,50], and has shown to be effective for increasing the yield of char according to the previous reports [51,52], which is in consistency with that observed in this study (Figure 4a). The higher yield of char means less production of volatile matters, and introduction of PO_4^{3-} may also strengthen the binding of Pb to the solid char, which will be investigated in-depth in the following Section 3.4. Therefore, PA pretreatment of the biomass and subsequent pyrolysis enhanced the retention of Pb in the char, which is beneficial for reducing the hazard of secondary pollution due to pyrolysis of heavy metal-contaminated biomass.

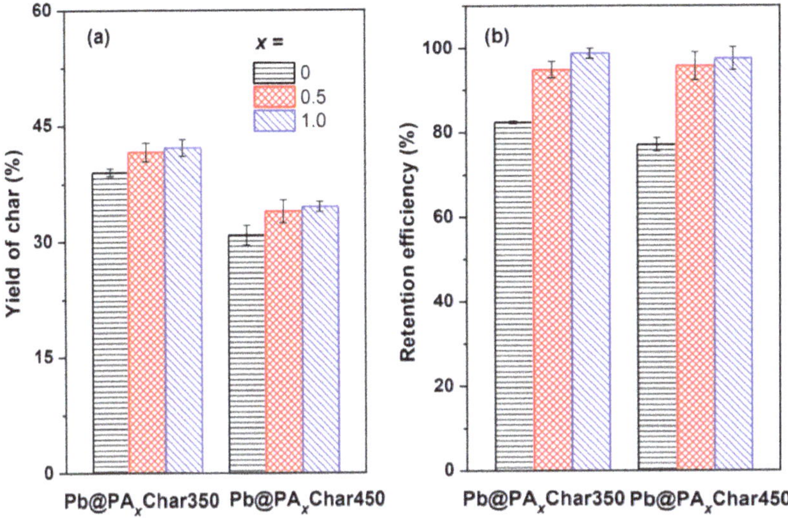

Figure 4. (**a**) Yield of chars, and (**b**) retention efficiency of Pb in the chars obtained by pyrolysis of Pb-contaminated biomass pretreated with various dosages of PA (0, 0.5 and 1.0 g/g mass).

3.3. Influence of Pyrolysis on Stability of Pb in the Char

The stability of Pb retained in the chars was first evaluated with the DTPA extraction method. Figure 5a shows the bioavailability of Pb in six chars, and the three Pb-contaminated biomass samples were also included for comparison. As can be seen, Pb adsorbed by biomass is readily bioavailable, whether the un-pretreated biomass or the PA-pretreated biomass was used as the sorbent. The DTPA- extractable Pb from the three biomass are all over 85%, so the sole PA pretreatment could not ensure the stability of Pb, although its sorption was enhanced as that shown in Figure 1b. Pyrolysis of the contaminated biomass sharply reduced the bioavailability of Pb in the char; in particular, the DTPA-extractable Pb is lower than 5% for the three chars obtained by pyrolysis at a lower temperature (Pb@Char350, Pb@PA$_{0.5}$Char350, and Pb@PA$_{1.0}$Char350). The DTPA-extractable Pb in the last two chars derived from the PA-pretreated biomass is close to zero. The results show that pyrolysis significantly enhanced the stability of Pb, and PA pretreatment can further reduce the bioavailability of Pb in the char. However, the DTPA-extractable Pb from the chars was lifted to >28% as the pyrolysis temperature increased to 450 °C, indicating that a higher pyrolysis temperature is unfavorable for immobilization of Pb. The similar results about the influence of pyrolysis temperature on stability of heavy metals in the char have been reported by Shi et al. [40]. The reason may be related to the different binding states between Pb and the chars obtained at different pyrolysis temperatures, which will be further discussed in the following Section 3.4.

The Pb immobilization by pyrolysis was further assessed with the TCLP test, and the results of leachable Pb in the biomass and chars are shown in Figure 5b. Interestingly, Pb adsorbed in the biomass was not readily leachable, with the highest leachable ratio of 5.34% observed on the biomass without PA pretreatment, although most of the Pb adsorbed in the biomass was bioavailable (Figure 5a). Such a difference should be related to the interactions between Pb and biomass, which are dominated by surface complexation and cation exchange, because there are plentiful oxygen-containing functional groups and exchangeable cations (e.g., K^+) on the biomass surface according to the IR and EDS analysis (Figures 2 and 3). These interactions are readily broken down by a chelating agent, such as DTPA, but insensitive to acetic acid used in the TCLP test. Furthermore, pyrolysis increased the amount of leachable Pb in the Pb@Char350 and Pb@Char450 chars, and higher pyrolysis temperatures resulted in more leachable Pb (Figure 5b). The reasons should be mainly related to the change of solid's properties during the pyrolysis. Pyrolysis makes biomass lose functional groups such as –OH and C–O that are essential for sorption of Pb through surface complexation. At the meantime, inorganic minerals in biomass will be transformed into alkaline matter, such as carbonates or (hydr)oxides [53]. Thus, the interactions between Pb and solid will be changed by formation of Pb oxide or carbonates that are readily dissolved in acid. Therefore, direct pyrolysis, especially at higher temperatures (450 °C), is not favorable for immobilization of Pb in an acidic environment, because nearly half of Pb in the Pb@Char450 sample is leachable (Figure 5b). Despite this result, the encouraging fact is that the two chars (Pb@PA$_{0.5}$Char350 and Pb@PA$_{1.0}$Char350) obtained at 350 °C and derived from the PA-pretreated biomass show good immobilization of Pb, with the ratio of leachable Pb accounting for less than 3%. In combination with the results shown in Figure 5a, we conclude that pyrolysis at 350 °C can immobilize Pb adsorbed by the PA-pretreated biomass. The similar immobilization of heavy metals in the char using phosphates as the additive for pyrolysis has been observed in our previous studies [39–41]. What makes this work differently is that the biomass was firstly pretreated with PA, which enhanced its efficiency for removal of Pb before it was immobilized by pyrolysis. In general, such a combined biosorption-pyrolysis will be a promising strategy for removal of heavy metals from aqueous solution and subsequent safe disposal of the contaminated biosorbents.

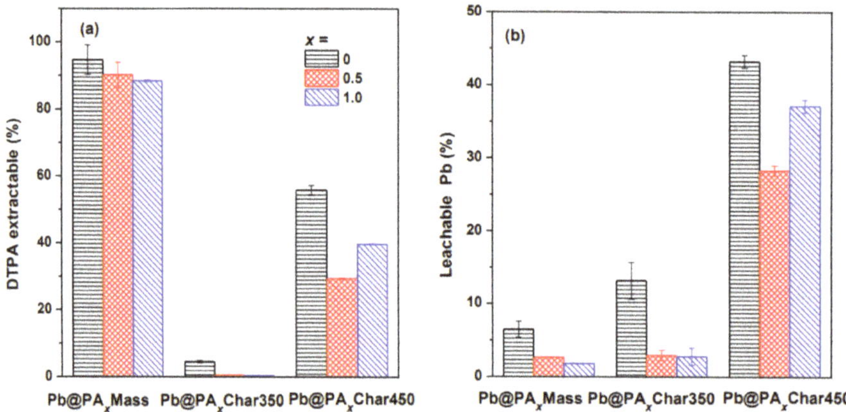

Figure 5. (**a**) DTPA-extractable Pb from, and (**b**) TCLP-leachable Pb from the biomass samples pretreated with various dosages of PA (0, 0.5, and 1.0 g/g mass) and the corresponding chars obtained at different pyrolysis temperatures (350 or 450 °C).

3.4. Investigations about the Immobilization Mechanisms

The X-ray diffraction (XRD) technique and IR spectra were used to examine the biomass and chars obtained by different treatments, so as to understand the mechanisms of Pb immobilization. The XRD patterns (Figure 6) indicate that there is no crystalline constituent other than cellulose in the Pb-contaminated biomass. However, the stable PbS crystal was found in all the chars, which should be one reason for immobilization of Pb. Furthermore, several strong peaks attributed to the lead phosphates (e.g., $Ca_8Pb_2(PO_4)_6(OH)_2$ at 2θ = 51.6°, and $Pb_9(PO_4)_6$ at 2θ = 26.8° and 33.2°) were observed in the Pb@PA$_{1.0}$Char350 sample. These lead phosphates contribute to the enhanced stability of Pb in this char, so the bioavailability and leachability of Pb were suppressed. However, as the pyrolysis temperature increased to 450 °C, the elemental Pb and unstable Pb compounds (PbO and $Pb_3(CO_3)(OH)_2$) were formed according to the XRD patterns (Figure 6), which can explain the results obtained in the TCLP test, in which more leachable Pb was observed in the chars obtained at a high pyrolysis temperature (450 °C) (Figure 5). Further, a new peak at 1156 cm^{-1} was found in the IR spectra (Figure 7) of two chars derived from the PA-pretreated biomass (Pb@PA$_{1.0}$Char350 and Pb@PA$_{1.0}$Char450), which could be assigned to C–O stretching vibrations in P–O–C (aromatic) linkage [49]. Thus, Pb phosphates would be bound to the carbon matrix through this linkage, which made Pb be stably anchored on the char [52].

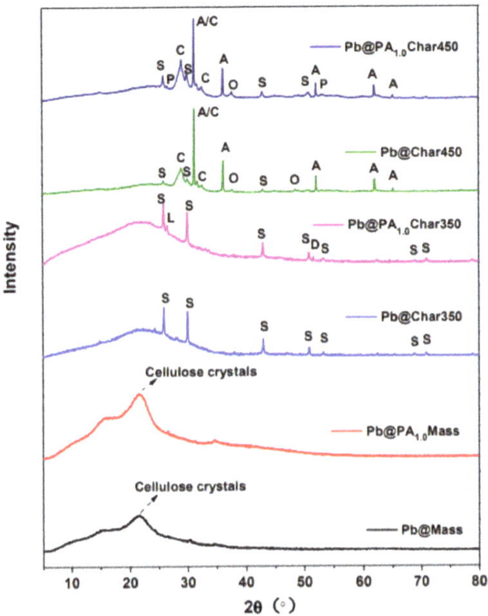

Figure 6. XRD patterns of two biomass and four char samples (A: Pb, C: $Pb_3(CO_3)(OH)_2$, D: $Ca_8Pb_2(PO_4)_6(OH)_2$, L: $Pb_9(PO_4)_6$, O: PbO, P: $Pb_5(PO_4)_3(OH)$, and S: PbS).

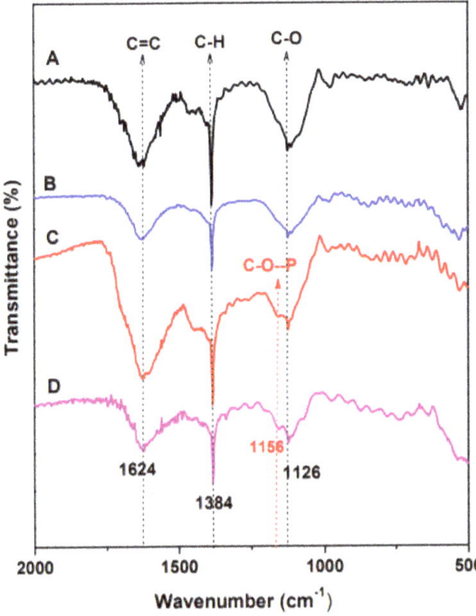

Figure 7. IR spectra of four char samples (A: Pb@Char350, B: Pb@Char450, C: Pb@$PA_{1.0}$Char350, and D: Pb@$PA_{1.0}$Char450).

4. Conclusions

The combined biosorption-pyrolysis process can be used for efficient removal of Pb from aqueous solution and then immobilization of Pb in the char, when the *Hydrocotyle* biomass is pretreated with PA. The PA-pretreatment increased the surface area and introduced more functional groups in the biomass, and enhanced the removal efficiency of Pb by biosorption. The subsequent pyrolysis of the Pb-contaminated biomass at relatively low temperatures (350 and 450 °C) retained most of the Pb in the char, and use of the PA-pretreatment further increased the retention efficiency to >95%. Furthermore, both the bioavailability and leaching potential of Pb in the char were significantly reduced by pyrolysis with the aid of PA-pretreatment, and pyrolysis at 350 °C is more favorable for the immobilization of Pb than that at 450 °C. The immobilization mechanisms are related to the formation of Pb phosphates and their linkage to the carbon matrix, according to the investigations with XRD and IR spectra. Despite the promising results obtained in this study, the long-time stability of Pb in the char deserves further investigation by considering more environmental factors, such as pH, temperature, humidity, and microbial effect, before the safe disposal of the char.

Supplementary Materials: The following are available online at http://www.mdpi.com/2073-4441/12/9/2381/s1, Figure S1: Change of the quantity of Pb adsorbed (Q_t) by the biosorbents with the contact time (t), Figure S2: Change of the quantity of Pb adsorbed after equilibrium (Q_e) with the initial pH, Figure S3: Change of the quantity of Pb adsorbed after equilibrium (Q_e) with the initial concentration (C_0), and Table S1: Parameters obtained by fitting the kinetic data (in Figure S1) with the pseudo-second order model.

Author Contributions: Conceptualization, J.L. (Jianfa Li); methodology, Y.W. and J.L. (Jinhong Lü); investigation, Y.W., D.F., S.G. and J.L. (Jinhong Lü); data curation, J.L. (Jinhong Lü); writing—original draft preparation, Y.W.; writing—review and editing, J.L. (Jinhong Lü); funding acquisition, J.L. (Jianfa Li); All authors have read and agreed to the published version of the manuscript.

Funding: This research was funded by the Natural Science Foundation of Zhejiang Province, China, grant number LY16B070003, and by the National Natural Science Foundation of China, grant number 41271475.

Conflicts of Interest: The authors declare no conflict of interest.

References

1. Ungureanu, G.; Santos, S.; Boaventura, R.; Botelho, C. Arsenic and antimony in water and wastewater: Overview of removal techniques with special reference to latest advances in adsorption. *J. Environ. Manag.* **2015**, *151*, 326–342. [CrossRef] [PubMed]
2. Ayangbenro, A.S.; Babalola, O.O. A new strategy for heavy metal polluted environments: A review of microbial biosorbents. *Int. J. Environ. Res. Public Health* **2017**, *14*, 94. [CrossRef] [PubMed]
3. Czikkely, M.; Neubauer, E.; Fekete, I.; Ymeri, P.; Fogarassy, C. Review of heavy metal adsorption processes by several organic matters from wastewaters. *Water* **2018**, *10*, 1377. [CrossRef]
4. Fernández-González, R.; Martín-Lara, M.Á.; Blázquez, G.; Pérez, A.; Calero, M. Recovering metals from aqueous solutions by biosorption onto hydrolyzed olive cake. *Water* **2019**, *11*, 2519. [CrossRef]
5. Filote, C.; Volf, I.; Santos, S.C.; Botelho, C.M. Bioadsorptive removal of Pb(II) from aqueous solution by the biorefinery waste of Fucusspiralis. *Sci. Total Environ.* **2019**, *648*, 1201–1209. [CrossRef]
6. Farooq, U.; Kozinski, J.A.; Khan, M.A.; Athar, M. Biosorption of heavy metal ions using wheat based biosorbents—A review of the recent literature. *Bioresour. Technol.* **2010**, *101*, 5043–5053. [CrossRef]
7. Abdolali, A.; Guo, W.S.; Ngo, H.H.; Chen, S.S.; Nguyen, N.C.; Tung, K.L. Typical lignocellulosic wastes and by-products for biosorption process in water and wastewater treatment: A critical review. *Bioresour. Technol.* **2014**, *160*, 57–66. [CrossRef]
8. Shaikh, R.B.; Saifullah, B.; Rehman, F.U. Greener method for the removal of toxic metal ions from the wastewater by application of agricultural waste as an adsorbent. *Water* **2018**, *10*, 1316. [CrossRef]
9. Salman, M.; Athar, M.; Farooq, U. Biosorption of heavy metals from aqueous solutions using indigenous and modified lignocellulosic materials. *Rev. Environ. Sci. BioTechnol.* **2015**, *14*, 211–228. [CrossRef]
10. Feng, N.; Guo, X.; Liang, S.; Zhu, Y.; Liu, J. Biosorption of heavy metals from aqueous solutions by chemically modified orange peel. *J. Hazard. Mater.* **2011**, *185*, 49–54. [CrossRef]

11. Abdolali, A.; Ngo, H.H.; Guo, W.; Zhou, J.L.; Du, B.; Wei, Q.; Wang, X.C.; Nguyen, P.D. Characterization of a multi-metal binding biosorbent: Chemical modification and desorption studies. *Bioresour. Technol.* **2015**, *193*, 477–487. [CrossRef]
12. Pintor, A.M.; Vieira, B.R.; Boaventura, R.A.; Botelho, C.M. Removal of antimony from water by iron-coated cork granulates. *Sep. Purif. Technol.* **2020**, *233*, 116020. [CrossRef]
13. Martín-Lara, M.A.; Pagnanelli, F.; Mainelli, S.; Calero, M.; Toro, L. Chemical treatment of olive pomace: Effect on acid-basic properties and metal biosorption capacity. *J. Hazard. Mater.* **2008**, *156*, 448–457. [CrossRef] [PubMed]
14. Buasri, A.; Chaiyut, N.; Tapang, K.; Jaroensin, S.; Panphrom, S. Equilibrium and kinetic studies of biosorption of Zn(II) ions from wastewater using modified corn cob. *APCBEE Procedia* **2012**, *3*, 60–64. [CrossRef]
15. Dada, A.O.; Olalekan, A.P.; Olatunya, A.M.; Dada, O.J.I.J.C. Langmuir, Freundlich, Temkin and Dubinin–Radushkevich isotherms studies of equilibrium sorption of Zn^{2+} unto phosphoric acid modified rice husk. *IOSR J. Appl. Chem.* **2012**, *3*, 38–45.
16. Xu, Y.L.; Song, S.Y.; Chen, J.D.; Chi, R.A.; Yu, J.X. Simultaneous recovery of Cu^{2+} and Pb^{2+} from metallurgical wastewater by two tandem columns fixed respectively with tetraethylenepentamine and phosphoric acid modified bagasse. *J. Taiwan Inst. Chem. Eng.* **2019**, *99*, 132–141. [CrossRef]
17. Naeem, M.A.; Imran, M.; Amjad, M.; Abbas, G.; Tahir, M.; Murtaza, B.; Zakir, A.; Shahid, M.; Bulgariu, L.; Ahmad, I. Batch and column scale removal of cadmium from water using raw and acid activated wheat straw biochar. *Water* **2019**, *11*, 1438. [CrossRef]
18. Nzihou, A.; Stanmore, B. The fate of heavy metals during combustion and gasification of contaminated biomass—A brief review. *J. Hazard. Mater.* **2013**, *256*, 56–66. [CrossRef]
19. Cao, X.; Ma, L.; Shiralipour, A.; Harris, W. Biomass reduction and arsenic transformation during composting of arsenic-rich hyperaccumulator *Pterisvittata* L. *Environ. Sci. Pollut. Res.* **2010**, *17*, 586–594. [CrossRef]
20. Bădescu, I.S.; Bulgariu, D.; Ahmad, I.; Bulgariu, L. Valorisation possibilities of exhausted biosorbents loaded with metal ions—A review. *J. Environ. Manag.* **2018**, *224*, 288–297. [CrossRef]
21. Liu, W.J.; Zeng, F.X.; Jiang, H.; Zhang, X.S.; Yu, H.Q. Techno-economic evaluation of the integrated biosorption–pyrolysis technology for lead (Pb) recovery from aqueous solution. *Bioresour. Technol.* **2011**, *102*, 6260–6265. [CrossRef] [PubMed]
22. Gonsalvesh, L.; Yperman, J.; Carleer, R.; Mench, M.; Herzig, R.; Vangronsveld, J. Valorisation of heavy metals enriched tobacco biomass through slow pyrolysis and steam activation. *J. Chem. Technol. Biotechnol.* **2016**, *91*, 1585–1595. [CrossRef]
23. Martín-Lara, M.Á.; Iáñez-Rodríguez, I.; Blázquez, G.; Quesada, L.; Pérez, A.; Calero, M. Kinetics of thermal decomposition of some biomasses in an inert environment: An investigation of the effect of lead loaded by biosorption. *Waste Manag.* **2017**, *70*, 101–113. [CrossRef]
24. Han, Z.; Guo, Z.; Zhang, Y.; Xiao, X.; Peng, C. Potential of pyrolysis for the recovery of heavy metals and bioenergy from contaminated *Broussonetia papyrifera* biomass. *BioResources* **2018**, *13*, 2932–2944. [CrossRef]
25. Gong, X.; Huang, D.; Liu, Y.; Zeng, G.; Wang, R.; Wei, J.; Huang, C.; Xu, P.; Wan, J.; Zhang, C. Pyrolysis and reutilization of plant residues after phytoremediation of heavy metals contaminated sediments: For heavy metals stabilization and dye adsorption. *Bioresour. Technol.* **2018**, *253*, 64–71. [CrossRef] [PubMed]
26. He, J.; Strezov, V.; Zhou, X.; Kumar, R.; Kan, T. Pyrolysis of heavy metal contaminated biomass pre-treated with ferric salts: Product characterisation and heavy metal deportment. *Bioresour. Technol.* **2020**, *313*, 123641. [CrossRef] [PubMed]
27. Stals, M.; Thijssen, E.; Vangronsveld, J.; Carleer, R.; Schreurs, S.; Yperman, J. Flash pyrolysis of heavy metal contaminated biomass from phytoremediation: Influence of temperature, entrained flow and wood/leaves blended pyrolysis on the behaviour of heavy metals. *J. Anal. Appl. Pyrol.* **2010**, *87*, 1–7. [CrossRef]
28. Dastyar, W.; Raheem, A.; He, J.; Zhao, M. Biofuel production using thermochemical conversion of heavy metal-contaminated biomass (HMCB) harvested from phytoextraction process. *Chem. Eng. J.* **2018**, *358*, 759–785. [CrossRef]
29. Chen, B.; Chen, Z.; Lv, S. A novel magnetic biochar efficiently sorbs organic pollutants and phosphate. *Bioresour. Technol.* **2011**, *102*, 716–723. [CrossRef]
30. Li, L.; Zou, D.; Xiao, Z.; Zeng, X.; Zhang, L.; Jiang, L.; Wang, A.; Ge, D.; Zhang, G.; Liu, F. Biochar as a sorbent for emerging contaminants enables improvements in waste management and sustainable resource use. *J. Clean. Prod.* **2019**, *210*, 1324–1342. [CrossRef]

31. Kim, H.; Ko, R.-A.; Lee, S.; Chon, K. Removal efficiencies of manganese and iron using pristine and phosphoric acid pre-treated biochars made from banana peels. *Water* **2020**, *12*, 1173. [CrossRef]
32. Stals, M.; Carleer, R.; Reggers, G.; Schreurs, S.; Yperman, J. Flash pyrolysis of heavy metal contaminated hardwoods from phytoremediation: Characterisation of biomass, pyrolysis oil and char/ash fraction. *J. Anal. Appl. Pyrol.* **2010**, *89*, 22–29. [CrossRef]
33. He, J.; Strezov, V.; Kan, T.; Weldekidan, H.; Asumadu-Sarkodie, S.; Kumar, R. Effect of temperature on heavy metal (loid) deportment during pyrolysis of Avicennia marina biomass obtained from phytoremediation. *Bioresour. Technol.* **2019**, *278*, 214–222. [CrossRef]
34. Li, S.; Zou, D.; Li, L.; Wu, L.; Liu, F.; Zeng, X.; Wang, H.; Zhu, Y.; Xiao, Z. Evolution of heavy metals during thermal treatment of manure: A critical review and outlooks. *Chemosphere* **2020**, *247*, 125962. [CrossRef] [PubMed]
35. Han, H.; Hu, S.; Syed-Hassan, S.S.A.; Xiao, Y.; Wang, Y.; Xu, J.; Xiang, J. Effects of reaction conditions on the emission behaviors of arsenic, cadmium and lead during sewage sludge pyrolysis. *Bioresour. Technol.* **2017**, *236*, 138–145. [CrossRef] [PubMed]
36. Al Chami, Z.; Amer, N.; Smets, K.; Yperman, J.; Carleer, R.; Dumontet, S.; Vangronsveld, J. Evaluation of flash and slow pyrolysis applied on heavy metal contaminated Sorghum bicolor shoots resulting from phytoremediation. *Biomass Bioenergy* **2014**, *63*, 268–279. [CrossRef]
37. He, J.; Strezov, V.; Kumar, R.; Weldekidan, H.; Jahan, S.; Dastjerdi, B.H.; Zhou, X.; Kan, T. Pyrolysis of heavy metal contaminated *Avicennia marina* biomass from phytoremediation: Characterisation of biomass and pyrolysis products. *J. Clean. Prod.* **2019**, *234*, 1235–1245. [CrossRef]
38. Lievens, C.; Yperman, J.; Vangronsveld, J.; Carleer, R. Study of the potentialvalorization of heavy metal contaminated biomass via phytoremediation by fast pyrolysis: Part I. Influence of temperature, biomass species and solid heat carrier on the behaviour of heavy metals. *Fuel* **2008**, *87*, 1894–1905. [CrossRef]
39. Li, S.; Zhang, T.; Li, J.; Shi, L.; Zhu, X.; Lü, J.; Li, Y. Stabilization of Pb(II) accumulated in biomass through phosphate-pretreated pyrolysis at low temperatures. *J. Hazard. Mater.* **2017**, *324*, 464–471. [CrossRef]
40. Shi, L.; Wang, L.; Zhang, T.; Li, J.; Huang, X.; Cai, J.; Wang, Y. Reducing the bioavailability and leaching potential of lead in contaminated water hyacinth biomass by phosphate-assisted pyrolysis. *Bioresour. Technol.* **2017**, *241*, 908–914. [CrossRef]
41. Zhang, T.; Wang, Y.; Liu, X.; Lü, J.; Li, J. Functions of phosphorus additives on immobilizing heavy metal cadmium in the char through pyrolysis of contaminated biomass. *J. Anal. Appl. Pyrol.* **2019**, *144*, 104721. [CrossRef]
42. Zeng, X.; Xiao, Z.; Zhang, G.; Wang, A.; Li, Z.; Liu, Y.; Wang, H.; Zeng, Q.; Liang, Y.; Zou, D. Speciation and bioavailability of heavy metals in pyrolytic biochar of swine and goat manures. *J. Anal. Appl. Pyrol.* **2018**, *132*, 82–93. [CrossRef]
43. Lu, H.; Zhang, W.; Yang, Y.; Huang, X.; Wang, S.; Qiu, R. Relative distribution of Pb^{2+} sorption mechanisms by sludge-derived biochar. *Water Res.* **2012**, *46*, 854–862. [CrossRef] [PubMed]
44. Xu, X.; Cao, X.; Zhao, L. Comparison of rice husk-and dairy manure-derived biochars for simultaneously removing heavy metals from aqueous solutions: Role of mineral components in biochars. *Chemosphere* **2013**, *92*, 955–961. [CrossRef]
45. Zhang, T.; Zhu, X.; Shi, L.; Li, J.; Li, S.; Lü, J.; Li, Y. Efficient removal of lead from solution by celery-derived biochars rich in alkaline minerals. *Bioresour. Technol.* **2017**, *235*, 185–192. [CrossRef]
46. Cantrell, K.B.; Hunt, P.G.; Uchimiya, M.; Novak, J.M.; Ro, K.S. Impact of pyrolysis temperature and manure source on physicochemical characteristics of biochar. *Bioresour. Technol.* **2012**, *107*, 419–428. [CrossRef]
47. Blázquez, G.; Ronda, A.; Martín-Lara, M.A.; Pérez, A.; Calero, M. Comparative study of isotherm parameters of lead biosorption by two wastes of olive-oil production. *Water Sci. Technol.* **2015**, *72*, 711–720. [CrossRef]
48. Patra, C.; Shahnaz, T.; Subbiah, S.; Narayanasamy, S. Comparative assessment of raw and acid-activated preparations of novel Pongamiapinnata shells for adsorption of hexavalent chromium from simulated wastewater. *Environ. Sci. Pollut. R.* **2020**, *27*, 14836–14851. [CrossRef]
49. Puziy, A.M.; Poddubnaya, O.I.; Martınez-Alonso, A.; Suárez-Garcıa, F.; Tascón, J.M.D. Synthetic carbons activated with phosphoric acid: I. Surface chemistry and ion binding properties. *Carbon* **2002**, *40*, 1493–1505. [CrossRef]

50. Shu, Y.; Tang, C.; Hu, X.; Jiang, L.; Hu, X.; Zhao, Y. H_3PO_4-activated cattail carbon production and application in chromium removal from aqueous solution: Process optimization and removal mechanism. *Water* **2018**, *10*, 754. [CrossRef]
51. Zhao, L.; Cao, X.; Zheng, W.; Kan, Y. Phosphorus-assisted biomass thermal conversion: Reducing carbon loss and improving biochar stability. *PLoS ONE* **2014**, *9*, e115373. [CrossRef] [PubMed]
52. Zhao, L.; Cao, X.; Zheng, W.; Scott, J.W.; Sharma, B.K.; Chen, X. Copyrolysis of biomass with phosphate fertilizers to improve biochar carbon retention, slow nutrient release, and stabilize heavy metals in soil. *ACS Sustain. Chem. Eng.* **2016**, *4*, 1630–1636. [CrossRef]
53. Yuan, J.H.; Xu, R.K.; Zhang, H. The forms of alkalis in the biochar produced from crop residues at different temperatures. *Bioresour. Technol.* **2011**, *102*, 3488–3497. [CrossRef] [PubMed]

Article

Use of Chemically Treated Human Hair Wastes for the Removal of Heavy Metal Ions from Water

Helan Zhang [1], Fernando Carrillo-Navarrete [2,3], Montserrat López-Mesas [1] and Cristina Palet [1,*]

[1] Centre Grup de Tècniques de Separació en Química, Unitat de Química Analítica, Departament de Química, Universitat Autònoma de Barcelona, 08193 Bellaterra, Spain; hlzhang1126@163.com (H.Z.); montserrat.lopez.mesas@uab.cat (M.L.-M.)

[2] Institut d'Investigació Tèxtil i Cooperació Industrial de Terrassa (INTEXTER), Universitat Politècnica de Catalunya (UPC), Colom 15, 08222 Terrassa, Spain; fernando.carrillo@upc.edu

[3] Departament d'Enginyeria Química, ESEIAAT—Universitat Politècnica de Catalunya (UPC), Colom 1, 08222 Terrassa, Spain

* Correspondence: cristina.palet@uab.cat; Tel.: +34-93-581-3475; Fax: +34-93-581-1985

Received: 1 April 2020; Accepted: 25 April 2020; Published: 29 April 2020

Abstract: Human hair is considered a ubiquitous waste product and its accumulation can cause environmental problems. Hence, the search for alternatives that take advantage of this waste as a new raw material is of interest, and contributes to the idea of the circular economy. In this study, chemically modified human hair was used as a low cost biosorbent for the removal of heavy metal ions from aqueous solutions. The effect of the contact time, the pH, and the biosorbent concentration on the biosorption process were investigated. Kinetic modeling indicated that the pseudo-second order kinetic equation fitted well with $R^2 > 0.999$. Furthermore, the equilibrium data fitted the Langmuir adsorption isotherm at 295 K resulting in saturation concentrations of 9.47×10^{-5}, 5.57×10^{-5}, 3.77×10^{-5}, and 3.61×10^{-5} mol/g for the sorption of Cr(III), Cu(II), Cd(II), and Pb(II), respectively. The biosorption process did not change the chemical structure and morphology of the hair, which was shown by FTIR and SEM. In addition, desorption experiments prove that 0.1 mol/L EDTA solution is an efficient eluent for the recovery of Pb(II) from the treated human hair. To summarize, treated human hair showed satisfactory biosorption capacity and can be considered as an effective biosorbent for the treatment of water with a low concentration of heavy metal ions.

Keywords: human hair; heavy metal; kinetics; isotherms; biosorption

1. Introduction

Heavy metal ions are considered extremely harmful to humans, aquatic organisms, and other life forms because of their toxicity, accumulation, and non-biodegradable nature, causing various diseases and disorders [1]. Hence, the removal of heavy metal ions from wastewater has attracted attention for the protection of public health and the environment [2]. Conventional methods for removing heavy metal ions, including chemical precipitation, flotation, ion exchange, evaporation, and membrane processes are practical and cost-effective only with high strength wastewater (which contains high concentration levels of pollutants), and they are ineffective when applied to low strength aqueous effluents with heavy metal ion concentrations less than 100 ppm [3]. Adsorption techniques currently play an important role in the removal of heavy metal ions from wastewater, offering considerable advantages, such as low-cost, availability, profitability, ease of operation, and efficiency [4,5]. Various materials have been developed as adsorbents for the removal of heavy metal ions. In particular, activated carbon is frequently used as an adsorbent due to its high surface area, high adsorption capacity, and high degree of surface reactivity [6]. However, activated carbon is relatively expensive and is difficult to recycle by eluting the heavy metal ions because of the strong interaction between activated carbon and heavy metal ions. Waste biogenic materials are considered ideal alternative biosorbents for the removal of heavy metal ions from low strength wastewater due to their relatively good cost-effective adsorption capacity [7]. Accordingly, various biogenic materials, including chitosan derivatives [8], agricultural waste materials [9], chicken feathers [10,11], cork waste [12,13], rubber leaf powder [14], chemically modified plant waste [15], and soybean stalks [16], among others, have been proposed and applied as biosorbents to effectively remove heavy metal ions. The good biosorption properties of these biogenic materials are attributed to the presence of abundant metal binding functional groups of these materials, such as carbonyl, carboxyl, hydroxyl, sulphate, sulfur, phosphate, and amido and amino groups [17].

Among natural resources, keratinous materials can be used as biosorbents, either directly or after activation, to remove heavy metal ions due to their intricate networks characterized by high stability, insolubility in water, and high surface area containing many carboxyl, amido, and sulfur functional groups [18]. In addition, keratin is an abundant nonfood protein; in fact, it is the major component of wool, hair, horns, nails, and feathers. Moreover, keratin wastes, such as feathers, horns, nails from butchery [19], human hair from hairdressers, poor quality raw wools from sheep breeding, and some by-products from the textile industry, amount globally to more than four millions tons per year [20]. Several examples of the use of keratinous materials have been already reported, especially using modified keratinous materials. Al-Asheh et al. compared adsorption capacities between inactivated and chemically activated chicken feathers as a biosorbent for removing heavy metal ions (i.e., Cu(II) and Zn(II)) from wastewater [21]. Park et al. prepared wool and silk blend nanofibrous membranes by electrospinning, which exhibited an excellent performance as an adsorbent of heavy metal ions [22]. The Aluigi research group successfully prepared keratin-rich nanofiber mats by electrospinning wool keratin/polyamide blends. This material shows good adsorption capacity for Cu(II) ions from water, with the adsorption capacity increasing with the increase of the specific surface area of the nanofiber mats [23].

Keratinous-composed human hair is considered a ubiquitous waste product and its accumulation can cause environmental problems. Hence, the search for alternatives that take advantage of this waste as a new raw material is of interest, and contributes to the idea of circular economy. In this sense, human hair can contribute significantly in many critical areas of public importance, such as agriculture, medicine, construction materials, and pollution control [24]. In particular, the presence of carboxyl, amido, and disulfide groups in human hair suggest this waste product could be a good biosorbent of several chemicals, including heavy metals, although it has been rarely studied for this application [11]. In this regard, one of the major drawbacks is that its hydrophobic nature in native form limits the diffusion of heavy metal ions from the solution to the external surface of the human hair [22]. To overcome this issue, disulfide bonds present in human hair can be readily oxidized to yield the corresponding cysteic acid residues, which increase the hydrophilic properties of the human hair and subsequently improve its capability to bind positively charged metal ions [25]. For this reason, oxidized human hair was chosen in the present work as a biosorbent for the removal of heavy metal ions from aqueous solutions. Environmental parameters affecting the biosorption process, such the pH value, biosorbent concentration, and contact time were studied. In addition, FT-IR and SEM analysis were conducted for the structural and morphological characterization of the biosorbent after the oxidation pretreatment and after the subsequent heavy metal biosorption process. The kinetic and isotherm data experimentally obtained were correlated with the established kinetic models (pseudo-first order, pseudo-second order, and Weber–Morris intraparticle diffusion model) and with well-known thermodynamic models (Freundlich and Langmuir). A comparison between these was performed. Finally, a desorption/regeneration test was performed in order to study the reusability of the biosorbent.

2. Experimental

2.1. Chemicals

All the chemicals used in this work were of analytical grade. Stock solutions of separate heavy metal ions, such as Cr(III), Mn(II), Ni(II), Co(II), Cu(II), Zn(II), Cd(II), and Pb(II) were prepared by dissolving their nitric salts (>99%, all from Panreac, Spain) in deionized water. A 1000 ppm stock solution of metal ions was first prepared, which was then diluted to the initial heavy metal concentration for each experiment. Sodium hydroxide (>98%, Panreac, Spain) and nitric acid (>70%, JT-Baker, Spain) were alternatively used for the pH adjustment of the initial aqueous solution prior to commencing the biosorption experiments. In all the experiments, the initial pH was measured, and usually the final pH was also checked, using an Omega 300 pH meter (Crison instruments, S.A., Spain).

2.2. Human Hair

Human hair waste (from different male individuals of approximately 13 years of age) was collected from local barbershops. The human hair samples were mixed together, washed with common laboratory detergent, rinsed several times with deionized water (purified with a milli-Q Gradient system from Millipore Corporation) and then left to dry at room temperature (22 ± 1 °C). The hair was cut to an approximate length of 1–2 mm using scissors.

2.3. Chemical Treatment of Human Hair

The treatment process of the human hair was carried out as follows: 20.0 g of the untreated human hair was weighed and soaked in 400 mL of the pretreatment reagent of known concentration (10% H_2O_2, originally at 35% in water, from Sigma-Aldrich, Germany) and at adjusted pH of 9 (pH 9.0 yields better biosorption results in comparison with others, when pH 5.0, pH 7.0 and pH 9.0 are assayed) [26]. After a given soaking time (5 h), the solution was filtered. The hair separated from the solution was washed many times with deionized water. To minimize any loss of the treated hair, at each washing step, the hair was separated by centrifugation, and the liquid was then decanted. Finally, the treated and cleaned hair was dried at room temperature.

2.4. Characterization of Human Hair

Structural characterization of the human hair was carried out to analyze any chemical change produced in the samples after the oxidative pretreatment and/or after the biosorption of heavy metals. The identification of the functional groups in the untreated and treated human hair was performed using a Fourier transform infrared (FT-IR) spectrometer (Tensor 27, Bruker, Germany). The spectra were recorded in the range of 600–4000 cm^{-1} with 16 scans and a resolution of 4 cm^{-1}. The surface morphology of the human hair samples (untreated and treated) was observed by scanning electron microscope (SEM; ZEISS EVO® MA 10, Oberkochen, Germany). The samples used the sputter-coating arrangement.

2.5. Heavy Metal Ions Biosorption Experiments

The uptake of heavy metal ions onto the hair systems was carried out by batch experiments at a constant temperature (22 ± 1 °C) on a rotary mixer (CE 2000 ABT-4, SBS Instruments SA, Barcelona, Spain) at 25 rpm. In all sets of experiments, 0.1 g of human hair (untreated or treated) was weighed in 50 mL plastic extraction tubes; the 10 mL of heavy metal ion aqueous solution was added and the final system was shaken for a certain period of time. Usually, the concentration of each heavy metal ion was 0.10 mmol/L (for multiple heavy metal system containing eight ions) and 0.18 mmol/L (in the multiple-metal system containing four ions, and also in the single-metal systems). The pH was adjusted to 4.0, unless otherwise specified. The initial pH of the multiple heavy metal aqueous solution was varied within the range 1.0 to 6.0 (higher pH values were not considered to avoid precipitation of metal hydroxides). To study the effect of the biosorbent concentration on metal uptake, its mass was varied from 1 to 20 g/L. The effect of the initial metal ion concentration on biosorption isotherms was studied in single-metal systems with 0.1 g of treated human hair. A range of initial metal ion concentrations from 0.5×10^{-3} to 2.0 mmol/L was used. In all cases, after agitation, the two phases were separated by decantation and the liquid was filtered through 0.22 μm syringe Millipore filters (Millex-GS, Millipore, Ireland). Then, the heavy metal concentration in the remaining aqueous solution was determined by an inductively coupled plasma optical emission spectrophotometer with mass detector, ICP-MS (XSERIES 2 ICP-MS, Thermo Scientific, Bremen, Germany).

The uptake of the metal ions by human hair was calculated using Equation (1), which quantifies the biosorption efficacy:

$$\%\ biosorption = \frac{C_i - C_f}{C_i} \times 100 \quad (1)$$

where C_i and C_f are the initial and the final concentration of heavy metal in the aqueous phase solution, respectively (in mmol/L).

The amount of metal sorbed per unit of mass of biosorbent at time t (q_t in mmol/g) was calculated using Equation (2):

$$q_t(\text{mmol/g}) = \frac{(C_i - C_f) \times V}{W} \quad (2)$$

where V is the total volume of the solution (in L), W is the amount of biosorbent (in g), and C_i and C_f are the initial and the final concentrations of heavy metal in the aqueous solution (each given in units of mmol/L), respectively.

The amount of metal sorbed at the equilibrium per unit of mass of biosorbent (q_e in mmol/g) was obtained as follows:

$$q_e(\text{mmol/g}) = \frac{(C_i - C_e) \times V}{W} \quad (3)$$

where V is the total volume of the solution (in L), W is the amount of biosorbent (in g), and C_i and C_e are the initial and equilibrium concentrations of heavy metal in the aqueous solution (each given in units of mmol/L), respectively.

All batch biosorption experiments were carried out in duplicate and the results are reported as their average in the corresponding figures (experimental errors found were less than 2.5% and 0.0025 mmol/g in the biosorption percentage and the biosorption capacity, respectively).

2.6. Desorption, Regeneration and Reuse

Desorption experiments were performed only for the removal of Pb(II) from treated human hair samples as biosorbent. Each hair sample containing the adsorbed Pb(II) was contacted and stirred with 10 mL of 0.1 mol/L HNO_3 or 10 mL of 0.1 mol/L EDTA. After 24 h of mixing (with the rotary mixer) at room temperature (22 ± 1 °C), the aqueous and solid phases were separated by centrifugation and subsequent filtration, and the Pb(II) content of the final solution was analyzed by ICP-MS, as indicated in Section 2.5. Desorption percentage was calculated using Equation (4):

$$\%\ desorption = \frac{amount \quad of \quad Pb(II) \quad desorbed}{amount \quad of \quad Pb(II) \quad adsorbed} \times 100 \quad (4)$$

The reuse of the treated human hair in a second biosorption step, after elution of the adsorbed metal ions (with nitric acid or EDTA), requires the cleaning of the remaining eluting solution from the surface of the biomaterial. The treated human hair was washed with deionized water and dried in an oven at 40 °C overnight. These regenerated human hair samples were employed to adsorb heavy metals again.

All batch biosorption, desorption, and regeneration experiments were carried out in duplicate and the results are reported as their average in the corresponding figures (experimental errors found were less than 2% and 0.002 mmol/g, in the biosorption and desorption percentages and the biosorption capacity, respectively).

3. Results and Discussion

3.1. Comparison of Biosorption Efficacy between Untreated and Treated Human Hair

The oxidation of human hair usually takes place with hydrogen peroxide in an acid or alkaline medium by attacking the disulfide bond of keratin. As a result of this reaction, sulfonic acid groups are formed, and hair samples are functionalized. In an alkaline medium, the oxidation process is much more effective [26]. Moreover, during oxidation, other proteins in the human hair are also oxidized, which leads to cell membrane damage causing the cortex and the cuticle to open and separate. The objective of this pretreatment is to increase the hydrophilicity of the human hair surface and also to increase its specific surface area. The biosorption capacities of untreated and treated human hair samples for recovering eight metal ions (Cr(III), Mn(II), Ni(II), Co(II), Cu(II), Zn(II), Cd(II), and Pb(II)) were determined. The obtained results are shown in Figure 1. As seen from the figure, the metal biosorption capacity of chemically treated human hair is significantly better than that of untreated human hair. Moreover, the affinity of both types of hair for Cr(III), Cu(II), and Pb(II) is greater than that for the other metal ions, which can be explained by the stronger interactions between the functional groups of the biosorbent and these three metal ions. The metal biosorption onto the treated human hair follows in the order of Cr(III) > Pb(II) > Cu(II) > Cd(II) > Ni(II) > Co(II) > Mn(II) > Zn(II). Finally, four metal ions, namely, Cr(III), Cu(II), Cd(II), and Pb(II), were selected from those metals to study the biosorption mechanism of treated human hair in subsequent experiments.

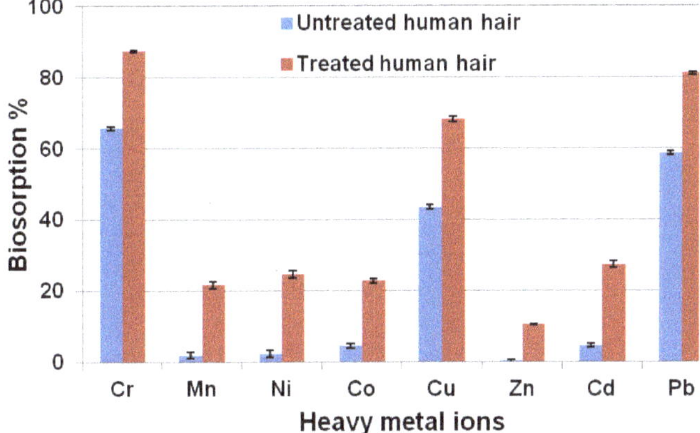

Figure 1. Comparison of biosorption between untreated and treated human hair in a multiple-metal system. The initial metal ion concentration was 0.1 mmol/L, the contact time was 24 h, the pH was 4.0, and the biosorbent was 0.1 g in 10 mL of the initial solution.

3.2. FT-IR and SEM Characterization

In order to understand how metal ions bind to the biosorbent, it is essential to identify the functional groups of its surface as these could be responsible for the metal binding. Thus, the untreated, treated, and metal loaded-treated human hairs were discriminated by FT-IR, as can be seen from the infrared spectra collected in Figure 2. The full-scan spectra of human hair (Figure 2a) display their corresponding infrared peaks. The broad and medium intensity band ranging from 3000 to 3600 cm^{-1} is indicative of the stretches of the bonds belonging to the carboxylic acid (-COOH), alcohol (-OH), and amino acid (-NH_2) groups. The peaks located at 1632 cm^{-1} (amide I), 1520 cm^{-1} (amide II), and 1241 cm^{-1} (amide III) are related to typical human hair amino acids. The peaks at 1041, 1075, 1180, and 1229 cm^{-1} all correspond to different products of cystine oxidation in human hair, and their peak assignment belongs to sulfonate (S-O sym. stretch), cystine monoxide (R-SO-S-R), sulfonate (S-O asym. stretch), and cystine dioxide (R-SO2-S-R), respectively. Carefully comparing the spectra of the three different hair samples, some differences can be seen between them, as expected, due first to the oxidation process (treated human hair), and, later, to the metal biosorption (metal loaded-treated human hair), particularly in the region from 850 to 1750 cm^{-1} (see Figure 2b). The intensity of the peaks at 1041 and 1180 cm^{-1} increased after chemical pretreatment, which means that conversion of cystine to cysteic acid, cystine monoxide, and cystine dioxide, as well as to sulfonates, occurred during this treatment process. The weak broad shoulder between approximately 1000 and 1130 cm^{-1} in the untreated human hair infrared spectra is probably due to environmental factors, such as sunlight, chlorinated water, and frequent shampooing causing partial oxidation of the hair surface [27]. The FT-IR spectra from treated human hair before and after the metal biosorption are very similar, indicating that the main functional groups on treated human hair did not change during the metal biosorption process (which can be an indication of a possible reuse of such biomaterial). However, the slight differences found around 1400 cm^{-1}, and some red shift of the emission spectra (from 3277.2 cm^{-1} to 3274.8 cm^{-1}, from 1526.8 cm^{-1} to 1519.7 cm^{-1}, and from 1078.4 cm^{-1} to 1074.9 cm^{-1}), before and after the biosorption process, is probably related to the presence of the heavy metal ions on the hair surface. Based on the FT-IR spectra changes, as seen in Figure 2, some hair surface chemical functional groups (including hydroxyl, amino, carboxyl, and sulfonic acid) could act as important biosorption sites for heavy metal ions.

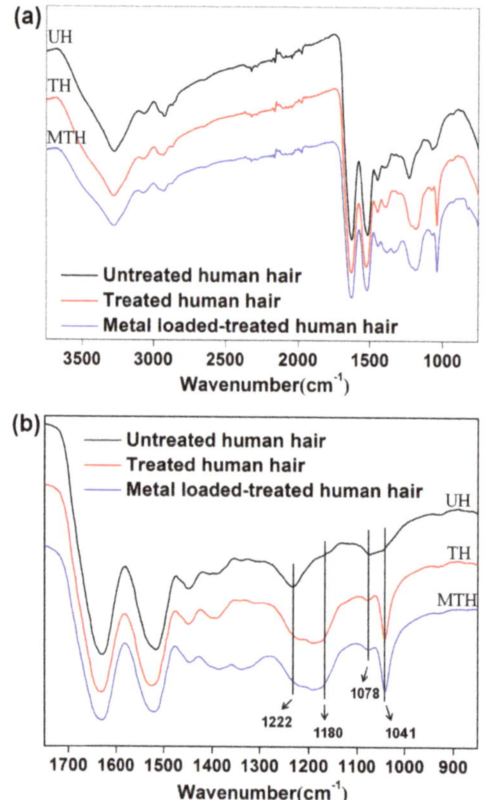

Figure 2. FT-IR spectra of human hair. UH: untreated human hair, TH: treated human hair, MTH: metal loaded-treated human hair. (**a**) Full-scan spectra, and (**b**) spectra in the range from 850 to 1750 cm^{-1}.

The scanning electron microscopy (SEM) technique was applied to address concern about the alteration of the human hair surface morphologies in the different cases of the study. Figure 3 shows the SEM micrographs of the untreated, treated, and metal-loaded treated human hair. It is observed that each cuticle scale of the human hair is uniquely shaped. Some have smooth rounded edges and others have jagged edges, overlapping each other as they ascend along the length of the fiber towards the tip (Figure 3). The surface topographies of the untreated and treated human hair are different (see Figure 3a,b for comparison): the majority of the cuticle scales of the treated human hair fibers represent a more jagged appearance, probably due to the oxidation treatment. After metal biosorption, the surface appears to be somewhat smoother compared with the hair prior to its use, suggesting that the cuticle scales are closed through biosorption, probably due to the acidic water media (see Figure 3b,c).

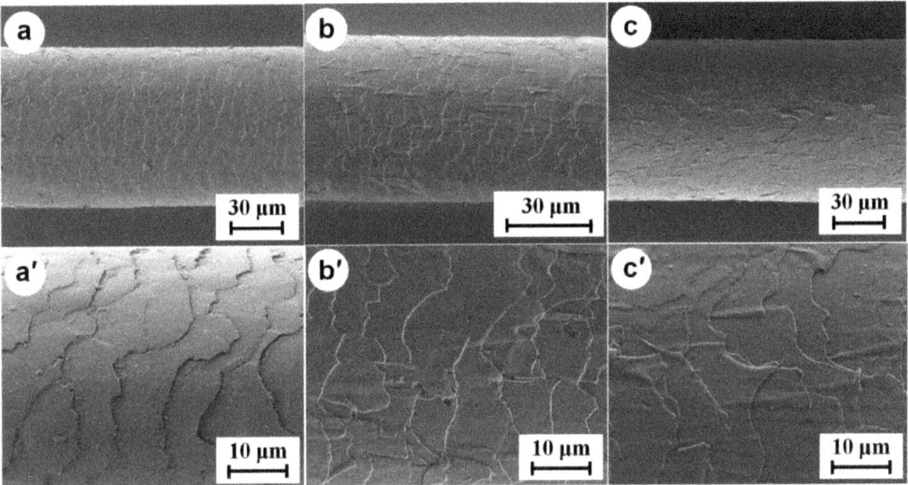

Figure 3. Scanning electron microscopy (SEM) micrographs of the human hair: (**a**) and (**a'**) correspond to the untreated human hair; (**b**) and (**b'**) correspond to the treated human hair; and (**c**) and (**c'**) correspond to the metal loaded-treated human hair.

3.3. Effect of the pH in Multiple Metal System

The pH level of the aqueous solution is an important variable for the biosorption of metal ions onto the biosorbents, due to the metal speciation and the speciation of the chemical functional groups present on the biosorbent's surface. The pH was controlled at the beginning and end of the experiments in order to evaluate any differences. The changes observed were lower than 0.3 units, and thus considered not significant.

The effect of the pH solution on the removal efficacy of the treated human hair for Cr(III), Cu(II), Cd(II), and Pb(II) was studied between pH 1.0 and 6.0 in the multiple-metal system (Figure 4). As observed from the results in Figure 4, the biosorption of metal ions increases significantly with increasing the pH. This behavior can be explained by the competition between the protons and the metal ions for the same binding site on the surface of the treated human hair. At low pH values, the surface of the biosorbent would also be surrounded by H^+ ions, which decrease the Cr(III), Cu(II), Cd(II), and Pb(II) ions interaction with binding sites of the treated human hair. As the pH increases, the basic forms of the chemical functional groups on the hair surface predominate, increasing negative charge, so metal cation biosorption increases. However, when the pH is around 5, the partial hydrolysis of metal ions (particularly for Cu; the remaining metals could occur at pH higher than 5) results in the formation of $M(OH)_n^{(m-n)}$ species affecting the biosorption hair capability. The biosorption percentage found could be related to the precipitation of the metals. Therefore, pH 4.0 was selected as the optimal condition in the subsequent experiments.

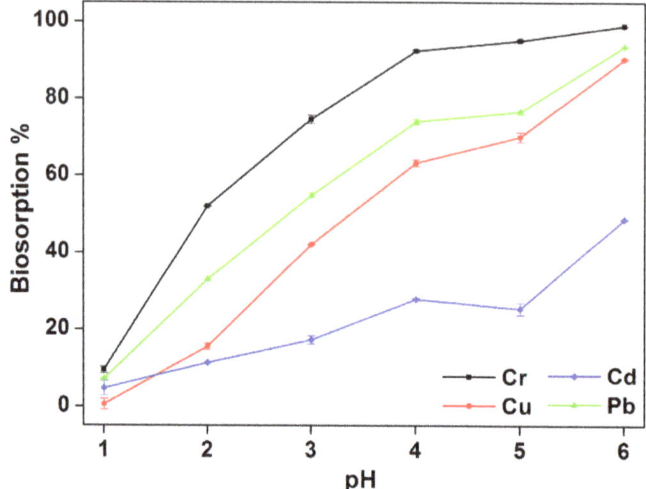

Figure 4. Effect of pH on the biosorption of the treated human hair for Cr(III), Cu(II), Cd(II), and Pb(II) in the multiple-metal system. The initial concentration was 0.18 mmol/L, the contact time was 24 h, and the biosorbent was 0.1 g in 10 mL initial solution.

3.4. *Effect of Biosorbent Concentration*

The effect of the biosorbent concentration on the removal efficacy of Cr(III), Cu(II), Cd(II), and Pb(II) ions was studied in the range of 1–20 g/L in a multiple-metal system (Figure 5). It was observed that the removal efficacy of the treated human hair for Cr(III), Cu(II), Cd(II), and Pb(II) ions increased with the increase of biosorbent concentration (Figure 5a). This can be explained by the increase in surface area of the biosorbent when increasing its amount, which in turn increases the binding sites. For Cr(III) and Pb(II), the sorbed metal ion (mmol) per unit weight of biosorbent significantly decreased by increasing the biosorbent concentration (Figure 5b). This can be explained due to the fact that at high biosorbent concentration, the available metal ions in the aqueous solution are insufficient to cover all the biosorbent sites due to the corroborated high affinity of these two metals (as can be seen from results collected in Figures 1 and 4). Furthermore, the metal uptakes (mmol/g) for Cu(II) and Cd(II) are basically stable with the increase of biosorbent concentration; this means the biosorption quantity of Cu(II) and Cd(II) increases through increasing the biosorbent concentration. Thus, functional groups induced on the biosorbent hair surface have stronger affinity for Cr(III) and Pb(II) than for Cu(II) and Cd(II).

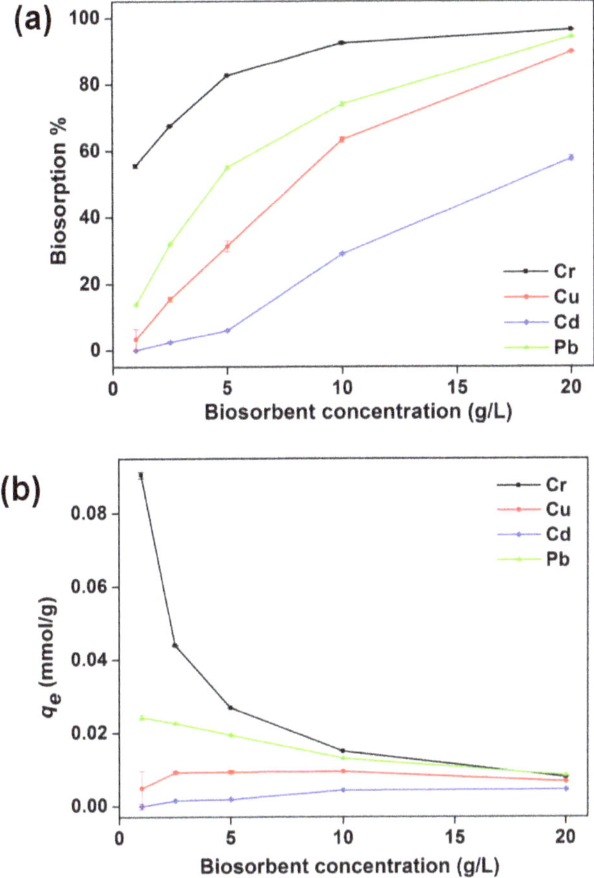

Figure 5. Effect of the biosorbent concentration on the percentage of biosorption (**a**) and on the amount of sorbed metal ion per unit weight of biosorbent (**b**) of the treated human hair for Cr(III), Cu(II), Cd(II), and Pb(II) in the multiple-metal system. The initial metal ions concentration was 0.18 mmol/L, the contact time was 24 h, and the pH of the 10 mL initial aqueous solution was 4.0.

3.5. Effect of Contact Time

Contact time with aqueous contaminated samples is an important parameter for successful usage of biosorbents in practice. Multiple- and single-metal aqueous systems (at pH = 4) of Cr(III), Cu(II), Cd(II), and Pb(II) were placed in contact with treated human hair (0.1 g) for periods of 5, 10, 20, 30, and 45 min, and 1, 2, 3, 4, 6, 12, 24, 48, and 72 h. Results plotted in Figure 6 show the biosorption capacity of treated human hair for removing Cr(III), Cu(II), Cd(II), and Pb(II) ions. Three steps can be differentiated during biosorption: the initial step with fast metal biosorption, the second step with gradual biosorption, and the third step, which can be related to the equilibrium uptake. The first step can be related to the diffusion of metal species from the solution to the external surface of the hair, which occurred instantaneously. The second step corresponds to a gradual biosorption uptake of heavy metal ions until reaching an equilibrium (the third stage). For each metal ion, the biosorption efficacy is higher for the single-metal system than for the multiple-metal system. Among these, the percentage of biosorption for Cd(II) is outstanding, with an increase from 29% to 86%. Furthermore, the single-metal system reached the biosorption equilibrium more rapidly than the multiple-metal system (around only 30 min in the former case), which is attributed to the effect of the competition between the heavy metal ions. Therefore, the selectivity order is Cr(III) > Pb(II) > Cu(II) > Cd(II), which corresponds to biosorption efficacy in the single-metal system of 98%, 96%, 95%, and 86%, respectively.

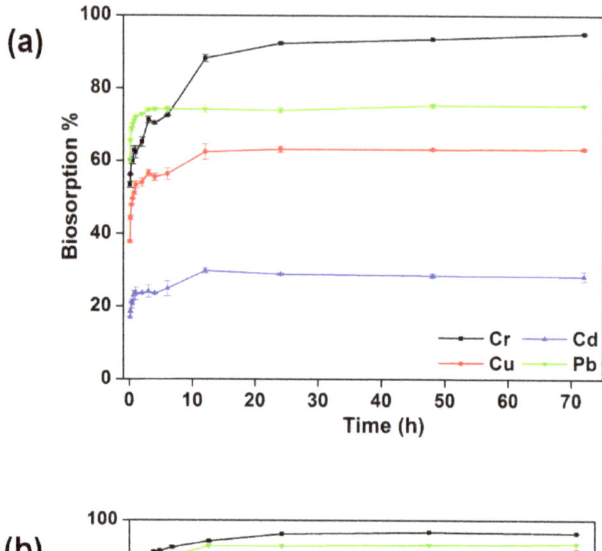

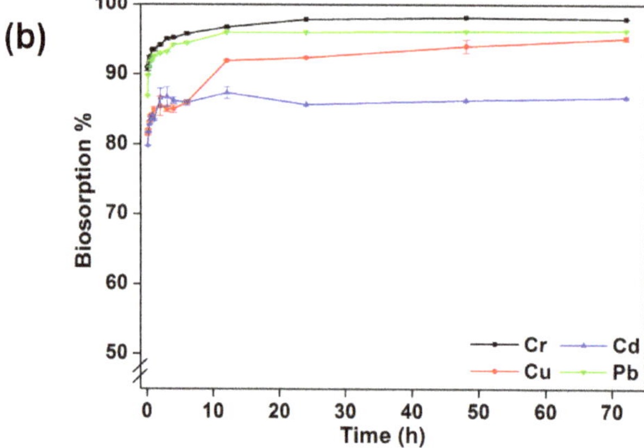

Figure 6. Percentage of biosorption of the treated human hair for Cr(III), Cu(II), Cd(II), and Pb(II) at different contact times. (**a**) Multiple-metal system, and (**b**) single-metal system. The initial metal ion concentration was 0.18mmol/L, the pH was 4.0, and the sorbent concentration was 10 g/L.

3.6. Kinetic Studies

Kinetic models have been used to model experimental data in order to investigate the mechanism of biosorption. Furthermore, it is important to determine the potential rate controlling steps, such as mass transport, chemical reaction, and intraparticle diffusion processes, in such systems. Many attempts have been made to formulate a general expression describing the kinetics of liquid–solid phase sorption systems [28]. In the present case, the kinetic models applied to the treated human hair as a biosorbent of heavy metals in solution were the pseudo-first order equation [29], the pseudo-second order equation, and the Weber–Morris intraparticle diffusion model. These are given by Equations (5)–(7), respectively:

$$\log(q_e - q_t) = \log q_e - \frac{k_1}{2.303}t \quad (5)$$

$$\frac{t}{q_t} = \frac{1}{k_2 q_e^2} + \frac{t}{q_e} \tag{6}$$

$$q_t = k_3 t^{1/2} + k_d \tag{7}$$

where, q_e and q_t are the amount of biosorbed metal ions per unit of mass biosorbent (in mmol/g) at the equilibrium and at time t (min), respectively; k_1 (in min^{-1}) is the rate constant of the pseudo-first order equation; k_2 (in g/(mmol·min)) is the rate constant of the pseudo-second order equation; k_3 is the intraparticle diffusion rate constant (in mmol/(g·$min^{1/2}$)); and k_d is the intercept that relates to the thickness of the boundary layer.

Experimental data were fitted to pseudo-first and pseudo-second order kinetic models and the rate corresponding constants (k), correlation coefficients (R^2), and q_e were estimated (values shown in Table 1). It is noteworthy that the pseudo-first order equation does not fit well for the whole range of time, which is generally applicable only over the initial time of the sorption processes, i.e., 30 and 60 min for the multiple- and single- metal biosorption systems, respectively [30]. Moreover, for the four metal ions in single and multiple systems, the calculated q_e did not match well with experimental data, which suggests the insufficiency of the pseudo-first order model to fit the experimental data.

The pseudo-second order model is more likely to predict kinetic behavior for the whole range of time studied, which indicates that chemical sorption is the rate-controlling step [31]. Correlation coefficients were always greater than 0.999, and the values of the predicted equilibrium biosorption capacities showed a good correlation with the experimental q_e values for all four metal ions in both systems. This shows that the biosorption process perfectly complies with the pseudo-second order model. In other words, the chemical sorption due to the formation of chemical bonds between the metal and sorbent in a monolayer onto the surface is the rate-controlling step [32]. The equilibrium biosorption capacities for Cr(III), Cu(II), Cd(II), and Pb(II) were 0.0166, 0.0174, 0.0165, and 0.0174 mmol/g, respectively, in the single-metal system, and 0.0155, 0.00945, 0.00429, and 0.0133 mmol/g, respectively, in the multiple-metal system.

Table 1. Sorption kinetic constants in the multiple- and single-metal systems for Cr(III), Cu(II), Cd(II), and Pb(II) for both pseudo-first and pseudo-second order models.

Metal		Cr		Cu		Cd		Pb	
System		Single	Multiple	Single	Multiple	Single	Multiple	Single	Multiple
Pseudo first order	$k_1 \times 10^3$ (min^{-1})	9.21 [a]	4.19 [b]	6.91 [a]	15.0 [b]	26.3 [a]	12.9 [b]	26.8 [a]	27.3 [b]
	q_e (mmol/g)	0.00132	0.00655	0.00263	0.00346	0.00151	0.00190	0.00173	0.00233
	R^2	0.9327	0.7343	0.9905	0.8794	0.9136	0.8592	0.8381	0.9030
Pseudo second order	k_2 (g/mmol min)	15.4	1.51	3.83	6.75	38.2	16.2	17.8	16.6
	q_e (mmol/g)	0.0166	0.0155	0.0174	0.00945	0.0165	0.00429	0.0174	0.0133
	R^2	1.000	0.9993	0.9999	0.9999	1.000	0.9997	1.000	1.000
	k_d (mmol/g)	0.0151	0.00831	0.0145	0.00517	0.0148	0.00223	0.0157	0.0103

[a]: 30 min, [b]: 1 h.

For the single-metal system, the kinetic constant values found for the pseudo-second order model (k_2) decreased in the following order: Cd (II) > Pb(II) > Cr(III) > Cu(III). This indicates that Cd (II) was more easily and rapidly adsorbed by treated human hair than Pb(II), Cr (III), and Cu (II). In this case, the sorption rate was lower for the heavy metal with the smallest ionic radius since Cu(II) has an ionic radius of 0.069 nm compared to 0.097 nm for Cd(II), 0.119 nm for Pb(II), and 0.070 nm for Cr(III).

The pseudo-first and pseudo-second order models cannot provide information about the diffusion mechanism controlling biosorption. Thus, the Weber–Morris intraparticle diffusion model was adjusted [33]. The plots of q_t versus $t^{1/2}$ are shown in the Figure 7. Unlike some simple cases, mathematical formulations representing the diffusion and biosorption are generally solvable analytically. In this case, these plots can be divided into multi-linear correlations, which indicates that the biosorption process take place in three steps and is not controlled solely by the intraparticle diffusion mechanism. The first stage corresponds to the sharper stage, where the metal ions move from the solution to the external surface of the biosorbent, through film diffusion, or the boundary layer diffusion of the metal species [34]. The second step describes the gradual biosorption onto the surface of the treated hair, where the intraparticle diffusion is the rate-limiting [35]. The third stage corresponds to the final biosorption equilibrium where the intraparticle diffusion starts to slow down due to extremely low metal ion concentration left in the solution. The presence of these three stages in the plots (Figure 7) suggests that the film diffusion and intraparticle diffusion were simultaneously controlling the biosorption process and both are enhanced with the increase of the initial metal concentration.

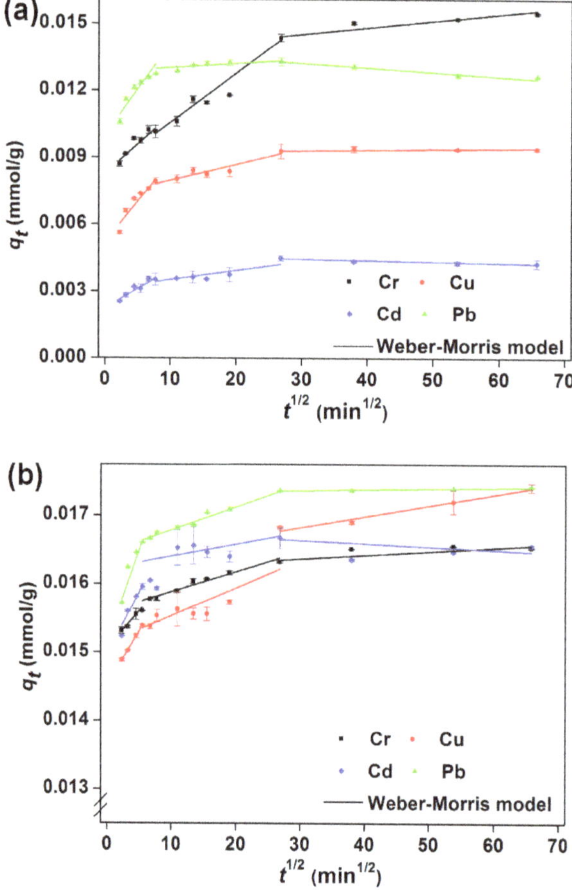

Figure 7. Weber–Morris intraparticle diffusion kinetic model applied for the metal ion biosorption onto the treated human hair, with time: (**a**) in the multiple-metal system, and (**b**) in the single-metal system.

3.7. Thermodynamic Isotherm Characterization

Sorption isotherms at equilibrium are very important data to understand the mechanism of each sorption system from a physicochemical perspective. The sorption capacity of a sorbent can be also described by the equilibrium sorption isotherm, which is characterized by some specific constants whose values provide information about the affinity between the liquid–solid sorption systems.

In the present study, two isotherm models were selected to fit the experimental data, namely, the Freundlich and Langmuir isotherm models [36]. The linear forms of the Freundlich and Langmuir isotherms are presented by Equations (8) and (9), respectively:

$$\log q_e = \log k_F + \frac{1}{n} \log C_e \tag{8}$$

$$\frac{C_e}{q_e} = \frac{b}{K_L} C_e + \frac{1}{K_L} \tag{9}$$

where C_e is the equilibrium concentration of the metal ion in the residual solution (in mol/L); q_e is the amount of the sorbed metal at the equilibrium per unit of mass of sorbent (in mol/g); k_F and n are Freundlich constants; and $K_L = Q_0 b$, where Q_0 and b are the Langmuir constants, corresponding to the saturation concentration of the sorbed metal ion per unit of mass of sorbent (in mol/g) and the ratio of sorption/desorption rates (in L/mol), respectively.

For both models and following Equations (8) and (9), log q_e versus log C_e and C_e/q_e versus C_e are calculated and compared with the experimental data, respectively. In addition, all the constants and correlation coefficients obtained for each model are summarized in Table 2.

Table 2. Freundlich and Langmuir isotherm constants for the biosorption of Cr(III), Cu(II), Cd(II), and Pb(II) by the treated human hair.

	Constant	Cr	Cu	Cd	Pb
Freundlich	$K_F \times 10^3$	1.56	2.87	0.546	0.247
	n	2.30	1.90	2.86	3.63
	R^2	0.8646	0.8402	0.9291	0.8607
Langmuir	$Q_0 \times 10^5$ (mol/g)	9.47	5.57	3.77	3.61
	$B \times 10^{-4}$ (L/mol)	1.07	2.06	8.64	8.04
	K_L (L/g)	1.01	1.15	3.26	2.90
	R^2	0.9912	0.9905	0.9952	1.000
	$-\Delta G^0$ (kJ/mol)	22.8	24.4	27.9	27.7

From the correlation coefficient values of both isotherm equations, it was observed that the Langmuir isotherm fitted the data better than the Freundlich isotherm, showing that the biosorption process relies on a specific site's sorption mechanism where adsorbate molecules occupy specific sites on the biosorbent. In Figure 8, experimental and calculated data for the Langmuir isotherm model are represented showing good correlation between the data. Taking on board the Langmuir equation, the saturated biosorption capacities of the treated human hair at 295 K for Cr(III), Cu(II), Cd(II), and Pb(II) were 9.47×10^{-5}, 5.57×10^{-5}, 3.77×10^{-5}, and 3.61×10^{-5} mol/g, respectively.

It is worth noting that the theoretical maximum values of adsorption capacity given by the Langmuir equation (Q_o) for the treated human hair were found to decrease in the order Cr (III) > Cu(II) > Cd(II) ~Pb(II). The metals with highest absorption capacities are those with lowest ionic radius, i.e., Cu(II) and Cr(III), while Pb(II), which has the largest ionic radius (0.119 nm), shows the smallest sorption capacity. This observed trend, based on the ionic radius, may be caused by the quick saturation of adsorption sites induced by the largest ions. This behavior is in agreement with that observed for the absorption of Pb(II), Cd (II), Ni(II), and Zn(II) using natural zeolite as a sorbent, where the adsorbed amount decreased as the ionic radius increased [37].

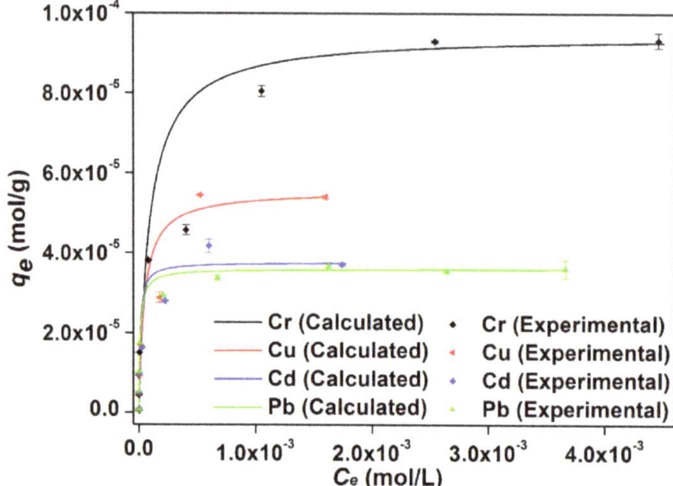

Figure 8. Experimental (.) and calculated (-) values adjusting using the Langmuir isotherm model for the sorption of Cr(III), Cu(II), Cd(II), and Pb(II) using the treated human hair.

In addition, from the estimated Langmuir sorption/desorption constant, the standard Gibb's free energy (ΔG^0) of the biosorption process can be evaluated by using Equation (10):

$$\Delta G^0 = -RT \ln b \tag{10}$$

where b is the Langmuir equilibrium constant shown in Equation (8), R is the universal gas constant (8.314 J/mol K), and T is the absolute temperature (K). The standard Gibb's free energy (ΔG^0) values are shown in Table 2. The negative ΔG^0 values indicate that the biosorption of metals into human hair is thermodynamically feasible and of spontaneous nature.

3.8. Desorption, Regeneration, and Reuse Studies

Recovery of the adsorbed heavy metals and reuse of the biosorbent are of significance from the viewpoint of practical application. As indicated previously, two eluent solutions, EDTA and HNO_3, were screened for their potential to desorb Pb(II) ions from metal-adsorbed treated human hair (98% of biosorption). Both eluents can effectively desorb the heavy metal ions from the metal loaded-treated human hair, with the elution efficiency of EDTA solution being slightly better than that of the HNO_3 solution, 89% ± 1% and 85% ± 1%, respectively. EDTA may combine both acidic and complexing effects, while nitric acid only has the acidic effect to liberate the adsorbed metals, which explains the higher ability of EDTA for removing the metal. In addition, the ability of EDTA to complex heavy metals, such as lead, due to the high complex constant, is well known.

The reuse of regenerated human hair for the possible continuous removal of heavy metals was investigated. After the desorption process, the hair samples were washed several times (following two different methods of cleaning), dried in an oven at 40 °C overnight, and their performance in a second biosorption step of fresh Pb(II) aqueous sample was then checked. The results show that the metal removal percentages (in the second biosorption step) of the regenerated hair samples are dependent on the elution methods. When using EDTA as eluent and deionized water for rinsing the used biomaterials, the Pb(II) removal percentage was 87 ± 2 %, while for samples eluted with HNO_3 the metal removal was 38 ± 4 %. Consequently, although HNO_3 is a very powerful metal eluent, it shows negative effects for the reuse of hair samples and results in a decrease of metal uptake capacity during the second application. This is probably because the acidic environment after the desorption step with nitric acid leads to the ionization state of functional groups on the biomaterial's surface, thus becoming a competitive medium for the next metal biosorption step. To confirm this concept, another set of samples were first eluted using EDTA, then rinsed with HNO_3 solution, and later washed with deionized water and dried. The biosorption capacities of the regenerated human hair samples by this method also decreased the second time (75 ± 1 %). It is clear that the desorption of metal adsorbed onto the biomaterials using HNO_3 negatively affects their reuse.

4. Conclusions

Chemically treated human hair behaved better than untreated hair in the process of removing metals from aqueous effluent. In particular, treated human hair was demonstrated to be an effective biosorbent for the removal of Cr(III), Cu(II), Cd(II), and Pb(II), and showed less effectiveness for metals such as Ni(II), Co(II), Mn(II), and Zn(II). It was observed that the operating parameters controlling the biosorption process, such as the pH of the aqueous heavy metal solution, the biosorbent concentration, and the contact time, had a significant influence on the metal uptake. In addition, treated human hair showed higher biosorption capacity when metals were applied in the single-system compared with the multiple-system solution, due to the induced competition between metal ions for the biosorbent sites. In the single system, removal efficacy of the treated human hair was found to be 86% for Cd(II), 92% for Cu(II), 96% for Pb(II), and 98% for Cr(III) when working with 10 g/L of biosorbent concentration at pH = 4.0.

According to the kinetic study, the biosorption of metal ions onto the treated human hair followed well the pseudo-second order kinetic model. Hence, physico-chemical interaction between induced functional groups (i.e., sulfonic acid groups as demonstrated by FTIR) in treated hair and metal ions is the fundamental mechanism controlling biosorption, with the film diffusion being the rate limiting step. Biosorption at equilibrium was better correlated with the Langmuir isotherm model compared to the Freundlich model, corroborating the finding that the mechanism of sorption is based on site-specific molecular sorption. The calculated standard Gibb's free energy (ΔG^0) indicated the thermodynamically feasible and spontaneous nature of the biosorption process.

Preliminary desorption experiments proved that EDTA and HNO_3 solutions were efficient eluents for the recovery of Pb(II) from the treated human hair. In particular, the treated human hair regenerated with EDTA showed the best biosorption efficiency when reused.

Taking into consideration the present findings, it can be stated that treated human hair could potentially be used as an effective and low-cost biosorbent for the removal of heavy metal ions from aqueous solution.

Author Contributions: Conceptualization, C.P. and M.L.-M.; methodology, C.P. and F.C.-N.; formal analysis, H.Z. and F.C.-N.; investigation, H.Z.; resources, C.P. and F.C.-N.; writing—original draft preparation, H.Z.; writing—review and editing, F.C.-N.; supervision, C.P.; project administration, C.P. and M.L.-M.; funding acquisition, C.P. and M.L.-M. All authors have read and agreed to the published version of the manuscript.

Funding: This research was funded by *Ministerio de Ciencia e Innovación del Gobierno de España*, grant numbers CTM2012-30970 and CTM2015-65414-C2-1-R, by the Interreg European Union, grant number ORQUE-SUDOE SOE3-P2-F591. Helan Zhang thanks to the China Sholarship Council for the grant [2001]3005.

Acknowledgments: The authors are grateful to the UAB Microscopy Service (*Servei de Microscopia Electrònica*, from UAB, Catalunya, Spain) for the SEM analysis and the Service of Analytical Chemistry (*Servei d'Anàlisi Química*, SCQ, from UAB, Catalunya, Spain) for the analysis of FTIR. Authors thanks to M. Resina who helped perform the analysis of heavy metals by ICP-MS.

Conflicts of Interest: The authors declare no competing interest.

References

1. Jaishankar, M.; Tseten, T.; Anbalagan, N.; Mathew, B.B.; Beer, K.N. Toxicity, mechanism and health effects of some heavy metals. *Interdiscip. Toxicol.* **2014**, *7*, 60–72. [CrossRef]
2. Vilela, D.; Parmar, J.; Zeng, Y.; Zhao, Y.; Sánchez, S. Graphene-Based Microbots for Toxic Heavy Metal Removal and Recovery from Water. *Nano Lett.* **2016**, *16*, 2860–2866. [CrossRef]

3. Saeed, A.; Iqbal, M.; Akhtar, M.W. Removal and recovery of lead(II) from single and multimetal (Cd, Cu, Ni, Zn) solutions by crop milling waste (black gram husk). *J. Hazard. Mater.* **2005**, *117*, 65–73. [CrossRef]
4. Ostovaritalab, M.; Hayati-Ashtiani, M. Investigation of Cs(I) and Sr(II) removal using nanoporous bentonite. *Part. Sci. Technol.* **2019**, *37*, 877–885. [CrossRef]
5. Shiferaw, Y.; Yassin, J.M.; Tedla, A. Removal of organic dye and toxic hexavalent chromium ions by natural clay adsorption. *Desalin. Water Treat.* **2019**, *165*, 222–231. [CrossRef]
6. Koohzad, E.; Jafari, D.; Esmaeili, H. Adsorption of lead and arsenic ions from aqueous solution by activated carbon prepared from tamarix leaves. *ChemistrySelect* **2019**, *4*, 12356–12367. [CrossRef]
7. De Freitas, G.R.; da Silva, M.G.C.; Vieira, M.G.A. Biosorption technology for removal of toxic metals: A review of commercial biosorbents and patents. *Environ. Sci. Pollut. Res.* **2019**, *26*, 19097–19118. [CrossRef]
8. Ngah, W.S.W.; Fatinathan, S. Pb(II) biosorption using chitosan and chitosan derivatives beads: Equilibrium, ion exchange and mechanism studies. *J. Environ. Sci.* **2010**, *22*, 338–346. [CrossRef]
9. Sud, D.; Mahajan, G.; Kaur, M.P. Agricultural waste material as potential adsorbent for sequestering heavy metal ions from aqueous solutions—A review. *Bioresour. Technol.* **2008**, *99*, 6017–6027. [CrossRef]
10. Kar, P.; Misra, M. Use of keratin fiber for separation of heavy metals from water. *J. Chem. Technol. Biotechnol.* **2004**, *79*, 1313–1319. [CrossRef]
11. Zhang, H.; Carrillo, F.; López-Mesas, M.; Palet, C. Valorization of keratin biofibers for removing heavy metals from aqueous solutions. *Text. Res. J.* **2019**, *89*, 1153–1165. [CrossRef]
12. López-Mesas, M.; Navarrete, E.R.; Carrillo, F.; Palet, C. Bioseparation of Pb(II) and Cd(II) from aqueous solution using cork waste biomass. Modeling and optimization of the parameters of the biosorption step. *Chem. Eng. J.* **2011**, *174*, 9–17. [CrossRef]
13. Sfaksi, Z.; Azzouz, N.; Abdelwahab, A. Removalof cr(VI) from water by cork waste. *Arab. J. Chem.* **2014**, *7*, 37–42. [CrossRef]
14. Kamal, M.H.M.A.; Azira, W.M.K.W.K.; Kasmawati, M.; Haslizaidi, Z.; Saime, W.N.W. Sequestration of toxic Pb(II) ions by chemically treated rubber (Hevea brasiliensis) leaf powder. *J. Environ. Sci.* **2010**, *22*, 248–256. [CrossRef]
15. Ngah, W.S.W.; Hanafiah, M.A.K.M. Removal of heavy metal ions from wastewater by chemically modified plant wastes as adsorbents: A review. *Bioresour. Technol.* **2008**, *99*, 3935–3948. [CrossRef] [PubMed]
16. Kong, H.; He, J.; Gao, Y.; Wu, H.; Zhu, X. Cosorption of Phenanthrene and Mercury(II) from Aqueous Solution by Soybean Stalk-Based Biochar. *J. Agric. Food Chem.* **2011**, *59*, 12116–12123. [CrossRef]
17. Lindholm-Lehto, P.C. Biosorption of heavy metals by lignocellulosic biomass and chemical analysis. *BioResources* **2019**, *14*, 4952–4995.
18. Saha, S.; Zubair, M.; Khosa, M.A.; Song, S.; Ullah, A. Keratin and chitosan biosorbents for wastewater treatment: A review. *J. Polym. Environ.* **2019**, *27*, 1389–1403. [CrossRef]
19. Salminen, E.; Rintala, J. Anaerobic digestion of organic solid poultry slaughterhouse waste—A review. *Bioresour. Technol.* **2002**, *83*, 13–26. [CrossRef]
20. Yin, J.; Rastogi, S.; Terry, A.E.; Popescu, C. Self-organization of Oligopeptides Obtained on Dissolution of Feather Keratins in Superheated Water. *Biomacromolecules* **2007**, *8*, 800–806. [CrossRef]
21. Al-Asheh, S.; Banat, F.; Al-Rousan, D. Beneficial reuse of chicken feathers in removal of heavy metals from wastewater. *J. Clean. Prod.* **2003**, *11*, 321–326. [CrossRef]
22. Ki, C.S.; Gang, E.H.; Um, I.C.; Park, Y.H. Nanofibrous membrane of wool keratose/silk fibroin blend for heavy metal ion adsorption. *J. Membr. Sci.* **2007**, *302*, 20–26. [CrossRef]
23. Aluigi, A.; Tonetti, C.; Vineis, C.; Tonin, C.; Mazzuchetti, G. Adsorption of copper(II) ions by keratin/PA6 blend nanofibres. *Eur. Polym. J.* **2011**, *47*, 1756–1764. [CrossRef]
24. Gupta, A. Human hair "waste" and its utilization: Gaps and possibilities. *J. Waste Manag.* **2014**, *2014*, 498018. [CrossRef]
25. Freddi, G.; Arai, T.; Colonna, G.M.; Boschi, A.; Tsukada, M. Binding of metal cations to chemically modified wool and antimicrobial properties of the wool–metal complexes. *J. Appl. Polym. Sci.* **2001**, *82*, 3513–3519. [CrossRef]
26. Brenner, L.; Squires, P.L.; Garry, M.; Tumosa, C.S. A measurement of human hair oxidation by Fourier transform infrared spectroscopy. *J. Forensic Sci.* **1985**, *30*, 420. [CrossRef]
27. Akhtar, W.; Edwards, H.G.M.; Farwell, D.W.; Nutbrown, M. Fourier-transform Raman spectroscopic study of human hair. *Spectrochim. Acta Part A Mol. Biomol. Spectrosc.* **1997**, *53*, 1021–1031. [CrossRef]

28. Febrianto, J.; Kosasih, A.N.; Sunarso, J.; Ju, Y.-H.; Indraswati, N.; Ismadji, S. Equilibrium and kinetic studies in adsorption of heavy metals using biosorbent: A summary of recent studies. *J. Hazard. Mater.* **2009**, *162*, 616–645. [CrossRef]
29. Lagergren, S. About the theory of so-called adsorption of soluble substances, Kungliga Svenska Vetenskapsakademiens. *Handlingar* **1898**, *24*, 1–39.
30. Ho, Y.S.; McKay, G. A comparison of chemisorption kinetic models applied to pollutant removal on various sorbents. *Process Saf. Environ. Prot.* **1998**, *76*, 332–340. [CrossRef]
31. Ho, Y.S.; McKay, G. Pseudo-second order model for sorption processes. *Process Biochem.* **1999**, *34*, 451–465. [CrossRef]
32. Ho, Y.S.; McKay, G. The kinetics of sorption of divalent metal ions onto sphagnum moss peat. *Water Res.* **2000**, *34*, 735–742. [CrossRef]
33. Weber, W.J.; Morris, J.C. Kinetics of adsorption of carbon from solution, Journal of the Sanitary Engineering Division. *Am. Soc. Civ. Eng.* **1963**, *89*, 31–60.
34. Onal, Y.; Akmil-Başar, C.; Sarici-Ozdemir, C. Investigation kinetics mechanisms of adsorption malachite green onto activated carbon. *J. Hazard. Mater.* **2007**, *146*, 194–203. [CrossRef] [PubMed]
35. Wu, F.-C.; Tseng, R.-L.; Juang, R.-S. Initial behavior of intraparticle diffusion model used in the description of adsorption kinetics. *Chem. Eng. J.* **2009**, *153*, 1–8. [CrossRef]
36. Guechi, E.; Benabdesselam, S. Removal of cadmium and copper from aqueous media by biosorption on cattail (typha angustifolia) leaves: Kinetic and isotherm studies. *Desalin. Water Treat.* **2020**, *173*, 367–382. [CrossRef]
37. Virgen, M.D.R.M.; Vázquez, O.F.G.; Montoya, V.H.; Gómez, R.T. Removal of heavy metals using adsorption processes subject to an external magnetic field. In *Heavy Metals*; Intechopen: London, UK, 2018; pp. 253–280. Available online: https://www.intechopen.com/books/heavy-metals/removal-of-heavy-metals-using-adsorption-processes-subject-to-an-external-magnetic-field (accessed on 28 April 2020).

Article

Removal of Aquatic Cadmium Ions Using Thiourea Modified Poplar Biochar

Yanfeng Zhu [1], Huageng Liang [2], Ruilian Yu [1], Gongren Hu [1,*] and Fu Chen [2,*]

[1] College of Chemical Engineering, Huaqiao University, Xiamen 361021, China; 18014087032@stu.hqu.edu.cn (Y.Z.); ruiliany@hqu.edu.cn (R.Y.)
[2] Low Carbon Energy Institute, China University of Mining and Technology, Xuzhou 221008, China; lianghg@cumt.edu.cn
* Correspondence: grhu@hqu.edu.cn (G.H.); chenfu@cumt.edu.cn (F.C.); Tel.: +86-516-8388-3501 (F.C.)

Received: 13 March 2020; Accepted: 11 April 2020; Published: 14 April 2020

Abstract: Removal of aquatic cadmium ions using biochar is a low-cost method, but the results are usually not satisfactory. Modified biochar, which can be a low-cost and efficient material, is urgently required for Cd-polluted water and soil remediation. Herein, poplar bark (SB) and poplar sawdust (MB) were used as raw materials to prepare modified biochar, which is rich in N- and S- containing groups, i.e., TSBC-600 and TMBC-600, using a co-pyrolysis method with thiourea. The adsorption characteristics of Cd^{2+} in simulated wastewater were explored. The results indicated that the modification optimized the surface structure of biochar, Cd^{2+} adsorption process by both TSBC-600 and TMBC-600 was mainly influenced by the initial pH, biochar dosage, and contact time, sthe TSBC-600 showed a higher adsorption capacity compared to TMBC-600 under different conditions. The Langmuir adsorption isotherm model and pseudo-second-order kinetic model were more consistent with the adsorption behavior of TSBC-600 and TMBC-600 to Cd^{2+}, the maximum adsorption capacity of TSBC-600 and TMBC-600 calculated by the Langmuir adsorption isotherm model was 19.998 mg/g and 9.631 mg/g, respectively. The modification method for introducing N and S into biochar by the co-pyrolysis of biomass and thiourea enhanced the removal rate of aquatic cadmium ions by biochar.

Keywords: biochar; thiourea; cadmium pollution; adsorption characteristics; water treatment

1. Introduction

Cadmium (Cd) has extreme biological toxicity, long half-life, and low elimination efficiency [1]. It is one of the "ten chemicals of major public health concern" listed by the World Health Organization [2]. Cd can readily cause damage to the internal organs and systems of the human body, and is the most toxic transition metal element that poses global human health risks [3]. In 2014, the fraction of points that exceeded the standard quantity of Cd in cultivated land in China was 7% [4], and the area of farmland polluted by Cd reached 20 million hectares, which was mainly caused by irrigation due to industrial wastewater [5]. With the improvement in food safety standards and environmental awareness of the public, the prevention of Cd contamination of the food chain has become a common concern [6].

Irrigation using wastewater is the main source of Cd pollution in farmland soil [7]. Therefore, eliminating Cd from wastewater is the most effective remediation method [8]. Currently, the methods for the removal of Cd^{2+} in wastewater are mainly chemical precipitation, ion exchange, membrane separation, coagulation, and adsorption [9]. Among them, the adsorption method is widely used due to the low energy consumption, high removal rate, and simple operation. Biochar is a type of carbon-rich porous material [10], which is accessible by high-temperature conversion of biomass in anoxic or anaerobic environments. It has been widely used in sewage treatment. Li et al. used vinegar-residue biochar to adsorb Cd^{2+} in water, while the maximum adsorption capacity for Cd^{2+} was only 2.91 mg/g [11]. Yakkala et al. prepared Buffal weed biochar to adsorb Cd^{2+} and Pb^{2+} in water, and the adsorption capacity was 11.63 mg/g and 333.33 mg/g, respectively [12]. Jing et al. prepared biochar to adsorb Cd^{2+} in soil, and the adsorption capacity was 5.00 mg/g [13]. Although biochar can adsorb cadmium ion to some extent, it is limited by the surface pore structure and functional groups, and the single biochar adsorption capacity is very limited [14]. To maximize the adsorption of biochar, modification by doping becomes an important method to optimize the pore structure of biochar and increase the functional groups and specific adsorption sites. Currently, the main methods include chemical modification of the surface, co-pyrolysis, and catalytic esterification [15–17]. Previous studies have shown that the type of modifier, as well as the modification method can affect the quality of the modified biochar [18–20]. Ma et al. synthesized modified biochar rich in $-NH_2$ by cross-linking using polyethyleneimine as the modifier, and the maximum adsorption capacity of the modified biochar to Cr^{6+} was 435 mg/g, much higher than the 23.09 mg/g before modification [21]. Park et al. prepared sulfur-modified biochar by the co-pyrolysis of sulfur and wood chips at 600 °C; after impregnation, the maximum adsorption capacity of the modified biochar for mercury was 107.5 mg/g, which was higher than that of the raw materials by 86%. This modification method enables sulfur to be doped into the biochar structure in the form of groups and enhanced the specific adsorption capacity of biochar [22]. At present, sulfur and nitrogen-modified biochar are the most common technologies, which have several disadvantages such as low practicability, low utilization of modifiers, and one modifier that can only be completed via modification [23].

Therefore, the specific objectives of this study are: (1) to prepare biochar with stable performance, rich functions, strong adsorption capacity, and multiple modifications as one step pyrolysis; (2) to compare the microstructure and physical characteristics of raw biochar and modified biochar by multiple techniques; (3) to evaluate the adsorption characteristics of thiourea-modified biochar on cadmium ions under different conditions, and experimental data were investigated with isotherm model and kinetic sorption model. The main aims of this study include understanding the role of the modified biochar produced from co-pyrolysed thiourea in the removal of Cd^{2+}, and providing technical support for its application in environmental remediation.

2. Materials and Methods

2.1. Preparation of Thiourea Modified Biochar

The collected sawdust and bark (from Wenchang Campus of China University of Mining and Technology, Xuzhou, Jiangsu, China) were thoroughly washed with deionized water and dried in an oven at 60 °C for 24 h. The biomass raw material and thiourea were weighed at a mass ratio of 1:1, respectively, and placed in a tubular furnace under an argon atmosphere after even mixing. The furnace was heated up to 300 to 700 °C at a heating rate of 5 °C/min and kept at the same temperature for 2 h to prepare thiourea-modified biochar (denoted as TSBC-600 and TMBC-600, where T stands for the pyrolysis temperature). The samples were crushed, passed through a 60-mesh sieve, and stored for future use (preparation is shown in Figure S1).

For comparison, unmodified biochar materials (MBC-600, SBC-600) were prepared by heating to 600 °C at 5 °C/min in an argon atmosphere without adding thiourea.

2.2. Physicochemical Properties and Structural Characterization of Biochar

Deionized water and biochar were mixed at a mass ratio of 20:1, and the pH value of the aqueous solution was measured with a pH meter to determine the acidity and basicity of the biochar. The ash content was measured by the combustion method [24]. The specific surface area and pore size distribution of biochar were investigated using N_2-isothermal adsorption and desorption experiments. The specific surface area of the material was calculated using the Brauner-Emmett-Teller (BET) equation, and the pore size distribution of the materials was determined by using the Density functional theory (DFT) model.

Field emission scanning electron microscope/energy-dispersive X-ray spectroscopy (SEM-EDX) (FEI QuantaTM 250, FEI, Boston, MA, USA) was used to observe the surface morphology; Fourier transform infrared spectroscopy (FTIR) (VERTEX 80V, Bruker, Karlsruhe, Germany) was used to characterize the surface properties of biochar. Typically, approximately 0.02 g of the dry sample was thoroughly mixed with KBr and pressed into semi-transparent pellets using a manual hydraulic press, and all spectra were acquired with a resolution of 2 cm^{-1} for 28 scans, over a wavenumber range between 400 and 4000 cm^{-1}. X-ray photoelectron spectroscopy (XPS) (ESCALAB 250Xi, Thermo Fisher, Boston, MA, USA) was used to characterize the chemical state of the biomass carbon material, and the binding energy values were all calibrated based on the hydrocarbon contamination using the C 1s peak at 284.8 eV.

2.3. Batch Adsorption Experiment

2.3.1. Determining the Relationship between Pyrolysis Temperature and Biochar Adsorption Performance

An amount of 0.02 g of the modified biochar prepared at five different temperatures was placed into a centrifuge tube, 20 mL of 10 mg/L Cd^{2+} solution was added and then shaken well. The mixture was oscillated at 25 °C for 24 h. Finally, the optimal pyrolysis temperature of biochar was determined.

2.3.2. Determining the Relationship between pH and the Amount of Cd^{2+} Adsorbed by Biochar

Amounts of 0.01 mol/L HCl and 0.01 mol/L NaOH were used to modulate the pH of 20 mL of 10 mg/L Cd^{2+} solution to 2, 3, 4, 5, 6, and 7 for minimizing the effect of Cd-chloride hydrolysis, then weremplaced one hour and there was no observed precipitation by the transillumination test, and then 0.02 g modified biochar were added. The mixture was mixed well and oscillated at 25 °C for 24 h.

2.3.3. Determination of the Relationship between the Dosage and the Capacity of Cd^{2+} Adsorbed by Biochar

The concentration of modified biochar was controlled to be 1, 2, 3, 4, 5, 6, 7, 8, 9, 10 g/L, respectively. The pH of the TMBC solution and TSBC solution was modulated to be 7, respectively. The initial concentration of Cd^{2+} in each solution was 10 mg/L. Finally, the optimal dose of biochar was determined.

2.3.4. Adsorption Kinetics

0.10 g of TMBC-600 was added into a 50 mL centrifuge tube, and subsequently, 20 mL of 100 mg/L Cd^{2+} solution was added. The pH was adjusted to be 7, and the samples were taken after oscillation for 5, 10, 30, 60, 120, 240, 360, 720, and 1440 min, respectively. For TSBC-600, the mass was 0.08 g, and the other procedures were the same as those of TMBC-600.

The adsorption kinetic equation can reflect the change in the adsorption capacity of the adsorbent to the solute as a function of time, i.e., the speed of the adsorption rate [25]. In this study, the more commonly used pseudo-first-order and pseudo-second-order kinetic model equations were used to fit the adsorption data.

Adsorption capacity:

$$q_e = \frac{(C_0 - C_e) \times V}{m} \quad (1)$$

$$q_t = \frac{(C_0 - C_t) \times V}{m} \quad (2)$$

pseudo-first-order kinetic model equation:

$$\ln(1-\frac{q_t}{q_e}) = -k_1 t \tag{3}$$

pseudo-second-order kinetic model equation:

$$\frac{t}{q_t} = \frac{1}{k_2 q_e^2} + \frac{t}{q_e} \tag{4}$$

where t is the adsorption time (min); C_0 is the initial concentration of Cd^{2+} in the solution (mg/L); C_e is concentration of Cd^{2+} in the solution when the adsorption reaches equilibrium (mg/L); C_t is the concentration of Cd^{2+} in the solution at t (mg/L); V is the volume of the solution participation in the adsorption reaction (L); m is the mass of the added biochar (g); q_e is the equilibrium adsorption capacity for biochar when the adsorption reaches equilibrium(mg/g); q_t is the adsorption capacity for biochar at t (mg/g); k_1 is the rate constant of the pseudo-first-order kinetic equation (min^{-1}); k_2 is the rate constant of the pseudo-first-order kinetic equation (g/mg/min).

2.3.5. Isothermal Adsorption

An amount of 0.10 g of TMBC-600 was added into a 50 mL centrifuge tube, and then 20 mL of 5, 10, 30, 100, 250, and 500 mg/L Cd^{2+} solution was added, respectively. The pH was adjusted to be 7 and subsequently oscillated for 24 h at room temperature. For TSBC, the mass was 0.08 g, and the other procedures were the same as those of TMBC-600.

In this study, Langmuir and Freundlich isothermal adsorption models were used to analyze the adsorption of the two modified biochars. For the Langmuir model, it is assumed that the surface of the adsorbent has the same adsorption active sites and that the adsorption occurs in the monolayer [26].

Langmuir equation:

$$Q_e = \frac{Q_m \times K_L \times C_e}{1 + K_L C_e} \tag{5}$$

Freundlich equation:

$$Q_e = K_f C_e^{1/n} \tag{6}$$

where Q_m is the maximum adsorption capacity (mg/g); K_L (L/g) is the affinity constant of the interaction between the adsorbate and the adsorbent; C_e is the concentration of the adsorbate when the adsorption reaches equilibrium (mg/L); K_f is the Freundlich adsorption capacity parameter; n is the empirical parameter of the adsorption strength (determined by the heterogeneity of the material).

3. Results

3.1. *Effects of Modification on the Physicochemical Properties of Biochar*

The physicochemical properties of biochar before and after modification are shown in Table 1. It can be seen from Table 1 that the biochar is alkaline due to the release of alkaline salts during the pyrolysis, and the pH of the biochar is increased after modification. This indicates that the change in pH and the ash content are positively correlated, as the ash of biochar contains K, Ca, Na, Mg, S, and other elements. These elements are attached to the surface of the biochar in the form of oxides and carbonates, and are alkaline in solution [27]. In addition, these inorganic components can co-precipitate with cadmium ions to remove [28]. The N_2 adsorption was carried out for the TMBC-600 and TSBC-600 (Figure S2). TMBC-600 and TSBC-600 exhibit reversible type I isotherm, a microporous filling at low P/P_0 and a H4 typed hysteresis loop within intermediate and high P/P_0, indicating the coexistence of both micropores and meso-/macropores in TMBC-600 and TSBC-600 [29]. From the DFT pore width distribution, TMBC-600 demonstrates low micropores and abundant mesopores, and TSBC-600 shows more micropores. Compared with the samples before modification, the BET specific surface area of

TMBC-600 and TSBC-600 increased by 3.08 m^2/g and 2.93 m^2/g, respectively, and the internal pore diameter increased by 0.17% and 1.48%, indicating that the surface and pore structure of the modified biochar has changed significantly. This is because, during the slow pyrolysis process, the steam generated by the modifier at high temperature penetrates the internal structure of the raw biochar, dredging the pores in it. This is consistent with the research results reported by O'Connor et al. [16].

Table 1. Physicochemical properties of MBC-600, SBC-600, TMBC-600, and TSBC-600. BET, Brauner-Emmett-Teller.

Samples	Yield (%)	Ash (%)	pH	BET Surface Area (m^2/g)	Average Pore Diameter (nm)
MBC-600	24.26	46.75	7.07	2.46	4.66
SBC-600	31.42	41.68	9.92	2.77	1.94
TMBC-600	—	57.46	8.94	5.54	4.83
TSBC-600	—	47.71	10.85	5.70	3.42

Notes: MBC-600, poplar sawdust biochar; SBC-600, poplar bark biochar; TMBC-600, thiourea modified poplar sawdust biochar; TSBC-600, thiourea modified Poplar bark biochar.

3.2. Effect of Modification on Morphology and Structure of Biochar

Figure 1 shows the SEM-EDS images of two materials before and after modification. From Figure 1, it is observed that before the modification, both SBC-600 and MBC-600 have a lamellar structure, which is more regularly arranged. The internal and external structures are well-formed and the surface is rough. The lamellar stack structure is obvious with an obvious pore structure. As proposed by Wu et al. [30] also, after modification, the granular and block structure of TSBC-600 and TMBC-600 increased, presumably due to the introduction of sulfur-containing groups that increased the number of biochar particles and the surface became smooth. Part of the pore structure was blocked, the impurities on the biochar surface reduced, and the outline of the structure was clear. EDS analysis indicated that the nitrogen content of the TSBC-600 increased from 3.57% to 15.43%, and the sulfur content increased from 0.30% to 3.68%; the nitrogen content of the TMBC-600 increased from 3.30% to 13.61%, and the sulfur content increased from 0.17% to 2.89%, which fully confirmed that N and S were successfully doped into the biochar during the modification process.

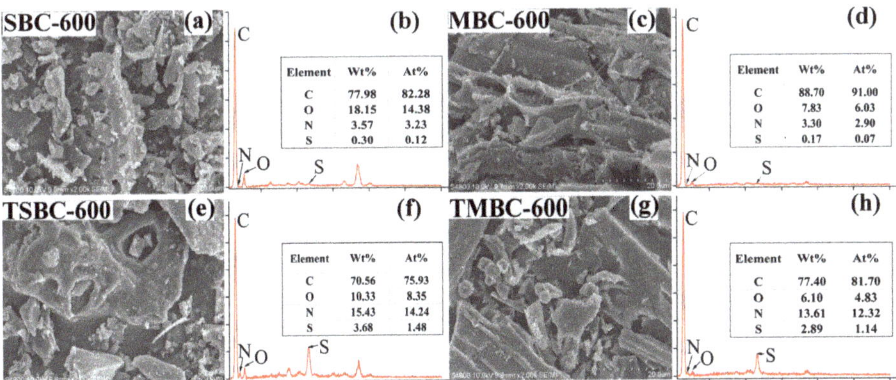

Figure 1. SEM image (**a**,**c**,**e**,**g**) and EDS spectrum (**b**,**d**,**f**,**h**) of MBC-600, SBC-600, TMBC-600 and TSBC-600.

3.3. Effects of Modification on Functional Groups on the Biochar Surface

The FTIR spectra of the two biochars (TMBC-600 and TSBC-600) before and after modification and before and after adsorption are shown in Figure 2. From Figure 2, the absorption peak at approximately

3300 cm^{-1} is assigned to the stretching vibration of -OH [31], and the strong absorption peaks appearing near 2988 cm^{-1} at TMBC-600 and TSBC-600 are the stretching vibration of $-NH_2$ [32], indicating that the nitrogen in thiourea participated in the amination of the surface of the biochar during the modification process. The absorption peak at 2901 cm^{-1} was assigned to the stretching vibration of $-CH_2$ or -CH [33]. The characteristic peak appearing at 2358 cm^{-1} after modification may be caused by the stretching vibration of the -C≡N triple bond, and the two strong absorption peaks, i.e., 1066 cm^{-1} and 1244 cm^{-1} appearing in the frequency band of 1050–1250 cm^{-1} are the stretching vibration absorption peaks of C-S and C=S [30], respectively, indicating the presence of sulfurization during the modification of biochar. The characteristic absorption peak at 1439 cm^{-1} is attributed to the stretching vibration of C-N [34], and the absorption peak at 1585 cm^{-1} may be the absorption characteristic peak of C=N or C=O [14]. Generally, C-H stretching vibrational absorption peaks of methyl and methylene appear in the vicinity of 2960 cm^{-1}, and the adsorption frequency of methylene will shift to the low-frequency region and the absorption intensity will become weaker after the introduction of the S atoms [35]. The peak at 2901 cm^{-1} evidences this and confirms that the S-containing groups were involved in the modification. The absorption peak in the 1400–1600 cm^{-1} band shifted after the reaction and the peak intensity weakened, indicating that Cd^{2+} and the functional group on the surface of the biochar were bound to the adsorption site by complexation [36]. The characteristic peaks of $-NH_2$ and sulfur-containing groups weakened after the adsorption reaction, indicating that the newly introduced functional groups participated in the adsorption of Cd^{2+}.

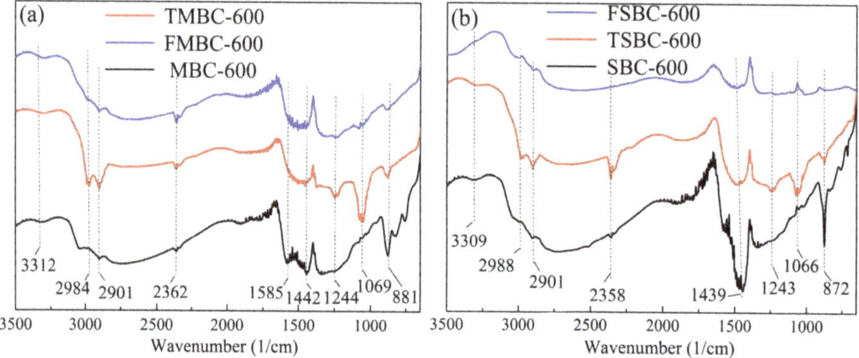

Figure 2. FTIR spectra of MBC-600, TMBC-600 and FMBC-600 (after Cd^{2+} adsorption) (**a**), and SBC-600, TSBC-600 and FSBC-600 (after Cd^{2+} adsorption) (**b**), (F stands for adsorption reaction).

3.4. XPS Analysis of Biochar before and after Modification

To further verify whether this modification method successfully doped N and S into the biochar carbon lattice, X-ray photoelectron spectroscopy analysis was used to analyze the change in the binding energy of C, N, and S in the biochar before and after modification. Figure 3a shows the XPS survey spectrum of MBC-600 and TMBC-600, in which the nitrogen content increased by 1.43% and the sulfur content increased by 2.03%. Figure 3b shows the XPS survey spectrum of SBC-600 and TSBC-600, in which the nitrogen content increased by 6.53% and the sulfur content increased by 3.39%. After modification, the extent of doping with N and S has improved significantly. The change in the mass fraction of the surface element is closely related to the functional group type and quantity of biochar before and after modification, which indicates that the modified material has better electrochemical performance [37]. Figure S3 (A1) and (A2) show the C 1s spectra of the two materials before and after modi3fication, in which the carbon was present in three forms, i.e., C-C (284.7 eV), C-O/C-N/C-S (286.2 eV), and C=O (288.6 eV) [38]. It can be seen from (A1) that, after the modification, the content of the three types of carbon in TMBC-600 changed significantly, while TSBC-600 in (A2) has not changed much. The changes in C-N and C-S indicate that N and S substitute C and become part of the carbon skeleton in the form of structural nitrogen and structural sulfur [39]. Figure S3 (B1) and (B2) show the N 1s spectra of the materials before and after modification. The nitrogen is mainly present in four forms, i.e., N-6 (pyridine nitrogen, 398.5 eV), N-5 (pyrrole nitrogen, 399.5 eV), and N-Q (graphite nitrogen, 400.7 eV) and NOx (nitrogen oxide, 402.2 eV) [40,41]. The N 1s spectra of MBC-600 and SBC-600 in (A1) and (A2) show that the nitrogen content of unmodified biochar is very small. Compared with MBC-600 and SBC-600, the number of different forms of nitrogen in TMBC-600 and TSBC-600 increased significantly and mainly existed in the form of pyridine nitrogen and graphite nitrogen, indicating that the nitrogen atom was successfully introduced into the carbon lattice and formed a bond with the carbon atom. The number of pyridine nitrogen was higher than that of graphite nitrogen, which has been confirmed in studies with thiourea as an additional doped nitrogen source [15]. Pyridine nitrogen can promote the coordination of metal ions, and thus, is more favorable for the removal of metal ions [42]. Figure S3 (C1) is the S 2p spectrum of MBC-600 and TMBC-600. It can be seen from the figure that the sulfur content of MBC-600 is small, while that in TMBC-600 is significantly increased, existing mainly in three forms, i.e., C=S (163.9 eV), C-S (165.1 eV), and S(=O)2 (168.3 eV). Figure S3 (C2) is the S 2p spectrum of SBC-600 and TSBC-600. Compared with SBC-600, the sulfur content of TSBC-600 is significantly increased, and the sulfur exists in five main forms, i.e., C=S (163.9 eV), C-S (165.1 eV), S(=O)2 (168.3 eV), Sul S (sulfide sulfur, 161.8 eV), and S(-II) (160.6 eV) [35,43]. This confirms that the sulfur was successfully doped into the biochar carbon skeleton structure after modification. The doping of N and S elements has given biochar materials some specific properties, such as easier electrochemical reaction and adsorption.

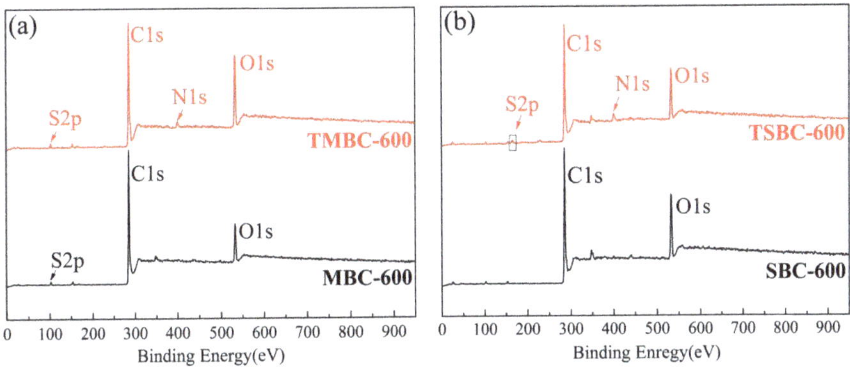

Figure 3. XPS spectra of MBC-600, SBC-600, TMBC-600 and TSBC-600 (**a**,**b**).

3.5. Adsorption Experiment Results

3.5.1. Effect of Pyrolysis Temperature on Cd^{2+} Adsorption by Modified Biochar

Cd^{2+} was removed by modified biochar at different pyrolysis temperatures and the results are shown in Figure 4. It can be seen from Figure 4 that when the dosage of biochar is 1 g/L and the initial pH of the solution is 7, the removal rate of Cd^{2+} by TSBC increases as the pyrolysis temperature increases. The removal rate of Cd^{2+} by TMBC in the range of 300–600 °C showed an upward trend, and the removal rate decreased above 600 °C. The removal rate of Cd^{2+} by TSBC was higher than that of TMBC. The removal rate of TMBC-600 and TSBC-600 were the highest at 43.83% and 87.29%, respectively. The removal rate associated with TMBC-600 increased by 36.53% compared with TMBC-300, and that of TSBC-600 increased by 81.07% compared with TSBC-300. Considering the energy consumption and raw material utilization in the process of preparing biochar, biochar was prepared at a pyrolysis temperature of 600 °C in this study.

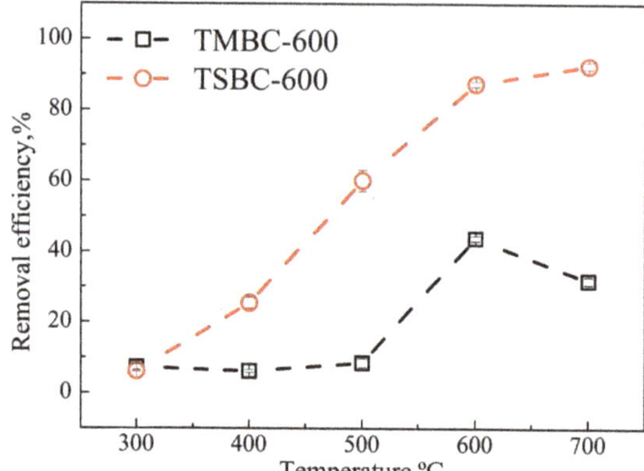

Figure 4. Effect of different pyrolysis temperatures on Cd^{2+} adsorption by TMBC-600 and TSBC-600.

3.5.2. Effect of Initial pH on Cd^{2+} Adsorption by Biochar

As shown in Figure 5, when the dosage of modified biochar was 1 g/L and the initial concentration of Cd^{2+} is 10 mg/L, as the pH increases, the removal rate of Cd^{2+} by both the biochars increase. Under strongly acidic conditions, the removal rate of Cd^{2+} by TSBC-600 is higher and that of TMBC-600 is lower. When the pH is in the range of 4–7 and 6–7, respectively, the adsorption of TSBC-600 and TMBC-600 is better, and the adsorption efficiency of TSBC-600 was higher than that of TMBC-600. When the pH was 4–7, the removal rate of Cd^{2+} in the solution can reach 93%; when pH = 7, the removal rate of Cd^{2+} by TSBC-600 became the highest, which was 96.05%. For TMBC-600, when the pH was at 2–6, the removal rate of Cd^{2+} increased slowly. When pH = 7, TMBC-600 had the highest removal rate of Cd^{2+}, which was 58.41%. Due to the difference in biomass materials, the alkalinity of the two modified biochars differed significantly. From Figure 6, it can be seen that the equilibrium pH of the biochar solution has increased to varying degrees after Cd^{2+} adsorption. The equilibrium pH of TSBC-600 was higher than TMBC-600, indicating that the basicity of TSBC-600 is greater, which is consistent with the pH measurement results in 2.1.

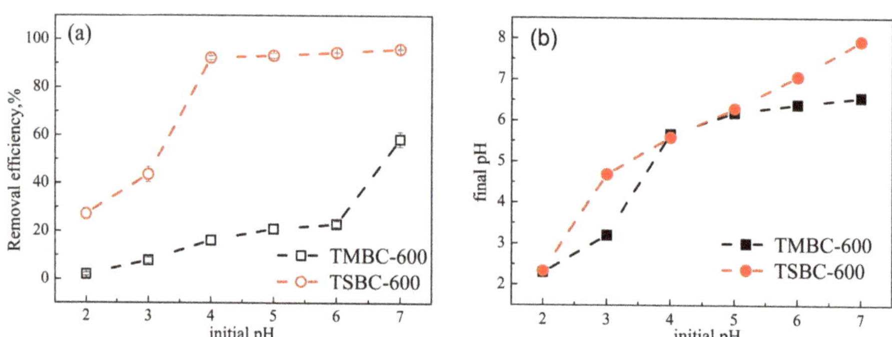

Figure 5. Effect of different pH on Cd^{2+} removal efficiency by TMBC-600 and TSBC-600 (**a**), comparison between the initial and final pH after the reaction (**b**).

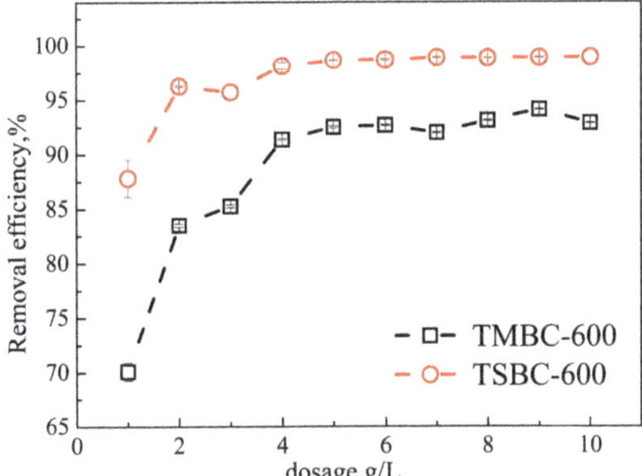

Figure 6. Effect of different dosage on adsorption of Cd^{2+} by TMBC-600 and TSBC-600.

Experiments have shown that TSBC-600 has good acid resistance. In the set pH range, TSBC-600 showed a strong removal rate of Cd^{2+}, which might be related to the role of the nitrogen-sulfur functional groups and cadmium ions. This has important implications for the research on the removal of cadmium in typical acidic wastewater such as mining wastewater.

3.5.3. Effect of Dosage of Biochar on Cd^{2+} Adsorption

It can be seen from Figure 6 that as the initial pH of the solution was 7, and the initial c (Cd^{2+}) of the solution was 10 mg/L, the removal rate of Cd^{2+} increased with the increase in the two kinds of biochar. When the dosage of TMBC-600 and TSBC-600 was 5 g/L and 4 g/L, the adsorption of Cd^{2+} by biochar reached saturation, and the removal rate reached 92.50% and 98.15%, respectively. As the dosage was increased continuously, the removal rate of Cd^{2+} did not vary much. This is possibly because the binding of cadmium ions to the limited adsorption sites on the surface of biochar reaches saturation when the dosage is small. With the increase in the dosage, the remaining cadmium ions in the solution bind to the new adsorption sites. Finally, the adsorption reached equilibrium, and further increase in dosage had little effect on the removal rate.

3.5.4. Analysis of Adsorption Kinetics

The adsorption capacity of Cd^{2+} by biochar as a function of time is shown in Figure 7. As shown in Figure 7, the adsorption capacity of TSBC-600, SBC-600, TMBC-600, and MBC-600 all increase with the increase of the adsorption time, and eventually reach equilibrium. TSBC-600, SBC-600, TMBC-600, and MBC-600 reached adsorption equilibrium at 240, 960, 360, and 960 min, respectively. In the initial stage of adsorption, the adsorption efficiency of each biochar is high and it gradually decreases with the increase in adsorption time. This is because there are more sites on the biochar surface that can bind to Cd^{2+} at the beginning of adsorption, and Cd^{2+} can be quickly adsorbed by the biochar. At this time, there is no competitive adsorption. As the adsorption time increases, the effective adsorption sites on the surface of the biochar gradually decrease, and Cd^{2+} needs to diffuse into the inside of porous medium. At this time, the mass transfer rate becomes slower, and the competitive adsorption becomes more and more obvious [44]. TSBC-600 showed a higher adsorption rate in the initial stages of adsorption, indicating that the surface of TSBC-600 has more binding sites.

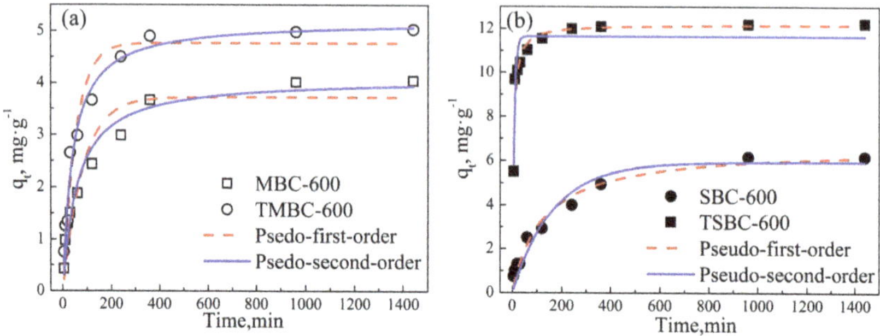

Figure 7. Fitting curves of the kinetics of TMBC-600 and MBC-600 (**a**) and TSBC-600 and SBC-600 (**b**) (C_0 = 100 mg/L, pH = 7, $doage_{TMBC-600}$ = 5 g/L and $doage_{TSBC-600}$ = 4 g/L, 25 °C).

3.5.5. Analysis of Isothermal Adsorption

The adsorption isotherm can reflect the affinity of the adsorbate for the adsorbent and can be used to describe the interaction between the adsorbate and the adsorbent [45]. The adsorption isotherms of Cd^{2+} solutions after adding biochar with different initial concentrations are shown in Figure 8. The adsorption capacity of TSBC-600, SBC-600, TMBC-600, and MBC-600 all increased with the initial concentration of Cd^{2+} and eventually reached equilibrium. At low concentrations of Cd^{2+}, biochar showed a higher adsorption rate, and then gradually flattened as the concentration increased. It can be seen from Figure 8 that the adsorption capacity of the four types of biochar is TSBC-600 > SBC-600 > TMBC-600 > MBC-600, among which TSBC-600 has the strongest adsorption capacity.

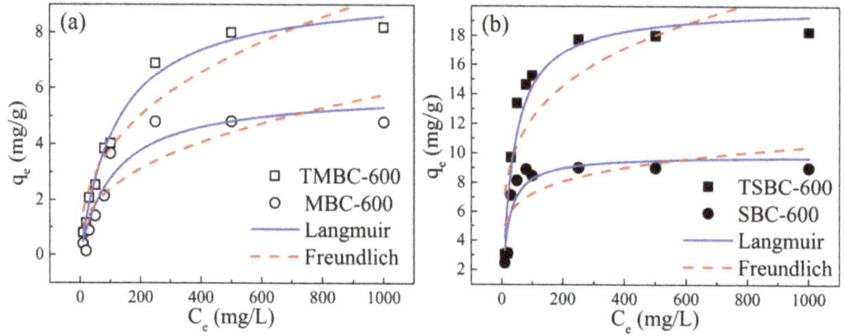

Figure 8. Fitting curves of isothermal adsorption of TMBC-600 and MBC-600 (**a**) and TSBC-600 and SBC-600 (**b**) (pH = 7, $doage_{TMBC-600}$ = 5 g/L and $doage_{TSBC-600}$ = 4 g/L, 25 °C).

4. Discussion

4.1. Effects of Pyrolysis Temperature, Dosage, and Initial pH on Cd^{2+} Removal

Biomass pyrolysis is a fairly complex thermochemical reaction process, in which temperature is a key factor that determines the performance of biochar [46]. The effect of modified biochar prepared at different pyrolysis temperatures on the removal rate of Cd^{2+} in aqueous solution is shown in Figure 4. When the initial concentration of Cd^{2+} in the solution was 10 mg/L, TMBC and TSBC had a high removal rate of Cd^{2+} at 600 °C. This is because, as the temperature increases, the moisture and volatile matter of the biomass is gradually released, and a large number of pore structures are formed on the biochar surface, the specific surface area increases, and the number of adsorption sites on the surface increases [47]. The reason for the lower growth may be that the introduction of sulfur and nitrogen enhances the ability of primary biochar to adsorb Cd^{2+}. At the same pyrolysis temperature, the difference in adsorption capacity between TMBC-600 and TSBC is remarkable. This may be because MB has volatile components up to 81.1% volatile components and its structure is loose. When the temperature is relatively low, organic solids in MB did not fully decompose and generate abundant biomass oil [48]. When the temperature was relatively high, the high temperature destroyed the pore structure of the biochar, which caused the TMBC pore structure to collapse, and the adsorption sites on the biochar surface decreased [49].

The dosage of adsorbent added is one of the important parameters that determine the adsorption capacity of the adsorbent for Cd^{2+}. The removal rate of TMBC-600 and TSBC-600 for Cd^{2+} in aqueous solution is shown in Figure 6. When the initial Cd^{2+} concentration of the solution was 10 mg/L, the dosage of TMBC-600 and TSBC-600 increased from 1 g/L to 10 g/L, and the removal rates of Cd^{2+} by TMBC and TSBC increased from 70.1% to 94.1% and from 87.8% to 99.0%, respectively. When the dosage increased to 4 g/L, the increase in removal rate did not vary much, similar results also appeared during the adsorption of transition metal element by sugarcane biochar and sulfurized biochar [22,50,51].

The initial pH of the solution is an important factor that affects the adsorption process. Cd-chloride hydrolysis have resulted in sacrificial dissolution of adsorbent and cannot be ignored, therefore, the neutralization method was used to minimize the effect of Cd-chloride hydrolysis on adsorption, so as to ensure that the adsorption of Cd^{2+} only affected by the initial pH of solution. Changes in pH can affect the surface charge of the adsorbent and the degree of ionization of the adsorbed cadmium ions [52]. The effect of initial pH on the adsorption of Cd^{2+} by modified biochar is shown in Figure 5. When the initial Cd^{2+} concentration is 20 mg/L and the amount of modified biochar is 1 g/L, TSBC-600 can still remove 93% of the Cd^{2+} in the solution at low pH, and TMBC-600 can obtain a removal rate of approximately 60%. In the same situation, Tang et al. found that the removal rate of Cd^{2+} by single amino-modified biochar was only 28%, and the removal rate of Cd^{2+} by single thiol-modified biochar was 48% [34]. Li et al. reported that when the dosage of nitrogen-added biochar was 8 g/L, the removal rate of Cd^{2+} could reach 98% at pH 4.0 [53]. Compared with single sulfur modification and nitrogen modification, the thiourea-modified biochar materials examined in this study have a higher tolerance to pH. The modified biochar mainly achieves the removal of Cd^{2+} through the specific binding of nitrogen and sulfur groups with Cd^{2+} and the non-specific complexation or ion exchange of oxygen-containing functional groups such as hydroxyl and carboxyl groups on the surface of biochar with Cd^{2+}. When the content of H^+ in the solution increases, a large amount of H^+ will dissociate the functional groups that are beneficial to the adsorption of Cd^{2+} [54,55], thereby reducing the removal rate of Cd^{2+}. Meanwhile, H^+ preferentially occupies the adsorption sites when competing with Cd^{2+}, causing a decrease in the number of sites bound to Cd^{2+}. In addition, the lower the pH of the solution, the higher the density of positive charges on the surface of biochar, and the stronger the mutual electrostatic repulsion between the positively charged metal cations and the surface of biochar [56–58], which will reduce the removal rate of Cd^{2+}.

4.2. Adsorption Kinetics of Cd^{2+} Adsorbed by Biochar before and after Modification

The pseudo-first-order kinetic equation and pseudo-second-order kinetic equation were used to fit the biochar adsorption kinetic data, and the fitting parameters are shown in Table 2. It can be seen from Table 2 that the R^2 values fitted by the pseudo-second-order kinetic equation are greater than those fitted by the pseudo-first-order kinetic equation, indicating that the adsorption rate of biochar are mainly controlled by the chemisorption mechanism, and Cd^{2+} adsorbed on the surface of biochar via the interaction with biochar, such as that shared by the electron-electron pairs, ion exchange, and surface complexation [59]. The adsorption process mainly includes external liquid film diffusion, surface adsorption, and intra-particle diffusion. The relationship between the equilibrium adsorption capacity of biochar for Cd^{2+} is TSBC-600 > SBC-600 > TMBC-600 > MBC-600. The adsorption of modified biochar was enhanced, indicating the important role of N- and S- containing groups on the adsorption of Cd^{2+}. Comparing the adsorption rate constant k^2 of the four biochars, 0.0184 (TSBC-600)

> 0.0049 (TMBC-600) > 0.0042 (MBC-600) > 0.0013 (SBC-600), indicating that the adsorption rate of TSBC-600 is faster, which is consistent with the analysis of k_1 in the pseudo-first order kinetic equation. This demonstrates that the affinity of SB to thiourea is greater than that of MB. In summary, thiourea-modified biochar can effectively increase the adsorption rate of Cd^{2+} by biochar.

Table 2. Adsorption kinetic parameters of MBC-600, SBC-600, TMBC-600, and TSBC-600 (C_0 = 100 mg/L, pH = 7, $doage_{TMBC\text{-}600}$ = 5 g/L and $doage_{TSBC\text{-}600}$ = 4 g/L, 25 °C).

Samples	Pseudo First Order			Pseudo Second Order		
	Q_e (mg/g)	k_1 (1/min)	R^2	Q_e (mg/g)	k_2 (g/mg/min)	R^2
TSBC-600	11.65	0.1391	0.88	12.19	0.0184	0.92
SBC-600	5.93	0.0062	0.94	6.63	0.0013	0.97
TMBC-600	4.76	0.0189	0.94	5.20	0.0049	0.98
MBC-600	3.72	0.0124	0.90	4.09	0.0042	0.96

Note: Q_m is the maximum adsorption capacity; k_1 is the rate constant of the pseudo-first-order kinetic equation; k_2 is the rate constant of the pseudo-first-order kinetic equation.

4.3. Isothermal Adsorption Curves before and after Biochar Modification

According to the shape of the isotherms, the isotherms of the four types of biochar can be designated as type I isotherms as per the BDDT (Brunauer Deming Deming Teller) classification [60]. Therefore, the commonly used Langmuir and Freundlich equations are used for fitting. The detailed fitting parameters and correlation coefficients are listed in Table 3. By comparing the R^2 fit using the two models, it is found that the Langmuir equation is more suitable than the Freundlich equation. This indicates that the adsorption of Cd^{2+} by the four biochars is monolayer adsorption. The functional groups on the absorbent surface are the main adsorption sites during the adsorption process, and the process takes place under the electrostatic attraction and hydrogen bonding [61]. There was no interaction between the homogeneity of the adsorbent surface and adsorption sites. The maximum adsorption capacity Q_m calculated by the Langmuir equation is not significantly different from the actual adsorption amount q_e. The maximum adsorption capacity of each biochar is TSBC-600 (19.998 mg/g) > SBC-600 (9.880 mg/g) > TMBC-600 (9.631 mg/g) > MBC-600 (5.898 mg/g). The adsorption capacity of thiourea modified TSBC-600 and TMBC-600 increased by 2.02 and 1.63 times, respectively. The adsorption capacities reported by other different adsorbents varied greatly among previous studies (Table S1). The adsorption capacity of TSBC-600 and TMBC-600 is nearly five times than biochar through pyrolysis with thiourea impregnation [15], the reason for this differentiation may be different activation of raw biochar in one-step co-pyrolysis. Moreover, the different feedstock and modifier made the variation of different surface characteristics, minerals, and functional groups of biochar.

Table 3. The fitting parameters for isothermal adsorption of MBC-600, SBC-600, TMBC-600, and TSBC-600 (pH = 7, $doage_{TMBC\text{-}600}$ = 5 g/L and $doage_{TSBC\text{-}600}$ = 4 g/L, 25 °C).

Samples	Langmuir Model			Freundlich Model		
	Q_m (mg/g)	K_L (L/mg)	R^2	K_F (L/g)	1/n	R^2
TSBC-600	19.998	0.026	0.91	4.230	0.233	0.70
SBC-600	9.880	0.050	0.83	3.506	0.158	0.55
TMBC-600	9.631	0.008	0.99	0.652	0.385	0.90
MBC-600	5.898	0.009	0.91	0.417	0.380	0.78

Notes: Q_m is the maximum adsorption capacity; K_L is the affinity constant of the interaction between the adsorbate and the adsorbent; K_f is the Freundlich adsorption capacity parameter; n is the empirical parameter of the adsorption strength.

For the separation factor R_L ($R_L = 1/(1 + R_LC_0)$, when $R_L > 1$, the adsorption is unfavorable; when $0 < R_L < 1$, the adsorption is favorable; when $R_L = 1$, the adsorption is linear [62]. In the Langmuir model, the R_L of biochar for Cd^{2+} is between zero and one, indicating that the adsorption behavior of biochar to Cd^{2+} is favorable. Similarly, the value of n in the Freundlich model can also confirm this viewpoint. The value of n is related to the properties of the adsorption material and the adsorption system. The larger the value of n, the better the adsorption performance of the material. When $0.1 < 1/n < 0.5$, it means that the adsorption is easier to perform; when $1/n > 2$, it means that the adsorption is difficult to carry out [53]. The 1/n values of the four biochars are all in the range of 0.1–0.5, which indicates that the N and S groups on the surface of the biochar have a higher modification efficiency. This demonstrates that the full adsorption process is preferential adsorption. The adsorption performance is relatively good, and the adsorption process is non-linear isothermal adsorption. The adsorption of Cd^{2+} by the four biochars is completed by a variety of mechanisms [14,50,63].

5. Conclusions

In this study, an N, S modified biochar was prepared by the co-pyrolysis of thiourea and poplar bark (SB) or poplar sawdust (MB), and used for the removal of aquatic cadmium ions. The results showed that the performance of thiourea-modified biochar was stable. The introduction of S and N optimized the surface structure of biochar, which was beneficial for enhancing the adsorption efficiency of biochar for Cd^{2+}. The optimal modification temperature of TSBC and TMBC was 600 °C, and the optimal reaction dosage of TSBC-600 and TMBC-600 was 4 g/L and 5 g/L, respectively. The optimal reaction pH was 7. Langmuir model and the pseudo-second-order kinetic model can explain the adsorption behavior of TSBC-600 and TMBC-600 with regard to Cd^{2+}. The adsorption behavior of TSBC-600 and TMBC-600 to Cd^{2+} was mainly chemical adsorption, which can be attributed to the surface functional groups. The maximum adsorption capacity of TSBC-600 and TMBC-600 calculated by the Langmuir model was 19.998 mg/g and 9.631 mg/g, respectively. These results indicated that the thiourea-modified biochar has strong adsorption capacity and good acid resistance, which is promising in the remediation of Cd-polluted water.

Supplementary Materials: The following are available online at http://www.mdpi.com/2073-4441/12/4/1117/s1: Figure S1: Schematic of the preparation of modified biochar, Figure S2: N_2 adsorption/desorption isotherms and pore size distribution of TSBC-600 (a) and TMBC-600 (b), Figure S3: High-resolution spectrum of C 1s (A1,A2), N 1s (B1,B2), S 2p (C1,C2) for MBC-600, SBC-600, TMBC-600 and TSBC-600, Table S1: Comparison of metal ion sorption capacity (mg/g) of different adsorbents.

Author Contributions: Conceptualization, F.C. and H.L.; methodology, F.C.; validation, R.Y., G.H., and F.C.; formal analysis, Y.Z.; investigation, Y.Z., H.L., R.Y., G.H., and F.C.; resources, Y.Z. and H.L.; writing—original draft preparation, Y.Z.; writing—review and editing, R.Y., G.H., and F.C.; supervision, F.C.; project administration, F.C.; funding acquisition, F.C. All authors contributed critical revisions during the editing process. All authors have read and agreed to the published version of the manuscript.

Funding: This research was funded by the National Natural Science Foundation of China (51974313, 41907405) and the Natural Science Foundation of Jiangsu Province (BK20180641).

Acknowledgments: The authors would like to thank Mapletrans English Service for providing English language, grammar, punctuation, and spelling help during the preparation of this manuscript.

Conflicts of Interest: The authors declare no conflicts of interest.

References

1. Jin, Z.; Deng, S.; Wen, Y.; Jin, Y.; Pan, L.; Zhang, Y.; Black, T.; Jones, K.C.; Zhang, H.; Zhang, D. Application of Simplicillium chinense for Cd and Pb biosorption and enhancing heavy metal phytoremediation of soils. *Sci. Total Environ.* **2019**, *697*, 134148. [CrossRef] [PubMed]
2. World Health Organization. *Ten Chemicals of Major Public Health Concern*; World Health Organization: Geneva, Switzerland, 2010; pp. 1–4.

3. Asara, Y.; Marchal, J.; Carrasco, E.; Boulaiz, H.; Solinas, G.; Bandiera, P.; Garcia, M.; Farace, C.; Montella, A.; Madeddu, R. Cadmium modifies the cell cycle and apoptotic profiles of human breast cancer cells treated with 5-fluorouracil. *Int. J. Mol. Sci.* **2013**, *14*, 16600–16616. [CrossRef] [PubMed]
4. Shen, Z.; Zhang, J.; Hou, D.; Tsang, D.C.; Ok, Y.S.; Alessi, D.S. Synthesis of MgO-coated corncob biochar and its application in lead stabilization in a soil washing residue. *Environ. Int.* **2019**, *122*, 357–362. [CrossRef] [PubMed]
5. Liu, N.; Huang, X.; Sun, L.; Li, S.; Chen, Y.; Cao, X.; Wang, W.; Dai, J.; Rinnan, R. Screening stably low cadmium and moderately high micronutrients wheat cultivars under three different agricultural environments of China. *Chemosphere* **2020**, *241*, 125065. [CrossRef] [PubMed]
6. Zeng, S.; Ma, J.; Yang, Y.; Zhang, S.; Liu, G.J.; Chen, F. Spatial assessment of farmland soil pollution and its potential human health risks in China. *Sci. Total Environ.* **2019**, *687*, 642–653. [CrossRef] [PubMed]
7. Feizi, M.; Jalali, M.; Antoniadis, V.; Shaheen, S.M.; Ok, Y.S.; Rinklebe, J. Geo-and nano-materials affect the mono-metal and competitive sorption of Cd, Cu, Ni, and Zn in a sewage sludge-treated alkaline soil. *J. Hazard. Mater.* **2019**, *379*, 120567. [CrossRef]
8. Li, H.; Xiong, J.; Zhang, G.; Liang, A.; Long, J.; Xiao, T.; Chen, Y.; Zhang, P.; Liao, D.; Lin, L. Enhanced thallium (I) removal from wastewater using hypochlorite oxidation coupled with magnetite-based biochar adsorption. *Sci. Total Environ.* **2020**, *698*, 134166. [CrossRef]
9. Peng, H.; Gao, P.; Chu, G.; Pan, B.; Peng, J.; Xing, B. Enhanced adsorption of Cu(II) and Cd(II) by phosphoric acid-modified biochars. *Environ. Pollut.* **2017**, *229*, 846–853. [CrossRef]
10. Rajendran, M.; Shi, L.; Wu, C.; Li, W.; An, W.; Liu, Z.; Xue, S. Effect of sulfur and sulfur-iron modified biochar on cadmium availability and transfer in the soil-rice system. *Chemosphere* **2019**, *222*, 314–322. [CrossRef]
11. Li, Y.; Pei, G.; Qiao, X.; Zhu, Y.; Li, H. Remediation of cadmium contaminated water and soil using vinegar residue biochar. *Environ. Sci. Pollut. Res. Int.* **2018**, *25*, 15754–15764. [CrossRef]
12. Yakkala, K.; Yu, M.-R.; Roh, H.; Yang, J.-K.; Chang, Y.-Y. Buffalo weed (Ambrosia trifida L. var. trifida) biochar for cadmium (II) and lead (II) adsorption in single and mixed system. *Desalin. Water Treat.* **2013**, *51*, 7732–7745. [CrossRef]
13. Jing, F.; Chen, C.; Chen, X.; Liu, W.; Wen, X.; Hu, S.; Yang, Z.; Guo, B.; Xu, Y.; Yu, Q. Effects of wheat straw derived biochar on cadmium availability in a paddy soil and its accumulation in rice. *Environ. Pollut.* **2019**, *257*, 113592. [CrossRef] [PubMed]
14. Zhu, L.; Tong, L.; Zhao, N.; Wang, X.; Yang, X.; Lv, Y. Key factors and microscopic mechanisms controlling adsorption of cadmium by surface oxidized and aminated biochars. *J. Hazard. Mater.* **2020**, *382*, 121002. [CrossRef] [PubMed]
15. Gholami, L.; Rahimi, G.; Khademi Jolgeh Nezhad, A. Effect of thiourea-modified biochar on adsorption and fractionation of cadmium and lead in contaminated acidic soil. *Int. J. Phytoremed.* **2019**, *22*, 468–481. [CrossRef]
16. O'Connor, D.; Peng, T.; Li, G.; Wang, S.; Duan, L.; Mulder, J.; Cornelissen, G.; Cheng, Z.; Yang, S.; Hou, D. Sulfur-modified rice husk biochar: A green method for the remediation of mercury contaminated soil. *Sci. Total Environ.* **2018**, *621*, 819–826. [CrossRef]
17. Zhang, Y.; Fan, J.; Fu, M.; Ok, Y.S.; Hou, Y.; Cai, C. Adsorption antagonism and synergy of arsenate(V) and cadmium(II) onto Fe-modified rice straw biochars. *Environ. Geochem. Health* **2019**, *41*, 1755–1766. [CrossRef]
18. Zhang, R.-L.; Xu, J.; Gao, L.; Wang, Z.; Wang, B.; Qin, S.-Y. Performance and Mechanism for Fluoride Removal in Groundwater with Calcium Modified Biochar from Peanut Shell. *Sci. Adv. Mater.* **2020**, *12*, 492–501. [CrossRef]
19. Li, Z.; Sun, Y.; Yang, Y.; Han, Y.; Wang, T.; Chen, J.; Tsang, D.C. Biochar-supported nanoscale zero-valent iron as an efficient catalyst for organic degradation in groundwater. *J. Hazard. Mater.* **2020**, *383*, 121240. [CrossRef]
20. Shi, S.; Yang, J.; Liang, S.; Li, M.; Gan, Q.; Xiao, K.; Hu, J. Enhanced Cr (VI) removal from acidic solutions using biochar modified by Fe3O4@ SiO2-NH2 particles. *Sci. Total Environ.* **2018**, *628*, 499–508. [CrossRef] [PubMed]
21. Ma, Y.; Liu, W.J.; Zhang, N.; Li, Y.S.; Jiang, H.; Sheng, G.P. Polyethylenimine modified biochar adsorbent for hexavalent chromium removal from the aqueous solution. *Bioresour. Technol.* **2014**, *169*, 403–408. [CrossRef] [PubMed]

22. Park, J.H.; Wang, J.J.; Zhou, B.; Mikhael, J.E.R.; DeLaune, R.D. Removing mercury from aqueous solution using sulfurized biochar and associated mechanisms. *Environ. Pollut.* **2019**, *244*, 627–635. [CrossRef] [PubMed]
23. Liu, W.J.; Jiang, H.; Yu, H.Q. Development of Biochar-Based Functional Materials: Toward a Sustainable Platform Carbon Material. *Chem. Rev.* **2015**, *115*, 12251–12285. [CrossRef] [PubMed]
24. Leksungnoen, P.; Wisawapipat, W.; Ketrot, D.; Aramrak, S.; Nookabkaew, S.; Rangkadilok, N.; Satayavivad, J. Biochar and ash derived from silicon-rich rice husk decrease inorganic arsenic species in rice grain. *Sci. Total Environ.* **2019**, *684*, 360–370. [CrossRef] [PubMed]
25. Wei, Y.; Wei, S.; Liu, C.; Chen, T.; Tang, Y.; Ma, J.; Yin, K.; Luo, S. Efficient removal of arsenic from groundwater using iron oxide nanoneedle array-decorated biochar fibers with high Fe utilization and fast adsorption kinetics. *Water Res.* **2019**, *167*, 115107. [CrossRef] [PubMed]
26. Zhao, D.; Yang, X.; Zhang, H.; Chen, C.; Wang, X. Effect of environmental conditions on Pb (II) adsorption on β-MnO2. *Chem. Eng. J.* **2010**, *164*, 49–55. [CrossRef]
27. Gondek, K.; Mierzwa-Hersztek, M.; Kopeć, M.; Mróz, T. The Influence of Biochar Enriched with Magnesium and Sulfur on the Amount of Perennial Ryegrass Biomass and Selected Chemical Properties and Biological of Sandy Soil. *Commun. Soil. Sci. Plan Anal.* **2018**, *49*, 1257–1265. [CrossRef]
28. Liu, Q.; Fang, Z.; Liu, Y.; Liu, Y.; Xu, Y.; Ruan, X.; Zhang, X.; Cao, W. Phosphorus speciation and bioavailability of sewage sludge derived biochar amended with CaO. *Waste Manag.* **2019**, *87*, 71–77. [CrossRef]
29. Gao, Z.; Chen, C.; Chang, J.; Chen, L.; Wang, P.; Wu, D.; Xu, F.; Jiang, K. Porous Co3S4@Ni3S4 heterostructure arrays electrode with vertical electrons and ions channels for efficient hybrid supercapacitor. *Chem. Eng. J.* **2018**, *343*, 572–582. [CrossRef]
30. Wu, C.; Shi, L.; Xue, S.; Li, W.; Jiang, X.; Rajendran, M.; Qian, Z. Effect of sulfur-iron modified biochar on the available cadmium and bacterial community structure in contaminated soils. *Sci. Total Environ.* **2019**, *647*, 1158–1168. [CrossRef]
31. Huang, Y.; Lee, X.; Grattieri, M.; Yuan, M.; Cai, R.; Macazo, F.C.; Minteer, S.D. Modified biochar for phosphate adsorption in environmentally relevant conditions. *Chem. Eng. J.* **2020**, *380*, 122375. [CrossRef]
32. Yang, G.X.; Jiang, H. Amino modification of biochar for enhanced adsorption of copper ions from synthetic wastewater. *Water Res.* **2014**, *48*, 396–405. [CrossRef] [PubMed]
33. Amin, M.T.; Alazba, A.A.; Shafiq, M. Removal of Copper and Lead using Banana Biochar in Batch Adsorption Systems: Isotherms and Kinetic Studies. *Arab. J. Sci. Eng.* **2017**, *43*, 5711–5722. [CrossRef]
34. Tang, N.; Niu, C.G.; Li, X.T.; Liang, C.; Guo, H.; Lin, L.S.; Zheng, C.W.; Zeng, G.M. Efficient removal of Cd(2+) and Pb(2+) from aqueous solution with amino- and thiol-functionalized activated carbon: Isotherm and kinetics modeling. *Sci. Total Environ.* **2018**, *635*, 1331–1344. [CrossRef] [PubMed]
35. Lyu, H.; Tang, J.; Huang, Y.; Gai, L.; Zeng, E.Y.; Liber, K.; Gong, Y. Removal of hexavalent chromium from aqueous solutions by a novel biochar supported nanoscale iron sulfide composite. *Chem. Eng. J.* **2017**, *322*, 516–524. [CrossRef]
36. Wu, H.; Wei, W.; Xu, C.; Meng, Y.; Bai, W.; Yang, W.; Lin, A. Polyethylene glycol-stabilized nano zero-valent iron supported by biochar for highly efficient removal of Cr (VI). *Ecotox. Environ. Saf.* **2020**, *188*, 109902. [CrossRef]
37. Liu, L.; Fan, S. Removal of cadmium in aqueous solution using wheat straw biochar: Effect of minerals and mechanism. *Environ. Sci. Pollut. Res.* **2018**, *25*, 8688–8700. [CrossRef]
38. You, F.; Dalal, R.; Huang, L. Biochar and biomass organic amendments shaped different dominance of lithoautotrophs and organoheterotrophs in microbial communities colonizing neutral copper(Cu)-molybdenum(Mo)-gold(Au) tailings. *Geoderma* **2018**, *309*, 100–110. [CrossRef]
39. Chang, Y.; Zhang, G.; Han, B.; Li, H.; Hu, C.; Pang, Y.; Chang, Z.; Sun, X. Polymer dehalogenation-enabled fast fabrication of N, S-codoped carbon materials for superior supercapacitor and deionization applications. *ACS Appl. Mater. Interfaces* **2017**, *9*, 29753–29759. [CrossRef]
40. Mian, M.M.; Liu, G.; Yousaf, B.; Fu, B.; Ullah, H.; Ali, M.U.; Abbas, Q.; Mujtaba Munir, M.A.; Ruijia, L. Simultaneous functionalization and magnetization of biochar via NH3 ambiance pyrolysis for efficient removal of Cr (VI). *Chemosphere* **2018**, *208*, 712–721. [CrossRef]

41. Song, H.; Liu, Z.; Gai, H.; Wang, Y.; Qiao, L.; Zhong, C.; Yin, X.; Xiao, M. Nitrogen-Dopped Ordered Mesoporous Carbon Anchored Pd Nanoparticles for Solvent Free Selective Oxidation of Benzyl Alcohol to Benzaldehyde by Using O2. *Front. Chem.* **2019**, *7*, 458. [CrossRef]
42. Wu, Z.; Chen, X.; Yuan, B.; Fu, M.-L. A facile foaming-polymerization strategy to prepare 3D MnO2 modified biochar-based porous hydrogels for efficient removal of Cd (II) and Pb (II). *Chemosphere* **2020**, *239*, 124745. [CrossRef] [PubMed]
43. Wei, Y.; Yan, Y.; Zou, Y.; Shi, M.; Deng, Q.; Zhao, N.; Wang, J.; You, C.; Yang, R.; Xu, Y. Sulfonated polyaniline coated bamboo-derived biochar/sulfur cathode for Li-S batteries with excellent dual conductivity and polysulfides affinity. *Electrochim. Acta* **2019**, *310*, 45–57. [CrossRef]
44. Mohan, D.; Pittman Jr, C.U.; Bricka, M.; Smith, F.; Yancey, B.; Mohammad, J.; Steele, P.H.; Alexandre-Franco, M.F.; Gómez-Serrano, V.; Gong, H. Sorption of arsenic, cadmium, and lead by chars produced from fast pyrolysis of wood and bark during bio-oil production. *J. Colloid Interface Sci.* **2007**, *310*, 57–73. [CrossRef] [PubMed]
45. Xia, Y.; Yang, T.; Zhu, N.; Li, D.; Chen, Z.; Lang, Q.; Liu, Z.; Jiao, W. Enhanced adsorption of Pb(II) onto modified hydrochar: Modeling and mechanism analysis. *Bioresour. Technol.* **2019**, *288*, 121593. [CrossRef]
46. Mahdi, Z.; El Hanandeh, A.; Yu, Q. Influence of pyrolysis conditions on surface characteristics and methylene blue adsorption of biochar derived from date seed biomass. *Waste Biomass Valorization* **2017**, *8*, 2061–2073. [CrossRef]
47. Hernandez-Soriano, M.C.; Kerré, B.; Kopittke, P.M.; Horemans, B.; Smolders, E. Biochar affects carbon composition and stability in soil: A combined spectroscopy-microscopy study. *Sci. Rep.* **2016**, *6*, 25127. [CrossRef]
48. Lai, F.; Chang, Y.; Huang, H.; Wu, G.; Xiong, J.; Pan, Z.; Zhou, C. Liquefaction of sewage sludge in ethanol-water mixed solvents for bio-oil and biochar products. *Energy* **2018**, *148*, 629–641. [CrossRef]
49. Zhang, X.; Zhang, P.; Yuan, X.; Li, Y.; Han, L. Effect of pyrolysis temperature and correlation analysis on the yield and physicochemical properties of crop residue biochar. *Bioresour. Technol.* **2020**, *296*, 122318. [CrossRef]
50. Cui, X.; Fang, S.; Yao, Y.; Li, T.; Ni, Q.; Yang, X.; He, Z. Potential mechanisms of cadmium removal from aqueous solution by Canna indica derived biochar. *Sci. Total Environ.* **2016**, *562*, 517–525. [CrossRef]
51. Park, J.H.; Ok, Y.S.; Kim, S.H.; Cho, J.S.; Heo, J.S.; Delaune, R.D.; Seo, D.C. Competitive adsorption of heavy metals onto sesame straw biochar in aqueous solutions. *Chemosphere* **2016**, *142*, 77–83. [CrossRef]
52. Huang, P.; Ge, C.; Feng, D.; Yu, H.; Luo, J.; Li, J.; Strong, P.; Sarmah, A.K.; Bolan, N.S.; Wang, H. Effects of metal ions and pH on ofloxacin sorption to cassava residue-derived biochar. *Sci. Total Environ.* **2018**, *616*, 1384–1391. [CrossRef] [PubMed]
53. Li, D.; Yang, R.; Luo, H.; Liu, S.; Liu, Y.; Peng, O.; Tie, B. Effect of adsorption of cadmium from aqueous solution by hexadecyl trimethyl ammonium bromide modified biochar. *Chin. J. Environ. Eng.* **2019**, *13*, 1809–1821.
54. Enders, A.; Hanley, K.; Whitman, T.; Joseph, S.; Lehmann, J. Characterization of biochars to evaluate recalcitrance and agronomic performance. *Bioresour. Technol.* **2012**, *114*, 644–653. [CrossRef] [PubMed]
55. Lam, Y.Y.; Lau, S.S.; Wong, J.W. Removal of Cd (II) from aqueous solutions using plant-derived biochar: Kinetics, isotherm and characterization. *Bioresour. Technol. Rep.* **2019**, *8*, 100323. [CrossRef]
56. Cui, H.-J.; Wang, M.K.; Fu, M.-L.; Ci, E. Enhancing phosphorus availability in phosphorus-fertilized zones by reducing phosphate adsorbed on ferrihydrite using rice straw-derived biochar. *J. Soil Sediment* **2011**, *11*, 1135. [CrossRef]
57. Jiang, T.-Y.; Jiang, J.; Xu, R.-K.; Li, Z. Adsorption of Pb (II) on variable charge soils amended with rice-straw derived biochar. *Chemosphere* **2012**, *89*, 249–256. [CrossRef]
58. Yin, W.; Zhao, C.; Xu, J. Enhanced adsorption of Cd (II) from aqueous solution by a shrimp bran modified Typha orientalis biochar. *Environ. Sci. Pollut. Res.* **2019**, *26*, 37092–37100. [CrossRef]
59. Wang, S.; Guo, W.; Gao, F.; Yang, R. Characterization and Pb (II) removal potential of corn straw-and municipal sludge-derived biochars. *R. Soc. Open Sci.* **2017**, *4*, 170402. [CrossRef]
60. Caurie, M. The derivation of the GAB adsorption equation from the BDDT adsorption theory. *Int. J. Food Sci. Technol.* **2006**, *41*, 173–179. [CrossRef]

61. Yao, L.; Zhao, Z.; Zhao, L.; Zhang, Z.; Shi, M.; Xu, M.M. Evaluation of Adsorption of Cadmium onto Ferrihydrite-Humic Acid Coprecipitation. *Sci. Adv. Mater.* **2019**, *11*, 1232–1240. [CrossRef]
62. Ao, H.; Cao, W.; Hong, Y.; Wu, J.; Wei, L. Adsorption of Sulfate Ion from Water by Zirconium Oxide-Modified Biochar Derived from Pomelo Peel. *Sci. Total Environ.* **2019**, 135092. [CrossRef] [PubMed]
63. Chen, F.; Luo, Z.; Liu, G.; Yang, Y.; Zhang, S.; Ma, J. Remediation of electronic waste polluted soil using a combination of persulfate oxidation and chemical washing. *J. Environ. Manag.* **2017**, *204*, 170–178. [CrossRef] [PubMed]

Article

Upcycling of Electroplating Sludge to Prepare Erdite-Bearing Nanorods for the Adsorption of Heavy Metals from Electroplating Wastewater Effluent

Yanwen Liu [1], Asghar Khan [1], Zhihua Wang [1], Yu Chen [2], Suiyi Zhu [1,*], Tong Sun [1], Dongxu Liang [1] and Hongbin Yu [1,*]

1 Science and Technology Innovation Center for Municipal Wastewater Treatment and Water Quality Protection, Northeast Normal University, Changchun 130117, China; liuyw401@nenu.edu.cn (Y.L.); asje962@nenu.edu.cn (A.K.); wangzh807@nenu.edu.cn (Z.W.); sunt281@nenu.edu.cn (T.S.); liangdx891@nenu.edu.cn (D.L.)

2 Jilin Institute of Forestry Survey and Design, Changchun 130022, China; papermanuscript2@126.com

* Correspondence: zhusy812@nenu.edu.cn (S.Z.); papermanuscript@126.com (H.Y.)

Received: 8 February 2020; Accepted: 31 March 2020; Published: 3 April 2020

Abstract: Electroplating sludge is a hazardous waste produced in plating and metallurgical processes which is commonly disposed of in safety landfills. In this work, electroplating sludge containing 25.6% Fe and 5.5% Co (named S1) and another containing 36.8% Fe and 7.8% Cr (S2) were recycled for the preparation of erdite-bearing particles via a facile hydrothermal route with only the addition of $Na_2S{\cdot}9H_2O$. In the sludges, Fe-containing compounds were weakly crystallized and spontaneously converted to short rod-like erdite particles (SP1) in the presence of Co or long nanorod (SP2) particles with a diameter of 100 nm and length of 0.5–1.5 μm in the presence of Cr. The two products, SP1 and SP2, were applied in electroplating wastewater treatment, in which a small portion of Co in SP1 was released in wastewater, whereas Cr in SP2 was not. Adding 0.3 g/L SP2 resulted in the removal of 99.7% of Zn, 99.4% of Cu, 37.9% of Ni and 53.3% of Co in the electroplating wastewater, with residues at concentrations of 0.007, 0.003, 0.33, 0.09 and 0.002 mg/L, respectively. Thus, the treated electroplating wastewater met the discharge standard for electroplating wastewater in China. These removal efficiencies were higher than those achieved using powdered activated carbon, polyaluminum chloride, polyferric sulfate or pure $Na_2S{\cdot}9H_2O$ reagent. With the method, waste electroplating sludge was recycled as nanorod erdite-bearing particles which showed superior efficiency in electroplating wastewater treatment.

Keywords: electroplating sludge; hydrothermal process; heavy metal; electroplating wastewater; upcycling

---

1. Introduction

Electroplating is a basic process in the machine manufacturing industry and uses various heavy metals to protect plating pieces [1–4]. High volumes of electroplating wastewater are generated, which is an extremely hazardous type of wastewater containing multiple heavy metals [5]. Electroplating wastewater can be categorized into acidic or alkaline wastewater in accordance with the electroplating technology used. Acidic wastewater is easily treated through pH adjustment and coagulation, which transfer heavy metals from wastewater to sludges [2,3]. By contrast, alkaline wastewater comprises complex agents, such as citric acid, ethylene diamine tetra-acetic acid (EDTA) and tartaric acid [6,7], which react with heavy metals to form heavy metal–organic complexes. Thus, heavy metals cannot be hydrolyzed even after adjusting wastewater pH to values above 12. Therefore, heavy metals are difficult to precipitate in alkaline wastewater through pH adjustment and are commonly treated using time-consuming processes, including pH adjustment, cationic exchange, extraction and/or precipitation with special agents [8–12]. Although the amount of heavy metals detected from the effluents of an electroplating workplace is usually low [10,13], heavy metals should be further removed until the discharge standard for electroplating wastewater is met.

The removal of heavy metals from alkaline electroplating wastewater generally involves two methods [14–21]. One is the decomposition of organic complex reagents by the Fenton reaction and/or wet oxidation [15–17]. For instance, Shin et al. [15] investigated the removal of a citrate–Ni complex in alkaline electroplating wastewater and found that, after the wastewater was adjusted to pH 3, 95% citrate removal was achieved by Fenton oxidation with the addition of 20 mM Fe^{2+} and 1080 mM H_2O_2, and subsequently 99.9% of Ni precipitation occurred when the wastewater was adjusted back to pH 10. Yong et al. also reported that more than 90% of Cu was precipitated after the decomposition of complex gluconic acid in a micro-electrolysis system [17]. Similarly, the derivatives of gluconic acid in the molasses-based distillery wastewater were effectively broken down by wet oxidation at a temperature above 150 °C [16]. Another method is the removal of heavy metals by adding special agents and/or functionalized resin [18–21]. For example, Li et al. [18] found that at a sodium diethyldithiocarbamate/Cu molar ratio of 1, approximately 99.6% of complex Cu (with EDTA as the coordination agent) is trapped by sodium diethyldithiocarbamate and precipitates during coagulation after the addition of polyferric sulfate and polyacrylamide. A few aminopolycarboxylic acids, such as iminodiacetic acid, nitrilotriacetic acid, and diethylenetriaminepentaacetic acid, include the necessary functionalized groups (e.g., carboxyl and amino) to chelate heavy metals [19]. Such organics were also grifted on the resin surface to considerably improve its affinity for heavy metals recycling from electroplating wastewater [20,21]. Although these methods can efficiently remove heavy metals from alkaline wastewater, they require expensive agents and complicated devices. Thus, their applications are limited. By contrast, adsorption is a low-cost process and has a simple operating method. Many adsorbents, such as natural minerals, carbon materials and artificial composites, exhibit a variety of surface groups (such as –OH and –COOH) to adsorb free heavy metals. These surface groups show relatively low affinity to heavy metals compared with complex agents and have low removal efficiencies in alkaline electroplating wastewater treatment [22]. Therefore, a novel adsorbent that can remove complex heavy metals should be urgently developed.

The resource reutilization of electroplating sludge for the preparation of novel adsorbents for electroplating wastewater treatment is a green route involving the "waste to treat wastewater" approach. This type of sludge is composed of heavy metals, precipitant reagent and hydrolyzed flocculant [23–25]. Fe/Al oxyhydroxide from hydrolyzed flocculants are usually used to adsorb free heavy metals in the absence of complex agents [24,25]. However, the reutilization of heavy metal-bearing sludges in alkaline electroplating wastewater has not yet been reported.

The aim of this study is to recycle Co/Cr-bearing sludge for the preparation of a novel erdite material that can be used in the advanced treatment of electroplating wastewater effluent. In contrast to conventional adsorbents, the novel erdite material can be spontaneously hydrolyzed in neutral solutions, thereby generating Fe/S-bearing oxyhydroxide with plenty of –SH groups for heavy-metal coordination. The release of Co and Cr in the prepared erdite materials was investigated during wastewater treatment.

2. Materials and Methods

2.1. Electroplating Sludge Pre-Treatment

The Co-bearing sludge, denoted as S1, was precipitated from the wastewater of a rolling-anode plant (Sanhe company, Changchun, China). The wastewater was first treated with a resin filter (CH-90, Kaiping company, Shanghai, China) to recycle Co, and then coagulated with the addition of polyferric sulfate and polyacrylamide. A yellowish precipitate was generated after coagulation treatment and pumped to a plate and frame filter (XAMY6/450-30U, Runnan company, Shanghai, China) to perform mechanical dewatering. Thus, the yellowish cake of S1 was generated and stored in the northwestern corner of the waste yard before transport and landfilling. The Cr-bearing sludge, named S2, was generated from the wastewater of the electroplating workshop (Sanhe company, Changchun, China). For Cr recovery, a resin filter (RS10, Kaiping company, Shanghai, China) was also employed to treat the wastewater, and then the effluent was further treated with the coagulation and dewatering process with the abovementioned procedures. The generated Cr-bearing sludge cake was placed at the south side of the waste yard.

The two sludges, S1 and S2, were vacuum-dried at 50 °C overnight and ground to pass through a 1 mm mesh. The powder of each sludge was analyzed by X-ray fluorescence (XRF, S4-Explorer, Bruker, Karlsruhe, Germany); only the diffraction intensity of metallic elements was recorded, except for the nonmetal elements C, H and O. For S1, the percentage of total metallic elements was about 36.31%, whilst the residual in S1 was affiliated with the coordinated groups (e.g., oxide/oxyhydroxide, sulfate and chloride) and the added polymeric flocculant. S2 showed a similar composition to S1, and the major elements are summarized in Table 1. S1 contains 25.6% Fe and 5.5% Co, demonstrating that S1 is a Fe/Co-rich sludge, whereas S2 is a Fe/Cr-rich sludge with 36.8% Fe and 7.8% Cr.

Table 1. Sludge composition.

Element	Relative Weight Percentage (wt.%)	
	S1	S2
Fe	25.6	36.8
Cr	0.06	7.8
Co	5.5	0.04
Ca	0.5	1.3
Si	0.9	1.1
Al	2.05	2.78
Na	1.7	1.4

2.2. Hydrothermal Conversion of Sludge

S1 and S2 were hydrothermally treated in accordance with our previous method [26,27] with the replacement of NaOH by Na_2S. In brief, S1 (1 g), $Na_2S{\cdot}9H_2O$ (2.4 g) and deionized water (30 mL) were mixed in a 50 mL Teflon vessel. Then, the vessel was sealed and heated at 180 °C for 10 h in a drying oven (DHG-9037A, Jinghong company, Shanghai, China) and water-cooled to below 25 °C. Finally, the blackish particles at the vessel bottom were collected, freeze-dried at −80 °C in a freeze dryer (FDU-2110, EYELA, Tokyo, Japan) overnight and denoted as SP1. SP2 was hydrothermally treated in accordance with the abovementioned steps, and the obtained product was named SP2.

2.3. Heavy Metal Release

In the Sanhe company, heavy-metal-bearing wastewaters were mixed in a storage tank to generated a comprehensive electroplating wastewater, in which Zn, Cu, Ni, Co and Cr values were 711.5, 36.8, 131.2, 2.7 and 0.9 mg/L. The comprehensive electroplating wastewater was alkaline at pH 13.5, and adjusted to pH 7–7.5 by adding hydrochloric acid (9.8 M, Binghai chemical Group, Jinhua, China), in accordance with the optimal pH range of the resin filter (KP752, Kaiping company, Shanghai, China) operation. The wastewater was then treated with resin (KP752, Kaiping company, Shanghai, China) for the recycling of heavy metals and precipitated with polyferric sulfate. After treatment, the pH of the electroplating wastewater effluent was 7.3, and the concentrations of Zn, Cu, Ni, Co and Cr were 3.03, 0.51, 0.56, 0.19 and 0.003 mg/L, respectively. In the effluent, the concentrations of Zn, Cu and Ni exceeded the emission standard of pollutants for electroplating in China (GB21900-2008). Thus, the effluent needed to be further treated before discharge.

In total, 0.8 g/L of SP1 and SP2 adsorbents were added to 100 mL of electroplating wastewater, and the mixture was magnetically stirred at 90 rpm for 2 h; then, SP1 and SP2 were collected and freeze-dried after the reaction. The dried SP1 and SP2 were respectively put into 50 mL deionized water until the concentration was 0.3 g/L and stirred magnetically for 24 h. Then, the concentration of heavy metals in the supernatant was determined by an inductively coupled plasma optical emission spectrometer (ICP-OES, AVIO-200, PerkinElmer, Waltham, MA, USA).

2.4. Electroplating Wastewater Treatment

The effluent was treated by placing SP1 in 50 mL of effluent until the SP1 concentration was 0.05 g/L. The mixture was magnetically stirred at 90 rpm for 2 h, and SP1 was separated by centrifuging at 6000 rpm for 5 min. Subsequently, the supernatant was collected, and the residual heavy metals in the supernatant were determined. According to the aforementioned steps, heavy metal removal was optimized by varying the SP1 dosage from 0.05 g/L to 0.8 g/L. Then, the dosage of SP2 on the effluent treatment was also investigated according to the abovementioned steps. Powdered activated carbon is a commercial adsorbent and is used in mass production in the project-scale advanced treatment of electroplating wastewater [28,29]; thus, it was targeted as the control in the experiment. After that, several common chemical reagents and the raw sludge were used to treat the electroplating wastewater, including polyaluminum chloride (PAC) [30], polyferric sulfate (PFS) [31] and $Na_2S{\cdot}9H_2O$ [32]. The experimental conditions are similar to the above, but the dose of the chemical reagent was 0.8 g/L, and the mixture was magnetically stirred at 90 rpm for 2 h.

2.5. Kinetic Experiment

A dose of 0.05 g/L SP2 was added to 50 mL electroplating wastewater, and the mixture was magnetically stirred at 90 rpm. The heavy metal concentration in the supernatant was measured at 0.25 h, 0.5 h, 1 h, 1.5 h and 2 h, respectively.

2.6. Adsorbent Regeneration

Adsorbent regeneration was carried out by treating with 15% NaCl of pH 5 for 48 h or calcining at 450 °C for 2 h. The adsorbent after regeneration treatment was put into 50 mL electroplating wastewater with a dose of 0.8 g/L and a reaction time of 2 h. After the reaction, the residual heavy metal concentration in the supernatant was measured.

2.7. Zeta Potential Measurement

In total, 0.5 g/L of SP2 was added to deionized water at pH 5 and magnetic stirring was performed, and the zeta potential of SP2 was measured by a zeta potential analyzer (Zetasizer Nano ZSP, Malvern, UK) at 30 min, 60 min, 180 min and 300 min, respectively.

2.8. Characterisation

The conversion mechanism was explained by characterizing the sludge and products by scanning electron microscopy (SEM, JSM-6400, Jeol, Tokyo, Japan), X-ray diffraction (XRD, Rint2200, Rigaku Corporation, Tokyo, Japan) and X-ray photoelectron spectroscopy (XPS, ADES-400, VG Scientific, Birmingham, Britain).

3. Results and Discussion

3.1. Conversion of the Two Types of Sludge to Erdite-Bearing Particles

The morphology and compositions of the sludges are shown in Figures 1 and 2. S1 was an irregular block (Figure 1a) and did not show the obvious peaks of Fe/Co-bearing compounds (Figure 2, S1). The Fe/Co-bearing compounds in S1 exhibited weakly crystallized forms. S2 showed fine particles (Figure 1b), and its XRD patterns were similar to those of S1 (Figure 2, S2) and corresponded to weakly crystallized Fe/Cr-bearing compounds. After hydrothermal treatment, SP1 synthesized using S1 as raw sludge was characterized by short rod-like particles (Figure 1c), with obvious peaks of erdite and sulfur (Figure 2, SP1). In comparison with SP1, SP2 from S2 showed long nanorods with diameters of 100 nm and lengths of 0.5–1.5 μm (Figure 1d). These products corresponded to the sharp peaks of erdite in the curve of SP2 (Figure 2, SP2). These findings demonstrated that Fe was involved in the formation of amorphous Fe-bearing compounds in S1 and S2. These Fe-bearing compounds were converted to well-crystallized erdite particles, which were short in SP1 and lengthened in SP2.

Figure 1. Scanning electron microscope (SEM) images of (**a**) S1, (**b**) S2, (**c**) SP1 and (**d**) SP2.

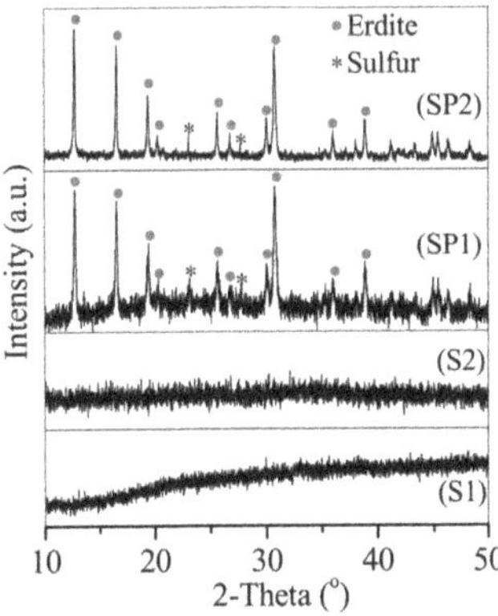

Figure 2. X-ray diffraction (XRD) patterns of S1, S2, SP1 and SP2.

The two types of sludge and the prepared SPs (SP1 and SP2) were further characterized by XPS. The Fe 2p spectra of S1 and S2 showed that the indicative peak at the binding energy of 710.5 eV (Figure 3a, S1 and S2) corresponded to Fe^{3+} in the Fe–O bond and was similar to that in ferrihydrite [33]. After the hydrothermal process, SP1 showed a new peak at the binding energy of 707.8 eV (Figure 3a, SP1). The peak belongs to Fe^{3+} in the Fe–S bond, in accordance with erdite formation. Compared with SP1, SP2 showed an intensified peak of Fe^{3+} in Fe–S bond (Figure 3a, SP2), in agreement with the long rod-sharped erdite. For the S 2p spectra, SP1 and SP2 showed the four major peaks at binding energies of 160.4, 161.3, 163.2 and 167.4 (Figure 3b) corresponding to the S in the Fe–S bond of the $(FeS2)_n{}^{n-}$ structure, and S^{2-}, sulfur and S in sulfate, respectively. In the Co 2p spectra, a major peak at 782.1 eV with a satellite peak was observed at the curve of S1 (Figure 3c, S1). The peak was related to the Co–O bond [34]. After hydrothermal treatment, new peaks related to Co in the Co–S bond appeared in SP1 at 778.6 eV (Figure 3c, SP1) [34], indicating the involvement of Co in CoS and/or CoS_2 after the addition of Na_2S. In the Cr 2p spectra, S2 exhibited two major peaks at 577.3 and 579.8 eV (Figure 3d,

S2). The peaks were attributed to the Cr(III) and Cr(VI) [35], respectively. However, the peak of Cr(VI) disappeared during the hydrothermal process, and only the peak of Cr(III) remained in SP2 (Figure 3d, SP2), indicating the reduction of Cr(VI) to Cr(III) by $Na_2S{\cdot}9H_2O$.

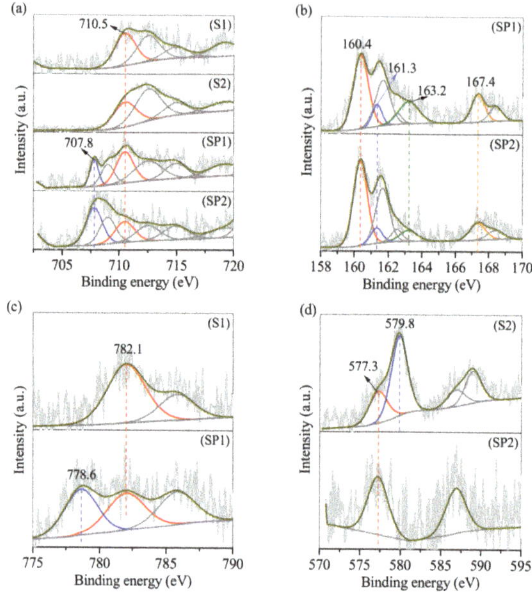

Figure 3. High resolution (**a**) Fe 2p, (**b**) S 2p, (**c**) Co 2p and (**d**) Cr 2p X-ray photoelectron spectroscopy (XPS) curves of S1, S2, SP1 and SP2.

Fe-bearing compounds were rich in S1 and S2 and converted into erdite in four steps. Firstly, the added Na_2S was hydrolyzed to release OH^- and HS^- to the solution, increasing the solution pH to above 13.6. Thus, many OH^- ions in the solution attacked the surface Fe at the Fe-bearing mineral and generated $Fe(OH)_4{}^-$ (Equation (1)) in the solution [36]. This result indicated the presence of residual Fe (approximately 15 mg/L) in the solution after the hydrothermal process (Figure 4). Secondly, a replacement reaction between free HS^- and the hydroxyl group of $Fe(OH)_4{}^-$ occurred, thereby generating $Fe(OH)_3HS^-$ (Equation (2)), followed by the conjunction reaction between two adjacent $Fe(OH)_3HS^-$ compounds. Thus, $Fe_2S_2(OH)_4{}^{2-}$ was generated with the dewatering of two water molecules (Equation (3)) [37]. Finally, the conjunction reaction continued in the presence of sufficient $Fe(OH)_4{}^-$ to form the final product $(FeS_2)_n{}^{n-}$. Na^+ neutralized the free charge of $(FeS_2)_n{}^{n-}$, and free water molecules occupied the free channels in the structure of $(FeS_2)_n{}^{n-}$, resulting in erdite nanorod formation. Cr(VI) was reduced by free HS^- in SP2, thereby generating Cr(III) with the generation of OH^-. This reaction employed plenty of OH^- during the $Fe(OH)_4{}^-$ formation, thereby increasing the length of the erdite nanorods in SP2. Conversely, the reaction between Co and HS^- in SP2 also occurred without OH^- generation, the contribution of which in erdite formation was negligible. During the hydrothermal process, Si/Al-bearing compounds were dissolved [38], thereby forming $Si(OH)_4$ and $Al(OH)_4{}^-$ in accordance with the high concentration of Si/Al in the supernatant (Figure 4). The dissolved Si/Al was not involved in erdite formation.

$$Fe(OH)_3 + OH^- \rightarrow Fe(OH)_4^- \tag{1}$$

$$Fe(OH)_4^- + HS^- \rightarrow Fe(OH)_3HS^- + OH^- \tag{2}$$

$$2Fe(OH)_3HS^- \rightarrow 2H_2O + (Fe_2S_2(OH)_4)^{2-} \tag{3}$$

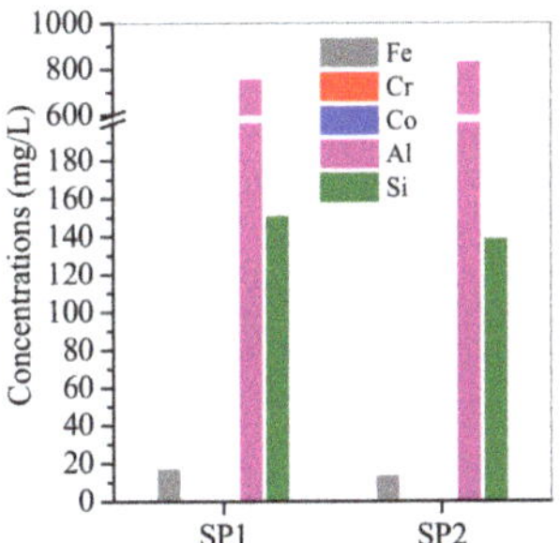

Figure 4. Concentrations of heavy metals in the supernatant after the hydrothermal process.

3.2. Adsorption of Cu, Zn, Ni and Co from Electroplating Wastewater Effluent

In order to study the adsorption stability of SP1 and SP2, the release of heavy metals in used SP1/SP2 after electroplating wastewater treatment was investigated. After stirring at 90 rpm for 24 h, Co was apparently released from used SP1, and its concentration was close to 0.7 mg/L, whist only 0.003 mg/L Co was released from used SP2. Other heavy metals—e.g., Zn, Cu, Ni and Cr—were at a level lower than 0.01 mg/L (Figure 5). This demonstrated that the used SP2 showed a stable adsorption performance in the electroplating wastewater treatment.

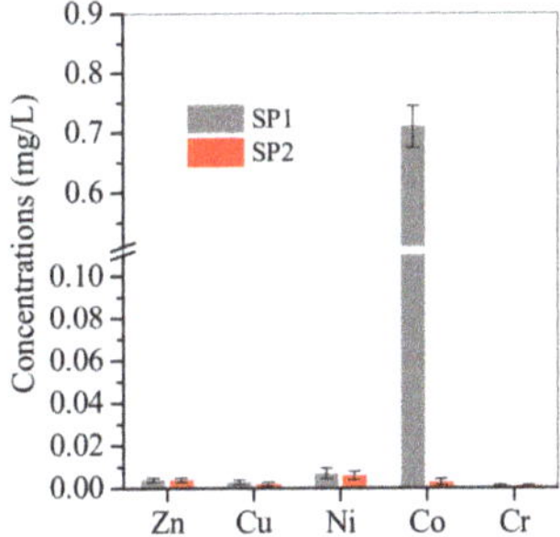

Figure 5. Heavy metals released from SP1/SP2 after electroplating wastewater treatment.

The electroplating wastewater contained 3.03 mg/L Zn, 0.51 mg/L Cu, 0.56 mg/L Ni and 0.19 mg/L Co. The first three heavy metals needed to be treated further because their concentrations exceeded the discharge standard for electroplating wastewater in China (GB21900-2008), and the concentration of Co met the discharge standard. The prepared adsorbents, SP1 and SP2, were employed to remove the first three heavy metals form electroplating wastewater, with powdered activated carbon as reference. At a dose of 0.05 g/L, the removal efficiency of Zn was 77.8% for SP2, 69.1% for SP1 and 29.1% for powdered activated carbon (Figure 6a). SP2 showed a high Zn removal efficiency in comparison with SP1 due to the formation of a well crystallized erdite nanorod in SP2. The mechanism for heavy metals removal by the erdite nanorod will be detailed in the next section. With the dosage increased from 0.05 to 0.3 g/L, the removal efficiency of Zn was elevated to nearly 100% for both SP1 and SP2, but only to 56.5% for powdered activated carbon (Figure 6a). In addition, a comparison experiment between SP2 and other materials were also performed, and their effects on Zn removal were sorted in the following order: SP2 > polyaluminum chloride (PAC) > $Na_2S{\cdot}9H_2O$ > Polymeric ferric sulfate (PFS) > S2 > S1. As mentioned in Section 1, Zn was complexed with organic reagents (e.g., citric acid, EDTA and tartaric acid) in the alkaline electroplating process, and thus was not spontaneously precipitated in the generated alkaline wastewater. For the advanced treatment of electroplating wastewater, powdered activated carbon was widely used because it has plenty of surface functional groups; e.g., –COOH, –C=O and –OH, for Zn coordination. However, it showed a low Zn removal efficiency (<64.7%), even though its dosage was increased to 0.8 g/L, in comparison with SP1 and SP2 (Figure 6a). This demonstrated that these surface groups on powdered activated carbon had a normal affinity to adsorbing Zn in comparison with the complexed organic reagents in wastewater. Ibrado et al. reported that the adsorption capacity of Zn on coconut-derived carbon was nearly 5 mg/g, but this decreased rapidly to 1.7 mg/g in the presence of cyano due to the formation of cyano complexes of Zn [39]. Therefore, the residual Zn was stable in the Zn-complex ligands and did not react with the surface groups of powdered activated carbon [40]. In electroplating wastewater treatment, PAC and PFS were common coagulants, and spontaneously hydrolysed to generate Al/Fe-bearing flocs. The raw sludge, S1 and S2, was rich in weakly crystallized Fe oxyhydroxides. Such Al/Fe-bearing flocs and Fe oxyhydroxides contained abundant of surface hydroxyl groups, similar to powdered activated carbon, for heavy metals adsorption [41]. In addition, $Na_2S{\cdot}9H_2O$ was an industrial chemical, and decomposed to HS^- in water, followed by reacting with free Zn to form ZnS precipitate [42], which was in accordance with the removal efficiency of 52.7% for Zn. In summary, the surface functional groups on powdered activated carbon, hydrolysed PAC and PFS, S1 and S2, along with free HS^- from Na_2S, were not particularly efficient in the removal of Zn from electroplating wastewater.

In the electroplating wastewater, Cu was at a low level in comparison with Zn, and was removed at a rate of nearly 100% with the addition of SP1 and SP2 at the dosage of 0.3 g/L. However, only 25.5% Cu removal was achieved by powdered activated carbon in the presence of chelating organics in wastewater (Figure 6b). Chu et al. employed coal-based carbon for Cu-bearing wastewater treatment and found that by adding EDTA at the M_{EDTA}/M_{Cu} ratio of 10, the Cu removal efficiency dropped dramatically from 83.7% to 16.5% [43]. The abovementioned materials—e.g., PAC, PFS, $Na_2S{\cdot}9H_2O$, S2 and S1—were also employed for Cu removal, showing similar values to those achieved for Zn removal. The atomic radius of Cu is 1.28 pm, which is close to that of Zn (1.39 pm), and thus Cu showed similar complex performance and removal efficiency to that of Zn in wastewater.

Although Ni was at a low level in wastewater—similar to Cu—its removal efficiency was apparently lower than that of Cu. As shown in Figure 6c, at the maximum dosage of 0.8 g/L, the removal efficiency of Ni was 49.6% for SP2, 44.3% for SP1 and 27.4% for powdered activated carbon. In the comparison experiment, the removal efficiency of Ni was ranked as follows: SP2 > PFS > PAC > $Na_2S{\cdot}9H_2O$ > S2 > S1. Compared with Cu and Zn, Ni has a smaller atomic radius (1.24 pm) and easily reacts with chelating organics to form more stable Ni-complexed ligands. For instance, in citric acid, the stability constant of Ni in citric acid is $\log K^{Ni}_{NiH_3A}$ 1.75, which is higher than that of Zn (1.25) [44]. In terms of Ni removal efficiency, $Na_2S{\cdot}9H_2O$ only achieved 7.5%, which was lower than that of PAC and PFS and close to that of S1 and S2. This is because NiS was metastable in aqueous solution and easily converted to an Ni_2S_3 and Ni-S-bearing mixture to redissolve in wastewater [45]. S1 and S2 showed low removal efficiencies for Ni in comparison with PAC and PFS due to the inadequacy of surface hydrogen groups.

The Co concentration was less than 0.5 mg/L, which met the discharge standard. Approximately 60% of Co was removed by adding SP2, which is higher than the value for powdered activated carbon (nearly 20%) (Figure 6d). However, by adding 0.8 g/L SP1, the Co concentration in the treated wastewater was 0.77 mg/L (Figure 6e), which was apparently higher than that in the raw wastewater (0.19 mg/L), suggesting the release of Co from SP1 to wastewater, probably from the dissolution of CoS and/or CoS_2. Thus, SP1 was not an ideal adsorbent for the effluent treatment. The abovementioned materials were also used for Co removal, as shown in Figure 6f, and showed a similar value as Ni due to the similar radius and chelating performance of Co and Ni [44].

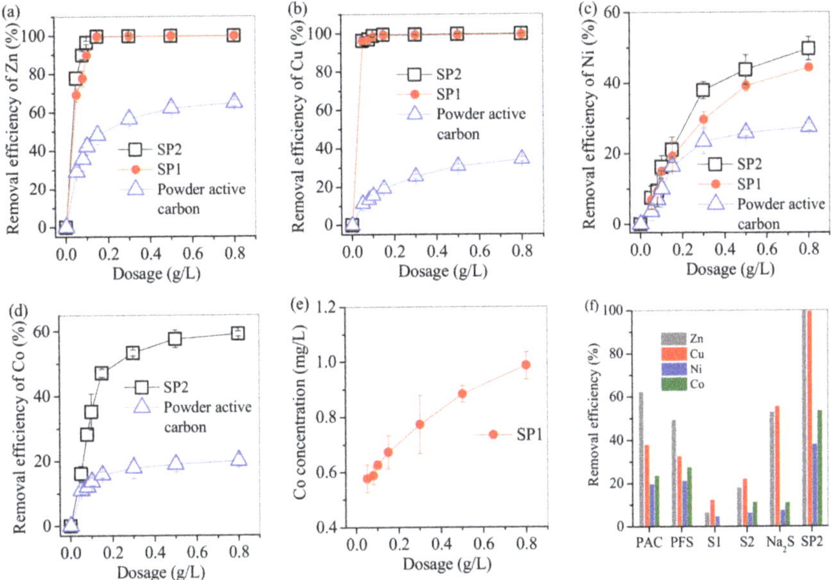

Figure 6. Application of SP1 and SP2 in electroplating wastewater treatment. (**a–d**) represent the removal efficiencies of Zn, Cu, Ni and Co, respectively, by SP1 and SP2, whilst (**e**) shows the concentration of Co in the supernatant with the addition of SP1; (**f**) shows the removal of heavy metals by adding SP2 in comparison with polyaluminum chloride (PAC), polyferric sulfate (PFS), the raw sludge S1 and S2 and $Na_2S{\cdot}9H_2O$ of chemically pure grade.

In comparison with the abovementioned materials, SP2 exhibited desirable removal efficiencies of Zn, Cu, Ni and Co. By adding 0.3 g/L SP2, the residual concentrations of Zn, Cu, Ni, Co and Cr in the wastewater were 0.007, 0.003, 0.33, 0.09 and 0.002 mg/L, respectively, which met the discharge standard [13]. This demonstrated that SP2 was an efficient reagent in smelting wastewater treatment.

The kinetic experiment of SP2 was investigated as shown in Figure 7, and the adsorption of heavy metal on SP2 was simulated using a pseudo-second-order model (Equation (4)). The kinetic model was expressed as follows:

$$q_t = \frac{kq_e^2 t}{1 + kq_e t} \tag{4}$$

where q_e and q_t are the adsorption capacity (mg/g) of the heavy metal at equilibrium and at any time, t, respectively; and k is the pseudo-second-order adsorption rate constant (g/mg h).

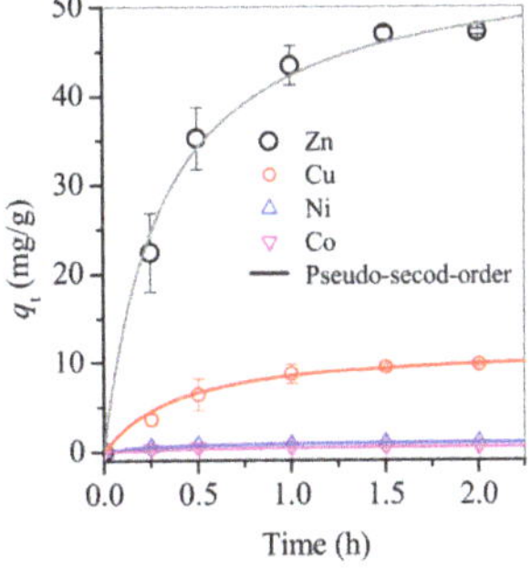

Figure 7. Non-linear plots of the pseudo-second-order model for the adsorption of Zn/Cu on SP2.

The adsorption data of Zn, Cu, Ni and Co on SP2 fitted well with the pseudo-second-order model (Figure 7) with correlation coefficients (R^2) of 0.995, 0.997, 0.991 and 0.989, respectively. This finding indicates the importance of heavy metals chemisorption on SP2. However, the calculated q_e values were in the following order: Zn > Cu > Ni > Co. The value of Zn was 3.03 mg/L in the electroplating wastewater, which was about six times the value of Cu and Ni, and thus showed the highest q_t. In comparison with Cu, Ni and Co have small ionic radii and are easily complexed with organics (e.g., ethylene diamine tetra-acetic acid) to form a stable complex product [6], resulting in a low q_t in comparison with Cu.

The used SP2 was regenerated with 15% NaCl solution at pH 5 for 48 h or calcinated at 450 °C for 2 h. The results showed that SP2 was regenerated easily using NaCl solution as the desorption agent. However, the removal efficiency of Zn, Cu, Ni and Co was dramatically decreased to 33.2%, 27.5%, 11.3% and 18.6% (Figure 8). After being calcinated at 450 °C, the regenerated SP2 also showed a similar low removal efficiency of heavy metals. This indicated that SP2 cannot feasibly be reused.

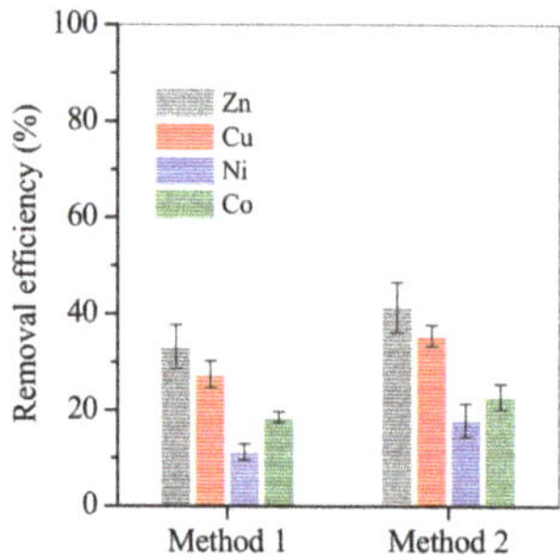

Figure 8. Reuse of the precipitate of hydrolyzed SP2 for electroplating wastewater treatment. In the figure, Methods 1 and 2 represent NaCl solution elution and high-temperature calcination, respectively.

The zeta potential measurement of SP2 showed that SP2 had a negative charge on the surface after hydrolysis (Figure 9), which showed a strong adsorption capacity for positively charged heavy metals. In addition, SP2 can compete with chelates for binding heavy metals and then remove heavy metals from wastewater.

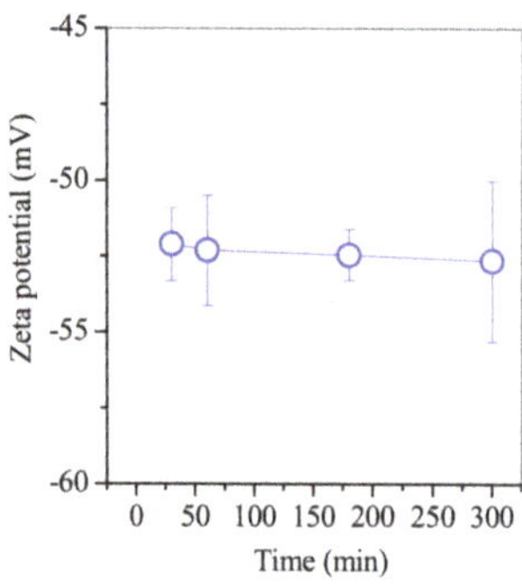

Figure 9. Zeta potential of SP2 in deionized water.

3.3. *SP1 and SP2 Characterisation after Electroplating Wastewater Treatment*

SP1 and SP2 were collected after wastewater effluent treatment and characterized by SEM, XRD and XPS, and the removal mechanism of heavy metals was investigated. The used SP1 and SP2 were irregular aggregates (Figure 10a,b); the peaks of erdite disappeared, and only peaks of element sulfur remained (Figure 11). Thus, erdite was spontaneously hydrolyzed. SP1 and SP2 spectra showed that the peaks of Fe^{3+} in the Fe–S bond of erdite were absent; instead, a new peak with a binding energy of 710.2 eV appeared (Figure 12), corresponding to the Fe^{2+} generated from the redox reaction of Fe^{3+}–S [46]. The typical peak of structural S in erdite also disappeared, and the peaks of S^{2-} and S were recorded in the curve of SP1 and SP2, in agreement with erdite hydrolysis (Figure 12). In addition, the peaks of Co and Cr did not change obviously in SP1 and SP2, respectively, indicating that Co/Cr oxidation did not occur (Figure 13).

Figure 10. SEM images of (**a**) SP1 and (**b**) SP2 after smelting wastewater treatment.

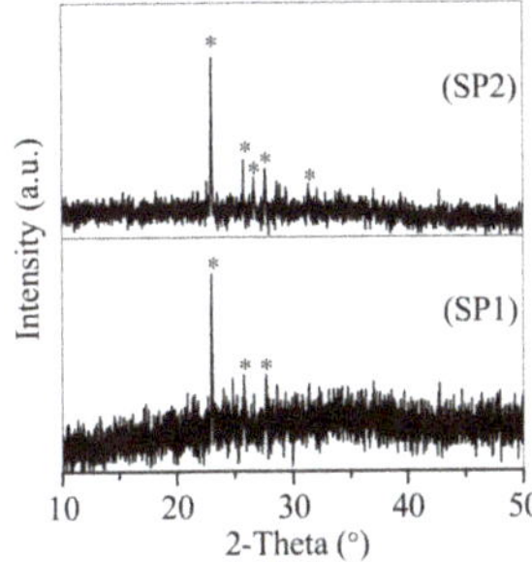

Figure 11. XRD patterns of SP1 and SP2 after electroplating wastewater treatment.

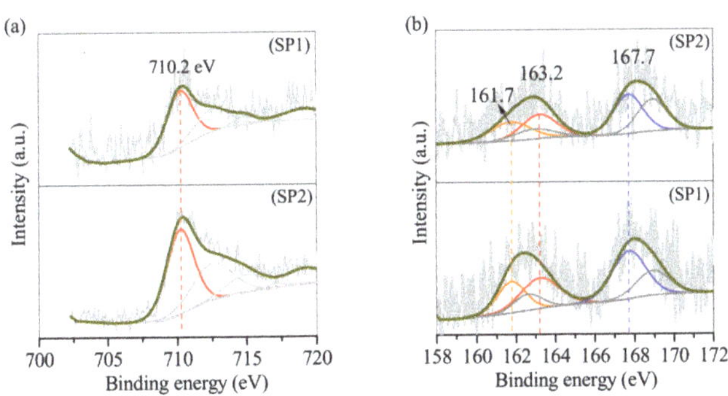

Figure 12. High resolution (**a**) Fe 2p, (**b**) S 2p XPS spectrum of SP1 and SP2 after smelting wastewater treatment.

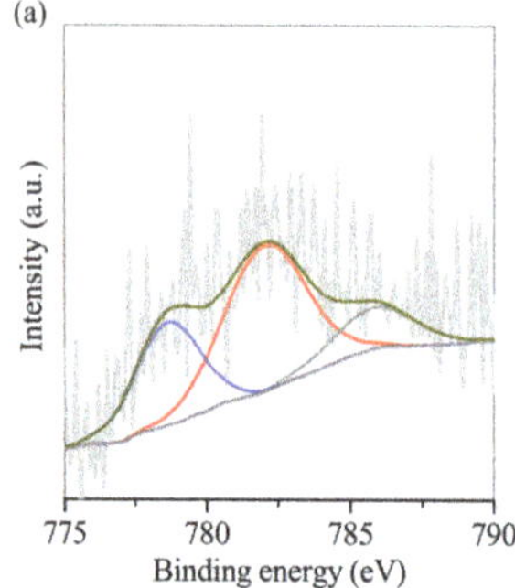

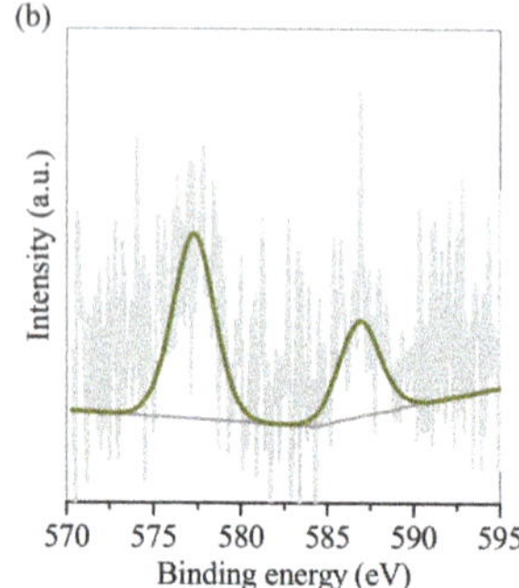

Figure 13. High resolution (**a**) Co 2p, (**b**) Cr 2p XPS spectrum of SP1 and SP2 after smelting wastewater treatment.

In the electroplating wastewater effluent, the residual heavy metals included Zn, Cu, Ni and Co and were chelated with organic matters, such as nitrilotriacetic acid and citric acid [6,7]; as such, they did not precipitate even at an effluent pH of >7.3. When the prepared SPs were introduced to the effluent, erdite was rich in SPs and spontaneously hydrolyzed and generated various Fe/S-bearing complexes and clusters, such as $=Fe(SH)_2$, =Fe(OH)(SH) and $=Fe(SH)^+$ [47]. These products, which are metastable, were further converted to Fe/S-bearing oxyhydroxide (such as =Fe–SH and =Fe–OH) through the homolytic cleavage of Fe–S–Fe bonds [48] after the release of OH^- and HS^- to the effluent (Figure 14). This phenomenon corresponds to the increase in the treated effluent pH from 7.3 to 8.6. During erdite hydrolysis, a redox reaction between surface Fe^{3+} and the adjacent free SH^- occurred [47], and thus surface-associated Fe^{2+}, which stabilized the new Fe/S-bearing oxyhydroxide, was generated. Surface functional groups (Fe–SH and Fe–OH) are rich in new Fe/S-bearing oxyhydroxide, and H^+ ions on their sides were dissociated with vacuum surface sites ($Fe–S^-$ and $Fe–O^-$) under alkaline conditions [26]. The adjacent vacuum sites exhibited high affinity to complex Cu and Zn (Figure 14) and competed with chelates, such as EDTA [49], leading to nearly 100% removal of Cu and Zn from the wastewater. However, Ni and Co have small ionic radii and easily react with complex agents to form stable complex products with high stability constants [6]; these products only break with difficulty in the presence of HS^- and hydrolyzed Fe/S-bearing oxyhydroxide, leading to the production of residual Co and Ni in treated effluents. Polyaluminum chloride and polyferric sulfate were hydrolyzed and formed Al/Fe flocs, exhibiting plenty of hydroxyl groups for heavy metal coordination; however, they showed low Zn, Cu, Ni and Co removal efficiencies (Figure 6f) given the absence of an –SH bond on the surface of the hydrolyzed Fe/Al flocs. In addition, the released HS^- from the hydrolyzed erdite can react with heavy metals, such as Zn and Cu, to form S-bearing precipitates; however, they play a minor role in heavy metal removal. For instance, with the addition of Na_2S, approximately 50% of Zn and Cu was removed, whereas lower than 10% of Ni and Co was removed (Figure 6f). Thus, the role of HS^- in the removal of heavy metals is minimal. The Fe/S-bearing oxyhydroxide of hydrolyzed erdite plays a key role in the removal of residual heavy metals.

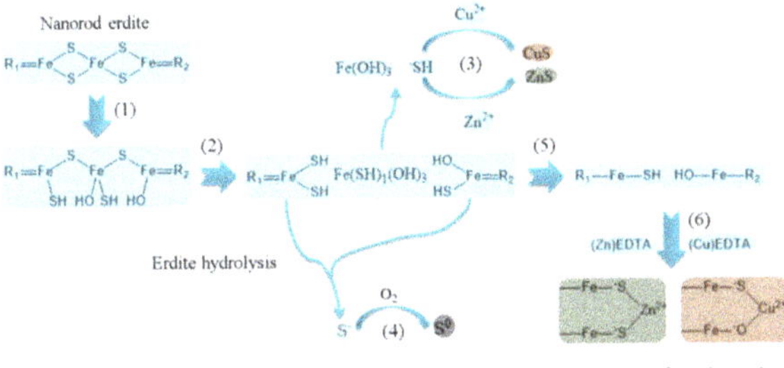

Figure 14. Illustration graph of SP2 for heavy metals removal from electroplating wastewater.

4. Conclusions

Two types of electroplating sludge, namely S1 with 25.6% Fe and 5.5% Co and S2 with 36.8% Fe and 7.8% Cr, were recycled as erdite-bearing particles for the advanced treatment of electroplating wastewater effluent. The weakly crystallized Fe-bearing minerals were rich in sludge and hydrothermally converted to well-crystallized erdite particles through the addition of Na_2S. In S1, Co was involved in CoS and/or CoS_2 formation, and the corresponding product SP1 was characterized by short rod-like particles. However, Cr had two valance states in S2, namely Cr(VI) and Cr(III). Cr(VI) was reduced to Cr(III) by Na_2S, and abundant OH^- was generated, which steadily promoted the conversion of Fe-bearing minerals to erdite. Consequently, the lengths of the erdite nanorods increased, and the corresponding product SP2 was characterized by long nanorods with a diameter of 100 nm and length of 0.5–1.5 μm.

When SP1 was added to the electroplating wastewater effluent, a fraction of Co was released to the effluent, and Zn and Cu were removed efficiently. Compared with SP1, SP2 showed ideal Zn, Cu, Ni and Co removal efficiencies without releasing Co and Cr. When the SP2 dosage was 0.3 g/L, 100% Zn and Cu removal efficiencies were nearly achieved, and 37.9% Ni and 53.3% Co removal efficiencies were observed—higher than those of powdered activated carbon, polyaluminum chloride, polyferric sulfate and Na_2S. After treatment, the concentrations of Zn, Cu, Ni and Co met the emission standard, and the treated effluent was dischargeable. Through the proposed method, the Cr-bearing electroplating sludge was recycled as erdite nanorod particles, and the product was efficient in the advanced treatment of electroplating wastewater.

Author Contributions: Data curation, A.K. and Z.W.; Project administration, Y.C.; Software, T.S.; Supervision, H.Y.; Validation, D.L.; Writing—original Draft, Y.L. and S.Z. All authors have read and agreed to the published version of the manuscript.

Funding: This work was supported by the National Natural Science Foundation of China (Grant Nos. 51578118, 51678273, 51878134, and 51878133), the Fundamental Research Funds for the Central Universities (Grant No. 2412017QD021) and the Science and Technology Program of Jilin Province (Grant No. 20190303001SF).

Acknowledgments: Yanwen Liu and Asghar Khan are co-first authors of the article.

Conflicts of Interest: The authors declare no conflict of interest.

References

1. Mizushima, I.; Tang, P.T.; Hansen, H.N.; Somers, M.A. Development of a new electroplating process for Ni–W alloy deposits. *Electrochim. Acta* **2005**, *51*, 888–896. [CrossRef]
2. Wang, G.; Sui, J.; Wang, C.; Cai, G. Treatment of electroplating synthetical wastewater by combined process of oxidation and reduction, coagulation and sedimentation. *China Water Wastewater* **2007**, *23*, 57.
3. Zhang, Z.; Li, L.; Zhu, H.; Wang, F.; Hua, J. Removing chromium from electroplating wastewater by chemical precipitation. *Environ. Sci. Technol.* **2008**, *31*, 96–97.
4. Lin, Y.; Yen, S. Effects of additives and chelating agents on electroless copper plating. *Appl. Surf. Sci.* **2001**, *178*, 116–126. [CrossRef]

5. Celary, P.; Sobik-Szołtysek, J. Vitrification as an alternative to landfilling of tannery sewage sludge. *Waste Manag.* **2014**, *34*, 2520–2527. [CrossRef]
6. Andrus, M.E. A review of metal precipitation chemicals for metal-finishing applications. *Met. Finish.* **2000**, *98*, 20–23. [CrossRef]
7. Guo, X.; Liu, J.; Tian, Q. Elemental behavior of multi-component metal powders from waste printed circuit board during low-temperature alkaline smelting. *Chin. J. Nonferrous Met.* **2013**, *23*, 1757–1763.
8. Sze, Y.K.P.; Xue, L.Z.; Lee, M.Y. Treatment of alloy-electroplating wastewater by an automated solvent extraction technique. *Environ. Technol.* **2001**, *22*, 979–990. [CrossRef]
9. Ajmal, M.; Rao, R.A.K.; Ahmad, R.; Ahmad, J.; Rao, L.A.K. Removal and recovery of heavy metals from electroplating wastewater by using kynaite as an adsorbent. *J. Hazard. Mater.* **2001**, *87*, 127–137. [CrossRef]
10. Zhonghe, M.; Zhumei, W.; Ming, J. Treatment of low concentration complex nickel Chromium(Cr^{6+}) COpper mixed electroplating wastewater. *Mod. Chem. Res.* **2018**, *11*, 26–27.
11. Zhang, H.; Sun, M.; Song, L.; Guo, J.; Zhang, L. Fate of NaClO and membrane foulants during in-situ cleaning of membrane bioreactors: Combined effect on thermodynamic properties of sludge. *Biochem. Eng. J.* **2019**, *147*, 146–152. [CrossRef]
12. Sun, M.; Yan, L.; Zhang, L.; Song, L.; Guo, J.; Zhang, H. New insights into the rapid formation of initial membrane fouling after in-situ cleaning in a membrane bioreactor. *Process Biochem.* **2019**, *78*, 108–113. [CrossRef]
13. Xu, T.; Lei, X.; Sun, B.; Yu, G.; Zeng, Y. Highly efficient and energy-conserved flocculation of copper in wastewater by pulse-alternating current. *Environ. Sci. Pollut. Res. Int.* **2017**, *24*, 20577–20586. [CrossRef] [PubMed]
14. Wang, Z.; Ye, G.; Yang, Y. Low-concentration electroplating wastewater treatment by heavy metal chelator. *J. Zhejiang Univ. (Sci. Ed.)* **2010**, *37*, 665–669. [CrossRef]
15. Shih, Y.; Lin, C.; Huang, Y. Application of Fered-Fenton and chemical precipitation process for the treatment of electroless nickel plating wastewater. *Sep. Purif. Technol.* **2013**, *104*, 100–105. [CrossRef]
16. Tembhekar, P.D.; Padoley, K.V.; Mudliar, S.L.; Mudliar, S.N. Kinetics of wet air oxidation pretreatment and biodegradability enhancement of a complex industrial wastewater. *J. Environ. Chem. Eng.* **2015**, *3*, 339–348. [CrossRef]
17. Yong, L. Progress of electroplating wastewater treatment with micro-electrolysis method. *Guangdong Chem. Ind.* **2008**, *35*, 56.
18. Li, Y.; Zeng, X.; Liu, Y.; Yan, S.; Hu, Z.; Ni, Y. Study on the treatment of copper-electroplating wastewater by chemical trapping and flocculation. *Sep. Purif. Technol.* **2003**, *31*, 91–95. [CrossRef]
19. Repo, E.; Warchoł, J.K.; Bhatnagar, A.; Mudhoo, A.; Sillanpää, M. Aminopolycarboxylic acid functionalized adsorbents for heavy metals removal from water. *Water Res.* **2013**, *47*, 4812–4832. [CrossRef]
20. Fadel, D.; El-Bahy, S.; Abdelaziz, Y. Heavy metals removal using iminodiacetate chelating resin by batch and column techniques. *Desalin. Water Treat.* **2016**, *57*, 25718–25728. [CrossRef]
21. Benouali, D.; Kherici, S.; Belabbassi, M.; Belkandouci, M.; Bennemra, A. Preliminary study of Zinc removal from cyanide-free alkaline electroplating effluent by precipitation using oxalis plants. *Orient. J. Chem.* **2014**, *30*, 515–519. [CrossRef]
22. Šćiban, M.; Radetić, B.; Kevrešan, Ž.; Klašnja, M. Adsorption of heavy metals from electroplating wastewater by wood sawdust. *Bioresour. Technol.* **2007**, *98*, 402–409. [CrossRef] [PubMed]
23. Li, C.; Xie, F.; Ma, Y.; Cai, T.; Li, H.; Huang, Z.; Yuan, G. Multiple heavy metals extraction and recovery from hazardous electroplating sludge waste via ultrasonically enhanced two-stage acid leaching. *J. Hazard. Mater.* **2010**, *178*, 823–833. [CrossRef] [PubMed]
24. Espinosa, D.C.R.; Tenório, J.A.S. Thermal behavior of chromium electroplating sludge. *Waste Manag.* **2001**, *21*, 405–410. [CrossRef]
25. McCafferty, E.; Wightman, J. Determination of the concentration of surface hydroxyl groups on metal oxide films by a quantitative XPS method. *Surf. Interface Anal.* **1998**, *26*, 549–564. [CrossRef]
26. Qu, Z.; Wu, Y.; Zhu, S.; Yu, Y.; Huo, M.; Zhang, L.; Yang, J.; Bian, D.; Wang, Y. Green synthesis of magnetic adsorbent using groundwater treatment sludge for tetracycline adsorption. *Engineering* **2019**, *5*, 880–887. [CrossRef]

27. Zhu, S.; Wu, Y.; Qu, Z.; Zhang, L.; Yu, Y.; Xie, X.; Huo, M.; Yang, J.; Bian, D.; Zhang, H. Green synthesis of magnetic sodalite sphere by using groundwater treatment sludge for tetracycline adsorption. *J. Clean. Prod.* **2020**, *247*, 119140. [CrossRef]
28. Kołodyńska, D.; Krukowska, J.; Thomas, P. Comparison of sorption and desorption studies of heavy metal ions from biochar and commercial active carbon. *Chem. Eng. J.* **2016**, *307*, 353–363. [CrossRef]
29. Luo, Z.; Yang, C.; Zeng, G.; He, H.; Luo, S. Treatment of vanadium smelting wastewater by means of coagulation-sand filtration-activated carbon filter-microfiltration-RO integration technology. *Chin. J. Environ. Eng.* **2014**, *8*, 2257–2261. [CrossRef]
30. Xu, Z. Application and progress of PAC to industrial wastewater treatment. *Shanghai Environ. Sci.* **2003**, *22*, 353–357. [CrossRef]
31. Zhang, L.; Wang, K.; Liu, X.; Liu, P.; Huang, J. Experimental study of advanced treatment of coking wastewater using PFS coagulation-photocatalytic oxidation technology. *Mod. Chem. Ind.* **2015**, *34*, 487–494. [CrossRef]
32. Chen, W.S.; Lin, H.S.; Lin, J.F. Study on treatment technology of waste water with copper in complex system. *Adv. Mater. Res.* **2011**, *317*, 1847–1851. [CrossRef]
33. Glasauer, S.M.; Hug, P.; Weidler, P.G.; Gehring, A.U. Inhibition of sintering by Si during the conversion of Si-rich ferrihydrite to hematite. *Clays Clay. Miner.* **2000**, *48*, 51–56. [CrossRef]
34. Ouyang, C.; Wang, X.; Wang, S. Phosphorus-doped CoS_2 nanosheet arrays as ultra-efficient electrocatalysts for the hydrogen evolution reaction. *Chem. Commun. (Camb.)* **2015**, *51*, 14160–14163. [CrossRef]
35. Ji, M.; Su, X.; Zhao, Y.; Qi, W.; Wang, Y.; Chen, G.; Zhang, Z. Effective adsorption of Cr (VI) on mesoporous Fe-functionalized Akadama clay: Optimization, selectivity, and mechanism. *Appl. Surf. Sci.* **2015**, *344*, 128–136. [CrossRef]
36. Diakonov, I.I.; Schott, J.; Martin, F.; Harrichourry, J.-C.; Escalier, J. Iron (III) solubility and speciation in aqueous solutions. Experimental study and modelling: Part 1. hematite solubility from 60 to 300 °C in NaOH–NaCl solutions and thermodynamic properties of $Fe(OH)_4^-$(aq). *Geochim. Cosmochim. Acta* **1999**, *63*, 2247–2261. [CrossRef]
37. Zhu, S.; Liu, Y.; Huo, Y.; Chen, Y.; Qu, Z.; Yu, Y.; Wang, Z.; Fan, W.; Peng, J.; Wang, Z. Addition of MnO_2 in synthesis of nano-rod erdite promoted tetracycline adsorption. *Sci. Rep.* **2019**, *9*, 1–12. [CrossRef]
38. Zhu, S.; Lin, X.; Dong, G.; Yu, Y.; Yu, H.; Bian, D.; Zhang, L.; Yang, J.; Wang, X.; Huo, M. Valorization of manganese-containing groundwater treatment sludge by preparing magnetic adsorbent for Cu (II) adsorption. *J. Environ. Manag.* **2019**, *236*, 446–454. [CrossRef]
39. Ibrado, A.S.; Fuerstenau, D.W. Adsorption of the cyano complexes of Ag(I), Cu(I), Hg(II), Cd(II) and Zn(II) on activated carbon. *Min. Metall. Explor.* **1989**, *6*, 23–28. [CrossRef]
40. Shi, T.; Tang, B. Adsorption characteristic study of Zn(II)and Zn(II)—EDTA at the activated carbon solution interface. *Ind. Water Treat.* **2000**, *20*, 12–15. [CrossRef]
41. Liu, H.; Yang, Y.; Kang, J.; Fan, M.; Qu, J. Removal of tetracycline from water by Fe-Mn binary oxide. *J. Environ. Sci.* **2012**, *24*, 242–247. [CrossRef]
42. Lewis, A.E. ChemInform abstract: Review of metal sulfide precipitation. *Cheminform* **2012**, *43*, 222–234. [CrossRef]
43. Chu, K.H.; Hashim, M.A. Adsorption of copper (II) and EDTA-chelated copper (II) onto granular activated carbons. *J. Chem. Technol. Biotechnol.* **2000**, *75*, 1054–1060. [CrossRef]
44. Campi, E.; Ostacoli, G.; Meirone, M.; Saini, G. Stability of the complexes of tricarballylic and citric acids with bivalent metal ions in aqueous solution. *J. Inorg. Nucl. Chem.* **1964**, *26*, 553–564. [CrossRef]
45. Legrand, D.L.; Nesbitt, H.W.; Bancroft, G.M. X-ray photoelectron spectroscopic study of a pristine millerite (NiS) surface and the effect of air and water oxidation. *Am. Min.* **1998**, *83*, 1256–1265. [CrossRef]
46. Sun, J.; Zhou, J.; Shang, C.; Kikkert, G.A. Removal of aqueous hydrogen sulfide by granular ferric hydroxide—Kinetics, capacity and reuse. *Chemosphere* **2014**, *117*, 324–329. [CrossRef]
47. Schlosser, C.; Streu, P.; Frank, M.; Lavik, G.; Croot, P.L.; Dengler, M.; Achterberg, E.P. H_2S events in the Peruvian oxygen minimum zone facilitate enhanced dissolved Fe concentrations. *Sci. Rep.* **2018**, *8*, 12642. [CrossRef]

48. Arantes, G.M.; Field, M.J. Ferric–thiolate bond dissociation studied with electronic structure calculations. *J. Phys. Chem. A* **2015**, *119*, 10084–10090. [CrossRef]
49. Yang, X.; Wang, J.-N.; Cheng, C. Preparation of new spongy adsorbent for removal of EDTA–Cu(II) and EDTA–Ni(II) from water. *Chin. Chem. Lett.* **2013**, *24*, 383–385. [CrossRef]

Article

Adsorptive Behavior of an Activated Carbon for Bisphenol A Removal in Single and Binary (Bisphenol A—Heavy Metal) Solutions

M.A. Martín-Lara [1,*], M. Calero [1,*], A. Ronda [2], I. Iáñez-Rodríguez [1] and C. Escudero [3]

1 Departamento de Ingeniería Química, Universidad de Granada, Avda, Fuentenueva, s/n, 18071 Granada, Spain; ireneir@ugr.es

2 Departamento de Ingeniería Química y Ambiental, Universidad de Sevilla, Camino de los descubrimientos, s/n, 41092 Sevilla, Spain; aronda@us.es

3 Departamento de Tecnología de Alimentos, Instituto Nacional de Investigación y Tecnología Agraria y Alimentaria (INIA), Ctra. de A Coruña, km 7.5, 28040 Madrid, Spain; carlos.escudero@inia.es

* Correspondence: marianml@ugr.es (M.A.M.-L.); mcaleroh@ugr.es (M.C.); Tel.: +34-958-240-445 (M.A.M.-L.); +34-958-243-315 (M.C.)

Received: 29 June 2020; Accepted: 28 July 2020; Published: 30 July 2020

Abstract: Bisphenol A (BPA) is an extensively produced and consumed chemical in the world. Due to its widespread use, contamination by this pollutant has increased in recent years, reaching a critical environmental point. This work investigates the feasibility of bisphenol A adsorption from industrial wastewater solutions, testing the reduction of bisphenol A in synthetic solutions by a commercial activated carbon, AC-40, in batch mode. Besides, mixtures of bisphenol A and different heavy metal cations were also studied. So far, no works have reported a complete study about bisphenol A removal by this activated carbon including the use of this material to remove BPA in the presence of metal cations. First, adsorption experiments were performed in batch changing pH, dose of adsorbent, initial bisphenol A concentration and contact time. Results showed greater retention of bisphenol A by increasing the acidity of the medium. Further, the percentage of bisphenol A adsorbed increased with increasing contact time. The selected conditions for the rest of the experiments were pH 5 and a contact time of 48 h. In addition, an increase in retention of bisphenol A when the dose of adsorbent increased was observed. Then, specific experiments were carried out to define the kinetics and the adsorption isotherm. Equilibrium data were adequately fitted to a Langmuir isotherm and the kinetics data fitted well to the pseudo-second-order model. The maximum adsorption capacity provided by Langmuir model was 94.34 mg/g. Finally, the effect of the presence of other heavy metals in water solution on the adsorption of bisphenol A was analyzed. Binary tests revealed competition between the adsorbates and a significant selectivity toward bisphenol A. Finally, the study of the adsorption performance in three consecutive adsorption–desorption cycles showed efficiencies higher than 90% in all cycles, indicating that the activated carbon has good reusability.

Keywords: activated carbon; adsorption; bisphenol A; equilibrium; heavy metals

1. Introduction

Bisphenol A (BPA) is recognized as a dangerous endocrine disruptor. It causes a number of hazardous effects on living beings such as fertility damage and fetal development problems [1–3]. It has been extensively used as a crude material in the industrial production of epoxy resins and polycarbonate plastics, which accounted for nearly 64% of BPA demand in 2018. These polycarbonate resins are mainly consumed in the electronics, construction and automotive industries. In terms of country distribution, high demand is concentrated in emerging consumers such as China, India,

Southeast Asia and Brazil. China is the largest player in the BPA industry, accounting for half of BPA's global consumption. Western Europe and the United States are the other important players in the BPA market [4].

As a consequence, concentrations of BPA are present both in surface waters as well as industrial wastewaters [5–7]. For all those reasons, novelty alternatives have been proposed for its removal from wastewater during recent years [8]. BPA is present in various types of water at different concentrations. For instance, it has a concentration of 17.2 mg/L in hazardous waste landfill leachate [9,10], 12 μg/L in stream water [11], between 3.5 and 59.8 ng/L in drinking water [7] and around 100 mg/L in effluents from polycarbonate factories [6,12].

Due to its high industrial yield and toxicological effect, the elimination of BPA from water solutions is of great importance and it has been listed as a priority pollutant in water treatment [13]. Several separation techniques such as adsorption, advanced oxidation, membrane separation, phytoremediation and photocatalysis, among others, have been applied to efficiently remove BPA from water effluents [14,15]. Particularly, among wastewater treatment technologies, phytoremediation is arousing attention because it is a promising ecologically friendly alternative to remove BPA from contaminated soil and water systems. For example, Phouthavong-Murphy et al. [9] investigated the potential of two United States native switchgrass varieties for BPA removal and found good BPA removals over approximately three months. Further, advanced oxidation processes are interesting due to their good performance in the degradation of BPA and other micropollutants. Recently, Luo et al. [16] studied the rapid removal of BPA and other organic micropollutants by heterogeneous peroxymonosulfate catalysis and reported that BPA had the highest removal efficiency at pH 9.0, almost 100% removal. Other researchers [17] integrated advanced oxidation processes with membrane filtration to form a catalytic membrane that demonstrated exceptional efficiency for instantaneous degradation of BPA. Moreover, the development of cheap and eco-friendly, effective and rapid remediation technologies for the removal of BPA has increased the investigations based on the use of transition metal-based nanomaterials for photocatalysis. For example, Rani et al. [18] synthesized and examined green zinc and cobalt ferrites nanoparticles for the degradation of BPA from water under direct sunlight. Results showed very good efficiencies, similar to those found with ZnO, Co_3O_4 and Fe_2O_3 nanoparticles.

Even novel technologies are being investigated with very good results: one of the most favorable methods is traditional adsorption onto carbonaceous materials [19]. Particularly, activated carbons have good characteristics to work as an adsorbent: high porosity and surface area (around 3000 m^2/g), high degree of surface reactivity and diverse characteristics of surface chemistry [20,21]. Due to these characteristics, activated carbons have been used for many different applications: as adsorbents, catalysts or catalyst supports. They have been tested as a cleaning material for removing pollutants from gaseous or liquid phases and the purification or recovery of chemicals [22,23]. Adsorption on activated carbon is one of the first water treatments to have been used [24,25] and it is recognized by the United States Environmental Protection Agency (US EPA) as one of the best methods for removing organic and inorganic compounds from water for human consumption. The aqueous-phase adsorption of both organic and inorganic compounds has been a very important application of activated carbons. In fact, around 80% of the world production of activated carbons is used in liquid-phase applications [26,27].

The adsorption mechanism of activated carbons is due to interactions (electrostatic or non-electrostatic) between the carbon surface and the adsorbate. The nature of these interactions, which can be attractive or repulsive, depends on the: (a) charge density of the carbon surface, (b) chemical characteristics of the adsorbate and (c) ionic strength of the solution. Non-electrostatic interactions are always attractive and can include: (a) van der Waals forces, (b) hydrophobic interactions and (c) hydrogen bonding.

Activated carbons can be manufactured by physical or chemical activation processes using a wide range of raw materials. Physical activation uses high temperature for the carbonization of the original material in absence of oxygen (usually CO_2 or steam atmosphere). During chemical activation, the material is impregnated with a chemical agent prior to its carbonization. One of the most frequent chemical agents extensively used for the production of activated carbons in industry is H_3PO_4 [28,29].

Adsorptive removal of BPA from wastewater solutions by carbon-based materials, such as activated carbon, has been extensively studied [30–35]. For example, Juhola et al. [31] prepared a biomass-based adsorbent with a BPA adsorption capacity of 41.5 mg/g. Pan et al. [33] investigated the BPA adsorption on carbon nanomaterials that showed a very good potential for the removal of BPA of waters. Redding et al. [34] studied the adsorption of BPA on various modified lignite carbons. Specifically, the work demonstrated that these modified carbons prepared using high-temperature steam or methane/steam offered better behavior than conventional activated carbons. However, there are few works available focusing on the BPA removal from water when other types of contaminants are also present in the water [36–39]. BPA is a high-volume industrial chemical used in the production of epoxy resins and polycarbonate plastics. Further, different additives are added in these materials to improve the performance of the plastics. The chemical substances used as additives can be very different depending on the objective. The additives can include metal-containing additives such as heavy metals as cadmium, copper, lead or zinc, among others [40]. Heavy metals are another important type of pollutants that have been found in waters. Some researchers have analyzed the simultaneous adsorption/biosorption of these priority pollutants (heavy metals and organic pollutants) [41–43]. The influence of heavy metals could enhance the removal of BPA or perhaps they could compete with BPA for the available active sites and decrease the removal of BPA when the removal of BPA and heavy metals is carried simultaneously.

Hence, one of the objectives of this investigation is to study the interactions between some of the most common heavy metals in wastewater environments and BPA during adsorption onto activated carbon (AC-40 commercial sample). First, the single adsorption of BPA on AC-40 activated carbon was investigated. The main factors influencing the adsorption process were analyzed. Then, the simultaneous adsorption of BPA and heavy metal cations on AC-40 activated carbon was performed in order to know the type of adsorption (competitive or cooperative) which occurs between BPA and these metal cations. Finally, desorption of BPA in three consecutive cycles was analyzed. Although the study has been performed using a commercial activated carbon, so no novel findings about the material are given, there are not a wide range of research works in the literature about the use of AC-40 for the removal of BPA from wastewaters including (1) the effect of the main parameters, (2) a kinetics study, (3) an isotherms study, (4) simultaneous adsorption of BPA and heavy metals and (5) desorption. The authors consider that specific complete studies about the removal of emergent pollutants with the current technology (adsorption on activated carbons) can be interesting since they increase the knowledge in this field and can be applied in the industrial wastewater sector.

2. Materials and Methods

2.1. Adsorbent

Activated carbon trademark CECA S.L., known as AC-40, was used as an adsorbent in all the experiments. It is an extruded, thermally activated carbon from bituminous coal with a high surface area (1200 m^2/g). The main features of this adsorbent have been determined by Méndez Díaz et al. [44], Abdel Daiem et al. [45] and Velo-Gala et al. [46]. Table 1 shows a summary of the properties.

Table 1. Main physic-chemical characteristics of activated carbon AC-40.

Median particle diameter (MPD) (mm)	3
Density (g/L)	450
Moisture content (%)	4
Specific surface area (m^2/g)	1201
Pore volumes (mean widths of micropores 0.71 nm) (cm^3/g)	0.406
Pore volumes (diameter between 6.6 and 50 nm) (cm^3/g)	0.046
Pore volumes (diameter > 50 nm) (cm^3/g)	0.409
Concentration of acidic groups (μeq/g)	907.7
Concentration of basic groups (μeq/g)	96.0
Point of zero charge (pH_{pzc})	7.5
%C	68.97
%H	1.03
%N	0.45
%S	1.02
%O	20.23
%Ash	8.30

2.2. Reagents

BPA, $C_{15}H_{16}O_{12}$, with 100% purity of Sigma-Aldrich company was used. Solutions of BPA were prepared by dissolving BPA in ultrapure water obtained from Milli-Q® equipment (Millipore). In adsorption tests conducted with mixtures of BPA and metal cations, copper (II) nitrate, $Cu(NO_3)_2{\cdot}3H_2O$, lead (II) nitrate, $Pb(NO_3)_2$, nickel (II) nitrate, $Ni(NO_3)_2{\cdot}6H_2O$, and cadmium (II) nitrate, $Cd(NO_3)_2{\cdot}4H_2O$, were used. All of them had purity higher than 98% and they were supplied by Panreac Quimica S.A. company.

2.3. Methodology

All adsorption experiments were conducted in batch mode in a stirred tank. BPA solutions were prepared in ultrapure water using a mild ultrasonic bath with a temperature of approximately 50 °C (323 K).

In the preparation of mixtures of BPA and metal cations, the proper amount of metal reagent was added to the previously prepared solution of BPA.

Each solution was located in a jacketed reactor with a capacity of 250 mL and maintained at a constant temperature (298 K) with stirring throughout the test by using a magnetic stirrer. When the operation time finished, a liquid sample was taken from the reactor by using an automated pipette. Moreover, a sample of the original solution was taken. The concentrations of initial and final contaminants were determined. Consequently, it was possible to determine the percentage of contaminant removed by the activated carbon.

The effect of main experimental conditions such as adsorbent dose, pH, contact time and initial BPA concentration was studied changing the variables between the following ranges: adsorbent dose (0.1–1.0 g/L), pH (3–10), contact time (0.5–48 h) and initial BPA concentration (1–160 mg/L) at a constant temperature of 25 °C. The desired value of pH was obtained by adding HCl solution (0.1 M) or NaOH solution (0.1 M) to the BPA solution. The real initial concentrations of BPA obtained in a pH range of 3–10 were measured in absence of the adsorbent.

To test the effect of pH on the adsorption process of BPA with activated carbon AC-40, an initial concentration of BPA of 20 mg/L, a contact time of 48 h and an adsorbent concentration of 0.5 g/L were selected. The pH of the medium was changed from 3 to 10. The temperature remained constant at 25 °C. Then, experiments were performed at pH values between 4 and 8 at different contact times in order to analyze the influence of contact time on the BPA removal efficiency.

To determine the minimum amount of activated carbon necessary to achieve maximum removal of BPA, experiments were carried out with a BPA initial concentration of 20 mg/L, three contact times

5.5, 22 and 48 h and a pH of 5, changing the concentration of the adsorbent from 0.1 to 1 g/L. The temperature remained constant during the tests at 25 °C (298 K).

The initial concentration of BPA is one of the most influential parameters in determining the contaminant removal on activated carbon from the aqueous solution. Experiments were carried out to test the effect of the BPA concentration in the adsorption process. A pH value of 5 with an adsorbent concentration of 0.5 g/L, and a contact time of 48 h were selected. The initial concentration of BPA was changed from 1 to 160 mg/L.

To analyze the kinetics of BPA adsorption, experiments were performed with different contact times (until a contact time of 48 h), at pH = 5 and a BPA initial concentration of 20 mg/L, with an adsorbent dose of 0.5 g/L and at a temperature of 298 K.

BPA adsorption isotherms of the activated carbon AC-40 were carried out at pH 5 and a temperature of 298 K. For this purpose, 100 mL of solutions containing different concentrations (1–160 mg/L) of BPA and 0.05 g of activated carbon were introduced into jacketed reactors, which were capped with flexible polyethylene (LDPE) film to avoid evaporation, prevent contamination and protect the solid–liquid system inside the reactor. Adsorption equilibrium was established when the concentration of BPA did not change (for 3 days). The BPA removal efficiency and the amount of BPA adsorbed by gram of activated carbon were calculated by the Equations (1) and (2), respectively:

$$\%\ \text{Removal efficiency} = \frac{(C_i - C_f)}{C_i} \times 100 \tag{1}$$

$$q = \frac{(C_i - C_f)}{m} \times V \tag{2}$$

where C_i and C_f are the initial and final concentrations of BPA in solution, respectively, mg/L, q is the amount of BPA adsorbed by gram of adsorbent, mg/g, m is the mass of adsorbent, g, and V is the volume of the solution, L.

The procedure to obtain the data of adsorption of BPA on the activated carbon in the presence of metal cations was similar to methods to obtain adsorption results of single BPA systems. The only difference was the addition of a specific content of some metal cations to each BPA solution. Tests were performed with a concentration of 20 mg/L for both BPA and each metal cation, with an activated carbon dose of 0.5 g/L, during a contact time of 48 h and at pH value of 5. When equilibrium was reached, the BPA and different metal cations concentrations were analyzed.

The determination of the content of BPA was performed by ultraviolet–visible (UV–Vis) spectrophotometry, using the spectrophotometer model Genesys 6. The BPA concentration was determined by absorbance measurements at 277.5 nm. All the samples were collected in triplicate to further ensure precision. Samples used filled three quarters of the 1 cm cell used in the UV–Vis spectrophotometer. This amounted to approximately 1.5 mL of sample. In binary tests, the determination of metal content was performed by atomic absorption spectrophotometry, using the spectrophotometer model Perkin-Elmer AAnalyst 200.

Desorption of the BPA adsorbed onto the AC-40 was studied in a batch system. First, adsorption was performed at pH 5 and an initial BPA concentration of 20 mg/L with a dose of AC-40 of 0.5 g/L. Then, desorption was performed under the following desorption conditions: use of a mixture of methanol/acetic acid of 4:1 (v/v), a dose of AC-40 of 0.5 g/L, operation time of 48 h and temperature of 25 °C, and when the operation time was finished, the supernatant was separated and analyzed by ultraviolet–visible spectrophotometry at 277.5 nm.

3. Results and Discussions

3.1. *Influence of pH and Contact Time*

The pH is one of the main parameters that control the removal of compounds present in aqueous solutions using solid adsorbents. According to several authors, pH can change the availability and characteristics of the pollutants in solution and it can modify the chemical state of the functional groups that are responsible for adsorption [47–49].

Some experiments were performed to test the effect of pH on the adsorption process of BPA with activated carbon AC-40. Figure 1 shows the results.

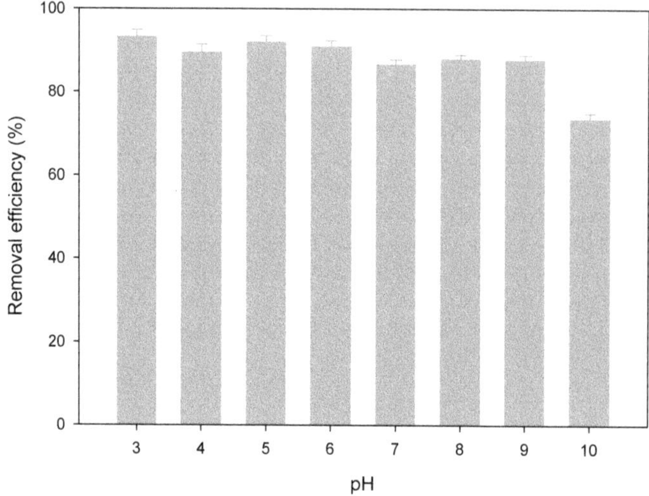

Figure 1. Bisphenol A (BPA) removal efficiency of activated carbon AC-40 according to the pH of the medium.

Data showed that pH did not significantly affect BPA adsorption. However, a slight tendency to obtain greater retention by increasing the acidity of the medium was observed. It could be because the zero point of charge was 7.5. The zero point of charge is the pH at which the surface of the adsorbent is globally neutral. Besides, it is a fundamental parameter of a material surface, governing the adsorbent–adsorbate interactions. Below this point, the surface is positively charged, otherwise it is negatively charged. Therefore, the CA surface was predominantly positive at pH values lower than 7.5. In contrast, negative charges appeared on the surface at pH values higher than 7.5 due to the dissociation of functional groups. In addition, BPA showed different states of equilibrium in an aqueous solution depending on the pH. BPA could be found in its molecular form at a pH less than 8. Deprotonation of the bisphenolate monoanion occurred at a pH of 8 [49], reaching a considerable concentration at pH 9 [50]. Therefore, electrostatic interactions between the negative charge of the surface of the activated carbon and bisphenolate anion could lead to a decrease in the overall adsorption of BPA. However, other interactions could also influence the adsorption rather than electrostatic forces, making the effect of surface charge less intense.

These results were in good agreement with those reported by other researchers. Thus, Wang and Xiao [51] indicated that the adsorption of BPA with four different activated carbons decreased at values of pH higher than 9. Chang et al. [52] found that the pH had an important role in the retention process of BPA with an activated carbon prepared from rice straw. Soni and Padmaja [53] also studied the adsorption of BPA on activated carbon prepared from palm shell. They found that at pH values below 9, the adsorption of BPA was not influenced by the pH of the medium. Bohdziewicz and Liszczyk [54] found a decrease in the adsorption of BPA by commercial activated carbon at pH values higher than or equal to 10. However, at pH values below 10, the adsorption capacity was not affected.

Figure 2 shows the BPA removal efficiency of activated carbon with different contact times. The removal efficiency increased with increasing contact time. At pH 5, the amount adsorbed increased quickly in the first 0.5 h of contact. Then, at 22 h, an adsorption percentage of 80% was reached. If the process continued until 48 h, the adsorbed percentage would increase approximately up to 91%. Finally, at 72 h of operating time, the removal of BPA reached a value of 92% (a very low change with respect to 48 h). At other pH values, as pH 7 or pH 8, the adsorption was performed in a more progressive way. However, the change between 48 and 72 h was also low.

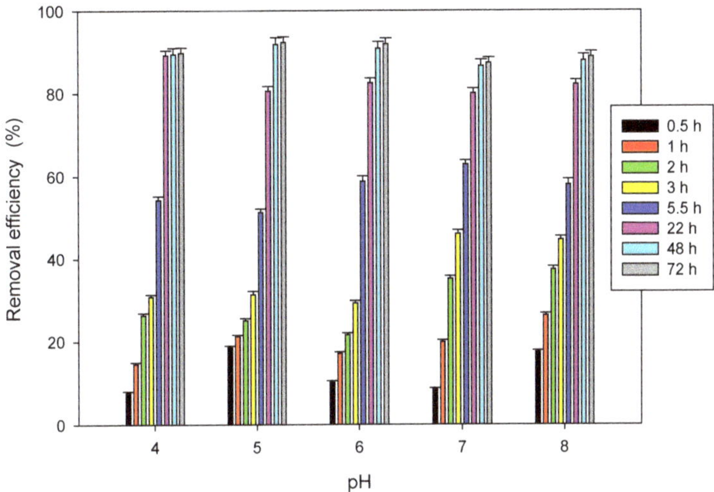

Figure 2. BPA removal efficiency of activated carbon AC-40 depending on pH at different contact times.

Considering the results obtained, from the removal efficiency point of view, a contact time of 48 h and a pH of 5 were selected for the rest of the experiments. There were not important differences in the pH range of 4 and 8. An adjustment by addition of acidic/basic chemicals could represent an important part of the cost of the wastewater treatment process and for that reason, an economic evaluation could indicate other pH values as a better option.

3.2. *Influence of Adsorbent Dosage*

Figure 3 shows the results of tests performed to determine the minimum dose of activated carbon necessary to achieve maximum removal of BPA.

As the dosage of activated carbon increased, the BPA removal efficiency from solutions increased for all the contact times. However, differences between 0.5, 0.75 and 1 g/L were very slight for long contacts times. For example, BPA removals (%) of 92%, 93% and 90% were achieved with adsorbent dosages of 0.5, 0.75 and 1 g/L, respectively, for a contact time of 48 h. However, great differences were observed at very short times. For example, very high adsorption attained at 5.5 h when using 1 g of adsorbent per L. With regard to the amount of BPA adsorbed by gram of activated carbon, in general, it was higher for a dose of 0.5 g/L. Above this adsorbent dose, a decrease in the uptake of the BPA was observed. In conclusion, if time was not considered, the use of an adsorbent dose of 0.5 g/L was recommended since at 22 h of operating time, very good BPA removals were achieved.

These results agreed with those obtained by numerous authors. Thus, Bautista-Toledo et al. [37] found an increase in the retention of BPA when 0.05 g of activated carbon was used for a period of 24 h in a BPA solution of 175 mg/L. Liu et al. [11] reported an increase in retention of BPA with two carbons activated with nitric acid that were selectively modified. They obtained maximum adsorption of BPA (432 mg/g) when changing the amount of activated carbon used. Tsai et al. [55] detected an increase in retention of BPA when the dose of adsorbent increased. This occurred because the number of adsorption sites increased in parallel with the increase in the dose of the adsorbent, when the concentration of BPA was small (<20 mg/L). If the concentration was too high, the adsorbent would reach saturation quickly. As a result, it would stop contaminant retention.

According to the results, a concentration of 0.5 g/L of activated carbon was selected, since at this concentration the minimum amount of adsorbent is used with good adsorption removal percentages (if enough contact time is maintained).

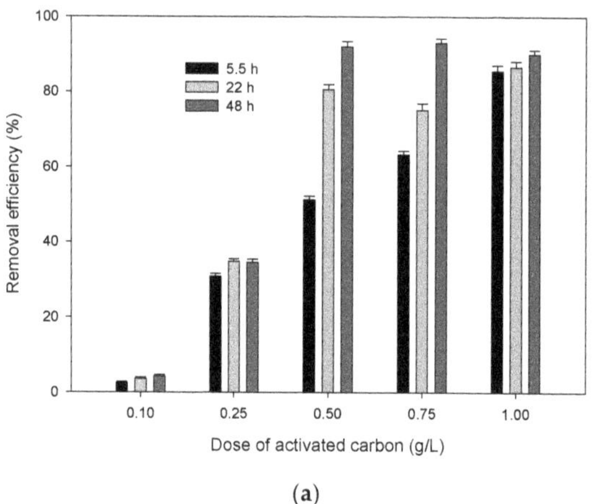

(**a**)

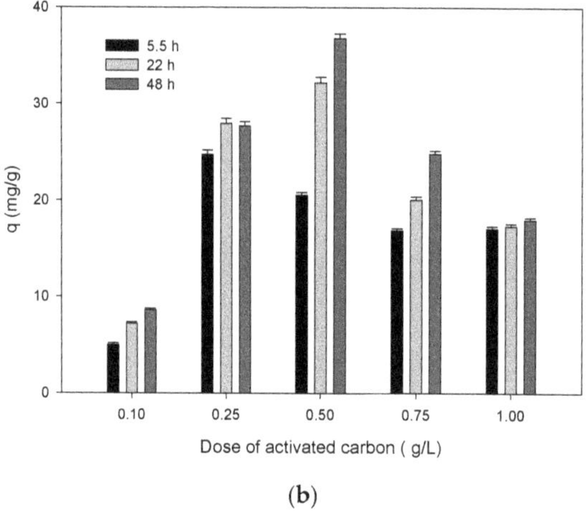

(**b**)

Figure 3. (**a**) BPA removal efficiency and (**b**) amount of BPA adsorbed by gram of activated carbon depending on adsorbent dosage at different contact times.

3.3. *Influence of the Initial Concentration of BPA*

Figure 4 shows the results of the influence of the initial concentration of BPA on its removal from the aqueous solution by AC-40.

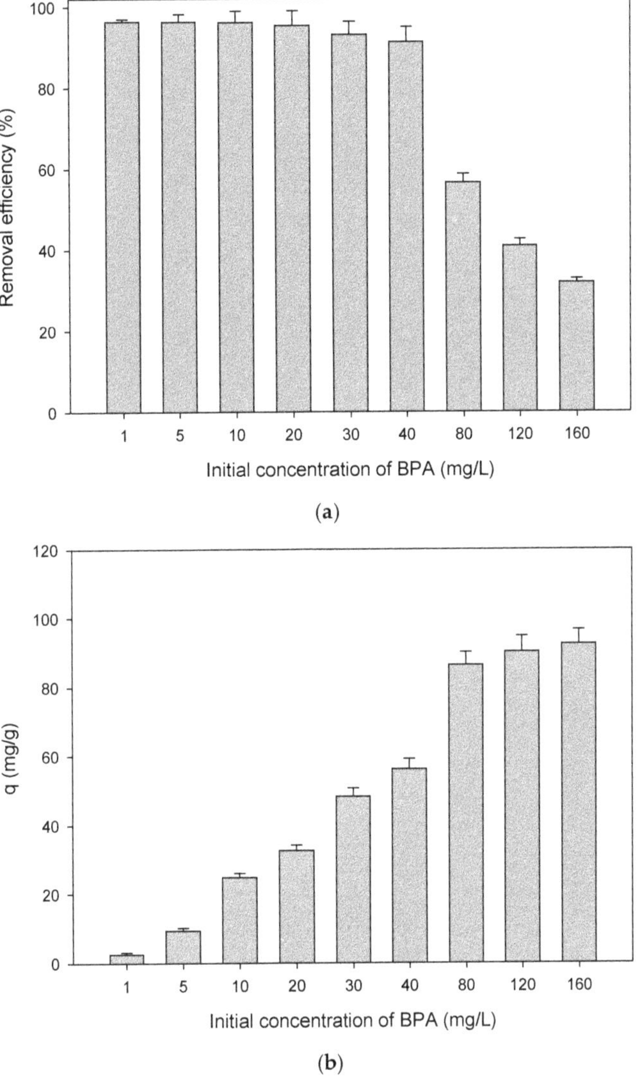

Figure 4. (**a**) BPA removal efficiency and (**b**) amount of BPA adsorbed by gram of activated carbon depending on initial concentration of BPA.

It should be noted that BPA concentration did not influence BPA removal (%) for concentration values below 40 mg/L because the active sites on the adsorbent surface are not saturated. However, when the concentration was higher than 40 mg/L, a significant decrease was observed in the adsorption percentage. In this sense, at concentrations of BPA below 80 mg/L, the retention was higher because the pores of the carbon surface were occupied almost entirely. However, when the concentration of BPA was too high, the carbon surface underwent a supersaturation of its pores. As a consequence, a fraction of BPA was not retained by the adsorbent. These results were in agreement with those obtained by numerous authors. Thus, Bautista-Toledo et al. [37] obtained the same results in several studies with different concentrations of BPA and activated carbon. With regard to the amount of BPA removed expressed in mg/g of adsorbent (Figure 4b), it increased from 2.6 to 92.4 mg/g with an increase in the initial concentration from 1 to 160 mg/L. The main reason for this important increase is the increase in the mass transfer driving force [56].

Based on the results obtained, an initial concentration of 20 mg/L was selected as the initial concentration of BPA for the rest of the experiments.

3.4. Kinetic Study

A kinetics study in adsorption allows the determination of the BPA removal rate from the aqueous medium. This study provides a basis for understanding the mechanism that controls the process and it is essential to select the optimum operating conditions in the system designed for the effluent treatment with this type of pollutant [57,58].

Experiments were performed with a total contact time of 48 h, pH = 5, a BPA initial concentration of 20 mg/L, an adsorbent concentration of 0.5 g/L and a temperature of 298 K. Figure 5 shows the values of q (mg retained/g sorbent) versus time.

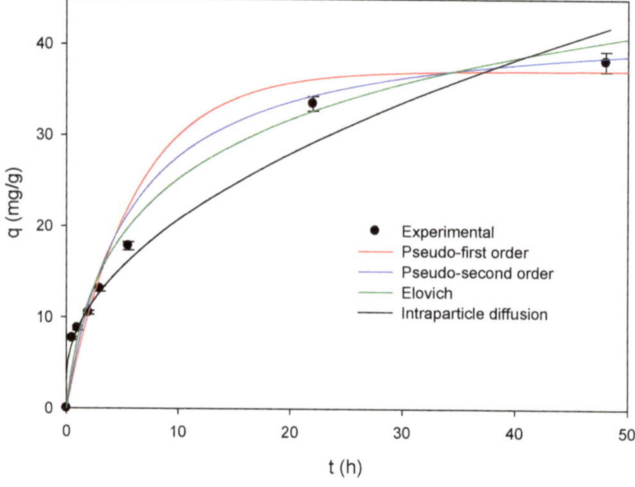

Figure 5. Variation in the adsorption capacity of BPA by activated carbon depending on the contact time and adjustment results of the four kinetic models.

These experimental results were fitted by a nonlinear regression to four kinetic models: pseudo-first-order, pseudo-second-order, Elovich and intraparticle diffusion models. Table 2 shows the representative equations of the models and their parameters. Likewise, Table 3 shows the results obtained when experimental data were fitted to the four selected kinetics models.

The pseudo-first- and pseudo-second-order models reproduced very well the experimental results. However, the intraparticle diffusion model showed a low R^2 value. Specially, the good correlation coefficients (R^2) of the pseudo-second-order model (it presented the highest R^2 value) and the similarity of the experimental and predicted adsorption capacity of this model indicated that the adsorption of BPA by activated carbon could be chemical adsorption, which is consistent with the results of Section 3.1. previously presented. In addition, the retention process was faster at first, but the equilibrium was not reached until 48 h of contact time. This indicates that, although adsorption happened quickly, the process subsequently proceeded more slowly until equilibrium was reached.

Other authors found similar results in their studies about BPA removal by activated carbons. Chang et al. [52] used an activated carbon obtained from rice straw agricultural waste. They found that the pseudo-second-order model fitted the experimental adsorption data better. Wang and Xiao [51] studied the adsorption of BPA by activated carbon reporting that the adsorption process followed the first-order kinetics. Soni and Padmaja [53] reported that the kinetic data of adsorption of BPA onto palm shell activated carbon could be fitted well by a pseudo-second-order kinetic model. Tang et al. [59] showed that the kinetics of BPA removal onto activated carbon-alginate beads with cetyltrimethyl ammonium bromide fitted good to the pseudo-first-order, pseudo-second-order and Elovich models.

Table 2. Equations and parameters representative of the four selected kinetic models.

Model	Equation	Parameters
Pseudo-first-order	$q = q_e\left(1 - e^{-k_1 \cdot t}\right)$	q_e adsorption capacity, mg/g k_1 kinetic constant of the pseudo-first-order, h^{-1}
Pseudo-second-order	$q = \frac{t}{\frac{1}{k_2 \times q_e^2} + \frac{t}{q_e}}$	q_e adsorption capacity, mg/g k_2 kinetic constant of the pseudo-second-order, g/mg·h
Elovich	$q = \frac{1}{b}\ln(a \times b) + \frac{1}{b}\ln(t)$	a velocity of initial adsorption, mg/g·h b area of the occupied surface, g/mg
Intraparticle	$q = k_p \times t^{\frac{1}{2}} + C$	k_p the intraparticle diffusion rate constant (mg/(g·min$^{1/2}$)) C constant of the intraparticle diffusion model (mg/g)

Table 3. Adjustment results of the experimental data to the four selected kinetic models.

Model	Parameters		
Pseudo-first-order	$k_1 = 0.164$	$q_e = 37.17$	$r^2 = 0.995$
Pseudo-second-order	$k_2 = 0.004$	$q_e = 43.06$	$r^2 = 1.000$
Elovich	$a = 11.325$	$B = 0.099$	$r^2 = 0.992$
Intraparticle diffusion	$k_p = 5.577$	$C = 3.082$	$r^2 = 0.964$

3.5. Equilibrium Study

Figure 6 shows the retention capacity of BPA, q_e (mg retained/g of sorbent when equilibrium is reached), versus the equilibrium concentration of BPA in the liquid phase, C_e.

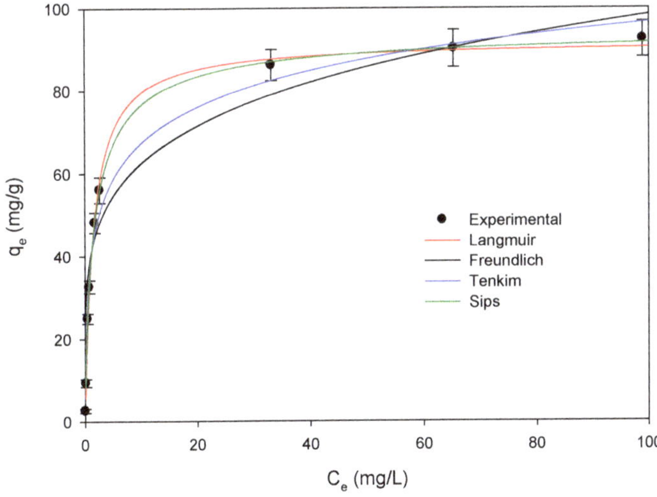

Figure 6. Equilibrium adsorption of BPA with activated carbon. Experimental results and adjustment to the selected models.

Different types of adsorption isotherms were found in the literature [51,60,61]. The isotherm obtained in this work was type I. The type I isotherm is monolayer adsorption easily explained using a Langmuir isotherm. Examples of type I adsorption isotherms are given by microporous solids such as activated carbon (material used in this study). It is also observed that the maximum adsorption capacity was close to 92 mg/g, and it was reached when the initial concentration of BPA was 160 mg/L.

In order to study the equilibrium of the process, these results have been adjusted to different models which are shown in Table 4. Further, Table 5 shows the fitting results and the parameters of the models obtained. Figure 7 represents the isotherm obtained with the models used.

The Langmuir and Sips models were the best in reproducing the experimental results, as the value of the parameter n of the Sips isotherm was equal to the unit, which indicated that this model tended to be a Langmuir isotherm. This result indicates that adsorption is monolayer on the active sites of the adsorbent that has a homogeneous surface and molecules do not interact with each other. Likewise, the value of the maximum adsorption obtained with the Langmuir and Sips models was very close to the result obtained experimentally. These findings were similar to those obtained by other authors [52–54]. However, maximum capacity values of adsorption of AC-40 activated carbon were higher than those reported in the literature [62–65].

Finally, although adsorption processes performed in batch systems applied to the decontamination of wastewater are more frequent to determine the maximum adsorption capacity as the maximum amount of chemical adsorbed onto the adsorbent by mass of the adsorbent, in this work, the partition coefficients at the studied concentrations were also determined as the ratio of the concentration of BPA in solid phase (AC-40) to the concentration in liquid phase when the two concentrations are at equilibrium. Table 6 shows the partition coefficient as a function of the initial concentration of BPA. Data showed that the degree of partitioning varied with concentration. The linear partition coefficient was about 50–52 L/g (the amount adsorbed versus its solution concentration) at initial BPA concentrations lower or equal than 10 mg/L. This can be observed in Figure 6 since the first experimental points were plotted in a linear trend. However, as the initial concentration of BPA increased, the linear partition coefficient decreased until approximately 1 L/g at 160 mg/L. It also can be observed in the change of the slope of the isotherm (Figure 6) as the BPA concentration was increased. In adsorption studies, when an L-type plot (isotherm) is obtained, the q_m coefficient is a convenient and reliable parameter to be applied for adsorption capacities assessment better than the partition coefficient.

Table 4. Representative equations and parameters of the four selected models for the study of the equilibrium process.

Model	Equation	Parameters
Langmuir	$q_e = \frac{bq_m C_e}{1+b C_e}$	q_m maximum adsorption capacity, mg/g b constant related to the affinity of the adsorbent for the adsorbate
Freundlich	$q_e = K_F \times C_e^{\frac{1}{n}}$	K_F equilibrium constant, (mg/g)·(L/mg)$^{1/n}$ n constant related to the affinity between the adsorbent and the adsorbate
Sips	$q_e = \frac{q_m bC_e^{1/n}}{1+bC_e^{1/n}}$	q_m maximum adsorption capacity, mg/g b constant related to the affinity of the adsorbent for the adsorbate n parameter characterizing the system's heterogeneity
Temkin	$q_e = \frac{RT}{b}\ln(A_T C_e)$	A_T constant of union of the equilibrium (L/g), b Temkin constant, B constant related to the heat of adsorption (J/mol), $B = \frac{RT}{b}$

Table 5. Adjustment results of the experimental data to the four selected isotherm models.

Model	Parameters			
Langmuir	q_m = 91.90	b = 0.64		r^2 = 0.996
Freundlich	K_F = 39.49	n = 5.04		r^2 = 0.941
Sips	q_m = 95.06	b = 0.632	n = 1.23	r^2 = 1.000
Temkin	b = 194.47	A_T = 19.65		r^2 = 0.981

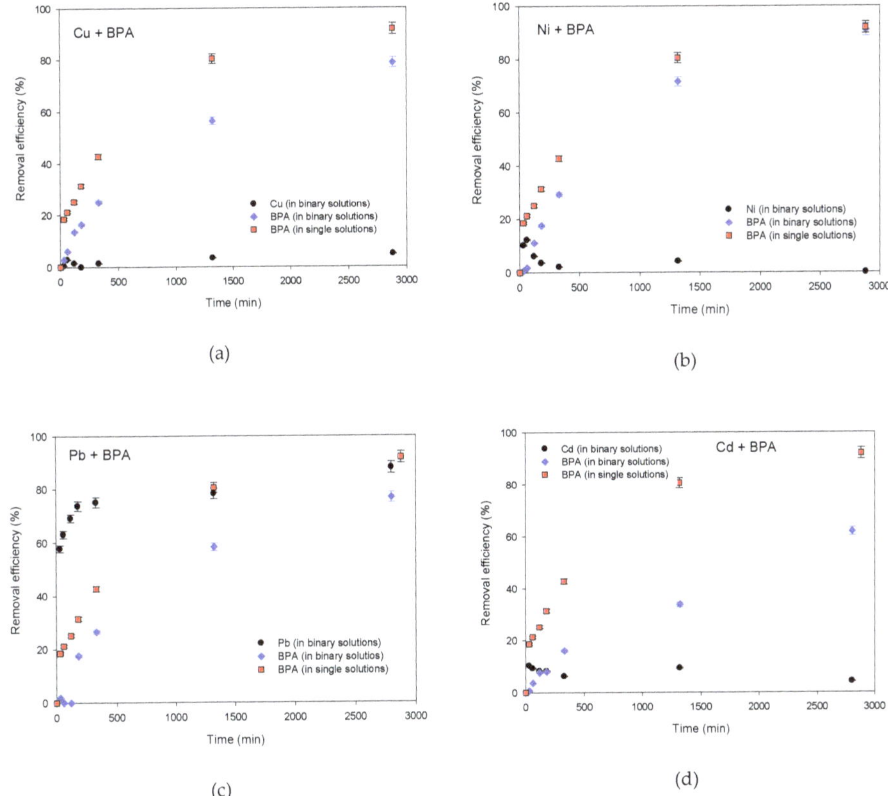

Figure 7. BPA removal (%) in single solutions (red marker) and in binary solutions with (**a**) copper, (**b**) nickel, (**c**) lead and (**d**) cadmium (black marker for metal and blue one for BPA) using activated carbon AC-40 as the adsorbent.

Table 6. Partition coefficient in function of concentration.

Initial Concentration of BPA, mg/L	Partition Coefficients, L/g
1	52.00
5	51.40
10	49.80
20	40.75
30	26.78
40	20.78
80	2.60
120	1.38
160	0.93

3.6. Adsorption of BPA-Metals Cations Mixtures

Different contaminants such as BPA and a wide variety of metal ions could be present in wastewater. The presence of these ions in solution can lead to competition between metals and the contaminant for the union with the adsorptive sites but also could enhance the BPA adsorption. For example, Han et al. [39] analyzed the influence of copper (II) and lead (II) ions on the adsorption of BPA onto lignin. These researchers found that the presence of the metal cations modified the surface of lignin, converting it on a less negatively charged surface. It was an advantage for the adsorption of anion species of BPA. Further, Liu et al. [38] studied the simultaneous adsorption of BPA and cadmium (II) ions on activated montmorillonite and found that the maximum adsorption capacity of BPA changed from 74.55 to 80.77 mg/g when cadmium was presented in the solution. Similarly, Bautista-Toledo et al. [37] reported that the presence of chromium (III) enhanced the adsorption of BPA on different activated carbons. The authors attributed the improvement in BPA adsorption to the in situ formation of complex compounds formed by Cr (III) and BPA (Cr (III)-BPA coordination compounds).

In this section, the interaction of some metal cations present in water with BPA was analyzed. The aim was to observe the possible interactions between the metal and the contaminant and/or metal and adsorbent, especially if they had synergistic properties. In all cases, the experiments of mixtures of metals and BPA have been compared with the initial experiment using only BPA. Cadmium, copper, lead and nickel were selected as the representatives of heavy metals because they are very common in wastewater. Figures 7 and 8 show the results.

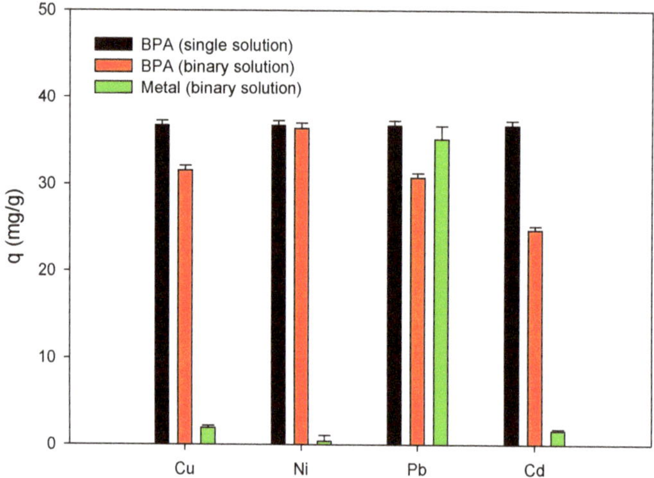

Figure 8. BPA adsorption capacity in single solutions (black color) and in binary solutions with different heavy metals (green marker for metal and red one for BPA) using activated carbon AC-40 as the adsorbent.

Contrary to the referenced studies of Han et al. [39], Lui et al. [38] and Bautista-Toledo et al. [37], the results showed that none of the metal cations improved the efficiency or adsorption capacity of BPA, suggesting a competition with BPA to occupy the porous surface of carbon and consequently decreasing its effectiveness. This is consistent with investigations carried out by Schiewer and Volesky [66], who systematically studied the effect of ionic strength on the adsorption of cations such as Zn^{2+}, Cd^{2+}, Cu^{2+} and Na^{+}. They reported that an increase in ionic strength led to a decrease in the adsorption achieved due to the increase in the electrostatic charge. Niu and Volesky [67] studied the removal of anionic metal complexes ($Au(CN)^{2-}$, $CrO_4{}^{2-}$, $SeO_4{}^{2-}$ and $VO_4{}^{3-}$), concluding that an increase in ionic strength decreases the removal.

In the case of the mixture of Cd (II) and BPA, retention of BPA on activated carbon decreased. The difference in adsorption of BPA in the mixture solution (metal-BPA) and in the BPA solution was significant. BPA adsorption capacity changed from 36.8 (single solution) to 24.7 mg/g (binary solution). However, Cd was not greatly adsorbed (adsorption capacity of 1.6 mg/g). This also occurred in mixtures of Cu (II) and BPA, although to a lesser extent. If the predominant binding force of BPA is electrostatic attraction, a decrease in the positively charged sites on the carbon surface could negatively influence the BPA adsorption. A change in the charge of the carbon surface and pH of the solution during the process due to the adsorption of metals could explain the decrease in the BPA adsorption. Earlier studies reported that the ionic strength of the solution notably influenced the adsorption of different species [37,67,68]. Further, Meng et al. [69] studied the adsorption of phenol and cadmium onto soil and found that the adsorption of mixed pollutants (Cd^{2+} and phenol) showed an antagonistic effect on adsorption. Chemical adsorption was the main mechanism of adsorption of cadmium ions while physical adsorption was the primary mechanism for phenol adsorption (very influenced by the physic-chemical properties of the carbon surface). This indication was consistent with the pH evolution of the system (Table 7) and the point of zero charge reported in Table 1. The net charge of activated carbon decreases with the increase in pH values. Yang et al. [70] noted that the electrostatic surface charge was smaller with the pH of the adsorption system closer to the pH_{pzc} of the activated carbon. In addition, BPA exists in its molecular form under pH < 8 but begins to deprotonate to a negatively charged form at around pH 8 [13,51,70]. Thus, when pH was increased above 7.5, a repulsive electrostatic interaction may be established between the negatively charged surface of the activated carbon and bisphenolate anion. In the mixture of Ni (II) and BPA, the metal was not adsorbed and BPA removal in the mixture solution was very similar to the BPA removal in the single BPA solution. This means that the presence of Ni (II) did not affect the adsorption of BPA. Finally, in the case of the mixture of Pb (II) and BPA, the results suggested that both Pb (II) and BPA were notably adsorbed by activated carbon, probably because they formed a complex in situ with activated carbon, adsorbing and synergistically removing both contaminants present in the solution, or perhaps they distributed the available active sites of the adsorbent for Pb and BPA adsorption.

Table 7. pH evolution on simultaneous adsorption experiments.

Time, min	Cd (II) + BPA	Cu (II) + BPA	Ni (II) + BPA	Pb (II) + BPA
0	5	5	5	5
30	6.9	6.3	6.2	6.4
60	7.5	6.7	6.5	6.8
120	7.8	6.8	6.6	7.1
240	8.1	7.2	7.0	7.2
480	8.0	7.5	7.0	7.3
1440	8.0	7.4	7.3	7.3
2880	7.9	7.6	7.5	7.5

In conclusion, the metal ions may affect the sorption of BPA on AC-40 from different aspects: (1) the metal ions can reduce the BPA adsorption due to the decrease in sorption sites on the AC surface; (2) the metal ions can change the ionic strength of the solution and influence the adsorption of the different species; (3) the metal ions can promote the hydrophobic interaction, which may enhance the BPA adsorption on AC; and (4) the metal ions can promote the formation of complex compounds (metal–BPA coordination compounds) which increase the BPA adsorption.

3.7. Cycles of BPA Adsorption/Desorption

The reuse of AC-40 was investigated in three consecutive cycles of adsorption/desorption, without significant loss in the adsorption capacity. The adsorption capacity in the three consecutive adsorption–desorption cycles is reported in the Table 8. Figure 9 shows the desorption percentage of BPA. The removal efficiency was higher than 90% in all cases, indicating that the activated carbon has good reusability. Further, desorption of BPA from the AC-40 was about 86.3% in the first desorption cycle and only decreased slightly after three adsorption–desorption cycles. These results indicated that the interaction of the BPA molecule to the binding sites on the adsorbent surface can be disrupted without any effect on the properties of the adsorbent. Similar results were found by other authors in their studies of the desorption of BPA by mixtures of methanol-acetic acid [71–73].

Table 8. Removal efficiency and adsorption capacity of BPA adsorption in three consecutive cycles.

Cycle	% Removal	q, mg/g
1	92.5	37.00
2	91.4	36.56
3	90.2	36.08

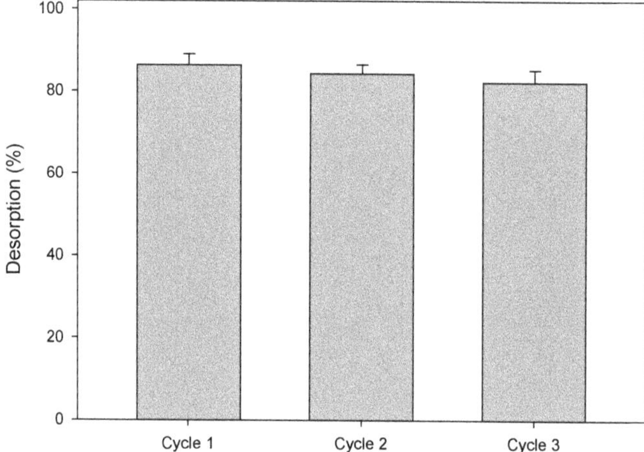

Figure 9. BPA desorption (%) in each adsorption–desorption cycle.

4. Conclusions

The aim of this work was to investigate the application of a commercial activated carbon (AC-40) to remove BPA in single and simultaneous adsorption systems. The results showed that pH did not significantly affect the BPA adsorption at low pH values. However, when the acidity of the medium was increased, a slight improvement in the removal efficiencies was observed. Further, as the dose of activated carbon increased, the BPA removal efficiency increased at all contact times. Additionally, the concentration of BPA did not influence the adsorption, for the values of initial BPA concentration analyzed in this work (similar to those than can be found in some industrial wastewaters).

The isotherm obtained for BPA adsorption onto AC-40 was type I and the maximum adsorption capacity was close to 90 mg/g. The Langmuir and Sips models were the best to reproduce the experimental results.

The experiments with mixtures of metals and BPA showed that none of the metal cations analyzed improved the efficiency of the adsorption process. In the case of the mixtures of Pb (II) and BPA, the results seemed to indicate that both Pb (II) and BPA can be efficiently removed from wastewater using AC-40.

Desorption tests showed good desorption efficiency (higher than 80%) and also BPA adsorption capacities were very similar after three adsorption–desorption cycles, indicating that the activated carbon has good reusability. In conclusion, results of the work showed good adsorptive behavior of AC-40 for BPA removal in batch systems in single- and binary-solute adsorption. However, from a practical point of view and as a potential research line, complete experimental research on packed-bed columns including desorption studies should be carried out.

Author Contributions: Conceptualization, M.C.; formal analysis, M.A.M.-L. and C.E.; investigation, M.A.M.-L. and C.E.; methodology, M.A.M.-L., A.R. and M.C.; software, I.I.-R. and C.E.; supervision, M.A.M.-L.; validation, C.E.; resources, M.C.; writing—original draft, M.A.M.-L., A.R., I.I.-R., and C.E.; writing—review and editing, M.A.M.-L. and M.C. All authors have read and agreed to the published version of the manuscript.

Funding: This research received no external funding.

Acknowledgments: Authors thank M.I. Bautista-Toledo for the activated carbon donation.

Conflicts of Interest: The authors declare no conflict of interest.

References

1. Gonsioroski, A.; Mourikes, V.E.; Flaws, J.A. Endocrine disruptors in water and their effects on the reproductive system. *Int. J. Mol. Sci.* **2020**, *21*, 1929. [CrossRef] [PubMed]
2. Rubin, B.S. Bisphenol A: An endocrine disruptor with widespread exposure and multiple effects. *J. Steroid Biochem. Mol. Biol.* **2011**, *127*, 27–34. [CrossRef] [PubMed]
3. Orimolade, B.O.; Adekola, F.A.; Adebayo, G.B. Adsorptive removal of bisphenol A using synthesized magnetite nanoparticles. *Appl. Water Sci.* **2018**, *8*, 46. [CrossRef]
4. ISH Markit. *Bisphenol A. Chemical Economics Handbook*; HIS Markit: London, UK, 2018.
5. Inoue, M.; Masuda, Y.; Okada, F.; Sakurai, A.; Takahashi, I.; Sakakibara, M. Degradation of bisphenol A using sonochemical reactions. *Water Res.* **2008**, *42*, 1379–1386. [CrossRef]
6. Ivanov, A.E.; Halthur, T.; Ljunggren, L. Flow permeable composites of lignin and poly(vinyl alcohol): Towards removal of bisphenol A and erythromycin from water. *J. Environ. Chem. Eng.* **2016**, *4*, 1432–1441. [CrossRef]
7. Santhi, V.A.; Sakai, N.; Ahmad, E.D.; Mustafa, A.M. Occurrence of bisphenol A in surface water, drinking water and plasma from Malaysia with exposure assessment from consumption of drinking water. *Sci. Total Environ.* **2012**, *427*, 332–338. [CrossRef]
8. Vandenberg, L.N.; Hauser, R.; Marcus, M.; Olea, N.; Welshons, W.V. Human exposure to bisphenol A (BPA). *Reprod. Toxicol.* **2007**, *24*, 139–177. [CrossRef]
9. Yamamoto, T.; Yasuhara, A.; Shiraishi, H.; Nakasugi, O. Bisphenol A in hazardous waste landfill leachates. *Chemosphere* **2001**, *42*, 415–418. [CrossRef]
10. Phouthavong-Murphy, J.C.; Merrill, A.K.; Zamule, S.; Giacherio, D.; Brown, B.; Roote, C.; Das, P. Phytoremediation potential of switchgrass (*Panicum virgatum*), two United States native varieties, to remove bisphenol-A (BPA) from aqueous media. *Sci. Rep.* **2020**, *10*, 835. [CrossRef]
11. Liu, G.; Ma, J.; Li, X.; Qin, Q. Adsorption of bisphenol A from aqueous solution onto activated carbons with different modification treatments. *J. Hazard. Mater.* **2009**, *164*, 1275–1280. [CrossRef]
12. Yeo, M.K.; Kang, M. Photodecomposition of bisphenol A on nanometer-sized TiO_2 thin film and the associated biological toxicity to zebrafish (*Danio rerio*) during and after photocatalysis. *Water Res.* **2006**, *40*, 1906–1914. [CrossRef] [PubMed]
13. Wang, J.; Zhang, M. Adsorption Characteristics and mechanism of bisphenol A by magnetic biochar. *Int. J. Environ. Res. Public Health* **2020**, *17*, 1075. [CrossRef] [PubMed]
14. Ahmed, M.B.; Zhou, J.L.; Ngo, H.H.; Guo, W.S.; Thomaidis, N.S.; Xu, J. Progress in the biological and chemical treatment technologies for emerging contaminant removal from wastewater: A critical review. *J. Hazard. Mater.* **2017**, *323*, 274–298. [CrossRef] [PubMed]
15. Esplugas, S.; Bila, D.M.; Krause, L.G.T.; Dezotti, M. Ozonation and advanced oxidation technologies to remove endocrine disrupting chemicals (EDCs) and pharmaceuticals and personal care products (PPCPs) in water effluents. *J. Hazard. Mater.* **2007**, *149*, 631–642. [CrossRef] [PubMed]
16. Luo, H.; Cheng, Y.; Zeng, Y.; Luo, K.; He, D.; Pan, X. Rapid removal of organic micropollutants by heterogeneous peroxymonosulfate catalysis over a wide pH range: Performance, mechanism and economic analysis. *Sep. Purif. Technol.* **2020**, *248*, 117023. [CrossRef]
17. Lin, H.; Fang, Q.; Wang, W.; Li, G.; Guan, J.; Shen, Y.; Ye, J.; Liu, F. Prussian blue/PVDF catalytic membrane with exceptional and stable Fenton oxidation performance for organic pollutants removal. *Appl. Catal. B Environ.* **2020**, *273*, 119047. [CrossRef]
18. Rani, M.; Rachna; Shanker, U. Efficient photocatalytic degradation of Bisphenol A by metal ferrites nanoparticles under sunlight. *Environ. Technol. Innov.* **2020**, *19*, 100792. [CrossRef]
19. Nakanishi, A.; Tamai, M.; Kawasaki, N.; Nakamura, T.; Tanada, S. Adsorption characteristics of bisphenol A onto carbonaceous materials produced from wood chips as organic waste. *J. Colloid Interface Sci.* **2002**, *252*, 393–396. [CrossRef]
20. Marsh, H.; Heintz, E.A.; Rodriguez-Reinoso, F. *Introduction to Carbon Technologies*; University of Alicante: Alicante, Spain, 1997.
21. Bansal, R.C.; Donnet, J.-B.; Stoeckli, F. *Active Carbon*; Marcel Dekker: New York, NY, USA, 1988.

22. Derbyshire, F.; Jagtoyen, M.; Andrews, R.; Rao, A.; Martin-Gullon, I.; Grulke, E. Carbon materials in environmental applications. *Chem. Phys. Carbon* **2001**, *27*, 1–66.
23. Musmarra, D.; Karatza, D.; Lancia, A.; Prisciandaro, M.; Mazziotti di Celso, G. Adsorption of elemental mercury vapors from synthetic exhaust combustion gas onto HGR carbon. *J. Air Waste Manag. Assoc.* **2016**, *66*, 698–706. [CrossRef]
24. Rivera-Utrilla, J.; Sánchez-Polo, M.; Gómez-Serrano, V. Activated carbon modifications to enhance its water treatment applications. An overview. *J. Hazard. Mater.* **2011**, *187*, 1–23. [CrossRef] [PubMed]
25. Dias, J.M.; Alvim-Ferraz, M.C.M.; Almeida, M.F.; Rivera-Utrilla, J.; Sánchez-Polo, M. Waste materials for activated carbon preparation and its use in aqueous-phase treatment: A review. *J. Environ. Manag.* **2007**, *85*, 833–846. [CrossRef]
26. Moreno-Castillo, C.; Rivera-Utrilla, J. Carbon materials as adsorbents for the removal of pollutants from the aqueous phase. *Mater. Res. Bull.* **2001**, *26*, 890–894. [CrossRef]
27. Radovic, L.R.; Moreno-Castilla, C.; Rivera-Utrilla, J. *Chemistry and Physics of Carbon*; Radovic, L.R., Ed.; Marcel Dekker: New York, NY, USA, 2000; Volume 27, p. 227.
28. Chen, J.; Zhang, L.; Yang, G.; Wang, Q.; Li, R.; Lucia, L.A. Preparation and characterization of activated carbon from hydrochar by phosphoric acid activation and its adsorption performance in prehydrolysis liquor. *BioResources* **2017**, *12*, 5928–5941. [CrossRef]
29. Jagtoyen, M.; Thwaites, M.; Stencel, J.; McEnaney, B.; Derbyshire, F. Adsorbent carbon synthesis from coals by phosphoric acid activation. *Carbon* **1992**, *30*, 1089–1096. [CrossRef]
30. Bhatnagar, A.; Anastopoulos, I. Adsorptive removal of bisphenol A (BPA) from aqueous solution: A review. *Chemosphere* **2017**, *168*, 885–902. [CrossRef] [PubMed]
31. Juhola, R.; Runtti, H.; Kangas, T.; Hu, T.; Romar, H.; Tuomikoski, S. Bisphenol A removal from water by biomass-based carbon: Isotherms, kinetics and thermodynamics studies. *Environ. Technol.* **2020**, *41*, 971–980. [CrossRef]
32. Lu, J.; Zhang, C.; Wu, J.; Luo, Y. Adsorptive removal of bisphenol A using N-doped biochar made of *Ulva prolifera*. *Water Air Soil Pollut.* **2017**, *228*, 327. [CrossRef]
33. Pan, B.; Lin, D.; Mashayekhi, H.; Xing, B. Adsorption and Hysteresis of Bisphenol A and 17-Ethinyl Estradiol on Carbon Nanomaterials. *Environ. Sci. Technol.* **2008**, *42*, 5480–5485. [CrossRef]
34. Redding, A.M.; Cannon, F.S.; Snyder, S.A.; Vanderford, B.J. A QSAR-like analysis of the adsorption of endocrine disrupting compounds, pharmaceuticals, and personal care products on modified activated carbons. *Water Res.* **2009**, *43*, 3849–3861. [CrossRef]
35. Solak, S.; Vakondios, N.; Tzatzimaki, I.; Diamadopoulos, E.; Arda, M.; Kabay, N.; Yüksel, M. A comparative study of removal of endocrine disrupting compounds (EDCs) from treated wastewater using highly crosslinked polymeric adsorbents and activated carbon. *J. Chem. Technol. Biotechnol.* **2014**, *89*, 819–824. [CrossRef]
36. Wu, Z.; Wei, X.; Xue, Y.; He, X.; Yang, X. Removal effect of atrazine in co-solution with bisphenol A or humic acid by different activated carbons. *Materials* **2018**, *11*, 2558. [CrossRef] [PubMed]
37. Bautista-Toledo, M.I.; Rivera-Utrilla, J.; Ocampo-Pérez, R.; Carrasco-Marín, F.; Sánchez-Polo, M. Cooperative adsorption of bisphenol-A and chromium(III) ions from water on activated carbons prepared from olive-mill waste. *Chem. Eng. J.* **2014**, *73*, 338–350. [CrossRef]
38. Liu, C.; Wu, P.; Zhu, Y.; Tran, L. Simultaneous adsorption of Cd^{2+} and BPA on amphoteric surfactant activated montmorillonite. *Chemosphere* **2016**, *144*, 1026–1032. [CrossRef]
39. Han, W.; Luo, L.; Zhang, S. Adsorption of bisphenol A on lignin: Effects of solution chemistry. *Int. J. Environ. Sci. Technol.* **2012**, *9*, 543–548. [CrossRef]
40. Hanladakis, J.N.; Velis, C.A.; Weber, R.; Iacovidou, E.; Purnell, P. An overview of chemical additives present in plastics: Migration, release, fate and environmental impact during their use, disposal and recycling. *J. Hazard. Mater.* **2018**, *344*, 179–199. [CrossRef] [PubMed]
41. Kyzas, G.Z.; Siafaka, P.I.; Pavlidou, E.G.; Chrissafis, K.J.; Bikiaris, D.N. Synthesis and adsorption application of succinyl-grafted chitosan for the simultaneous removal of zinc and cationic dye from binary hazardous mixtures. *Chem. Eng. J.* **2015**, *259*, 438–448. [CrossRef]
42. Stawinski, W.; Wegrzyn, A.; Freitas, O.; Chmielarz, L.; Mordarski, G.; Figueiredo, S. Simultaneous removal of dyes and metal cations using an acid, acid-base and base modified vermiculite as a sustainable and recyclable adsorbent. *Sci. Total Environ.* **2017**, *576*, 398–408. [CrossRef] [PubMed]

43. Morosanu, I.; Teodosiu, C.; Fighir, D.; Paduraru, C. Simultaneous Biosorption of Micropollutants from Aqueous Effluents by Rapeseed Waste. *Process Saf. Environ. Prot.* **2019**, *132*, 231–239. [CrossRef]
44. Méndez-Díaz, J.D.; Abdel Daiem, M.M.; Rivera-Utrilla, J.; Sánchez-Polo, M.; Bautista-Toledo, I. Adsorption/bioadsorption of phthalic acid, an organic micropollutant present in landfill leachates, on activated carbons. *J. Colloid Interface Sci.* **2012**, *369*, 358–365. [CrossRef]
45. Abdel Daiem, M.M.; Rivera-Utrilla, J.; Sánchez-Polo, M.; Ocampo-Pérez, R. Single, competitive and dynamic adsorption on activated carbon of compounds used as plasticizers and herbicides. *Sci. Total Environ.* **2015**, *537*, 335–342. [CrossRef] [PubMed]
46. Velo-Gala, I.; López-Peñalver, J.J.; Sánchez-Polo, M.; Rivera-Utrilla, J. Role of activated carbon on micropollutants degradation by ionizing radiation. *Carbon* **2014**, *67*, 288–299. [CrossRef]
47. Bernal, V.; Erto, A.; Giraldo, L.; Moreno-Pijarán, J.C. Effect of solution pH on the adsorption of paracetamol on chemically modified activated carbons. *Molecules* **2017**, *22*, 1032. [CrossRef] [PubMed]
48. Pereira, R.C.; Anizelli, P.R.; Di Mauro, E.; Valezi, D.F.; da Costa, A.C.S.; Zaia, C.T.B.V.; Zaia, D.A.M. The effect of pH and ionic strength on the adsorption of glyphosate onto ferrihydrite. *Geochem. Trans.* **2019**, *20*, 3. [CrossRef] [PubMed]
49. Bautista-Toledo, M.I.; Ferro-García, M.A.; Rivera-Utrilla, J.; Moreno-Castilla, C.; Vegas, F.J. Bisphenol A removal from water by activated carbon. Effects of carbon characteristics and solution chemistry. *Environ. Sci. Tecnhol.* **2005**, *39*, 6246–6250. [CrossRef]
50. Tursi, A.; Chatzisymeon, E.; Chidichimo, F.; Beneduci, A.; Chidichimo, G. Removal of endocrine disrupting chemicals from water: Adsorption of bisphenol-A by biobased hydrophobic functionalized cellulose. *Int. J. Environ. Res. Public Health* **2018**, *15*, 2419. [CrossRef]
51. Wang, X.; Xiao, Y. Removal of the endocrine disrupting chemical Bisphenol A from water by activated carbon. *Adv. Mater. Res.* **2013**, *671*, 2726–2731. [CrossRef]
52. Chang, K.L.; Hsieh, J.F.; Ou, B.M.; Chang, M.H.; Hseih, W.Y.; Lin, J.F.; Huang, P.J.; Wong, K.F.; Chen, S.T. Adsorption studies on the removal of an endocrine-disrupting compound (Bisphenol A) using activated carbon from rice straw agricultural waste. *Sep. Sci. Technol.* **2012**, *47*, 1514–1521. [CrossRef]
53. Soni, H.; Padmaja, P. Palm shell based activated carbon for removal of bisphenol A: An equilibrium, kinetic and thermodynamic study. *J. Hazard. Mater.* **2014**, *21*, 275–284. [CrossRef]
54. Bohdziewicz, J.; Liszczyk, G. Evaluation of effectiveness of bisphenol a removal on domestic and foreign activated carbons. *Ecol. Chem. Eng. Soc.* **2013**, *20*, 371–379. [CrossRef]
55. Tsai, W.-T.; Lai, C.-W.; Su, T.-Y. Adsorption of bisphenol-A from aqueous solution onto minerals and carbon adsorbents. *J. Hazard. Mater.* **2006**, *134*, 169–175. [CrossRef] [PubMed]
56. Yu, J.; Shen, H.; Liu, B. Adsorption Properties of polyethersulfone-modified attapulgite hybrid microspheres for bisphenol A and sulfamethoxazole. *Int. J. Environ. Res. Public Health* **2020**, *17*, 473. [CrossRef]
57. Ho, Y.S.; Mckay, G. A comparison of chemisorption kinetic models applied to pollutant removal on various sorbents. *Process Saf. Environ. Prot.* **1998**, *76*, 332–340. [CrossRef]
58. King, P.; Srinivas, P.; Prasanna Kumar, Y.; Prasad, V.S.R.K. Sorption of copper(II) ion from aqueous solution by *Tectona grandis* l.f. (teak leaves powder). *J. Hazard. Mater.* **2006**, *136*, 560–566. [CrossRef] [PubMed]
59. Tang, Z.; Peng, S.; Hu, S.; Hong, S. Enhanced removal of bisphenol-AF by activated carbon-alginate beads with cetyltrimethyl ammonium bromide. *J. Colloid Interface Sci.* **2017**, *495*, 191–199. [CrossRef]
60. Rouquerol, F.; Rouquerol, J.; Sing, K.S.W.; Llewellyn, P.; Maurin, G. *Adsorption by Powders and Porous Solids. Principles, Methodology and Applications*, 2nd ed.; Elsevier: New York, NY, USA, 2012. [CrossRef]
61. Brunauer, S.; Deming, L.S.; Deming, W.E.; Teller, E. On a theory of the Van der Waals adsorption of gases. *J. Am. Chem. Soc.* **1940**, *62*, 1723–1732. [CrossRef]
62. Asada, T.; Oikawa, K.; Kawata, K.; Ishihara, S.; Iyobe, T.; Yamada, A. Study of removal effect of bisphenol A and β-estradiol by porous carbon. *J. Health Sci.* **2004**, *50*, 588–593. [CrossRef]
63. Kuo, C.-Y. Comparison with as-grown and microwave modified carbon nanotubes to removal aqueous bisphenol A. *Desalination* **2009**, *249*, 976–982. [CrossRef]
64. Li, S.; Gong, Y.; Yang, Y.; He, C.; Hu, L.; Zhu, L.; Sun, L.; Shu, D. Recyclable CNTs/Fe_3O_4 magnetic nanocomposites as adsorbents to remove bisphenol A from water and their regeneration. *Chem. Eng. J.* **2015**, *260*, 231–239. [CrossRef]
65. Xu, J.; Zhu, Y.F. Elimination of bisphenol A from water via graphene oxide adsorption. *Acta Phys. Chim. Sinica* **2013**, *29*, 829–836.

66. Schiewer, S.; Volesky, B. Modeling multi-metal ion exchange in biosorption. *Environ. Sci. Technol.* **1996**, *30*, 2921–2927. [CrossRef]
67. Niu, H.; Volesky, B. Biosorption mechanism for anionic metal species with waste crab shells. *Eur. J. Miner. Proc. Environ. Prot.* **2003**, *3*, 75–87.
68. López-Ramón, V.; Moreno-Castilla, C.; Rivera-Utrilla, J.; Radovic, L.R. Ionic strength effects in aqueous phase adsorption of metals ion son activated carbons. *Carbon* **2003**, *41*, 2020. [CrossRef]
69. Meng, Z.-F.; Zhang, Y.-P.; Zhang, Z.-Q. Simultaneous adsorption of phenol and cadmium on amphoteric modified soil. *J. Hazard. Mater.* **2008**, *159*, 492–498. [CrossRef] [PubMed]
70. Yan, L.; Huang, X.; Shi, H.; Zhang, G. Adsorption characteristics of bisfenol A on tailored activated carbon in aqueous solutions. *Water Sci. Technol.* **2016**, *74*, 1744–1751. [CrossRef] [PubMed]
71. Guo, W.; Hu, W.; Pan, J.; Zhou, H.; Guan, W.; Wang, X.; Dai, J.; Xu, L. Selective adsorption and separation of BPA from aqueous solution using novel molecularly imprinted polymers based on kaolinite/Fe_3O_4 composites. *Chem. Eng. J.* **2011**, *171*, 603–611. [CrossRef]
72. Bayramoglu, G.; Arica, M.Y.; Liman, G.; Celikbicak, O.; Salih, B. Removal of bisphenol A from aqueous medium using molecularly surface imprinted microbeads. *Chemosphere* **2016**, *150*, 275–284. [CrossRef]
73. Yu, C.; Shan, J.; Chen, Y.; Shao, J.; Zhang, F. Preparation and adsorption properties of rosin-based bisphenol A molecularly imprinted microspheres. *Mater. Today Commun.* **2020**, *24*, 101076. [CrossRef]

 water

Article

Removal of Diclofenac in Wastewater Using Biosorption and Advanced Oxidation Techniques: Comparative Results

José M. Angosto, María J. Roca and José A. Fernández-López *

Department of Chemical and Environmental Engineering, Technical University of Cartagena, Paseo Alfonso XIII, 52, 30203 Cartagena, Murcia, Spain; jm.angosto@upct.es (J.M.A.); mjose.roca@upct.es (M.J.R.)
* Correspondence: josea.fernandez@upct.es

Received: 25 November 2020; Accepted: 17 December 2020; Published: 19 December 2020

Abstract: Wastewater treatment is a topic of primary interest with regard to the environment. Diclofenac is a common analgesic drug often detected in wastewater and surface water. In this paper, three commonly available agrifood waste types (artichoke agrowaste, olive-mill residues, and citrus waste) were reused as sorbents of diclofenac present in aqueous effluents. Citrus-waste biomass for a dose of 2 g·L^{-1} allowed for removing 99.7% of diclofenac present in the initial sample, with a sorption capacity of 9 mg of adsorbed diclofenac for each gram of used biomass. The respective values obtained for olive-mill residues and artichoke agrowaste were around 4.15 mg·g^{-1}. Advanced oxidation processes with UV/H_2O_2 and UV/HOCl were shown to be effective treatments for the elimination of diclofenac. A significant reduction in chemical oxygen demand (COD; 40–48%) was also achieved with these oxidation treatments. Despite the lesser effectiveness of the sorption process, it should be considered that the reuse and valorization of these lignocellulosic agrifood residues would facilitate the fostering of a circular economy.

Keywords: agrowaste biomass; biosorption; diclofenac removal; advanced oxidation treatments; low-cost sorbents

---

1. Introduction

Water is a renewable but limited natural resource, which forces us to manage it, especially in arid and semiarid areas where water resources are limited. The presence of some pollutants can jeopardize water recovery, and the reuse of urban and industrial effluents [1]. In addition to pollution from municipal discharges, water bodies are also contaminated by industry, mining, and pollution from agrochemicals used on irrigated land, which in many cases are discharged directly into lakes and rivers without prior treatment [2].

The reuse of urban and industrial wastewater with some pollutants can contribute to soil contamination, as well as its passage into the tissue of cultivated plants. This can produce adverse effects on the ecosystem [3–6].

While the international scientific community continues to make considerable efforts to minimize or alleviate the problem of what we may call traditional pollutants (heavy metals, organic matter, inorganic nutrients, etc.), other environmental pollutants, called emerging pollutants (ECs), are of growing concern [7] and are leading us to adapt our strategies to respond to their threats [8,9].

Aqueous effluents that arrive at urban wastewater-treatment plants show a complex composition associated with current human and industrial development. Therefore, the presence of certain pollutants can, in many cases, interfere with the proper functioning of biological wastewater-treatment systems. Emerging contaminants are considered among these pollutants [9]. Emerging contaminants are compounds of different origin and chemical nature of which the presence in the environment has gone largely unnoticed [10].

The lack of international monitoring programs for ECs makes it impossible to fully know their effects, problems, and behavior from an ecotoxicological point of view [2]. They can be a danger to the environment and human health [11,12]. Another particularity of this type of pollutant is that, due to its high production and consumption, and the constant and continuous introduction of these pollutants into the environment, they do not need to be persistent to cause negative effects.

Emerging contaminants can be classified as pharmaceutical and personal-care products, plasticizers, food additives, wood preservatives, contrast media, detergents, surfactants, fire retardants, biocides, hormones, and some disinfection byproducts. Advances in analytical technologies have helped to identify this group of contaminants [13,14].

These compounds have high transformation/disposal rates that can compensate for their continued introduction into the environment. For this reason, it is necessary to raise awareness of their origin and transformation, and the effects that this new generation of pollutants can have in order to propose modifications or improvements in water-treatment mechanisms. Some authors suggested the use of toxicity tests with chemical analysis to evaluate the efficiency of treatment methods [15–18].

The ongoing processing of aqueous effluents in conventional wastewater-treatment systems does not provide an adequate approach for the removal of emerging contaminants. These can arrive in their original form and/or be metabolized, making it difficult to operate biological wastewater treatment, sludge treatment, and the subsequent reuse of wastewater; there is also a lack of knowledge about the possible impact of these pollutants on the environment [19]. In many cases, products resulting from the transformation of the original emerging contaminants produce compounds that turn out to be more toxic than the original ones are [20].

Diclofenac (2-(2,6-dichloroanilino) phenylacetic acid) is one of the best-selling nonsteroidal anti-inflammatory drugs [21]. The high consumption level of this compound means that it is one of the most detected pharmaceuticals in aqueous effluents, and according to reported data, the measured values of diclofenac in municipal wastewater can be up 7.1 $\mu g \cdot L^{-1}$ [22,23]. A distinctive feature of diclofenac is that it undergoes low biodegradability and high persistence in wastewater-treatment plants, leading to its bioaccumulation in surface waters, sediments, and sludges [24,25]. Therefore, it is necessary to implement specific treatments to reduce or eliminate its presence in effluents.

Different sorption studies of diclofenac are reported in the literature. The highest bioremoval efficacies were obtained with carbon nanotubes [26], clays [27], and biochars [28] as sorbents. Raw lignocellulosic biomass was also used as a low-cost biosorbent, but the achieved efficiency was not as high as that with previous sorbents [29,30]. Despite this, biosorption has lately been presented as a very interesting technology, in line with the concept of the circular economy [31,32]. Efficient biowaste management is among the most important challenges in agrifood industries. Suggested uses for these lignocellulosic residues include their utilization as sorbents of pollutants in aqueous effluents [33–35] since they present very interesting physicochemical properties, such as their high surface area and the presence of a variety of functional groups that facilitate the sorption process [36–38]. Bioadsorptive techniques are the best ecofriendly solutions for removing contaminants because they are economical, efficient, highly selective on pollutants, and easily operable.

The most commonly used treatments for the removal of emerging contaminants, such as diclofenac, are advanced oxidation processes [39–43], which utilize free radical reactions to directly degrade chemical contaminants as an alternative to traditional treatments. For these oxidative treatments, it is necessary to carry out research studies that examine what the most appropriate combination of oxidizing reagents is to degrade a specific pollutant.

The main objective of this work was to carry out a comparative study of diclofenac elimination in aqueous effluents through biosorption with agrifood residues and through different advanced oxidation techniques (ultraviolet radiation (UV), sodium hypochlorite, and combinations of ultraviolet radiation with hypochlorite and with hydrogen peroxide).

2. Materials and Methods

2.1. Preparation and Characterization of Agrowaste Biomass

In preliminary studies, different agricultural wastes residues were processed and examined for their ability to remove diclofenac from a test solution. Of these, the three most interesting agrifood waste types that are abundant in the Murcia region (Spain) were chosen to be investigated in this study: artichoke agrowaste, olive-mill residues, and citrus waste from agrifood industries. Biomass samples were washed with distilled water and dried at 70 °C for 24 h before being ground and passed through a sieve with a number 18 mesh (1 mm). The different surface morphologies of each biomass were explored, and a qualitative determination of each was made using a field-emission scanning electron microscope and an energy-dispersive X-ray analyzer (SEM) (Hitachi S-3500N). FTIR measurements of the sorbents were performed in the range of 4000 to 400 cm^{-1} in a Thermo Nicolet 5700 (Karlsruhe, Germany). Moreover, the pH point of zero charge (pH_{ZPC}) was determined by the mass-titration method [44].

2.2. Preparation of Diclofenac Solutions

A diclofenac stock solution (1 $g{\cdot}L^{-1}$) was prepared using diclofenac sodium ($C_{14}H_{10}Cl_2NNaO_2$, ≥99%; Sigma-Aldrich, Madrid, Spain) in deionized water. The working solution (20 L) was achieved by diluting the stock solution in treated and filtered wastewater from a wastewater-treatment plant. The pH value was adjusted to 6.6 with a 0.1 $mol{\cdot}L^{-1}$ HCl or 0.1 $mol{\cdot}L^{-1}$ NaOH solution using a Metrohm 654 pH meter with a pH combination electrode. All chemicals were of analytical grade. The analytical characterization of the working solution was performed by American Public Health Association (APHA) methods [45].

2.3. Batch-Biomass Tests

Batch-biosorption experiments were performed at 25 °C under stirring in a reciprocal contact agitator with 100 mL of the diclofenac working solution at a fixed concentration (C_o = 37.6 $mg{\cdot}L^{-1}$) and a known amount of sorbent for 24 h in a conical flask at a constant stirring speed (150 rpm). Lastly, the solutions were filtered and quantified obtaining the final concentration (C_e). The initial (C_o) and final (C_e) concentrations were quantified by high-performance liquid chromatography (HPLC).

The amounts of diclofenac adsorbed at equilibrium (q_e), also called diclofenac removal efficiency and expressed as mg diclofenac/g dry biomass, were determined by mass-balance equation (Equation (1)) on the basis of values of diclofenac concentration in the solution at the beginning (C_o) and end (C_e) of the test:

$$q_e = \frac{(C_o - C_e) \cdot V}{m}, \tag{1}$$

where V is the solution volume (L) and m the sorbent dry weight (g).

The percentage of diclofenac elimination in each case was calculated by Equation (2):

$$\%\ Removal = \frac{(C_o - C_e)}{C_o} \times 100, \tag{2}$$

where C_o is the initial concentration ($mg{\cdot}L^{-1}$) and C_e is the equilibrium concentration ($mg{\cdot}L^{-1}$) of diclofenac.

2.4. Diclofenac Determination

Diclofenac concentration in the water samples was determined by high-performance liquid chromatography (HPLC). A Waters modular liquid chromatography system (Waters, Milford, MA, USA) equipped with two M510 pumps, a 717 plus autosampler, and an M996 photodiode array detector (PDA) was used. HPLC was run by the data system of the Millenium 2010 Chromatography Manager.

An Atlantis C18 5 μm, 25 cm × 4.6 mm i.d. (Waters) column was used. Elution was performed using a gradient between 175 mM phosphoric acid in water and 175 mM acetic acid in acetonitrile as the mobile phase. Flow rate was 0.6 mL/min. Samples, before being analyzed, were filtered using a 0.45 μm pore size cellulose acetate membrane, and the filtrate was analyzed for HPLC. Diclofenac was monitored at 270 nm. The retention time of diclofenac was about 6.4 min.

2.5. Advanced Oxidation Treatments

2.5.1. Sodium Hypochlorite

Treatment with sodium hypochlorite was performed with 1000 mL of filtered diclofenac solution. Four different amounts of available chlorine, namely, 1.0, 2.0, 3.0, and 4.0 mL of a 5% (*w*/*v*) sodium hypochlorite solution (Panreac, Barcelona, Spain) were added to each sample. Experiments were carried out in sealed glass bottles. They were incubated for 2 h until the moment of their quantification in a dark isothermal chamber at 20 °C, with slow agitation. Lastly, samples were filtered and analyzed by HPLC.

2.5.2. UV Light Treatments

UV radiation was applied using an 8 watt low-pressure UV lamp (UV-C 8 watt Strahler–Messner, Kalletal, Germany), with a flow range of 0.5 to 1.1 $L{\cdot}h^{-1}$ and a wavelength of 254 nm. The facility was designed to be used with 500 mL of a sample, irradiated for short periods of time of 2, 5, 10, and 15 min. According to these irradiation durations, the provided doses were 15, 37, 73, and 110 $mJ{\cdot}cm^{-2}$, respectively. Before each treatment, the lamp was cleaned with deionized water. After irradiation, samples were filtered with a 0.45 μm filter and analyzed by HPLC.

2.5.3. UV Radiation Combined with Sodium Hypochlorite (UV/HOCl) and Hydrogen Peroxide (UV/H_2O_2)

UV/HOCl and UV/H_2O_2 assays were conducted as batch experiments using a concentration of 5 $mg{\cdot}L^{-1}$ sodium hypochlorite and 5 $mg{\cdot}L^{-1}$ of hydrogen peroxide, respectively, which were added into the initial sample and subjected to different UV exposure times (2, 5, 10, and 15 min) before filtration and analysis of diclofenac concentration was performed by HPLC.

2.6. Chemical Oxygen Demand

Chemical oxygen demand (COD) was determined using the methods described in APHA [45]. For COD measurements, a Spectroquant Nova 30 spectrophotometer (Merck, Darmstadt, Germany) was used.

3. Results and Discussion

3.1. Characterization of Biomass Samples

The surface characteristics of the different biomass samples used in diclofenac bioadsorption studies were investigated using scanning electron microscopy (SEM) and energy-dispersive X-ray spectroscopy (EDS), as shown in Figures 1–3. SEM is considered to be one of the most widely used tools in the scientific field for studying the surface characteristics of biomass used as biosorbent [46]. In the different micrographs that were made in the SEM, important differences in the surface characteristics of the different studied agrifood waste types were observed, presenting a majority composition of carbon and oxygen in all cases.

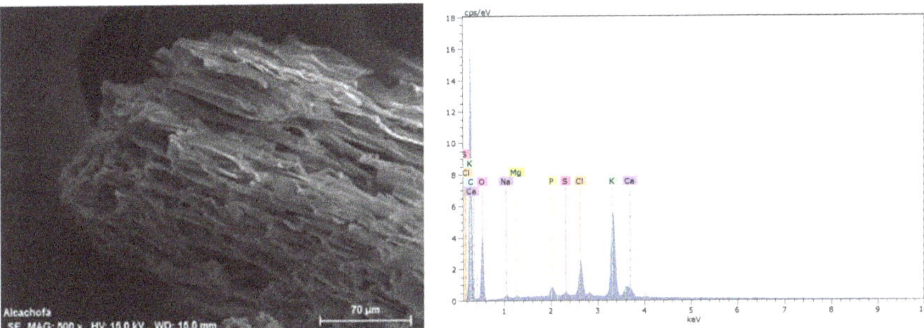

Figure 1. SEM image and EDS analysis of artichoke-agrowaste biomass.

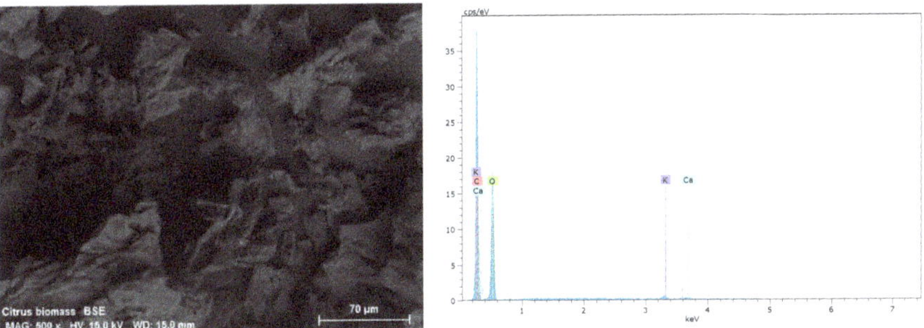

Figure 2. SEM image and EDS analysis of citrus-waste biomass.

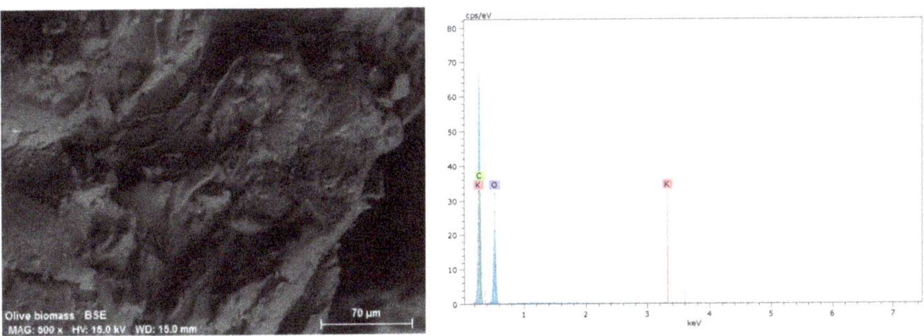

Figure 3. SEM image and EDS analysis of olive-mill residue biomass.

Figure 1 shows that artichoke biomass had filamentous structures made up of fiber cells that were longitudinally aligned. This fibrous structure provides consistency, resistance, and a large surface area [47] that, together with the small channels, allows for the rapid movement of fluids through the fibers, increasing the possibility of retaining diclofenac molecules. SEM analysis also showed that it was a rough surface, with the presence of holes of irregular shape and size [47] that could improve the retention of diclofenac.

Citrus-waste biomass (Figure 2) presented an irregular surface, but with a much more compact appearance and with fewer cavities than in the previous case, although with cracks that provided a large surface for the retention of diclofenac. Olive-mill residues (Figure 3) also presented an irregular surface with a fairly compact appearance, bit with few cavities. In all cases, the presence of cavities of irregular shape and dimensions would enhance the fixation of contaminants to be retained.

FTIR analysis allowed for inferring the structure of the sorbents on the basis of the locations and shapes of the bands in the spectra (Figure 4). The three sorbents showed a broad band at 3600–3100 cm^{-1}

that was assigned to –OH stretching vibrations of hydroxyl and carboxyl groups. A –CH aliphatic stretching vibration band was also observed at 2900–2850 cm^{-1} in all sorbents. Around 1630 cm^{-1}, a band attributed to C=O stretching vibrations and typical skeletal vibrations of aromatic rings was presented. In the region of 1030–1000 cm^{-1}, a broad band, an indication of C–O and C=C vibrational stretching, was revealed. These are, in fact, typical FTIR spectra, corresponding to nonmodified plant biomass samples, where the lignocellulosic constituents are particularly outstanding [32,47–49].

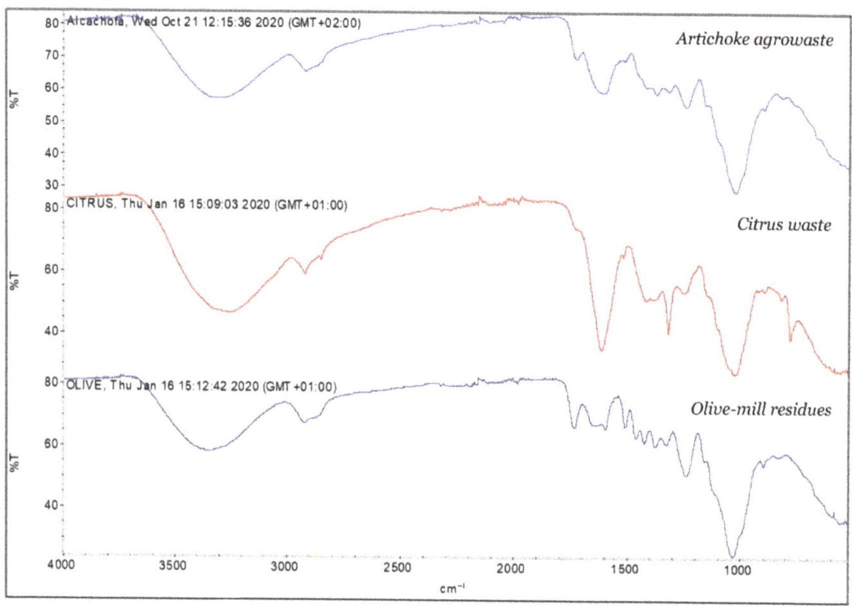

Figure 4. FTIR spectra between 4000 and 400 cm^{-1} for sorbent samples.

3.2. Diclofenac Biosorption

Table 1 shows the characterization of the working water prepared from diclofenac stock solution and filtered effluent from a treatment plant. The initial concentration (C_o) of diclofenac for the different experiments was 37.6 mg·L^{-1} and an initial COD value of 282 mg O_2·L^{-1}.

Table 1. Initial composition of working solution. COD, chemical oxygen demand.

Parameter	Value
pH	6.6 ± 0.2 *
Conductivity (μS·cm^{-1})	12.5 ± 2.5
TSS (mg/L)	98.2 ± 8.6
Turbidity (NTU)	4.7 ± 0.3
NO_3^- (mg·L^{-1})	16.8 ± 2.7
PO_4^{3-} (mg·L^{-1})	5.6 ± 1.1
BOD (mg O_2·L^{-1})	113.3 ± 10.2
COD (mg O_2·L^{-1})	282.0 ± 15.0
Diclofenac initial (mg·L^{-1})	37.6

* mean ± SD of 3 determinations.

The sorption capacities and percentages of diclofenac removal in the experiments carried out for the different used sorbents and doses (Table 2) revealed that citrus-waste biomass was the most effective. Citrus residues, with a zero point of charge of 4.2, reduced the pH of the solution during the sorption process to 5.7 by the end of the experiment. These pH reductions were less pronounced in the olive-mill-residue (pH_{ZPC} = 5.4) and in the artichoke-agrowaste (pH_{ZPC} = 4.8) sorption tests, of which the final pH values were 6.4 and 6.1, respectively. In all cases, the modifications in pH were very slight, hardly affecting the solubility of the adsorbate and sorption-capacity values. The biomass of artichoke agrowaste showed the least capacity to retain diclofenac. In the case of citrus waste, for a dose of 2 g/L, all the diclofenac is retained, obtaining a sorption capacity (q_e) of 9 mg·g^{-1}. Olive-mill residues had intermediate and very similar retention values regardless of the dose used in the experiments.

Table 2. Elimination of diclofenac for different biomass samples.

Biomass	Sorbent Dosage (g·L^{-1})	Removal %	q_e (mg·g^{-1})
Artichoke agrowaste	1	18.7 ± 2.4 *	3.9 ± 0.5
	2	19.9 ± 3.1	4.2 ± 0.4
Olive-mill residues	1	46.4 ± 4.4	4.2 ± 0.6
	2	46.3 ± 5.8	4.2 ± 0.8
Citrus-waste	1	87.1 ± 5.7	7.8 ± 1.1
	2	99.7 ± 0.5	9.0 ± 0.3

* mean ± SD of 3 determinations.

The results of this study revealed that the reuse of raw lignocellulosic biomass from agrifood waste for the sorption of diclofenac may be an efficient and common method of wastewater treatment. The sorption capacities of diclofenac obtained in this study were lower than those reported for microalgae biomass (20–28 mg·g^{-1}) [29] and grape bagasse (24 mg·g^{-1}) [49], but comparable for that shown with activated carbon from olive stones (11 mg·g^{-1}) [50]. In the present investigation, the agrifood waste used as sorbent was not previously modified or subjected to any thermal treatment. The obtained results in the present investigation reveal that this is a new line of research that is worth further exploring, and it may be of interest to test some cost-effective activation mechanism that enhances the bioadsorptive capacity of waste to specific contaminants. In this context, future studies will be made in application in real systems, where fixed-bed citrus-waste adsorbers can be implemented for the treatment of wastewater with diclofenac.

3.3. Advanced Oxidation Treatments

As summarized in Table 3, when the working solution (C_o = 37.6 mg·L^{-1}) was exposed to different exposure times of UV radiation ranging from 2 to 15 min, significant reductions in the initial diclofenac concentration were achieved, ranging from 9.78% removal for 2 min of exposure and 88.49% removal for 15 min of exposure, without being able to achieve total removal. A significant reduction in COD values was also achieved, in a similar way to that in other reported investigations [51].

Table 3. Treatment with ultraviolet (UV) radiation.

Contact Time (min)	Diclofenac (mg·L^{-1})	Removed Diclofenac %	COD Reduction %
2	33.9 ± 2.8 *	9.8 ± 2.4	44.3 ± 3.6
5	16.6 ± 2.1	55.9 ± 3.9	62.1 ± 3.5
10	6.1 ± 1.1	83.8 ± 4.6	62.4 ± 4.1
15	4.3 ± 0.9	88.5 ± 3.8	66.3 ± 3.9

* mean ± SD of 3 determinations.

In the treatment with sodium hypochlorite and a contact time of 2 h (Table 4), for a dose of 4 mg·L^{-1}, all diclofenac present in the starting sample was eliminated. As the dose of hypochlorite was increased, greater elimination was achieved. The reduction in COD was also quite significant.

Table 4. Treatment with sodium hypochlorite (contact time 2 h).

Hypochlorite Concentration (mg·L^{-1})	Diclofenac (mg·L^{-1})	Removed Diclofenac %	COD Reduction %
1	14.4 ± 3.3 *	61.7 ± 4.7	24.1 ± 2.3
2	4.4 ± 1.2	88.4 ± 5.1	38.3 ± 2.8
3	3.3 ± 0.9	91.3 ± 4.4	44.0 ± 2.6
4	n.d. †	100	44.7 ± 1.8

* mean ± SD of 3 determinations; † not detected.

In the treatment that combined UV radiation with the addition of sodium hypochlorite (Table 5), practically all the diclofenac (93.59%) was eliminated for a radiation time of 10 min. When exposure time was increased (15 min), it was possible to eliminate it completely. In addition, the unreacted chlorine in this process provided residual protection for water reuse purposes [52]. Comparable reductions in COD were achieved as those with previous treatments. For the treatment in which UV radiation was combined with hydrogen peroxide (Table 6), the initial diclofenac was eliminated for all treatments. For an exposure time of 2 min of UV radiation, 99.20% of the initial diclofenac was degraded. The achieved elimination of COD was quite comparable in all performed treatments.

Table 5. Ultraviolet radiation treatment combined with sodium hypochlorite (5 mg·L^{-1}).

Contact Time (min)	Diclofenac (mg·L^{-1})	Removed Diclofenac %	COD Reduction %
2	29.1 ± 2.2 *	22.6 ± 2.8	14.2 ± 1.9
5	13.5 ± 1.8	64.2 ± 3.1	22.0 ± 2.4
10	2.4 ± 0.5	93.6 ± 4.8	45.0 ± 3.8
15	n.d. †	100	47.9 ± 3.4

* mean ± SD of 3 determinations; † not detected.

Table 6. UV radiation treatment combined with H_2O_2 (5 mg·L^{-1}).

Contact Time (min)	Diclofenac (mg·L^{-1})	Removed Diclofenac %	COD Reduction %
2	0.3 ± 0.1 *	99.2 ± 4.4	23.4 ± 1.7
5	n.d. †	100	25.9 ± 2.3
10	n.d.	100	36.5 ± 2.0
15	n.d.	100	40.1 ± 3.1

* mean ± SD of 3 determinations; † not detected.

It is scientifically accepted that advanced oxidation processes are highly effective novel methods that accelerate the oxidation and the degradation of a wide range of organic and inorganic pollutants that are resistant to conventional treatment methods [53]. In our experiments, diclofenac was 88% by UV-C degraded irradiation; when the wastewater was treated for a time as short as 15 min, this result verified that this compound absorbs UV irradiation well at 254 nm and can be easily photolyzed [54]. When UV-C irradiation treatment was combined with 5 mg·L^{-1} of sodium hypochlorite, the degradation of diclofenac was increased to 100%. Nevertheless, UV/H_2O_2 was the most efficient oxidation process, combining the immediate UV effect and the action of the HO$^{\cdot}$ radicals produced from the homolytic disruption of H_2O_2. The levels of diclofenac degradation under the adopted experimental conditions were higher than the values reported in the literature for oxidation processes, such as ozonation (32%) and UV/H_2O_2 (39%), after 39 min treatment [55]. The use of hypochlorite to degrade diclofenac in aqueous effluents was also reported with successful results [56]. Recent research in a Fenton-like system with FeCeOx–H_2O_2 achieved 84% degradation of diclofenac within 40 min [57]. All these results are in accordance with those obtained in our experimental design and confirm the feasibility of the tested methods. A significant reduction in chemical oxygen demand (COD) (40–48%) was also achieved with these oxidation treatments, although there was no direct relationship between the percentage of diclofenac degradation and the decrease in COD.

4. Conclusions

This paper offers a comparative study between the application of biosorption and advanced oxidation techniques to remove diclofenac from wastewater at high concentrations.

Artichoke agrowaste, olive-mill residues, and citrus waste were used as low-cost biosorbents to remove diclofenac from aqueous effluents. Citrus-waste biomass, at a sorbent dosage of 2 $g{\cdot}L^{-1}$, showed the highest sorption capacity (9 $mg{\cdot}g^{-1}$). This value was achieved with raw biomass without being subjected to any chemical or thermal activation process. The obtained results highlight the sustainable reutilization of agrifood waste for producing cost-effective sorbents, providing additional income for the agroindustrial sector. Waste valorization can efficiently help in reducing environmental stress by decreasing unwarranted pollution and promoting economy circularization.

UV-driven advanced oxidation treatments were effective in the degradation of diclofenac in wastewater. The UV/H_2O_2 (5 $mg{\cdot}L^{-1}$) oxidation process was faster than UV/HOCl (5 $mg{\cdot}L^{-1}$) to reach the same degradation rate. Chemical oxygen demand (COD) was considerably reduced (40–48%) with these oxidation treatments.

The obtained results in this study confirm biosorption as a low-cost and environmentally sustainable technique, and the high efficiency demonstrated by advanced oxidation techniques, may lead to applying specifically designed techniques that combine biosorption and advanced oxidation to eliminate specific emerging contaminants in aqueous effluents.

Author Contributions: Conceptualization, J.M.A. and J.A.F.-L.; methodology, J.M.A. and M.J.R.; formal analysis, J.M.A. and J.A.F.-L.; investigation, M.J.R.; writing—original-draft preparation, J.M.A.; writing—review and editing, J.A.F.-L. All authors have read and agreed to the published version of the manuscript.

Funding: This research was funded by the Technical University of Cartagena (UPCT), grant number ACI B.

Acknowledgments: This research is part of the Environmental Engineering R&D group. The excellent technical assistance of the Technical Research Support Service of the UPCT is greatly acknowledged.

Conflicts of Interest: The authors declare no conflict of interest.

References

1. Dalecka, B.; Oskarsson, C.; Juhna, T.; Rajarao, G.K. Isolation of fungal strains from municipal wastewater for the removal of pharmaceutical substances. *Water* **2020**, *12*, 524. [CrossRef]
2. Geisen, V.; Mol, H.; Klumpp, E.; Umlauf, G.; Nadal, M.; Van der Ploeg, M.; van de Zee, S.E.A.T.M.; Ritsema, C.J. Emerging pollutants in the environment: A challenge for water resource management. *Int. Soil Water Conserv. Res.* **2015**, *3*, 57–65. [CrossRef]
3. Sauvé, S.; Desrosiers, M. A review of what is an emerging contaminant. *Chem. Cent. J.* **2014**, *8*, 15. [CrossRef] [PubMed]
4. López-Doval, J.C.; Montagner, C.C.; de Alburquerque, A.F.; Moschini-Carlos, V.; Umbuzeiro, G.; Pompeo, M. Nutrients, emerging pollutants and pesticides in a tropical urban reservoir: Spatial distributions and risk assessment. *Sci. Total Environ.* **2017**, *575*, 1307–1324. [CrossRef] [PubMed]
5. Margenat, A.; Matamoros, V.; Díez, S.; Cañameras, N.; Comas, J.; Bayona, J.M. Ocurrence of chemical contaminants in peri-urban agricultural irrigation waters and assesment of their phytotoxicity and crop productivity. *Sci. Total Environ.* **2017**, *599*, 1140–1148. [CrossRef] [PubMed]
6. Tran, N.; Drogui, P.; Brar, S.K. Sonochemical techniques to degrade pharmaceutical organic pollutants. *Environ. Chem. Lett.* **2015**, *13*, 251–268. [CrossRef]

7. Carmalin, A.; Lima, E. Removal of emerging contaminants from the environment by adsorption. *Ecotoxicol. Environ.* **2018**, *150*, 1–17.
8. Munthe, J.; Brorstrom-Lunden, E.; Rahmberg, M.; Posthuma, L.; Alterburger, R.; Brack, W.; Bunke, D.; Engelen, G.; Gawlik, B.; Gils, J.; et al. An expanded conceptual framework for solution-focused management of chemical pollution in European waters. *Environ. Sci. Eur.* **2017**, *29*, 1072. [CrossRef]
9. Richardson, S.D.; Kimura, S.Y. Emerging environmental contaminants: Challenges facing our next generation and potential engineering solutions. *Environ. Technol. Innov.* **2017**, *8*, 40–56. [CrossRef]
10. Rodriguez-Narvaez, O.M.; Peralta-Hernandez, J.M.; Goonetilleke, A.; Bandala, E.R. Treatment technologies for emerging contaminants in water: A review. *Chem. Eng. J.* **2017**, *323*, 361–380. [CrossRef]
11. Petrovic, M.; Solé, M.; López, M.; Barceló, D. Endocrine disruptors in sewage treatment plants, receiving river waters, and sediments: Integration of chemical analysis and biological effects of feral carp. *Environ. Toxic. Chem.* **2002**, *21*, 2146–2156. [CrossRef]
12. Esplugas, S.; Bila, D.M.; Krause, L.G.T.; Dezotti, M. Ozonation and advanced oxidation technologies to remove endocrine disrupting chemicals (EDCs) and pharmaceuticals and personal care products (PPCPs) in water effluents. *J. Hazard. Mater.* **2007**, *149*, 631–642. [CrossRef] [PubMed]
13. Richardson, S.D. Environmental mass spectrometry: Emerging contaminants and current issues. *Anal. Chem.* **2012**, *84*, 747–778. [CrossRef] [PubMed]
14. Richardson, S.D.; Kimura, S.Y. Water analysis: Emerging contaminants and current issues. *Anal. Chem.* **2016**, *88*, 546–582. [CrossRef]
15. Blanco, M.; Pérez-Albadalejo, E.; Piña, B.; Kuspilic, G.; Milun, V.; Lille-Langoy, R. Assessing the environmental quality of sediments from Split coastal area (Croatia) with a battery of cell-based bioassays. *Sci. Total Environ.* **2018**, *624*, 1640–1648.
16. Boehler, S.; Strecker, R.; Heinrich, P.; Prochazka, E.; Northcott, G.L.; Ataria, J.M. Assessment of urban stream sediment pollutants entering estuaries using chemical analysis and multiple bioassays to characterize biological activities. *Sci. Total Environ.* **2017**, *593*, 498–507. [CrossRef]
17. Abbas, M.; Adil, M.; Ehtisham-ul-Haque, S.; Munir, B.; Yameen, M.; Ghaffar, A.; Abbas Shar, G.; Asif Tahir, M.; Iqbal, M. *Vibrio fischeri* bioluminescence inhibition assay for ecotoxicity assessment: A review. *Sci. Total Environ.* **2018**, *626*, 1295–1309.
18. Schereiber, B.; Fischer, J.; Schiwy, S.; Hollert, H.; Schulz, R. Towards more ecological relevance in sediment toxicity testing with fish: Evaluation of multiple bioassays with embryos of the benthic weatherfish (*Misgurnus fossilis*). *Sci. Total Environ.* **2018**, *619*, 391–400. [CrossRef]
19. Sang, Z.; Jiang, Y.; Tsoi, Y.-K.; Leung, K.S. Evaluating the environmental impact of artificial sweeteners: A study of their distributions, photodegradation ant toxicities. *Water Res.* **2014**, *52*, 260–274. [CrossRef]
20. Funke, J.; Prasse, C.; Eversloh, C.L.; Ternes, T.A. Oxypurinol-A novel marker for wastewater contamination of the aquatic environment. *Water Res.* **2015**, *74*, 257–265.
21. Lonappan, L.; Brar, S.K.; Das, R.K.; Verma, M.; Rao, Y. Diclofenac and its transformation products: Environmental occurrence and toxicity. *Environ. Int.* **2016**, *96*, 127–138. [CrossRef] [PubMed]
22. Grandclément, C.; Piram, A.; Petit, M.E.; Seyssiecq, I.; Laffont-Schwob, I.; Vanot, G.; Tiliacos, N.; Roche, N.; Doumenq, P. Biological removal and fate assessment of diclofenac using *Bacillus subtilis* and *Brevibacillus laterosporus* strains and ecotoxicological effects of diclofenac and 4′-hydroxy-diclofenac. *J. Chem.* **2020**, *2020*, 9789420. [CrossRef]
23. Vieno, N.; Silanpää, M. Fate of diclofenac in municipal wastewater treatment plants—A review. *Environ. Int.* **2014**, *69*, 28–39. [CrossRef] [PubMed]
24. Langford, K.H.; Reid, M.; Thomas, K.V. Multi-residue screening of prioritised human pharmaceuticals, illicit drugs and bactericides in sediments and sludge. *J. Environ. Monit.* **2011**, *13*, 2284–2291. [CrossRef] [PubMed]
25. Kunkel, U.; Radke, M. Fate of pharmaceuticals in rivers: Deriving a benchmark dataset at favorable attenuation conditions. *Water Res.* **2012**, *46*, 5551–5565. [CrossRef]
26. Gil, A.; Santamaría, L.; Korili, S.A. Removal of caffeine and diclofenac from aqueous solution by adsorption on multiwalled carbon nanotubes. *Coll. Interf. Sci. Commun.* **2018**, *22*, 25–28. [CrossRef]
27. Mabrouki, H.; Akretche, D.E. Diclofenac potassium removal from water by adsorption on natural and pillared clay. *Des. Water Treatm.* **2016**, *57*, 6033–6043. [CrossRef]

28. Thi Minh Tam, N.; Liu, Y.; Bashir, H.; Yin, Z.; He, Y.; Zhou, X. Efficient removal of diclofenac from aqueous solution by potassium ferrate-activated porous graphitic biochar: Ambient condition influences and adsorption mechanism. *Int. J. Environ. Res. Public Health* **2020**, *17*, 291. [CrossRef]
29. Coimbra, R.N.; Escapa, C.; Vázquez, N.C.; Noriega-Hevia, G.; Otero, M. Utilization of non-living microalgae biomass from two different strains for the adsorptive removal of diclofenac from water. *Water* **2018**, *10*, 1401. [CrossRef]
30. Li, Y.; Taggart, M.A.; McKenzie, C.; Zhang, Z.; Lu, Y.; Pap, S.; Gibb, S. Utilizing low-cost natural waste for the removal of pharmaceuticals from water: Mechanisms, isotherms and kinetics at low concentrations. *J. Clean. Prod.* **2019**, *227*, 88–97. [CrossRef]
31. Yusuf, M.; Elfghi, F.M.; Zaidi, S.A.; Abdullah, E.C.; Khan, M.A. Applications of graphene and its derivatives as an adsorbent for heavy metal and dye removal: A systematic and comprehensive overview. *RSC Adv.* **2015**, *5*, 50392–50420. [CrossRef]
32. Rosique, M.; Angosto, J.M.; Guibal, E.; Roca, M.J.; Fernández-López, J.A. Factorial design methodological approach for enhanced cadmium ions bioremoval by *Opuntia* biomass. *CLEAN Soil Air Water* **2016**, *44*, 959–966. [CrossRef]
33. Ben-Othman, S.; Jõudu, I.; Bhat, R. Bioactives from agri-food wastes: Present insights and future challenges. *Molecules* **2020**, *25*, 510. [CrossRef] [PubMed]
34. Nowicki, P.; Kazmierczak-Razna, J.; Pietrzak, R. Physico-chemical and adsorption properties of carbonaceous sorbents prepared by activation of tropical fruit skins with potassium carbonate. *Mater. Des.* **2016**, *90*, 579–585. [CrossRef]
35. Castrica, M.; Rebucci, R.; Giromini, C.; Tretola, M.; Cattaneo, D.; Baldi, A. Total phenolic content and antioxidant capacity of agri-food waste and by-products. *Ital. J. Anim. Sci.* **2019**, *18*, 336–341. [CrossRef]
36. Manna, S.; Toy, D.; Adhikari, B.; Thomas, S.; Das, P. Biomass for water defluoridation and current understanding on biosorption mechanisms: A review. *Environ. Prog. Sustain. Energy* **2018**, *37*, 1560–1572. [CrossRef]
37. Ummartyotin, S.; Pechyen, C. Strategies for development and implementation of bio-based materials as effective renewable resources of energy: A comprehensive review on adsorbent technology. *Renew. Sustain. Energy Rev.* **2016**, *62*, 654–664. [CrossRef]
38. Arief, V.O.; Trilestari, K.; Sunarso, J.; Indraswati, N.; Ismadji, S. Recent progress on biosorption of heavy metals from liquids using low cost biosorbents: Characterization, biosorption parameters and mechanism studies. *CLEAN Soil Air Water* **2008**, *36*, 937–962. [CrossRef]
39. Ternes, T.; Stüber, J.; Herrmann, N.; McDowell, D.; Ried, A.; Kampmann, M.; Teiser, B. Ozonation: A tool for removal of pharmaceuticals, contrast media and musk fragrances from wastewater? *Water Res.* **2003**, *37*, 1976–1982. [CrossRef]
40. Cerreta, G.; Roccamante, M.A.; Oller, I.; Malato, S.; Rizzo, L. Contaminants of emerging concern removal from real wastewater by UV/free chlorine process: A comparison with solar/free chlorine and UV/H_2O_2 at pilot scale. *Chemosphere* **2019**, *236*, 124354.
41. Lee, Y.; von Gunten, U. Oxidative transformation of micropollutants during municipal wastewaters treatment: Comparison of kinetic aspects of selective (chlorine, chlorine dioxide, ferrate VI and ozone) and non-selective oxidants (hydroxyl radicals). *Water Res.* **2010**, *39*, 555–566. [CrossRef] [PubMed]
42. Kanakaraju, D.; Glass, B.D.; Oelgemöller, M. Titanium dioxide photocatalysis for pharmaceutical wastewater treatment. *Environ. Chem. Lett.* **2014**, *12*, 27–47. [CrossRef]
43. Jiang, J.; Pang, S.Y.; Ma, J.; Liu, H. Oxidation of phenolic endocrine disrupting chemicals by potassium permanganate in synthetic and real waters. *Environ. Sci. Technol.* **2012**, *46*, 1774–1781. [CrossRef]
44. Fiol, N.; Villaescusa, I. Determination of sorbent point zero charge: Usefulness in sorption studies. *Environ. Chem. Lett.* **2009**, *7*, 79–84. [CrossRef]
45. Eaton, A.D.; American Public Health Association; American Water Works Association; Water Environment Federation. *Standard Methods for the Examination of Water and Wastewater*, 21st ed.; APHA-AWWA-WEF: Washington, DC, USA, 2005.
46. Amiri, H.; Karimi, K. Improvement of acetone, butanol, and ethanol production from woody biomass using organosolv pretreatment. *Bioproc. Biosyst. Eng.* **2015**, *38*, 1959–1972. [CrossRef] [PubMed]

47. Saavedra, M.I.; Doval Miñarro, M.; Angosto, J.M.; Fernández-López, J.A. Reuse potential of residues of artichoke (*Cynara scolymus* L.) from industrial canning processing as sorbent of heavy metals in multimetallic effluents. *Ind. Crops Prod.* **2019**, *141*, 111751. [CrossRef]
48. Mahato, N.; Sharma, K.; Sinha, M.; Baral, E.R.; Koteswararao, R.; Dhyani, A.; Cho, M.H.; Cho, S. Bio-sorbents, industrially important chemicals and novel materials from citrus processing waste as a sustainable and renewable bioresource: A review. *J. Adv. Res.* **2020**, *23*, 61–82. [CrossRef] [PubMed]
49. Antunes, M.; Esteves, V.I.; Guégan, R.; Crespo, J.S.; Fernandes, A.N.; Giovanela, M. Removal of diclofenac sodium from aqueous solution by Isabel grape bagasse. *Chem. Eng. J.* **2012**, *192*, 114–121. [CrossRef]
50. Larous, S.; Meniai, A.-H. Adsorption of diclofenac from aqueous solution using activated carbon prepared from olive stones. *Int. J. Hydrog. Energy* **2016**, *41*, 10380–10390. [CrossRef]
51. Plauta, M.; Tisler, T.; Toman, M.J.; Pintar, A. Efficiency of advanced oxidation processes in lowering bisphenol A toxicity and oestrogenic activity in aqueous samples. *Arch. Ind. Hyg. Toxicol.* **2014**, *65*, 77–87.
52. Zhao, Q.; Shang, C.; Zhang, X.; Ding, G.; Yang, X. Formation of halogenated organic byproducts during medium-pressure UV and chlorine coexposure of model compounds, NOM and bromide. *Water Res.* **2011**, *45*, 6545–6554. [CrossRef] [PubMed]
53. Stasinakis, A. Use of selected advanced oxidation processes (AOPs) for wastewater treatment—A mini review. *Glob. NEST J.* **2008**, *10*, 376–385.
54. De la Cruz, N.; Giménez, J.; Esplugas, S.; Grandjean, D.; de Alencastro, L.F.; Pulgarín, C. Degradation of 32 emergent contaminants by UV and neutral photo-fenton in domestic wastewater effluent previously treated by activated sludge. *Water Res.* **2012**, *46*, 1947–1957. [CrossRef] [PubMed]
55. Vogna, D.; Marotta, R.; Napolitano, A.; Andreozzi, R.; d'Ischia, M. Advanced oxidation of the pharmaceutical drug diclofenac with UV/H_2O_2 and ozone. *Water Res.* **2004**, *38*, 414–422. [CrossRef] [PubMed]
56. Luongo, G.; Guida, M.; Siciliano, A.; Libralato, G.; Saviano, L.; Amoresano, A.; Previtera, L.; Di Fabio, G.; Zarrelli, A. Oxidation of diclofenac in water by sodium hypochlorite: Identification of new degradation by-products and their ecotoxicological evaluation. *J. Pharmac. Biomed. Anal.* **2020**, 113762. [CrossRef] [PubMed]
57. Chong, S.; Zhang, G.; Zhang, N.; Liu, Y.; Huang, T.; Chang, H. Diclofenac degradation in water by $FeCeO_x$ catalyzed H_2O_2: Influencing factors, mechanism and pathways. *J. Hazard. Mat.* **2017**, *334*, 150–159. [CrossRef]

Article

Removal of Phenolic Compounds from Olive Mill Wastewater by a Polydimethylsiloxane/oxMWCNTs Porous Nanocomposite

Antonio Turco [1,2,*] **and Cosimino Malitesta** [1,*]

1 Department of Biological and Environmental Sciences and Technologies (Di.S.Te.B.A.), University of Salento, Via Monteroni, 73100 Lecce, Italy

2 CNR-NANOTEC Institute of Nanotechnology, Via Monteroni, 73100 Lecce, Italy

* Correspondence: antonio.turco@nanotec.cnr.it (A.T.); cosimino.malitesta@unisalento.it (C.M.)

Received: 13 November 2020; Accepted: 7 December 2020; Published: 10 December 2020

Abstract: User-friendly and energy-efficient methods able to work in noncontinuous mode for *in situ* purification of olive mill wastewater (OMW) are necessary. Herein we determined the potential of oxidized multiwalled carbon nanotubes entrapped in a microporous polymeric matrix of polydimethylsiloxane in the removal and recovery of phenolic compounds (PCs) from OMW. The fabrication of the nanocomposite materials was straightforward and evidenced good adsorption capacity. The adsorption process is influenced by the pH of the OMW. Thermodynamic parameters evidenced the good affinity of the entrapped nanomaterial towards phenols. Furthermore, the kinetics and adsorption isotherms are studied in detail. The presence of oil inside the OMW can speed up the uptake process in batch adsorption experiments with respect to standard aqueous solutions, suggesting a possible use of the nanocomposite for fast processing of OMW directly in the tank where they are stored. Moreover, the prepared nanocomposite is safe and can be easily handled and disposed of, thus avoiding the presence of specialized personnel. After the adsorption process the surface of the nanomaterial can be easily regenerated by mild treatments with diluted acetic acid, thus permitting both the recyclability of the nanomaterial and the recovery of phenolic compounds for a possible use as additives in food and nutraceutical industries and the recovery of OMW for fertirrigation.

Keywords: olive mill wastewater; carbon nanotubes; polydimethilsiloxane; waste treatment; phenolic compounds; resources recovery

---

1. Introduction

Olive mill wastewater (OMW) is an acid waste derived from olive pressing, which has a production range from 10 to 30 million of m^3 per year [1]. OMW is composed of water, oil, and solids and exhibits ecotoxic and phytotoxic properties due to its high content of phenols [2]. For that reason, OMW has been considered as a matter of treatment and minimization [3]. However, it could represent a cheap source of components that can be recovered and used as natural food additives [4,5]. For instance, at low concentrations, phenols of olive have antioxidant properties with potential benefits for the health [6–8]. Moreover, OMW purified from phenols can be a valuable source for fertirrigation [9].

Different methods have been developed to purify OMW from phenols such as electrochemical oxidation [10], physical methods [11], solvent extraction [12], chemical treatments [13–15], filtration [16], and bioremediation [17,18]. However, these techniques could require high energy, the use of chemicals, and could generate secondary pollutants during the remediation process [19]. Moreover, phenolic compounds (PCs) could be degraded during the treatment thus not permitting their recovery. For all these reasons, there is an urgent need to develop alternative methods for the removal and recovery of PCs from OMW. In this view, adsorption techniques can be an attractive alternative [20–25]. Different adsorbents have been developed, such as granular activated carbon [20], zeolites [26], agricultural wastes [3], and amberlite [27]. One of the major issues is that these materials are in powder form requiring large centrifuges or filtration systems for their management during the treatment. Another approach could be to pack the powder in a column for a continuous flow separation in a plant [21]. However, the seasonally production of OMW and the necessity to collect and transport OMW from the large number of olive mills to the plant can make such systems expensive [19]. Recently we have developed a nanocomposite material in which oxidized carbon nanotubes (oxCNTs) were physically entrapped on the surface of porous polydimethylsiloxane (PDMS) for the removal of phenolic compounds from aqueous solutions with good adsorption capacity. The nanocomposite can be easily handled and disposed of, making easier its recovery after the adsorption process [28]. Although the use of the material to remove phenolic compounds for OMW treatment was suggested a complete characterization of the adsorption process was not performed. In the present work we tested the ability of PDMS/oxidized multiwalled carbon nanotubes (oxMWCNTs)—spongeous materials to remove phenolic compounds from complex matrices such as OMW. The adsorption mechanisms, thermodynamic parameters, and kinetics were studied with different theoretical models. A good and fast adsorption capacity was observed. The system was demonstrated to be effective for the purification of complex OMW matrices in batch samples, suggesting their possible use for *in situ* purification of OMW, being able to work directly in the tank where the waste is stored. It has been demonstrated that the different phenols present in OMW can affect the adsorption process with respect to our previous observations. Moreover, the presence of oil in OMW can speed up the uptake process, probably due to the swelling of the pores inside the adsorbent phase. The reusability of the nanocomposite and the possibility of recovering adsorbed phenols was also demonstrated.

2. Materials and Methods

2.1. Materials

Multiwalled carbon nanotubes with a diameter of 25.4 ± 4 nm were provided by Nanostructured & Amorphous Materials, Inc., Los Alamos, NM, USA. The PDMS polymerization kit (Sylgard 184), comprising monomer and curing agent, were purchased from Dow Corning, Midland, MI, USA. All the other reagents were analytical grade and purchased from VWR International srl, Milano, Italy, and used as received.

2.2. OMW Origin and Composition

OMW was obtained from a three-phase continuous extraction unit in Miggiano, Italy. The pH at 25 °C was equal to 4.8 and the density 1.08 ± 0.02 g/L. Total suspended solids were 2.57 ± 0.05 g/L, the total solids were 25.12 ± 0.8 g/L, the mineral matter was 4.02 ± 0.07 g/L. The phenolic amount equal to 2.089 ± 0.01 g/L was calculated by Folin–Ciocalteau assay [29].

2.3. Preparation of the Spongeous Adsorbent

A sponge of polydimethilsiloxane (PDMS), in which oxidized multiwalled carbon nanotubes (oxMWCNTs) were stably entrapped and homogenously dispersed on pores' surfaces, was prepared following our well-developed procedure [28]. Briefly, microparticles of glucose crystals with an average dimension of 290 ± 170 μm were mixed with a shaker overnight with pristine MWCNTs at 3% *w/w*. The obtained mixture was packed in a centrifuge tube and an appropriate amount of PDMS prepolymer mixed with curing agent in the ratio of 10:1 and diluted with 40% in *wt.*% of hexane was added on the top. The composites were centrifuged at 8000 rpm for 20 min to allow the packing of the mixture and the permeation of the prepolymerization solutions between the sugar particles. The composite was then cured at 60 °C overnight to accomplish the polymerization. Finally, glucose was removed by firstly soaking the nanocomposite in boiling water under continuous stirring and then by sonication in warm water and ethanol.

MWCNTs entrapped in the nanocomposites were then oxidized. The sponges were placed in water under reduced pressure to allow the permeation of the solution in the pores of the hydrophobic nanocomposite, then nitric acid was added until a concentration of 3 M was reached and left for two hours under continuous stirring. At the end of the process the as-obtained nanocomposites were washed repetitively with water until the pH of the washing solution remained stable. The nanocomposites were dried under vacuum and dipped in H_2O_2 30% *w/w* solution under stirring for 2 h. Finally, the obtained materials (PDMS/oxMWCNTs) were repetitively washed in water and dried at 100 °C overnight.

2.4. Determination of PCs in OMW

The PCs in the OMW were quantified by Folin–Ciocalteu assay [29]. 240 μL of water, 50 μL of OMW, and 250 μL of Folin–Ciocalteau reagent were added in a flask. After 120 s, 2.7 mL of sodium carbonate (20% *w/v*) was added. The mixture was left at 25 °C for 2 h and then centrifuged at 8000 rpm for 5 min. The absorbance of the supernatant was read at 765 nm with a Cary UV 50 spectrophotometer. Gallic acid was used as a standard for the calibration of the method.

2.5. Adsorption Experiments

A proper amount of the PDMS/oxMWCNTs nanocomposite (6 g/100 mL) was immersed in 5 mL of OMW diluted with water (pH = 4.8) with a known concentration of phenols (1.251 g/L) and shacked at 25 °C. Studies at different pH were performed, adjusting the pH with 0.1 M HCl or 0.1 M NaOH. The adsorption rates of phenols were monitored at different times (namely 0, 0.5, 1, 1.5, 2, 2.5, 3, 4, 22, and 24 h) collecting small aliquots from the solution for further spectrophotometric analysis. Removal efficiency (Q%) and equilibrium adsorption capacity (q_e) of the sponges were calculated by Equations (1) and (2) respectively:

$$removal\quad efficiency\quad (\%) = \left[\frac{(C_0 - C_t)}{C_0}\right] \times 100 \tag{1}$$

$$adsorption\quad capacity,\quad q_e\left(\frac{mg}{g}\right) = \left[\frac{(C_0 - C_e)}{W}\right] \times V \tag{2}$$

where C_0 and C_t are respectively the concentration of PCs at the beginning of the experiment and at a given time (hours) in ppm (mg/L), C_e is the PCs concentrations (ppm) at the equilibrium (24 h), W is the weight of the sponge in grams, and V is the volume of the solution in liters.

2.6. Desorption Experiments

The adsorbent was separated from the OMW solution and washed with water. Desorption experiments were conducted by dipping the PDMS/oxMWCNTs sponge in 10% acetic acid, which was vortexed for 2 h.

3. Results and Discussion

3.1. Adsorption of OMW PCs on PDMS/oxMWCNTs Sponges

Our developed fabrication route described in experimental methods can easily allow the synthesis of black porous PDMS/oxMWCNTs sponges (Figure 1a) in which the pores dimensions are comparable to that of the used hard template. The carbon nanotubes are well-dispersed in the polymeric matrices thanks to the mechanical destroying of π-π stacking during the fabrication steps. Moreover, the oxidation of the nanomaterials in the sponges occurred after the synthesis of the 3D nanocomposites thus reducing the need of complex apparatus thanks to the easy handle of the material [28,30]. Although it is known that treatment with strong acid can degrade PDMS matrices [31], it is interesting to note as the oxidation procedure did not significantly affect the mechanical stability of the material.

This is probably due to the low concentration of nitric acid and short time of incubation used for the synthetic procedure. The prepared sponges were dipped in an OMW solution at pH 4.8 and the adsorption of PCs was monitored at different times. As reported in Figure 1c, at the beginning of the process we observed a fast phenols adsorption. After 4 h the adsorption process become slower due to the decrease in the number of easily accessible sites on oxMWCNTs [3,28,32]. Finally, for times higher than 20 h an equilibrium between the adsorbent and adsorbate is achieved. Moreover, we observed that the adsorption of PCs did not occur in 24 h on porous PDMS prepared following the procedure in Section 2.3, but in the absence of MWCNTs, and is very low on a spongeous nanocomposite in which MWCNTs were not oxidized (removal efficiency approximately 2%) (data not shown).

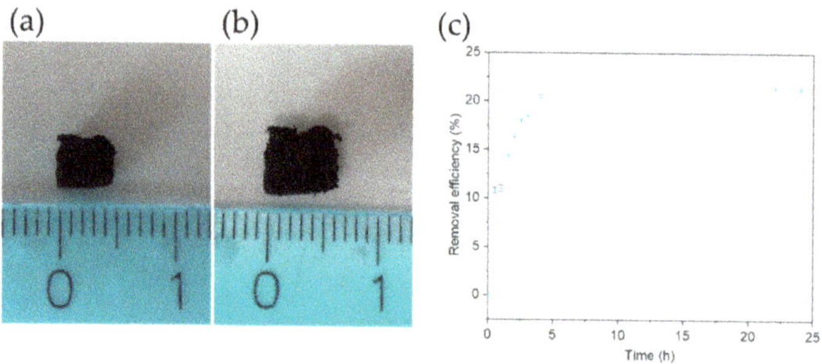

Figure 1. A piece of polydimethylsiloxane/oxidized multiwalled carbon nanotubes (PDMS/oxMWCNTs) sponge (**a**) before and (**b**) after dipping in an olive mill wastewater (OMW) solution. (**c**) Effect of contact time on adsorption of OMW phenolic compounds (PCs) on the PDMS/oxMWCNTs sponge.

Interestingly, most of the uptake process (around 95%) was completed in 4 h, evidencing a faster uptake with respect to what we observed in aqueous solutions [28].

We hypothesized that this can be due to two different reasons: on one hand it could be due to an increased affinity of phenols contained in the OMW with the adsorbent; on the other hand, it could be due to the presence of oil in the OMW, which could cause the swelling of the PDMS/oxMWCNTs sponge [30,33] thus favoring the diffusion of the mixture inside the adsorbent phase. To verify the first hypothesis the PDMS/oxMWCNTs sponge was dipped in two different solutions containing two different PCs, namely 4-nitrophenol and phenol, for which the nanomaterial has evidenced different affinities at the same concentration (i.e., 0.18 mM) [28]. We observed a significant increment in the removal efficiency at equilibrium, but not in the rate of adsorption. In fact, 84% of the process is completed in 4 h in both cases (Figure S1). Therefore, the increment in the adsorption rate was attributed to the evident swelling of the nanocomposite in OMW (Figure 1b).

3.2. *Effect of pH and Adsorbent Amount on PCs Adsorption*

pH can affect both the adsorption mechanisms and the nature of soluble species' interactions with the adsorbents [34].

We tested the pH effect on the PCs removal from OMW with our porous nanocomposite. As visible in Figure 2, the removal efficiency (%) increases from pH 2 to 4.8, then slightly decreases for pH values up to 6.5 and finally decreases dramatically at higher pH, reaching a removal efficiency (%) close to zero at pH over 10.5.

The increment in removal efficiency observed until pH of around 6.5 is mainly due to π-π interactions. Although this mechanism is still not clear, it is well known that higher pH values can alter the π donation strength of PCs thus causing an increase of their adsorption on the oxMWCNTs' surface [28,35]. However, the removal efficiency decreased very fast at pHs over 6.5 with an apparent different behavior to that observed in adsorption of PCs on oxMWCNTs in aqueous solutions [28]. A similar behavior was also reported for some PCs adsorbed by MWCNTs [35], demonstrating that the decreased removal efficiency of phenols for pH over their pKa could be due to the increased electrostatic repulsion between the dissociated phenols and negatively charged oxMWCNTs. Moreover, the dissociation of the PCs would increase their hydrophilicity, thus decreasing their adsorption. Consequently, the observed trend can be explained by the fact that most of the PC constituents of OMW are deprotonated at higher pH, due to their pKa lower than 5 (Figure 3) [36]. This suggests also that the adsorption of different type of phenols could be achieved at different pHs.

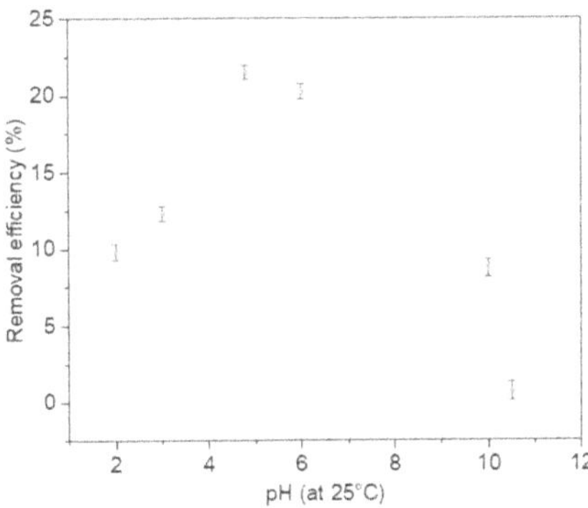

Figure 2. Effect of pH on adsorption of PCs by the PDMS/oxMWCNTs sponge.

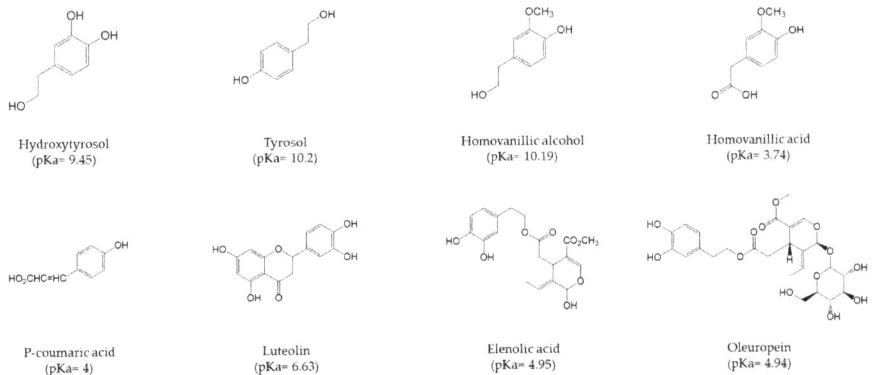

Figure 3. The major constituents of OMW. The common name and the pKa value are reported under each compound.

Figure 4 shows the removal efficiency (%) and adsorption capacity (q_e) of phenols in OMW as a function of the PDMS/oxMWCNTs sponge amount in the given conditions. It is evident that an increase in sponge amount in the mixture results in an obvious increase in the phenol adsorption percent. With the increase in PDMS/oxMWCNTs amount from 4 to 10 g/100 mL, the phenol removal efficiency increased rapidly from 20.2 to 35.5%. This was intuitively due to the increase in the number of adsorption sites with the increase in the sponge amount. The results indicated that it was possible to remove phenols completely from OMW when there was a high enough PDMS/oxMWCNTs sponge amount in the mixture. On the other hand, the q_e was high at low doses and reduced at high doses, thus suggesting that some adsorption sites remain unsaturated during the adsorption process [3,37–40].

The results of this section also indicated that, in order to obtain the optimal adsorbent dosage, higher initial phenol concentrations should be tested in conjunction with appropriate adsorbent dosage depending on the concentration of phenolic compound in OMW [37,38].

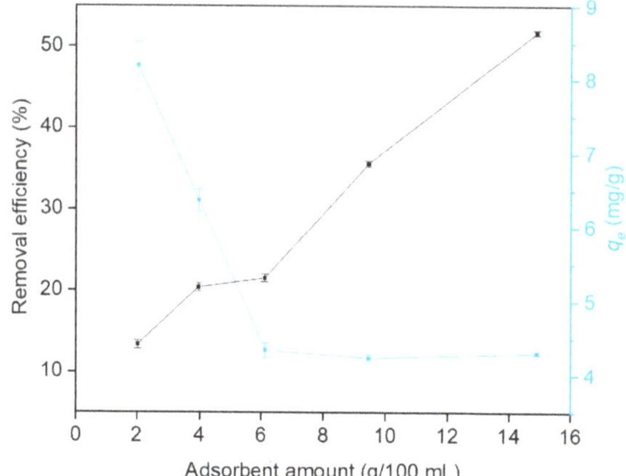

Figure 4. Influence of foam concentration on removal efficiency and adsorption capacity of phenolic compounds in OMW.

3.3. Adsorption Isotherms and Thermodynamic Parameters

Several models have been used to describe adsorption equilibrium, among which the Freundlich and the Langmuir models are the most frequently used.

The Freundlich isotherm [41] describes a reversible process in which the adsorption can occur through homogeneous and/or heterogenous interactions.

The linear form of the equation is:

$$\log q_e = \log K_F + \frac{1}{n} \log C_e \tag{3}$$

where K_F is the Freundlich constant representative of the adsorption capacity of adsorbent and n describes the strength of adsorption. An R^2 of 0.80 was obtained by linear fitting (Figure S2), evidencing that the model did not describe well the experimental data, thus being in contrast with what was observed in aqueous solutions [28]. We ascribe this behavior to the complexity of OMW in which other interferences species can influence the adsorption mechanisms.

Experimental data were also fitted with the Langmuir theory [42], which is valid for monolayer adsorption onto surfaces with homogenous binding sites.

The linearized form of the isotherm is:

$$\frac{C_e}{q_e} = \frac{1}{K_L q_{max}} + \frac{C_e}{q_{max}} \tag{4}$$

where K_L (L/mg) is the Langmuir constant and is representative of the affinity of the sorbate for the sorbent and q_{max} (mg/g) is the maximum adsorption capacity.

The R^2 value higher than 0.99 suggests that the model is more appropriate to describe the adsorption process. From the fitting, a q_{max} of 4.39 mg/g was found. The dimensionless equilibrium parameter (R_L) was calculated as in Equation (5) at different concentrations of PCs.

$$R_L = \frac{1}{1 + K_L C_0} \tag{5}$$

The results summarized in the table of Figure 5 evidenced that the R_L value is always comprised between 0 and 1, confirming the high affinity of PDMS/oxMWCNTs sponge for phenols contained in OMW [3].

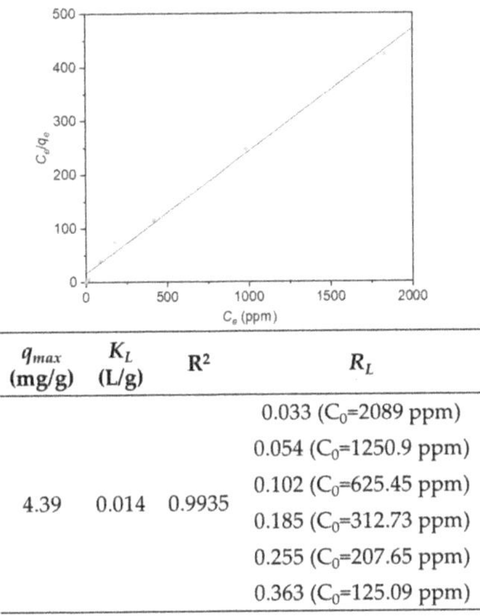

q_{max} (mg/g)	K_L (L/g)	R^2	R_L
4.39	0.014	0.9935	0.033 (C_0=2089 ppm)
			0.054 (C_0=1250.9 ppm)
			0.102 (C_0=625.45 ppm)
			0.185 (C_0=312.73 ppm)
			0.255 (C_0=207.65 ppm)
			0.363 (C_0=125.09 ppm)

Figure 5. Fitting of experimental data with the linearized Langmuir isotherm model for phenols in OMW. The table reports calculated values from Equations (6) and (7).

The obtained results were compared with other sorbents used for the same purpose (Table 1). The highest values were reached with wheat bran and banana peel [3,37]. However, it should be pointed out that in this work the adsorption capacities are calculated per gram of sponges and not per gram of oxMWCNTs. This is important since the adsorption of PCs is exclusively due to the oxMWCNTs. Therefore, calculating the q_{max} considering only the grams of oxMWCNTs a value of 454.55 mg/g is obtained, which is comparable with the most efficient materials. Moreover, the PDMS/oxMWCNTs sponge has the advantages of being user-friendly and easy to manipulate during all the steps of the waste treatment.

Table 1. Langmuir constants for PCs adsorption from various absorbents reported in literature.

Adsorbent	q_{max} (mg/g)	K_L (L/g)	References
PDMS/oxMWCNTs	4.39 (454.55)	0.014	This work
Banana peel	688.9	0.24	[3]
Wheat bran	487.3	0.13	[37]
Olive pomace	11.40	0.005	[43]
Activated carbon coated with milk protein	246.45	9.1	[44]
Activated carbon	268.17	0.14	[45]

From the variation of K_L values with temperature we calculate the Gibbs free energy ($\Delta G°$) of adsorption, enthalpy ($\Delta H°$), and entropy ($\Delta S°$) using the following equation:

$$\Delta G^{\circ} = -RT\ln(K_{LD}) \tag{6}$$

and the van't Hoff equation:

$$\ln(K_{LD}) = (\frac{\Delta S^{\circ}}{R}) - (\frac{\Delta H^{\circ}}{RT}) \tag{7}$$

in which R is the gas constant (8.314 J/(mol K)), T is the temperature expressed in Kelvin, and K_{LD} is obtained by multiplying K_L by 1000 [46,47].

The van't Hoff plot of $\ln(K_{LD})$ against 1/T evidenced a good linearity with an R^2 = 0.96 and both $\Delta H°$ and $\Delta S°$ were calculated (Figure 6). The negative values of $\Delta G°$ suggests that the process occurs spontaneously. The positive value of $\Delta H°$ represents an endothermic reaction while values of $\Delta S°$ higher than zero evidenced that the randomness increased at the solid liquid interface due to the high affinity of the sorbent for the PCs [46].

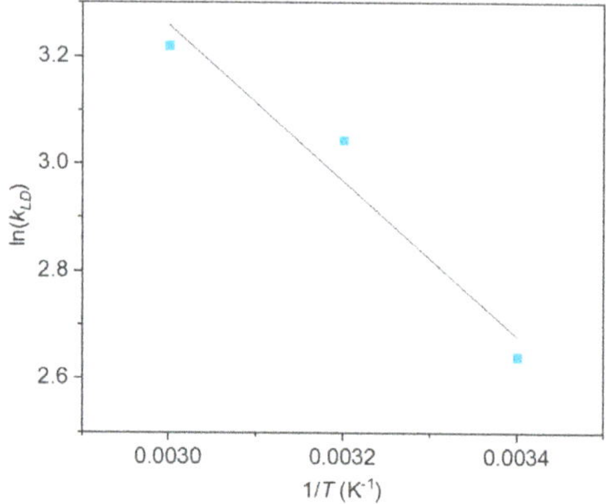

Temperature (K)	$\Delta G°$ (KJ/mol)	$\Delta H°$ (KJ/mol)	$\Delta S°$ (J/K)
298.15	-65.44		
313.15	-79.25	12.06	63.25
333.15	-89.19		

Figure 6. Van't Hoff plot for phenols adsorption from OMW. The calculated values from Equation (7) are reported in the table.

3.4. Kinetic of the Adsorption Process

To elucidate the adsorption mechanisms in OMW, adsorption kinetics have been evaluated. The mechanism of the adsorption strongly depends on the physical and chemical characteristics of the adsorbent as well as on the mass transport process. Pseudo-first-order and pseudo-second-order equations were examined in this study.

The pseudo-first-order equation [48] is represented by the following equation:

$$\ln(q_e - q_t) = \ln q_e - k_1 t \tag{8}$$

in which q_t is relative to the number of PCs adsorbed (mg/g) at any time t (h), and k_1 (h^{-1}) is the equilibrium rate constant of pseudo-first-order sorption. By plotting $\ln(q_e - q_t)$ against t a straight line should be obtained with slope $-K_1$ and intercept $\ln q_e$.

The pseudo-second-order equation [49] is expressed in the form:

$$\frac{t}{q_t} = \frac{1}{k_2 q_e^2} + \frac{t}{q_e} \tag{9}$$

in which k_2 is the rate constant of the pseudo-second-order equation (g/g h). The rate constant (k_2) and the equilibrium adsorption capacity (q_e) can be obtained from the slope and the intercept of the plot of t/q_t versus t. The experimental data fitted with both the models are reported in Figure 7.

It is evident that pseudo-second-order kinetic model better describes the experimental data ($R^2 > 0.99$), thus suggesting that chemical sorption, mainly due to π-π interactions [28], occurs between the PCs and PDMS/oxMWCNTs sponge [3]

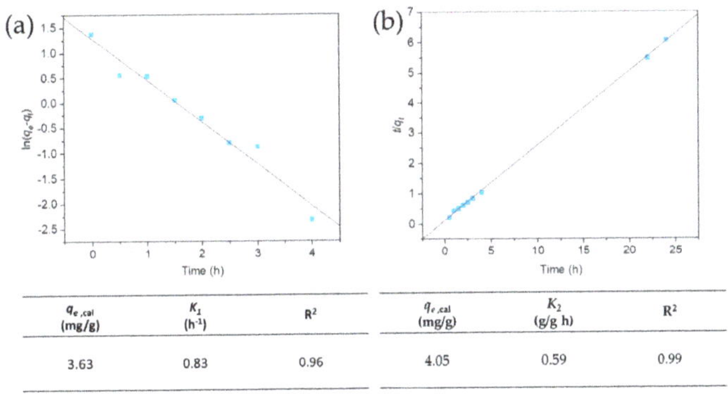

$q_{e,cal}$ (mg/g)	K_1 (h^{-1})	R^2
3.63	0.83	0.96

$q_{e,cal}$ (mg/g)	K_2 (g/g h)	R^2
4.05	0.59	0.99

Figure 7. Application of (**a**) pseudo-first-order adsorption model and (**b**) pseudo-second-order adsorption model. The calculated values from Equations (8) and (9) are reported in the tables under graphs (**a**) and (**b**) respectively.

3.5. Intraparticle Diffusion Model

The pseudo-first-order and the pseudo-second-order models can explain the adsorption process, but are not useful to identify the diffusion mechanisms.

$$q_t = k_p t^{1/2} + C \tag{10}$$

k_p is the rate constant of intraparticle diffusion model and C is a constant for any experiment (mg/g). By plotting q_t versus $t^{1/2}$ (Figure 8) two linear ranges were observed and ascribed to at least two different diffusion mechanisms of adsorption.

The lower value of k_{p2} with respect to k_{p1} (with k_{p1} and k_{p2} representing k_p values for step I and II, respectively) indicated that the free path available for diffusion of PCs inside the sponge became smaller, thus causing the reduction of the diffusion rate [50]. It can be hypothesized that firstly the adsorption occurs on the most accessible sites on the oxMWCNTs surface. Once these sites are saturated, the PCs entered into smaller pores and/or reached the binding sites in the interstitial space between oxMWCNTs, causing a decrease in the diffusion rate [28,32,51].

Interestingly, more than 95% of the total adsorption process occurs at the faster step. Moreover, a higher K_p value in the first step was obtained with respect to that obtained in aqueous solutions [28], thus confirming that the swelling of the sponge in OMW can speed up the entire adsorption process [30,33].

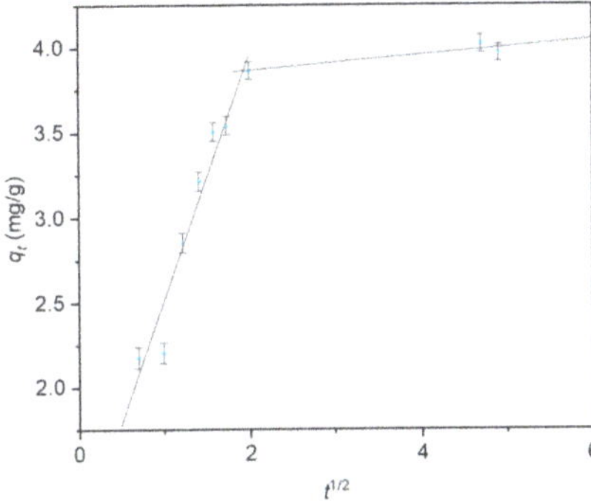

Figure 8. Application of intraparticle diffusion model for the adsorption of phenols in OMW onto PDMS/oxMWCNTs sponges.

3.6. Desorption and Reusability Studies

Since the chemisorption process occurs between phenols and oxMWCNTs, adsorbed compounds can be removed with an acidic treatment [28]. After the first adsorption process, the PDMS/oxMWCNTs sponges were washed in 10% acetic acid at 60 °C to break the π-π interactions. The washing procedure permits the solubilization of most of the adsorbed PCs (~99%) (Figure 9, blue columns), which can thus be used in phenolic-enriched foods after simple further purification steps. Furthermore, the sponge can be reused without losing its adsorption capacity, confirming the high stability of the nanocomposite despite the swelling process. This suggests the possibility to use the nanocomposite for an higher number of adsorption/desorption cycles, thus permitting the decrease of costs for the treatment of OMW with higher phenols concentrations and the complete purification of the waste that can thus be used for fertirrigation [9].

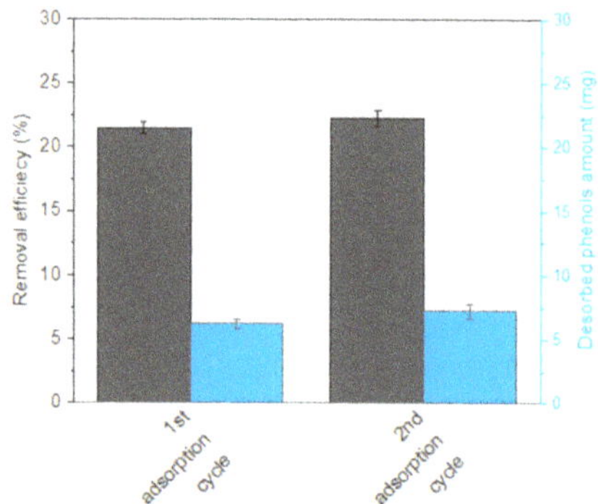

Figure 9. Reusability tests for PDMS/oxMWCNTs sponges reporting removal efficiency % (gray column) and the amount of desorbed phenols (blue column).

4. Conclusions

The present work describes the application of spongeous nanocomposites made of PDMS and oxidized MWCNTs for the adsorption of phenols from olive mill wastewater. The MWCNTs loading was performed with a straightforward method without the use of complex procedures. The oxidation of the nanomaterial was performed directly on the sponge, simplifying the post-treatment and speeding up the whole fabrication process. The entrapped nanomaterials were stable and evidenced good adsorption capacity compared to other systems. The pH of the OMW and pKa of the phenolic compounds can influence the removal efficiency of the nanocomposites. By the evaluation of thermodynamic parameters, we observed that the adsorption process is spontaneous and with a high affinity for the phenolic compounds. The adsorption process in OMW is described by the Langmuir isotherm, suggesting the formation of a monolayer on the nanomaterial surface and evidencing a different behavior of the nanocomposites with respect to what happens in standard aqueous solutions in which the formation of a PCs heterogeneous multilayer on an adsorbent surface was observed. It is also interesting to note that the presence of complex matrices such as OMW can speed up the entire adsorption process (more than 95% of the adsorption is completed in less than 4 h) with respect to standard aqueous solution. We ascribed this behavior to the presence of a small amount of oil inside the OMW that can promote the swelling of the sponge and the diffusion of the mixture inside the polymeric matrices. This is interesting since the oil seasonal production required the fast processing of OMW.

The nanocomposite can thus be easily used to work in batch conditions. This could be useful for a real application of the system. After its production, OMW is stored in a big tank in which the spongeous nanocomposites can be added for the adsorption process. Thus, the nanocomposite can be used directly *in situ*. In this view, the entrapment of oxMWCNTs adsorbent in a porous polymeric matrix represents a significant advantage; in this way, post-treatment processes, such as filtration and/or centrifugation, are not required to remove the adsorbent phase after the adsorption process, decreasing the time and the costs of the treatment. Moreover, the nanomaterial can be easily handled and disposed in a safe way without the use of specialized personnel. The surface of the nanomaterial can be regenerated with a mild treatment with diluted acetic acid. On one hand this permits us to further decrease the costs of production in wastewater treatment. On the other hand, the adsorbed phenolic compound can be easily desorbed from the sponge and can be used, after small further purification, to produce phenolic-enriched food due to its health benefits ranging from reduced incidence of cardiovascular disease, diabetes, and cancers. Furthermore, the treated OMW can be a useful source for fertirrigation. This could permit us to transform a waste product to a resource, especially for the regions of the Mediterranean area in which is concentrated most of the world's production of olive oil. Moreover, the reusability of the material can be useful for OMW with high concentration of phenols that needs repetitive cycles of purification.

Supplementary Materials: The following are available online at http://www.mdpi.com/2073-4441/12/12/3471/s1, Figure S1: Normalized removal efficiency % data to 0–1 range on the maximal value after 4 h of the uptake process and removal efficiency % at equilibrium for an aqueous solution of phenol and 4-nitrophenol. Figure S2: Fitting of experimental data with linearized Freundlich isotherm model for phenols in OMW ($R^2 = 0.8$).

Author Contributions: Conceptualization, A.T.; methodology, A.T. and C.M.; validation, A.T. and C.M.; formal analysis, A.T. and C.M.; investigation, A.T.; resources, A.T. and C.M.; data curation, A.T. and C.M.; writing—original draft preparation, A.T.; writing—review and editing, A.T. and C.M.; supervision, A.T. and C.M.; funding acquisition, A.T. and C.M. All authors have read and agreed to the published version of the manuscript.

Funding: This research was funded by Fondazione CARIPUGLIA, project: "Materiali innovativi porosi nanocompositi per la rimozione e il recupero di composti fenolici da acque di vegetazione olearie" and Cohesion fund 2007–2013–APQ Ricerca Regione Puglia "Programma regionale a sostegno della specializzazione intelligente e della sostenibilità sociale ed ambientale-FutureInResearch" under Grant No. 9EC1495 (Ultrasensitive sensor for food analysis).

Conflicts of Interest: The authors declare no conflicts of interest. The funders had no role in the design of the study; in the collection, analyses, or interpretation of data; in the writing of the manuscript, or in the decision to publish the results.

References

1. El-Abbassi, A.; Hafidi, A.; García-Payo, M.C.; Khayet, M. Concentration of olive mill wastewater by membrane distillation for polyphenols recovery. *Desalination* **2009**, *245*, 670–674. [CrossRef]
2. Della Greca, M.; Monaco, P.; Pinto, G.; Pollio, A.; Previtera, L.; Temussi, F. Phytotoxicity of Low-Molecular-Weight Phenols from Olive Mill Waste Waters. *Bull. Environ. Contam. Toxicol.* **2001**, *67*, 352–359. [CrossRef] [PubMed]
3. Achak, M.; Hafidi, A.; Ouazzani, N.; Sayadi, S.; Mandi, L. Low cost biosorbent "banana peel" for the removal of phenolic compounds from olive mill wastewater: Kinetic and equilibrium studies. *J. Hazard. Mater.* **2009**, *166*, 117–125. [CrossRef] [PubMed]
4. Sklavos, S.; Gatidou, G.; Stasinakis, A.S.; Haralambopoulos, D. Use of solar distillation for olive mill wastewater drying and recovery of polyphenolic compounds. *J. Environ. Manag.* **2015**, *162*, 46–52. [CrossRef] [PubMed]
5. Dutournié, P.; Jeguirim, M.; Khiari, B.; Goddard, M.-L.; Jellali, S. Olive Mill Wastewater: From a Pollutant to Green Fuels, Agricultural Water Source, and Bio-Fertilizer. Part 2: Water Recovery. *Water* **2019**, *11*, 768. [CrossRef]
6. Schaffer, S.; Podstawa, M.; Visioli, F.; Bogani, P.; Müller, W.E.; Eckert, G.P. Hydroxytyrosol-Rich Olive Mill Wastewater Extract Protects Brain Cells in Vitro and ex Vivo. *J. Agric. Food Chem.* **2007**, *55*, 5043–5049. [CrossRef]
7. Visioli, F.; Galli, C. The Effect of Minor Constituents of Olive Oil on Cardiovascular Disease: New Findings. *Nutr. Rev.* **2009**, *56*, 142–147. [CrossRef]
8. Bulotta, S.; Celano, M.; Lepore, S.M.; Montalcini, T.; Pujia, A.; Russo, D. Beneficial effects of the olive oil phenolic components oleuropein and hydroxytyrosol: Focus on protection against cardiovascular and metabolic diseases. *J. Transl. Med.* **2014**, *12*, 219. [CrossRef]

9. Colarieti, M.L.; Toscano, G.; Greco, G. Toxicity attenuation of olive mill wastewater in soil slurries. *Environ. Chem. Lett.* **2006**, *4*, 115–118. [CrossRef]
10. Abdelwahab, O.; Nassef, E.M. Treatment of Petrochemical Wastewater Containing Phenolic Compounds by Electrocoagulation Using a Fixed Bed Electrochemical Reactor. *Int. J. Electrochem. Sci.* **2013**, *8*, 1534–1550.
11. Jerman Klen, T.; Mozetič Vodopivec, B. Ultrasonic Extraction of Phenols from Olive Mill Wastewater: Comparison with Conventional Methods. *J. Agric. Food Chem.* **2011**, *59*, 12725–12731. [CrossRef] [PubMed]
12. Pelendridou, K.; Michailides, M.K.; Zagklis, D.P.; Tekerlekopoulou, A.G.; Paraskeva, C.A.; Vayenas, D.V. Treatment of olive mill wastewater using a coagulation–flocculation process either as a single step or as post-treatment after aerobic biological treatment. *J. Chem. Technol. Biotechnol.* **2014**, *89*, 1866–1874. [CrossRef]
13. Hosseini, S.A.; Davodian, M.; Abbasian, A.R. Remediation of phenol and phenolic derivatives by catalytic wet peroxide oxidation over Co-Ni layered double nano hydroxides. *J. Taiwan Inst. Chem. Eng.* **2017**, *75*, 97–104. [CrossRef]
14. Domingues, E.; Assunção, N.; Gomes, J.; Lopes, D.V.; Frade, J.R.; Quina, M.J.; Quinta-Ferreira, R.M.; Martins, R.C. Catalytic Efficiency of Red Mud for the Degradation of Olive Mill Wastewater through Heterogeneous Fenton's Process. *Water* **2019**, *11*, 1183. [CrossRef]
15. Amor, C.; Marchão, L.; Lucas, M.S.; Peres, J.A. Application of advanced oxidation processes for the treatment of recalcitrant agro-industrial wastewater: A review. *Water* **2019**, *11*, 205. [CrossRef]
16. Garcia-Segura, S.; Bellotindos, L.M.; Huang, Y.-H.; Brillas, E.; Lu, M.-C. Fluidized-bed Fenton process as alternative wastewater treatment technology—A review. *J. Taiwan Inst. Chem. Eng.* **2016**, *67*, 211–225. [CrossRef]
17. Ran, N.; Gilron, J.; Sharon-Gojman, R.; Herzberg, M. Powdered Activated Carbon Exacerbates Fouling in MBR Treating Olive Mill Wastewater. *Water* **2019**, *11*, 2498. [CrossRef]
18. Kapellakis, I.; Tzanakakis, V.A.; Angelakis, A.N. Land Application-Based Olive Mill Wastewater Management. *Water* **2015**, *7*, 362–376. [CrossRef]
19. Paraskeva, P.; Diamadopoulos, E. Technologies for olive mill wastewater (OMW) treatment: A review. *J. Chem. Technol. Biotechnol.* **2006**, *81*, 1475–1485. [CrossRef]
20. Bernal, V.; Giraldo, L.; Moreno-Piraján, J.C. Insight into adsorbate–adsorbent interactions between aromatic pharmaceutical compounds and activated carbon: Equilibrium isotherms and thermodynamic analysis. *Adsorption* **2020**, *26*, 153–163. [CrossRef]
21. Takahashi, K.; Yoshida, S.; Urkasame, K.; Iwamura, S.; Ogino, I.; Mukai, S.R. Carbon gel monoliths with introduced straight microchannels for phenol adsorption. *Adsorption* **2019**, *25*, 1241–1249. [CrossRef]
22. Duy Nguyen, H.; Nguyen Tran, H.; Chao, H.-P.; Lin, C.-C. Activated Carbons Derived from Teak Sawdust-Hydrochars for Efficient Removal of Methylene Blue, Copper, and Cadmium from Aqueous Solution. *Water* **2019**, *11*, 2581. [CrossRef]
23. Nikić, J.; Tubić, A.; Watson, M.; Maletić, S.; Šolić, M.; Majkić, T.; Agbaba, J. Arsenic Removal from Water by Green Synthesized Magnetic Nanoparticles. *Water* **2019**, *11*, 2520. [CrossRef]
24. Nabeela Nasreen, S.A.A.; Sundarrajan, S.; Syed Nizar, S.A.; Ramakrishna, S. Nanomaterials: Solutions to water-concomitant challenges. *Membranes* **2019**, *9*, 40. [CrossRef] [PubMed]
25. Nasreen, S.A.A.N.; Sundarrajan, S.; Nizar, S.A.S.; Balamurugan, R.; Ramakrishna, S. Advancement in electrospun nanofibrous membranes modification and their application in water treatment. *Membranes* **2013**, *3*, 266–284. [CrossRef] [PubMed]
26. Aly, A.A.; Hasan, Y.N.Y.; Al-Farraj, A.S. Olive mill wastewater treatment using a simple zeolite-based low-cost method. *J. Environ. Manag.* **2014**, *145*, 341–348. [CrossRef] [PubMed]
27. Frascari, D.; Bacca, A.E.M.; Zama, F.; Bertin, L.; Fava, F.; Pinelli, D. Olive mill wastewater valorisation through phenolic compounds adsorption in a continuous flow column. *Chem. Eng. J.* **2016**, *283*, 293–303. [CrossRef]
28. Turco, A.; Monteduro, A.; Mazzotta, E.; Maruccio, G.; Malitesta, C. An Innovative Porous Nanocomposite Material for the Removal of Phenolic Compounds from Aqueous Solutions. *Nanomaterials* **2018**, *8*, 334. [CrossRef]
29. Singleton, V.L.; Orthofer, R.; Lamuela-Raventós, R.M. Analysis of total phenols and other oxidation substrates and antioxidants by means of folin-ciocalteu reagent. *Methods Enzymol.* **1999**, *299*, 152–178. [CrossRef]
30. Turco, A.; Malitesta, C.; Barillaro, G.; Greco, A.; Maffezzoli, A.; Mazzotta, E. A magnetic and highly reusable macroporous superhydrophobic/superoleophilic PDMS/MWNT nanocomposite for oil sorption from water. *J. Mater. Chem. A* **2015**, *3*, 17685–17696. [CrossRef]
31. Delor-Jestin, F.; Tomer, N.S.; Singh, R.P.; Lacoste, J. Durability of crosslinked polydimethylsyloxanes: The case of composite insulators. *Sci. Technol. Adv. Mater.* **2008**. [CrossRef] [PubMed]

32. Agnihotri, S.; Mota, J.P.B.; Rostam-Abadi, M.; Rood, M.J. Theoretical and experimental investigation of morphology and temperature effects on adsorption of organic vapors in single-walled carbon nanotubes. *J. Phys. Chem. B* **2006**, *110*, 7640–7647. [CrossRef] [PubMed]
33. Turco, A.; Primiceri, E.; Frigione, M.; Maruccio, G.; Malitesta, C. An innovative, fast and facile soft-template approach for the fabrication of porous PDMS for oil–water separation. *J. Mater. Chem. A* **2017**, *5*, 23785–23793. [CrossRef]
34. Turco, A.; Pennetta, A.; Caroli, A.; Mazzotta, E.; Monteduro, A.G.; Primiceri, E.; de Benedetto, G.; Malitesta, C. Easy fabrication of mussel inspired coated foam and its optimization for the facile removal of copper from aqueous solutions. *J. Colloid Interface Sci.* **2019**, *552*, 401–411. [CrossRef] [PubMed]
35. Lin, D.; Xing, B. Adsorption of phenolic compounds by carbon nanotubes: Role of aromaticity and substitution of hydroxyl groups. *Environ. Sci. Technol.* **2008**, *42*, 7254–7259. [CrossRef]
36. Daâssi, D.; Lozano-Sánchez, J.; Borrás-Linares, I.; Belbahri, L.; Woodward, S.; Zouari-Mechichi, H.; Mechichi, T.; Nasri, M.; Segura-Carretero, A. Olive oil mill wastewaters: Phenolic content characterization during degradation by *Coriolopsis gallica*. *Chemosphere* **2014**, *113*, 62–70. [CrossRef] [PubMed]
37. Achak, M.; Hafidi, A.; Mandi, L.; Ouazzani, N. Removal of phenolic compounds from olive mill wastewater by adsorption onto wheat bran. *Desalin. Water Treat.* **2014**, *52*, 2875–2885. [CrossRef]
38. Lin, K.; Pan, J.; Chen, Y.; Cheng, R.; Xu, X. Study the adsorption of phenol from aqueous solution on hydroxyapatite nanopowders. *J. Hazard. Mater.* **2009**, *161*, 231–240. [CrossRef]
39. Han, R.; Zou, W.; Li, H.; Li, Y.; Shi, J. Copper (II) and lead (II) removal from aqueous solution in fixed-bed columns by manganese oxide coated zeolite. *J. Hazard. Mater.* **2006**, *137*, 934–942. [CrossRef]
40. Ho, Y.S.; Chiang, C.C. Sorption studies of acid dye by mixed sorbents. *Adsorption* **2001**, *7*, 139–147. [CrossRef]
41. Freundlich, H.M. Over the adsorption in solution. *J. Physicochem.* **1906**, *57*, 385.
42. Langmuir, I. The constitution and fundamental properties of solids and liquids. Part I. Solids. *J. Am. Chem. Soc.* **1916**, *38*, 2221–2295. [CrossRef]
43. Stasinakis, A.S.; Elia, I.; Petalas, A.V.; Halvadakis, C.P. Removal of total phenols from olive-mill wastewater using an agricultural by-product, olive pomace. *J. Hazard. Mater.* **2008**, *160*, 408–413. [CrossRef] [PubMed]
44. Yangui, A.; Abderrabba, M. Towards a high yield recovery of polyphenols from olive mill wastewater on activated carbon coated with milk proteins: Experimental design and antioxidant activity. *Food Chem.* **2018**, *262*, 102–109. [CrossRef]
45. Senol, A.; Hasdemir, İ.M.; Hasdemir, B.; Kurdaş, İ. Adsorptive removal of biophenols from olive mill wastewaters (OMW) by activated carbon: Mass transfer, equilibrium and kinetic studies. *Asia-Pacific J. Chem. Eng.* **2017**, *12*, 128–146. [CrossRef]
46. Ghosal, P.S.; Gupta, A.K. Determination of thermodynamic parameters from Langmuir isotherm constant-revisited. *J. Mol. Liq.* **2017**, *225*, 137–146. [CrossRef]
47. Milonjić, S.K. A consideration of the correct calculation of thermodynamic parameters of adsorption. *J. Serb. Chem. Soc.* **2007**, *72*, 1363–1367. [CrossRef]
48. Lagergren, S. Zur theorie der sogenannten adsorption geloster stoffe, Kungliga Svenska Vetenskapsakademiens. *Handlingar* **1898**, *24*, 1–39.
49. Blanchard, G.; Maunaye, M.; Martin, G. Removal of heavy metals from waters by means of natural zeolites. *Water Res.* **1984**, *18*, 1501–1507. [CrossRef]
50. Cheng, C.S.; Deng, J.; Lei, B.; He, A.; Zhang, X.; Ma, L.; Li, S.; Zhao, C. Toward 3D graphene oxide gels-based adsorbents for high-efficient water treatment via the promotion of biopolymers. *J. Hazard. Mater.* **2013**, *263*, 467–478. [CrossRef]
51. Pham, X.-H.; Li, C.A.; Han, K.N.; Huynh-Nguyen, B.-C.; Le, T.-H.; Ko, E.; Kim, J.H.; Seong, G.H. Electrochemical detection of nitrite using urchin-like palladium nanostructures on carbon nanotube thin film electrodes. *Sens. Actuators B Chem.* **2014**, *193*, 815–822. [CrossRef]

Publisher's Note: MDPI stays neutral with regard to jurisdictional claims in published maps and institutional affiliations.

Article

Adsorption of Mixed Dye System with Cetyltrimethylammonium Bromide Modified Sepiolite: Characterization, Performance, Kinetics and Thermodynamics

Jian Yu [1], Aiyi Zou [1], Wenting He [1] and Bin Liu [1,2,*]

[1] Department of Water Engineering and Science, College of Civil Engineering, Hunan University, Changsha 410082, China; yujian@hnu.edu.cn (J.Y.); zouaiyi@hhu.edu.cn (A.Z.); lgc@hnu.edu.cn (W.H.)

[2] Department of Chemical Engineering, Process Engineering for Sustainable Systems (ProcESS), KU Leuven, Celestijnenlaan 200F, B-3001 Leuven, Belgium

* Correspondence: ahxclb@163.com

Received: 2 March 2020; Accepted: 28 March 2020; Published: 30 March 2020

Abstract: In this study, sepiolite was modified by calcination (200 °C) and cetyltrimethylammonium bromide (CTMAB) treatment. Though the specific surface area sharply declined, the adsorption amount of Acid Orange II (AO), Reactive Blue (RB), Acid Fuchsin (AR) and their mixed solution were improved. The morphology of modified sepiolite showed a better dispersibility and looser structure. The adsorption performance was highly impacted by the pH condition and adsorbent dosage. The electrostatic attraction of positively charged adsorption sites on the adsorbent surface and the negatively charged anionic dye could enhance the adsorption amount especially under acid condition. The order of preferentially adsorbed dye was AO > RB > AR. The adsorption process was much correlated to the quasi-second-order reaction kinetics. The adsorption amount and equilibrium amount of single dye system, as well as in the mixed system were in accordance with the Langmuir model and extended Langmuir isotherm.

Keywords: anionic dye; cetyltrimethylammonium bromide; sepiolite; two-step modification; adsorption

---

1. Introduction

Synthetic dyes are common pollutants in industrial wastewater. They are stable, highly soluble and when enter in the water bodies present environmental hazards and potential threats to health [1]. Some of the azo dyes have been found to be carcinogenic, sensitizing and reproductive to humans; some of the triarylmethane dyes are also carcinogenic and have long-term adverse effects in the aquatic environment; some of the anthraquinone dyes are chemically stable and refractory to biodegradable [2–4]. Hence, Acid Orange II (azo dye), Reactive Blue (anthraquinone dye) and Acid Fuchsin (triarylmethane dye) were employed in this study to investigate the various kinds of dye.

The treatment methods for dye-bearing wastewater include physical methods, chemical methods and biological methods [5–11]. Among them, adsorption decolorization is the most effective method [12]. According to the different interaction forces between the adsorbate and the adsorbent, the adsorption can be divided into physical adsorption and chemical adsorption [13]. The physical adsorption is caused by the intermolecular force (Van der Waals force) [14]. Chemical adsorption was caused by the formation of chemical bonds or surface coordination compounds by adsorbate molecules and adsorbents by means of ion exchange, electron transfer and electron pair sharing [15,16].

Sepiolite is a hydrous magnesium silicate with a fibrous cross-section [17,18]. The merits of sepiolite were low thermal conductivity, high salt resistance, non-polluting, environmentally friendly [19]. It has been widely used in various fields due to the special structural properties and low cost. In China, the sepiolite was presented with two types, hydrothermal type and clay type. The clay type is mainly distributed in Hunan province [17]. The water in sepiolite mainly exists with three forms, adsorbing water, coordinating water and hydroxyl water, in which the adsorption water enters the sepiolite pores, the coordinating water is mainly bound by Mg^{2+} and the hydroxyl water exists as OH group [20]. The structure of sepiolite was greatly impacted by the bound water, especially after heated [21]. The three forms of water are gradually lost and the structure of sepiolite will change differently at various temperatures. Serna et al found that the adsorbed water in the sepiolite pores was mainly lost and the specific surface area of the pores was increased at 25~250 °C, but the structure of the sepiolite is folded when the temperature was higher than 300 °C [22].

The adsorbent generally has the following characteristics, large specific surface area, suitable pore structure and surface structure, high adsorption capacity, no chemically react with the medium, good mechanical strength, etc. [19]. The sepiolite, a fibrous hydrous magnesium silicate, has been widely explored as an adsorbent by two reasons: suitable specific surface area; low price, only 20–30 dollars per ton in China. There are three kinds of adsorption active sites from sepiolite: oxygen atom in silicon tetrahedron, water molecules coordinated with magnesium ions, which can form hydrogen bonds with adsorbate and Si-OH combination, formed after the destruction of Si-O-Si.

In order to further improve the adsorption performance of sepiolite, modifications such as thermal acid treatment, surfactant organic and inorganic modifications have been employed [11]. Mahir Alkan et al found thermally modification at 200, the highest adsorption amount was obtained [23]. A wider temperature range of the sepiolite modification was reported by Jiquan Wang et al [24]. The surfactant modification also attracts researcher's attention. Bulent Armagan et al used an organic modification method of natural sepiolite with cetyltrimethylammonium bromide to investigate its adsorption property for anionic dyes (active black, reactive red, active yellow) [25].

In this study, to improve the adsorption property of sepiolite, a combination of heat and surfactant modification method was applied. To confirm the success of this combined method, unmodified and modified forms of sepiolite were characterized by BET, XRD, XPS, FT-IR and SEM. On the other hand, this study also focused on the adsorption performance and mechanism of mixed dye system with modified sepiolite since the kinetics and thermodynamics of two-component and three-component solution were much more complicated. Hence, the three dyes, Acid Orange II, Reactive Blue and Acid Fuchsin were employed as target adsorbate with single, two-component and three-component solution type.

2. Method and Material

2.1. Materials

In this study, the original sepiolite was obtained from Guangda sepiolite company (Hunan province, China). The three dyes, Acid Orange II (AO), Reactive Blue (RB) and Acid Fuchsin (AR) was purchased from Sigma company and the detail information can be found in previous work [26–28]. The Cetyltrimethylammonium bromide (CTMAB), with the formula of $C_{16}H_{33}(CH_3)_3NBr$, was purchased from Tianjin Bodi Chemical company. The chemical agents were all above analytical grade in this work. To compare the adsorption amount and specific surface area, commercial powder activated carbon (granularity of 180–220 μm) purchased from Xingyuan company (Hunan province, China) was utilized.

2.2. Preparation of Organically Modified Sepiolite

Before the modification process, the sepiolite was first settled and purified, then dried in an oven at 120 °C and sieved with a 100 mesh. The sepiolite was calcined under 200 °C for 2 h in a muffle. Then, the sepiolite was immersed into CTMAB-saturated solution for 8 h. To remove the CTMAB solution, the modified sepiolite was repeatedly cleaned and settled. Finally, the modified sepiolite was placed into a 60 °C vacuum oven for 24 h.

2.3. Dye Concentration Determination and Adsorption Test

In this study, a multiple linear regression method was used to process the data. When the photometric system satisfied the linear sum of absorbance, the concentration of each component can be obtained by solving linear equations. Suppose dye wastewater contains dyes with n different absorbance components without mutual react. The sum of the absorbance of each single-component dye is the total absorbance of the mixed dye. It is measured at different wavelengths according to Beer-Lambert law. The absorbance of the mixed dye can be expressed by Equation (1).

$$\begin{vmatrix} A_1 \\ A_2 \\ \cdots \\ A_n \end{vmatrix} = \begin{vmatrix} k_{11}k_{12}\ldots k_{1m} \\ k_{21}k_{22}\ldots k_{2m} \\ \cdots \\ k_{n1}k_{n2}\ldots k_{nm} \end{vmatrix} \begin{vmatrix} C_1 \\ C_2 \\ \cdots \\ C_m \end{vmatrix} \quad (1)$$

where A_i and a_i presents the absorbance at λ_i, k_{ij} was the absorption coefficient of the jth component at wavelength λ_i and C_i is the concentration of the ith component in the sample. To determine the concentration of individual dye, the mixed solution was scanned in the wavelength region of 594–474 nm; the results of dye concentration were calculated using MATLAB software.

A lab scale static adsorption experiment was utilized to investigate the adsorption performance of modified sepiolite with the dye system. A beaker with 300 mL target solution was employed in the experiment. The adsorption test was conducted with 2 g/L sepiolite and 200 mg/L dye solution for 240 min under pH value of 1, oscillation rate of 180 r/min and 25 °C. Considering the effect of pH and adsorbent dosage on adsorption performance, pH range from 1 to 9, dosage range from 0.5 to 5 g/L were utilized. For the mixed dye system, for instance, the RB+AO+AR represents mixed dye of 200 mg/L RB, 200 mg/L AO and 200 mg/L AR, the RB+AO represents mixed dye of 200 mg/L RB and 200 mg/L AO. AO (RB + AO + AR) means the concentration of AO in the three-component system.

After the determination of fixed dye, the decolorization percentage and adsorption amount were calculated by Equations (2) and (3).

$$\eta = \frac{(C_0 - C_e)}{C_0} \times 100\% \quad (2)$$

$$q = \frac{(C_0 - C_e)}{m} V \quad (3)$$

where η and q are the percentage of decolorization and adsorption amount of fixed dye, respectively, C_0 and C_e are the initial and final concentration of fixed dye, V is the dye solution volume and m is the weight of the sepiolite. In order to ensure the accuracy of the test data, each group of experiment was repeated at least 3 times.

2.4. Adsorption Kinetics and Thermodynamics

The kinetic characteristics of mixed dye removal with sepiolite adsorption were discussed by using the quasi-first-order and quasi-secondary kinetics models. The mathematical expressions of quasi-first-order and quasi-secondary kinetic models were listed in Equations (4) and (5), respectively [29].

quasi-first-order model

$$\ln(q_e - q_t) = \ln q_e - k_1 t \quad (4)$$

quasi-secondary-order model

$$(\frac{1}{q_t}) = \frac{1}{k_2 q^2{}_e} + \frac{1}{q_e} t \quad (5)$$

where q_e and q_t represent the amount of fixed dye adsorbed by the unit mass adsorbent after equilibrium and at the moment t, k_1 and k_2 represent the adsorption rate constant of the quasi-first-order reaction kinetic equation and the quasi-second-order reaction kinetic equation.

The thermodynamics characteristics of dye solution adsorption were discussed by Langmuir linear fitting model. The mathematical expression of the thermodynamics model was listed in Equation (6).

Langmuir linear fitting model

$$\frac{C_e}{q_e} = \frac{1}{q_{\max}K_L} + \frac{C_e}{q_{\max}} \tag{6}$$

where q_e is the balanced amount of adsorption, $q_{\max}$ is the saturated adsorption capacity of a single layer, c_e is the balanced mass concentration of the dye, K_L is the adsorption equilibrium constant in Langmuir linear fitting model.

In the two-component system, the Langmuir fitting model was transformed to Equations (7) and (8) [30,31]. Where the subscript 1 and 2 present component 1 and component 2 in the mixed system. In this extended Langmuir model, the $\frac{c_{e1}}{q_{e1}}$ shows linear correlation with c_{e1} and $\frac{q_{e2}c_{e1}}{q_{e1}}$.

$$\frac{C_{e1}}{q_{e1}} = \frac{1}{q_{\max1}K_{L1}} + \frac{C_{e1}}{q_{\max1}} + \frac{q_{e2}C_{e1}}{q_{e1}q_{\max2}} \tag{7}$$

$$\frac{C_{e2}}{q_{e2}} = \frac{1}{q_{\max2}K_{L2}} + \frac{C_{e2}}{q_{\max2}} + \frac{q_{e1}C_{e2}}{q_{e1}q_{\max2}} \tag{8}$$

In the three-component system, the Langmuir fitting model was transformed to Equations (9)–(11) [30,31]. Where the subscript 1, 2 and 3 present component 1, component 2 and component 3 in the mixed system. In this extended Langmuir model, the $\frac{C_{e1}}{q_{e1}}$ shows linear correlation with C_{e1}, $\frac{q_{e2}C_{e1}}{q_{e1}}$ and $\frac{q_{e3}C_{e1}}{q_{e1}}$.

$$\frac{C_{e1}}{q_{e1}} = \frac{1}{q_{\max1}K_{L1}} + \frac{C_{e1}}{q_{\max1}} + \frac{q_{e2}C_{e1}}{q_{e1}q_{\max2}} + \frac{q_{e3}C_{e1}}{q_{e1}q_{\max3}} \tag{9}$$

$$\frac{C_{e2}}{q_{e2}} = \frac{1}{q_{\max2}K_{L2}} + \frac{C_{e2}}{q_{\max2}} + \frac{q_{e1}C_{e2}}{q_{e2}q_{\max1}} + \frac{q_{e3}C_{e2}}{q_{e2}q_{\max3}} \tag{10}$$

$$\frac{C_{e3}}{q_{e3}} = \frac{1}{q_{\max3}K_{L3}} + \frac{C_{e3}}{q_{\max3}} + \frac{q_{e1}C_{e3}}{q_{e3}q_{\max1}} + \frac{q_{e2}C_{e3}}{q_{e3}q_{\max2}} \tag{11}$$

2.5. Characterization

A Brunner–Emmet–Teller (BET, QuadraSorb SI, Quantachrome, USA) was utilized for the measurement of specific surface area [32]. An X-ray diffraction spectrometer (D8-Advance, Burke, Germany) was employed for X-ray diffraction spectroscopy (XRD) test [33]. The tube pressure was 20 KV, the scanning range was 5–60° and the scanning speed was 0.02°/0.2 s [34]. A Fourier-transform infrared spectrometer (Spectrum One B, Perkin Elmer, USA) was used to detect the Fourier-transform infrared spectroscopy (FT-IR) [33]. A scanning electron microscopy (Quanta 200, FEI, Hillsboro, OR, USA) was employed to observe the feature of modified sepiolite [35].

3. Result and Discussion

3.1. Characterization of Modified Sepiolite

Figure 1a shows that the adsorption capacity of calcined sepiolite and modified sepiolite were much higher than that of sepiolite ore. The adsorption capacity obviously improved from 45.28 mg/g to 74.82 mg/g and 78.66 mg/g after calcination and CTMAB modification, respectively. The previous study had reported that the calcination process at a suitable temperature range could enhance the adsorption amount of sepiolite, but the enhancement was undermined at too high temperature. Moreover organic modification could further improve the adsorption capacity.

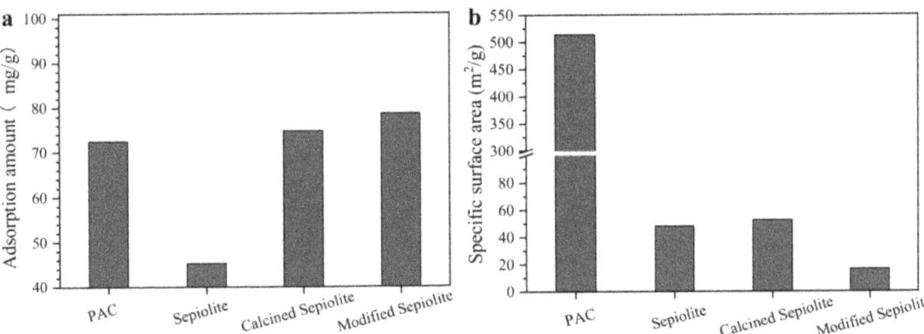

Figure 1. Comparison of Acid Orange II (AO) adsorption amount (**a**) and specific surface area (**b**) of different adsorbents.

Figure 1b shows that the specific surface area of calcined sepiolite was slightly increased, but significantly declined after surfactant modification. Though the specific surface area of CTMAB modified sepiolite highly decreased, the adsorption amount was nearly twice when compared to the sepiolite ore. The decrease of electrostatic repulsion of the adsorption site by CTMAB was the reason for adsorption enhancement. Another interesting point was the adsorption amount of organically modified sepiolite was comparable with the powdered activated carbon (PAC), but the specific surface area of PAC was forty times large than the modified sepiolite [36]. This suggests that the specific surface area was not in line with the adsorption capacity, the property of valid adsorption site was the key factor for adsorption performance.

Figure 2a shows that the characteristic dispersion peak of sepiolite at 2θ value of 9.4/9.44 was significantly strengthened after CTMAB modification and no significant changes occurred at other characteristic peak positions. On the other hand, the diffraction peak of the quartz also enhanced after the surfactant modification, which was different with previous report [34]. The d_{001} value of modified sepiolite was slightly increased to 0.9365 nm, which indicate that the CTMAB intercalated into the sepiolite.

The FT-IR spectrums of original sepiolite, modified sepiolite and modified sepiolite after adsorption saturated are shown in Figure 2b. The 3670 cm^{-1}, 3622 cm^{-1}, 3411 cm^{-1} and 1796 cm^{-1} peaks corresponding to the stretching vibration of O-H bond existed, which reflects the structure of the sepiolite itself. The peak at 3670 cm^{-1} is the stretching vibration of OH (belonging to Mg-OH) which coordinated by the Mg ion octahedron in the sepiolite pore channel. The peak at 3622 cm^{-1} is the crystal water with weak hydrogen bonding and peak at 3411 cm^{-1} is the stretching vibration of the hydroxyl group adsorbing water. The vibration at 1030 cm^{-1} is the stretching vibration of the Si-O group and the absorption band near 796 cm^{-1}, 669 cm^{-1} and 471 cm^{-1} belong to the bending expansion vibration of Si-O and Mg-O in the octahedron. When comparing the spectra of original sepiolite and modified sepiolite, the Si-O and Mg-O stretching vibration peaks of the organically modified sepiolite were enhanced and the peaks at 2920 cm^{-1}, 2856 cm^{-1} and 1420 cm^{-1} appeared [37]. The peaks at 2920 cm^{-1} and 2856 cm^{-1} were corresponding to the stretching vibration of CH_3 and CH_2 from CTMAB, while 1420 cm^{-1} represented the vibration of the C-N. This indicates that cetyltrimethylammonium bromide was electrostatically coated on the surface of sepiolite, which could change the surface of sepiolite from hydrophilic to lipophilic [38]. When comparing the spectra before and after adsorption, the characteristic peak (1500–1600 cm^{-1}) of AO appeared on the sepiolite after adsorption, which indicates the surface adsorption distribution occurs during the adsorption process. The XPS spectrum (Figure 2c) shows that the characteristic peaks of Mg and Si decreased while the characteristic peak of C enhanced after CTMAB modification.

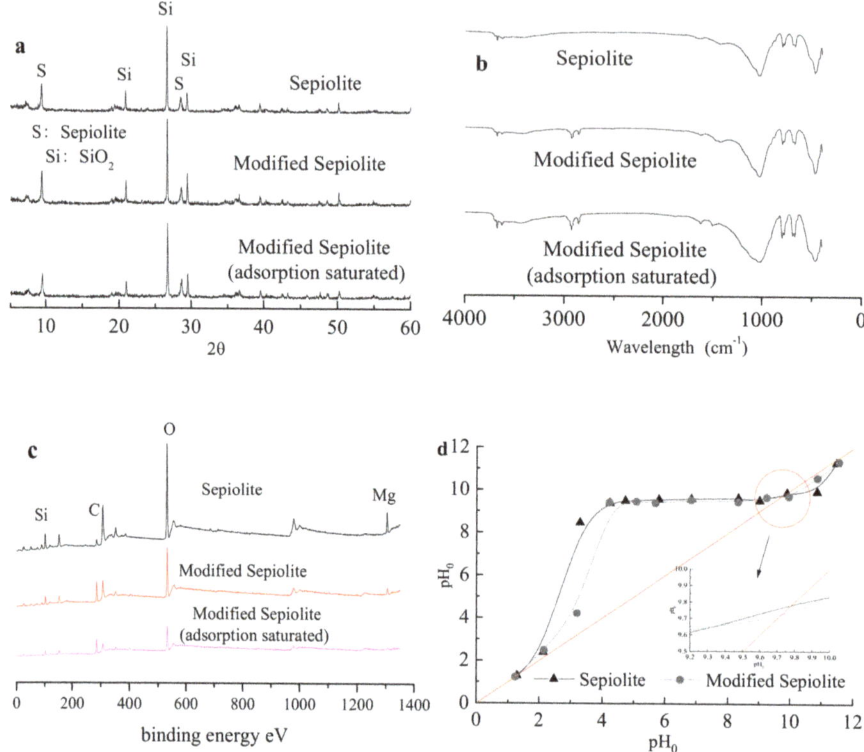

Figure 2. (**a**) X-ray diffraction spectrum of original sepiolite, modified sepiolite and modified sepiolite after adsorption saturated; (**b**) Fourier-transform infrared spectrum of original sepiolite, modified sepiolite and modified sepiolite after adsorption saturated; (**c**) X-ray photoelectron spectrum of original sepiolite, modified sepiolite and modified sepiolite after adsorption saturated; (**d**) Isoelectric point of original sepiolite and modified sepiolite.

The isoelectric point of the sepiolite before and after the modification was shown in Figure 2d. The isoelectric point changed from 9.62 to 9.87 after modification, which indicates that the isoelectric point value of the surface increased and the surface was positive charged. The isoelectric point is important when investigating the effect of pH on the adsorption effect. When the solution pH is lower than the isoelectric point, the surface of the adsorbent was positively charged. At lower pH condition, the surface of the sepiolite was combined with H^+ and the surface was positively charged, while at higher pH condition, the surface of the sepiolite was combined with OH^- and the surface was negatively charged.

$$M - OH + H^+ \rightarrow MOH_2{}^+$$

$$M - OH + OH^- \rightarrow MO^- H_2O$$

The morphology images confirm that the original crystal structure of the sepiolite was retained after the modification process. Better dispersibility and looser morphology were obtained after modification. The diameter of the fiber increased and the gap between the layered structure was cleaner and smoother (Figure 3a,b).

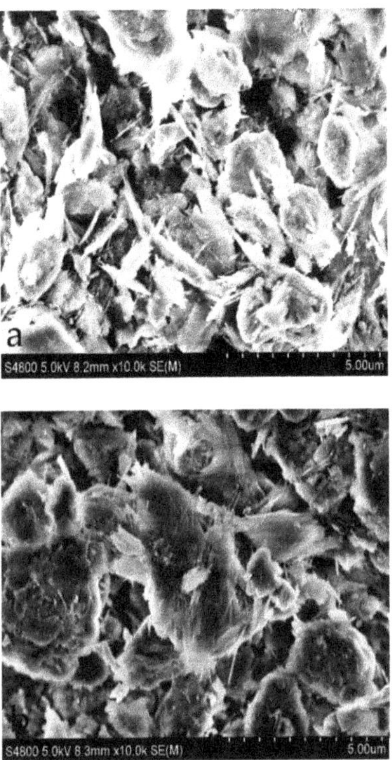

Figure 3. SEM image of original sepiolite (**a**) and modified sepiolite (**b**).

3.2. Effect of the Operation Condition on Adsorption Performance

It can be seen from Figure 4a that the removal efficiency of AO and AR by modified sepiolite declined with the increase of pH, which probably because the anionic dye is more protonated under acidic condition [23]. The negatively charged anionic dye could attracted on the positively charged adsorption sites by electrostatic attraction [39]. When OH^- increased, the anionic dye and OH^- had competitive adsorption [40]. That is also in accordance with the result from Figure 2. In the two-component and three-component mixed solution, the removal efficiency of each dye was basically as same as that of single dye system. The decolorization efficiency of the single component dye in the multi-component system was lower than that of the single dye system. The reduction degree of removal efficiency was AR > AO > RB. For instance, the removal efficiency of AR in a single dye system was 96.5%, while reduced to less than 30% in the meta-mixing system and the three-component mixed system.

It can be seen from Figure 4c,d that the removal efficiency of each dye was basically the same during the fixed adsorption condition in single dye system. When the dosage was 3 g/L, the removal efficiency of the three dyes were all 100%. In the mixed dye system, the decolorization behavior of the three dyes was distinct compared to the single dye system. When the dosage of adsorbent was 2 g/L, the decolorization efficiency of AO, RB and AR decreased from 97.48%, 94.49% and 93.6% in the single dye system, to 40.09%, 40.62%, and 15.36% in the three-component system. The decolorization efficiency of AR was rapidly decreased compared with the RB and AO. In the three-component system, when the dosage was lower than 2 g/L, the decolorization efficiency of AO and RB was maintained at the same level and much higher than AR. While the removal efficiency of AO and AR increased rapidly when increasing the adsorbent dosage. This indicates that the AO and RB were preferentially adsorbed.

Moreover, the AO was preferentially adsorbed in the two-component system of AO and RB. The above results indicate that AO had the characteristics of preferential adsorption in the two-component system.

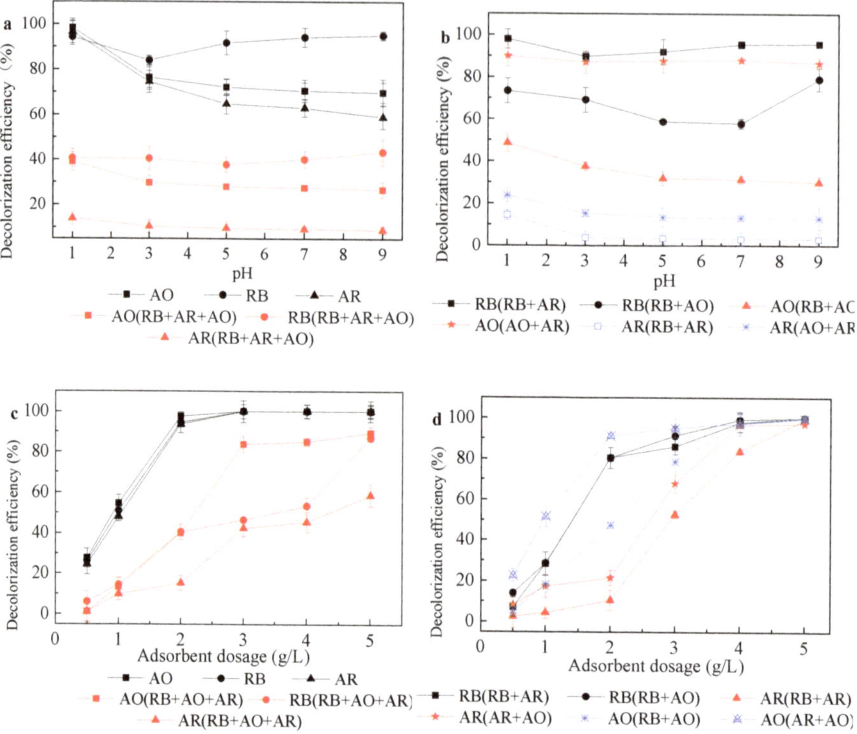

Figure 4. Effect of pH and adsorbent dosage on the Acid Orange II (AO), Reactive Blue (RB) and Acid Red (AR) adsorption performance. In (**a**) and (**b**), the adsorption test was conducted with 2 g/L modified sepiolite; In (**c**) and (**d**), the adsorption test was conducted under initial pH value of 1.

3.3. Adsorption Kinetics and Thermodynamics

Figure 5a is the kinetic process of the one-component and three-component mixed system. In the single dye system, the adsorption of dyes by the organic modified sepiolite reached equilibrium at 120 min; the adsorption amount of the single dye is sorted as: AO > RB > AR. In the three-component system, the adsorption reaction was also quickly balanced. The rapid reaction of dye and adsorbent may be due to the fact that the total concentration of the anionic dye in the three-component system was larger than the one-component system. On the other hand, the adsorption amount of the three anionic dyes in the three-component mixed system was reduced. Figure 5b is the kinetic curve of the two-component mixed system; the adsorption also reached equilibrium in a short time. The AR appears desorption in the mixed system, which may because the decrease of the valid sites of the adsorbent after the adsorption reached saturation, while the RB and AO had a competitive advantage and the "substitution effect" occurred [23]. That is, the RB and AO replaced AR from the adsorption site during competitive adsorption.

Since the adsorption kinetics is very fast, most of the experimental data points belong to the steady state, which may skew the mathematical modelling of the adsorption dynamics. It can be seen from Figure 6 that the correlation of the quasi-second-order reaction kinetics fitting was much better than the first-order reaction kinetics fitting; the data in Table 1 also verified. The coefficient of determination of quasi-second-order fitting was between 0.97–0.99, while the coefficient of determination of the quasi-first-order reaction kinetics was lower than 0.85. The adsorption amount is also consistent with the rate constant, which indicates that the rate constant may have a certain relationship with the adsorption amount. The adsorption rate of the single dye was sorted as AO > RB > AR.

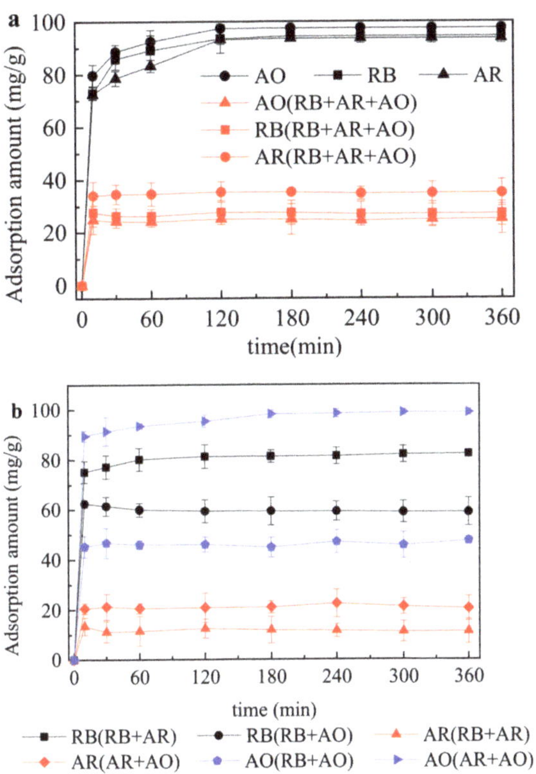

Figure 5. Effect of adsorption duration on the Acid Orange II (AO), Reactive Blue (RB) and Acid Red (AR) adsorption performance. (**a**) one component and three component system; (**b**) two component system.

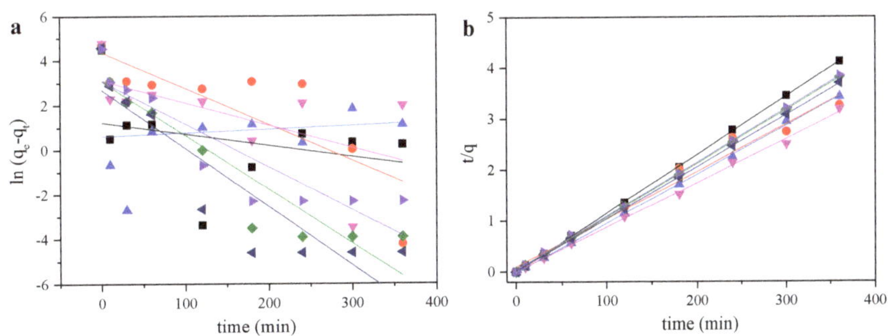

Figure 6. (**a**) fitting of experimental data with quasi-first-order reaction kinetics; (**b**) fitting of experimental data with the quasi-second-order model. The adsorption test was conducted with 2 g/L modified sepiolite and 200 mg/L dye solution for 240 min under oscillation rate of 180 r/min and 25 °C.

Table 1. The adsorption kinetics model with various dyes, the adsorption test was conducted with 200 mg/L Acid Orange II and 2 g/L modified sepiolite under oscillation rate of 180 r/min and initial pH value of 1.

Solution	Quasi-First-Order Reaction Kinetics Model			Quasi-Secondary Reaction Kinetics Model		
	q_{1e} (mg/g)	k_1 (1/min)	R_1^2	$q2_e$ (mg/g)	k_2 (1/min)	R_1^2
RB	21.62	2.42×10^{-2}	0.85	95.24	4.53×10^{-3}	0.99
AO	14.65	2.61×10^{-2}	0.79	98.14	5.36×10^{-3}	0.99
AR	22.01	1.93×10^{-2}	0.80	94.70	2.91×10^{-3}	0.99
RB + AR	79.46	1.61×10^{-2}	0.61	107.30	8.6×10^{-2}	0.97
RB + AO	1.84	1.59×10^{-3}	0.23	104.17	1.10×10^{-2}	0.99
AO + AR	22.53	1.00×10^{-2}	0.28	115.61	7.09×10^{-2}	0.99
RB + AO + AR	3.39	5.09×10^{-3}	0.56	87.18	1.42×10^{-1}	0.99

Figure 7 shows the adsorption isotherms of three anionic dyes in single dye systems, two-component and three-component mixed system. In the single dye system, the amount of adsorption increased rapidly as the initial concentration increased until reached saturation and equilibrates. According to the shape of the adsorption isotherm, it is preliminarily determined that the adsorption of the single dye conforms to the Langmuir model; the adsorption of the dye on the sepiolite is single layer adsorption. In the two-component and three-component system, the adsorption isotherm of each dye is not followed a regular Langmuir model [24]. For example, in the three-component mixed system, the adsorption amount of AR and RB decreased with the increase of initial concentration. This trend mainly because of the competitive effect in the adsorption process.

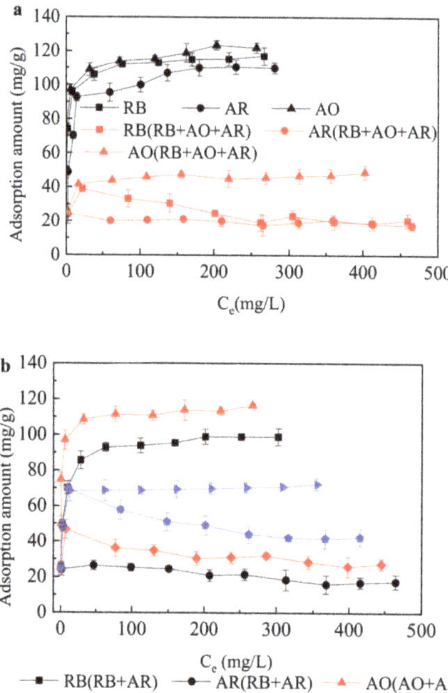

Figure 7. Adsorption isotherm with various dyes. To ensure the adsorbed saturation, the adsorption test was conducted at least 4 h at a certain temperature in a constant temperature oscillating box. (**a**) one component and three component system; (**b**) two component system.

Though Langmuir, Freundlich linear fitting model and intraparticle diffusion are the means to discuss the adsorption isotherm, previous work provided that monolayer adsorption was the main adsorption mechanism during and mixed dye adsorption process [30,31]. Hence, the extended Langmuir model was utilized in this study. As can be seen from Table 2, the saturated adsorption capacities of AO, RB and AR were 119 mg/g, 115 mg/g and 110 mg/g, respectively. The adsorption amount of sepiolite to each dye in the mixed system was lower than that from the single system. For instance, when the AO was mixed with RB or AR, its adsorption capacity decreased by 41.2% and 2.1%, and decreased by 58.8% in the three-component mixed system. In the two-component mixed system, the presence of AR had little effect on the adsorption amount of other dyes, and the amount of adsorption itself was minimized in the competitive adsorption process. The decrease of adsorption capacity in the mixed system may have the following factors: the interaction force between the dye solutions, the change of the surface charge of the adsorbent during the adsorption process, the competition of the dye on the active adsorption sites. The adsorption amount and equilibrium amount of single dye system, as well as in the mixed system, were in accordance with the Langmuir model; the coefficient of determination were all over 0.99.

Table 2. Langmuir model fitting results of modified sepiolite adsorption.

Solution	Langmuir Model		
	q_{max} (mg/g)	K_L (L/mg)	R_1^2
RB	115.0	0.32	0.99
AO	119.0	0.50	0.99
AR	110.0	0.16	0.99
RB + AR	129.9	0.34	0.99
RB in RB + AR	107.5	0.86	0.99
AR in RB + AR	26.5	0.28	0.99
RB + AO	134.9	1.55	0.99
RB in RB + AO	72.5	2.19	0.99
AO in RB + AO	69.9	1.99	0.99
AO + AR	144.9	1.00	0.99
AO in AO + AR	116.5	3.73	0.99
AR in AO + AR	46.7	3.70	0.99
RB + AO + AR	96.2	1.85	0.99
RB in RB + AO + AR	39.0	0.05	0.99
AO in RB + AO + AR	49.0	0.17	0.99
AR in RB + AO + AR	22.7	0.04	0.99

4. Conclusions

In this study, the sepiolite was modified by a two-step method, heat and surfactant modification. The adsorbent was characterized by BET, X-ray diffraction spectroscopy, Fourier-transform infrared spectroscopy and SEM methods. Then the impact of pH and adsorbent dosage on the one-component, two-component and three-component dye system were comprehensively investigated. At last, the kinetics and thermodynamics analysis were studied. The conclusions were listed as follow:

1. The adsorption amounts of Acid Orange II, Reactive Blue and Acid Fuchsin improved after CTMAB modification process, which confirmed the applicability of the modified sepiolite in industrial dye wastewater treatment.
2. The specific surface area of modified sepiolite was obviously decreased, but the adsorption capacity was enhanced. The SEM image shows that the modified sepiolite has a dispersible morphology and the gaps were clean and smooth. The characterization indicate the modification does not deform the sepiolite structure and the CTMAB was successfully loaded.
3. Acid Orange II had the characteristics of preferential adsorption in the two-component system. The electrostatic attraction of positively charged adsorption sites on the adsorbent surface with the negatively charged anionic dye could enhance the adsorption amount under acid condition.

4. The adsorption performance of one-component, two-component and three-component dye system was in accordance with the quasi-second-order reaction kinetics and the adsorption equilibrium time was all around 120 min.
5. The adsorption equilibrium were fitted very well to the Langmuir model and extended Langmuir isotherm.

Author Contributions: Data curation, A.Z. and W.H.; investigation, J.Y.; writing—original draft, B.L. All authors have read and agreed to the published version of the manuscript.

Funding: This research was funded by the special funding for Hunan's innovative province construction (No. 2019SK2111).

Conflicts of Interest: The authors declare no conflict of interest.

References

1. Kolmakov, K.; Hebisch, E.; Wolfram, T.; Nordwig, L.A.; Wurm, C.A.; Ta, H.; Westphal, V.; Belov, V.N.; Hell, S.W. Far-Red Emitting Fluorescent Dyes for Optical Nanoscopy: Fluorinated Silicon-Rhodamines (SiRF Dyes) and Phosphorylated Oxazines. *Chem. A Eur. J.* **2015**, *21*, 13344–13356. [CrossRef]
2. Singh, R.L.; Singh, P.K.; Singh, R.P. Enzymatic decolorization and degradation of azo dyes–A review. *Int. Biodeterior. Biodegrad.* **2015**, *104*, 21–31. [CrossRef]
3. Oladipo, A.A.; Gazi, M.; Yilmaz, E. Single and binary adsorption of azo and anthraquinone dyes by chitosan-based hydrogel: Selectivity factor and Box-Behnken process design. *Chem. Eng. Res. Des.* **2015**, *104*, 264–279. [CrossRef]
4. Baptista, M.S.; Indig, G.L. Effect of BSA Binding on Photophysical and Photochemical Properties of Triarylmethane Dyes. *J. Phys. Chem. B* **1998**, *102*, 4678–4688. [CrossRef]
5. Robinson, T.; McMullan, G.; Marchant, R.; Nigam, P.; McMullan, G. Remediation of dyes in textile effluent: A critical review on current treatment technologies with a proposed alternative. *Bioresour. Technol.* **2001**, *77*, 247–255. [CrossRef]
6. Ghoreishi, S.; Haghighi, R. Chemical catalytic reaction and biological oxidation for treatment of non-biodegradable textile effluent. *Chem. Eng. J.* **2003**, *95*, 163–169. [CrossRef]
7. Demirbas, A. Agricultural based activated carbons for the removal of dyes from aqueous solutions: A review. *J. Hazard. Mater.* **2009**, *167*, 1–9. [CrossRef] [PubMed]
8. Shao, S.; Fu, W.; Li, X.; Shi, D.; Jiang, Y.; Li, J.; Gong, T.; Li, X. Membrane fouling by the aggregations formed from oppositely charged organic foulants. *Water Res.* **2019**, *159*, 95–101. [CrossRef] [PubMed]
9. Yu, J.; Zhang, D.; Ren, W.; Liu, B. Transport of Enterococcus faecalis in granular activated carbon column: Potential energy, migration, and release. *Colloids Surf. B Biointerfaces* **2019**, *183*, 110415. [CrossRef] [PubMed]
10. Li, G.; Zhou, S.; Shi, Z.; Meng, X.; Li, L.; Liu, B. Electrochemical degradation of ciprofloxacin on BDD anode using a differential column batch reactor: Mechanisms, kinetics and pathways. *Environ. Sci. Pollut. Res.* **2019**, *26*, 17740–17750. [CrossRef] [PubMed]
11. Li, G.; Liu, B.; Bai, L.; Shi, Z.; Tang, X.; Wang, J.; Liang, H.; Zhang, Y.; Van Der Bruggen, B. Improving the performance of loose nanofiltration membranes by poly-dopamine/zwitterionic polymer coating with hydroxyl radical activation. *Sep. Purif. Technol.* **2020**, *238*, 116412. [CrossRef]
12. Garg, V.; Gupta, R.; Yadav, A.B.; Kumar, R. Dye removal from aqueous solution by adsorption on treated sawdust. *Bioresour. Technol.* **2003**, *89*, 121–124. [CrossRef]
13. Abbas, A.; Al-Amer, A.M.; Laoui, T.; Al-Marri, M.J.; Nasser, M.S.; Khraisheh, M.; Atieh, M. Heavy metal removal from aqueous solution by advanced carbon nanotubes: Critical review of adsorption applications. *Sep. Purif. Technol.* **2016**, *157*, 141–161.
14. Tillotson, M.J.; Brett, P.; Bennett, R.; Grau-Crespo, R. Adsorption of organic molecules at the TiO2(110) surface: The effect of van der Waals interactions. *Surf. Sci.* **2015**, *632*, 142–153. [CrossRef]
15. Song, J.; Xu, T.; Gordin, M.L.; Wang, D. Nitrogen-Doped Mesoporous Carbon Promoted Chemical Adsorption of Sulfur and Fabrication of High-Areal-Capacity Sulfur Cathode with Exceptional Cycling Stability for Lithium-Sulfur Batteries. *Adv. Funct. Mater.* **2014**, *24*, 1243–1250. [CrossRef]

16. Grassi, M.; Rizzo, L.; Farina, A. Endocrine disruptors compounds, pharmaceuticals and personal care products in urban wastewater: Implications for agricultural reuse and their removal by adsorption process. *Environ. Sci. Pollut. Res.* **2013**, *20*, 3616–3628. [CrossRef]
17. Wang, Z.; Liao, L.; Hursthouse, A.S.; Song, N.; Ren, B. Sepiolite-Based Adsorbents for the Removal of Potentially Toxic Elements from Water: A Strategic Review for the Case of Environmental Contamination in Hunan, China. *Int. J. Environ. Res. Public Health* **2018**, *15*, 1653. [CrossRef]
18. Fayazi, M.; Afzali, D.; Ghanei-Motlagh, R.; Iraji, A. Synthesis of novel sepiolite-iron oxide-manganese dioxide nanocomposite and application for lead(II) removal from aqueous solutions. *Environ. Sci. Pollut. Res.* **2019**, *26*, 18893–18903. [CrossRef]
19. Zaini, N.A.M.; Ismail, H.; Rusli, A. Short Review on Sepiolite-Filled Polymer Nanocomposites. *Polym. Plast. Technol. Eng.* **2017**, *56*, 1665–1679. [CrossRef]
20. Kara, M.; Yuzer, H.; Sabah, E.; Çelik, M.S. Adsorption of cobalt from aqueous solutions onto sepiolite. *Water Res.* **2003**, *37*, 224–232. [CrossRef]
21. Chen, Q.; Zhu, R.; Liu, S.; Wu, D.; Fu, H.; Zhu, J.; He, H. Self-templating synthesis of silicon nanorods from natural sepiolite for high-performance lithium-ion battery anodes. *J. Mater. Chem. A* **2018**, *6*, 6356–6362. [CrossRef]
22. Serna, C.; Vanscoyoc, G. Infrared study of sepiolite and palygorskite surfaces. In *Developments in Sedimentology*; Elsevier: Amsterdam, The Netherlands, 1979; pp. 197–206.
23. Alkan, M.; Demirbas, O.; Çelikçapa, S.; Doğan, M. Sorption of acid red 57 from aqueous solution onto sepiolite. *J. Hazard. Mater.* **2004**, *116*, 135–145. [CrossRef] [PubMed]
24. Wang, J.; Wang, D.; Zhang, G.; Guo, Y.; Liu, J. Adsorption of rhodamine B from aqueous solution onto heat-activated sepiolite. *Wuhan Univ. J. Nat. Sci.* **2013**, *18*, 219–225. [CrossRef]
25. Armağan, B.; Ozdemir, O.; Turan, M.; Celik, M.S. Adsorption of Negatively Charged Azo Dyes onto Surfactant-Modified Sepiolite. *J. Environ. Eng.* **2003**, *129*, 709–715. [CrossRef]
26. Yu, J.; He, W.; Liu, B. Adsorption of Acid Orange II with Two Step Modified Sepiolite: Optimization, Adsorption Performance, Kinetics, Thermodynamics and Regeneration. *Int. J. Environ. Res. Public Health* **2020**, *17*, 1732. [CrossRef]
27. Zhang, L.; Zhou, X.; Guo, X.; Song, X.; Liu, X. Investigation on the degradation of acid fuchsin induced oxidation by MgFe2O4 under microwave irradiation. *J. Mol. Catal. A Chem.* **2011**, *335*, 31–37. [CrossRef]
28. Özcan, A.; Ömeroğlu, Ç.; Erdoğan, Y.; Özcan, A.S. Modification of bentonite with a cationic surfactant: An adsorption study of textile dye Reactive Blue 19. *J. Hazard. Mater.* **2007**, *140*, 173–179. [CrossRef]
29. Yang, X.; Yi, H.; Tang, X.; Zhao, S.; Yang, Z.; Ma, Y.; Feng, T.; Cui, X. Behaviors and kinetics of toluene adsorption-desorption on activated carbons with varying pore structure. *J. Environ. Sci.* **2018**, *67*, 104–114. [CrossRef]
30. Kurniawan, A.; Sutiono, H.; Indraswati, N.; Ismadji, S. Removal of basic dyes in binary system by adsorption using rarasaponin–bentonite: Revisited of extended Langmuir model. *Chem. Eng. J.* **2012**, *189–190*, 264–274. [CrossRef]
31. Zhang, Y.-Z.; Li, J.; Zhao, J.; Bian, W.; Li, Y.; Wang, X.-J. Adsorption behavior of modified Iron stick yam skin with Polyethyleneimine as a potential biosorbent for the removal of anionic dyes in single and ternary systems at low temperature. *Bioresour. Technol.* **2016**, *222*, 285–293. [CrossRef]
32. Oliveira, L.C.; Petkowicz, D.I.; Smaniotto, A.; Pergher, S. Magnetic zeolites: A new adsorbent for removal of metallic contaminants from water. *Water Res.* **2004**, *38*, 3699–3704. [CrossRef] [PubMed]
33. Nayak, P.S.; Singh, B.K. Instrumental characterization of clay by XRF, XRD and FTIR. *Bull. Mater. Sci.* **2007**, *30*, 235–238. [CrossRef]
34. Yu, J.; Zhang, L.; Liu, B. Adsorption of Malachite Green with Sodium Dodecylbenzene Sulfonate Modified Sepiolite: Characterization, Adsorption Performance and Regeneration. *Int. J. Environ. Res. Public Health* **2019**, *16*, 3297. [CrossRef] [PubMed]
35. Liu, B.; Qu, F.; Yu, H.; Tian, J.; Chen, W.; Liang, H.; Li, G.; Van Der Bruggen, B. Membrane Fouling and Rejection of Organics during Algae-Laden Water Treatment Using Ultrafiltration: A Comparison between in Situ Pretreatment with Fe(II)/Persulfate and Ozone. *Environ. Sci. Technol.* **2018**, *52*, 765–774. [CrossRef]
36. Shao, S.; Liang, H.; Qu, F.; Li, K.; Chang, H.; Yu, H.; Li, G. Combined influence by humic acid (HA) and powdered activated carbon (PAC) particles on ultrafiltration membrane fouling. *J. Membr. Sci.* **2016**, *500*, 99–105. [CrossRef]

37. Tunç, S.; Duman, O.; Çetinkaya, A. Electrokinetic and rheological properties of sepiolite suspensions in the presence of hexadecyltrimethylammonium bromide. *Colloids Surf. A Physicochem. Eng. Asp.* **2011**, *377*, 123–129. [CrossRef]
38. Yan, L.; Qin, L.-L.; Yu, H.-Q.; Li, S.; Shan, R.-R.; Du, B. Adsorption of acid dyes from aqueous solution by CTMAB modified bentonite: Kinetic and isotherm modeling. *J. Mol. Liq.* **2015**, *211*, 1074–1081. [CrossRef]
39. Habish, A.J.; Lazarević, S.; Janković-Častvan, I.; Jokić, B.; Kovač, J.; Rogan, J.; Janaćković, Đ.; Petrović, R. Nanoscale zerovalent iron (nZVI) supported by natural and acid-activated sepiolites: The effect of the nZVI/support ratio on the composite properties and Cd^{2+} adsorption. *Environ. Sci. Pollut. Res.* **2016**, *24*, 628–643. [CrossRef]
40. Dashamiri, S.; Ghaedi, M.; Asfaram, A.; Zare, F.; Wang, S. Multi-response optimization of ultrasound assisted competitive adsorption of dyes onto Cu (OH)2-nanoparticle loaded activated carbon: Central composite design. *Ultrason. Sonochem.* **2017**, *34*, 343–353. [CrossRef]

Article

Adsorption of Methylene Blue in Water onto Activated Carbon by Surfactant Modification

Yu Kuang [1], Xiaoping Zhang [1,2,3,4,*] and Shaoqi Zhou [1]

1 School of Environment & Energy, Guangzhou Higher Education Mega Centre, South China University of Technology, Guangzhou 510006, China; eskuangyu@mail.scut.edu.cn (Y.K.); fesqzhou@yeah.net (S.Z.)

2 The Key Laboratory of Pollution Control and Ecosystem Restoration in Industry Clusters of Ministry of Education, Guangzhou 510006, China

3 Guangdong Environmental Protection Key Laboratory of Solid Waste Treatment and Recycling, Guangzhou 510006, China

4 Guangdong Provincial Engineering and Technology Research Center for Environmental Risk Prevention and Emergency Disposal, Guangzhou 510006, China

* Correspondence: xpzhang@scut.edu.cn; Tel.: +86-1367-892-0429

Received: 19 January 2020; Accepted: 18 February 2020; Published: 21 February 2020

Abstract: In this paper, the enhanced adsorption of methylene blue (MB) dye ion on the activated carbon (AC) modified by three surfactants in aqueous solution was researched. Anionic surfactants—sodium lauryl sulfate (SLS) and sodium dodecyl sulfonate (SDS)—and cationic surfactant—hexadecyl trimethyl ammonium bromide (CTAB)—were used for the modification of AC. This work showed that the adsorption performance of cationic dye by activated carbon modified by anionic surfactants (SLS) was significantly improved, whereas the adsorption performance of cationic dye by activated carbon modified by cationic surfactant (CTAB) was reduced. In addition, the effects of initial MB concentration, AC dosage, pH, reaction time, temperature, real water samples, and additive salts on the adsorption were studied. When Na^+, K^+, Ca^{2+}, NH^{4+}, and Mg^{2+} were present in the MB dye solution, the effect of these cations was negligible on the adsorption (<5%). The presence of NO^{2-} improved the adsorption performance significantly, whereas the removal rate of MB was reduced in the presence of competitive cation (Fe^{2+}). It was found that the isotherm data had a good correlation with the Langmuir isotherm through analyzing the experimental data by various models. The dynamics of adsorption were better described by the pseudo-second-order model and the adsorption process was endothermic and spontaneous. The results showed that AC modified by anionic surfactant was effective for the adsorption of MB dye in both modeling water and real water.

Keywords: activated carbon; modification; surfactant; adsorption; methylene blue; ions effect

---

1. Introduction

As a cationic dye, methylene blue ($C_{16}H_{18}ClN_3S$, MB) is widely used in chemical indicators, dyes and biological dyes. A large amount of organic dye wastewater is produced in the processes of the printing and dyeing industries. The dye wastewater has characteristics such as large discharge, high chromaticity, high organic matter concentration, and poor biodegradability, and greatly affects the water body health and the photosynthesis of microorganisms in the water environment [1,2].

At present, many researchers have used different methods to treat the dye wastewater [3]. Typical treatment methods include physical, chemical, and biological methods, such as flocculation [4], membrane filtration [5,6], advanced oxidation [7], ozonation, photocatalytic degradation [8], and biodegradation. These traditional methods have inherent limitations [9] such as the complex and uneconomical of nature of the technology, and thus it is necessary to seek efficient and simple dye wastewater treatment methods [10].

Compared with other treatment methods, the adsorption method is considered as prevailing over other dye wastewater treatment technology due to its advantages such as high efficiency, low cost, simple operation, and insensitive of toxic substances [11]. Activated carbon is the most commonly used adsorbent, and is widely used to remove the organic and inorganic pollutants in water phase.

Adsorption capacity is an important index to evaluate the adsorption effect of adsorbent. Various low-cost alternative adsorbents from agricultural solid waste, industrial solid waste, agricultural by-products, and biomass are used in wastewater treatment. For example, clay [12], sludge [13], montmorillonite [14], flax fiber [15], zeolite [16,17], and biochar (rice husk [18,19], pinewood [20], wheat [21], sugarcane bagasse [22], switchgrass [23], *Ficus carica* bast [24]) as adsorbents have been used for adsorption treatment of dye wastewater. However, common activated carbon is not widely used to adsorb and remove pollutants due to its low adsorption capacity, which is caused by small specific surface area and poor adsorption selection performance, as well as limitations of the surface functional groups and electrochemical properties.

In many cases, it is required to modify activated carbon surface to enhance its affinity with target pollutants and increase its adsorption capacity and improve the removal effect of pollutants in different types of industrial wastewater. The methods used to modify activated carbon include physical, chemical, and biological methods. As a chemical modification technology of activated carbon, surfactant modification shows great significance. Surfactant modification for activated carbon can improve the hydrophilicity and the dispersion of activated carbon in water due to an increase in the affinity between activated carbon and water and the decrease of the attractive energy between particles. Surfactant has the advantages of low cost [25] and less damage to the structure of activated carbon. Moreover, the surfactant can change the surface charge characteristics of activated carbon and provide more ionic adsorption sites for pollutants, which can not only enhance the adsorption capacity of ionic pollutants on activated carbon [26], but also can promote selective adsorption.

In wastewater treatment, activated carbon modified by surfactant is used as an adsorbent to remove various pollutants, including organic pollutants [26,27], reactive dyes [15,21,28,29], and heavy metals [30,31]. Surfactants are classified into anionic, cationic, non-ionic, and amphoteric ions according to the type and performance of the hydrophilic group [22]. Sodium lauryl sulfate (SLS) is a nontoxic anionic surfactant. Much research into the effect on inorganic metal ions by activated carbon modified by surfactant has been reported, however, the study of organic dyes adsorption by SLS-C has been rarely reported.

Previous research has shown that activated carbon modified by different surfactants have different effects on various water quality dye wastewater. Therefore, it is necessary to determine how the coexistence of other ions would affect the adsorption. Meanwhile, more information is still required in order to better understand of the adsorption behavior of methylene blue cationic dyes on modified-activated carbon.

The feasibility of removing MB from aqueous solution by surfactant-modified activated carbon was investigated. The purposes of the study were to (1) confirm the adsorption rate and capacity through the adsorption models; (2) determine the various parameters that affect sorption, such as initial dye concentration, temperature, pH, adsorbent dose, and additive salts; and (3) obtain the adsorption effect of adsorbent in actual water samples.

2. Materials and Methods

2.1. Materials

Methylene blue (Figure 1, $C_{16}H_{18}ClN_3S$) was used as adsorbent. The molar mass of MB molecule is 319.85 g·mol^{-1}. The surfactants used in this study were sodium lauryl sulfate (SLS, $C_{12}H_{25}SO_4Na$), sodium dodecyl sulfonate (SDS, $C_{12}H_{25}SO_3Na$), and hexadecyl trimethyl ammonium bromide (CTAB). Other chemicals used were calcium chloride ($CaCl_2$), magnesium sulfate heptahydrate ($MgSO_4{\cdot}7H_2O$), sodium sulfate (Na_2SO_4), potassium chloride (KCl), sodium chloride (NaCl), sodium nitrite ($NaNO_2$)

and ferrous sulfate ($FeSO_4$). All chemicals were purchased from Shanghai Aladdin Bio-Chem Technology Co., LTD. The powdered activated carbon (AC) was purchased from Tianjin Fuchen Chemical Reagent Co., Ltd. All the chemicals were used without any further purification.

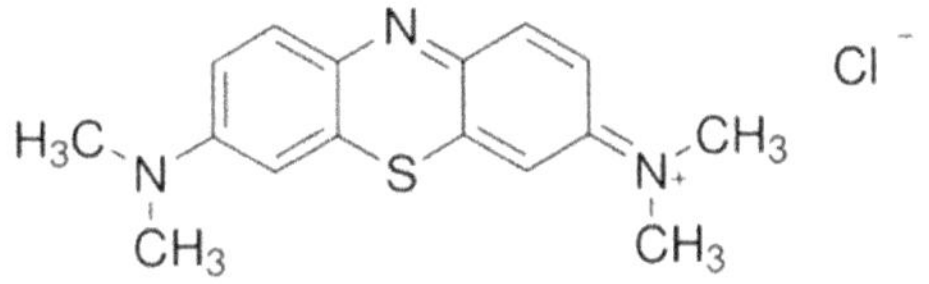

Figure 1. The chemical structure of methylene blue (MB).

2.2. Preparation of Surfactant-modified Activated Carbon

A total of 5 g of AC and 8.60 mM anionic surfactant SLS were added to 100 mL of solution. The mixture was oscillated in a shaker at 301 K for 6 h, and after that, the AC was filtrated and washed with deionized water. The filtrated AC was dried in an air-dry oven at 313 K for 24 h, and then was stored in a sealed and dry environment. Studies have shown that the concentration of surfactant corresponding to critical micelle concentration (CMC) is the best for adsorption [32]. It has been reported that the critical micelle concentration (CMC) of SLS is 8.60 mM at 28°C. The raw AC and surfactant-modified AC were called Virgin-C and SLS-C, respectively. The surfactant sodium dodecyl sulfonate (SDS) and hexadecyl trimethyl ammonium bromide (CTAB) were used to modify AC in the same method at the concentration of 1 CMC, called SDS-C and CTAB-C, respectively.

2.3. Adsorption Experiments

The adsorption capacity of MB on modified AC carried out in the batch adsorption experiment by investigating the effect of experimental variables such as pH (1–12), initial MB concentration (10, 30, 50 $mg{\cdot}L^{-1}$), contact time, adsorbent dosage, temperature (298, 308, 318, 328K), and ionic species. The initial pH of the solution was adjusted with 0.1 M hydrochloric acid (HCl) and sodium hydroxide (NaOH) solution and determined with pH meter. A certain amount of modified AC was added to a 250 mL conical flask solution containing 100 mL of solution at certain concentrations of MB. Each mixture was oscillated at 170 $r{\cdot}min^{-1}$ at 25 °C at the given time interval. After the adsorption process, the samples were filtered and analyzed. The reserve solution of MB (1 $g{\cdot}L^{-1}$) was prepared by dissolving the suitable amount MB in distilled water. All the experiments were done in triplicate.

2.4. Analytical Methods

According to the Standard Methods of Water and Wastewater Monitoring and Analysis Method, the absorbance of MB solution was measured by UV-visible spectrophotometer (INESA Scientific Instrument Co., Ltd. Shanghai, China) at 664 nm, which is the maximum absorption peak of MB. The concentration and removal rate of MB was obtained through the curve of relationship between solution concentration and absorbance of MB (Figure S1). The pH was measured using a DZS-706A multi-parameter meter (INESA Scientific Instrument Co., Ltd. Shanghai, China).

By measuring the absorbance of dye solution before and after treatment, the removal of MB on SLS-C adsorbent was studied with the batch equilibrium method. On the basis of Equations (1) and (2), the removal rate of dye in equilibrium state was calculated.

$$Q_e = \frac{(C_0 - C_e)V}{m} \tag{1}$$

$$Removel\ percentage = \frac{(C_0 - C_e)}{C_0} \times 100, \tag{2}$$

where Q_e represents the quantity of dye per unit mass of adsorbent (mg·g^{-1}); C_0 and C_e are the initial and equilibrium concentrations, respectively; V is the volume of MB dye solution in liters; and m is the weight of SLS-C adsorbent in grams.

3. Results

3.1. Effect of Different Surfactants

Figure 2 showed the extent of adsorption of MB on AC modified with different surfactants in aqueous solution at the initial MB concentration of 50 mg·L^{-1}. The results showed that SLS-C represents the highest MB adsorption capacity, followed by Virgin-C, SDS-C, and CTAB-C.

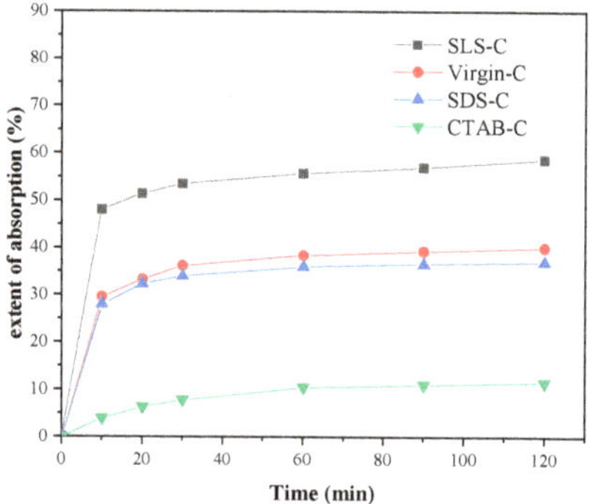

Figure 2. MB removal by using different activated carbons (ACs) (initial MB concentration = 50 mg·L^{-1}, S/L = 0.15 g·L^{-1}, T = 298 K, pH = 5.0).

Compared with SLS-C and SDS-C, CTAB-C showed the weakest adsorption for MB dye. As a cationic surfactant, CTAB had a repulsive force with dye cations and occupied the adsorption sites on activated carbon, resulting in a low adsorption capacity of cationic MB dye.

Compared with the unmodified AC, surfactants loaded on activated carbon may clog the pores of AC, which reduces its ability to adsorb MB. On the other hand, the ionic functional groups of loaded surfactants can provide ion exchange sites with a higher affinity for MB than the unmodified AC surface. The higher MB adsorption on SLS-C compared with Virgin-C indicated that the positive effect of the functional group of SLS-C exceeded the negative effect of the pore blocking, whereas the lower MB removal by SDS-C compared with Virgin-C indicated that the negative effect of the pore blocking exceeded the positive effect of the functional group of SDS-C.

AC modified by anionic surfactants showed a strong adsorption capacity for MB dye due to the strong binding between the anionic surfactant and the cationic MB dye. The chemical properties of the functional groups of surfactant play an important role in adsorption. The cations attached to the strong acid conjugate base of SLS, such as sodium ions (e.g., R-SO_3^- Na^+) and protons (e.g., R-SO_3^- H^+), can be easily dissociated in aqueous solution and exchanged with an MB ion [27]. Therefore the SLS-C showed high MB removal effect. The affinity between the functional group on SDS and MB dye was weaker than that of SLS due to it being without strong acid conjugate base, which made the adsorption effect on MB lower than SLS-C.

The possible chemical adsorption processes of MB on SLS-C are shown in Figure 3. The reactions between MB dye cation (M^+) with RSO_3Na on SLS-C occurred regarding Equations (3)–(5). RSO_3Na is the active hydrophilic group of SLS and ROH represents the carboxyl, phenolic hydroxyl on AC.

$$RSO_3Na + H^+ \leftrightarrow RSO_3H + Na^+, \tag{3}$$

$$RSO_3Na + M^+ \leftrightarrow RSO_3M + Na^+, \tag{4}$$

$$\text{ROH} + \text{M}^{+} \leftrightarrow \text{ROM} + \text{H}^{+}, \tag{5}$$

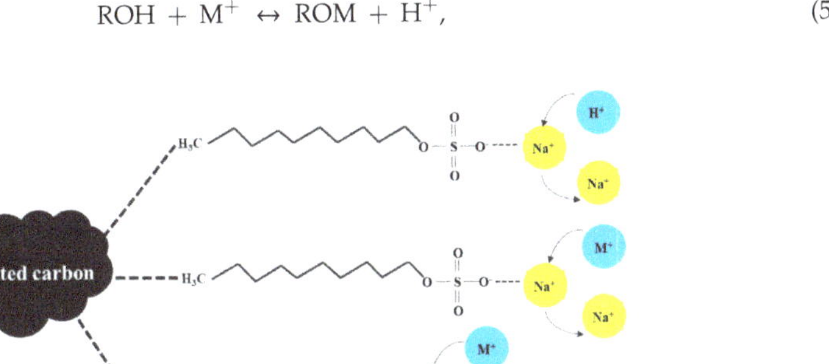

Figure 3. The chemical adsorption between MB dye and the surface functional groups on sodium lauryl sulfate (SLS)-C.

3.2. *Characterization of SLS-C*

FE-SEM (Field Emission Scanning Electron Microscope) was used to observe the surface of the adsorbent. Figure 4 shows the detailed surface characteristics of the adsorbent. Figure 4a shows the surface of Virgin-C with a small amount of carbon debris and micropore of different sizes. Figure 4b shows the surface of SLS-C with a great quantity of pores and a large number of material debris occupying the pores compared with untreated AC. This was due to the electrostatic interaction and adhesion of SLS molecules. A large number of material debris occupied the pores, which could lead to a reduction in surface area, but could provide adsorption and ion exchange sites for MB.

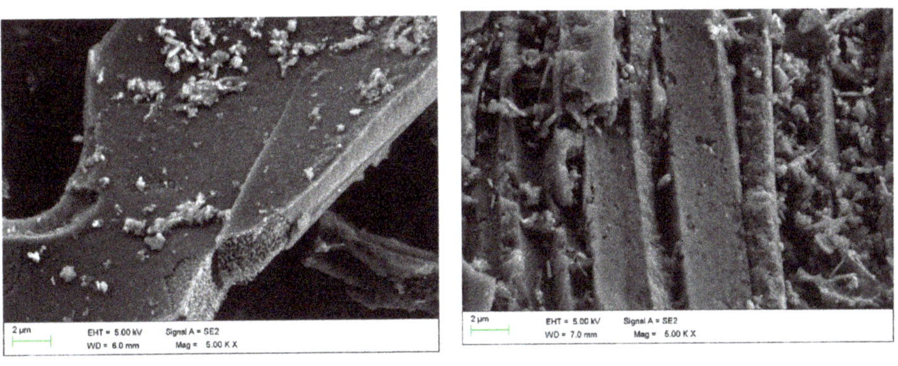

(**a**) FE-SEM-image of Virgin-C (**b**) FE-SEM-image of SLS-C

Figure 4. FE-SEM pictures of (**a**) Virgin-C and (**b**) SLS-C.

3.3. *Effects of Initial Solution pH*

Figure 5 shows the variation of removal rate and adsorption capacities of MB on SLS-C at various pH values. As can be seen from Figure 5a, the alkaline condition was favorable for the adsorption of MB on SLS-C. According to Figure 5b, when the initial concentration of MB was 10, 30, and 50 mg·L^{-1}, the adsorption capacities of MB were 62.67, 168.39, and 193.33 mg·g^{-1}, respectively, at a natural pH value of 4.8, and the adsorption capacities of MB were 62.90, 179.78, and 220.49 mg·g^{-1} at a pH of 12.00. Thus, both the MB adsorption removal rate and adsorption capacities were increased with the increase of pH value.

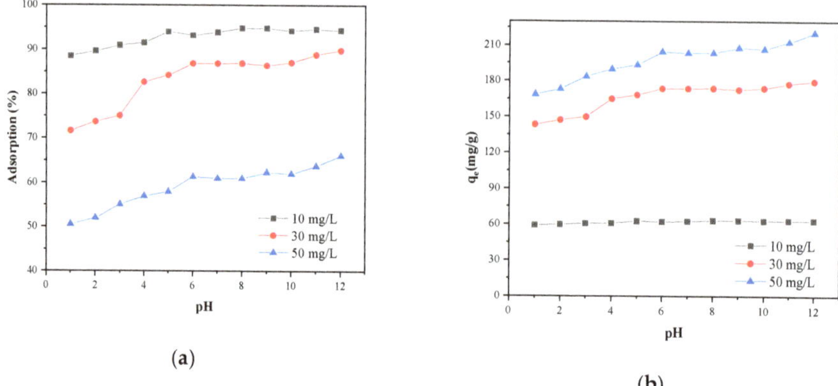

Figure 5. The adsorption rate (**a**) and the adsorption capacity (q_e) (**b**) of MB adsorption on the SLS-C sample at various pH values (S/L = 0.15 g·L^{-1}, contact time = 120 min, 25 °C).

The pH values of dye solution control the extent of ionization of the acidic and basic compounds and affect the surface charge of SLS-C [7,33]. At lower pH, the dissociation of hydrogen ions (H^+) by oxygen-containing functional groups on SLS-C would be inhibited, and the electronegativity of SLS-C and electrostatic attraction force between dye cation and SLS-C were relatively weak. Moreover, the fact that the free hydrogen ions inhibited the adsorption reaction of dye cation onto AC site by competing adsorption could lead to a reduction in MB removal rate. The increase in the concentration of hydroxide ions in the solution made the dissociation degree of MB small, and thus the removal rate of MB was improved as the pH value increased [34]. In addition, the dissociation degree of H^+ by the oxygen-containing functional groups on the surface of the SLS-C increased with the increase of pH, which increased the electronegativity of SLS-C and the electrostatic attractive force between the dye cation and SLS-C [25]. Furthermore, hydroxyl groups (O-H) and carbonyl groups (C=O) on the surface of the adsorbent can also attract the cationic dye molecules under the condition of high pH value [35]. Therefore, SLS-C has a good adsorption capacity of MB in an alkaline environment.

The results have resemblance to previous studies by Karaca [11]. The study by Yagub et al. [36] showed that the adsorption rate of cationic dye MB by raw pine leaf biochar at high pH values was better than that at low pH values. The adsorption rate had a noticeable increase at pH from 2 to 7, and slightly increased at the range of pH 7–9. The effect of pH on MB adsorption by various ACs was studied by Kannan N et al. [37], showing that the adsorption capacity of dye increased with the increase of initial pH.

Figure 6 shows the zeta potential variation of AC with pH before and after modification. The pH_{PZC} of SLS-C and Virgin-C were 3.37 and 4.10. The result demonstrated that SLS-C has a higher surface electronegativity than that of Virgin-C, which was attribute to the hydrophobic alkyl end of surfactant attached to the nonpolar surface of activated carbon by van der Waals force. The AC surface oxygen-containing functional groups such as carboxyl and phenolic hydroxyl were covered with SLS to lead the amount of dissociated H^+ to decrease. The enhanced electronegativity of SLS-C made it have a better electrostatic attraction and adsorption capacity to MB than Virgin-C. When pH was around 4 to 8, the adsorption rate of MB on SLS-C was relatively stable in this experiment. Therefore, a pH value of 5 was chosen for the adsorption research.

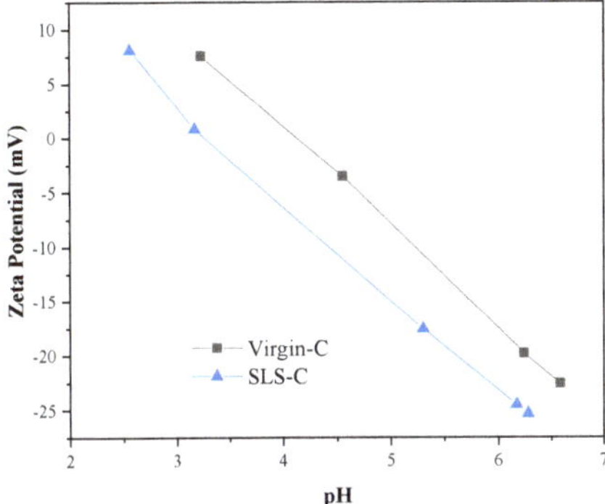

Figure 6. The zeta potential of Virgin-C and SLS-C.

3.4. Effect of Adsorbent Dose

Adsorbent dose is an important factor that affects the adsorption performance. The influence of adsorbent dose in adsorption of MB was studied to obtain a most appropriate amount of adsorbent at various MB concentrations [38]. The effect of adsorbent dose was studied by 100 mL of three different MB concentrations (10, 30, and 50 mg·L^{-1}) under different adsorbent doses (5, 7.5, 10, 15, 20, 30, 40, 50, 75, and 100 mg), as shown in the diagram below (Figure 7). A similar trend in adsorption behavior of MB on SLS-C under various MB concentrations was observed.

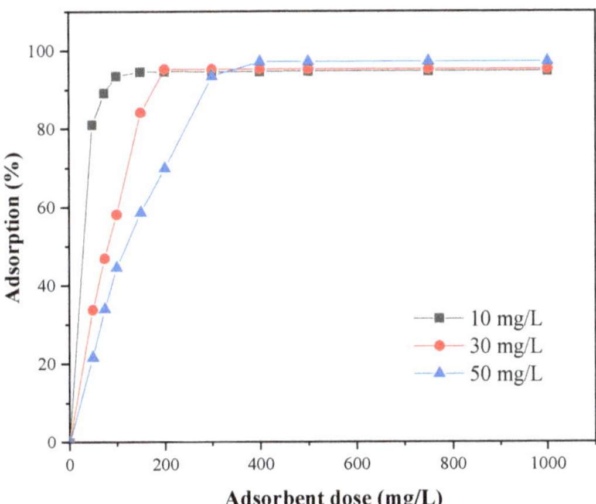

Figure 7. Effect of adsorbent dose on MB adsorption on SLS-C sample (contact time = 120 min).

As the mass of adsorbents increased, the removal rate of MB gradually increased due to the increases of the number of adsorbent pores and adsorption sites. The adsorption would tend to an equilibrium when the mass of adsorbent reached a certain value. The removal rate of MB reached the saturated value at adsorbent mass of 15, 20, and 30 mg corresponding to the initial MB concentration of 10, 30, and 50 mg·L^{-1}, respectively. At high adsorbent dosages, the available number of MB dye molecules in solution was not enough to completely combine with all effective adsorption sites on the adsorbent, resulting in a surface equilibrium state and a reduction in the adsorption capacity per unit mass of adsorbent.

3.5. Effects of Contact Time

Figure 8 showed the relationship between the adsorption rate and the adsorption capacity of MB on SLS-C and time at three initial MB concentrations (10, 30, and 50 mg·L^{-1}).

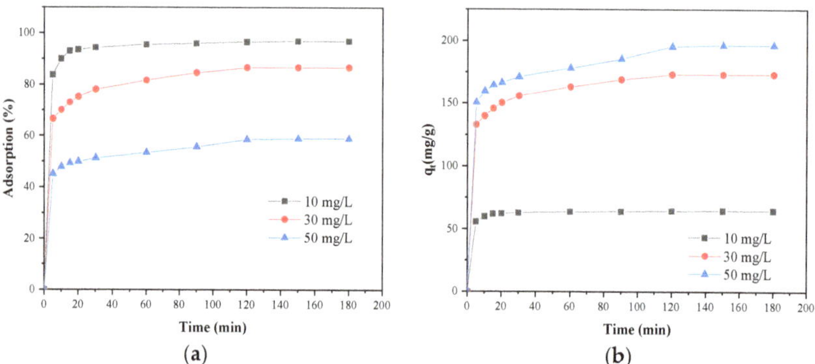

Figure 8. Effect of contact time on adsorption of MB on SLS-C at pH = 5.0, S/L = 0.15 g·L^{-1}. (**a**) The adsorption rate; (**b**) the adsorption capacity.

It can be seen in Figure 8 that the removal efficiency and adsorption capacity of MB by SLS-C was increased with an increase in contact time and then reached a maximum value. The process was divided into two phases. The first step took 5–30 min to reach the relative adsorption equilibrium state called fast adsorption. This performance was due to the binding process between MB dye and the adsorption active sites, and functional groups on the SLS-C adsorbent were fully and efficiently completed. The absorption rate of the dye was controlled by the rate of the dye transported from the solution to the surface of the adsorbent particles. The second step was called slow adsorption process. After 30 min of contact time, the relative increase in the removal extent of MB was not significant, and with the increase of time, the adsorption rate decreased and gradually stabilized. This performance was due to the binding process between MB dye and the adsorption active sites, and functional groups on the SLS-C adsorbent were gradually saturated. The absorption rate of the dye was controlled by the rate of the dye transported from the exterior to the interior pore sites of the adsorbent particles [39].

Moreover, the lower the dye initial concentration, the shorter the time to achieve the adsorption equilibrium state. The results were basically consistent with previous studies on the removal rate of dyes [29]. Considering the relationship between the contact time and the MB removal, the contact time for subsequent experiments was selected as 120 min.

3.6. Effects of Initial MB Concentration

The influence of different initial dye concentrations (10, 20, 30, 40, 50 mg·L^{-1}) on the decolorization efficiency of MB dye was studied using the same mass of SLS-C and adjusting the dye solution pH to 5. A 15 mg adsorbent sample was added to 100 mL dye solution and adsorbed for 120 min at 298K. The experimental data are given in Table 1.

According to the listed data (Table 1), the decolorization rate (%) of MB by 15 mg of the SLS-C adsorbent decreased from 96.6% to 58.7% when the initial dye concentration increased from 10 to 50 mg·L^{-1}. However, as the initial dye concentration increased from 10 to 50 mg·L^{-1}, the adsorption capacity at equilibrium (Q_e) increased from 64.4 to 195.8 mg·g^{-1}. Moreover, the maximum adsorption capacity at equilibrium (Q_e) was calculated up to value of 195.8 mg·g^{-1} when the dye concentration was 40 mg·L^{-1}. It was found that when the MB concentration was greater than 40 mg·L^{-1}, the adsorption sites on the adsorbent were completely adsorbed, and thus the adsorption amount of activated carbon at a MB concentration of 40 mg·L^{-1} was similar to that at MB concentration of 50 mg·L^{-1}.

Table 1. Effect of initial MB concentration on the percentage extraction of dye.

MB Concentration ($mg{\cdot}L^{-1}$)	SLS-C		Virgin-C	
	Percent Adsorption (%)	Adsorption Capacity ($mg{\cdot}g^{-1}$)	Percent Adsorption (%)	Adsorption Capacity ($mg{\cdot}g^{-1}$)
10	96.6	64.4	80.9	53.9
20	90.6	120.8	72.7	96.0
30	86.7	173.4	66.3	132.6
40	73.4	195.8	51.9	138.4
50	58.7	195.7	41.3	137.6

The removal extent of dyes was decreased with an increase in the initial MB concentration due to the lack of available active sites under high concentration condition of MB [40], whereas the adsorption capacity of MB on SLS-C increased with the increase of initial MB concentration. The sulfate functional group of SLS provided ion exchange sites conducive to the MB ion adsorption. Compared with untreated AC, activated carbon modified by anionic surfactant has more positively charged adsorption sites and high adsorptive capacity for removing cationic dye [12,27].

3.7. *Adsorption Kinetics and Isotherm*

3.7.1. Adsorption Kinetics

Adsorption kinetics mainly studies the reaction rate between adsorbents and adsorbates and the factors affecting the reaction rate. The adsorption data were fitted by kinetic models and the adsorption mechanism was discussed according to the fitting results. The adsorption kinetics of MB on SLS-C were studied by using four kinetic equation models to confirm the most effective equation. The consistency between the experimental results and the theoretical values of the model was assessed on the basis of the coefficient of determination (R^2 value was close to or equal to 1).

The principle of the pseudo-first-order kinetic model is that the reaction rate is proportional to the number of ions remaining in the solution, assuming that adsorption is controlled by diffusion steps [41]. It is assumed that the adsorption rate is proportional to the difference between the saturated concentration and the adsorption amount of SLS-C with time. The integral equation is shown below in Equation (6).

$$\ln(Q_e - Q_t) = \ln Q_e - k_1 t \tag{6}$$

where k_1 is the rate constant of adsorption (min^{-1}), Q_e is the quantity of dye adsorbed at equilibrium ($mg{\cdot}g^{-1}$), and Q_t is the equilibrium concentration at various times t (in $mg{\cdot}L^{-1}$). The rate constant in this model was determined by the slope of the plot of $\ln(Q_e - Q_t)$ over time (t). The results of R^2 calculated were in the range of 0.644–0.974. In addition, the large differences between the value of $Q_{e,exp}$ and $Q_{e,cal}$ indicated that the pseudo-first order model was not an effective model to explain the adsorption process.

The pseudo-second-order kinetic model, which can be expressed as the reaction rate being proportional to the concentrations of the two reactants, assuming the adsorption is controlled by chemical adsorption steps, can be expressed by Equation (7) [42].

$$\frac{t}{Q_t} = \frac{1}{k_2 Q_e^2} + \frac{t}{Q_e} \tag{7}$$

where k_2 is the second-order-rate constant ($g{\cdot}mg^{-1}{\cdot}min^{-1}$) that can be determined for different MB concentrations according to the linear plots of t/Q_t versus t, as shown in Figure 9. The calculated correlation coefficients (R^2) were found to be greater than 0.999, and the experimental Q_e differing from the calculated ones are listed in Table 2. Obviously, from the values calculated by pseudo-second-order kinetic model, $Q_{e,cal}$ had good agreement with the experimental $Q_{e,exp}$. This behavior indicated that the pseudo-second order model was the best model to explain the adsorption process of MB on SLS-C.

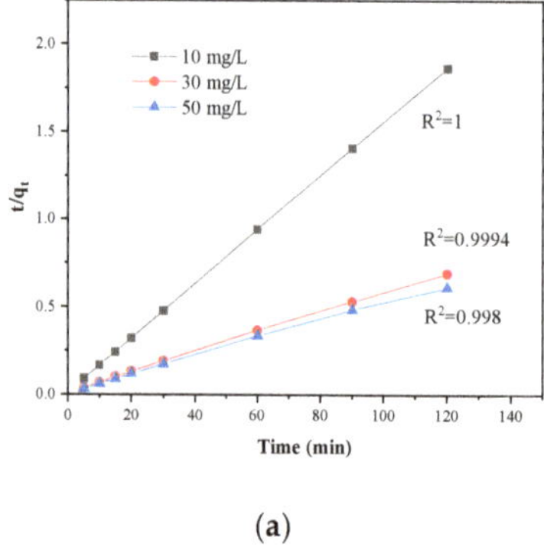

(a)

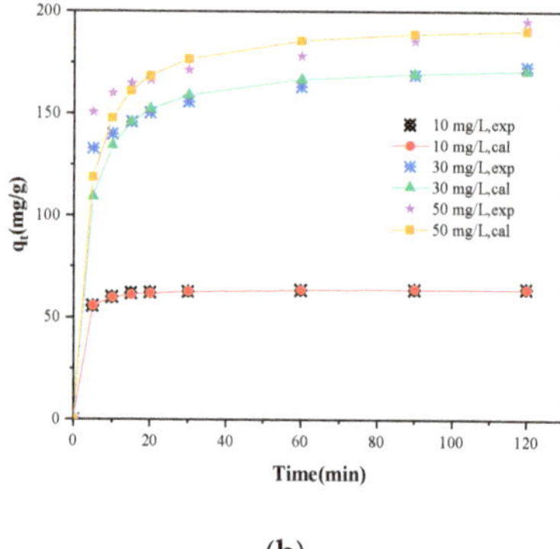

(b)

Figure 9. (**a**) Pseudo-second order plots and (**b**) the adsorption capacity (q_e) for the adsorption of MB on SLS-C at various initial concentrations, pH = 5.0, S/L = 0.15 g·L^{-1}, T = 25 °C.

Table 2. Pseudo-First-Order model and Pseudo-Second-Order model form removal of MB by SLS-C.

C_0 (mg.L^{-1})	$Q_{e,cal}$ (mg.g^{-1})	Pseudo-First-Order Model			Pseudo-Second-Order Model		
		k_1 (min^{-1})	Q_e (mg.g^{-1}) (exp.)	R^2	k_2 (g.mg^{-1}. min^{-1})	Q_e (mg.g^{-1}) (exp.)	R^2
10	64.4	0.0090	4.4	0.644	0.00195	64.1	1
30	173.4	0.0107	40.9	0.966	0.00188	171.1	0.999
50	195.7	0.0069	42.1	0.974	0.00157	190.9	0.998

The intra-particle diffusion model divides the adsorption process into two main steps including the migration of solute molecules from aqueous solution to the surface of adsorbent particles and the diffusion of adsorbate molecules into the interior pores of the adsorbent [43].

The adsorption process was studied Weber–Morris intra-particle diffusion model which is expressed as Equation (8).

$$Q_t = k_{ip}t^{\frac{1}{2}} + C_i \quad (8)$$

where k_{ip} (mg·g^{-1}·min$^{-1/2}$) is the rate constant of the intra-particle diffusion model, and C_i is the constants related to boundary layer thickness expressed in milligrams per gram (mg·g^{-1}), which can be calculated according to slope and intercept of the plot of Q_t versus the square root of time $t^{1/2}$.

The straight line fitted by the intra-particle model (as shown in Figure S2) without passing through the origin indicated that the adsorption process was controlled not only by the intra-particle diffusion but also by other adsorption processes. The constant C_i (57.64, 127.73, 144.74) was found to increase with the increase of MB dye concentration, which may have been due to the increase of the boundary layer thickness (as shown in Table 3). The high values of C_i indicated that the external diffusion of MB molecule on SLS-C was very important in the initial adsorption period. The values of R^2 were close to 1 and show a good application of the intra-particle model in the sorption process.

Table 3. Intra-particle diffusion model and Elovich model from removal of MB by SLS-C.

C_0 ($mg \cdot L^{-1}$)	Intra-Particle Diffusion Model			Elovich Model		
	k_{ip} ($mg \cdot g^{-1} \cdot min^{-1/2}$)	C_i	R^2	α ($mg \cdot g^{-1} \cdot min^{-1}$)	β ($mg \cdot g^{-1}$)	R^2
10	K_{ip1} = 3.800 K_{ip2} = 0.303	57.64	0.985 0.974	2.395	0.436	0.816
30	4.426	127.73	0.957	14.373	0.078	0.997
50	4.542	144.74	0.974	14.197	0.078	0.970

The Elovich model of adsorption process was expressed as Equation (9), which describes adsorption in a non-ideal state. The adsorption process is divided into the fast adsorption and slow adsorption process.

$$Q_t = \frac{1}{\beta} \ln \alpha \, \beta + \frac{1}{\beta} \ln t \tag{9}$$

where α represents the initial adsorption rate ($mg \cdot g^{-1} \cdot min^{-1}$) and β is the desorption coefficient ($mg \cdot g^{-1}$). The graph of Q_t versus $\ln t$ is a line with a slope of $1/\beta$ and an intercept of $1/\beta \ \ln(\alpha \, \beta)$. The correlation coefficient (R^2) at MB concentrations of 10, 30, and 50 $mg \cdot L^{-1}$ were identified as being 0.816, 0.997, and 0.970, respectively. The fitting data calculated from Elovich kinetic model showed that when the initial concentration of MB was 10 $mg \cdot L^{-1}$, the models described the observations not well. Nevertheless, that the experimental results could be well described by the Elovich model at the initial concentration of MB higher than 10 $mg \cdot L^{-1}$ indicated that the rate-limiting step was the intraparticle diffusion process but not the only process [22,44].

The results showed that the pseudo-second-order kinetic model can be used to describe the adsorption of MB on SLS-C and proved that chemical adsorption dominated the adsorption process. The adsorption process not only included the processes of the liquid film diffusion and the internal diffusion of micropores, but also the chemical adsorption process with electrons shared and gained or loss of electrons between cationic dyes and functional groups on the SLS-C surface [30].

3.7.2. Adsorption Isotherm

The adsorption process is divided into two parts, adsorbent adsorption pollutants and pollutants desorption from the adsorbent. When the rate of the two processes is the same, the adsorption will enter a dynamic equilibrium state. Adsorption isotherms are used to study the relationship between the equilibrium adsorption capacity and the equilibrium concentration of pollutants under certain conditions (temperature and pH remain unchanged). The adsorption isotherms were studied to provide a basis for revealing adsorption behavior, indicating possible adsorption mechanism, and estimating adsorption capacity. Three models, the Langmuir model, Freundlich model, and Temkin model, were used to describe adsorption isothermal behavior.

The Langmuir model assumes that adsorption is localized on a monolayer, and all adsorption sites on the adsorbent are homogeneous and have the same adsorption capacity [45]. The Langmuir isotherm equation is Equation (10):

$$\frac{C_e}{Q_e} = \frac{1}{Q_{max} \, K_L} + \frac{C_e}{Q_{max}} \tag{10}$$

where C_e is the equilibrium concentration ($mg \cdot L^{-1}$), Q_e is the amount of adsorbed dye at equilibrium ($mg \cdot g^{-1}$), Q_{max} ($mg \cdot g^{-1}$) is the Langmuir constants that are related to the adsorption capacity, and K_L ($L \cdot mg^{-1}$) is the adsorption rate. The value of R^2 obtained from the linear graph of C_e/Q_e versus C_e was 0.993, which demonstrated that the adsorption process meets the Langmuir isotherm model, as exhibited in the diagram (Figure 10). The value of Q_{max} and K_L are shown in Table 4.

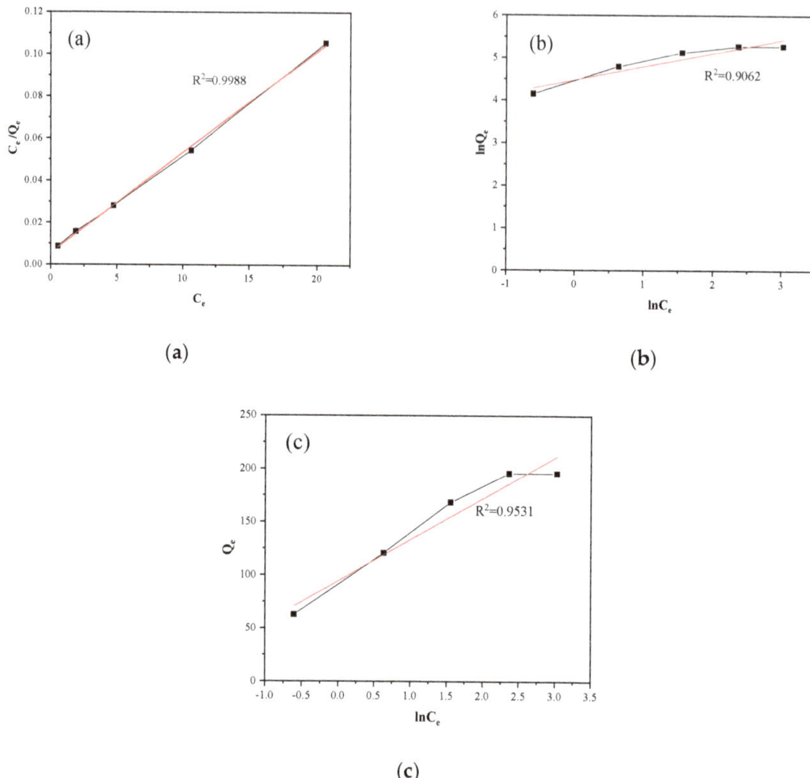

(a) (b) (c)

Figure 10. (**a**) Langmuir, (**b**) Freundlich, and (**c**) Temkin isotherm plots for adsorption of MB on SLS-C.

Table 4. Isotherm parameters of the Langmuir, Freundlich, and Temkin models.

Adsorption Model	Isotherm Parameters	R^2
Langmuir	Q_{max} = 232.5 mg·g^{-1} K_L = 0.842 L·m^{-1}·g^{-1} R_L= 0.106	0.999
Freundlich	K_f = 88.235 mg·g^{-1} $1/n$ = 0.318	0.906
Temkin	B = 38.73 J·mol^{-1} A = 11.53 L·g^{-1} b = 63.98 L·g^{-1}	0.953

The feasibility of adsorbent adsorption was evaluated by R_L. It can be defined by Equation (11) [46]:

$$R_L = \frac{1}{1 + K_L C_0} \tag{11}$$

where C_0 (mg·L^{-1}) is initial dye concentration. The value of R_L indicates the type of the isotherm: irreversible (R_L = 0), favorable (0 < R_L< 1), linear (R_L = 1), or unfavorable (R_L > 1). The R_L value of MB adsorption on SLS-C was in the range of 0.661–0.842. It can be testified that the adsorption process was favorable.

The Freundlich isotherm model is based on the assumption that multi-layer adsorption processes occur on heterogeneous surfaces. The Freundlich isotherm linear equation is shown in Equation (12):

$$\ln Q_e = \ln K_f + \frac{1}{n} \ln C_e \quad (12)$$

where K_f ($mg \cdot g^{-1}$) is a constant related to the adsorption energy. The value of R^2 obtained from the linear graph of $\ln Q_e$ versus $\ln C_e$ was 0.906. Compared the values of R^2 of two isotherms models, the Langmuir model was more suitable for the experimental equilibrium adsorption data than the Freundlich model, as shown in Figure 10. The values of $1/n$ and K_f were calculated and listed in Table 4. The slope ($1/n$) values reflected adsorption intensity or surface heterogeneity. It is generally considered as a sign of a good adsorption process when the value of $1/n$ is in the range of 0.1 to 1.0. In this experiment, the index value was within the range, indicating that the adsorption process was favorable. This was in line with the conclusion of the Langmuir model.

The Temkin model considers the interaction between adsorbent and contaminant as a chemical adsorption process. The Temkin isotherm equation are shown in Equations (13) and (14):

$$Q_e = \frac{RT}{b} \ln A + \frac{RT}{b} \ln C_e \quad (13)$$

$$\mathrm{B} = \frac{RT}{b} \quad (14)$$

where b ($mg \cdot L^{-1}$) is the Temkin isotherm constant, A ($L \cdot g^{-1}$) is the Temkin isotherm equilibrium binding constant, and B is the constant related to the heat of adsorption ($J \cdot mol^{-1}$). The calculated values of A, B, and b are shown in Table 4.

The isotherm data were linearized using the Langmuir equation, as shown in Figure 10. The parameters of Langmuir isotherm are shown in Table 4. The high value of R^2 indicated that there was a good agreement between the model parameters. The theoretical maximum adsorption capacity of MB on SLS-C, obtained by the Langmuir adsorption isothermal equation fitting, was up to 232.5 $mg \cdot g^{-1}$, whereas the theoretical maximum adsorption capacity of Virgin-C was 153.8 $mg \cdot g^{-1}$. The Freundlich equation and Temkin equation were also used to fit the same data, as shown in Figure 10. The relevant coefficients can be calculated from three isotherms, such as K_L, Q_{max}, R_L, K_f, n, A, B, and b. The values are listed in Table 4.

The adsorption isotherm analysis results were in good agreement with the Langmuir, Freundlich, and Temkin models, but the Langmuir model had better consistency. These results indicated that the MB adsorption sites of the adsorbent were homogeneous, which was consistent with the assumption of the Langmuir model. For SLS-C, the surfactant molecules covered the AC surface uniformly, resulting in the homogeneous adsorption sites. Therefore, it can be inferred that the adsorption mechanism of MB adsorption by SLS-C was the physical and chemical monolayer adsorption.

3.8. Temperature Effect and Thermodynamic Parameters

The influence of temperature was studied by three MB concentrations (10, 30, and 50 $mg.L^{-1}$) at different temperature values (25, 35, 45, and 55°C) and the results are graphed in Figure 11. The higher the temperature, the higher the adsorption ability of MB on SLS-C at all concentrations. This indicated that the adsorption was a spontaneous endothermic process.

The thermal motion, the solubility, and the chemical potential of dye molecules increased with the increase of temperature [47]. Moreover, the pore structure of AC was closely related to temperature. The pore structure and number of active adsorption sites of AC increased with the increase of temperature due to thermal expansion. These reasons led to the increase of adsorption capacity of MB on SLS-C with the increase of temperature.

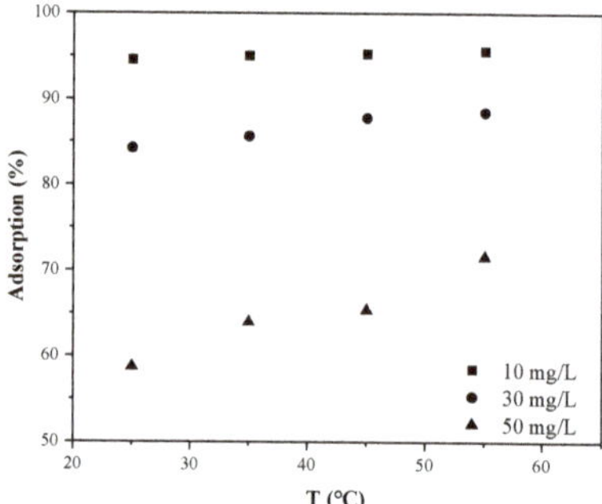

Figure 11. Influence of temperature on MB removal efficiency by SLS-C (adsorbent dosage S/L = 0.15 g·L^{-1}, pH = 5.0, contact time = 120 min).

The temperature effect and sorption mechanism were further discussed in the study of adsorption thermodynamics. The thermodynamics of adsorption of SLS-C were analyzed from the view of energy. Studying on the driving force of adsorption by the adsorption thermodynamics method determined whether the adsorption process was spontaneous or not. The values of thermodynamic parameters, such as free energy change (ΔG), enthalpy change (ΔH), and entropy change (ΔS), are usually calculated on the basis of the thermodynamic formulas shown in Equations (15)–(18) [48]:

$$\Delta G = -RT \ln K_C, \tag{15}$$

$$K_c = \frac{Q_e}{C_e}, \tag{16}$$

$$\ln K_C = \frac{\Delta S}{R} - \frac{\Delta H}{RT} \tag{17}$$

$$\Delta G = \Delta H - T\Delta S \tag{18}$$

where K_C is the thermodynamic constant, R is the universal gas constant (8.314 J·mol^{-1}·K^{-1}), and T is the absolute temperature (K). The ΔG is the Gibbs free energy change (kJ·mol^{-1}), ΔS is the entropy change (kJ·mol^{-1}·K^{-1}), and ΔH is the enthalpy change (kJ·mol^{-1}). The slope and intercept of the linear graph about ln K_C versus 1/T, as shown in Figure 12, can be respectively obtained from the values of ΔH and ΔS, as listed in Table 5.

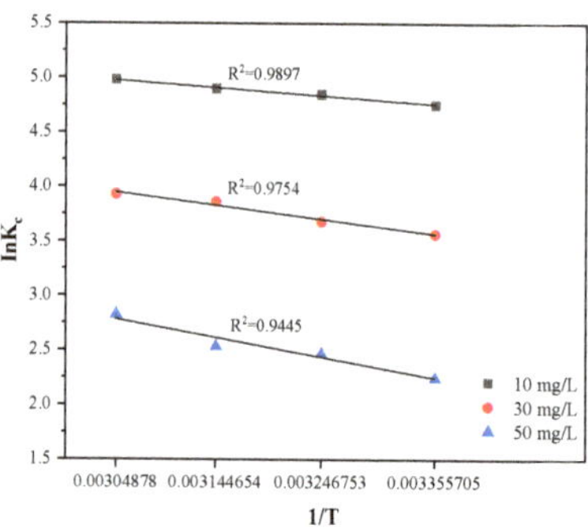

Figure 12. Thermodynamics parameters on the adsorption of MB by SLS-C.

Table 5. Thermodynamic parameters for removal efficiency of MB on SLS-C.

		Thermodynamics		
c/mg·L^{-1}	T/K	ΔG/kJ·mol^{-1}	ΔS/kJ·mol^{-1}·K^{-1}	ΔH/kJ·mol^{-1}
10	298	−11.78	0.059	5.90
	308	−12.38		
	318	−12.97		
	328	−13.56		
30	298	−8.84	0.064	10.28
	308	−9.48		
	318	−10.12		
	328	−10.77		
50	298	−5.57	0.067	14.46
	308	−6.24		
	318	−6.91		
	328	−7.58		

The negative ΔG values, which can be seen from Table 5, demonstrated the discolored process of MB on SLS-C in terms of feasibility and spontaneity [49]. The positive ΔH values of 5.90, 10.28, and 14.46 KJ·mol^{-1} at different temperatures prove that the adsorption was an endothermic process at initial dye concentrations of 10 mg·L^{-1}, 30 mg·L^{-1}, 50 mg·L^{-1}, respectively [50]. In the adsorption of different MB dye concentrations on SLS-C, ΔS values were 0.059, 0.064, and 0.067 kJ·mol^{-1}·K^{-1}, respectively. The positive values of ΔS indicated that the randomness of the solid/solution interface increased during the adsorption process [13].

3.9. Effect of Additive Salts

The effects of different salts such as NaCl, KCl, $CaCl_2$, NH_4Cl, $MgSO_4.7H_2O$, $NaNO_2$, and $FeSO_4$ on adsorption were studied in 100.0 mL MB solution under the different concentration conditions of 10, 30, and 50 mg·L^{-1}. A total of 200.0 mg of additive salt was added into the dye solution, adjusting the pH value to 5.0 and oscillating for 30 min. The effect of salt species on the adsorption of MB on SLS-C are listed in Table 6.

Table 6. Effect of salt species on the removal efficiency (%) of MB at 25°C, t = 30 min.

MB Concentration (mg·L^{-1})	Removal Efficiency (%) in Presence of Salt Species							
	None	NaCl	KCl	$CaCl_2$	NH_4Cl	$MgSO_4$	$NaNO_2$	$FeSO_4$
10	94.3	94.6	94.6	94.6	94.6	94.6	94.6	90.4
30	78.1	79.7	77.0	78.3	78.3	73.7	85.3	60.5
50	53.2	55.5	52.7	53.9	53.9	52.0	56.9	44.2

The presence of coexisting ions (Na^+, K^+, Ca^{2+}, NH^{4+}, Mg^{2+}) had little effect on the adsorption of MB on SLS-C. When the concentration of MB was 30 mg·L^{-1}, the removal rates of MB in the presence of coexisting ions were 73.7%-79.7%, which was similar to the removal rates of 78.1% in the absence of ions. When the dye concentration was 50 mg·L^{-1}, the removal rate of MB in the presence of coexisting ions (52.0%-55.5%) was similar to that in the absence of ions (53.2%). This result is similar to the study on effect of the cations (Na^+, K^+, Ca^{2+}) on adsorption of ammonium on SLS-C by Wooram Lee [27]. It indicated that the adsorption process was as insensitive to ionic as Na^+, K^+, Ca^{2+}, NH^{4+}, and Mg^{2+} [51].

The adsorption capacity of MB on SLS-C increased slightly when NaCl, $CaCl_2$, or NH_4Cl were present in the solution system. It may have been due to the increase of dimerization reaction of MB in solution in the presence of forces such as ion dipole, dipole–dipole interaction, and van-der-Waals force among dye molecules. The result agreed well with the article of Mohamed E. Mahmoud that additive salts (NaCl, NaAc, KCl, $MgSO_4$, NH_4Cl, and $CaCl_2$) improve the removal rate of dye by surfactant-modified AC [22]. When the ion (NO^{2-}) was present, it had a greater impact on the adsorption of MB by SLS-C. When MB concentration was 30 mg·L^{-1}, the removal rate of MB (85.3%) in the presence of coexisting ions (NO^{2-}) was higher than that in the absence of ions (78.1%). The reason may be that NO^{2-} has strong oxidation capacity in dilute solutions. Nitrite such as sodium nitrite and potassium nitrate are widely used in dye production. At a MB concentration of 50 mg·L^{-1}, the removal rate of MB (44.2%) by SLS-C in the presence of competitive cation (Fe^{2+}) was lower than that in the absence of ions (53.2%). The rate of adsorption was reduced because the strong reducibility ferrous ions occupied the adsorption site of MB on SLS-C.

3.10. Decolorization of MB in Real Water Samples by SLS-C

The adsorption rate of MB by SLS-C in three different real water samples of tap water, raw water, and waste samples are listed in Table 7. The tap water sample was collected from laboratory faucets, raw water sample was collected from the back of the Pearl River, and waste water sample was collected from a nameless river in the campus. The parameters of the water samples are shown in Table S1. A certain amount of MB was added into 100.0 mL water samples and prepared at concentrations of 10.0, 30.0, and 50.0 mg·L^{-1}, respectively. All the samples were adjusted to pH of 5.0 and shaken for 30min in a shaker. Distilled water samples were treated as blank water samples for comparison. The experimental results expressed that the adsorption rate of MB by SLS-C in real water samples was slightly improved relative to modeling dye wastewater.

Table 7. Adsorptive removal of MB dye from real water samples' SLS-C adsorbent.

Water Sample	MB Concentration (mg·L^{-1})	Removal (%)
Distilled water	10	94.3
	30	78.1
	50	53.4
Tap water	10	94.3
	30	80.1
	50	54.1
Raw water	10	94.3
	30	81.0
	50	55.3
waste water	10	94.3
	30	81.1
	50	55.9

4. Conclusions

This study showed that AC modified by anionic surfactants could significantly improve the adsorption properties of MB on AC. The solution initial pH, adsorbent dosage, contact time, initial MB concentration, temperature, and additive salts had a great impact on adsorption properties. The high adsorption capacity of SLS-C can be attributed to the hydrophobic group of the surfactant, which is expected to bind to the hydrophobic surface of AC. A specific binding site with the anionic functional group on SLS-C provided an efficient sorption field for the MB target dye cations. The SLS-C has advantages such as a strong affinity of acid conjugate base, a number of functional groups per unit mass adsorbent, and good dispersibility. The study of adsorption of dyes in aqueous solutions at pH values in the range of 1.0–12.0 confirmed that the adsorption rate increased with the increase of pH. The adsorption rate of MB by SLS-C increased with increased temperature. The adsorption kinetics conformed to the pseudo-second-order reaction model and the adsorption isotherm conformed to the Langmuir adsorption isotherm. The negative ΔG values and positive ΔH values proved that the adsorption process was an endothermic and spontaneous process. The theoretical maximum adsorption capacity of MB on SLS-C was 232.5 mg·g^{-1}, whereas the theoretical maximum adsorption capacity of Virgin-C was 153.8 mg·g^{-1}. The adsorption equilibrium time was about 120 min. The presence of cations such as Na^+, K^+, Ca^{2+}, NH^{4+}, and Mg^{2+} had negligible impact on the adsorption of MB on SLS-C (< 5%). The adsorption capacity was significantly improved in the presence of NO^{2-} and decreased in the presence of cation Fe^{2+}. The application of SLS-C in the decolorization of MB in tap water, raw water, and waste water proved that the existence of ions has little influence in the MB adsorption in real water samples.

Supplementary Materials: The following are available online at http://www.mdpi.com/2073-4441/12/2/587/s1, Figure S1: title, Table S1: title, Video S1: title. Figure S1: MB standard curve (at pH of 5). Figure S2: The intraparticle diffusion model plots for the adsorption of MB on SLS-C (a) at MB concentration of 10 mg/L, and (b) at MB concentrations of 10, 30, and 50 mg/L. Table S1: The water quality parameters of different water samples.

Author Contributions: Y.K. conceived the study. She also conducted the experiment, data collection, and analysis of data, and prepared the first edition of the manuscript. X.Z. participated in the design of the study and helped to draft and edit the manuscript. S.Z. helped to edit the manuscript. All authors read and approved the final manuscript.

Funding: This research was funded by National Key R&D Plan Project of China (2016YFC0400702-2), and the Research Funds for the National Natural Science Foundation (no. 21377041).

Conflicts of Interest: The authors declare no conflict of interest.

References

1. Wong, Y.C.; Szeto, Y.S.; Cheung, W.H.; McKay, G. Adsorption of acid dyes on chitosan—equilibrium isotherm analyses. *Process. Biochem.* **2004**, *39*, 695–704. [CrossRef]
2. Tan, I.A.W.; Ahmad, A.L.; Hameed, B.H. Adsorption of basic dye on high-surface area activated carbon prepared from coconut husk: Equilibrium, kinetic and thermodynamic studies. *J. Hazard. Mater.* **2008**, *154*, 337–346. [CrossRef]
3. Pavithra, K.G.; Kumar, P.S.; Jaikumar, V.; Rajan, P.S. Removal of colorants from wastewater: A review on sources and treatment strategies. *J. Ind. Eng. Chem.* **2019**, *75*, 1–19. [CrossRef]
4. Verma, A.K.; Dash, R.R.; Bhunia, P. A review on chemical coagulation/flocculation technologies for removal of colour from textile wastewaters. *J. Environ. Manag.* **2012**, *93*, 154–168. [CrossRef] [PubMed]
5. Yu, S.; Liu, M.; Ma, M.; Qi, M.; Lü, Z.; Gao, C. Impacts of membrane properties on reactive dye removal from dye/salt mixtures by asymmetric cellulose acetate and composite polyamide nanofiltration membranes. *J. Membr. Sci.* **2010**, *350*, 83–91. [CrossRef]
6. Alventosa-deLara, E.; Barredo-Damas, S.; Alcaina-Miranda, M.I.; Iborra-Clar, M.I. Ultrafiltration technology with a ceramic membrane for reactive dye removal: Optimization of membrane performance. *J. Hazard. Mater.* **2012**, *209–210*, 492–500. [CrossRef] [PubMed]
7. Asghar, A.; Abdul Raman, A.A.; Wan Daud, W.M.A. Advanced oxidation processes for in-situ production of hydrogen peroxide/hydroxyl radical for textile wastewater treatment: A review. *J. Clean. Prod.* **2015**, *87*, 826–838. [CrossRef]
8. Kordouli, E.; Bourikas, K.; Lycourghiotis, A.; Kordulis, C. The mechanism of azo-dyes adsorption on the titanium dioxide surface and their photocatalytic degradation over samples with various anatase/rutile ratios. *Catal. Today* **2015**, *252*, 128–135. [CrossRef]
9. Yagub, M.T.; Sen, T.K.; Afroze, S.; Ang, H.M. Dye and its removal from aqueous solution by adsorption: A review. *Adv. Colloid Interface Sci.* **2014**, *209*, 172–184. [CrossRef]
10. Sun, D.; Zhang, X.; Wu, Y.; Liu, X. Adsorption of anionic dyes from aqueous solution on fly ash. *J. Hazard. Mater.* **2010**, *181*, 335–342. [CrossRef]
11. Karaca, S.; Gürses, A.; Açıkyıldız, M.; Ejder, M.K. Adsorption of cationic dye from aqueous solutions by activated carbon. *Microporous Mesoporous Mater.* **2008**, *115*, 376–382. [CrossRef]
12. Xiang, Y.; Gao, M.; Shen, T.; Cao, G.; Zhao, B.; Guo, S. Comparative study of three novel organo-clays modified with imidazolium-based gemini surfactant on adsorption for bromophenol blue. *J. Mol. liquids* **2019**, 286. [CrossRef]
13. Fan, S.; Wang, Y.; Wang, Z.; Tang, J.; Li, X. Removal of methylene blue from aqueous solution by sewage sludge-derived biochar: Adsorption kinetics, equilibrium, thermodynamics and mechanism. *J. Environ. Chem. Eng.* **2017**, *5*, 601–611. [CrossRef]

14. Wang, G.; Wang, S.; Sun, Z.; Zheng, S.; Xi, Y. Structures of nonionic surfactant modified montmorillonites and their enhanced adsorption capacities towards a cationic organic dye. *Appl. Clay Sci.* **2017**, *148*, 1–10. [CrossRef]
15. Wang, W.; Huang, G.; An, C.; Zhao, S.; Chen, X.; Zhang, P. Adsorption of anionic azo dyes from aqueous solution on cationic gemini surfactant-modified flax shives: Synchrotron infrared, optimization and modeling studies. *J. Clean. Prod.* **2018**, *172*, 1986–1997. [CrossRef]
16. Hailu, S.L.; Nair, B.U.; Mesfin, R.A.; Diaz, I.; Tessema, M. Preparation and characterization of cationic surfactant modified zeolite adsorbent material for adsorption of organic and inorganic industrial pollutants. *J. Environ. Chem. Eng.* **2017**, *5*, 3319–3329. [CrossRef]
17. Jiménez-Castañeda, M.; Medina, D. Use of Surfactant-Modified Zeolites and Clays for the Removal of Heavy Metals from Water. *Water* **2017**, *9*, 235. [CrossRef]
18. Leng, L.; Yuan, X.; Zeng, G.; Shao, J.; Chen, X.; Wu, Z.; Wang, H.; Peng, X. Surface characterization of rice husk bio-char produced by liquefaction and application for cationic dye (Malachite green) adsorption. *Fuel* **2015**, *155*, 77–85. [CrossRef]
19. Mohamed, M.M. Acid dye removal: Comparison of surfactant-modified mesoporous FSM-16 with activated carbon derived from rice husk. *J. Colloid Interface Sci.* **2004**, *272*, 28–34. [CrossRef]
20. Abdel-Fattah, T.M.; Mohamed, T.M.; Mahmoud, E.; Somia, B.; Matthew, A.; Huff, D.; James, W.L.; Sandeep, K. Biochar from woody biomass for removing metal contaminants and carbon sequestration. *J. Ind. Eng. Chem.* **2015**, *22*, 103–109. [CrossRef]
21. Mubarik, S.; Saeeda, A.; Athar, M.M.; Iqbal, M. Characterization and mechanism of the adsorptive removal of 2,4,6-trichlorophenol by biochar prepared from sugarcane baggase. *J. Ind. Eng. Chem.* **2016**, *33*, 115–121. [CrossRef]
22. Mahmoud, M.E.; Nabil, G.M.; El-Mallah, N.M.; Bassiouny, H.I.; Kumar, S.; Abdel-Fattah, T.M. Kinetics, isotherm and thermodynamic studies of the adsorption of reactive red 195 A dye from water by modified Switchgrass Biochar adsorbent. *J. Ind. Eng. Chem.* **2016**, *37*, 156–167. [CrossRef]
23. Pathania, D.; Sharma, S.; Singh, P. Removal of methylene blue by adsorption onto activated carbon developed from Ficus carica bast. *Arab. J. Chem.* **2017**, *10*, S1445–S1451. [CrossRef]
24. Rafatullah, M.; Sulaiman, O.; Hashim, R.; Ahmad, A. Adsorption of methylene blue on low-cost adsorbents: A review. *J. Hazard. Mater.* **2010**, *177*, 70–80. [CrossRef] [PubMed]
25. Zhang, R. Adsorption of Dye by Modified Activated Carbon and Heavy and Heavy Metals by Rice Husk-Based Activated Carbon. Master's Thesis, Nanjing Agricultural University, Nanjing, China, 2011. (In Chinese).
26. Choi, H.D.; Jung, W.S.; Cho, J.M.; Ryu, B.G.; Yang, J.S.; Baek, K. Adsorption of Cr(VI) onto cationic surfactant-modified activated carbon. *J. Hazard. Mater.* **2009**, *166*, 642–646. [CrossRef]
27. Lee, W.; Yoon, S.; Choe, J.K.; Lee, M.; Choi, Y. Anionic surfactant modification of activated carbon for enhancing adsorption of ammonium ion from aqueous solution. *Sci. Total Environ.* **2018**, *639*, 1432–1439. [CrossRef]
28. Lin, S.Y.; Chen, W.f.; Cheng, M.T.; Li, Q. Investigation of factors that affect cationic surfactant loading on activated carbon and perchlorate adsorption. *Colloids Surf. A Physicochem. Eng. Asp.* **2013**, *434*, 236–242. [CrossRef]
29. Choi, H.D.; Shin, M.C.; Kim, D.H.; Jeon, C.S.; Baek, K. Removal characteristics of reactive black 5 using surfactant-modified activated carbon. *Desalination* **2008**, *223*, 290–298. [CrossRef]
30. Namasivayam, C.; Suresh Kumar, M.V. Removal of chromium (VI) from water and wastewater using surfactant modified coconut coir pith as a biosorbent. *Bioresour. Technol.* **2008**, *99*, 2218–2225. [CrossRef]
31. Zhou, Y.; Wang, Z.; Andrew, H.; Ren, B. Gemini Surfactant-Modified Activated Carbon for Remediation of Hexavalent Chromium from Water. *Water* **2018**, *10*, 91. [CrossRef]
32. Akhter, M.S. Effect of acetamide on the critical micelle concentration of aqueous solutions of some surfactants. *Colloids Surf. A Physicochem. Eng. Asp.* **1997**, *121*, 103–109. [CrossRef]
33. Fu, J.; Chen, Z.; Wang, M.; Liu, S.; Zhang, J.; Zhang, J.; Han, R.; Xu, Q. Adsorption of methylene blue by a high-efficiency adsorbent (polydopamine microspheres): Kinetics, isotherm, thermodynamics and mechanism analysis. *Chem. Eng. J.* **2015**, *259*, 53–61. [CrossRef]
34. Wu, S.H.; Pendleton, P. Adsorption of anionic surfactant by activated carbon: Effect of surface chemistry, ionic strength, and hydrophobicity. *J. Colloid Interface Sci.* **2001**, *243*, 306–315. [CrossRef]

35. Wang, C. Fabrication of Polyacrylonitrile-Based Activated Carbon Fibers Functionalized Sodium Dodecyl Sulfate for the Adsorptive Removal of Organic Dye from Aqueous Solution. Master's Thesis, Hunan University, Changsha, China, 2017. (In Chinese).
36. Yagub, M.T.; Sen, T.K.; Ang, H.M. Equilibrium, Kinetics, and Thermodynamics of Methylene Blue Adsorption by Pine Tree Leaves. *Water Air Soil Pollut.* **2012**, *223*, 5267–5282. [CrossRef]
37. Kannan, N.; Sundaram, M.M. Kinetics and mechanism of removal of methylene blue by adsorption on various carbons—a comparative study. *Dye Pigment* **2001**, *51*, 25–40. [CrossRef]
38. Salleh, M.A.M.; Mahmoud, D.K.; Karim, W.A.; Idris, A. Cationic and anionic dye adsorption by agricultural solid wastes: A comprehensive review. *Desalination* **2011**, *280*, 1–13. [CrossRef]
39. Amin, N.K. Removal of reactive dye from aqueous solutions by adsorption onto activated carbons prepared from sugarcane bagasse pith. *Desalination* **2008**, *223*, 152–161. [CrossRef]
40. Eren, Z.; Acar, F.N. Adsorption of Reactive Black 5 from an aqueous solution: Equilibrium and kinetic studies. *Desalination* **2006**, *194*, 1–10. [CrossRef]
41. Plazinski, W.; Rudzinski, W.; Plazinska, A. Theoretical models of sorption kinetics including a surface reaction mechanism: A review. *Adv. Colloid Interface Sci.* **2009**, *152*, 2–13. [CrossRef]
42. Ho, Y.S.; Mckay, G. Pseudo-second order model for sorption processes. *Process Biochem.* **1999**, *34*, 451–465. [CrossRef]
43. Feng, M.; You, W.; Wu, Z.; Chen, Q.; Zhan, H. Mildly alkaline preparation and methylene blue adsorption capacity of hierarchical flower-like sodium titanate. *ACS Appl. Mater. Interfaces* **2013**, *5*, 12654–12662. [CrossRef] [PubMed]
44. Huang, X.; Bu, H.; Jiang, G.; Zeng, M. Cross-linked succinyl chitosan as an adsorbent for the removal of Methylene Blue from aqueous solution. *Int. J. Biol. Macromol.* **2011**, *49*, 643–651. [CrossRef] [PubMed]
45. Mall, I.D.; Srivastava, V.C.; Agarwal, N.K. Removal of Orange-G and Methyl Violet dyes by adsorption onto bagasse fly ash—kinetic study and equilibrium isotherm analyses. *Dyes Pigment* **2006**, *69*, 210–223. [CrossRef]
46. Yao, Y.; Xu, F.; Chen, M.; Xu, Z.; Zhu, Z. Adsorption behavior of methylene blue on carbon nanotubes. *Bioresour. Technol.* **2010**, *101*, 3040–3046. [CrossRef]
47. Mouni, L.; Belkhiri, L.; Bollinger, J.C.; Bouzaza, A.; Assadi, A.; Tirri, A.; Dahmoune, F.; Madani, K.; Reminie, H. Removal of Methylene Blue from aqueous solutions by adsorption on Kaolin: Kinetic and equilibrium studies. *Appl. Clay Sci.* **2018**, *153*, 38–45. [CrossRef]
48. Gobi, K.; Mashitah, M.D.; Vadivelu, V.M. Adsorptive removal of methylene blue using novel adsorbent from palm oil mill effluent waste activated sludge: Equilibrium, thermodynamics and kinetic studies. *Chem. Eng. J.* **2011**, *171*, 1246–1252. [CrossRef]
49. Auta, M.B.H. Hameed, Chitosan–clay composite as highly effective and low cost adsorbent for batch and fixed-bed adsorption of methylene blue. *Chem. Eng. J.* **2014**, *237*, 350–361. [CrossRef]
50. Bulut, Y.; Aydın, H. A kinetics and thermodynamics study of methylene blue adsorption on wheat shells. *Desalination* **2006**, *194*, 259–267. [CrossRef]
51. Que, W.; Jiang, L.H.; Wang, C.; Liu, Y.G.; Zeng, Z.W.; Wang, X.H.; Ning, Q.M.; Liu, S.H.; Zhang, P.; Liu, S.B. Influence of sodium dodecyl sulfate coating on adsorption of methylene blue by biochar from aqueous solution. *J. Environ. Sci.* **2018**, *70*, 166–174. [CrossRef]

MDPI
St. Alban-Anlage 66
4052 Basel
Switzerland
Tel. +41 61 683 77 34
Fax +41 61 302 89 18
www.mdpi.com

Water Editorial Office
E-mail: water@mdpi.com
www.mdpi.com/journal/water

www.ingramcontent.com/pod-product-compliance
Lightning Source LLC
LaVergne TN
LVHW071613170726
843515LV00009B/2382